Salmonella in Domestic Animals

———————————————

To Patricia, who despite suffering from verbal salmonellosis for many years has given us every support, and also to the many friendships that have resulted from mutual interest in this fascinating bacterium

Salmonella in Domestic Animals

_____

Edited by

C. Wray

Formerly of
Veterinary Laboratories Agency (Weybridge), Addlestone, Kent, UK

and

A. Wray

*Department of Oral Biology, College of Dentistry, University of Florida,
Gainesville, Florida, USA*

CABI *Publishing*

CABI *Publishing* is a division of CAB *International*

CABI Publishing
CAB International
Wallingford
Oxon OX10 8DE
UK

Tel: +44 (0)1491 832111
Fax: +44 (0)1491 833508
Email: cabi@cabi.org

CABI Publishing
10 E 40th Street
Suite 3203
New York, NY 10016
USA

Tel: +1 212 481 7018
Fax: +1 212 686 7993
Email: cabi-nao@cabi.org

A catalogue record for this book is available from the British Library, London, UK.

Library of Congress Cataloging-in-Publication Data
Salmonella in domestic animals / edited by C. Wray and A. Wray.
        p. cm.
    Includes bibliographical references.
    ISBN 0–85199–261–7 (alk. paper)
        1. Salmonellosis in animals.   I. Wray, C. (Clifford).   II. Wray, A.
    SF809.S24 S25   2000
    636.089'6927--dc21
                                                                        99–043803

ISBN 0 85199 261 7

Typeset by Columns Design Ltd, Reading.
Printed and bound in the UK by Biddles Ltd, Guildford and King's Lynn.

Contents

Contributors

P.A. Barrow, Institute for Animal Health, Compton Laboratory, Compton, Newbury, Berkshire RG20 7NN, UK

A.J. Bäumler, Department of Medical Microbiology and Immunology, Texas A&M University Health Science Center, 407 Reynolds Medical Building, College Station, TX 77843-1114, USA

P. Bouvet, Centre National de Référence des Salmonella et Shigella, Unité des Entérobactéries, INSERM Unit 389, Institut Pasteur, 75724 Paris Cedex 15, France

M.E. Carter, Department of Veterinary Microbiology and Parasitology, Faculty of Veterinary Medicine, University College Dublin, Ballsbridge, Dublin 4, Eire

R.H. Davies, Veterinary Laboratories Agency (Weybridge), New Haw, Addlestone, Surrey KT15 3NB, UK

P.J. Fedorka-Cray, Richard Russell Research Center, 950 College Station Road, Athens, GA 30605-2720, USA

J.T. Gray, University of Nebraska, Lincoln Veterinary Diagnostic Center, Lincoln, Nebraska, USA

F. Grimont, Centre National de Référence pour le Typage Moléculaire des Entérobactéries, Unité des Entérobactéries, INSERM Unit 389, Institut Pasteur, 75724 Paris Cedex 15, France

P.A.D. Grimont, Centre National de Référence des Salmonella et Shigella, Unité des Entérobactéries, INSERM Unit 389, Institut Pasteur, 75724 Paris Cedex 15, France

H.M. Hafez, Free University of Berlin, Faculty of Veterinary Medicine, Institute of Poultry Diseases, Koserstrasse 21, 14195 Berlin, Germany

F. Heffron, Department of Molecular Microbiology and Immunology, Oregon Health Sciences University, Portland, OR 97201-3098, USA

R. Helmuth, Federal Institute for Health Protection of Consumers and Veterinary Medicine BgVV, Laboratory for Molecular Biology and National Salmonella Reference Laboratory, Diedersdorfer Weg 1, 12277 Berlin, Germany

R.R. Henry, Cherry Valley Farms, North Kelsey Moor, Lincoln LN7 6HH, UK

M.H. Hinton, Home Office, 50 Queen Anne's Gate, London SW1H 9AT, UK

K. Hollinger, FDA CVM, Division of Epidemiology and Surveillance, 7500 Standish Place, Rockville, MD 20855, USA

P.S. Holt, US Department of Agriculture, Agricultural Research Service, Southeast Poultry Laboratory, 934 College Station Road, Athens, GA 30605, USA

J.K. House, Department of Medicine and Epidemiology, School of Veterinary Medicine, University of California – Davis, Davis, CA 95616, USA

J.H.J. Huis in't Veld, Department of Food of Animal Origin, Utrecht University, PO Box 80.175, 3508 TD Utrecht, The Netherlands

T. Humphrey, PHLS Food Microbiology Research Unit, Church Lane, Heavitree, Exeter EX2 5AD, UK

S. Jodas, Staatliches Tierarztliches Untersuchungasmt, Azenbergestrasse 16, 70174 Stuttgart, Germany

Y.E. Jones, Veterinary Laboratories Agency (Weybridge), New Haw, Addlestone, Surrey KT15 3NB, UK

K.A. Linklater, SAC, West Mains Road, Edinburgh EH9 3JG, UK

I.M. McLaren, Veterinary Laboratories Agency (Weybridge), New Haw, Addlestone, Surrey KT15 3NB, UK

G. Mead, Royal Veterinary College, University of London, Boltons Park, Hawkshead Road, Potters Bar, Hertfordshire EN6 1NB, UK

C.J. Murray, Institute of Medical and Veterinary Science, Box 14, Rundle Mall, Adelaide, SA 5000, Australia

J.E. Olsen, Department of Veterinary Microbiology, The Royal Veterinary and Agricultural University, Stigbøjlen 4, DK-1870 Frederiksberg C, Denmark

C. Poppe, Health Canada, 110 Stone Road West, Guelph, Ontario N1G 3W4, Canada

P.J. Quinn, Department of Veterinary Microbiology and Parasitology, Faculty of Veterinary Medicine, University College Dublin, Ballsbridge, Dublin 4, Eire

A.N. Rycroft, Royal Veterinary College, Hawkshead Lane, North Mymms, Hatfield, Hertfordshire AL9 7TA, UK

C. Schneitz, Orion Corporation, Animal Health, PO Box 425, FIN-20101 Turku, Finland

B.P. Smith, Department of Medicine and Epidemiology, School of Veterinary Medicine, University of California – Davis, Davis, CA 95616, USA

C.J. Thorns, Bacteriology Department, Veterinary Laboratories Agency (Weybridge), New Haw, Addlestone, Surrey KT15 3NB, UK

R.M. Tsolis, Department of Veterinary Pathobiology, Texas A&M University, College Station, TX 77843-4467, USA

H. van der Zee, Inspectorate for Health Protection, Regional Service East, De Stoven 22, 7206 AX Zutphen, The Netherlands

T.S. Wallis, Institute for Animal Health, Compton Laboratory, Compton, Newbury, Berkshire RG20 7NN, UK

W.D. Waltman, Georgia Poultry Laboratory, Oakwood, GA 30566, USA

M.J. Woodward, Bacteriology Department, Veterinary Laboratories Agency (Weybridge), New Haw, Addlestone, Surrey KT15 3NB, UK

A. Wray, Department of Oral Biology, College of Dentistry, University of Florida, PO Box 100424, Gainesville, FL 32610-0424, USA

C. Wray, Veterinary Laboratories Agency (Weybridge), New Haw, Addlestone, Surrey KT15 3NB, UK

Preface

It is now more than 40 years since the publication of 'Salmonellosis in Animals' by Professor Buxton and, although there have been books regarding diseases in domestic animals which contain a chapter on salmonellosis, there has been no English-language book that brings together all aspects of Salmonella infection in domestic animals.

Professor Buxton pointed out 'salmonella are a large group of bacteria, which does not recognise international frontiers, shows little host specificity and from which there seems to be derived an ever increasing list of antigenic combinations by which these organisms are classified'. Since the publication of this book, the situation has gradually worsened and the prevalence of Salmonella infection has increased markedly in both humans and domestic animals. There are many factors for this increase, including the rapid changes in animal husbandry and production that have taken place. In the UK, the number of farms has decreased, as has the number of farm-workers; at the same time, the number of animals on the farm has increased through the adoption of intensive systems of animal production. To feed the increased number of farm animals, protein and vegetable by-products are imported on a large scale, which has resulted in widespread international outbreaks of salmonellosis in animals and subsequently humans, e.g. S. agona. These developments have been accompanied by marked changes in food distribution and the eating habits of the human population;

chicken is now the cheapest source of animal protein in Western Europe and North America.

Over 2400 different Salmonella serovars have been described, but only a few predominate in an animal population or a country at any one time. There are currently global pandemics of S. enteritidis and S. typhimurium DT104 and we have little knowledge as to the factors that have resulted in these current pandemics and why the predominant phage types vary in different countries. A consequence of the S. enteritidis outbreak was political concern and legislation has been enacted in many countries to control the prevalence of Salmonella infections in farm animals in order to prevent food-borne infection. Likewise, the rise of the penta-resistant S. typhimurium DT104 has reopened the debate on the agricultural use of antibacterial drugs and the use of alternate methods of control.

Some Salmonella serovars are host-adapted; thus S. choleraesuis is associated with pigs. The reasons for the host adaptation are largely unknown. In contrast, S. typhimurium may infect most animal species. Infection is primarily by ingestion of the organism and large doses of Salmonella are usually required to cause experimental infections; yet epidemiological evidence suggests that the infective dose must be much smaller. Although Salmonella may multiply in the small intestine, disease is not an inevitable consequence, and most infections in pigs and poultry are asymptomatic.

Fimbriae were first identified in Salmonella

more than 40 years ago, when it was suggested that they might assist *Salmonella* to attach to epithelial cells. However, although many studies claim to have demonstrated attachment, other studies have failed to detect any measurable effect. In recent years, other distinct fimbriae have been detected on some *Salmonella* serovars, but their role in the pathogenesis of disease requires clarification.

Although *Salmonella* may colonize the intestine without causing disease, close association and penetration of the intestinal mucosa are necessary for the induction of diarrhoea and systemic disease. Many possible virulence mechanisms have been identified in tissue culture systems and in experimental animals but there is still much to understand about the *in vivo* role of various *Salmonella* genes that may be involved in mucosal penetration. Likewise, evidence for the role of toxins is confusing and often contradictory.

Immunity to *Salmonella* infection has been studied predominantly in mice, but to what extent these studies are applicable to farm animals is not known and the relative importance of humoral and cellular immunity in the resistance of farm animals to *Salmonella* infections has not yet been established. Despite our lack of knowledge on immunity and the pathogenesis of infection, vaccines of varying degrees of efficacy have been used for many years. More recently, rationally attenuated vaccines have been developed and some are undergoing field trials. The feeding of faeces from adult hens to young chickens has been shown to prevent *Salmonella* colonization and this competitive exclusion is being increasingly used in the poultry industry. However, despite many years of research, little is known about the ecology of *Salmonella* in the intestine and the bacteria that may prevent colonization.

The wider aspects of the ecology and epidemiology of *Salmonella* are, however, of impor-tance to all those involved in *Salmonella* investigations. *Salmonella* are widespread in the environment and in recent years their prolonged persistence has been demonstrated on many calf and pig units. Indeed, some consider *Salmonella* to be primarily an environmental organism that is pathogenic for animals. Other survival mechanisms have been demonstrated and further studies on the epidemiology and persistence of the organism are desirable, using the molecular techniques that have been used to study the pathogenesis.

The emphasis of the book is on the role of *Salmonella* in animal disease and it is written in five sections. In the first part, the characteristics of the microorganism are discussed. The following section considers its virulence, effect on the host and antibacterial resistance. The third part reviews current knowledge of *Salmonella* infection in farm and companion animals. Each of the chapters in this section is intended to provide a comprehensive account, although more detailed information on some of the topics will be found in other sections of the book. Subsequent sections discuss the epidemiology and prevention, and laboratory methods.

One problem that confronts all involved with *Salmonella* is the taxonomy and nomenclature of the bacteria. As discussed in Chapter 1 the full nomenclature of each serovar is complicated and, while taxonomically correct, it is cumbersome. As a consequence, the binomial nomenclature of the different serovars has been used.

It is hoped that the book will provide information for all those whose work involves *Salmonella* and that the book will assist in the control of *Salmonella* infections in animals to facilitate the economic production of food that is free from infection and safe for human consumption.

Chapter 1

Taxonomy of the Genus *Salmonella*

Patrick A.D. Grimont,[1,2] **Francine Grimont**[2] **and Philippe Bouvet**[1]

[1]*Centre National de Référence des Salmonella et Shigella;* [2]*Centre National de Référence pour le Typage Moléculaire des Entérobactéries, Unité des Entérobactéries, INSERM Unit 389, Institut Pasteur, 75724 Paris Cedex 15, France*

Introduction

The habitat of the genus *Salmonella* seems to be limited to the digestive tract of humans and animals. Thus, the presence of *Salmonella* in other habitats (water, food, natural environment) is explained by faecal contamination. Some serovars (serotypes) have a habitat limited to a host species, such as humans (serovars Typhi, Paratyphi A), sheep (serovar Abortusovis) or fowl (Gallinarum). Different infectious syndromes can be caused by *Salmonella* serovars, e.g. serovar Typhi causes typhoid in humans, serovar Typhimurium causes diarrhoea in humans and other animal species and a typhoid-like syndrome in mice, serovar Abortusovis is responsible for abortion in ewes and serovar Dublin has been associated with different extra-intestinal infections in AIDS patients. The genetics of this pathogenic diversity is only beginning to be uncovered. Because no tools were available to identify virulence factors associated with the diverse salmonellosis syndromes, the genus *Salmonella* was subdivided into subspecific taxa (types), which could more or less be associated with a host species or syndrome. Furthermore, prevention of salmonellosis implies local (industry, hospital, district), national (national reference centre) or international surveillance based on the systematic typing of strains.

History of *Salmonella* Taxonomy and Nomenclature

In 1884, Gaffky cultivated the typhoid bacillus (Kauffmann, 1978), which Eberth had observed in 1880 in spleen sections and mesenteric lymph nodes from a patient who died from typhoid (Le Minor, 1994). The organism now known as *S. choleraesuis* was first isolated from pigs by Salmon and Smith (1886), when they considered the organism to be the cause of swine fever (hog cholera). Later, Pfeiffer and Kolle (1896) and Gruber and Durham (1896) discovered that the serum of an animal immunized with the typhoid bacillus agglutinated the typhoid bacillus. At the same time, Widal (1896) and Grunbaum (1896) found that the serum of a typhoid patient agglutinated the typhoid bacillus. This new test was called 'serodiagnostic' by Widal (1896). The same year, two isolates were recovered from patients with clinical symptoms of typhoid and negative Widal serodiagnosis (Achard and Bensaude, 1896). The organism was called 'bacille paratyphique'.

This was only the beginning of an ongoing story and new serovars of what is now known as *Salmonella* are described each year.

In an early stage, *Salmonella* strains isolated from different clinical conditions or hosts were considered to be different species. This gave names such as '*Eberthella typhosa*' (*S. typhi*), *S. enteritidis*, '*S. abortusovis*', '*S. gallinarum*', '*S. bovismorbificans*', *S. choleraesuis* or *S. typhimurium*. It was soon realized that a number of these so-called species were ubiquitous.

Analysis of O and H antigens, initiated by White (1926) and extended by Kauffmann (1941), resulted in the description of a great number of serovars. The species was defined by Kauffmann (1961) as 'a group of related sero-fermentative phage types', with the result that each serovar was considered as a species. Names were given to more than 2000 serovars. These names were generally derived from the geographical location where the first strain was isolated (e.g. 'S. *london*').

This one serovar–one species concept was later found to be untenable since most serovars cannot be separated by biochemical tests. Proposals were made to reduce the number of species. Borman *et al.* (1944) proposed that only three species ('S. *typhosa*', S. *choleraesuis* and 'S. *kauffmannii*') should be recognized. Kauffmann and Edwards (1952) also proposed three species ('S. *typhosa*', S. *choleraesuis* and 'S. *enterica*'). Ewing (1963) proposed S. *typhi*, S. *choleraesuis* and S. *enteritidis*. All these proposals had in common that, apart from S. *typhi* and S. *choleraesuis*, all serovars were placed into one species ('S. *kauffmannii*', 'S. *enterica*' or S. *enteritidis*). The latter was confusing as S. *enteritidis* meant either a precise serovar or a very large set of serovars.

Strains able to liquefy gelatin slowly and to ferment lactose were considered to form a separate genus, *Arizona* (Kauffmann and Edwards, 1952). After some nomenclature confusion, the name was validly published by Ewing with one species, *Arizona hinshawii* (Ewing, 1969).

Kauffmann (1966a,b) divided the genus *Salmonella* into four subgenera on the basis of biochemical reactions. These subgenera were designated by Roman numerals (I–IV) without formal nomenclature. The genus *Arizona* constituted subgenus III. Later, Le Minor *et al.* (1970) considered Kauffmann's subgenera to represent species named 'S. *kauffmannii*' (subgenus I), 'S. *salamae*' (subgenus II), S. *arizonae* (subgenus III) and 'S. *houtenae*' (subgenus IV).

A landmark in bacterial nomenclature was the publication of the Approved Lists of Bacterial Names (Skerman *et al.*, 1980). Names which did not appear in the Approved Lists lost standing in the nomenclature (when cited, these names should be printed with quotation marks). All new names proposed after 1 January 1980 can only be validated by publication or announcement in the *International Journal of Systematic Bacteriology*. The Approved Lists included five *Salmonella* species: S. *arizonae*, S. *choleraesuis*, S. *enteritidis*, S. *typhi* and S. *typhimurium*.

DNA-relatedness studies showed that the so-called subgenera I–IV constituted a single DNA hybridization group with five subgroups delineated by studies of the thermal stability of hybridized DNA (Crosa *et al.*, 1973; Stoleru *et al.*, 1976; Le Minor *et al.*, 1982, 1986). The subgroups corresponded to the former subgenera except that subgenus III was split into DNA subgroups IIIa and IIIb. Later, an additional subgroup (subgroup VI) was identified and a few rare serovars (Bongor group) were found to constitute a second DNA hybridization group (Le Minor *et al.*, 1982, 1986).

However, in the absence of rules for delineating bacterial species, Le Minor *et al.* (1982) considered all *Salmonella* serovars to constitute a single species, which was named S. *choleraesuis*, since this is the name of the type species of the genus *Salmonella*. The species contained six subspecies: S. *choleraesuis* subsp. *choleraesuis*, S. *choleraesuis* subsp. *salamae*, S. *choleraesuis* subsp. *arizonae*, S. *choleraesuis* subsp. *diarizonae*, S. *choleraesuis* subsp. *houtenae* and S. *choleraesuis* subsp. *bongori*. A new subspecies, S. *choleraesuis* subsp. *indica*, was added subsequently (Le Minor *et al.*, 1986). This nomenclature, which strictly followed the rules of the International Code of Nomenclature of Bacteria (Rules Revision Committee, 1975) had a serious drawback, since the specific name (S. *choleraesuis*) was also the name of a serovar. To overcome this, Le Minor and Popoff (1987) proposed the name S. *enterica* for the single *Salmonella* species, with the following subspecies, S. *enterica* subsp. *enterica*, S. *enterica* subsp. *salamae*, S. *enterica* subsp. *arizonae*, S. *enterica* subsp. *diarizonae*, S. *enterica* subsp. *houtenae*, S. *enterica* subsp. *bongori* and S. *enterica* subsp. *indica*. This proposal requested an opinion from the Judicial Commission of the International Committee of Systematic Bacteriology. Unfortunately, the opinion has not yet been awarded, probably because the request was not limited to nomenclature (the only scope of the Judicial Commission) and included the recognition of a single species in the genus *Salmonella* (a taxonomic proposal).

From a taxonomic standpoint, a genomic species is now defined as a set of strains more than 70% related by DNA–DNA hybridization with ΔT_m values below 5°C (Wayne *et al.*, 1987). Application of these guidelines allowed the recognition of two species in the genus *Salmonella* – *S. enterica* and *S. bongori* (Reeves *et al.*, 1989) – with six subspecies – *S. enterica* subsp. *enterica*, *S. enterica* subsp. *salamae*, *S. enterica* subsp. *arizonae*, *S. enterica* subsp. *diarizonae*, *S. enterica* subsp. *houtenae* and *S. enterica* subsp. *indica*. Although this nomenclature is not yet validated, it is widely used, since it is scientifically based and less confusing than the *S. choleraesuis* proposal.

Serovar names are no longer considered as species names and therefore should not be printed in italics. *S. typhimurium* becomes *S. enterica* subsp. *enterica* serovar Typhimurium, or simply *Salmonella* serovar Typhimurium. Only serovars of *S. enterica* subsp. *enterica* are given names (usually geographical names). Serovars of other subspecies are designated by their O : H formula.

Phylogenetic Position of the Genus *Salmonella*

Bacterial classification is now based on phylogenetic grounds. A phylogenetic tree can be derived from the comparison of 16S rRNA or other gene sequences. The two *Salmonella* species (*S. enterica* and *S. bongori*) were separated by 16S rRNA sequence analysis. Within *S. enterica*, the diphasic subspecies *enterica* and *indica* were separated from the monophasic subspecies *arizonae* and *houtenae* by 23S rRNA comparison. The genus *Salmonella* was found to be related to the *Escherichia coli/Shigella* genomic species and to *Citrobacter freundii* by both 16S and 23S rRNA sequence comparison (Christensen *et al.*, 1998). Divergence within the genus *Salmonella* and proximity with *E. coli* and *C. freundii* makes the choice of a *Salmonella*-specific oligonucleotide probe difficult (Lane and Collins, 1991). It was discovered that 23S rRNA is fragmented in several *Salmonella* serovars (Winkler, 1979). This fragmentation is due to the presence of non-transcribed intervening sequences inserted in genes coding for 23S rRNA (Burgin *et al.*, 1990).

More gene sequences are now used for phylogenetic studies. A combined comparison of five gene sequences (proline permease, glyceralde-hyde-3-phosphate dehydrogenase, malate dehydrogenase, 6-phosphogluconate dehydrogenase and isocitrate dehydrogenase kinase/phosphatase) yielded a phylogenetic tree consistent with DNA hybridization data (Barker *et al.*, 1988; Beltran *et al.*, 1988, 1991; Selander *et al.*, 1990a,b). It is interesting that *S. enterica* subspecies *enterica*, *salamae*, *indica* and *diarizonae*, which are predominantly diphasic in flagellar expression, cluster apart from monophasic subspecies *arizonae* and *houtenae*, whereas *S. bongori* branches apart.

From these sequence data, the following phylogenetic hypothesis has been drawn (Selander *et al.*, 1996). The genera *Salmonella* and *E. coli* might have diverged from a common ancestor 120–160 million years ago, coincident with the origin of mammals. *E. coli* evolved as a commensal and opportunistic pathogen of mammals and birds. The lineage of the *Salmonella* remained associated with reptiles (which are still the primary hosts of the monophasic subspecies of *S. enterica*) and evolved as intracellular pathogens through acquisition of genes that mediate invasion of host epithelial cells (*inv/spa* genes and others). Building a mechanism of flagellar antigen phase shifting (diphasic serovars) has permitted an extension of ecological range to mammals and birds as a pathogen (*S. enterica* subsp. *enterica*, *salamae*, *diarizonae* and *indica*). *S. enterica* subsp. *enterica* became highly specialized for mammals and birds with some serovars adapting to single species.

S. enterica subsp. *enterica* serovar Typhi might have appeared when humans were available as a host (3 million years ago). It has been hypothesized that serovar Typhi could have appeared first in Indonesia, where diphasic strains of this serovar (a supposedly ancestral form of the serovar) can still be found (Frankel *et al.*, 1989a,b).

DNA Relatedness within the Genus *Salmonella*

A bacterial species can be defined as a DNA hybridization group. Strains within a species are generally more than 70% related and the thermal instability of reassociated DNA (ΔT_m, divergence) does not exceed 5°C (Wayne *et al.*, 1987). DNA hybridization studies (Crosa *et al.*,

1973; Stoleru *et al.*, 1976; Le Minor *et al.*, 1982, 1986), have shown the genus *Salmonella* to be composed of only two genomic species, *S. enterica* and *S. bongori* (Le Minor and Popoff, 1987; Reeves *et al.*, 1989). *S. enterica* (the most common *Salmonella* species) has been subdivided into six subspecies (Le Minor and Popoff, 1987). The subspecific epithets are *enterica, salamae, arizonae, diarizonae, houtenae* and *indica*. The habitat of *S. enterica* subsp. *enterica* is the intestinal tract of humans and warm-blooded animals.

Population Genetics

The multilocus enzyme electrophoresis (MLEE) method has been used to assess allelic variation in multiple genes in a collection of isolates. Electromorphs of an enzyme are equated with alleles of the corresponding structural gene. Distinctive allele profiles are designated as electrophoretic types (ETs). They represent multilocus enzyme genotypes. MLEE analysis indicates that *S. bongori* is the most divergent group of *Salmonella*. The other *Salmonella* show clusters corresponding to subspecies (Selander *et al.*, 1990a,b).

Within *S. enterica* subsp. *enterica*, MLEE analysis shows serovar Typhi as a single clone, distinct from all other serovars studied. Serovars Paratyphi A and Sendai constitute a group, whereas serovars Typhimurium, Paratyphi B, Saintpaul, Heidelberg and Muenchen form a loose cluster.

MLEE analysis has identified serovar Enteritidis as a close relative of the non-motile serovar Gallinarum (Selander *et al.*, 1996).

For *Salmonella*, a basic clonal population structure is evidenced by the presence of strong linkage disequilibrium among alleles at enzyme loci, the association of specific O and H serovars with only one or a small number of multilocus enzyme genotypes and the global distribution of certain genotypes (Selander *et al.*, 1996).

Phenotypic Characteristics

Strains belonging to the genus *Salmonella* comply with the definition of the family *Enterobacteriaceae*: straight rods, generally motile with peritrichous flagella, grow on nutrient agar, aeroanaerobes, ferment glucose, often with production of gas, reduce nitrate into nitrite and the oxidase test is negative. Some serovars have peculiarities: the avian serovar Gallinarum is regularly non-motile and non-motile mutants of normally motile serovars are occasionally observed. Most *Salmonella* strains are prototrophic, i.e. they have no growth-factor requirement and can grow in a minimal medium with glucose as sole carbon and energy source and ammonium ion as nitrogen source. Some host-adapted serovars (e.g. Typhi, Paratyphi A, Sendai, Abortusovis, Gallinarum) are auxotrophic and require one or more growth factors. Some serovars (e.g. Typhi) never produce gas from glucose.

The following characteristics are used for *Salmonella* identification: urea not hydrolysed; tryptophan and phenylalanine not deaminated; acetoin not produced; lactose, adonitol, sucrose, salicin and 2-ketogluconate not fermented; hydrogen sulphide (H_2S) produced from thiosulphate; lysine and ornithine decarboxylated; growth on Simmons citrate agar; 4-methylumbelliferyl caprylate (MUCAP) hydrolysed. Some serovars behave differently. Typhi never decarboxylates ornithine and fails to grow on Simmons citrate agar. Paratyphi A fails to produce H_2S, to decarboxylate lysine and to grow on Simmons citrate agar. Subspecies other than *S. enterica* subsp. *enterica* may ferment lactose. Tests allowing identification of these subspecies are shown in Table 1.1.

Some phenotypic properties of *Salmonella* are so specific that they have been used for enrichment, selective isolation or colony differentiation. *Salmonella* and other genera of *Enterobacteriaceae* are more resistant to novobiocin, selenite, tergitol and bile salts, especially desoxycholate, than other bacteria. *Salmonella* are more resistant to brilliant green and malachite green than other genera of *Enterobacteriaceae*. However, these properties are not sufficient for a true selective isolation and no medium is at present available with the ability to isolate only *Salmonella* and no other bacteria. As 'selective' media are insufficiently selective, most of these media need to be more differential. Lactose, sucrose, salicin, cellobiose or glycerol is often included with pH indicators in 'selective' media, since most *Salmonella* fail to produce acid from these substrates. Alternatively, a chromogenic substrate (e.g. 5-bromo-4-chloro-3-β-D-

Table 1.1. Phenotypic differentiation of *Salmonella* species and subspecies.

Trait	Salmonella enterica subsp.						Salmonella bongori
	enterica	salamae	arizonae	diarizonae	houtenae	indica	bongori
ONPG test	−	−	+	+	−	d	+
β-Glucuronidase	d	d	−	+	−	d	−
α-Glutamyl transferase	d	+	−	+	+	+	+
Acid from dulcitol	+	+	−	−	−	d	+
Acid from sorbitol	+	+	+	+	+	+	−
Acid from galacturonate	−	+	−	+	+	+	+
Malonate alkalinized	−	+	+	+	−	−	−
L(+) Tartrate utilized	+	−	−	−	−	−	−
Gelatin hydrolysed	−	+	+	+	+	+	−
Growth on KCN	−	−	−	−	+	−	+
Phage O1 susceptible	+	+	−	+	−	+	+

+, More than 90% strains positive; −, less than 10% strains positive; d, 10–90% strains positive; ONPG, *ortho*-nitrophenyl-β-D-galactopyranoside.

galactopyranoside or X-gal) is used to detect β-galactosidase production. Thiosulphate and iron salts allow the production and detection of H_2S unless the pH is acid. Thus, in *Salmonella–Shigella* (SS) agar, selective agents are bile salts and brilliant green, substrates of interest are lactose and sodium thiosulphate and indicators are neutral red and ferric citrate. Typically, *Salmonella* strains give colourless colonies with black centres. *Proteus* strains have a lower efficiency of plating than *Salmonella* strains but their colonies are very similar. *E. coli* strains give red colonies and *Citrobacter* strains give red, pink or colourless colonies with or without black centres. Hektoen agar contains bile salts (selective agents), lactose, sucrose, salicin and sodium thiosulphate (substrates) and bromthymol blue, acid fuchsin and ferric ammonium citrate (indicators). *Salmonella* strains give green, black-centred colonies. XLT4 agar contains tergitol-4 (selective agent), xylose, lactose, sucrose, lysine and thiosulphate (substrates), a pH indicator and ferric ion. *Salmonella* strains produce acid from xylose. The low pH triggers lysine decarboxylation, which alkalinizes the medium around *Salmonella* colonies. Ferric sulphide accumulates, giving black colonies on a green background. Occasional sucrose- or lactose-fermenting *Salmonella* strains would give black colonies on a yellow background. Chromagar (Rambach-agar) contains desoxycholate (selective agent), propylene glycol (substrate) and X-gal (chromogenic substrate). Most *Salmonella* colonies are red or fuchsia. However,

serovars Typhi and Paratyphi A give colourless colonies. (The many different plating media are considered more fully in Chapter 21.)

Colonies that are suspected to be *Salmonella* can be submitted to the MUCAP test. *Salmonella* regularly produce caprylate esterase.

Strains of *Citrobacter* (when H_2S-positive and lactose-negative) or *Hafnia* (lactose-negative at 37°C) are often misidentified as *Salmonella*. However, *Citrobacter* strains fail to decarboxylate lysine and often ornithine and *Hafnia* strains develop a characteristic reaction at 20–30°C [*ortho*-nitrophenyl-β-D-galactopyranoside (ONPG) hydrolysed, acetoin produced] and never produce H_2S or grow on Simmons citrate agar at any temperature. Furthermore, most *S. enterica* subsp. *enterica* strains are susceptible to phage O1 (Felix and Callow, 1943) and all are resistant to *Hafnia* phage (Guinée and Valkenburg, 1968), whereas *Hafnia* strains are often susceptible to *Hafnia* phage and all are resistant to phage O1.

Antigenic Diversity

Classically, three sorts of antigens are considered: somatic (O), flagellar (H) and (mostly for serovar Typhi) surface (Vi) antigens. The antigenic structure of *Salmonella* has been revealed mostly by cross absorption of antisera, which subdivided antigens into different factors. The typing system, which was built up over more than 70 years

by White, Kauffmann and Le Minor, is a model
of its kind.

O antigens

The chemical structure of diverse O factors (i.e.
the specific part of the bacterial lipopolysaccharide) has been determined and the genes
involved in the production of essential enzymes
for the assembly of some O factors have been
located, cloned or sequenced. A new nomenclature has been proposed for the genes coding for
those enzymes involved in polysaccharide synthesis (Reeves *et al.*, 1996). A bacterial polysaccharide database is available on the Internet
(http://www.microbio.usyd.edu.au/BPGD/default.
htm).

A core contains, in addition to 3-deoxy-D-
manno-octulosonic   acid   and   lipid   A,
L-glycero-D-manno-heptose, D-glucose, D-galactose, N-acetylglucosamine and ethanolamine
pyrophosphate. From this core, a poly-O side-
chain extends to the bacterial surface. The poly-
O side-chain is made of repeated monomers
containing D-galactose, L-rhamnose, D-mannose
and, for some serogroups, abequose (factor O4 in
group B), paratose (group A) or tyvelose (factor
9 in group D) branched in position 1–3 on D-
mannose (Rick, 1987; Jiang *et al.*, 1991; Raetz,
1996; Fig. 1.1).

A genetic locus, *rfa*, located between genes
cysE and *pyrE* at 79 minutes on the genetic map
of strain LT2 (serovar Typhimurium), contains
the structural genes coding for the glycosyltrans-
ferases involved in core synthesis. Locus *rfb*,
located in the vicinity of gene *his* (at 42 minutes), contains the genes necessary for synthesis
of an oligosaccharide monomer. Thus, from glu-
cose-1-phosphate, CDP-4-keto-6-deoxyglucose is
obtained by the action of glucose-1-phosphate
cytidylyltransferase (coded by gene *ddhA*, formerly *rfbF*) and CDP-glucose-4,6-dehydratase
(coded by gene *ddhB*, formerly *rfbG*). Then, after
action of enzymes coded by genes *ddhC* (formerly
rfbH) and *ddhD* (formerly *rfbI*), CDP-4-keto-3,6-
dideoxyglucose is obtained, and finally abequose
synthase (gene *abe*, formerly *rfbJ*) produces CDP-
abequose. In groups A and D, gene *prt* (formerly
rfbS) codes for the final step yielding CDP-
paratose and, in group D, the product of gene *tyv*

(formerly *rfbE*) turns CDP-paratose into CDP-
tyvelose. A close examination (local G+C content) of sequences in region *rfb* suggests that
diverse parts could have been inserted or
exchanged in the course of bacterial evolution.
Genes *ddhC* (*rfbH*), *ddhD* (*rfbI*) and *abe* (*rfbJ*),
which control the last steps in abequose synthesis, could originate from genetic exchange with
another species (Jiang *et al.*, 1991). In region *rfb*
are also located *manB*, formerly *rfbK* (phospho-
mannomutase), and *wbaN*, formerly *rfbN* (rham-
nosyl transferase). However, the genes, which are
also involved in other syntheses (*galE* for UDP-
galactose-4-epimerase and *pmi* for phosphoman-
nose isomerase), are located elsewhere on the
chromosomal map (Rick, 1987; Jiang *et al.*,
1991).

The oligosaccharide monomer is built by
sequential transfer of galactose-1-phosphate,
rhamnose, mannose and abequose moieties from
UDP-galactose, dTDP-L-rhamnose, GDP-man-
nose and CDP-abequose, on a lipid carrier, undecaprenyl phosphate.

Oligosaccharide monomers are polymerized
(action of gene *wzy*, formerly *rfc*, located in the
vicinity of *trp* at 32 minutes) and then transferred from undecaprenyl phosphate to the independently synthesized core. The lipopolysaccharide is then translocated from the inner
membrane to the surface of the outer membrane
(Rick, 1987).

Mutations in regions *rfa* and *rfb* cause rough
phenotypes, whereas a mutation in *wzy* (*rfc*) prevents the monomer polymerization (semi-rough
mutants). When abequose is acetylated (effect of
gene *oafA* located at 46 minutes on the genetic
map), factor O4 becomes O5. Gene *oafR* causes
the α1–4 branching of a glycosyl residue on
galactose, thus yielding factor 12_2.

Converting phages can modify O factor
structure. Phage P22 changes the 1–4 link
between glucose and galactose into a 1–6 link,
thus yielding factor O1. Phage Φ27 changes the
1–2 link between monomers into a 1–6 link, thus
yielding factor O27. Phages ε15 and ε34 alter
several O factors in group E1 (Rick, 1987).
Plasmids can also change O factors.

A 7.5 kb plasmid has been found to determine factor O54 (Popoff and Le Minor, 1985).
The lipopolyoside also carries receptors for phage
binding (Ackermann and Dubow, 1987).

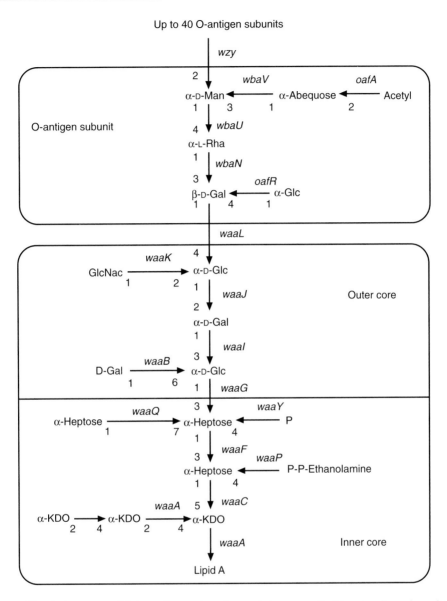

Fig. 1.1. Chemical structure of *Salmonella enterica* subsp. *enterica* serovar Typhimurium lipopolysaccharide (adapted from Raetz, 1996, and the Bacterial Polysaccharide Gene Database). Gal, galactose; Glc, glucose; GlcNac, *N*-acetyl-D-glucosamine; Heptose, L-glycero-D-manno-heptose; KDO, 3-deoxy-D-manno-octulosonic acid; Man, mannose; P, phosphate; P-P-Ethanolamine, ethanolamine pyrophosphate; Rha, rhamnose. Genes are indicated in italics following the new nomenclature (Reeves *et al.*, 1996).

H antigens

H antigens are carried by flagella. These are composed of protein subunits called flagellin. H antigens are typically diphasic in *Salmonella*. The availability of two genetic systems (genes distantly located on the chromosome) expressing different flagellins could help the organism to survive the host's defences (Macnab, 1987). The genes coding for the two sorts of flagellin are somewhat similar although not identical, thus suggesting that they may have resulted from the

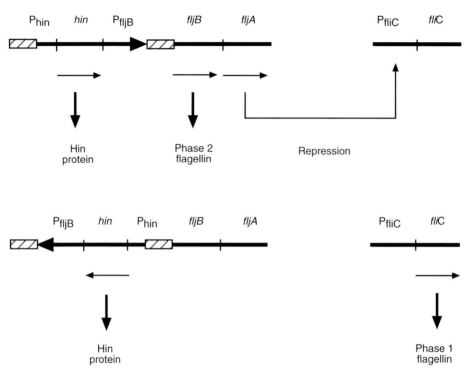

Fig. 1.2. Flagellar phase inversion system in *Salmonella*. Promoters are indicated with capital P. Hatched zones indicate inverted repeats. Arrows indicate transcription directions.

duplication of an ancestral gene. In a random fashion and after 1000–10,000 generations, the formerly silent gene is expressed and the expression of the other gene is turned off (Jones and Aizawa, 1991). Gene *fliC* (formerly *H1*) is repressed by the product of gene *fljA* (formerly *rh1*) which is part of operon *flj* (Fig. 1.2). The expression of operon *flj* prevents expression of gene *fliC*. The alternating gene expression is due to the inversion of a 750 nucleotide-pair DNA fragment located upstream of gene *fljB* and which includes the promoter of *fljB*. In a given orientation, the promoter can initiate transcription; in the other orientation, it cannot. The inversion region is flanked by inverted repeats, enabling homologous recombination. It also contains the promoter and the coding sequence of the gene *hin* (formerly *vh2*), whose product is necessary for the inversion process. It is noteworthy that the protein Hin (product of the gene *hin*) resembles protein TnpR of transposon Tn3. The phase inversion system could have evolved from a transposon occurring in *Salmonella* (Simon *et al.*, 1980).

A non-motile *Salmonella* can have structural genes *fliC* and *fljB* (and possibly have a defect in flagella assembly) and a monophasic *Salmonella* can be defective in phase inversion and still have the genes corresponding to both phases (Jones and Aizawa, 1991). This should be considered when nucleic acid probes are devised for the identification of major serovars. Confirming the results of an early transduction experiment (Lederberg and Edwards, 1953), Kilger and Grimont (1993) found that a *fliC* gene is present in the non-motile serovar Gallinarum.

There are more than 50 different alleles of gene *fliC* and more than 30 of gene *fljB*. To understand the molecular bases of such a diversity, some of these genes have been sequenced (Frankel *et al.*, 1989a,b; Smith and Selander, 1990, 1991; Smith *et al.*, 1990). Gene *fliC* comprises three parts: a 5' part containing 300 nucleotide pairs, a 3' part of 200 nucleotide pairs and a middle part of 350 nucleotide pairs. It is remarkable that 5' and 3' distal parts have been largely conserved in their sequences across the diverse serovars. The sequence of the middle part is hypervariable, with less than 32% homology among serovars. The carboxy and amino distal parts of flagellin are essential for flagellin polymerization and secretion (hence the conserved sequences), whereas the middle part, which has no functional role, carries the major epitope of the H antigen. When comparing the global bacterial genotype (studied by MLEE), it appeared that strains with very similar global genotypes could have very different *fliC* genes. Strains with very different global genotypes could have identical *fliC* genes (Smith and Selander, 1991). This strongly suggests that flagellin gene exchange has occurred among strains (possibly by transduction). Interstrain exchange and recombination are major evolutionary mechanisms, generating both allelic variation in *fliC* and serovar diversity in natural populations of the *Salmonella* (Selander *et al.*, 1990a,b).

Atypical strains of serovar Typhi have been isolated from Indonesia (Guinée *et al.*, 1981). Strains of serovar Typhi are normally monophasic with H:d. Some strains from Indonesia have a factor H:j in place of H:d and others are diphasic with H:z66. The following flagellar formulae are observed, H:d:–, or j:–, or d:z66, or j:z66. In fact, gene *fliC*, which codes for factor j, is identical to that which codes for factor d except for a deletion of 261 nucleotide pairs, which occurs in the middle part of the gene (Frankel *et al.*, 1989a,b). Factor z66 is coded by gene *fljB*. The current hypothesis is that serovar Typhi, originally diphasic, appeared first in Indonesia (when humans had appeared as a species), where mutations or deletions would have occurred, thus yielding monophasic variants or H:j variants. Serovar Typhi would have spread throughout the world from a monophasic H:d clone (Frankel *et al.*, 1989a,b).

The *Salmonella* flagella can also carry a receptor for flagellotropic phage χ binding (Ackermann and Dubow, 1987).

Vi antigen

Vi antigen is a linear homopolymer of 2-acetamido-2-deoxy-D-galacturonic acid linked by α(1–4) bonds. This capsular polysaccharide is found in serovars Typhi, Paratyphi C and Dublin. Three loci (*viaA*, *viaB* and *ompB*) are involved in

the genetic control of Vi production. Loci *viaA* and *ompB* are not limited to Vi producing species and genera and are involved in the regulation of Vi synthesis. Locus *viaB* is found only in strains able to produce Vi antigen. It contains genes coding for proteins involved in polysaccharide synthesis (*wcdABCD*), polysaccharide exportation (*wza, wzm, wzt, wzf*) and anchoring to the bacterial surface (*wcdE*). A nucleic acid probe targeting *viaB* has been proposed for the detection of serovar Typhi (Rubin *et al.*, 1985). Such a probe also reacted with occasional other serovars which are able to produce Vi antigen. The Vi antigen is the receptor of phages ViI, ViII and ViIV (Craigie and Yen, 1938a).

Epidemiological Typing

Serotyping

The White–Kauffmann–Le Minor (WKL) (Popoff *et al.*, 1998) scheme is a practical summary of the antigenic structure of *Salmonella* serovars. O antigenic factors (numerals) which are easily modified by mutation are indicated in brackets and those that are determined by bacteriophages or plasmids are underlined. In 1998, 2449 serovars were listed in the WKL scheme, 1443 in *S. enterica* and 20 in *S. bongori*. Within *S. enterica*, 1443 serovars belonged to subspecies *enterica* and were given names, 488 corresponded to subspecies *salamae*, 94 to subspecies *arizonae*, 323 to subspecies *diarizonae*, 70 to subspecies *houtenae* and 11 to subspecies *indica* (Popoff *et al.*, 1998). However, only a limited number of serovars are encountered in practice. In the French National Reference Centre for *Salmonella* and *Shigella*, 19,174 strains isolated from humans were distributed into 194 serovars. Fifteen serovars represented 91% of the strains and serovars Typhimurium and Enteritidis represented 69% of all *Salmonella* isolates from human sources (P.A.D. Grimont and P. Bouvet, unpublished data).

Salmonella isolates should be sent to a national reference centre. In some countries, laboratories are requested to keep a minimum set of sera (anti-O4, 5 – O6, 7, 8 – O9), enabling the agglutination of 90% of isolates from human sources. Anti-Hi–Hb–Hd–HG (mixture) and Hr sera are also useful.

Phage typing

Phage typing evaluates the susceptibility (or resistance) of isolates to a set of selected bacteriophages. Most bacteriophages were wild phages, which were isolated from sewage. Bacteriophage can also be produced by lysogenic bacteria.

International phage-typing schemes
The most elaborate system had been described by Craigie and Yen (1938a,b) and Craigie and Felix (1947) for typing serovar Typhi. This system uses 87 variants derived from a single phage strain (ViII). These variants were 'adapted' by passage on different strains and thus submitted to different host restriction and modification systems. This set of phages allowed the subdivision of serovar Typhi into 110 phage types. Since the phage receptor is antigen Vi, isolates must produce Vi to be susceptible to any phage in the set. The specificity of typing phages is partly governed by the presence of prophages. This system has been standardized and, to avoid host-range drift, laboratories should not multiply these phages for production but rather request phage suspensions from the Central Public Health Laboratory, Colindale, UK. Unfortunately, 70% of isolates are distributed into only three phage types (A, E1 and C1). Vi-negative isolates (2.5%) and 'degraded Vi strains' (DVS) (9%) are untypable. A complementary set composed of 13 unadapted Vi phages can subdivide phage type A into ten phage subtypes (Nicolle *et al.*, 1954, 1958). Type A subtype Tananarive, which is not lysogenic, is considered as the precursor of all other phage types of *Salmonella* serovar Typhi.

The international scheme for serovar Paratyphi B uses a set of 12 adapted and unadapted phages and contains 48 phage types (Felix and Callow, 1943, 1951). Serovar Paratyphi B is split into biotype Paratyphi B (d-tartrate negative, associated with paratyphoid) and biotype Java (d-tartrate positive, associated with diarrhoea). Unfortunately, phage typing cannot differentiate these biotypes (Vieu *et al.*, 1988). In France, the most frequent phage types are 1var3 (35%) and 1010 (24%) among biotypes Paratyphi B and Java (F. Grimont, unpublished data).

The Paratyphi A scheme uses six phages and contains six phage types (Banker, 1955; Anderson, 1964). Isolates encountered in France (imported cases) often correspond to phage types 1 (56% of isolates) and 2 (18% of isolates).

Other phage-typing systems
The Colindale scheme for serovar Typhimurium uses 37 phages and contains more than 210 phage types (Felix and Callow, 1943; Callow, 1959; Anderson and Wilson, 1961; Anderson, 1964; Anderson *et al.*, 1977). Phage type DT104 shows multiple resistance to antibiotics in different European countries and is considered more fully in Chapter 6. Two other phage-typing systems have been published (Guinée and van Leeuwen, 1978).

Several phage-typing schemes have been published for serovar Enteritidis. The most widely used is that of Ward *et al.* (1987), which now differentiates 65 phage types by use of 16 phages. Phage type (PT) 4 has been associated with a pandemic in Europe and elsewhere. The scheme of Vieu *et al.* (1990) was first designed for serovar Dublin and 14 phages delineated 101 phage types. When applied to serovar Enteritidis, 85 phage types were differentiated. In this scheme, phage types correspond to either serovar Dublin or serovar Enteritidis, never to both. Enteritidis phage type 33 corresponds to PT4 of Ward *et al.* (1987). There might be a relationship between phage type and insertion position of transposon IS200 (Stanley *et al.*, 1991).

Other phage-typing schemes have been proposed for serovars Adelaide, Anatum, Bareilly, Blockley, Braenderup, Bovismorbificans, Gallinarum, Newport, Panama and Weltevreden. These have been reviewed (Guinée and van Leeuwen, 1978). More recently described schemes were for serovars Bareilly (Jayasheela *et al.*, 1987), Hadar (De Sa *et al.*, 1980), Infantis (Kasatiya *et al.*, 1978), Montevideo (Vieu *et al.*, 1981) and Virchow (Chambers *et al.*, 1987). Although *Salmonella* phage-typing is cheap and requires no expensive equipment, the method should be in the hands of well-trained personnel. This means that it is generally limited to reference laboratories.

Bacteriocin typing

Bacteriocin typing of serovar Typhimurium has been described (Barker, 1980). However, it is seldom used.

Biotyping

Biotypes have been described in serovar Typhimurium (Cordano *et al.*, 1971; Pohl *et al.*, 1973; Duguid *et al.*, 1975; Descamp *et al.*, 1982). However, the choice of tests for biotyping has often been empirical.

Utilization of d-tartrate (= L-tartrate) is used to separate two biotypes in serovar Paratyphi B. Biotype Paratyphi B cannot utilize d-tartrate whereas biotype Java can. Biotype Java is commonly associated with diarrhoea and isolated from stools, whereas biotype Paratyphi B is often associated with paratyphoid and isolated from blood or stools.

Plasmid profiling and plasmid restriction profiling

It is now easy to extract plasmid DNA and to separate plasmids of different sizes by agarose gel electrophoresis and ethidium bromide staining. Isolates derived from the same epidemic strain will have plasmids with identical sizes. However, this test requires the presence of at least one plasmid type. Some phage types in serovars Enteritidis (Threlfall *et al.*, 1989) and Typhimurium (Threlfall *et al.*, 1990) were subdivided by plasmid profiling.

More precise results are obtained when plasmid DNA is extracted, digested by a restriction endonuclease and the fragments separated by agarose gel electrophoresis. Identical plasmids should have the same restriction pattern (Tacket, 1989).

rRNA gene restriction patterns (ribotyping)

When bacterial DNA is extracted, purified, digested by a restriction endonuclease and the fragments separated by agarose gel electrophoresis, fragments in restriction patterns are too numerous for these patterns to be compared. Instead, the fragments in the gel are transferred to a nylon membrane, to maintain their relative positions, and hybridized with a labelled mixture of 16+23S rRNA. Revealing the label (autoradiography or immunoenzymatic reaction) yields simpler patterns, often referred to as ribotypes (Grimont and Grimont, 1986).

Ribotyping has uncovered a wide heterogeneity within serovar Typhi (Altweg *et al.*, 1989). A diversity of ribotypes has been found in the most frequent phage types, A and E1. Such diversity is in contrast with the homogeneity of MLEE (Reeves *et al.*, 1989) and could be explained by the presence of 'intervening sequences' inserted in 23S rRNA genes (Burgin *et al.*, 1990).

Ribotyping of other serovars gives very different results. Some serovars have almost a serovar-specific ribotype. However, for serovars of group B (e.g. Typhimurium, Paratyphi B), there is no relationship between ribotype and serovar (F. Grimont, unpublished data).

Pulse-field gel electrophoresis (PFGE)

PFGE uses restriction endonucleases, which have infrequently occurring restriction sites in a given bacterial DNA. A small number of fragments of a much larger size are produced. Conventional agarose gel electrophoresis cannot resolve DNA molecules larger than ~20 kb pairs. Thus, to separate large DNA molecules above 50 kbp, PFGE which allows resolution of DNA molecules millions of base pairs long is used. The most sophisticated configuration for this technique is called clamped homogeneous electric field (CHEF) electrophoresis, which uses an array of hexagonally arranged electrodes to generate uniform electric fields at an angle of 120°C to each other, thus ensuring that large DNA fragments migrate through the gel in a straight line. Although PFGE is highly discriminative for *Salmonella* with the endonucleases XbaI, BlnI or SpeI, it is expensive and time-consuming and standardization, analysis and comparison of restriction profiles require effort (Murase *et al.*, 1995).

IS200 typing

DNA of most *Salmonella* serovars contain several copies of a 708-base-pair insertion sequence, IS200. Stanley *et al.* (1991) found that isolates could be differentiated by comparing the restriction patterns of bacterial DNA after hybridization with an IS200 probe. Strains differed by the number of visualized fragments (IS200 number of copies) and the size of fragments.

Random amplification of polymorphic DNA (RAPD)

RAPD is a rapid genomic typing method of broad application. This technique uses the polymerase chain reaction (PCR), in which a single arbitrarily chosen primer (usually 10-mer) can be annealed at multiple sites throughout the genome. The profiles of amplified products are characteristic of the template DNA. An RAPD method has been developed to differentiate *S. enteritidis* isolates (Lin *et al.*, 1996), providing more discrimination than any other subtyping method. This method proved to discriminate between isolates of different *Salmonella* serovars (Hilton *et al.*, 1996). However, between-laboratory reproducibility is a problem (Meunier and Grimont, 1993).

ERIC-PCR

A repetitive element, highly conserved, called the enterobacterial repetitive intergenic consensus (ERIC) sequence, about 126 bp in length, was described by Hulton *et al.* (1991). The chromosome locations of ERIC sequences can differ in different species and strains. This element has been identified in *Salmonella* and can be used for typing.

Restriction of amplified flagellin genes

In an attempt to substitute molecular methods for serotyping, flagellin genes from 264 serovars were amplified by two phase-specific PCR systems (Dauga *et al.*, 1998). Amplification products were cleaved by endonucleases *Hha*I or *Hph*I. Restriction of 329 (195 phase 1 plus 134 phase 2) flagellin genes coding for 26 antigens yielded 64 *Hha*I profiles and 42 *Hph*I profiles. The phase 1 gene (*fli*C) showed 46 patterns with *Hha*I and 30 patterns with *Hph*I. The phase 2 gene (*flj*B) showed 23 patterns with *Hha*I and 17 patterns with *Hph*I. When the data from both enzymes were combined, 116 patterns were obtained: 74 for *fli*C, 47 for *flj*B and five shared by both genes. Of these combined patterns, 80% were specifically associated with one flagellar antigen and 20% were associated with more than one antigen. Each flagellar antigen was divided into two to 18 different combined patterns. The pattern corresponding to serovar Typhi H:d was specific and different from other H:d patterns. The pattern corresponding to serovar Typhimurium H:i was also unique and differed from other H:i patterns. Overall, the diversity uncovered by restriction of flagellin genes did not precisely match that evidenced by flagellar agglutination.

New Approaches for the Detection of *Salmonella*

The procedure of the colorimetric Gene-Trak *Salmonella* assay used nucleic acid probes to detect polynucleotide sequences that are uniquely conserved among *Salmonella* bacteria. This method was considered equivalent to the conventional *Bacteriological Analytical Manual*/AOAC culture method (Foster *et al.*, 1992).

PCR will amplify DNA molecules 1000-fold, but the presence of a specifically amplified product must be identified. Several PCR primers for specific detection of *Salmonella* have been published utilizing specific gene sequences for targeting. Kwang *et al.* (1996) described the usefulness of a primer set of oligonucleotides from the *omp*C gene of *Salmonella*. This primer set successfully amplified 40 *Salmonella* serovars, but not 24 non-*Salmonella* bacteria. The sensitivity for boiled whole bacteria was 400 cells.

The magnetic immuno-PCR assay utilizes magnetic particles coated with monoclonal antibodies to extract bacteria from a sample. Fluit *et al.* (1993) applied this method to *Salmonella* and detected the presence of 0.1 colony-forming units (cfu) of *Salmonella* g^{-1} of chicken meat after enrichment.

An alternative method for analysing PCR products utilizes the oligonucleotide ligation assay (OLA). The OLA procedure used two adjacent oligonucleotides. The first one (capture probe) is 5′-biotinylated with the 3′ end adjacent to the second probe. The second (reporter) probe is 5′-phosphorylated and 3′-end-labelled with a reporter substance, such as digoxigenin. If the two oligonucleotides are hybridized to the target DNA, DNA ligase covalently joins the two oligonucleotides. The capture of the biotinylated probe is accomplished by binding the biotin to immobilized streptavidin in a micro-titration plate. Stone *et al.* (1995) described a combined cultivation and PCR-hybridization procedure enzyme-linked immunosorbent assay (ELISA)-based oligonucleotide ligation assay, adapted to a 96-well microtitration PCR-OLA format for detection of the product from the *invE* and *invA* genes of *Salmonella* serovars. Sensitivity, specificity and predictive value were comparable to the conventional culture.

A digoxigenin-based ELISA (DIG-ELISA) following a PCR to detect the amplified lipopolysaccharide *rfbS* gene as a means for rapid screening of serogroup D *Salmonella* in stool specimens was described by Luk *et al.* (1997). In the presence of stool materials, the *Salmonella* were isolated by an immunomagnetic separation technique with an O9-specific monoclonal antibody, followed by PCR and DIG-ELISA. The sensitivity was 10–100 bacteria.

A recent technique for the separation/concentration of bacteria from background food is the immunomagnetic separation (IMS) procedure, which uses a novel biosorbent consisting of a *Salmonella*-specific bacteriophage immobilized on to a solid phase (Bennett *et al.*, 1997). This biosorbent could remove *Salmonella* from culture fluid and separate *Salmonella* from suspensions of other *Enterobacteriaceae*. The advantage is the ease of production of phage, the high affinity of phage–cell interaction and the ability of phage to infect host cells.

Chen *et al.* (1997) have developed a rapid, sensitive and automated fluorescence PCR method, the AG-9600 AmpliSensor assay, for the detection of *Salmonella* species in food samples. The AmpliSensor assay comprises two steps: an initial asymmetric amplification with normal primers to overproduce one strand of the target and subsequent semi-nested amplification and signal detection, in which one of the outer primers and the AmpliSensor primer direct the amplification. The AmpliSensor primer is a double-stranded signal probe; one strand is labelled with fluorescein isothiocyanate and the other with Texas red. During semi-nested amplification, one strand of the AmpliSensor duplex serves as a primer and the other as an 'energy sink'. The semi-nested amplification results in strand dissociation of the duplex and disruption of the fluorescence signal. The extent of signal disruption is proportional to the amount of the primer incorporation into the amplification product and can be measured cycle by cycle and used for quantification of the initial target.

Cocolin *et al.* (1998) described a primer set of oligonucleotides from the *invA* gene of *Salmonella* which amplified 33 *Salmonella* serovars but did not amplify 16 non-*Salmonella* bacteria. Moreover, after PCR amplification, it was possible to identify serovar Typhimurium by *Hin*fI restriction enzyme analysis.

A *Salmonella*-specific PCR system targeting a virulence gene with hybridization to a covalently immobilized oligonucleotide probe on a microplate (Chevrier *et al.*, 1995) is now marketed under the name Probelia.

Another PCR system, involving electrophoresis, has been devised for *Salmonella* and is marketed under the name BAX (Bailey, 1998).

References

Achard, C. and Bensaude, R. (1896) Infections paratyphoïdiques. *Bulletin et Mémoires de la Société de Médecine des Hôpitaux de Paris* 13, 820–833.

Ackermann, H.-W. and Dubow, M.S. (1987) *Viruses of Prokaryotes*, Vol. 1. CRC Press, Boca Raton, Florida, 202 pp.

Altwegg, M., Hickman-Brenner, F.W. and Farmer, J.J., III (1989) Ribosomal RNA gene restriction patterns provide increased sensitivity for typing *Salmonella typhi* strains. *Journal of Infectious Diseases* 160, 145–149.

Anderson, E.S. (1964) Phage typing of *Salmonella* other than *Salmonella typhi*. In: van Oye, W. (ed.) *The World Problem of Salmonellosis*. Junk Publishers, The Hague, pp. 89–110.

Anderson, E.S. and Wilson, E.M.J. (1961) Die Bedeutung der *Salmonella typhimurium*-Phagen-typisierung in der Human- und Veterinarmedizin. *Zentralblatt für Bakteriologie, Mikrobiologie und Hygiene Series I. Abteilung Orig.* 224, 368–373.

Anderson, E.S., Ward, L.R., De Saxe, M.J. and De Sa, J.D.H. (1977) Bacteriophage-typing designations of *Salmonella typhimurium*. *Journal of Hygiene, Cambridge* 78, 297–300.

Bailey, J.S. (1998) Detection of *Salmonella* cells within 24 to 26 hours in poultry samples with the polymerase chain reaction BAX^tm. *Journal of Food Protection* 61, 792–795.

Banker, D.D. (1955) Paratyphoid A phage typing. *Nature* 175, 309–310.

Barker, R.M. (1980) Colicinogeny in *Salmonella typhimurium*. *Journal of General Microbiology* 120, 21–26.

Barker, R.M., Kearney, G.M., Nicholson, P., Blair A.L., Porter, R.C. and Crichton, P.B. (1988) Types of *Salmonella paratyphi B* and their phylogenetic significance. *Journal of Medical Microbiology* 26, 285–293.

Beltran, P., Musser, J.M, Helmuth, R., Farmer, J.J., III, Frerichs, W.M., Wachsmuth, I.K., Ferris, K., McWhorter, A.C., Wells, J.G., Cravioto, A. and Selander, R.K. (1988) Toward a population genetic analysis of *Salmonella*: genetic diversity and relationships among strains of serovars *S. choleraesuis*, *S. derby*, *S. dublin*, *S. enteritidis*, *S. heidelberg*, *S. infantis*, *S. newport*, and *S. typhimurium*. *Proceedings of the National Academy of Sciences USA* 85, 7753–7757.

Beltran, P., Plock, S.A., Smith, N.H., Whittam, T.S., Old, D.C. and Selander, R.K. (1991) Reference collection of strains of the *Salmonella typhimurium* complex from natural populations. *Journal of General Microbiology* 137, 601–606.

Bennett, A.R., Davids, F.G.C, Vlahodimou, S., Banks, J.G. and Betts, R.P. (1997) The use of bacteriophage-based systems for the separation and concentration of *Salmonella*. *Journal of Applied Microbiology* 83, 259–265.

Borman, E.K., Stuart, C.A. and Wheeler, K. (1944) Taxonomy of the family *Enterobacteriaceae*. *Journal of Bacteriology* 48, 351–367.

Burgin, A.B., Parodos, K., Lane, D.J., Pace, N.R. (1990) The excision of intervening sequences from *Salmonella* 23S ribosomal RNA. *Cell* 60, 405–414.

Callow, B.R. (1959) A new phage typing scheme for *Salmonella typhimurium*. *Journal of Hygiene, Cambridge* 57, 346–559.

Chambers, R.M., McAdam, P., De Sa, J.D.H., Ward, L.R. and Rowe, B. (1987) A phage typing scheme for *Salmonella virchow*. *FEMS Microbiology Letters* 40, 155–157.

Chen, S., Yee, A., Griffiths, M., Wu, K.Y., Wang, C.-N., Rahn, K. and De Grandis, S.A. (1997) A rapid, sensitive and automated method for the detection *Salmonella* species in foods using AG-9600 AmpliSensor Analyser. *Journal of Applied Microbiology* 83, 314–321.

Chevrier, D., Popoff, M.Y., Dion, M.Y., Hermant, D. and Guesdon, J.L. (1995) Rapid detection of *Salmonella* subspecies I by PCR combined with non-radioactive hybridisation using covalently immobilised oligonucleotide on a microplate. *FEMS Microbiology Letters* 10, 245–252.

Christensen, H., Nordentoft, S. and Olsen, J.E. (1998) Phylogenetic relationships of *Salmonella* based on rRNA sequences. *International Journal of Systematic Bacteriology* 48, 605–610.

Cocolin, L., Manzano, M., Canton, C. and Comi, G. (1998) Use of polymerase chain reaction and restriction enzyme analysis to directly detect and identify *Salmonella typhimurium* in food. *Letters in Applied Microbiology* 85, 673–677.

Cordano, A.M., Richard, C. and Vieu, J.-F. (1971) Biotypes de *Salmonella typhimurium*. Enquête sur 513 souches isolées en France en 1969–1970. *Annales de l'Institut Pasteur* 121, 473–478.

Craigie, J. and Felix, A. (1947) Typing of typhoid bacilli with Vi bacteriophages. *Lancet* i, 823–827.

Craigie, J. and Yen, C.H. (1938a) The demonstration of types of *B. typhosus* by means of preparations of type II Vi phage. I. Principles and technique. *Canadian Public Health Journal* 29, 448–484.

Craigie, J. and Yen C.H. (1938b) The demonstration of types of *B. typhosus* by means of preparations of type II Vi phage. II. The stability and epidemiological significance of V form types of *B. typhosus*. *Canadian Public Health Journal* 29, 484–496.

Crosa, J.H., Brenner, D.J., Ewing, W. H. and Falkow, S. (1973) Molecular relationships among the salmonellae. *Journal of Bacteriology* 115, 307–315.

Dauga, C., Zabrovskaia, A. and Grimont, P.A.D. (1998) Restriction fragment length polymorphism analysis of some flagellin genes of *Salmonella enterica*. *Journal of Clinical Microbiology* 36, 2835–2843.

De Sa, J.D.H., Ward, L.R. and Rowe, B. (1980) A scheme for the phage typing of *Salmonella hadar*. *FEMS Microbiology Letters* 9, 175–177.

Descamp, P., Veron, M., Le Minor, S. and Bussière, J. (1982) Phénotypes et marqueurs épidémiologiques de *Salmonella typhimurium*. *Revue d'Epidémiologie et de Santé Publique* 30, 423–435.

Duguid, J.P., Anderson, E.S. and Alfresson, G.A. (1975) A new biotyping scheme for *Salmonella typhimurium* and its phylogenetic significance. *Journal of Medical Microbiology* 8, 149–166.

Ewing, W.H. (1963) An outline of nomenclature for the family *Enterobacteriaceae*. *International Bulletin of Bacteriological Nomenclature and Taxonomy* 13, 95–110.

Ewing, W.H. (1969) *Arizona hinshawii* com. nov. *International Journal of Systematic Bacteriology* 19, 1.

Felix, A. and Callow, B.R. (1943) Typing of paratyphoid B bacilli by means of Vi bacteriophage. *British Medical Journal* ii, 127–130.

Felix, A. and Callow, B.R. (1951) Paratyphoid B. Vi phage-typing. *Lancet* ii, 10–14.

Fluit A.D., Widjojoatmodjo, M.N., Box, A.T.A., Torensma, R. and Verhoef, J. (1993) Rapid detection of salmonellae in poultry with the magnetic immuno-polymerase chain reaction assay. *Applied and Environmental Microbiology* 59, 1342–1346.

Foster, K., Garramone, S., Ferraro, K. and Groody, E.P. (1992) Modified colorimetric DNA hybridization method and conventional culture method for detection of *Salmonella* in foods: comparison of methods. *Journal of AOAC International* 75, 685–692.

Frankel, G., Newton, S.M.C., Schoolnik, G.K. and Stocker, B.A.D. (1989a) Intragenic recombination in a flagellin gene: characterization of the H1-j gene of *Salmonella typhi*. *EMBO Journal* 8, 3149–3152.

Frankel, G., Newton, S.M.C., Schoolnik G.K. and Stocker, B.A.D. (1989b) Unique sequences in region VI of the flagellin gene of *Salmonella typhi*. *Molecular Microbiology* 3, 1379–1383.

Grimont, F. and Grimont, P.A.D. (1986) Ribosomal ribonucleic acid gene restriction patterns as potential taxonomic tools. *Annales de l'Institut Pasteur/Microbiologie* 137B, 165–175.

Gruber, M. and Durham, H.E. (1896) Eine neue Methode zur raschen Erkennung des choleras vibrio und des Typhusbacillus. *Münschauer Medizinische Wochenschrift* 43, 285–286.

Grunbaum, A.S. (1896) Preliminary note on the use of the agglutinative action of human action for the diagnostic of enteric fever. *Lancet* 16, 806–807.

Guinée, P.A.M. and Valkenburg, J. (1968) Diagnostic value of a *Hafnia* specific bacteriophage. *Journal of Bacteriology* 96, 564.

Guinée, P.A.M. and van Leeuwen, W.J. (1978) Phage typing of *Salmonella*. In: Bergan, T. and Norris, J.R. (eds) *Methods in Microbiology*, Vol. 2. Academic Press, New York, pp. 157–196.

Guinée, P.A.M., Jansen, W.H., Maas, H.M.E., Le Minor, L. and Beaud, R. (1981) An unusual H antigen (z66) in strains of *Salmonella typhi*. *Annales de Microbiologie* 132, 331–334.

Hilton, A.C., Banks, J.G. and Penn, C.W. (1996) Random amplification of polymorphic DNA (RAPD) of *Salmonella*: strain differentiation and characterization of amplified sequences. *Journal of Applied Bacteriology* 81, 575–584.

Hulton, C.S., Higgins, C.F. and Sharp, P.M. (1991) ERIC sequences: a novel family of repetitive elements in the genomes of *Escherichia coli, Salmonella typhimurium* and other enterobacteria. *Molecular Microbiology* 5, 825–834.

Jayasheela, M., Singh, G,. Sharma, N.C. and Saxena, S.N. (1987) A new scheme for phage typing *Salmonella bareilly* and characterization of typing phages. *Journal of Applied Bacteriology* 62, 429–432.

Jiang, X.-M., Neal, B., Santiago, F., Lee, S.J., Romana, L.K. and Reeves, P.R. (1991) Structure and sequence of the *rfb* (O antigen) gene cluster of *Salmonella* serovar *typhimurium* (strain LT2). *Molecular Microbiology* 5, 695–713.

Jones, C. and Aizawa, S.-I. (1991) The bacterial flagellum and flagellar motor: structure, assembly and function. *Advances in Microbiology Physiology* 32, 110–172.

Kasatiya, S., Caprioli, T. and Champoux, S. (1978) Bacteriophage typing scheme for *Salmonella infantis*. *Journal of Clinical Microbiology* 10, 637–640.

Kauffmann, F. (1941) Über mehrere neue *Salmonella* Typen. *Acta Pathologica Microbiologica Scandinavica* 18, 351–366.

Kauffmann, F. (1961) The species definition in the *Enterobacteriaceae*. *International Bulletin of Bacteriological Nomenclature and Taxonomy* 11, 5–6.

Kauffmann, F. (1966a) *The Bacteriology of* Enterobacteriaceae. Munksgaard, Copenhagen.

Kauffmann, F. (1966b) *Das* Salmonella *sub-genus IV*. *Annales Immunologiae Hungaricae* 9, 77–80.

Kauffmann, F. (1978) *Das Fundament*. Munksgaard, Copenhagen.

Kauffmann, F. and Edwards, P.R. (1952) Classification and nomenclature of *Enterobacteriaceae*. *International Bulletin of Bacteriological Nomenclature and Taxonomy* 2, 2–8.

Kilger, G. and Grimont, P.A.D. (1993) Differentiation of *Salmonella* phase 1 flagellar antigen types by restriction of the amplified *fliC* gene. *Journal of Clinical Microbiology* 31, 1108–1110.

Kwang, J., Littledike, E.T. and Keen, J.E. (1996) Use of the polymerase chain reaction for *Salmonella* detection. *Letters in Applied Microbiology* 22, 46–51.

Lane, D.J. and Collins, M.L. (1991) Current methods for detection of DNA/ribosomal RNA hybrids. In: Vaheri, A., Tilton, R.C. and Balows, A. (eds.) *Rapid Methods and Automation in Microbiology and Immunology*. Springer-Verlag, Berlin, pp. 55–75.

Lederberg, J. and Edwards, P.R. (1953) Serotypic recombination in *Salmonella*. *Journal of Immunology* 71, 232–240.

Le Minor, L. (1994) The genus *Salmonella* In: Ballows, A., Trüper, H.G., Dworkin, M., Harber, W. and Scheiffer, K.-H. (eds) *The Prokaryotes*. Springer, New York, pp. 2760–2774.

Le Minor, L. and Popoff, M.Y. (1987) Request for an opinion. Designation of *Salmonella enterica* sp. nov., nom. rev., as the type and only species of the genus *Salmonella*. *International Journal of Systematic Bacteriology* 37, 465–468.

Le Minor, L., Rohde, R. and Taylor, J. (1970) Nomenclature des *Salmonella. Annales de l'Institut Pasteur* 119, 206–210.

Le Minor, L., Veron, M. and Popoff, M.Y. (1982) Taxonomie des *Salmonella. Annales de Microbiologie* 133B, 223–243.

Le Minor, L., Popoff, M.Y., Laurent, B. and Hermant, D. (1986) Individualisation d'une septième sous-espèce de *Salmonella*: *S. choleraesuis* subsp. *indica* subsp. nov. *Annales de l'Institut Pasteur/Microbiologie* 137B, 211–217.

Lin, A.W., Usera, M.A., Barrett, T.J. and Goldsby, R.A. (1996) Application of random amplified polymorphic DNA analysis to differentiate strains of *Salmonella enteritidis. Journal of Clinical Microbiology* 34, 870–876.

Luk, J.M., Kongmuang, U., Tsang, R.S.W. and Lindberg, A.A. (1997) An enzyme-linked immunosorbent assay to detect PCR products of the rfbS gene from serogroup D salmonellae: a rapid screening prototype. *Journal of Clinical Microbiology* 35, 714–718.

Macnab, R.M. (1987) Flagella. In: Neidhardt, F.C., Ingraham, J.L., Low, K.B., Magasanik, B., Schaechter, M. and Umberger, H.E. (eds.) Escherichia coli *and* Salmonella typhimurium: *Cellular and Molecular Biology*, Vol. I. American Society for Microbiology, Washington, DC, pp. 70–83.

Meunier, J-R. and Grimont, P.A.D. (1993) Factors affecting reproducibility of random amplified polymorphic DNA fingerprinting. *Research in Microbiology* 144, 373–379.

Murase, T., Okitsu, T., Suzuki, R., Morozumi, H., Matsushima, A., Nakamura, A. and Yamai, S. (1995) Evaluation of DNA fingerprinting by PFGE as an epidemiological tool for *Salmonella* infections. *Microbiology and Immunology* 39, 673–676.

Nicolle, P., Pavlatou M. and Diverneau, G. (1954) Les lysotypies auxiliaires de *Salmonella typhi*. I. Subdivision du type A et du groupe I + IV par une nouvelle série de phages. *Annales de l'Institut Pasteur* 87, 493–509.

Nicolle, P., Diverneau, G. and Brault, J. (1958) Relation entre les divers états lysogènes du type A de *Salmonella typhi* et ses sous-types tels qu'ils sont mis en évidence par une lysotypie complémentaire. *Bulletin of the Research Council of Israel* 7E, 89–100.

Pfeiffer, R. and Kolle, W. (1896) Zur Differentialdiagnose des Typhus-bacillus vermittels Serum der gegen Typhus immunisierten Thiere. *Deutsche Medizinische Wochenschrift* 22, 185–186.

Pohl, P., Thomas, J. and Laub, R. (1973) Classification biochimique des *Salmonella typhimurium. Revue de Fermentation Industrielle et Alimentaire* 27, 239–242.

Popoff, M.Y. and Le Minor, L. (1985) Expression of antigenic factor O:54 is associated with the presence of a plasmid in *Salmonella. Annales de l'Institut Pasteur/Microbiologie* 136B, 169–179.

Popoff, M.Y., Bockemühl, J. and Brenner, F.W. (1998) Supplement 1997 (no. 41) to the Kauffmann–White scheme. *Research in Microbiology* 149, 601–604.

Raetz, C.R.H. (1996) Bacterial lipopolysaccharides: a remarkable family of bioactive macroamphiphiles. In: Neidhardt, F.C., Curtiss, R., III, Ingraham, J.L., Lin, E.C.C., Low, K.B., Magasanik, B., Reznikoff, W.S., Riley, M., Schaechter, M. and Umbarger, H.E. (eds) Escherichia coli *and* Salmonella, Vol. I, 2nd edn. American Society for Microbiology Press, Washington, DC, pp. 1035–1063.

Reeves, M.W., Evins, G.M., Heiba, A.A., Plikaytis, B.D. and Farmer, J.J., III (1989) Clonal nature of *Salmonella typhi* and its genetic relatedness to other salmonellae as shown by multilocus enzyme electrophoresis, and proposal of *Salmonella bongori* comb. nov. *Journal of Clinical Microbiology* 27, 313–320.

Reeves, P.R., Hobbs, M., Valvano, M.A., Skurnik, M., Whitfield, C., Coplin, D., Kido, N., Klena, J., Maskell, D., Raetz, C.R.H. and Rick, P.D. (1996) Bacterial polysaccharide synthesis and gene nomenclature. *Trends in Microbiology* 4, 495–503.

Rick, P.D. (1987) Lipopolysaccharide biosynthesis. In: Neidhardt, F.C., Ingraham, J.L., Low, K.B., Magasanik B., Schaechter, M. and Umbarger, H.E. (eds) Escherichia coli *and* Salmonella typhimurium: *Cellular and Molecular Biology*, Vol. I. American Society for Microbiology Press, Washington, DC, pp. 648–662.

Rubin, F.A., Kopecko, D.J., Noon, K.F. and Baron, L.S. (1985) Development of a DNA probe to detect *Salmonella typhi. Journal of Clinical Microbiology* 22, 600–605.

Rules Revision Committee, Judicial Commission (1975) Proposal to amend the International Code of Nomenclature of Bacteria. *International Journal of Systematic Bacteriology* 35, 123.

Salmon, D.E. and Smith, T. (1886) The bacterium of swine plague. *American Monthly Microbiology Journal* 7, 204.

Selander, R.K., Beltran, P., Smith, N.H., Barker, R.M., Crichton, P.B., Old, D., Musser, J.M. and Whittam, T.S. (1990a) Genetic population structure, clonal phylogeny, and pathogenicity of *Salmonella paratyphi* B. *Infection and Immunity* 58, 1891–1901.

Selander, R.K., Beltran, P., Smith, N.H., Helmuth, R., Rubin, F.A., Kopecko, D.J., Ferris, K., Tall, B.D., Cravioto, A. and Musser, J. M. (1990b) Evolutionary genetic relationships of clones of *Salmonella* serovars that cause human typhoid and other enteric fevers. *Infection and Immunity* 58, 2262–2275.

Selander, R.K., Li, J. and Nelson, K. (1996) Evolutionary genetics of *Salmonella enterica*. In: Neidhardt, F.C.,

Curtiss, R., III, Ingraham, J.L., Lin, E.C.C., Low, K.B., Magasanik, B., Reznikoff, W.S., Riley, M., Schaechter, M. and Umbarger, H.E. (eds) *Escherichia coli* and *Salmonella*, Vol. II, 2nd edn. American Society for Microbiology Press, Washington, DC, pp. 2691–2707.

Simon, M., Zieg, J., Silverman, M., Mandel, G., and Doolittle, R. (1980) Phase variation: evolution of a controlling element. *Science* 209, 1370–1374.

Skerman, V.B.D., McGowan, V. and Sneath, P.H.A. (1980) Approved list of bacterial names. *International Journal of Systematic Bacteriology* 30, 225–420.

Smith, N.H. and Selander, R.K. (1990) Sequence invariance of the antigen-coding central region of the phase 1 flagellar filament gene (fliC) among strains of *Salmonella typhimurium*. *Journal of Bacteriology* 172, 603–609.

Smith, N.H. and Selander, R.K. (1991) Molecular genetic basis for complex flagellar antigen expression in a triphasic serovar of *Salmonella*. *Proceedings of the National Academy of Sciences USA* 88, 956–960.

Smith, N.H., Beltran, P. and Selander, R.K. (1990) Recombination of *Salmonella* phase 1 flagellin genes generates new serovars. *Journal of Bacteriology* 172, 2209–2216.

Stanley, J., Jones, C.S., Threlfall, E.J. (1991) Evolutionary lines among *Salmonella enteritidis* phage types are identified by insertion sequence IS 200 distribution. *FEMS Microbiology Letters* 82, 83–90.

Stoleru, G.H., Le Minor, L. and L'Heritier, A.M. (1976) Polynucleotide sequence divergence among strains of *Salmonella* sub-genus IV and closely related organisms. *Annales de Microbiologie* 127A, 477–486.

Stone, G.G., Oberst, R.D., Hays, M.P., McVey, S. and Chengappa, M.M. (1995) Combined PCR–oligonucleotide ligation assay for rapid detection of *Salmonella* serovars. *Journal of Clinical Microbiology* 33, 2888–2893.

Tacket, C.O. (1989) Molecular epidemiology of *Salmonella*. *Epidemiologic Reviews* 11, 99–108.

Threlfall, E.J., Rowe, B. and Ward, L.R. (1989) Subdivision of *Salmonella enteritidis* phage types by plasmid profile typing. *Epidemiology and Infection* 102, 459–465.

Threlfall, E.J., Frost, J.A., Ward, L.R. and Rowe, B. (1990) Plasmid profile typing can be used to subdivide phagetype 49 of *Salmonella typhimurium* in outbreak investigations. *Epidemiology and Infection* 104, 243–251.

Vieu, J.F., Hassan-Massoud, B., Klein, B. and Leherissey, M. (1981) Bacteriophage-typing and biotyping of *Salmonella montevideo*. In: *FEMS Symposium on Salmonellae and Salmonellosis*. Istanbul, 15–17 September 1981.

Vieu, J.F., Binette, H. and Leherissey, M. (1988) *Salmonella paratyphi* B d-tartrate positif (var. java): lysotypie de 1200 souches isolées en France (1975–1985). *Zentralblatt fur Bakteriologie, Mikrobiologie und Hygiene Series A* 268, 424–432.

Vieu, J.F., Jeanjean, S., Tournier, B. and Klein, B. (1990) Application d'une série unique de bacteriophages à la lysotypie de *Salmonella* sérovar Dublin et de *Salmonella* sérovar Enteritidis. *Médecine et Maladies Infectieuses* 20, 229–233.

Ward, L.R., De Sa, J.D.H. and Rowe, B. (1987) A phage-typing scheme for *Salmonella enteritidis*. *Epidemiology and Infection* 99, 291–294.

Wayne, L.G., Brenner, D.J., Colwell, R.R., Grimont, P.A.D., Kandler, O., Krichevsky, M.I., Moore, H., Moore, W.E.C., Murray, R.G.E., Stackebrandt, E., Starr, M.P. and Trüper H.G. (1987) Report of the ad hoc committee on reconciliation of approaches to bacterial systematics. *International Journal of Systematic Bacteriology* 37, 463–464.

White, B. (1926) Further studies on the *Salmonella* group. *Medical Research Council Special Report* 103, 3–160.

Widal, F. (1896) Sérodiagnostic de la fièvre typhoïde. *Bulletin et Mémoires de la Société de Médecine des Hôpitaux de Paris* 13, 561–566.

Winkler, M.E. (1979) Ribosomal ribonucleic acid isolated from *Salmonella typhimurium*: absence of the intact 23S species. *Journal of Bacteriology* 139, 842–849.

Chapter 2

Structure, Function and Synthesis of Surface Polysaccharides in *Salmonella*

Andrew N. Rycroft

Royal Veterinary College, Hawkshead Lane, North Mymms, Hatfield, Hertfordshire AL9 7TA, UK

Introduction

The surface polysaccharides of *Salmonella* spp. form the outermost components of the bacterial cell. They are in direct contact with the immediate environment of the organism and are therefore of great significance in the interaction of the organism with its habitat. For a bacterial pathogen such as *Salmonella*, which is able to exist in different habitats as it passes from the dry, external environment, through the acidity of the stomach, the lumen of the gut, the extracellular space of host tissues and the inside of the macrophage, the surface components provide a protective and yet porous shield against the outside world. This chapter aims to review the chemical structure, biological function and biosynthesis of the surface polysaccharides found on the surface of *Salmonella*.

The Architecture of the Surface Structures of *Salmonella*

Much of what we understand about the nature of the outer envelope of Gram-negative bacteria has been derived from studies with *Escherichia coli* K-12 and *S. typhimurium* LT2. Essentially, there are three layers: the cytoplasmic membrane (inner membrane), the peptidoglycan (murein) and the outer membrane (Fig. 2.1). The compartment between the two membranes is referred to as the periplasmic space.

The cytoplasmic membrane in *Salmonella* is composed of phospholipids and proteins. As in other Gram-negative bacteria, it transports nutrients and it is the site of oxidative phosphorylation and the synthesis of phospholipid, peptidoglycan units and lipopolysaccharide (LPS). The cytoplasmic membrane is also the site of anchorage of the DNA during replication and has a role in the partitioning of daughter cells at cell division.

The peptidoglycan is a relatively thin layer in Gram-negative bacteria. It is composed of alternating residues of *N*-acetyl muramic acid and *N*-acetyl glucosamine, forming long glycan chains, which are covalently cross-linked by peptide bridges. This forms a single bag-like molecule surrounding the cell protoplast, which serves to stabilize it against osmotic lysis. Approximately 3.5 atmospheres pressure is thought to be exerted by the cytoplasm (Stock *et al.*, 1977).

The peptidoglycan confers rigidity and shape to the bacterial cell. Degrading the peptidoglycan, by first disrupting the outer membrane and then allowing lysozyme to penetrate and hydrolyse it, causes the cell to swell and lyse by the uptake of water through the cytoplasmic membrane into the cytosol. If this is prevented by immersion of the bacteria in a hypertonic medium, such as 8% sucrose, the rod-shaped cells will round up to form osmotically fragile spheroplasts.

The periplasm contains the peptidoglycan

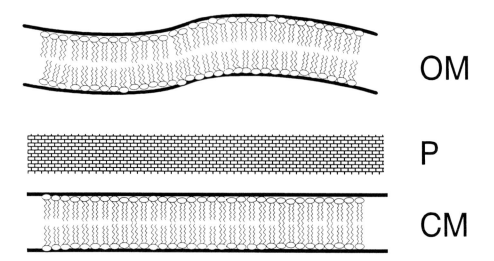

Fig. 2.1. Simplified representation of the envelope of *Salmonella*. Two lipid bilayers: the cytoplasmic membrane (CM) and the outer membrane (OM) are separated by the periplasmic space (P), in which the peptidoglycan provides tensile strength to support the cell against osmotic pressure.

and also numerous soluble proteins, which usually have one of three functions. These are: (i) catabolic functions, such as alkaline phosphatase, where solutes for which no transport system exists are converted to a form that can be transported though the cytoplasmic membrane; (ii) binding proteins, which fasten on to nutrients, such as amino acids, ions and sugars, and assist their transport; and (iii) proteins which degrade or modify harmful substances, such as antibiotics, e.g. β-lactamase.

The two membranes are connected at various points, known as zones of adhesion or Bayer bridges, first described by Bayer (1968). These sites of contact become visible by electron microscopy only when the inner membrane is plasmolysed or 'shrunk' away from the outer membrane by holding the organisms in hypertonic medium. These quasi-stable zones of adhesion are thought to facilitate the transport of hydrophobic materials, such as LPS, from the inner surface of the cytoplasmic membrane to the outer membrane and may be the site of synthesis of some outer-membrane proteins.

The outer membrane is a highly complex lipid-bilayer membrane structure, which surrounds the peptidoglycan layer and shields the periplasm from the external environment. It also prevents leakage of the periplasmic proteins away

from the immediate environment of the cytoplasmic membrane. Electron microscopy has determined the thickness of the outer membrane to be like that of the cytoplasmic membrane, 7.5 nm. However, in composition and function it is quite different from the cytoplasmic membrane. It is composed primarily of phospholipid and protein but also LPS and lipoprotein. LPS is found exclusively in the outer leaflet of the outer membrane and the lipoprotein is present in the inner leaflet, where it functions to anchor the outer membrane to the cell peptidoglycan. Enterobacterial common antigen is a minor component, contributing only 0.2% of the cell's dry weight.

Since the natural habitat of *Salmonella* is the lower intestinal tract of animals of all kinds, it is logical to assume that the outer membrane functions to protect or assist the cell in this environment. In order to compete effectively with other microorganisms in the anaerobic, nutritionally sparse conditions of the gut, *Salmonella* need to be able to take up limited nutrients effectively and to adapt to rapidly changing conditions. The outer membrane serves to allow the passive transport of selected molecules into the periplasm, where they can be held and transported across the cytoplasmic membrane. At the same time, the outer membrane must serve to protect the delicate components of the cytoplasmic membrane from the detergent-like action of bile salts, fatty acids and glycerides. The intestinal lumen is replete with proteases and lipases, and these must be prevented from gaining access to the vicinity of the cytoplasmic membrane, where they will cause damage to the membrane structures. Therefore, the outer membrane of *Salmonella* may be considered as a molecular sieve, whose purpose is to allow required nutrients to access the periplasm while resisting the penetration of dangerous substances from the external environment.

Many Gram-negative bacteria, such as *Escherichia coli*, *Klebsiella* or *Pasteurella* spp., possess a polysaccharide layer external to the outer membrane. *S. enterica* is unusual among the enteric Gram-negative bacteria of mammals in that it usually possesses no capsular polysaccharide. The polysaccharide exposed at the surface is primarily the O side-chain of the LPS. The one exception to this is the Vi polysaccharide carried on the external surface of a very few strains of *Salmonella* such as some strains of *S. typhi* and *S. dublin*.

Lipopolysaccharide

LPS is the molecule that is most closely associated with the surface of Gram-negative bacteria. It is also the immunodominant antigen of the majority of Gram-negative bacteria. During the 1950s and 1960s, a considerable research effort, due to the interest in LPS as mediators of biological activity known as endotoxin activity, led to a wealth of information on their structure and biosynthesis.

The structure of LPS has been elucidated over a number of years, investigations largely being conducted with *S. typhimurium* LT2 and *E. coli* K-12 using a variety of techniques, including biochemical analysis and examination of mutants deficient in LPS production. *Salmonella* has therefore come to be perceived as 'the norm' among Gram-negative bacteria, although there is considerable variety of structure even within the genus.

LPS is amphipathic, having both hydrophilic and hydrophobic components on the same molecule. Three regions of the molecule are recognized: the lipid A, the core oligosaccharide and the O side-chain repeating oligosaccharide. The core oligosaccharide is further subdivided into the inner and outer core regions (Fig. 2.2). The hydrophobic lipid A portion of the molecule resides within the outer leaflet of the outer membrane. The polysaccharide portion, which is hydrophilic, projects into the external environment.

Colonies of wild strains of *Salmonella* bacteria usually have a smooth appearance. This is associated with the presence of a full O side-chain, which is therefore termed the S form. Mutants that have lost, through natural occurrence or deliberate mutagenesis in the laboratory, their O side-chain often produce irregular-edged colonies with a dull surface. These are referred to as rough mutants and the LPS present in these bacteria as the R form.

Preparation and purification of *Salmonella* lipopolysaccharide

There are approximately 10^6 molecules of lipid A and 10^7 molecules of glycerophospholipid per bacterial cell (Goldman *et al.*, 1988).

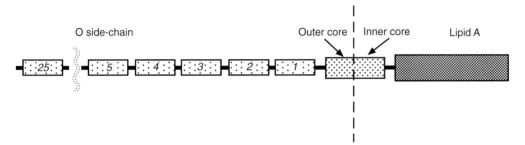

Fig. 2.2. Schematic representation of the structure of S-form lipopolysaccharide. Lipid A is joined, via the inner- and outer-core oligosaccharide, to the repeating oligosaccharide that forms the O side-chain. The length of the O side-chain is variable, but between 25 and 40 units is common in *Salmonella*.

LPSs from *Salmonella* are often purified from the bacterial cells by one of two methods. S-form LPS is prepared by the phenol–water extraction method of Westphal and Jann (1965). In this procedure, the bacteria are held at 65°C in 45% aqueous phenol. The S-form LPS, together with a proportion of the R-form LPS with a complete or nearly complete core oligosaccharide, is partitioned into the aqueous phase. For R-form LPS, the PCP method, devised by Galanos *et al.* (1969), is preferable. This uses a mixture of 90% phenol, chloroform and petroleum ether to extract the bacteria at room temperature. LPS from deep rough mutants is only extracted using this method, since it largely partitions into the hydrophobic phase. Other methods intended to prepare purified LPS in both forms have also been devised (Darveau and Hancock, 1983). Early studies with purified *Salmonella* LPS showed it to be highly variable in the O side-chain but conserved in structure in the core oligosaccharide and the lipid A.

Lipid A

Chemistry of lipid A

The structure of lipid A from *S. enterica* serovars is thought to be conserved. The lipid A component is embedded in the outer leaflet of the outer membrane of the organism, forming part of the lipid bilayer. It is responsible for the endotoxic properties of the LPS molecule (Raetz, 1993). While the general or approximate structure of lipid A was known for many years, the exact structure of lipid A was not finally established until 1983, in the laboratory of Ernst Rietschel, Otto Lüderitz and co-workers (Rietschel *et al.*, 1983). The delay in defining the structure was due to the complex structure of the molecule. This discovery opened the way to a fuller appreciation of the biosynthesis and pharmacology of lipid A endotoxin.

Enterobacteria such as *Salmonella* which are unable to synthesize lipid A are non-viable. The minimum requirement of the LPS molecule is that found in Re mutants. In these mutants, the LPS comprises lipid A and two 3-deoxy-D-manno-octulosonic acid (KDO) residues (Fig. 2.3). The reasons why these components are essential for the bacteria to be viable are not understood.

Lipid A can be prepared from LPS by mild acid hydrolysis. Diphosphoryl lipid A (DPL) is obtained by treatment with 0.02 M sodium acetate, pH 4.5 at 100°C for 30 min. If harsher conditions are used (0.1 M HCl at 100°C for 30 min) monophosphoryl lipid A (MPL), lacking the phosphate group at the 1 position, is released.

Lipid A from *S. typhimurium*, other *Salmonella* and *E. coli* consists of two glucosamine residues joined by a β(1–6) linkage, to which are substituted four fatty acid residues at positions 2 and 3 and 2' and 3'. Those attached to positions 2 and 2' are joined via amide links. The acyl moieties of lipid A are unusual because they are hydroxylated at carbon 3. They are also two to six carbons shorter than glycerophospholipids. Those hydroxylated fatty acids attached to positions 2' and 3' are further esterified, through the hydroxyls at C3, to additional fatty acyl residues, giving six fatty acid chains in all.

Biological activities of lipid A

Lipid A has potent biological activity. Since lipid A has been a component of the Gram-negative cell envelope probably throughout the evolution of eukaryotes, the immune system of animals is exceedingly sensitive to it as a marker of infection. Lipid A has been known to induce pathophysiological effects, such as endotoxic shock, pyrogenicity, activation of complement, coagulation and haemodynamic changes, for many years. The dose of Re LPS from *S. typhimurium* producing a febrile response in 50% of rabbits is between 0.1 and 0.3 µg kg^{-1} body weight. Long-recognized immunological effects include mitogenicity for B lymphocytes and activation of macrophages, but it is only during the last decade that it has been known that these effects are mediated through the induction and release of numerous cytokines of monocyte and macrophage origin: interferon, tumour necrosis factor, colony-stimulating factor and interleukin 1(IL-1) (Qureshi and Takayama, 1990). It is through these activities that lipid A (endotoxin) contributes to the pathogenic or toxic activity of Gram-negative bacteria, including *Salmonella*. As an example of a substance that can modify or modulate an immune response, LPS is now considered to be a potent modulin (Henderson *et al.*, 1996).

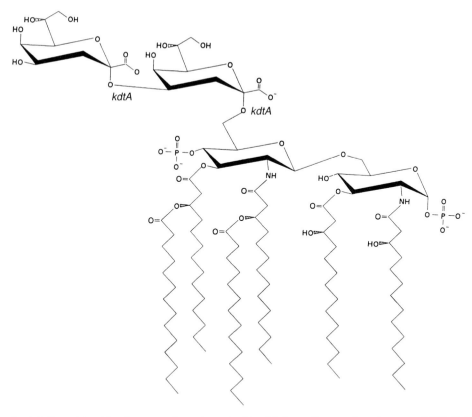

Fig. 2.3. The minimal Re chemotype lipopolysaccharide of *Salmonella*, sometimes known as KDO$_2$-lipid A. Two KDO residues are linked to the acetylated glucosamine disaccharide (lipid A). Linkage of the KDO to the glucosamine disaccharide involves the product of the *kdtA* gene.

Until the middle of the 1980s, endotoxin was thought, in simple terms, to act by somehow disrupting the cytoplasmic membrane of cells such as macrophages and neutrophils and releasing cell contents, including IL-1 (previously known as endogenous pyrogen), which led to fever. The mechanism by which lipid A interacts with host cells is now much clearer. Membrane blebs or LPS released from bacteria is bound by LPS-binding protein (LBP). This is an intermediate in the transfer of LPS to the CD14 protein on the surface of macrophages and certain other cell types. Cell types lacking CD14 are rendered more responsive to LPS when transfected with the gene for CD14 (Lee *et al.*, 1992). A second, transmembrane, lipid A receptor protein is thought to be involved in the generation of intracellular signals. This stimulates transcrip-

tional activity of cytokines, including IL-1 and tumour necrosis factor. Enhanced production of these cytokines is responsible for the symptoms of endotoxic shock.

Genetics of lipid A synthesis

The biosynthetic pathway of lipid A has been elucidated primarily in *E. coli* K-12. However, the genes and gene products required for this process seem to be identical for lipid A in *Salmonella*. Uridine diphosphate-N-acetyl glucosamine (UDP-GlcNAc) is a central precursor for the synthesis of both peptidoglycan and LPS. This leads to a group of phospholipids based not on glycerol, but on glucosamine. The only known function of these is as precursors of lipid

A. Only four genes are known to be involved in lipid A synthesis and all of these are derived from studies with *E. coli*. These are *lpxA*, *lpxB*, *lpxC* and *lpxD*. These genes encode the enzymes which: (i) transfer β-hydroxymyristic acid from acyl carrier protein (ACP) to the 3-hydroxy group of UDP-GlcNAc (*lpxA*); (ii) remove the N-acetyl group from UDP-3-hydroxymyristoyl-GlcNAc (*lpxC*); (iii) transfer the β-hydroxymyristic acid from ACP to the 2-amino group of deacylated GlcNAc to form UDP-2,3-dihydroxymyristoyl-glucosamine (*lpxD*); and (iv) form the β(1–6) bond between the glucosamines of one molecule of UDP-2,3-dihydroxymyristoyl-glucosamine and one of 2,3-dihydroxymyristoyl-glucosamine (lipid X) to form the lipid A disaccharide with four fatty acid residues and one phosphate group (*lpxB*) (Crowell *et al.*, 1987).

Core Oligosaccharide

The lipid A is linked through carbon 6' of the glucosamine disaccharide to the core oligosaccharide. This link (α2–6) is made through a unique eight-carbon sugar called KDO. A second, branch KDO (known as KDO II) is also added through an α(2–4) linkage. This structure is known as KDO_2-lipid A or Re endotoxin, because it is the form of LPS seen in mutants with the Re chemotype – the minimal LPS substructure for growth of the bacteria (Rick and Young, 1982). It is generally believed that a third KDO residue is added to KDO II at a late stage of core completion through a further α(2–4) linkage, although this is not essential for viability in *Salmonella* (Lehmann *et al.*, 1971).

The core oligosaccharide consists of a conserved, non-repeating group of six to eight sugars. In *Salmonella*, a single core oligosaccharide type (termed the Ra core) is conserved throughout the genus (Holst and Brade, 1992). It consists of eight sugar residues, in addition to the KDO residues, which link it to the lipid A. Two L-glycero-D-manno-heptose residues are attached to the KDO, forming the so-called inner core. A third L-glycero-D-manno-heptose branches from the outermost heptose. This in turn is linked to two D-glucose, two D-galactose residues and N-acetyl glucosamine forming the outer core (Fig. 2.4). In addition, O-pyrophosphorylethanolamine and O-phosphorylethanolamine are frequently substituted on to the L-glycero-D-manno-heptose and KDO residues, respectively.

Genetics of the LPS core biosynthesis

The genetics of core biosynthesis have been extensively reviewed by Schnaitman and Klena (1993). The genes required for the production of core LPS belong primarily to the *rfa* gene cluster. This name has recently been revised by Reeves *et al.* (1996), who have proposed a new system for the nomenclature of genes for polysaccharide biosynthesis. The *rfa* gene cluster has largely been renamed the *waa* cluster (Table 2.1). The two systems are currently coexisting, but in this chapter the new system will take priority. While many of the genes are known from work with *S. typhimurium*, others have only been identified in *E. coli* and their equivalent function in *Salmonella* is not known.

Two *kds* genes (*kdsA* and *kdsB*) are required for synthesis of KDO. The KDO is then added to the glucosamine of lipid A disaccharide by the

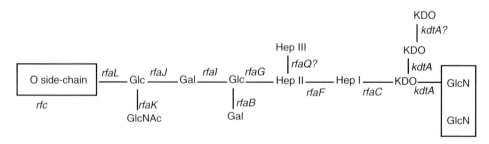

Fig. 2.4. The structure of the lipopolysaccharide core oligosaccharide of *Salmonella*. Genes known to be required for the synthesis of the core at various points are shown.

Table 2.1. Genes required for the synthesis of the lipopolysaccharide core in *Salmonella typhimurium*.

Gene	New designation	Function
kdsA		KDO-8-phosphate synthesis
kdsB		CMP–KDO synthesis
kdtA	(*waaA*)	KDO transferase (transfer to glucosamine)
rfaE	(*waaE*)	ADP-heptose synthesis
rfaD	(*gmhD*)	Epimerization of ADP heptose to L-glycero-D-manno-heptose
rfaC	(*waaC*)	Heptosyltransferase: addition of heptose
rfaF	(*waaF*)	Addition of heptose II to complete the inner core
rfaG	(*waaG*)	Glucosyltransferase: addition of glucose to Hep II
rfaB	(*waaB*)	Addition of branch D-galactose to glucose
rfaI	(*waaI*)	Galactosyltransferase: addition of D-galactose
rfaJ	(*waaJ*)	Addition of second D-glucose
rfaK	(*waaK*)	Transfer of branch *N*-acetyl glucosamine to some terminal glucose
rfaH		Positive regulation of *waa* (*rfa*)
rfaQ	(*waaQ*)	Addition of Hep III to core?
rfaL	(*waaL*)	Addition of O-antigen to core
rfaP	(*waaP*)	Phosphorylation of Hep I
rfaS	(*waaS*)	Synthesis of lipooligosaccharide; not in *S. typhimurium*
rfaZ	(*waaZ*)	Function unknown

product of the *waaA* gene (*kdtA*), KDO transferase. KDO II is also added by this enzyme, and indirect evidence suggests that KDO III is further added by this enzyme at a late stage in synthesis of the core. The final steps in the acylation of lipid A are thought to be coupled to the attachment of KDO (Brozek and Raetz, 1990); however, the genes required for terminal acylation of lipid A are not yet known.

The synthesis of heptose and its conversion to L-glycero-D-manno-heptose involve the *waaD* (*rfaD*) and *waaE* (*rfaE*) genes. Addition of the first heptose to KDO I requires a heptosyltransferase encoded by the *waaC* (*rfaC*) gene, and completion of the inner core, by addition of the second heptose, requires *waaF* (*rfaF*). Mutants lacking these genes are said to exhibit a deep-rough phenotype.

The first D-glucose of the outer core is attached by a glucosyl transferase, which is the product of the *waaG* (*rfaG*) gene. The substrate, UDP-glucose, is cleaved to add the glucose residue to the second heptose (Hep II). Interestingly, mutants with a defective *waaG* gene produce neither flagella nor type 1 fimbriae. This suggests either a regulatory role for the WaaG protein or that the synthesis of these surface appendages is dependent upon a certain degree of completion of the LPS core.

A D-galactose residue is then attached by an α(1–3) link to the D-glucose by a galactosyl transferase. This is the product of the *waaI* (*rfaI*) gene. A branch D-galactose residue is also attached to the glucose in an α(1–6) link by the product of the *waaB* (*rfaB*) gene. The second terminal α(1–2)-linked D-glucose is then attached by the *waaJ* (*rfaJ*) product and finally, completing the *Salmonella* core oligosaccharide, the *waaK* (*rfaK*) gene product transfers a branch *N*-acetyl glucosamine in α(1–2) linkage to a proportion of the terminal glucose residues.

The O-antigen

The serologically dominant and highly variable region of the LPS is the O side-chain. This is hydrophilic in nature and reaches out to the microenvironment of the bacterial cell (see Fig. 2.2). It is a repeated tetra- or pentasaccharide, characterized by the inclusion of deoxy- and dideoxyhexoses. The number of repeats in the O side-chain varies from strain to strain and is dependent upon the prevailing growth conditions. The repeating nature and quantum variation of the O side-chain size can be visualized by separation of S-form LPS on a polyacrylamide gel and silver staining (Hitchcock and Brown,

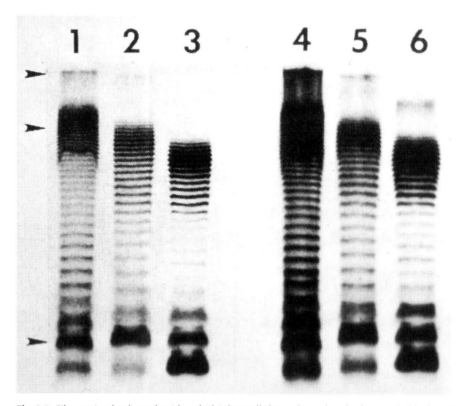

Fig. 2.5. Silver stained polyacrylamide gel of *Salmonella* lipopolysaccharide showing ladder banding from the presence of different lengths of O side-chain (from Peterson and McGroarty, 1985, with permission from the *Journal of Bacteriology*).

1983). Each band in the 'ladder' represents an LPS molecule with one single oligosaccharide unit more or less than the one next to it. Between 25 and 40 units of the oligosaccharide are a common finding (Fig. 2.5).

The O side-chain in serological classification

The system that is used to differentiate the *Salmonella* serovars (previously referred to as species) is based on the grouping of strains according to the structure of their O antigenic side-chain polysaccharide, coupled with determination of the serological specificity of the H (flagellar) antigen, which may be expressed in one of two phases. This is the White–Kauffmann–Le Minor scheme. Serological distinction of the O antigens of *Salmonella* is made on the detection of particular antigens, which are designated by a number. These factors are determined by the component sugars of

the O side-chain. *S. typhimurium* has the antigenic formula 1, 4, 5, 12. Antigens 4 and 12 are found in all group B serovars, although factor 12 is also found elsewhere. Factors 1 and 5 are additional factors that assist in the definition of serological specificity of *S. typhimurium*.

Synthesis and genetics of the O side-chain

The genes required for the biosynthesis of the O side-chain obviously vary from serovar to serovar, since these contain different sugars in the repeating oligosaccharide. Nevertheless, there are basic similarities in these genes from those serovars that have been examined. The group of genes involved in O side-chain synthesis is the 19 or so genes of the O-antigen gene cluster (previously known as the *rfb* locus (Fig. 2.6)), which maps near *his* at approximately 45 minutes on the *Salmonella* chromosome (Jiang *et al.*, 1991).

Among the enterobacteria, there are two systems of O-antigen synthesis based on the gene involved in attachment of the first sugar of the O side-chain to the undecaprenol phosphate antigen carrier lipid (ACL). Most *Salmonella* serovars use the product of the *wbaP* (*rfbP*) gene rather than that of the alternative, *wecA* (*rfe*) gene. These include O groups A, B, C2, D and E1. However, some serogroups – C1 and L – are *wecA*-dependent.

The O side-chain repeat unit of group B strains, such as *S. typhimurium*, has four or five sugars. The backbone sugars are the same in serogroups A, D and E1, comprising mannose, rhamnose and galactose, whereas in group C2 this is Rha–Man–Man–Gal. In group B, the mannose is substituted with abequose, a 3,6-dideoxygalactose (which confers the O4 antigen factor) and the galactose is partially substituted with glucose. The abequose is further O-acetylated, which confers the O5 antigen. In other groups, the dideoxyhexose component is different: paratose in group A, abequose in groups B and C2, tyvelose in group D and none in E1.

Synthesis of group B O antigen begins with the transfer of galactose phosphate to ACL by the product of the *wbaP* (*rfbP*) gene. The gene *wbaN* (*rfbN*) encodes the sugar transferase for addition of the rhamnose, while *wbaU* (*rfbU*) and *wbaV* (*rfbV*) encode the enzymes that attach rhamnose and abequose, respectively, using the sugar nucleotides as substrates. The glycosylation of galactose requires the *oafA* gene, which lies just outside the *rfb* cluster, and the O-acetylation of abequose involves the *oafC* and *oafR* genes, far away from *rfb*.

Mutants are known which produce the so-called semi-rough (SR) phenotype. The LPS produced by this class of mutant has a high proportion of core oligosaccharides, substituted with a single O side-chain unit, and the remainder have complete cores that carry no antigen. The proportion of LPS molecules with core only is approximately the same as that seen in strains with S-form LPS. Therefore it was recognized that these mutants can attach the first O-antigen unit but are defective in the polymerization step required to generate the repeating O side-chain. They are nevertheless able to ligate the first O unit to the core and efficiently transfer the resulting LPS to the surface of the outer membrane. The gene encoding this function is termed *wzy*

(*rfc*) and it is located in the O-antigen gene cluster in *Salmonella* strains of serogroups C1, C2 and E1. However, it is at a separate site in other serogroups, such as B. The Wzy protein is very hydrophobic, suggesting that it is a membrane protein (Table 2.2).

The mechanism by which O-antigen units are added to the growing O side-chain is still not entirely clear. The sugars of the oligosaccharide are transferred from sugar nucleotide phosphates to the carrier lipid molecule, the undecaprenol phosphate ACL. Galactose, rhamnose and mannose are sequentially transferred to the ACL while it is in the inner face of the cytoplasmic membrane. Branch sugars, such as abequose of *S. typhimurium*, are transferred to the mannose and galactose. This must then be transferred to the site of the lipid A core. As the ACL–trisaccharide reaches the outer leaflet of the cytoplasmic membrane, the galactosyl-phosphate bond of the ACL–trisaccharide is broken and the galactosyl bond is transferred to the terminal mannosyl residue of a second or acceptor lipid, forming an oligosaccharide lipid carrier. This is repeated several times, chain growth taking place by addition of oligosaccharide units to the reducing end of the polysaccharide chain. Therefore, the developing polysaccharide is always transferred to an ACL carrying an O-antigen unit. To recycle the ACL, a specific phosphorylase dephosphorylates the free ACL to the monophosphate derivative, which then returns to the inner cytoplasmic face to begin the addition of sugars to form an oligosaccharide unit (Fig. 2.7). In a few cases, e.g. *S. minneapolis* and *S. typhimurium*, certain branch sugars (glucose) are added not in the cytoplasm but after polymerization. Frequently when this is the case, substitution is incomplete. Similarly, O acetylation is carried out once polymerization has taken place.

The completed polysaccharide O side-chain–ACL then interacts with the lipid A core. The transfer of the O side-chain polysaccharide from the ACL to the glucose of the *Salmonella* core is carried out by O-antigen LPS ligase. Once transferred, the LPS molecule must be relocated to the outer membrane. The flip-flop transfer of the LPS to the outer membrane is still not fully understood, but seems to involve outer-membrane proteins for which LPS has a strong affinity, as well as the Bayer-bridge adhesion points linking the inner and outer membranes.

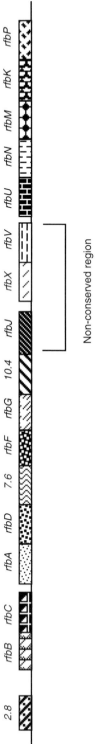

Fig. 2.6. The gene cluster of the *rfb* operon (now known as *wba*), which is responsible for the biosynthetic enzymes for O side-chain synthesis. These genes map at 45 minutes on the chromosome of *Salmonella typhimurium*.

Table 2.2. Genes involved in the synthesis of the O side-chain.

Gene	New designation	Function
rfbBDAC	(*wbaBDAC*)	Synthesis of TDP-rhamnose
rfbKM	(*manB*)	Synthesis of GDP-mannose
rfbFG	(*ddhAB*)	Synthesis of CDP-4-keto-3,6-dideoxyglucose precursor
rfbJ	(*abe*)	Synthesis of CDP-abequose (group B)
rfbS	(*prt*)	Synthesis of CDP-paratose (group A)
rfbE	(*tyv*)	Synthesis of CDP-tyvelose (group D)
rfbP	(*wbaP*)	Initial step of attachment of Gal-P to ACL
rfe	(*wecA*)	Initial step of attachment of GlcNAc-P to ACL
rfbN	(*wbaN*)	Rhamnose transfer to galactose (groups A, B, D and E1)
rfbU	(*wbaU*)	Mannose transfer to rhamnose (groups A, B, and D)
rfbV	(*wbaV*)	Abequose transfer to mannose (groups B and D)
rfc	(*wzy*)	Polymerization of O side-chain units
rfaL	(*waaL*)	Attaches O side-chain to core oligosaccharide
rol	(*wzz*)	Regulates length of O side-chain

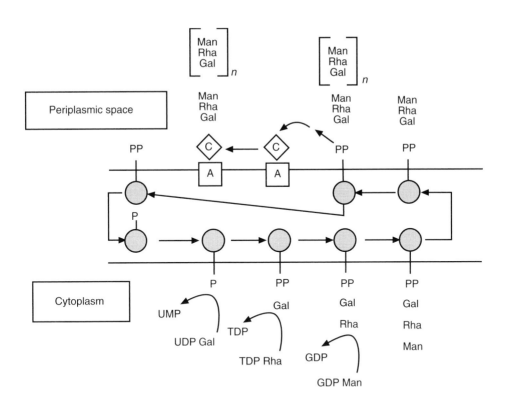

Fig. 2.7. Biosynthesis of O side-chain in *Salmonella*. At the inner face of the cytoplasmic membrane, the nucleotide sugars are transferred in sequence to the C_{55} polysioprenoid phosphate carrier lipid. Following the transfer of the mannosylrhamnosylgalactose trisaccharide-carrier lipid to the outer face of the membrane, the trisaccharides polymerize, forming the lipid-linked O-antigen polymer, the polysaccharide portion of which is then passed to the lipid A core (AC). The liberated carrier lipid is then dephosphorylated and returned to the cytoplasmic face of the membrane.

The distribution of size of molecules of LPS can be seen on the characteristic ladder pattern in silver-stained polyacrylamide gels (see Fig. 2.5). It is clear that, while there are a large number of molecules substituted with one, two or three O repeat units, this quickly decreases until there seem to be very few of intermediate size. As the size increases, the abundance once again increases to a maximum that is dependent upon the strain and its environmental conditions. Goldman and Hunt (1990) suggested that this distribution indicates a mechanism by which larger molecules are preferentially selected for ligation on to the LPS core. Mutation of the gene termed *wzz* (*rol*) caused an even distribution of chain lengths characteristic of random ligation and polymerization (Batchelor *et al.*, 1992).

The O side-chain in pathogenicity

The O-antigen component is also of importance in the interaction with the host in disease, particularly in evading the innate host defences. The structure and immunogenicity of the O side-chain may have profound effects on the ability of the humoral immune system to mount a response to infection and on the initial interaction with professional phagocytic cells.

Influence of chemical structure of LPS on the pathogenicity of *Salmonella*

Differences in the virulence of serovars and strains of *Salmonella* for animals are well known. In order to examine the importance and function of the O side-chain composition in an isogenic background, Valtonen and co-workers carried out genetic exchange experiments, using transduction to alter a strain of *S. typhimurium* (somatic antigens O4, 12) to carry the O antigen of *S. enteritidis* (O9, 12) or *S. montevideo* (O6, 7) (Fig. 2.8). They clearly showed that the strain possessing the O6, 7 antigen was least virulent, while the parent (O4, 12) showed the greatest virulence. The strain carrying the O9, 12 antigen was of intermediate virulence (Valtonen, 1970). These differences were shown to be manifest in immunosuppressed mice also, suggesting that the innate immunity of the animal was responsible for the differential response (Valtonen *et al.*, 1971). Subsequent experiments showed that, when the O9, 12 somatic antigen of a natural *S. enteritidis* strain was replaced with the O4, 12 somatic antigen of *S. typhimurium* by transduction, a statistically significant increase in virulence was observed (Valtonen *et al.*, 1975). Since these strains were shown to differ little in their uptake by phagocytes both *in vitro* and *in vivo*, an O6, 7 strain was constructed to be isogenic with the O4, 12 strain and this correlated well

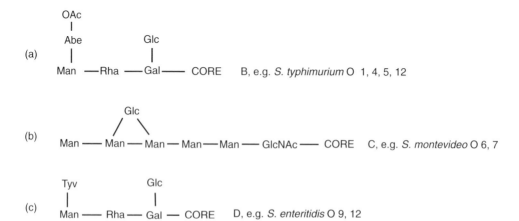

Fig. 2.8. Representative O side-chain repeat units from *Salmonella typhimurium* (group B), *S. montevideo* (group C1) and *S. enteritidis* (group D).

with the clearance rates *in vivo*. However, neither strain was killed in an *in vitro* phagocytosis assay (Valtonen, 1977).

This work was continued by Liang-Takasaki *et al.* (1982). They found that the rate of uptake of the isogenic strains by a macrophage-like cell line was inversely proportional to their mouse virulence. The differences in uptake were attributed to the differential affinity of the bacteria for the macrophages, rather than the rate of uptake following the interaction. This was proposed as evidence for a receptor-mediated process. In addition, the uptake was shown to be complement-dependent and direct activation of complement by LPS was postulated as one important factor determining virulence. It was concluded that virulent bacteria may evade both complement killing and opsonophagocytosis by evolving LPS modified so as to reduce activation of complement. Later studies confirmed the differential activation of C3 by these strains and showed it to proceed primarily via the alternative pathway (Liang-Takasaki *et al.*, 1983a). Finally, these studies reverted to examining the strains in the whole animal and concluded that the difference in complement-dependent phagocytosis noted *in vitro* was the primary factor responsible for the observed virulence differences *in vivo* (Liang-Takasaki *et al.*, 1983b).

Enterobacterial Common Antigen

The enterobacterial common antigen (ECA) is an acidic cell surface glycolipid shared by essentially all members of the *Enterobacteriaceae*, including *Salmonella*. It is of relatively little importance in the life of *Salmonella* and, as such, is frequently ignored. ECA is found in two forms: the haptenic form and the immunogenic form (Mayer and Schmidt, 1979). The immunogenic form is covalently linked to the LPS core region of rough mutants. The haptenic form is not covalently linked to the LPS and it is found in both smooth and rough strains, including those that produce the immunogenic form.

The serological specificity of ECA is governed by an amino sugar-containing heteropolysaccharide. This is primarily a trisaccharide repeating unit linear chain of 1–4 linked N-acetyl-D-glucosamine, N-acetyl-D-mannosaminuronic acid and N-acetylfucosamine residues.

The *wecA–E* (*rfe* and *rff*) gene cluster is involved in the biosynthesis of ECA. The *rml* gene products are known to be required for production of ECA, as well as O side-chains. The products of the *mnaAB* and *fcnAB* genes are also required for ECA synthesis (such as synthesis of UDP-N-acetyl-D-mannosaminuronic acid). Strains of some *Salmonella* harbouring *wecA* mutations (e.g. *S. montevideo* and *S. minnesota*) cannot synthesize the O side-chain serogroup antigen or ECA, while others (e.g. *S. typhimurium*) with similar mutations fail to produce ECA. Biosynthesis is thought to involve a lipid intermediate (undecaprenylphosphate) in a manner similar to that for O-antigen side-chain and peptidoglycan components (Rick *et al.*, 1985).

The function of ECA in the natural history of *Salmonella* is not known, although they may represent a primitive, redundant antigenic component on the cell surface, which predates the O side-chain.

The Vi Antigen

The Vi antigen is the only true capsular polysaccharide produced by *Salmonella* spp. It was discovered in 1934 and was termed the Vi antigen because of its association with virulence. It is only produced by strains of *S. typhi* and *S. paratyphi* C, together with a few strains of *S. dublin* and *Citrobacter freundii*. As recognized with other enterobacteria bearing capsular polysaccharide, *Salmonella* carrying the Vi antigen are not agglutinable with anti-O antiserum. They therefore appear inagglutinable in slide agglutination tests until the bacteria have been heated (100°C, 60 min) to remove the masking effect of the capsule, revealing the O-specific antigen beneath. Since few strains infecting animals carry the Vi antigen, inagglutinability due to Vi is encountered very infrequently in clinical veterinary microbiology.

The Vi antigen is an unbranched homopolymer of α-1,4–2-deoxy-2-N-acetylgalactosamine uronic acid. It is O-acetylated in approximately 60% of the residues at the C3 position (Heynes and Kiessling, 1967). The Vi antigen is recognized as a group I polysaccharide in the scheme of Jann and Jann (1990). This scheme is based on chemical, physical and genetic criteria, in which group I polysaccharides have high molecular mass and are thicker than type II polysaccharides, include hexuronic acid as the acidic component and are expressed along with LPS O antigens.

Two independent loci of genes are required for biosynthesis of Vi: viaA and viaB. The viaB locus, comprising 11 genes, is only found in those strains showing a Vi-positive phenotype. Investigation of the function of the viaB locus has identified genes responsible for Vi polymer

synthesis and others required for translocation of the polysaccharide to the bacterial cell surface (Virlogeux et al., 1995). The viaA region is also present in certain Vi-negative bacteria, such as E. coli. In this organism, a locus that is allelic to viaA is known by the designation rcsB. This is known to be a positive regulator of capsule biosynthesis, which is functional in E. coli. Therefore, transfer of the viaB locus to E. coli was shown to cause expression of Vi in E. coli (Johnson and Baron, 1969).

References

Batchelor, R.A., Alifano, P., Biffali, E., Hull, S.I. and Hull, R.A. (1992) Nucleotide sequence of the genes regulating O polysaccharide antigen chain length (rol) from Escherichia coli and Salmonella typhimurium: protein homology and functional complementation. Journal of Bacteriology 174, 5228–5236.

Bayer, M.E. (1968) Areas of adhesion between wall and membrane of Escherichia coli. Journal of General Microbiology, 53, 395–404.

Brozek, K.A. and Raetz, C.R.H. (1990) Biosynthesis of lipid A in Escherichia coli: acyl carrier protein-dependent incorporation of laurate and myristate. Journal of Biological Chemistry 265, 15410–15417.

Crowell, D.N., Reznikoff, W.S. and Raetz, C.R.H. (1987) Nucleotide sequence of the Escherichia coli gene for lipid A disaccharide synthetase. Journal of Bacteriology 169, 5727–5734.

Darveau, R.P. and Hancock, R.E.W. (1983) Procedure for isolation of bacterial lipopolysaccharides from both smooth and rough Pseudomonas aeruginosa and Salmonella typhimurium strains. Journal of Bacteriology 155, 831–838.

Galanos, C., Lüderitz, O. and Westphal, O. (1969) A new method for the extraction of R lipopolysaccharides. European Journal of Biochemistry 9, 245–249.

Goldman, R.C. and Hunt, F. (1990) Mechanism of O-antigen distribution in lipopolysaccharide. Journal of Bacteriology 172, 5352–5359.

Goldman, R.C., Doran, C.C. and Capobianco, J.O. (1988) Analysis of lipopolysaccharide biosynthesis in Salmonella typhimurium and Escherichia coli by using agents which specifically block incorporation of 3-deoxy-D-manno-octulosonate. Journal of Bacteriology 170, 2185–2192.

Henderson, B., Poole, S. and Wilson, M. (1996) Bacterial modulins: a novel class of virulence factors which cause host tissue pathology by inducing cytokine synthesis. Microbiological Reviews 60, 316–341.

Heynes, K. and Kiessling, G. (1967) Strukturaufklarung des Vi-antigens aus Cirobacter freundii (E. coli) 5396/38. Carbohydrate Research 3, 340–352.

Hitchcock, P.J. and Brown, T.M. (1983) Morphological heterogeneity among Salmonella lipopolysaccharide chemotypes in silver-stained polyacrylamide gels. Journal of Bacteriology 154, 269–277.

Holst, O. and Brade, H. (1992) Chemical structure of the core region of lipopolysaccharides. In: Morrison, D.C. and Ryan, J.L. (eds) Bacterial Endotoxic Lipopolysaccharides, Vol. 1, Molecular Biochemistry and Cellular Biology of Lipopolysaccharides. CRC Press, Boca Raton, Florida, pp. 135–170.

Jann, B. and Jann, K. (1990) Structure and biosynthesis of the capsular antigens of Escherichia coli. Current Topics in Microbiology and Immunology 150, 19–42.

Jiang, X.-M., Neal, B., Santiago, F., Lee, S.J., Romana, L.K. and Reeves, P.R. (1991) Structure and sequence of the rfb (O-antigen) gene cluster of Salmonella serovar typhimurium (strain LT2). Molecular Microbiology 5, 695–713.

Johnson, E.M. and Baron, L.S. (1969) Genetic transfer of the Vi antigen from Salmonella typhosa to Escherichia coli. Journal of Bacteriology 99, 355–359.

Lee, J.D., Kato, K., Tobias, P.S., Kirkland, T.N. and Ulevitch, R.J. (1992) Transfection of CD14 into 70Z/3 cells dramatically enhances the sensitivity to complexes of lipopolysaccharide (LPS) and LPS binding protein. Journal of Experimental Medicine, 175, 1697–1705.

Lehmann, V. Lüderitz, O. and Westphal, O. (1971) The linkage of pyrophosphorylethanolamine to heptose in the core of Salmonella minnesota lipopolysaccharides. European Journal of Biochemistry 21, 339–347.

Liang-Takasaki, C.-J., Makela, P.H. and Leive, L. (1982) Phagocytosis of bacteria by macrophages: changing the carbohydrate of lipopolysaccharide alters interaction with complement and macrophages. *Journal of Immunology* 128, 1229–1235.

Liang-Takasaki, C.-J., Grossman, N. and Leive, L. (1983a) Salmonellae activate complement differentially via the alternative pathway depending on the structure of their lipopolysaccharide O-antigen. *Journal of Immunology* 130, 1867–1870.

Liang-Takasaki, C.-J., Saxen, H., Makela, P.H. and Leive, L. (1983b) Complement activation by polysaccharide of lipopolysaccharide: an important virulence determinant of salmonellae. *Infection and Immunity* 41, 563–569.

Mayer, H. and Schmidt, G. (1979) Chemistry and biology of the enterobacterial common antigen (ECA). *Current Topics in Microbiology and Immunology* 85, 99–153.

Peterson, A.A. and McGroarty, E.J. (1985) High-molecular-weight components in lipopolysaccharides of *Salmonella typhimurium, Salmonella minnesota,* and *Escherichia coli. Journal of Bacteriology* 162, 738–745.

Qureshi, N. and Takayama, K. (1990) Structure and function of lipid A. In: Inglewski, B.H. and Clark, V.L. (eds) *The Bacteria, Vol. XI, Molecular Basis of Bacterial Pathogenesis.* Academic Press, London, pp. 319–338.

Raetz, C.R.H. (1993) Bacterial endotoxins: extraordinary lipids that activate eukaryotic signal transduction. *Journal of Bacteriology* 175, 5745–5753.

Reeves, P.R., Hobbs, M., Valvano, M.A., Skurnik, M., Whitfield, C., Coplin, D., Kido, N., Klena, J., Maskell, D., Raetz, R.H. and Rick, P.D. (1996) Bacterial polysaccharide synthesis and gene nomenclature. *Trends in Microbiology* 4, 495–503.

Rick, P.D. and Young, D.A. (1982) Isolation and characterization of a temperature-sensitive lethal mutant of *Salmonella typhimurium* that is conditionally defective in 3-deoxy-D-mannooctulosonate-8-synthetase. *Journal of Bacteriology* 150, 447–455.

Rick, P.D., Mayer, H., Neumeyer, B.A., Wolski, S. and Bitter-Suermann, D. (1985) Biosynthesis of enterobacterial common antigen. *Journal of Bacteriology* 162, 494–503.

Rietschel, E.T., Sidorczyk, Z., Zahringer, U., Wollen-Weber, H.-W. and Lüderitz, O. (1983) Analysis of the primary structure of lipid A. *ACS Symposium on Serology* 231, 214.

Schnaitman, C.A. and Klena, J.D. (1993) Genetics of lipopolysaccharide biosynthesis in enteric bacteria. *Microbiological Reviews* 57, 655–682.

Stock, J.B., Rauch, B. and Roseman, S. (1977) Periplasmic space in *Salmonella typhimurium* and *Escherichia coli. Journal of Biological Chemistry* 252, 7850–7861.

Valtonen, V.V. (1970) Mouse virulence of *Salmonella* strains: the effect of different smooth-type O side-chains. *Journal of General Microbiology* 64, 255–268.

Valtonen, V.V., Aird, J., Valtonen, M.V., Makela, O. and Makela, P.H. (1971) Mouse virulence of *Salmonella*: antigen dependent differences are demonstrable also after immunosuppression. *Acta Pathologica et Microbiologica Immunologica Scandinavica Section B* 79, 715–718.

Valtonen, M.V., Plosila, M., Valtonen, V.V. and Makela, P.H. (1975) Effect of the quality of the lipopolysaccharide on mouse virulence of *Salmonella enteritidis. Infection and Immunity* 12, 828–832.

Valtonen, M.V. (1977) Role of phagocytosis in mouse virulence of *Salmonella typhimurium* recombinants with O-antigen 6, 7 or 4, 12. *Infection and Immunity* 18, 574–582.

Virlogeux, I., Waxin, H., Ecobishon, C. and Popoff, M.Y. (1995) Role of the *viaB* locus in synthesis, transport and expression of *Salmonella typhi* Vi antigen. *Microbiology* 141, 3039–3047.

Westphal, O. and Jann, K. (1965) Bacterial lipopolysaccharide extraction with phenol water and further application of the procedure. *Methods in Carbohydrate Chemistry* 5, 83–91.

Chapter 3
Fimbriae of *Salmonella*

Christopher J. Thorns and Martin J. Woodward
Bacteriology Department, Veterinary Laboratories Agency (Weybridge), New Haw, Addlestone, Surrey KT15 3NB, UK

Introduction

Fimbriae are a family of polymeric proteinaceous surface organelles expressed by many bacteria, of which those of *Escherichia coli* have been most extensively studied (Smyth *et al.*, 1994). In particular, the fimbriae of enterotoxigenic *E. coli*, such as F4 and F5 (K88 and K99), which are responsible for the attachment of the bacterium to the villous epithelium of the small intestine of domestic animals, are now widely used in diagnostic tests and as active components in bacterins and subunit vaccines (Walker and Foster, 1983; Thorns *et al.*, 1989a,b).

Although the expression of fimbriae by certain strains of *Salmonella* was first described nearly 40 years ago (Duguid and Gillies, 1958), until recently there was little understanding of the variety of *Salmonella* fimbriae and their functions. However, in the last 10 years, there has been a renewed interest in the molecular and antigenic characterization and functions of *Salmonella* fimbriae, which has identified their potential as diagnostic and protective antigens (Thorns, 1995). Most of this work has focused on fimbriae expressed by *S. typhimurium* and, more recently, on *S. enteritidis*. In this review, we describe the different types of fimbriae expressed by *Salmonella* and methods of isolation and detection, molecular organization and regulation and current perspectives in the functions of these surface structures.

Classification of *Salmonella* Fimbriae

The association between *Salmonella* fimbriae and the ability of the bacteria to agglutinate certain species of erythrocytes was first noticed by Duguid and Gillies (1958). At this time, it was also discovered that this association between fimbriae and erythrocytes included mannose-containing carbohydrates in lectin-based interactions (Collier *et al.*, 1955). In a large study with over 1400 *Salmonella* strains, fimbriae were classified on the basis of their morphology and ability to mediate erythrocyte agglutination in the presence or absence of D-mannose (Duguid *et al.*, 1966). This classification has been used successfully for many years. However, it is now less appropriate, since a number of novel fimbriae are morphologically similar and do not appear to mediate haemagglutination reactions of any type. A great variety of *Salmonella* fimbriae have now been described (Table 3.1) and the important features of each are detailed below.

Type 1 fimbriae

Type 1 fimbriae comprise a family of rod-shaped organelles, 7–8 nm in diameter and up to 100 nm long (Fig. 3.1a). They have a peritrichous distribution, although only about 10% of cultured bacteria appear to express them at any one time (M. Sojka, unpublished observations).

Table 3.1. Characteristics of *Salmonella* fimbriae.

Fimbrial class	Fimbrial name	Morphology	Diameter (nm)	M_r subunit (kDa)	Haemagglutination reaction	*Salmonella* serotype	Genetic determinant Operon	Location
Type 1	F1	R	7–8	21	MS	*S. typhimurium*	*fim*	C
	SEF21	R	7–8	21	MS	*S. enteritidis*	*fim*	C
Type 2		R	7–8		NH	*S. pullorum,*		
					NH	*S. dublin*		
					NH	*S. gallinarum*		
Type 3		F	3–5	22	MRTE	*S. enteritidis*		
					MRTE	*S. typhimurium*		
Type 4		F	3		MRFE	*S. typhimurium*		
Type 4-like	Bundle-forming pili	F	7	18.5–21*	?	*S. dublin*		
NC	SEF14	F	3	14	NH	*S. enteritidis*	*sef*	C
					NH	*S. dublin*		
GVVPQ	SEF17	F	3	17	NH	*S. enteritidis*	*agf*	C
	Thin aggregative fimbriae	F	3	17	NH	*S. typhimurium*		
NC	Long polar fimbriae	R	7–8†		?	*S. typhimurium*	*lpf*	C
NC	Plasmid-encoded fimbriae				?	*S. enteritidis*‡	*pef*	P
						S. typhimurium		

*Deduced from subunits produced in *E. coli*.

†When expressed in *E. coli*.

‡Assumption based on specific antibodies produced following infection.

R, rigid and rod-shaped; F, fibrillar; MS, mannose-sensitive; MRTE, mannose-resistant with tanned erythrocytes; MRFE, mannose-resistant with fresh erythrocytes;
NC, not classified; NH, no haemagglutination; C, chromosome; P, plasmid.

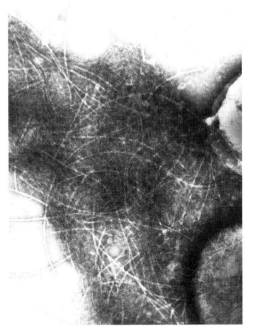

(b)

Fig. 3.1. Distinct fimbrial antigens expressed by *Salmonella enteritidis.* (a) Rigid SEF21 (type 1 class) fimbriae arranged peritrichously around the cell. (b) Fine and tightly coiled SEF17 (GVVPQ class) fimbriae closely associated with flagella and giving a dense matted appearance when viewed by electron microscopy. (c) Very thin and flexible SEF14 fimbriae expressed by the bacterium when grown in nutrient-limiting conditions and enhanced by immunostaining with an SEF14 monoclonal antibody.

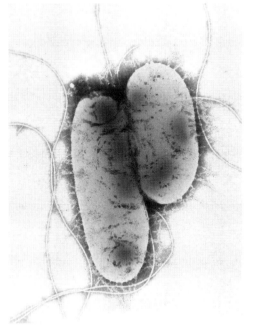

(a)

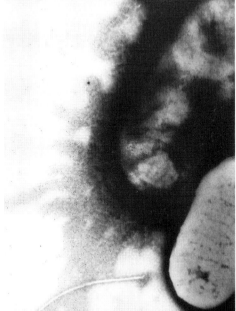

(c)

SEF21 fimbriae are members of the type 1 class of fimbriae and were first described on strains of *S. enteritidis* by Müller *et al.* (1991). Cultured *S. enteritidis* express SEF21 at 18°C and 37°C optimally in the late logarithmic to stationary growth phase. Like other type 1 fimbriae, they are composed of repeating major protein subunits, with a smaller number of minor proteins non-covalently linked around a hollow core, which gives a channelled appearance when viewed by electron microscopy. Major subunits of molecular mass between 20 and 22 kDa have been described for SEF21 and other type 1 fimbriae of *Salmonella* (Korhonen *et al.*, 1980; Kukkonnen *et al.*, 1993; Sojka *et al.*, 1996). There is considerable genetic and antigenic cross-reactivity between SEF21 and other type 1 fimbriae of *Enterobacteriaceae*, although SEF21-specific epitopes have also been described (Purcell *et al.*, 1987; Sojka *et al.*, 1996, 1998). Like other type 1 fimbriae of *Salmonella*, SEF21 mediate mannose-sensitive haemagglutination and lectin-based binding to mannoside glycoprotein receptors situated on a variety of epithelial cells.

Type 2 fimbriae

Type 2 fimbriae are morphologically indistinguishable from type 1 but are unable to agglutinate erythrocytes of any species. They were originally observed on strains of *S. gallinarum/pullorum* (Duguid and Gillies, 1958) and have subsequently been described on *S. paratyphi* B and *S. dublin* strains (Duguid *et al.*, 1966). There is a very close molecular and antigenic relationship between type 1 and 2 fimbriae (Swenson *et al.*, 1991; Sojka *et al.*, 1998), which indicates that the type 2 fimbriae may be non-haemagglutinating type 1 variants.

Type 3 fimbriae

Type 3 fimbriae are thin, flexible structures with diameters between 3 and 5 nm, arranged peritrichously around the bacterial cell when viewed by electron microscopy. They are able to mediate the agglutination of tannic-acid-treated erythrocytes in the presence of α-D-mannose (Duguid *et al.*, 1966). Like type 1 fimbriae, there is consider-able antigenic conservation between type 3 fimbriae expressed by *Salmonella*, *Klebsiella* and *Yersinia* spp. (Adegbola *et al.*, 1983; Old and Adegbola, 1985).

Type 4 fimbriae

Type 4 fimbriae were originally described by Duguid *et al.* (1966) as thin, flexible fimbriae 4 nm in diameter, with the ability to mediate agglutination of fresh erythrocytes in the presence of mannose (mannose-resistant (MR)), and until recently they had not been described on *Salmonella*. However, thin, flexible fimbriae 3 nm in diameter that mediate MR haemagglutination of pigeon erythrocytes have been described on a strain of *S. typhimurium* isolated from pigeons (Grund and Weber, 1988; Grund and Seiler, 1993). Unfortunately, 'type 4 fimbriae of enterobacteria' has also been used to refer to a recently described class of fimbriae with an unusual export system for the translocation of the fimbrin subunits across the cytoplasmic membrane (Strom and Lory, 1993). Morphologically, these are quite different from the type 4 fimbriae originally described by Duguid *et al.* (1966), being 7 nm in diameter and 10–20 nm long and having a polar arrangement around the cell (Girón *et al.*, 1994). Type 4-like fimbriae, termed bundle-forming pili (bfp), normally associated with enteropathogenic *E. coli*, have also been described as being expressed by strains of *S. dublin*, although the evidence is based only on indirect observation of *S. dublin* bacteria by electron microscopy (Sohel *et al.*, 1993).

GVVPQ fimbriae

This designation refers to the short amino acid sequence conserved at the N terminus of the major fimbrial subunit of certain fimbrial types. This conserved sequence occurs between the *Salmonella* fimbrial antigen SEF17 and the thin, coiled, fibrillar structures, called curli, that are expressed by the entero-aggregative *E. coli* associated with infantile diarrhoea in humans (Collinson *et al.*, 1992). SEF17 fimbriae were first described on a strain of *S. enteritidis* by Collinson *et al.* (1991). They have a similar morphology to curli fimbriae, appear tightly coiled in a peritrichous arrangement around the cell (Fig. 3.1b) and comprise protein subunits of molecular mass 17 kDa, which contain a receptor for the tissue-matrix protein fibronectin (Collinson *et al.*, 1993). Typically, SEF17 fimbriae are expressed by organisms cultured on solid medium such as colony factor antigen (CFA) agar, at temperatures up to 30°C, although a few strains of *S. enteritidis* express SEF17 constitutively to 42°C (Dibb-Fuller *et al.*, 1997).

SEF14 fimbriae

SEF14 fimbriae comprise a unique class of *Salmonella* fimbriae and were first described on strains of *S. enteritidis* (Thorns *et al.*, 1990; Müller *et al.*, 1991). Their structure comprises thin, filamentous organelles, < 3 nm in diameter, peritrichously arranged (Fig. 3.1c) and composed of repeating protein subunits of molecular mass 14.3 kDa. Under certain cultural conditions, *S. enteritidis* can co-express SEF14 and SEF21 fimbriae. SEF14 is unique to certain serovars in serogroup D, including all strains of *S. enteritidis* so far examined and a proportion of *S. dublin* strains (Thorns *et al.*, 1992). Cultured organisms express SEF14 at temperatures above 30°C and their expression is subject to catabolite repression.

The structural gene encoding SEF14 has been cloned and sequenced (*sef*A) and shown to be limited in distribution to serovars belonging to serogroup D. Interestingly, *S. gallinarum/pullorum* and *S. typhi* all possess the entire *sef*A gene but do not express SEF14 fimbriae (Turcotte and Woodward, 1993). The unique specificity of SEF14 to *S. enteritidis* has been successfully applied to the development of rapid antigen and serological tests for the specific detection of *S. enteritidis* in poultry (McLaren *et al.*, 1992; Hoorfar and Thorns, 1996; Thorns *et al.*, 1996a).

SEF18 fimbriae

The SEF14 fimbrial operon (see below) appears to possess two structural genes; *sef*A encodes the SEF14 fimbriae, whereas *sef*D encodes for the antigenically distinct SEF18 fimbriae, whose expression is also independent of growth temperature (Clouthier *et al.*, 1994). The same workers report that SEF18 fimbriae are highly conserved structures, ubiquitous among the *Enterobacteriaceae*, and speculate that they may contribute to cell-to-cell adherence. However, conclusive identification of SEF18 as fimbriae has not yet been made, since isolation, detection and purification of the assembled structure has not been reported.

Long polar fimbriae (LPF), plasmid-encoded fimbriae (PEF) and bovine colonization factor (BCF)

Long (10–20 μm) fimbriae, with a diameter of c. 7 nm and morphologically similar to enterobacterial type 4 fimbriae, have been observed on a non-fimbriated *E. coli* strain into which an *S. typhimurium* fimbrial operon had been introduced (Bäumler and Heffron, 1995). The fimbriae were expressed at the pole of the bacterium only and have been termed long polar fimbriae (LPF). Similarly, profuse peritrichous fimbriae were observed on *E. coli* harbouring the PEF operons from *S. typhimurium* and *S. enteritidis* (Friedrich *et al.*, 1993; Woodward *et al.*, 1996). However, neither native LPF nor PEF fimbriae have yet been detected on wild-type *Salmonella*, although PEF seroconversion has been reported in chickens infected with *S. enteritidis* (Woodward *et al.*, 1996). A newly identified fimbrial operon of *S. typhimurium* (BCF) is associated with specific adherence to bovine but not murine Peyer's patches (see Chapter 4).

Detection of Fimbriae and their Application as Diagnostic Reagents

Rapid immunoassays using specific monoclonal antibodies (Mabs) have now replaced the traditional methods of haemagglutination and electron microscopy for the detection of many types of fimbriae, such as type 1, GVVPQ and SEF14 fimbrial classes (Fig. 3.2). However, sugar-sensitive and sugar-resistant haemagglutination reactions will continue to aid the detection of novel fimbriae that mediate lectin-specific binding to cell receptors (Fig. 3.3).

The first diagnostic tests based on the detection of fimbrial antigens were developed for the detection of enterotoxigenic *E. coli* (ETEC)

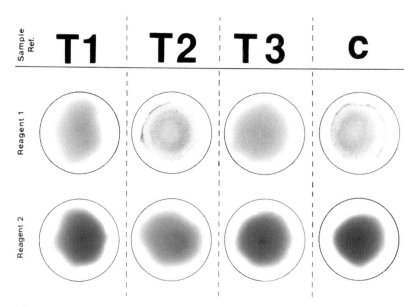

Fig. 3.2. Agglutination of coloured latex particles coated with an SEF14 monoclonal antibody in the presence of *S. enteritidis* expressing SEF14 fimbriae. For comparison, two non-producing SEF14 strains (T1 and T3) are also shown.

(Sojka, 1971). The stimulus for their development and use was the discovery that fimbriae such as F4 and F5 are essential for the *E. coli* to cause disease, and their detection on *E. coli* isolates from cases of diarrhoea is of diagnostic significance. A variety of tests that incorporate Mabs are now commonly used for the rapid detection of fimbriae expressed by ETECs (Thorns *et al.*, 1989a,b).

Although much recent effort has gone into the characterization of fimbriae expressed by *Salmonella*, their application for the detection of *Salmonella* infections has not been fully exploited to date. Two main reasons for this have been the concentration of research towards characterization of type 1 fimbriae and a lack of understanding of the role of these antigens in the life cycle of the bacterium, such that their detection is of uncertain clinical significance.

Detection of *Salmonella* Genus

Genus-specific tests, based mainly on the detection of surface antigens, such as lipopolysaccharide (LPS), flagellin and outer-membrane proteins, have been used with some success in enzyme-linked immunosorbent assays (ELISAs) and agglutination tests (Clark *et al.*, 1989; Feldsine *et al.*, 1992; Kerr *et al.*, 1992; Manafi and Sommer, 1992; Wyatt *et al.*, 1993). Recently, a DNA-based test that targets the *agfA* structural gene of SEF17 fimbriae (GVVPQ class) has been developed and shown to react strongly with 603 of 604 *Salmonella* isolates and only very weakly with 31 of 266 other members of the family *Enterobacteriaceae* (Doran *et al.*, 1993). Similar strategies have been developed by Cohen *et al.* (1996), who used *Salmonella*-specific polymerase chain reaction (PCR) probes to the type 1 *fimA* gene of *S. typhimurium*, and by Woodward and Kirwan (1996), who used a PCR assay based on the *sefA* gene to detect *S. enteritidis* in eggs. These are the first instances in which fimbrial gene probes have been used successfully as a genus-specific diagnostic tool and may, in the future, offer quicker and more sensitive approaches than existing culture methods.

Specific detection of *Salmonella enteritidis*

The rapid spread of *S. enteritidis* through the poultry population and the subsequent increase

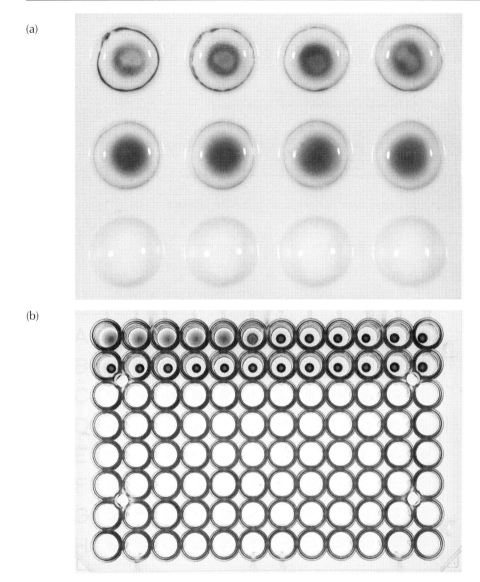

Fig. 3.3. Traditional techniques for the detection of type 1 fimbriae. (a) Agglutination of horse erythrocytes with SEF21 fimbriae in the rock-tile test. For comparison, absence of agglutination is shown in the second row. (b) Similar test carried out in microtitre plates with fimbriae diluted from left to right and incubated overnight. The lower row contains mannose, which inhibits the agglutination.

in human food poisoning cases caused by this serovar highlight the need for more rapid, specific tests to identify *S. enteritidis* infections in animal and human populations in order to apply appropriate control strategies. Recently, a simple latex-agglutination test has been developed for the specific identification of *S. enteritidis*, based on the detection of SEF14 (McLaren *et al.*, 1992; Thorns *et al.*, 1992, 1994). Extensive evaluation

by laboratories worldwide has demonstrated that this is an accurate, presumptive test for *S. enteritidis* and can be carried out in routine laboratories that would normally have to send all their *Salmonella* isolates to specialized laboratories for serotyping.

The increase in bacteriological monitoring of poultry flocks for *Salmonella* has required an evaluation of alternative screening procedures in order to reduce the time and costs involved. The most likely alternative is serological monitoring, especially as more specific ELISA-based assays become available (Barrow, 1994; see also Chapter 24). A recent development is the application of SEF14 fimbriae to the specific serodiagnosis of chicken flocks infected with *S. enteritidis*. Results indicate that birds infected with *S. enteritidis* readily seroconvert within 10 days of infection and the immunoglobulin G (IgG) response persists for at least 4 weeks thereafter (Thorns *et al.*, 1996a). The production of SEF14 antibodies following infection was the first demonstration of a specific anti-fimbrial response to *Salmonella* infection. Preliminary data indicated that the potential advantages of the detection of SEF14 antibodies are high specificity and sensitivity.

The Genetics of Fimbrial Biosynthesis and Regulation

Type 1 fimbriae

Clegg and colleagues (Clegg *et al.*, 1987; Swenson and Clegg, 1992) defined a 13.7 kb *SphI* chromosomal region of *S. typhimurium* that encoded the entire type 1 gene cluster (Fig. 3.4). Nucleotide sequence analysis identified nine open reading frames within a contiguous 9500 base pair region, flanked by genes that encoded tetrahydrofolate dehydrogenase and an arginine tRNA. This latter gene has been redefined as *fimU* and it is speculated that the tRNA function may be required for translation of some other *fim* genes (Swenson *et al.*, 1994). Indeed, Clouthier *et al.* (1998) suggested that *fimU* was involved in the co-regulation of both type 1 and SEF14 fimbriae of *S. enteritidis*. The structure of the operon is similar to that encoding the type 1 fimbriae of *E. coli*, but with notable exceptions with regard to the haemagglutinin and regulatory genes. The *fimA* gene encodes a 21 kDa protein, which is the major structural monomer of the fimbriae. The *fimA* gene of the genus *Salmonella* is highly conserved (Warner and M.J. Woodward, unpublished data) and the published *S. typhimurium* sequence has considerable similarity to that of *E. coli*. The *fimA* gene is transcribed monocistronically from its own promoter under complex phase-variation regulation,

involving three other gene products of this gene cluster. The adjacent *fimI* gene encodes a 16 kDa protein, which shares 65% identity with FimA, but the role of this protein is unclear and it may be a minor component of the mature fimbrial structure (Rossilini *et al.*, 1993). Genes *fimC* and *fimD* encode proteins of 25 and 82 kDa, respectively, which, due to their similarity to other defined fimbrial genes (Hultgren *et al.*, 1991), perform chaperone and usher functions for fimbrial synthesis, while *fimF* encodes a minor protein that may play a role in initiation of fimbrial synthesis (Russell and Orndorff, 1992). The *fimH* gene encodes a 34 kDa protein that is related to the SfaS protein of the *E. coli* S-fimbrial adhesin, but not the *E. coli* FimH haemagglutinin, and it is probable that this protein confers the mannose-sensitive adhesive properties of *Salmonella* (Krogfelt *et al.*, 1990). The *fimICDHF* genes are transcribed polycistronically from a single promoter (Swenson and Clegg, 1992), whereas three genes encoding regulatory functions, *fimZYW*, are transcribed polycistronically but in the opposite orientation (Yeh *et al.*, 1995).

Regulation of the type 1 fimbrial operon in *S. typhimurium* seems to have evolved differently from that of *E. coli*, in which an invertible genetic element encoding a promoter sequence is located immediately downstream of the *fimA* gene. Phase-on or phase-off orientation is regulated by the gene products of *fimB* and *fimE* genes, both located further downstream of the *fim* operon (for review, see Dorman *et al.*, 1997). No similar genes are found in *S. typhimurium* and, although the *fimA* gene is flanked by a 10 bp repeat sequence, Southern hybridization analysis of phase-on and phase-off bacteria indicates that no genetic rearrangement of the repeat sequences or the *fimA* region occurs (Clegg *et al.*, 1996). Unlike the regulation of the type 1 fimbriae of *E. coli*, relatively little is known about regulation in *Salmonella*, other than media and temperature effects (Sojka *et al.*, 1996). The roles of global regulators have yet to be established.

SEF14 and SEF18 fimbriae

The *sefABCD* gene cluster (Fig. 3.4) encodes SEF14 fimbriae (Clouthier *et al.*, 1993; Turcotte

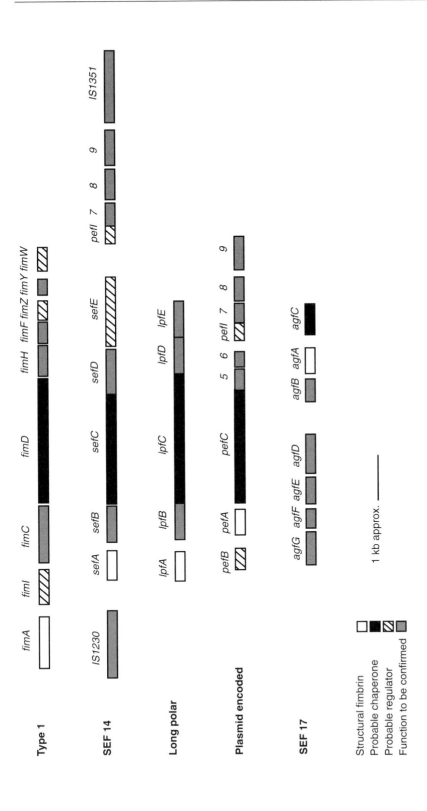

Fig. 3.4. Genetic maps of each of the fimbrial operons of *S. typhimurium* (type 1, LPF and PEF) and *S. enteritidis* (type 1, SEF14, SEF17 and PEF) drawn to scale and coded according to assignment of gene function, where known.

and Woodward, 1993). The major fimbrial subunit antigen is encoded by *sefA*, whilst putative usher and chaperone functions are encoded by *sefB* and *sefC*. These genes share identity with those encoding the P and CS31 fimbriae of *E. coli* and the MRK type 3 fimbriae of *Klebsiella pneumoniae*. Upstream, Clouthier *et al.* (1994) identified a further open reading frame, designated *sefD*, which overlapped *sefC* and shared identity with other major fimbrial subunit antigen genes, including *sefA* itself. Unlike the *sefABC* gene cluster, whose distribution is limited to group D *Salmonella* (Turcotte and Woodward, 1993), *sefD*, as determined by colony dot–blot hybridization, is ubiquitous amongst the *Enterobacteriaceae* (Clouthier *et al.*, 1994).

The expression of *sefD* occurs under a wide variety of environmental conditions, as demonstrated by Western blot of whole-cell extracts (Clouthier *et al.*, 1998). SEF14 fimbriae of *S. enteritidis* and *S. dublin* are expressed under specific environmental conditions *in vitro*, whereas elaboration by *S. typhi* or *S. gallinarum/pullorum* was not detected, although these *Salmonella* are genetically competent (Turcotte and Woodward, 1993). Intriguingly, Rodriguez-Pena *et al.* (1997) demonstrated co-location of *sef* and *pef* genes on the chromosome of *S. typhi*, a finding we have confirmed also for *S. enteritidis* (R.J. Collighan and M.J. Woodward, unpublished observations). Sequence analysis downstream of the cluster revealed potential promoter recognition sites, complex hairpin-loop structures (Turcotte and Woodward, 1993) and an IS3-like element (Collighan and Woodward, 1997), although no evidence was gained for regulation of expression by inversion of the element. Indeed, the element seemed to be inactive, with the putative transposase reading frame interrupted by three point mutations. Interestingly, the GC ratio of the IS3-like element and the SEF14 gene cluster, but with the notable exception of the *sefA* gene, is considerably lower than that of adjacent sequences. Thus, the limited distribution within group D *Salmonella* and the co-location of an IS element may indicate the origin of the gene cluster as recent and possibly associated with a pathogenicity island (for reviews, see Finlay and Falkow, 1997; Hacker *et al.*, 1997).

Perhaps the most intriguing observation is the role of *fimU*, which encodes a tRNA molecule encoding the rare arginine anticodon UCU, in co-regulation of both SEF21 (type 1) and SEF14 expression (Clouthier *et al.*, 1998). These authors examined a Tn10 insertion mutant of *fimU*, which abolished SEF14 expression under all conditions and severely modulated SEF21 expression. Interestingly, the rare arginine codon is not found in *sefA* but is found in *sefBCD* and *sefE*, the latter gene is reported by Clouthier *et al.* (1998) to be an AraC-like transcriptional activator. We have confirmed this finding by insertional inactivation studies and shown that PEF sequences are also co-located upstream of the *sef* gene cluster and bound by a further IS element and adjacent to *leuX*, a leucine-specific tRNA gene (R.J. Collighan and M.J. Woodward, unpublished observations). The question arises as to whether this constitutes a pathogenicity islet.

SEF17 fimbriae

The *agfBAC* gene cluster (Fig. 3.4) encodes SEF17 fimbriae of *S. enteritidis* (Collinson *et al.*, 1991, 1992, 1993, 1996a); it shares significant sequence identity with the *csgBAC* operon encoding the 'curli' fibronectin-binding fimbriae of *E. coli* (Arnquist *et al.*, 1992). The *agfA* gene encoding major fimbrial subunit antigen, and presumably the entire operon, is ubiquitous amongst *Salmonella* and has been developed as a probe for the specific detection of the microorganism (Doran *et al.*, 1993). Northern blot analysis confirmed transcription of polycistronic *agfBA* mRNA from a region immediately downstream of *agfB*, but transcription of *agfC* was not detected in these experiments (Collinson *et al.*, 1996a). The role of an intercistronic stem–loop structure between *agfA* and *agfC* is unclear. Hammar *et al.* (1995) demonstrated the role of two co-located but distinct gene clusters, *agfBAC* and *agfDEFG*, which were transcribed in opposite orientations for the regulation and elaboration of the 'curli' fimbriae of *E. coli*. A similar gene arrangement exists in *S. typhimurium* (Romling *et al.*, 1998). Significantly, 'curli' elaboration was shown to be influenced by environmental factors with a direct involvement of RpoS and HN-S factors for expression (Olsen *et al.*, 1993, 1994) and possibly OmpR (Romling *et al.*, 1998) in *S. typhimurium*. Humphrey *et al.* (1996) has described a naturally occurring avirulent *S. enteritidis* isolate, designated strain I, which is unable to elaborate SEF17 fimbriae due to a genetic lesion within the *rpoS* allele (Allen-Vercoe *et al.*, 1997).

Long polar fimbriae

Bäumler and Heffron (1995), using a novel approach to negative screening, identified a chromosomal region in *S. typhimurium* and related serovars, and some serovars of group D, which included *S. enteritidis* and *S. dublin*, that was absent from *S. typhi*, *S. arizonae*, *E. coli*, *Shigella* spp. and other *Enterobacteriaceae*. The nucleotide sequence of the region identified a putative fimbrial operon comprising five contiguous open reading frames (Fig. 3.4), located at 78 minutes on the genetic map of *S. typhimurium*; the operon had both sequence identity and gene organization similarity to the *S. typhimurium fim* cluster and, less so, to the MR/P fimbrial operon of *P. mirabilis* and the P fimbrial operon of *E. coli*. The *lpfA* gene, encoding the putative major fimbrial subunit antigen, is flanked by 14 bp repeat sequences and the promoter region contains four 9 bp direct repeat sequences. As yet, nothing is understood of the mode of regulation of this operon. However, the afimbriate *E. coli* K12 strain ORN172 harbouring a cosmid encoding the entire *lpf* region derived from *S. typhimurium* elaborated very thick, rod-like, polar fimbriae 2–10 μm long. In this laboratory, similar experiments with *S. enteritidis* have confirmed the presence of the *lpf* operon but have been unable to demonstrate expression of polar fimbriae under any of the conditions tested thus far (Allen-Vercoe and Woodward, 1999; Allen-Vercoe *et al.*, 1999).

Plasmid-encoded fimbriae

The serovar associated plasmid (SAP, virulence plasmid) has been shown to contribute to the systemic phase of *S. typhimurium* infection of mice (Pardon *et al.*, 1986) and of *S. dublin* infection of cattle (Wallis *et al.*, 1995). The *spv* region of the SAP has been well studied (for review, see Gulig *et al.*, 1993) but a Tn5 insertional mutation in the *S. typhimurium* SAP distant from the *spv* region resulted in attenuation

(Sizemore *et al.*, 1991). The nucleotide sequence of this region, a 13.9 kb contiguous segment of the SAP between the *repB* and *repC* genes, was determined by Friedrich *et al.* (1993). A putative fimbrial operon containing seven open reading frames (Fig. 3.4), which were, in terms of gene organization and deduced amino acid sequence, similar to the P and F4 (K88) adherence fimbriae of ETEC. Woodward *et al.* (1996) demonstrated the presence of *pef* sequences encoded by the SAPs of *S. enteritidis*, *S. choleraesuis* and *S. bovismobificans*. Interestingly, the deduced amino acid sequence of the *S. enteritidis* PefA, the presumed major fimbrial subunit antigen, shared only 76% identity with the *S. typhimurium* analogue, indicating discrete evolutionary pathways for the acquisition of these genes in *Salmonella* (A. Baumler, Texas, and D. Platt, Glasgow, 1999, personal communications). Nothing is known of the regulation of expression of PEF and elaboration of these fimbriae has yet to be observed on wild-type *S. typhimurium* and *S. enteritidis* grown under any of the *in vitro* conditions tested so far (unpublished observations). Interestingly, regions of the probable regulatory region of the *pef* operon are co-located with the *sef* operon on *S. enteritidis* (R.J. Collighan and M.J. Woodward, unpublished observations). *E. coli* K12 recombinants that harboured the entire *S. typhimurium* or *S. enteritidis pef* region elaborated profuse fimbriae (Fig. 3.5; Friedrich *et al.*, 1993; Woodward *et al.*, 1996), which indicated zygotic induction. It is possible that chromosomal genes are involved in their regulation and that specific *in vivo* conditions are required for induction, as for the *spv* genes (Rhen *et al.*, 1993).

Bundle-forming pili (BFP) and other newly described fimbriae

Enteropathogenic *E. coli* (EPEC) adhere to epithelial cells in microcolonies, to form the so-called localized adherence (LA) pattern, for which the EAF plasmid is essential (Baldini *et al.*, 1983). This plasmid also encodes the region where BFP were first identified (Sohel *et al.*, 1993). The nucleotide sequence of some 14 genes encoding the elaboration of the fimbriae was determined simultaneously by Stone *et al.* (1996) and Sohel *et al.* (1996) and a further three regulatory genes were identified by Tobe *et al.* (1996). Optimal expression of BFP was obtained during exponential growth at 37°C in the presence of calcium (Puente *et al.*, 1996). Sohel *et al.* (1993) used the *bfp* region to probe other enteropathogens and demonstrated hybridization with many *Salmonella* at low stringency, indicating the presence of analogous genes in some *Salmonella*, although this finding has yet to be confirmed by other laboratories.

Brocchi *et al.* (1999) studied a preferred site for mini-Tn5 integration in *S. typhimurium*. One such site mapped to a hitherto uncharacterized site on the SAP, and sequence analysis identified four genes in this region that shared sequence identities in the region of 40–46% homology with the type 1 fimbriae described previously by Purcell *et al.* (1987). The relationship between this and the BCF described by Bäumler *et al.* (Chapter 4) and Tsolis *et al.* (unpublished observations) remains to be determined. Recently, Folkesson *et al.* (1999) described a large region of DNA unique to *Salmonella* spp., a putative pathogenicity island, located at centisome 7 of the genome. This region encoded a deduced operon encoding four genes which shared a high degree

of identity with a number of other fimbrial operons and was designated the *Salmonella* atypical fimbria (SAF). Whilst knock-out mutants of this operon did not attenuate *S. typhimurium* in the mouse model, the question arises as to its significance in human and animal disease.

The Role of Fimbriae

Contemporary molecular approaches to the study of *Salmonella* pathogenesis have identified at least five pathogenicity islands (PIs) and, although these studies point to a significant role of secretory functions and secreted antigens in pathogenesis, genetic loci encoding fimbriae have not been identified within the PIs by these approaches. However, Takeuchi (1967) described fimbria-like structures involved in the interaction between *S. typhimurium* and murine gut epithelium *in vivo*. Type 1 fimbriae have been considered to play a role in early events of gut colonization, due to their ability to cause haemagglutination and the fact that they are ubiquitous amongst *Salmonella* (Duguid *et al.*, 1966, 1976; Darekar and Eyer, 1973). Many studies support this view

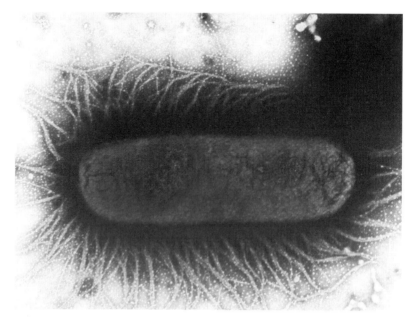

Fig. 3.5. An electron micrograph of an afimbriate *E. coli* K12 derivative harbouring a cosmid with genomic DNA that encoded the entire *pef* operon derived from *S. enteritidis*. Copious fine flexible fimbriae forming rope-like clusters were observed.

(Tanaka and Katsube, 1978; Korhonen *et al.*, 1981; Tanaka *et al.*, 1981) and other studies have indicated an additional adhesive role for type 3 fimbriae (Jones and Richardson, 1981; Jones *et al.*, 1982; Tavendale *et al.*, 1983) and also the SAP (Jones *et al.*, 1982). Campbell *et al.* (1987) noted reversible adherence to chicken connective muscle of all *Salmonella* tested while Stenstrom and Kjelleberg (1985) reported enhanced non-specific binding of fimbriate *S. typhimurium* to minerals. Adherence to tissue culture has also been used to assess the role of fimbriae, although the results have been conflicting (Mintz *et al.*, 1983; Old *et al.*, 1986) – probably because the key factors that influence adherence are the bacterial cultural conditions and the tissue types, which determine fimbrial elaboration and expression of different targets, respectively. Kukkonen *et al.* (1993, 1998) reported binding of type 1 fimbriae to laminin and plasminogen and Ghosh *et al.* (1996) described a 60 kDa glycoprotein receptor of the *S. typhimurium* type 1 fimbriae in rat intestinal brush-border membranes. Interestingly, Lockman and Curtiss (1992a,b) described an *S. typhimurium* mutant defective for the elaboration of type 1 fimbriae that was not attenuated in the murine typhoid model and a mutant defective for the elaboration of both type 1 fimbriae and flagella that was tenfold less virulent, as assessed by median lethal dose (LD_{50}), which supported the observation by Jones *et al.* (1981) that flagella play a role in attachment.

SEF17 production is required for the elaboration of the Congo-red-binding and autoaggregative phenotypic traits of *S. enteritidis* (Collinson *et al.*, 1991; Dibb-Fuller *et al.*, 1997) and *S. typhimurium* (Romling *et al.*, 1998). Recently, a particular 'lacy' colonial morphology has been observed with strains of *S. enteritidis*, which may be associated with increased virulence of the organism (Guard-Petter, 1993; Guard-Petter *et al.*, 1996; Humphrey *et al.*, 1996). Guard-Petter (1993) demonstrated that *S. enteritidis* isolated directly from infected hen's eggs generate a 'lacy' colony morphology, originally described as 'convoluted' by Jameson (1966), which was considered more virulent in the chick model (Guard-Petter *et al.*, 1995). The 'lacy' phenotype was dependent upon expression of SEF17 by the organism (Allen-Vercoe *et al.*, 1997; Romling *et al.*, 1998). The thin aggregative

fimbrial analogues, 'curli' of *E. coli* and SEF17 of *Salmonella*, both bind fibronectin, plasminogen and Congo red (Baloda, 1988; Collinson *et al.*, 1993; Sjöbring *et al.*, 1994; Hammar *et al.*, 1995).

Significantly, Ben Nasr *et al.* (1996) reported the role of 'curli' in the assembly of contact-phase factors, notably H-kininogen (HK), at the bacterial surface and the release of the proinflammatory and vasoactive peptide bradykinin. These authors discuss the likely impact that this facet of the biology of the microorganism has on the disease process. Both *E. coli* and *S. enteritidis* bound HK at 26°C, the optimal temperature for *in vitro* elaboration of both 'curli' and SEF17. Another facet of 'curli', and presumably SEF17, was the ability for fimbrial assembly to occur extracellularly through a self-assembly process dependent upon a nucleator protein, although the significance for colonization and/or adherence remains to be tested (Hammar *et al.*, 1996). Romling *et al.* (1998) postulated that curli fibres may form a network, such as a biofilm, because of the adhesive nature of the fibres, the colonial morphology developed and the expression of curli apparently limited to ambient temperatures only. However, whilst this is an attractive hypothesis, we have established *in vitro* growth conditions that induce *S. enteritidis* to express SEF17 at 37°C (S. Walker and M.J. Woodward, unpublished observations). This finding and the array of phenotypes associated with the curli fibres of *Salmonella* spp. pointed strongly to a role in pathogenesis of infection. Indeed, Sukupolvi (1997a, b) demonstrated that curli fibres of *S. typhimurium* mediated interaction between bacterium and immortalized proximal small intestinal epithelial cells in the mouse model. Likewise, Herwald *et al.* (1998) demonstrated that *S. typhimurium* strain SR11, which is constitutively curliated, when mixed with mouse plasma depletes contact-phase proteins and clotting factors, incorporates these into the fibrin network, disturbs clot formation, prolongs clotting and changes clot fibril morphology.

Reverse-genetics approaches have been used to determine the role of fimbriae in pathogenesis. Bäumler *et al.* (1996a) generated a *pefC* mutant of *S. typhimurium* by insertional inactivation to assess the role in adhesion to the murine small intestine. In organ culture, the mutant adhered in lower numbers than wild-type *Salmonella* and non-adherent *E. coli* harbouring the *pef* gene cluster became adherent. Intragastric inoculation of wild-type *S. typhimurium* in the infant murine model caused fluid accumulation, whereas the *pefC* mutant did not. *E. coli* harbouring and expressing the *pef* gene cluster failed to induce fluid accumulation, which raises doubts about the sole function of PEF. Bäumler *et al.* (1996b), using a similar approach, demonstrated the role of the *lpf* fimbrial operon in the attachment of *S. typhimurium* to Peyer's patches in the murine model and of *fim* and *pef* fimbrial operons in *in vitro* adhesion assays (Bäumler *et al.*, 1996a). Interestingly, *invA lpfC* double mutants of *S. typhimurium* were attenuated 150-fold, as compared with either mutation alone, when inoculated orally in the mouse model (Bäumler *et al.*, 1997a).

To date, the study of *S. enteritidis* wild-type and defined SEF14 fimbrial deletion mutants in various animal models, including the mouse, rat and chick, suggests that SEF14 fimbriae do not contribute significantly to pathogenicity of the organism (Thorns *et al.*, 1996b; Oggunniyi *et al.*, 1997). However, it has been reported that the attachment of *S. enteritidis* to chicken ovarian granulosa cells *in vitro* can be partially blocked by cell-free SEF14 fimbriae (Thiagarajan *et al.*, 1996). This provides some evidence that SEF14 fimbriae contribute to infection of the reproductive tissue in mature hens. The co-location of *fim*, *sef* and *agf* genes to a single region of 850 kb, inverted with respect to one another on the genetic maps of *S. typhimurium* and *S. enteritidis*, is intriguing (Collinson *et al.*, 1996b) and presents evidence for the discussion of the evolution of fimbrial patterns within *Salmonella* (Bäumler *et al.*, 1997b).

Thus, evidence is accumulating that suggests that fimbriae and flagella play a role in *Salmonella* pathogenesis. However, in our laboratory, multiply afimbriate isogenic mutants of *S. enteritidis* were constructed for assay in the *in vivo* chick model and the results suggested that, at least in this model, flagella and not fimbriae were significant virulence determinants (Allen-Vercoe and Woodward, 1999; Allen-Vercoe *et al.*, 1999; Dibb-Fuller and Woodward, 1999). Interestingly, Stein *et al.* (1996) described the production of fimbria-like structures encoded by the *sifA* gene (*Salmonella*-induced filament) by *S. typhimurium* within the phagolysosome. It is clear

that numerous fimbriae have yet to be identified and functionally characterized and it is interesting that the BCF may play a role in host restriction, a theme that deserves further analysis. Additionally, fimbriae are expressed under a variety of stress conditions, with, for example, SEF14 upregulated by contact with inanimate surfaces (Walker *et al.*, 1999), which raises the possibility of a role for fimbriae in environmental survival and transmission.

Vaccines

The major diseases caused by *E. coli* in domestic animals include neonatal and post-weaning diarrhoea, oedema disease in pigs, septicaemia and mastitis. In the first recorded use of fimbriae as vaccine, pregnant gilts vaccinated by the intramammary route with purified F4 (K88) fimbriae afforded protection to sucking neonates upon challenge with ETEC (Rutter and Jones, 1973; Rutter, 1975). Other ETEC fimbriae, such as F5 (K99), F6 (987P) and F41, have been developed subsequently as vaccines (for review, see Issacson, 1994), which are available commercially. The rationale for their development was the prevention of gut colonization by inhibiting adhesion mediated by fimbriae. Whether a similar approach will be successful for *Salmonella* vaccines has yet to be tested, but evidence is becoming available that the many fimbriae of *Salmonella* may be virulence determinants, possibly associated with tissue-specific adherence, and may be vaccine candidates. It is interesting that live, multivalent *Salmonella* vaccines have been engineered to elaborate the fimbriae of ETEC and which afford protection against the heterologous challenge (Bertram *et al.*, 1994; Morona *et al.*, 1994; Giron *et al.*, 1995). Both B- and T-cell immune responses to the SEF14 fimbrial antigen (Thorns *et al.*, 1996a; Ogunniyi *et al.*, 1997) and B-cell responses to the PEF fimbrial antigen (Woodward *et al.*, 1996) were detected in chicken infected with wild-type *S. enteritidis*. Similarly, responses to SEF14 were detected in *S. enteritidis aroA* vaccinated chickens (Cooper and Thorns, 1996) and passive immunity in mice was afforded against experimental *S. enteritidis* challenge by egg-yolk-derived antibodies specific for the SEF14 fimbrial antigen (Peralta *et al.*, 1994). Studies with *S. enteritidis aroA* vaccines in a chicken model demonstrated solid protection from oral challenge and a significant reduction in the extent of invasion, persistence and shedding of the challenge strain (Cooper *et al.*, 1994a,b). Similarly no invasion was detected after an aerosol-delivered challenge (Cooper *et al.*, 1996). Detection of secretory immunoglobulin A (sIgA) against a number of antigens in these models gave inconclusive results (G.L. Cooper, personal communication), so the contribution to protection made by the fimbriae remains unknown but worthy of further investigation.

Concluding Remarks

It is now evident that *Salmonella* can express several fimbriae during their life cycle. The expression of fimbriae is tightly regulated by numerous environmental and cell signals which are still poorly understood. It is assumed, however, that their expression in specific niches in the infected host or in the environment plays an important part in the organism's survival. Current evidence indicates that *Salmonella* can produce numerous fimbriae, which mediate a variety of functions and are important for the maintenance and survival of the organism in the host and its environment. These functions might include the following.

1. Initiation and/or stabilization of the organism to epithelial cells.

2. Colonization of and microcolony formation in tissues via site-specific binding to receptors on tissue-matrix proteins.
3. Maintenance of persistent infections in the host by mediating selective bacterial trapping by phagocytic cells, and subsequent survival in specific niches of the infected animal.
4. Evasion of the host's specific immunological defences by:
(a) presentation of self epitopes on the fimbriae;
(b) production of harmless immune responses to fimbriae, which are then rapidly shed or their expression switched off from the surface of the organism.
5. Increased survival in the environment by enveloping the organism with highly hydrophobic fimbriae, which are resistant to changes in temperature, pH and water availability.

Recent advances in the molecular and structural characterization of *Salmonella* fimbriae in our laboratory and elsewhere have provided a range of defined afimbrial mutants and specific immunological reagents, such as monoclonal antibodies. The use of these tools in the next few years will enable researchers to gain a far greater understanding of the precise functions of this fascinating group of bacterial organelles.

References

Adegbola, R.A., Old, D.C. and Aleksic, S. (1983) Rare MR/K-like haemagglutinins (and type 3-like fimbriae) of *Salmonella* strains. *FEMS Microbiological Letters* 19, 233–238.

Allen-Vercoe, E. and Woodward, M.J. (1999a) Adherence of *Salmonella enterica* serovar Enteritidis to chick gut explant: the role of flagella but not fimbriae. *Journal of Medical Microbiology* 48, 771–780.

Allen-Vercoe, E., Dibb-Fuller, M., Thorns, C.J. and Woodward, M.J. (1997) SEF17 fimbriae are essential for the convoluted colonial morphology of *Salmonella enteritidis*. *FEMS Microbiological Letters* 153, 33–42.

Allen-Vercoe, E., Sayers, A.R. and Woodward, M.J. (1999b) Virulence of *Salmonella enterica* serovar Enteritidis aflagellate and afimbriate mutants in a day-old chick model. *Epidemiology and Infection* 122, 395–402.

Arnquist, A., Olsen, A., Pfeifer, J., Russell, D.G. and Normark, S. (1992) The Crl protein activates cryptic genes for curli formation and fibronectin binding in *Escherichia coli* HB101. *Molecular Microbiology* 6, 2443–2452.

Baldini, M.M., Kaper, J.B., Levine, M.M., Candy, D.C. and Moon, H.W. (1983) Plasmid-mediated adhesion in enteropathogenic *Escherichia coli*. *Journal of Pediatrics, Gastroenterology and Nutrition* 2, 534–538.

Baloda, S.M. (1988) Cell surface properties of enterotoxogenic and cytotoxigenic *Salmonella enteritidis* and *Salmonella typhimurium*: studies on haemagglutination, cell surface hydrophobicity, attachment to human intestinal cells and fibronectin. *Microbiology and Immunology* 32, 447–459.

Barrow, P.A. (1994) Serological diagnosis of *Salmonella* serovar *enteritidis* infections in poultry by ELISA and other tests. *International Journal of Food Microbiology* 21, 55–68.

Bäumler, A.J. and Heffron, F. (1995) Identification and sequence analysis of *lpfABCDE*, a putative fimbrial operon of *Salmonella typhimurium. Journal of Bacteriology* 177, 2087–2097.

Bäumler, A.J., Tsolis, R.M. and Heffron, F. (1996a) Contribution to fimbrial operons to attachment to and invasion of epithelial cell lines by *Salmonella typhimurium. Infection and Immunity* 64, 1862–1865.

Bäumler, A.J., Tsolis, R.M. and Heffron, F. (1996b) The *lpf* operon mediates adhesion of *Salmonella typhimurium* to murine Peyer's patches. *Proceedings of the National Academy of Sciences USA* 93, 270–283.

Bäumler, A.J., Tsolis, R.M., Valentine, P.J., Ficht, T.A. and Heffron, F. (1997a) Synergistic effect of mutations in *invA* and *lpfC* on the ability of *Salmonella typhimurium* to cause murine typhoid. *Infection and Immunity* 65, 2254–2259.

Bäumler, A.J., Gilde, A.J., Tsolis, R.M., van der Velden, A.W. M., Ahmer, B.M.M. and Heffron, F. (1997b) Contribution of horizontal gene transfer and deletion events to the development of fimbrial operons during evolution of *Salmonella* serotypes. *Journal of Bacteriology* 179, 317–322.

Ben-Nasr, A., Olsén, A., Sjöbring, U., Muller-Esterl, W. and Bjorck, L. (1996) Assembly of human contact phase proteins and release of bradykinin at the surface of curli-expressing *Escherichia coli. Molecular Microbiology* 20, 927–935.

Bertram, E.M., Attridge, S.R. and Kotlarski, I. (1994) Immunogenicity of the *Escherichia coli* fimbrial antigen K99 when expressed by *Salmonella enteritidis* 11RX. *Vaccine* 12, 1372–1378.

Brocchi, M., Giuseppina Covone, M., Palla, E. and Galeotti, C.S. (1999) Integration of minitransposons for expression of the *Escherichia coli elt* genes at a preferred site in *Salmonella typhimurium* identifies a novel putative fimbrial locus. *Archives in Microbiology* 171, 122–126.

Campbell, S., Duckworth, S., Thomas, C.J. and McMeekin, T.A. (1987) A note on the adhesion of bacteria to chicken muscle connective tissue. *Journal of Applied Bacteriology* 63, 67–71.

Clark, C., Candlish, A.A.G. and Steell, W. (1989) Detection of *Salmonella* in foods using a novel coloured latex test. *Food and Agricultural Immunology* 1, 3–9.

Clegg, S., Purcell, B.K. and Pruckler, J. (1987) Characterisation of genes encoding type 1 fimbriae of *Klebsiella pneumoniae, Salmonella typhimurium,* and *Serratia marcescens. Infection and Immunity* 55, 281–287.

Clegg, S., Hancox, L.S. and Yeh, K.-S. (1996) *Salmonella typhimurium* fimbrial phase variation and FimA expression. *Journal of Bacteriology* 178, 542–545.

Clouthier, S.C., Müller, K.-H, Doran, J.L., Collinson, S.K. and Kay, W.W. (1993) Characterisation of three fimbrial genes, *sefABC,* of *Salmonella enteritidis. Journal of Bacteriology* 175, 2523–2533.

Clouthier, S.C., Collinson, S.K. and Kay, W.W. (1994) Unique fimbriae-like structures encoded by *sefD* of the SEF14 fimbrial gene cluster of *Salmonella enteritidis. Molecular Microbiology* 12, 893–903.

Clouthier, S.C., Collinson, S.K., White, A.P., Banser, P.A. and Kay, W.W. (1998) tRNA^Arg (*fimU*) and expression of SEF14 and SEF21 in *Salmonella enteritidis. Journal of Bacteriology* 180, 840–845.

Cohen, H.J., Mechanda, S.M. and Lin, W. (1996) PCR amplification of the *fimA* gene sequence of *Salmonella typhimurium,* a specific method for detection of *Salmonella* spp. *Applied and Environmental Microbiology* 62, 4303–4308.

Collier, W.A., Wong, S.T. and De Miranda, J.C. (1955) Bacterien-Haemagglutination. II. Unspecifische hemmung der Coli-Haemagglutination. *Antonie van Leeuwenhoek Journal of Microbiology and Serology, Netherlands* 21, 124–132.

Collighan, R.J. and Woodward, M.J. (1997) Sequence analysis and distribution of an IS3-like element isolated from *Salmonella enteritidis. FEMS Microbiology Letters* 154, 207–213.

Collinson, S.K., Emödy, L., Müller, K.-H., Trust, T.J. and Kay, W.W. (1991) Purification and characterisation of thin aggregative fimbriae from *Salmonella enteritidis. Journal of Bacteriology* 173, 4773–4781.

Collinson, S.K., Emödy, L., Trust, T.J. and Kay, W.W. (1992) Thin aggregative fimbriae from diarrheagenic *Escherichia coli. Journal of Bacteriology* 174, 4490–4495.

Collinson, S.K., Doig, P.C., Doran, J.L., Clouthier, S., Trust, T.J. and Kay, W.W. (1993) Thin aggregative fimbriae mediate binding of *Salmonella enteritidis* to fibronectin. *Journal of Bacteriology* 175, 12–18.

Collinson, S.K., Clouthier, S.C., Doran, J.L., Banser, P.A. and Kay, W.W. (1996a) *Salmonella enteritidis agfBAC* operon encoding thin, aggregative fimbriae. *Journal of Bacteriology* 178, 662–667.

Collinson, S.K., Liu, S.-L., Clouthier, S.C., Banser, P.A., Doran, J.L., Sanderson, K.E. and Kay, W.W. (1996b) The location of four fimbrin-encoding genes, *agfA, fimA, sefA* and *sefD,* on the *Salmonella enteritidis* and/or *S. typhimurium* XbaI-BlnI genomic restriction maps. *Gene* 169, 75–80.

Cooper, G.L. and Thorns, C.J. (1996) Evaluation of the SEF14 fimbrial dot-blot and flagella western blot tests as indicators of *Salmonella enteritidis* infections in chicks. *Veterinary Record* 138, 149–152.

Cooper, G.L., Venables, L.M., Woodward, M.J. and Hormaeche, C.E. (1994a) Invasiveness and persistence of *Salmonella enteritidis, Salmonella typhimurium,* and a genetically defined *S. enteritidis aroA* strain in young chickens. *Infection and Immunity* 62, 4739–4746.

Cooper, G.L., Venables, L.M., Woodward, M.J. and Hormaeche, C.E. (1994b) Vaccination of chickens with strain CVL30, a genetically defined *Salmonella enteritidis aroA* live oral vaccine candidate. *Infection and Immunity* 62, 4747–4754.

Cooper, G.C., Venables, L.M. and Lever, M.S. (1996) Airborne challenge of chickens vaccinated orally with the genetically-defined *Salmonella enteritidis aroA* strain CVL30. *Veterinary Record* 139, 447–448.

Darekar, M.R. and Eyer, H. (1973) The role of flagella, but not fimbriae, in the adherence of *Salmonella enterica* serovar enteriditis to chick gut explant. *Zentralblatt für Bakteriologie [A]* 225, 130–134.

Dibb-Fuller, M.P. and Woodward, M.J. (1999) The contribution of fimbriae and flagella of *Salmonella enterica* serovar Enteritidis in the colonisation, invasion, persistence and lateral transfer in chicks. *Avian Pathology* (in press).

Dibb-Fuller, M., Allen-Vercoe, E., Woodward, M.J. and Thorns, C.J. (1997) Expression of SEF17 fimbriae by *Salmonella enteritidis*. *Letters in Applied Microbiology* 25, 447–452.

Doran, J.L., Collinson, K.S., Burian, J., Sarlos, G., Todd, E.C.D., Munro, C.K., Kay, C.M., Bauser, P.A., Peterkin, P.I. and Kay, W.W. (1993) DNA-based diagnostic tests for *Salmonella* species targeting *agfA*, the structural gene for thin, aggregative fimbriae. *Journal of Clinical Microbiology* 31, 2263–2273.

Dorman, C.J., Nolan, N.C. and Smith, S.G.J. (1997) Control of type 1 fimbrial expression by a random genetic switch in *Escherichia coli*. *Society for General Microbiology Symposia* 54, 192–209.

Duguid, J.P. and Gillies, R.R. (1958) Fimbriae and haemagglutinating activity in *Salmonella, Klebsiella, Proteus* and *Chromobacterium*. *Journal of Pathology and Bacteriology* 75, 519–520.

Duguid, J.P., Anderson, E.S. and Campbell, I. (1966) Fimbriae and adhesive properties in *Salmonellae*. *Journal of Pathology and Bacteriology* 92, 107–138.

Duguid, J.P., Dareker, M.R. and Wheater, D.W. (1976) Fimbriae and infectivity in *Salmonella typhimurium*. *Journal of Medical Microbiology* 9, 459–473.

Feldsine, P.T., Falbo-Nelson, M.T. and Hustead, D.L. (1992) Polyclonal enzyme immunoassay method for detection of motile and non-motile *Salmonella* in foods: collaborative study. *Journal of AOAC International* 75, 1032–1044.

Finlay, B.B. and Falkow, S. (1997) Common themes in microbial pathogenicity revisited. *Microbiology and Molecular Biology Reviews* 61, 136–169.

Folkeson, A., Advani, A., Sukulpolvi, S., Pfeifer, J.D., Normark, S. and Lodfahl, S. (1999) Multiple insertions of fimbrial operons correlate with Salmonella serovars responsible for human disease. *Molecular Microbiology* 33, 612–622.

Friedrich, M.J., Kinsey, N.E., Vila, J. and Kadner, R.J. (1993) Nucleotide sequence of a 13.9 kb segment of the 90 kb virulence plasmid of *Salmonella typhimurium*: the presence of fimbrial biosynthetic genes. *Molecular Microbiology* 8, 543–548.

Ghosh, S., Mittal, A., Vohra, H. and Ganguly, N.K. (1996) Interaction of rat intestinal brush border membrane glycoprotein with type-1 fimbriae of *Salmonella typhimurium*. *Molecular and Cellular Biochemistry* 158, 125–131.

Girón, J.A., Levine, M.M. and Kaper, J.B. (1994) Longus: a long pilus ultrastructure produced by human enterotoxigenic *Escherichia coli*. *Molecular Microbiology* 12, 71–82.

Girón, J.A., Xu, J.G., Gonzalez, C.R., Hone, D., Kaper J.B. and Levine M.M. (1995) Simultaneous expression of CFA/I and CS3 colonisation factor antigens on enterotoxigenic *Escherichia coli* by delta *aroC* delta *aroA Salmonella typhi* vaccine strain CVD 908. *Vaccine* 13, 939–946.

Grund, S. and Seiler, A. (1993) Fimbriae and lectinophagocytosis of *Salmonella typhimurium* variatio copenhagen (STMVC) – an electron microscopic study. *Journal of Veterinary Medicine* B 40, 105–112.

Grund, S. and Weber, A. (1988) A new type of fimbriae on *Salmonella typhimurium*. *Journal of Veterinary Medicine* B 34, 779–782.

Guard-Petter, J. (1993) Detection of two smooth colony phenotypes in a *Salmonella enteritidis* isolate which vary in their ability to contaminate eggs. *Applied and Environmental Microbiology* 59, 2884–2890.

Guard-Petter, J., Lakshmi, B., Carlson, R. and Ingram, K. (1995) Characterisation of lipopolysaccharide heterogenity in *Salmonella enteritidis* by an improved gel electrophoresis method. *Applied and Environmental Microbiology* 61, 2845–2851.

Guard-Petter, J., Keller, L.H., Rahman, M.M., Carlson, R.W. and Silvers, S. (1996) A novel relationship between O-antigen variation, matrix formation and invasiveness of *Salmonella enteritidis*. *Epidemiology and Infection* 117, 219–231.

Gulig, P.A., Danbara, H., Guiney, D.G., Lax, A.J., Norel, F. and Rhen, M. (1993) Molecular analysis of *spv* virulence genes of the salmonella virulence plasmids. *Molecular Microbiology* 7, 825–830.

Hacker, J., Blum-Oehler, G., Muhldorfer, I. and Tschape, H. (1997) Pathogenicity islands of virulent bacteria: structure, function and impact on microbial evolution. *Molecular Microbiology* 23, 1089–1097.

Hammar, M., Arnquist, A., Bian, Z., Olsen, A. and Normark, S. (1995) Expression of two *csg* operons is required for production of fibronectin- and Congo red-binding curli polymers in *Escherichia coli*. *Molecular Microbiology* 18, 661–670.

Hammar, M., Bian, Z. and Normark, S. (1996) Nucleator-dependent intercellular assembly of adhesive curli organelles in *Escherichia coli*. *Proceedings of the National Academy of Sciences USA* 93, 6562–6566.

Herwald, H., Morgelin, M., Olsen, A., Rhen, M., Dahlback, B., Muller-Esterl, W. and Bjorck, L. (1998) Activation of the contact-phase system on bacterial surfaces – a clue to serious complications in infectious diseases. *Nature Medicine* 4, 298–302.

Hoorfar, J. and Thorns, C.J. (1996) Rapid identification and serotyping of *Salmonella enteritidis*. *Journal of Rapid Methods and Automation in Microbiology* 4, 317–320.

Hultgren, S.J., Normark, S. and Abraham, S.N. (1991) Chaperone assisted assembly and molecular architecture of adhesive pili. *Annual Review of Microbiology* 45, 383–415.

Humphrey, T.J., Williams, A., McAlpine, K., Lever, M.S., Guard-Petter, J. and Cox, J.M. (1996) Isolates of *Salmonella enterica* Enteritidis PT4 with enhanced heat and acid tolerance are more virulent in mice and more invasive in chickens. *Epidemiology and Infection* 117, 79–88.

Issacson, R.E. (1994) Vaccines against *E. coli* diseases. In: Gyles, C.L. (ed.) Escherichia coli *in Domestic Animals and Humans*. CAB International, Wallingford, UK, pp. 629–647.

Jameson, J.E. (1966) Differentiation of *Salmonella* strains by colonial morphology. *Journal of Pathology and Bacteriology* 91, 141–148.

Jones, G.W. and Richardson, L.A. (1981) The attachment to and invasion of HeLa cells by *Salmonella typhimurium*: the contribution of mannose-sensitive and mannose resistant haemagglutinating activities. *Journal of General Microbiology* 127, 361–370.

Jones, G.W., Richardson, L.A. and Uhlman, D. (1981) The invasion of HeLa cells by *Salmonella typhimurium*: reversible and irreversible bacterial attachment and the role of bacterial motility. *Journal of General Microbiology* 127, 351–360.

Jones, G.W., Rabert, D.K., Svinarich, D.M. and Whitfield, H.J. (1982) Association of adhesive, invasive and virulent phenotypes of *Salmonella typhimurium* with autonomous 60-megadalton plasmids. *Infection and Immunity* 38, 476–486.

Kerr, S., Ball, H.J., Mackie, D.P., Pollock, D.D. and Finlay, D.A. (1992) Diagnostic application of monoclonal antibodies to outer membrane protein for rapid detection of salmonella. *Journal of Applied Bacteriology* 72, 302–308.

Korhonen, T.K., Lounatmaa, K., Ranta, H. and Kuusi, N. (1980) Characterisation of type 1 pili of *Salmonella typhimurium* LT2. *Journal of Bacteriology* 144, 800–805.

Korhonen, T.K., Leffler, H. and Svanborg Eden, C. (1981) Binding specificity of piliated strains of *Escherichia coli* and *Salmonella typhimurium* to epithelial cells, *Saccharomyces cerevisiae* cells and erythrocytes. *Infection and Immunity* 32, 796–804.

Krogfelt, K.A., Bergmans, H. and Klemm, P. (1990) Direct evidence that the FimH protein is the mannose-specific adhesin of *Escherichia coli* type 1 fimbriae. *Infection and Immunity* 58, 1995–1999.

Kukkonen, M., Raunio, T., Virkola, R., Hynonen, U., Westerlund-Wikstrom, B., Rhen, M. and Korkhonen, T.K. (1993) Basement membrane carbohydrate as a target for bacterial adhesion: binding of type 1 fimbriae of *Salmonella enterica* and *Escherichia coli* to laminin. *Molecular Microbiology* 7, 229–237.

Kukkonen, M., Saarela, S., Lahteenmaki, K., Hyonen, U., Westerlund-Wikstrom, B., Rhen, M. and Korhonen T.K. (1998) Identification of two laminin-binding fimbriae, the type 1 fimbria of *Salmonella enterica* serovar Typhimurium and the G fimbria of *Escherichia coli*, as plasminogen receptors. *Infection and Immunity* 66, 4965–4970.

Lockman, H.A. and Curtiss, R. (1992a) Isolation and characterisation of conditionally adherent and non-type 1 fimbriated *Salmonella typhimurium* mutants. *Molecular Microbiology* 6, 933–945.

Lockman, H.A. and Curtiss, R. (1992b) Virulence of non-type 1-fimbriated and non-fimbriated non-flagellated *Salmonella typhimurium* in murine typhoid fever. *Infection and Immunity* 60, 491–496.

McLaren, I.M., Sojka, M.G., Thorns, C.J. and Wray, C. (1992) An interlaboratory trial of a latex agglutination kit for rapid identification of *Salmonella enteritidis*. *Veterinary Record* 131, 235–236.

Manafi, M. and Sommer, R. (1992) Comparison of three rapid screening methods for *Salmonella* spp.: 'MUCAP Test, Microscreen[R] Latex and Rambach Agar'. *Letters in Applied Microbiology* 14, 163–166.

Mintz, C.S., Cliver, D.O. and Deibel, R.H. (1983) Attachment of *Salmonella* to mammalian cells *in vitro*. *Canadian Journal of Microbiology* 29, 1731–1735.

Morona, R., Morona, J.K., Considine, A., Hackett, J.A., Van den Bosch, L., Boyer, L. and Attridge, S.R. (1994) Development of a live vaccine against colibacillosis in pigs: construction and immunogenicity of K88- and K99-expressing clones of *Salmonella typhimurium* G30. *Vaccine* 12, 513–521.

Müller, K.-H., Collinson, K.S., Trust, T.J. and Kay, W.W. (1991) Type 1 fimbriae of *Salmonella enteritidis*. *Journal of Bacteriology* 173, 4765–4772.

Ogunniyi, A.D., Kotlarski, I., Morona, R. and Manning, P. (1997) Role of SefA subunit protein of SEF14 fimbriae in the pathogenesis of *Salmonella enterica* serovar Enteritidis. *Infection and Immunity* 65, 708–717.

Old, D.C. and Adegbola, R.A. (1985) Antigenic relationships among type 3 fimbriae of *Enterobacteriaceae* revealed by immunoelectron microscopy. *Journal of Medical Microbiology* 20, 113–121.

Old, D.C., Roi, A.I. and Tavendale, A. (1986) Differences in adhesiveness among type 1 fimbriate strains of Enterobacteriaceae revealed by and *in vitro* Hep2 cell adhesion model. *Journal of Applied Bacteriology* 61, 563–568,

Olsen, A., Arnquist, A., Hammar, M., Sukupolvi, S. and Normark, S. (1993) The RpoS sigma factor relieves H-NS-mediated transcriptional repression of *csgA*, the subunit gene of fibronectin binding curli in *Escherichia coli*. *Molecular Microbiology* 7, 523–537.

Olsen, A., Arnquist, A., Hammar, M. and Normark, S. (1994) Environmental regulation of curli production in *Escherichia coli*. *Infectious Agents and Disease* 2, 272–274.

Pardon, P., Popoff, M.Y., Coynault, C., Marly, J. and Miras, I. (1986) Virulence associated plasmids of *Salmonella typhimurium* in experimental murine infection. *Annual Annales de l'Institut Pasteur Microbiologie* 137, 47–60.

Peralta, R.C., Yokoyama, H., Ikemori, M. and Kodama, Y. (1994) Passive immunisation against experimental salmonellosis in mice by orally administered hen egg-yolk antibodies specific for 14-kDa fimbriae of *Salmonella* enteritidis. *Journal of Medical Microbiology* 41, 29–35.

Puente, J.L., Bieber, D., Ramer, S.W., Murray, W. and Schoolnik, G.K. (1996) The bundle forming pili of enteropathogenic *Escherichia coli*: transcriptional regulation by environmental signals. *Molecular Microbiology* 20, 87–100.

Purcell, B.K., Pruckler, J. and Clegg, S. (1987) Nucleotide sequences of the genes encoding type 1 fimbrial subunits of *Klebsiella pneumoniae* and *Salmonella typhimurium*. *Journal of Bacteriology* 169, 5831–5834.

Rhen, M., Riikonen, P. and Taira, S. (1993) Transcriptional regulation of *Salmonella enterica* virulence plasmid genes in cultured macrophages. *Molecular Microbiology* 10, 45–56.

Rodriguez-Pena, J.M., Alvarez, I., Ibanez, M. and Rotger, R. (1997) Homologous regions of the *Salmonella enteritidis* virulence plasmid and the chromosome of *Salmonella typhi* encode thiol:disulphide oxidoreductactases belonging to the DsbA thioredoxin family. *Microbiology* 143, 1405–1413.

Romling, U., Bian, Z., Hammar, M., Sierralta, W.D. and Normark, S.J. (1998) Curli fibres are highly conserved between *Salmonella typhimurium* and *Escherichia coli* with respect to operon structure and regulation. *Journal of Bacteriology* 180, 722–731.

Rossilini, G.M., Muscas, P., Chiesurin, A. and Satta, G. (1993) Analysis of the *Salmonella fim* cluster: identification of a new gene (*fimI*) encoding a fimbrin-like protein and located down stream from the *fimA* gene. *FEMS Microbiology Letters* 114, 259–266.

Russell, P.W. and Orndorff, P.E. (1992) Lesions in two *Escherichia coli* type 1 pilus genes alter pilus number and length without affecting receptor binding. *Journal of Bacteriology* 174, 5923–5929.

Rutter, J.M. (1975) *Escherichia coli* infection in piglets: pathogenesis, virulence and vaccination. *Veterinary Record* 96, 171–175.

Rutter, J.M. and Jones, G.W. (1973) Protection against enteric disease caused by *Escherichia coli* – a model for vaccination with a virulence determinant. *Nature (London)* 242, 531–532.

Sizemore, D.R., Fink, P.S., Ou, J.T., Baron, L.S., Kopecko, D.J. and Warren, R.L. (1991) Tn5 mutagenesis of the *Salmonella typhimurium* 100kb plasmid: definition of new virulence regions. *Microbial Pathogenesis* 10, 493–498.

Sjöbring, U., Pohl, G. and Olsén, A. (1994) Plasminogen, absorbed by *Escherichia coli* expressing curli or by *Salmonella enteritidis* expressing thin, aggregative fimbriae, can be activated by simultaneously captured tissue-type plasminogen activator (t-PA). *Molecular Microbiology* 14, 443–452.

Smyth, C.J., Marrion, M. and Smith, S.G. (1994) Fimbriae of *Escherichia coli*. In: Gyles, C.L. (ed.) Escherichia coli *in Domestic Animals and Humans*. CAB International. Wallingford, UK, pp. 399–435.

Sohel, I., Puente, J.L., Murray, W.J., Vuopio-Varkila, J. and Schoolnik, G.K. (1993) Cloning and characterisation of the bundle-forming pilin gene of enteropathogenic *Escherichia coli* and its distribution in *Salmonella* serovars. *Molecular Microbiology* 7, 563–575.

Sohel, I., Puente, J.L., Ramer, S.W., Bieber, D., Wu, C.-Y. and Schoolnik, G. (1996) Enteropathogenic *Escherichia coli*: identification of a gene cluster coding for bundle-forming pilus morphogenesis. *Journal of Bacteriology* 178, 2613–2628.

Sojka, M.G., Dibb-Fuller, M. and Thorns, C.J. (1996) Characterisation of monoclonal antibodies specific to SEF21 fimbriae of *Salmonella enteritidis* and their reactivity with other salmonellae and enterobacteria. *Veterinary Microbiology* 48, 207–221.

Sojka, M.G., Carter, M.A. and Thorns, C.J. (1998) Characterisation of epitopes of type 1 fimbriae of Salmonella using monoclonal antibodies specific for SEF21 fimbriae of *Salmonella enteritidis*. *Veterinary Microbiology* 59, 157–174.

Sojka, W.J. (1971) Enteric diseases in newborn piglets, calves and lambs due to *Escherichia coli* infection. *Veterinary Bulletin* 41, 509–522.

Stein, M.A., Leung, K.Y., Zwick, M., Portillo, F.G. and Findlay, B.B. (1996) Identification of a *Salmonella* virulence gene required for formation of filamentous structures containing lysosomal membrane glycoproteins within epithelial cells. *Molecular Microbiology* 20, 151–165.

Stenstrom, T.A. and Kjelleberg, S. (1985) Fimbriae mediated nonspecific adhesion of *Salmonella typhimurium* to mineral particles. *Archives in Microbiology* 143, 6–10.

Stone, K.D., Zhang, H.-Z., Carlson, L.K. and Donnenberg, M.S. (1996) A cluster of fourteen genes from enteropathogenic *Escherichia coli* is sufficient for the biogenesis of a type IV pilus. *Molecular Microbiology* 20, 325–337.

Strom, M.S. and Lory, S. (1993) Structure–function and biogenesis of the type IV pili. *Annual Reviews of Microbiology* 47, 565–596.

Sukupolvi, S., Edelstein, A., Rhen, M., Normark, S.J. and Pfeifer, J.D. (1997a) Development of a murine model of chronic *Salmonella* infection. *Infection and Immunity* 65, 838–842.

Sukupolvi, S., Lorenz, R.G., Gordon, J.I., Bian, Z., Pfeifer, J.D., Normark, S.J. and Rhen, M. (1997b) Expression of thin aggregative fimbriae promotes interaction of *Salmonella typhimurium* SR-11 with mouse small intestinal epithelial cells. *Infection and Immunity* 65, 5320–5325.

Swenson, D.L. and Clegg, S. (1992) Identification of ancillary *fim* genes affecting *fimA* expression in *Salmonella typhimurium*. *Journal of Bacteriology* 174, 7697–7704.

Swenson, D.L., Clegg, S. and Old, D.C. (1991) The frequency of *fim* genes among *Salmonella* serovars. *Microbial Pathogenesis* 10, 487–492.

Swenson, D.L., Kim, K.-J., Six, E.W. and Clegg, S. (1994) The gene *fimU* affects expression of *Salmonella typhimurium* type 1 fimbriae and is related to *Escherichia coli* tRNA gene *argU*. *Molecular and General Genetics* 244, 216–218.

Takeuchi, A. (1967) Electron microscope studies of experimental *Salmonella* infections. *American Journal of Pathology* 50, 109–136.

Tanaka, Y. and Katsube, Y. (1978) Infectivity of *Salmonella typhimurium* for mice in relation to fimbriae. *Nippon Juigashi Zasshi* 40, 671–681.

Tanaka, Y., Katsube, Y., Mutoh, T. and Imaizumi, K. (1981) Fimbriae of *Salmonella typhimurium* and their role in mouse intestinal colonisation of the organism. *Japanese Journal of Veterinary Science* 43, 51–62.

Tavendale, A., Jardine, C.K., Old, D.C. and Duguid, J.P. (1983) Haemagglutinins and adhesion of *Salmonella typhimurium* to Hep2 and HeLa cells. *Journal of Medical Microbiology* 16, 371–380.

Thiagarajan, D., Thacker, H.L. and Saeed, A.M. (1996) Experimental infection of laying hens with *Salmonella enteritidis* that express different types of fimbriae. *Poultry Science* 75, 1365–1372.

Thorns, C. J. (1995) Salmonella fimbriae: novel antigens in the detection and control of salmonella infections. *British Veterinary Journal* 151, 643–658.

Thorns, C.J., Wells, G.A.H., Morris, J.A., Bridges, A. and Higgins, R. (1989a). Evaluation of monoclonal antibodies to K88, K99, F41 and 987P fimbrial adhesins for the detection of porcine enterotoxigenic *Escherichia coli* in paraffin-wax tissue sections. *Veterinary Microbiology* 20, 377–381.

Thorns, C.J., Sojka, M.G. and Roeder, P.L. (1989b) Detection of fimbrial adhesins of ETEC using monoclonal antibody-based latex reagents. *Veterinary Record* 125, 91–92.

Thorns, C.J., Sojka, M.G. and Chasey, D. (1990) Detection of a novel fimbrial structure on the surface of *Salmonella enteritidis* using a monoclonal antibody. *Journal of Clinical Microbiology* 28, 2409–2414.

Thorns, C.J., Sojka, M.G., McLaren, I.M. and Dibb-Fuller, M. (1992) Characterisation of monoclonal antibodies against a fimbrial structure of *Salmonella enteritidis* and certain other serogroup D salmonellae and their application as serotyping reagents. *Research in Veterinary Science* 53, 300–308.

Thorns, C.J., McLaren, I.M. and Sojka, M.G. (1994) The use of latex particle agglutination to specifically detect *Salmonella enteritidis*. *International Journal of Food Microbiology* 21, 47.

Thorns, C.J., Bell, M.M., Sojka, M.G. and Nicholas, R. A. (1996a) Development and application of enzyme-linked immunosorbent assay for specific detection of *Salmonella enteritidis* infections in chickens based on antibodies to SEF14 fimbrial antigen. *Journal of Clinical Microbiology* 34, 792–797.

Thorns, C.J., Turcotte, C., Gemmell, C.G. and Woodward, M.J. (1996b) Studies into the role of the SEF14 fimbrial antigen in the pathogenesis of *Salmonella enteritidis*. *Microbial Pathogenesis* 20, 235–246.

Tobe, T., Schoolnik, G.K., Sohel, I., Bustmente, V.H. and Puente, J.L. (1996) Cloning and characterisation of *bpfTVW* genes required for the transcriptional activation of *bfpA* in enteropathogenic *Escherichia coli*. *Molecular Microbiology* 21, 963–975.

Turcotte, C. and Woodward, M.J. (1993) Cloning, DNA nucleotide sequence and distribution of the gene encoding SEF14, a fimbrial antigen of *Salmonella enteritidis*. *Journal of General Microbiology* 139, 1477–1485.

Walker, P.D. and Foster, W. (1983) Bacterial vaccines. *Biologist* 30, 67–74.

Walker, S., Sojka, M., Dibb-Fuller, M. and Woodward, M.J. (1999) Effect of pH, temperature and surface contact on the elaboration of fimbriae and flagella by *Salmonella enterica* serovar Enteritidis. *Journal of Medical Microbiology* 48, 771–780.

Wallis, T.S., Paulin, S.M., Plested J.S., Watson., P.R. and Jones, P.W. (1995) The *Salmonella dublin* virulence plasmid mediates systemic but not enteric phases of salmonellosis in cattle. *Infection and Immunity* 63, 2755–2761.

Woodward, M.J. and Kirwan S.E.S. (1996) Detection of *Salmonella enteritidis* in eggs by the polymerase chain reaction. *Veterinary Record* 138, 411–413.

Woodward, M.J., Allen-Vercoe, E. and Redstone, J.S. (1996) Distribution, gene sequence and expression *in vivo* of the plasmid encoded fimbrial antigen of *Salmonella* serotype Enteritidis. *Epidemiology and Infection* 117, 17–28.

Wyatt, G.M., Langley, M.N., Lee, H.A. and Morgan, M.R.A. (1993) Further studies on the feasibility of one-day *Salmonella* detection by enzyme-linked immunosorbent assay. *Applied and Environmental Microbiology* 59, 1383–1390.

Yeh, K.-S., Hancox, L.S. and Clegg, S. (1995) Construction and characterisation of a *fimZ* mutant of *Salmonella typhimurium*. *Journal of Bacteriology* 177, 6861–6865.

Chapter 4

Virulence Mechanisms of *Salmonella* and their Genetic Basis

Andreas J. Bäumler,[1] Renée M. Tsolis[2] and Fred Heffron[3]

[1]*Department of Medical Microbiology and Immunology, Texas A&M University Health Science Center, 407 Reynolds Medical Building, College Station, TX 77843-1114, USA;* [2]*Department of Veterinary Pathobiology, Texas A&M University, College Station, TX 77843-4467, USA;* [3]*Department of Molecular Microbiology and Immunology, Oregon Health Sciences University, Portland, OR 97201-3098, USA*

Introduction

The basic virulence strategy common to *Salmonella* species is to invade the intestinal mucosa and multiply in the gut-associated lymphoid tissue (GALT). From the infected intestinal tissues the pathogens are drained to the regional lymph nodes, where macrophages that line the lymphatic sinuses form a first effective barrier to prevent further spread. If this host defence mechanism successfully limits bacterial expansion, the infection remains localized to the intestine and the GALT (Fig. 4.1). In humans, non-typhoidal *Salmonella* serovars typically cause a localized disease, which manifests itself as acute gastroenteritis. Similarly, *S. typhimurium* causes a localized disease in cattle and pigs, which is characterized by acute diarrhoea.

If, on the other hand, the macrophages located in the draining lymph nodes are unable to limit spread, *Salmonella* can cause a systemic disease. The systemic disease caused by the human-adapted serovar *S. typhi* is also known as typhoid fever or enteric fever, reflecting the initial focus of infection. Some *Salmonella* serovars cause typhoid fever-like diseases in animals, such as *S. gallinarum* in poultry (fowl typhoid), *S. choleraesuis* in pigs (porcine paratyphoid) or *S. typhimurium* and *S. enteritidis* in mice (murine typhoid). Diarrhoea is not a typical sign of disease during murine typhoid or pig paratyphoid. Similarly, diarrhoea develops in only about one-third of typhoid-fever patients (Miller *et al.*, 1995). During systemic infection, the pathogens spread from the GALT via the efferent lymphatics and the thoracic duct into the vena cava. The capillary systems of liver and spleen constitute an efficient filtering system, which focuses infection to liver and spleen, and these organs are usually enlarged during systemic infection (Fig. 4.1).

Distinction between *Salmonella* serovars causing localized or systemic illness is complicated by the fact that the disease outcome depends on the immune status of the host. For example, most (if not all) *Salmonella* serovars are able to cause systemic disease in immunocompromised patients, e.g. those at the extremes of age or those with underlying conditions, such as HIV infection. However, only a few host-adapted serovars (*S. typhi*, *S. paratyphi* A, *S. paratyphi* B, *S. paratyphi* C, and *S. sendai*) can cause enteric fever in an immunocompetent patient.

The ability to cause both localized and systemic disease relies on a repertoire of elaborate virulence determinants. As *Salmonella* encounters a variety of drastically different microenvironments and host defence mechanisms during its course of infection, adaptation to these conditions involves a large number of genes, which may be considered virulence determinants in a

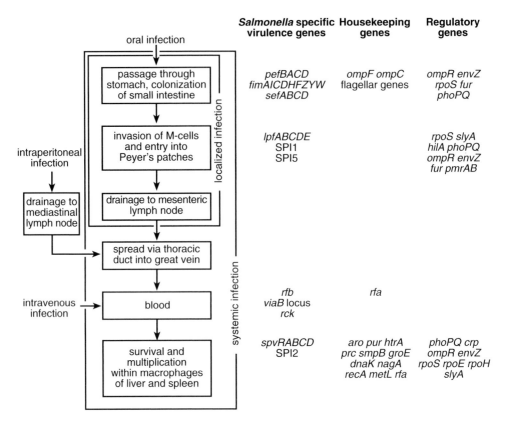

Fig. 4.1. Course of *Salmonella* infection in the mouse (left) after different routes of administration (oral, intraperitoneal and intravenous) and virulence genes required during colonization of various niches in the host (right). Virulence genes are categorized as described in the text.

broad sense. Among these genes are those involved in nutrient biosynthesis/uptake, stress response and repair of cell damage. Virulence determinants that belong in this category are also necessary for survival of stresses outside the host and may therefore be considered housekeeping genes. Housekeeping genes are typically present in other closely related bacteria, such as *Escherichia coli*. A second group of virulence determinants sets the genus *Salmonella* apart from its close relatives, because these genes are present only in members of this genus (Bäumler, 1997). These *Salmonella*-specific virulence genes encode adaptations that are specially designed to overcome host defence mechanisms and may therefore be considered bona fide virulence determinants. Finally, expression of both groups of virulence genes is tightly regulated in response to environmental signals encountered in the

host. The regulatory genes mediating this control may themselves be considered virulence determinants (Fig. 4.1).

Intestinal Phase of Infection

Stress conditions encountered in the lumen of the alimentary tract

The alimentary tract is a hostile environment, which imposes severe stress upon invading bacteria. The first host defence mechanism encountered after oral infection is the acid barrier of the stomach (Fig. 4.1). The passage through this environment induces expression of a number of genes whose products are involved in pH homoeostasis and repair of macromolecules (Bearson *et al.*, 1997). Although this acid-stress response confers increased resistance to low pH, it does not grant complete protection against the stomach acid. For example, studies on oral infection of S. *enteritidis* in mice have shown that only 1% of the inoculum survives the low pH during the passage through the stomach (Carter and Collins, 1974). The surviving bacteria then reach the small intestine, which contains bactericidal compounds, such as bile salts. Like other enterobacteria, *Salmonella* serovars are well adapted to cope with these stress conditions. A still greater challenge than survival in the lumen of small and large intestines appears to be resisting removal by peristalsis and gaining a foothold at the preferred niche in the intestinal wall, the GALT. This is illustrated by the finding that, during infection of mice with S. *enteritidis*, about 80% of the bacteria that survive the passage through the stomach are passed with the faeces within 6–10 h post-infection. Approximately 15% remain localized in the lumen of the caecum and large intestine, and only 5% manage to penetrate the intestinal wall of the small intestine and reach the GALT (Carter and Collins, 1974). An important factor that impedes colonization by *Salmonella* serovars is the normal gut flora. Disruption of the indigenous flora by streptomycin treatment results in a 100,000-fold reduction in the 50% implantation dose (ID_{50}) of S. *typhimurium* for mice (Que and Hentges, 1985). The phenomenon of the indigenous flora being able to prevent colonization by exogenous bacteria is known as bacterial interference. Several mechanisms of bacterial interference have been proposed. These include production of inhibitory substances, as well as competition for tissue adhesion sites and limiting nutrients. The main strategy used by *Salmonella* serovars to evade bacterial interference is to escape the competitive environment of the gut by penetrating the intestinal mucosa.

Tissue tropism in the intestine

One of the hallmarks of *Salmonella* infection is the preferential invasion of lymph follicles that are located in the intestinal wall of the ileum (Carter and Collins, 1974; Hohmann et al., 1978). In mammals, lymph follicles of the small intestine are clustered in organs known as Peyer's patches. The Peyer's patches serve as the main

port of entry for *Salmonella* serovars, and their colonization contributes to the development of disease during both localized and systemic infection (Fig. 4.1). In fact, intestinal perforations at areas of Peyer's patches are the most frequent cause of death during typhoid fever (Bitar and Tarpley, 1985). Despite the importance of this step during infection, little is known about factors involved in the tropism of *Salmonella* serovars for Peyer's patches. How can *Salmonella* distinguish between Peyer's patches and other areas of the alimentary tract?

The epithelium overlying lymph follicles in Peyer's patches exhibits features distinct from the epithelium of the ileal villi (Pappo and Owen, 1988; Giannasca et al., 1994). Whereas the villous epithelium contains an abundance of goblet cells, which produce the intestinal mucus layer, these cell types are not found in the follicle-associated epithelium (FAE; Fig. 4.2). The FAE contains a cell type that is not found in the villous epithelium, the M cell. The M cell content of the FAE varies among different species and can range from 5–10% in humans and mice to 100% in cattle. The apical membrane of M cells displays abundant glycoconjugates, which differ in their structure from those of adjacent enterocytes (Clark et al., 1993). While the presence of M cells is restricted to the FAE (Owen, 1977; Bye et al., 1984), enterocytes form the predominant cell type throughout the intestine. However, there is evidence that the surface of enterocytes located in the FAE differs in its properties from that of enterocytes in the villous epithelium. For example, receptors for polymeric immunoglobulin are absent from enterocytes of the FAE (Pappo and Owen, 1988). In addition, the presence of enterocytes in FAE with unique glycosylation patterns has been reported (Bye et al., 1984; Finzi et al., 1993). Any of these surface structures that distinguish FAE from the surrounding villous epithelium could play a role in targeting *Salmonella* serovars to Peyer's patches.

Since recognition and binding of surface structures in Peyer's patches is a necessary first step during colonization of these organs, it is likely that adhesins play an important role in this process. Consistent with this idea, long polar (LP) fimbriae have been shown to be necessary for adhesion of S. *typhimurium* to murine Peyer's patches in an intestinal organ culture model (Bäumler et al., 1996; Norris et al., 1998). LP

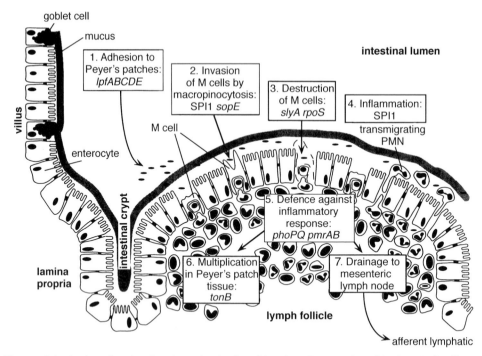

Fig. 4.2. Colonization of murine Peyer's patches by *S. typhimurium*. Cross-section of the intestinal wall at the area of a Peyer's patch showing a lymph follicle (centre) and part of an intestinal villus (left). Consecutive stages of Peyer's patch colonization and *S. enterica* virulence genes required during infection are described in boxes 1–6.

fimbriae are encoded by five genes organized in the *lpf* (long polar fimbrial) operon located at 80 centisomes on the physical map of serovar *S. typhimurium* (Fig. 4.3; Bäumler and Heffron, 1995; Bäumler *et al.*, 1997b). A survey on the distribution of *lpf* genes among *Salmonella* serovars revealed that, among 90 strains tested, only 44 contained this operon (Bäumler *et al.*, 1997a). Interestingly, all *Salmonella* serovars that can cause lethal systemic infection in mice possess the *lpf* operon (Bäumler *et al.*, 1997c). One factor that may determine the host range of adhesins, such as LP fimbriae, is the variation in glycosylation patterns found between epithelial surfaces of different host species. For example, M cells located in mouse and rabbit Peyer's patches differ in expression of glycoconjugates on their apical surfaces (Jepson *et al.*, 1995). The existing diversity in glycosylation patterns observed between mucosal surfaces of different animals may create a necessity to utilize alternate adhesion determinants during infection of distinct host species

(Bäumler *et al.*, 1998). Consistent with this idea, a mutation in a novel fimbrial operon of *S. typhimurium* (termed *bcf*, for bovine colonization factor) has been shown to reduce its ability to colonize bovine but not murine Peyer's patches (Tsolis *et al.*, 1999b). Thus fimbrial adhesins may function as host-range factors of *Salmonella* serovars.

Entering Peyer's patches

Penetration of the host epithelium appears to be essential for *Salmonella* virulence. For instance, secretion of monoclonal immunoglobulin A (IgA) antibody directed against the O4 antigen of the *S. typhimurium* lipopolysaccharide (LPS) results in avirulence (> 10,000-fold increased median lethal dose (LD_{50})) for mice upon oral challenge (Michetti *et al.*, 1992). The protection against oral challenge is due to immune exclusion at the mucosal surface, which prevents entry

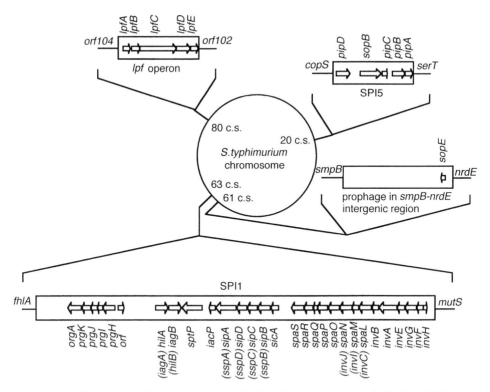

Fig. 4.3. *Salmonella*-specific virulence genes involved during the intestinal phase of infection. DNA regions that are present in *S. typhimurium* but absent from the *E. coli* chromosome are shown as open boxes. Arrows indicate position of *S. enterica* virulence genes. Map positions of virulence genes on the physical map (circle) of the *S. typhimurium* chromosome are given in centisomes (c.s.).

of *Salmonella* into epithelial cells (Michetti *et al.*, 1994). These data imply that blockage of mucosal invasion results in avirulence of *Salmonella* serovars for mice by the oral route of infection. However, mutational inactivation of a single virulence factor involved in mucosal invasion results only in a modest (five- to 50-fold) increase of the oral LD_{50} value for infection of mice (Galán and Curtiss, 1989; Bäumler *et al.*, 1996, 1997c). These data suggest that *S. typhimurium* possesses multiple pathways for intestinal penetration. Virulence factors involved in two distinct pathways of entry into murine Peyer's patches have been identified.

The major factor for intestinal penetration is encoded by genes that are clustered in a large (40 kb) area on the chromosome designated *Salmonella* pathogenicity island 1 (SPI1) (Fig. 4.3; Mills *et al.*, 1995). Studies with cultured epithelial cell lines revealed that the virulence

genes encoded on SPI1 allow *Salmonella* serovars to trigger macropinocytosis, a form of endocytosis that is accompanied by cell-surface ruffling (Frances *et al.*, 1993). In epithelial cells, macropinocytosis results in non-selective endocytosis of large particles, such as bacteria. *Salmonella* serovars are therefore able to use induction of macropinocytosis as a mechanism for epithelial-cell invasion. SPI1 encodes a secretion system that exports proteins in response to bacterial contact with epithelial cells (Ginocchio *et al.*, 1994). Such contact-dependent secretory pathways are known as type III export systems and are among the most sophisticated bacterial virulence determinants. Genes located on SPI1 fall into three categories: those encoding regulatory proteins, those involved in formation of the type III secretion apparatus and those encoding the secreted targets of the type III exporter. Additional secreted targets of the invasion-associated type III secretion system are encoded in areas located outside of SPI1. These include genes located on SPI5 (Wood *et al.*, 1998) and the *sopE* gene, which is carried by a prophage of *S. typhimurium* (Hardt *et al.*, 1998b; Mirold, 1999). This phage is located within the *smpB–nrdE* intergenic region of the *S. typhimurium* chromosome (Bäumler and Heffron, 1998). In *S. typhi*, a genomic inversion has placed part of this phage adjacent to the *nrdE* gene (A.J. Bäumler, unpublished data). The type III secretion machinery translocates SopE (*Salmonella* outer protein) through three biological membranes. SopE is first transported across both the cytoplasmic membrane and the outer membrane of the bacterial cell and subsequently traverses the cytoplasmic membrane of the cultured epithelial cell line (Wood *et al.*, 1996). Once inside the cytosol, SopE induces macropinocytosis in the host cell by inducing signalling events through direct interaction with small GTP-binding proteins, such as CDC42 and Rac-1 (Hardt *et al.*, 1998a).

Studies using murine ligated ileal loops revealed that the genes located on SPI1 are necessary for invasion of M cells in the FAE of Peyer's patches (Jones *et al.*, 1994; see Fig. 4.2). Since the presence of M cells is restricted to the FAE, selective invasion of these cells may play a role in preferential colonization of Peyer's patches. This idea is supported by the finding that *S. typhimurium* strains carrying mutations in SPI1 have a reduced ability to colonize murine Peyer's patches (Galán and Curtiss, 1989). Subsequent to their invasion, M cells are destroyed by *S. typhimurium* (Jones *et al.*, 1994). *Salmonella* strains that carry mutations in *slyA* are able to enter murine M cells by induction of cell-surface ruffling but are non-cytotoxic. These results indicate that *S. typhimurium* kills M cells using an SPI1-independent mechanism (Daniels *et al.*, 1996).

A second gene cluster involved in colonization of murine Peyer's patches is the *lpf* operon (Fig. 4.3; Bäumler *et al.*, 1996). Like strains carrying mutations in SPI1, *S. typhimurium lpfC* mutants have a reduced ability to enter Peyer's patches and are attenuated if mice are challenged orally, but not if the intestinal phase of infection is bypassed by intraperitoneal injection (Bäumler *et al.*, 1996, 1997c). Combining mutations in *invA* (located on SPI1) and *lpfC* (located in the *lpf* operon) results in a much greater attenuation of *S. typhimurium* ATCC 14028 for mice infected by the oral route than inactivation of either of these genes alone (Bäumler *et al.*, 1997c). These data are evidence that SPI1 and *lpf* determine independent pathways for intestinal penetration and that the encoded virulence factors act synergistically. However, an *lpfC invA* mutant is still able to cause mortality in mice when administered orally at a high dose, suggesting that *S. typhimurium* possesses yet another pathway to penetrate the intestinal wall.

Diarrhoea and inflammation

Although the onset of systemic and localized infection is fairly similar (see Fig. 4.1), the symptoms produced are not identical. Localized infection generally manifests itself as acute gastroenteritis. In contrast, only one-third of patients suffering from typhoid fever develop diarrhoea, usually several days after the onset of fever. Because most scientists study *S. typhimurium* infection of mice, a typhoid fever model, the aetiology of diarrhoeal disease is poorly understood. The contribution of toxins to the generation of diarrhoea during localized infection has never been conclusively demonstrated, although several toxic activities have been described in *Salmonella* (Stephen *et al.*, 1985). A putative *Salmonella* enterotoxin, Stn (Chopra *et al.*, 1994), was shown not to be involved in diarrhoea in calves (Watson *et al.*, 1998). Early studies performed by Ralph Giannella suggested that *Salmonella* serovars may cause an inflammatory diarrhoea, since fluid secretion in rabbit ileal loops can be markedly inhibited by treatment with nitrogen mustard, an agent that depletes the polymorphonuclear neutrophil (PMN) pool (Giannella *et al.*, 1973; Giannella, 1979). However, it was shown later that nitrogen mustard treatment also inhibits fluid secretion induced by cholera toxin, an agent that causes a secretory diarrhoea (Wallis *et al.*, 1990). To study processes leading to diarrhoea, a cell-culture model, consisting of a polarized T84 monolayer (a crypt-like intestinal epithelial cell line) and human PMN placed into the basolateral compartment, was developed. In this model, *S. typhimurium* is added to the apical surface, where it adheres to and invades the T84 epithelial cells. Bacterial attachment and entry result in polarized basolateral secretion by the T84 monolayer of interleukin 8 (IL-8), a potent PMN chemotaxin. Release of IL-8 by the T84 monolayer triggers transepithelial migration of PMN, a process associated with *Salmonella* infections, since PMN are present in stools of enteritis patients (McCormick *et al.*, 1993, 1995). Induction of IL-8 release by epithelial cells is dependent on the function of the invasion-associated type III secretion system of *S. typhimurium* (Hobbie *et al.*, 1997). More direct evidence for the involvement of the invasion-associated type III secretion system in diarrhoea comes from studies of *S. typhimurium* and *S. dublin* infections in calves. The product of *S. dublin sopB*, a gene located on SPI5, is secreted by an SPI1-dependent mechanism and is required for fluid accumulation in bovine ligated ileal loops (Jones *et al.*, 1998; Wood *et al.*, 1998). Furthermore, *S. typhimurium hilA*, *prgH* or *orgA* mutants are avirulent and do not cause diarrhoea in calves (Tsolis *et al.*, 1999a). These data suggest that the type III export system on SPI1 secretes one or more proteins with enterotoxic activity.

As the host mounts an inflammatory response at the site of mucosal invasion, *Salmonella* genes involved in defence against inflammation have to be expressed subsequently to bacterial entry into the epithelium. Coordinated expression of these virulence genes appears to be mediated by PhoPQ, a two-component regulatory system that changes gene expression in response to changes in the external Mg^{2+} and Ca^{2+} concentrations (Vescovi *et al.*, 1996, 1997). Ca^{2+} and Mg^{2+} cations stabilize the outer membrane by neutralizing the negative charge of phosphate groups and bridging adjacent LPS molecules (Vaara, 1992). The intracellular parasitophorous vacuole in which *Salmonella* resides was shown to be low in Mg^{2+} and Ca^{2+}. In such environments, PhoPQ activates *pmrAB*, two genes encoding a second two-component regulatory system (Gunn and Miller, 1996; Soncini and Groisman, 1996). Activation of PmrAB increases the substitution of phosphates in both the core oligosaccharide and the lipid A part of LPS with 4-amino-4-deoxy-L-arabinose, thereby compensating for the lack of Ca^{2+} and Mg^{2+} cations. These structural changes in LPS result in increased resistance to bactericidal/permeability-increasing protein (BPI), a cationic antibacterial

protein that is released by human PMN during inflammation (Helander *et al.*, 1994). Furthermore, in response to low Mg^{2+} and Ca^{2+} concentrations, PhoPQ activates a second, *pmrAB*-independent pathway, which results in increased resistance to defensins released by recruited PMN and cryptdins produced by Paneth cells located in the intestinal crypts (Selsted *et al.*, 1992; Gunn and Miller, 1996). Defensin/cryptdin resistance may be beneficial during penetration of the mucosa, as their production is increased in intestinal epithelial cells as part of an inflammatory response (Schonwetter *et al.*, 1995).

Systemic Phase of Infection

Growth at systemic sites of infection

Most of the experimental data on systemic *Salmonella* infection have been obtained using the mouse typhoid model. The course of *S. typhimurium* infection in the mouse is dependent on the dose administered and on the route of inoculation (O'Brien, 1986; see Fig. 4.1). By oral infection, approximately 180 genes are required for virulence in a murine infection model based on sampling random, independent bacteriophage Mu insertion mutations for virulence in mice. In contrast, it has been estimated that as few as 60 genes are required for infection by the intraperitoneal (i.p.) route, when graded doses of mutant bacteria were first identified as potential avirulent mutants and then false positives eliminated in a second screen (Bowe *et al.*, 1998). An alternative, more sensitive approach, utilizing tagged transposons, found that roughly 180 genes are required for i.p. infection (Hensel *et al.*, 1995) and 270 genes for oral infection of mice (Tsolis *et al.*, 1999b). The tagged-transposon approach identifies mutants based on competition within the host – even small increases in LD_{50} often correlate with large differences in the number of organisms recovered from the spleen and liver.

Hormaeche and co-workers have divided *Salmonella* infection of the mouse into four phases (Hormaeche and Maskell, 1989). Following oral inoculation of a mouse, the bacteria invade the intestinal mucosa as described above. Within 1 day post-infection, the bacteria can be found in the filtering organs of the reticuloendothelial system (RES) – the spleen and liver – while the blood remains sterile. The effectiveness of this filtering system becomes apparent upon intravenous injection, where large numbers of bacteria (10^6 colony-forming units (cfu)) are removed from the circulation within 2 h (Mackaness et al., 1966; Dunlap et al., 1991). The bacteria replicate in the liver and spleen at a net growth rate of $\log_{10}$ 0.5–1.5 day^{-1} and are predominantly located within cells, particularly phagocytes, of these organs (Maw and Meynell, 1968; Hormaeche, 1980). The assertion that the bacteria replicate intracellularly in this phase (the third according to Hormaeche's division) is borne out by several observations, including the effect of chemical agents (such as silica particles), which affect macrophage function, the effect of antibiotics that do not reach intracellular compartments (Dunlap et al., 1991; Gulig and Doyle, 1993) and the increased susceptibility to infection of Ity^s inbred mouse strains. Macrophages from Ity^s animals (e.g. BALB/c mice) are less microbicidal than those from resistant mice (O'Brien et al., 1982; Lissner et al., 1983). Salmonella have been directly identified within phagocytic cells of the spleen and liver after disruption and cell fractionation on a flow cytometer (Dunlap et al., 1992) and by laser confocal microscopy (Richter-Dahlfors et al., 1997). Furthermore, S. typhimurium mutants that have a reduced capacity to survive in murine macrophages in vitro are unable to cause systemic disease in mice (Fields et al., 1986). At the end of the third phase, the bacteria reappear in the blood and replicate rapidly there until the mouse dies. Death seems not to be caused by endotoxic shock (Khan et al., 1998); however, LPS-hyporesponsive mice are highly susceptible to Salmonella infection (O'Brien et al., 1980).

In order to be a successful pathogen, Salmonella must have strategies to overcome each of the deep-tissue defences, including resistance to phagocytic cells, complement, specific antibodies, cellular immunity and perhaps other components of the immune system that little is known about. In addition, it must reach its preferred site of replication, replicate and then somehow be able to spread to a new host. Although death is the usual result in the inbred mouse model of systemic disease, this may not be the outcome that maximizes spread. Instead, in a more resistant host, the ability to cause systemic infection may provide a selective advantage, because it facilitates the transformation of the host into a chronic carrier, which has the potential to transmit the disease (Bäumler et al., 1998). For example, in typhoid fever, the primary mechanism of spread appears to be the ability of S. typhi to be carried within the gall-bladder and shed in the bile.

Resistance to components of the non-specific immune defence

Salmonella is resistant to complement and can grow with little inhibition in sera containing all of the complement components. The resistance varies with LPS serovar, primarily because certain O side-chains stimulate complement deposition by the alternative pathway better than others (Grossman et al., 1986). The complement factors are deposited on the ends of the LPS, thereby reducing the amount of attack complex reaching the bacterial membrane. In addition, a plasmid-encoded protein, Rck, specifies resistance to complement. However, the level of resistance conferred varies widely with strain background. In smooth strains containing intact LPS, the product does not appear necessary. The gene rck encodes a protein that is related to PagC, Ail and Lom (Heffernan et al., 1992). It is a membrane protein, with three transmembrane loops that prevent insertion of the complement attack complex into the membrane. When transferred to E. coli, rck also confers increased resistance to complement and promotes low levels of cell invasion, similarly to ail (Heffernan et al., 1994). A single transmembrane loop of Rck has been associated with both of these characteristic phenotypes (Cirillo et al., 1996).

Professional phagocytic cells, which are a component of the body's innate immunity and crucial defence against many pathogens, are the preferred niches for Salmonella during growth in liver and spleen. Upon phagocytosis by macrophages, Salmonella is able to sense this hostile environment and respond appropriately. Analysis of the expression of bacterial proteins within macrophages suggests that at least 35 proteins, including the stress protein GroEL, are induced within macrophages, while many others are repressed (Buchmeier and Heffron, 1990). Many of the same proteins that are expressed in macrophages are also expressed in epithelial cells, but the two sets are by no means identical (Gahring *et al.*, 1990). The ability to sense and respond to the intracellular environment is governed by several regulators, including PhoP/PhoQ (Fields *et al.*, 1989) and OmpR/EnvZ (Lindgren *et al.*, 1996). PhoP/PhoQ responds to Mg^{2+} ions, which are apparently at low concentration in the phagocytic vesicle (Vescovi *et al.*, 1996), while OmpR/EnvZ appears to respond to both osmolarity and pH. Within the macrophage, most evidence suggests that the phagocytic vesicle containing *Salmonella* is acidified normally (Rathman *et al.*, 1996). Analysis of mutants defective in macrophage survival *in vitro* revealed that, in addition to regulators, the inactivated genes encode repair/defence mechanisms for macrophage-induced damage (*recA*, *htrA*, *prc*, *metL*), nutrient biosynthesis (*aro*, *pur*) and LPS biosynthesis (Buchmeier *et al.*, 1993; Bäumler *et al.*, 1994; DeGroote *et al.*, 1996).

The phagocytic vesicle passes through much the same pathway, as do endocytic vesicles – to an early and a late compartment and hence to the lysosome, where the pathogen is killed and degraded. Early studies suggested that *Salmonella* follows one of two fates – either the traditional route to the lysosome, as happens with killed bacteria, *E. coli* or latex beads, or along a pathway that avoids fusion with the lysosome, as discussed below (Buchmeier and Heffron, 1991). The number of bacteria that survive or are killed varies with the macrophage population and the strain of mouse infected (Buchmeier and Heffron, 1989). Recent studies using the confocal microscope suggest that regardless of its fate, the vesicle containing *Salmonella* does not contain the mannose-6-phosphate receptor – a major targeting signal to the late endosome (del Garcia and Finlay, 1995; Rathman *et al.*, 1997). Normally, mannose-6-phosphate is added in the Golgi or trans-Golgi network. Proteins that are tagged with this sugar are captured at the cell surface or in the Golgi by one of the two mannose-6-phosphate receptors. The receptor/protein complex moves by vesicular transport to the late endosome, where the acid pH causes dissociation. The receptor recycles to the cell surface and the Golgi, while the mannosylated protein moves to the lysosome. The implication is that many proteins that would normally be targeted to the late endosome and lysosome are not

present in *Salmonella*-containing vesicles, because they lack the mannose-6-phosphate receptor. Many of these same proteins are presumably toxic to bacteria and *Salmonella* must therefore avoid exposure to them.

Salmonella is cytotoxic for macrophages *in vitro* (Arai *et al.*, 1995; Chen, L.M. *et al.*, 1996; Guilloteau *et al.*, 1996; Lindgren *et al.*, 1996; Monack *et al.*, 1996). After infecting a macrophage in tissue culture, *Salmonella* will grow initially, but, after several hours, the macrophages begin to die. Killing of macrophages appears to take place by two pathways – the first early killing pathway uses the same component (*sipB*) as is used by *Shigella* to kill cells (Zychlinsky *et al.*, 1992; Hersh *et al.*, 1999). The product of the *Shigella ipaB* gene has been shown to bind to caspase-1 and causes it to be constitutively active (Chen, Y. *et al.*, 1996; Hilbi *et al.*, 1997; Thirumalai *et al.*, 1997). The SPI1-mediated pathway may be important during interaction with macrophages in the intestine, since this pathogenicity island is not required during the systemic phase of infection in mice. Some authors have reported that cell killing is entirely dependent on the type III export system encoded by SPI1 that exports SipB, as well as on SipB itself (Fig. 4.3; Chen, L.M. *et al.*, 1996). Our own results suggest that there is a second pathway of late killing, which is independent of SipB (Lindgren *et al.*, 1996). The late killing pathway probably enables *Salmonella* to remain within phagocytic cells and use components of the host cell for its own nutritional needs. Whether this pathway is responsible for the apoptosis of splenic macrophages that has been observed in mice infected with *S. typhimurium* by the intravenous route of infection (see Fig. 4.1) remains to be worked out (Richter-Dahlfors *et al.*, 1997).

The *Salmonella* virulence plasmid and *Salmonella* pathogenicity island 2 (SPI2) are required for systemic infection

A cluster of five genes (the *spv* operon) located on a large plasmid is essential for systemic infection in the mouse (Roudier *et al.*, 1990) and is present in strains of *Salmonella* that are most likely to cause a disseminated infection, except for *S. typhi* (Woodward *et al.*, 1989; Fierer *et al.*,

1992). The genes were identified in the Curtiss laboratory by complementation of the virulence defect in a plasmidless derivative of *Salmonella* (Gulig and Curtiss, 1987; Gulig *et al.*, 1992). A *Salmonella* library was constructed and transformed into a plasmid-cured strain, and the resulting library was fed to a mouse to select for positive clones in the spleen. With the exception of the first gene, *spvR*, which encodes a positive activator of *spvABCD*, the exact function of the encoded proteins is not yet known. These genes are induced by starvation, in stationary phase and within the macrophage (Fierer *et al.*, 1993; Rhen *et al.*, 1993). The *spv* genes appear to be essential for an early step in infection that involves phagocytic cells. Macrophages that are infected with *Salmonella* lacking the plasmid express heat shock proteins and attract gamma–delta T cells, unlike wild-type *Salmonella*. The *spv* genes increase the rate of intracellular replication in a murine infection model, even if this phenotype cannot be duplicated *in vitro* (Riikonen *et al.*, 1992; Gulig and Doyle, 1993). It is likely that the *spv* gene products affect some component of the eukaryotic signalling pathway that allows the body to detect and respond appropriately to infection.

A second *Salmonella*-specific DNA region that is essential for extra-intestinal growth is SPI2, a pathogenicity island located at 30 centisomes on the *S. typhimurium* physical map (Hensel *et al.*, 1995, 1997; Ochman *et al.*, 1996). Mutations in SPI2 render *S. typhimurium* unable to cause systemic disease in mice (Hensel *et al.*, 1995; Deiwick *et al.*, 1998) and result in reduced growth in murine macrophages *in vitro* (Ochman *et al.*, 1996; Cirillo *et al.*, 1998; Hensel *et al.*, 1998). As on SPI1, the virulence factor encoded on SPI2 is a type III secretion apparatus. Although *S. typhimurium* SPI2 mutants are more than 10,000-fold attenuated in mice (Hensel *et al.*, 1995, 1998; Ochman *et al.*, 1996; Shea *et al.*, 1996), they cause diarrhoea and are less than 15-fold attenuated in calves (Tsolis *et al.*, 1999). Thus, *Salmonella* serovars may use different virulence factors during localized disease (e.g. bovine enteritis) and systemic infection (e.g. murine typhoid).

Adaptive immunity and *Salmonella* infection

The immune response to *Salmonella* in the mouse is delayed compared with that of other pathogens, such as *Listeria*. This finding suggests that *Salmonella* encodes a mechanism to delay adaptive immunity and may be one reason for this infection being so lethal in mice. Depletion experiments suggest that *Salmonella*-infected macrophages are sufficient to inhibit the immune response (Lee *et al.*, 1985). Several studies suggest that production of NO by macrophages may contribute to this, although other factors were not ruled out (Eisenstein *et al.*, 1994; Schwacha and Eisenstein, 1997; Schwacha *et al.*, 1998). NO production has been shown to block T-cell proliferation at higher concentrations (al-Ramadi *et al.*, 1992). However, in studies in which NO production has been eliminated by treating with a specific inhibitor, no difference in course of infection, timing or strength of the immune response was noted.

Infection with *Salmonella* serovars elicits both humoral and cellular immunity (Mastroeni *et al.*, 1993; Eisenstein, 1998). The relative contribution of a cytotoxic T-cell and a B-cell response to protective immunity in mice appears to depend on their inherent susceptibility to *S. typhimurium*. Passive transfer of antibody protects inherently resistant mice but not inherently susceptible mouse strains (Eisenstein *et al.*, 1984). Since antibodies directed against killed vaccines or purified Vi capsular antigen confer protection against *S. typhi* infections in humans, it was concluded that the resistant mouse is the appropriate model to study immunity to typhoid fever (Eisenstein and Sultzer, 1983; Eisenstein, 1998).

Conclusions

The use of an inexpensive inbred animal model, the mouse, has allowed investigators to amass a large amount of data on the genetics of *Salmonella* pathogenesis. This has led to rapid progress in our understanding of *Salmonella*-mediated systemic disease and to the discovery of major virulence factors. The future study of these virulence factors promises to yield new insights into the mechanisms used by *Salmonella* to subvert the host's defences. However, infection of inbred mice may not model all aspects of disease caused in other animal hosts, such as the diarrhoeal illness of humans and livestock. In some of these host species, the disease resulting from infection with *Salmonella* differs markedly from the typhoid-like disease studied in the mouse model. There is little known about genes involved in determining host adaptations found among *Salmonella* serovars, such as *S. dublin* for cattle, *S. typhi* for humans, *S. abortusovis* for sheep, *S. abortusequi* for horses, *S. gallinarum* for poultry or *S. choleraesuis* for pigs. In addition, typhoid fever differs from other systemic *Salmonella* infections, including that caused by *S. typhimurium* in the mouse, in that the *spv* genes are not required. Consequently, some of the virulence factors necessary for infection of other host species may differ from those identified in the mouse. Much work remains to be done to fully understand *Salmonella*-mediated disease in these hosts.

Acknowledgements

Work in A.B.'s laboratory is supported by grant no. AI40124-01A1 from the National Institute of Allergy and Infectious Diseases. R.T. is supported by Postdoctoral Fellowship no. 9702568 from the US Department of Agriculture. Research in F.H.'s laboratory is funded by US Public Health Service grant no. AI22933 from the National Institutes of Health.

References

Arai, T., Hiromatsu, K., Nishimura, H., Kimura, Y., Kobayashi, N., Ishida, H., Nimura, Y. and Yoshikai, Y. (1995) Endogenous interleukin 10 prevents apoptosis in macrophages during *Salmonella* infection. *Biochemical and Biophysical Research Communications* 213, 600–607.

Bäumler, A.J. (1997) The record of horizontal gene transfer in *Salmonella*. *Trends in Microbiology* 5, 318–322.

Bäumler, A.J. and Heffron, F. (1995) Identification and sequence analysis of *lpfABCDE*, a putative fimbrial operon of *Salmonella typhimurium*. *Journal of Bacteriology* 177, 2087–2097.

Bäumler, A.J. and Heffron, F. (1998) Mosaic structure of the smpB–nrdE intergenic region of *Salmonella enterica*. *Journal of Bacteriology* 180, 2220–2223.

Bäumler, A.J., Kusters, J.G., Stojiljkovic, I. and Heffron, F. (1994) *Salmonella typhimurium* loci involved in survival within macrophages. *Infection and Immunity* 62, 1623–1630.

Bäumler, A.J., Tsolis, R.M. and Heffron, F. (1996) The *lpf* fimbrial operon mediates adhesion to murine Peyer's patches. *Proceedings of the National Academy Sciences USA* 93, 279–283.

Bäumler, A.J., Gilde, A.J., Tsolis, R.M., van der Velden, A.W.M., Ahmer, B.M.M. and Heffron, F. (1997a) Contribution of horizontal gene transfer and deletion events to the development of distinctive patterns of fimbrial operons during evolution of *Salmonella* serovars. *Journal of Bacteriology* 179, 317–322.

Bäumler, A.J., Tsolis, R.M. and Heffron, F. (1997b) Fimbrial adhesins of *Salmonella typhimurium* – role in bacterial interactions with epithelial cells. In: Paul, P.S., Francis, D.H. and Benfield, D. (eds) *Mechanisms in the Pathogenesis of Enteric Diseases*. Plenum Press, New York, pp. 149–158.

Bäumler, A.J., Tsolis, R.M., Valentine, P.J., Ficht, T.A. and Heffron, F. (1997c) Synergistic effect of mutations in *invA* and *lpfC* on the ability of *Salmonella typhimurium* to cause murine typhoid. *Infection and Immunity* 65, 2254–2259.

Bäumler, A.J., Tsolis, R.M., Ficht, T.A. and Adams, L.G. (1998) Evolution of host adaptation in *Salmonella enterica*. *Infection and Immunity* 66, 4579–4587.

Bearson, S., Bearson, B. and Foster, J.W. (1997) Acid stress responses in enterobacteria. *FEMS Microbiology Letters* 147, 173–180.

Bitar, R. and Tarpley, J. (1985) Intestinal perforation and typhoid fever: a historical and state-of-the-art review. *Review of Infectious Diseases* 7, 257.

Bowe, F., Lipps, C.J., Tsolis, R.M., Groisman, E.A., Kusters, J.G. and Heffron, F. (1998) At least four percent of the *Salmonella typhimurium* genome is required for fatal infection of mice. *Infection and Immunity* 66, 3372–3377.

Buchmeier, N.A. and Heffron, F. (1989) Intracellular survival of wild-type *Salmonella typhimurium* and macrophage-sensitive mutants in diverse populations of macrophages. *Infection and Immunity* 57, 1–7.

Buchmeier, N.A. and Heffron, F. (1990) Induction of *Salmonella* stress proteins upon infection of macrophages. *Science* 248, 730–732.

Buchmeier, N.A. and Heffron, F. (1991) Inhibition of macrophage phagosome–lysosome fusion by *Salmonella typhimurium*. *Infection and Immunity* 59, 2232–2238.

Buchmeier, N.A., Lipps, C.J., So, M.Y. and Heffron, F. (1993) Recombination-deficient mutants of *Salmonella typhimurium* are avirulent and sensitive to the oxidative burst of macrophages. *Molecular Microbiology* 7, 933–936.

Bye, W.A., Allen, C.H. and Trier, J.S. (1984) Structure, distribution, and origin of M cells in Peyer's patches of mouse ileum. *Gastroenterology* 86, 789–801.

Carter, P.B. and Collins, F.M. (1974) The route of enteric infection in normal mice. *Journal of Experimental Medicine* 139, 1189–1203.

Chen, L.M., Kaniga, K. and Galan, J.E. (1996) *Salmonella* spp. are cytotoxic for cultured macrophages. *Molecular Microbiology* 21, 1101–1115.

Chen, Y., Smith, M.R., Thirumalai, K. and Zychlinsky, A. (1996) A bacterial invasin induces macrophage apoptosis by binding directly to ICE. *EMBO Journal* 15, 3853–3860.

Chopra, A.K., Peterson, J.W., Chary, P. and Prasad, R. (1994) Molecular characterization of an enterotoxin from *Salmonella typhimurium*. *Microbial Pathogenesis* 16, 85–98.

Cirillo, D.M., Heffernan, E.J., Wu, L., Harwood, J., Fierer, J. and Guiney, D.G. (1996) Identification of a domain in Rck, a product of the *Salmonella typhimurium* virulence plasmid, required for both serum resistance and cell invasion. *Infection and Immunity* 64, 2019–2023.

Cirillo, D.M., Valdivia, R.H., Monack, D.M. and Falkow, S. (1998) Macrophage-dependent induction of the *Salmonella* pathogenicity island 2 type III secretion system and its role in intracellular survival. *Molecular Microbiology* 30, 175–188.

Clark, M.A., Jepson, M.A., Simmons, N.L., Booth, T.A. and Hirst, B. (1993) Differential expression of lectin-binding sites defines mouse intestinal M-cells. *Journal of Histology and Cytochemistry* 41, 1679–1687.

Daniels, J.J.D., Autenrieth, I.B., Ludwig, A. and Goebel, W. (1996) The gene *slyA* of *Salmonella typhimurium* is required for destruction of M cells and intracellular survival but not for invasion or colonization of the small intestine. *Infection and Immunity* 64, 5075–5084.

DeGroote, M.A., Testerman, T., Xu, Y., Stauffer, G. and Fang, F.C. (1996) Homocysteine antagonism of nitric oxide-related cytostasis in *Salmonella typhimurium*. *Science* 272, 414–417.

Deiwick, J., Nikolaus, T., Shea, J.E., Gleeson, C., Holden, D.W. and Hensel, M. (1998) Mutations in *Salmonella* pathogenicity island 2 (SPI2) genes affecting transcription of SPI1 genes and resistance to antimicrobial agents. *Journal of Bacteriology* 180, 4775–4780.

del Garcia, P.F. and Finlay, B.B. (1995) Targeting of *Salmonella typhimurium* to vesicles containing lysosomal membrane glycoproteins bypasses compartments with mannose 6-phosphate receptors. *Journal of Cell Biology* 129, 81–97.

Dunlap, N.E., Benjamin, W.H., Jr, McGall, R.D., Tilden, A.B. and Briles, D.E. (1991) A 'safe-site' for *Salmonella typhimurium* is within splenic cells during the early phase of infection in mice. *Microbial Pathogenesis* 10, 297–310.

Dunlap, N.E., Benjamin, W.J., Berry, A.K., Eldridge, J.H. and Briles, D.E. (1992) A 'safe-site' for *Salmonella typhimurium* is within splenic polymorphonuclear cells. *Microbial Pathogenesis* 13, 181–190.

Eisenstein, T.K. (1998) Intracellular pathogens: the role of antibody-mediated protection in *Salmonella* infection [letter; comment]. *Trends in Microbiology* 6, 135–136.

Eisenstein, T.K. and Sultzer, B.M. (1983) Immunity to *Salmonella* infection. *Advances in Experimental Medicine and Biology* 162, 261–296.

Eisenstein, T.K., Killar, L.M. and Sultzer, B.M. (1984) Immunity to infection with *Salmonella typhimurium*: mouse-strain differences in vaccine- and serum-mediated protection. *Journal of Infectious Diseases* 150, 425–435.

Eisenstein, T.K., Huang, D., Meissler, J.J. and al Ramadi, R.B. (1994) Macrophage nitric oxide mediates immunosuppression in infectious inflammation. *Immunobiology* 191, 493–502.

Fields, P.I., Swanson, R.V., Haidaris, C.G. and Heffron, F. (1986) Mutants of *Salmonella typhimurium* that cannot survive within the macrophage are avirulent. *Proceedings of the National Academy Sciences USA* 83, 5189–5193.

Fields, P.I., Groisman, E.G. and Heffron, F. (1989) A *Salmonella* locus that controls resistance to microbicidal proteins from phagocytic cells. *Science* 243, 1059–1062.

Fierer, J., Krause, M., Tauxe, R. and Guiney, D. (1992) *Salmonella typhimurium* bacteremia: association with the virulence plasmid. *Journal of Infectious Diseases* 166, 639–642.

Fierer, J., Eckmann, L., Fang, F., Pfeifer, C., Finley, B.B. and Guiney, D. (1993) Expression of the *Salmonella* virulence plasmid gene *spvB* in cultured macrophages and nonphagocytic cells. *Infection and Immunity* 61, 5231–5236.

Finzi, G., Cornaggia, M., Capella, C., Fiocca, R., Bosi, F. and Solcia, E. (1993) Cathepsin E in follicle associated epithelium of intestine and tonsils: localization to M cells and possible role in antigen processing. *Histochemistry* 99, 201–211.

Frances, C.L., Ryan, T.A., Jones, B.D., Smith, S.J. and Falkow, S. (1993) Ruffles induced by *Salmonella* and other stimuli direct macropinocytosis of bacteria. *Nature* 364, 639–642.

Gahring, L.C., Heffron, F., Finlay, B.B. and Falkow, S. (1990) Invasion and replication of *Salmonella typhimurium* in animal cells. *Infection and Immunity* 58, 443–448.

Galán, J.E. and Curtiss, R., III (1989) Cloning and molecular characterization of genes whose products allow *Salmonella typhimurium* to penetrate tissue culture cells. *Proceedings of the National Academy Sciences USA* 86, 6383–6387.

Giannasca, P.J., Giannasca, K.T., Falk, P., Gordon, J.I. and Neutra, M. (1994) Regional differences in glycoconjugates of intestinal M cells in mice: potential targets for mucosal vaccines. *American Journal of Physiology* 267, G1108–G1121.

Giannella, R.A. (1979) Importance of the intestinal inflammatory reaction in *Salmonella*-mediated intestinal secretion. *Infection and Immunity* 23, 140–145.

Giannella, R.A., Formal, S.B., Dammin, G.J. and Collins, H. (1973) Pathogenesis of salmonellosis: studies on fluid secretion, and morphologic reaction in the rabbit ileum. *Journal of Clinical Investigation* 52, 441–453.

Ginocchio, C.C., Olmsted, S.B., Wells, C.L. and Galán, J.E. (1994) Contact with epithelial cells induces the formation of surface appendages on *Salmonella typhimurium*. *Cell* 76, 717–724.

Grossman, N., Joiner, K.A., Frank, M.M. and Leive, L. (1986) C3b binding, but not its breakdown, is affected by the structure of the O-antigen polysaccharide in lipopolysaccharide from salmonellae. *Journal of Immunology* 136, 2208–2215.

Guilloteau, L.A., Wallis, T.S., Gautier, A.V., MacIntyre, S., Platt, D.J. and Lax, A.J. (1996) The *Salmonella* virulence plasmid enhances *Salmonella*-induced lysis of macrophages and influences inflammatory responses. *Infection and Immunity* 64, 3385–3393.

Gulig, P.A., Caldwell, A.L., and Chiodo, V.A. (1992) Identification, genetic analysis and DNA sequence of a 7.8-kb virulence region of the *Salmonella typhimurium* virulence plasmid. *Molecular Microbiology* 6, 1395–1411.

Gulig, P.A. and Curtiss, R. (1987) Plasmid-associated virulence of *Salmonella typhimurium*. *Infection and Immunity* 55, 2891–2901.

Gulig, P.A. and Doyle, T.J. (1993) The *Salmonella typhimurium* virulence plasmid increases the growth rate of salmonellae in mice. *Infection and Immunity* 61, 504–511.

Gunn, J.S. and Miller, S.I. (1996) PhoP–PhoQ activates transcription of pmrAB, encoding a two-component regulatory system involved in *Salmonella typhimurium* antimicrobial peptide resistance. *Journal of Bacteriology* 178, 6857–6864.

Hardt, W.D., Chen, L.M., Schuebel, K.E., Bustelo, X.R. and Galan, J.E. (1998a) *S. typhimurium* encodes an activator of Rho GTPases that induces membrane ruffling and nuclear responses in host cells. *Cell* 93, 815–826.

Hardt, W.D., Urlaub, H. and Galan, J.E. (1998b) A substrate of the centisome 63 type III protein secretion system of *Salmonella typhimurium* is encoded by a cryptic bacteriophage. *Proceedings of the National Academy of Sciences USA* 95, 2574–2579.

Heffernan, E.J., Harwood, J., Fierer, J. and Guiney, D. (1992) The *Salmonella typhimurium* virulence plasmid complement resistance gene rck is homologous to a family of virulence related outer membrane protein genes including pagC and ail. *Journal of Bacteriology* 174, 84–91.

Heffernan, E.J., Wu, L., Louie, J., Okamoto, S., Fierer, J. and Guiney, D.G. (1994) Specificity of the complement resistance and cell association phenotypes encoded by the outer membrane protein genes rck from *Salmonella typhimurium* and ail from *Yersinia enterocolitica*. *Infection and Immunity* 62, 5183–5186.

Helander, I.M., Kilpelainen, I. and Vaara, M. (1994) Increased substitution of phosphate groups in lipopolysaccharides and lipid A of the polymyxin-resistant pmrA mutants of *Salmonella typhimurium*: a 31P-NMR study. *Molecular Microbiology* 11, 481–487.

Hensel, M., Shea, J.E., Bäumler, A.J., Gleeson, C., Blattner, F. and Holden, D.W. (1997) Analysis of the boundaries of *Salmonella* pathogenicity island 2 and the corresponding chromosomal region of *Escherichia coli* K-12. *Journal of Bacteriology* 179, 1105–1111.

Hensel, M., Shea, J.E., Gleeson, C., Jones, M.D., Dalton, E. and Holden, D.W. (1995) Simultaneous identification of bacterial virulence genes by negative selection. *Science* 269, 400–403.

Hensel, M., Shea, J.E., Waterman, S.R., Mundy, R., Nikolaus, T., Banks, G., Vazquez-Torres, A., Gleeson, C., Fang, F.C. and Holden, D.W. (1998) Genes encoding putative effector proteins of the type III secretion system of *Salmonella* pathogenicity island 2 are required for bacterial virulence and proliferation in macrophages. *Molecular Microbiology* 30, 163–174.

Hersh, D., Monack, D.M., Smith, M.R., Ghori, N., Falkow, S. and Zychlinsky, A. (1999) The *Salmonella* invasin SipB induces macrophage apoptosis by binding to caspase-1. *Proceedings of the National Academy of Sciences USA* 96, 2396–2401.

Hilbi, H., Chen, Y., Thirumalai, K. and Zychlinsky, A. (1997) The interleukin 1beta-converting enzyme, caspase 1, is activated during *Shigella flexneri*-induced apoptosis in human monocyte-derived macrophages. *Infection and Immunity* 65, 5165–5170.

Hobbie, S., Chen, L.M., Davis, R.J. and Galan, J.E. (1997) Involvement of mitogen-activated protein kinase pathways in the nuclear responses and cytokine production induced by *Salmonella typhimurium* in cultured intestinal epithelial cells. *Journal of Immunology* 159, 5550–5559.

Hohmann, A.W., Schmidt, G. and Rowley, D. (1978) Intestinal colonization and virulence of *Salmonella* in mice. *Infection and Immunity* 22, 763–770.

Hormaeche, C.E. (1980) The *in vivo* division and death rates of *Salmonella typhimurium* in the spleens of naturally resistant and susceptible mice measured by the superinfecting phage technique of Meynell. *Immunology* 41, 973–979.

Hormaeche, C.E. and Maskell, D.J. (1989) Influence of the Ity gene on *Salmonella* infections. *Research in Immunology* 140, 791–793.

Jepson, M.A., Mason, C.M., Clark, M.A., Simmons, N.L. and Hirst, B.H. (1995) Variations in lectin binding properties of intestinal M cells. *Journal of Drug Targeting* 3, 75–77.

Jones, B.D., Ghori, N. and Falkow, S. (1994) *Salmonella typhimurium* initiates murine infection by penetrating and destroying the specialized epithelial M cells of the Peyer's patches. *Journal of Experimental Medicine* 180, 15–23.

Jones, M.A., Wood, M.W., Mullan, P.B., Watson, P.R., Wallis, T.S. and Galyov, E.E. (1998) Secreted effector proteins of *Salmonella dublin* act in concert to induce enteritis. *Infection and Immunity* 66, 5799–5804.

Khan, S.A., Everest, P., Servos, S., Foxwell, N., Zahringer, U., Brade, H., Rietschel, E.T., Dougan, G., Charles, I.G. and Maskell, D.J. (1998) A lethal role for lipid A in *Salmonella* infections. *Molecular Microbiology* 29, 571–579.

Lee, J.C., Gibson, C.W. and Eisenstein, T.K. (1985) Macrophage-mediated mitogenic suppression induced in mice of the C3H lineage by a vaccine strain of *Salmonella typhimurium*. *Cellular Immunology* 91, 75–91.

Lindgren, S.W., Stojiljkovic, I. and Heffron, F. (1996) Macrophage killing is an essential virulence mechanism of *Salmonella typhimurium*. *Proceedings of the National Academy of Sciences USA* 93, 4197–4201.

Lissner, C.R., Swanson, R.N. and O'Brien, A.D. (1983) Genetic control of the innate resistance of mice to *Salmonella typhimurium*: expression of the Ity gene in peritoneal and splenic macrophages isolated in vitro. *Journal of Immunology* 131, 3006–3013.

McCormick, B., Colgan, S.P., Delp-Archer, C., Miller, S.I. and Madara, J.L. (1993). *Salmonella typhimurium* attachment to human intestinal epithelial monolayers: transepithelial signalling to subepithelial neutrophils. *Journal of Cell Biology* 123, 895–907.

McCormick, B.A., Miller, S.I., Carnes, D. and Madara, J.L. (1995) Transepithelial signaling to neutrophils by salmonellae: a novel virulence mechanism for gastroenteritis. *Infection and Immunity* 63, 2302–2309.

Mackaness, G.B., Blandon, R.V. and Collins, F.M. (1966) Host parasite relationships in mouse typhoid. *Journal of Experimental Medicine* 124, 573–583.

Mastroeni, P., Villarreal-Ramos, B. and Hormaeche, C.E. (1993) Adoptive transfer of immunity to oral challenge with virulent salmonellae in innately susceptible BALB/c mice requires both immune serum and T cells. *Infection and Immunity* 61, 3981–3984.

Maw, J. and Meynell, G.G. (1968) The true division and death rates of *Salmonella typhimurium* in the mouse spleen determined with superinfecting phage P22. *British Journal of Experimental Pathology* 49, 597–613.

Michetti, P., Mahan, M.J., Slauch, J.M., Mekalanos, J.J. and Neutra, M.R. (1992) Monoclonal secretory immunoglobulin A protects mice against oral challenge with the invasive pathogen *Salmonella typhimurium*. *Infection and Immunity* 60, 1786–1792.

Michetti, P., Porta, N., Mahan, M.J., Slauch, J.M., Mekalanos, J.J., Blum, A.L., Kraehenbuhl, J.P. and Neutra, M.R. (1994) Monoclonal immunoglobulin A prevents adherence and invasion of polarized epithelial cell monolayers by *Salmonella typhimurium*. *Gastroenterology* 107, 915–923.

Miller, S.I., Hohmann, E.L. and Pegues, D.A. (1995) *Salmonella* (including *Salmonella typhi*). In: Mandell, G.L., Bennett, J.E. and Dolin, R (eds) *Principles and Practice of Infectious Diseases*, Vol. 2. Churchill Livingstone, New York, pp. 2013–2033.

Mills, D.M., Bajaj, V. and Lee, C.A. (1995) A 40kb chromosomal fragment encoding *Salmonella typhimurium* invasion genes is absent from the corresponding region of the *Escherichia coli* K-12 chromosome. *Molecular Microbiology* 15, 749–759.

Mirold, S., Rabsch, W., Rohde. M., Stender, S., Tschape, H., Russmann, H., Igwe, E. and Hardt, W.D. (1999) Isolation of a temperate bacteriophage encoding the type III effector protein SopE from an Epidemic *Salmonella typhimurium* strain. *Proceedings of the National Academy of Sciences USA* 96, 9845–9850.

Monack, D.M., Raupach, B., Hromockyj, A.E. and Falkow, S. (1996) *Salmonella typhimurium* invasion induces apoptosis in infected macrophages. *Proceedings of the National Academy of Sciences USA* 93, 9833–9838.

Norris, T.L., Kingsley, R.A. and Bäumler, A.J. (1998) Expression and transcriptional control of the *Salmonella typhimurium lpf* fimbrial operon by phase variation. *Molecular Microbiology* 29, 311–320.

O'Brien, A.D. (1986) Influence of host genes on resistance of inbred mice to lethal infection with *Salmonella typhimurium*. *Current Topics in Microbiology and Immunology* 124, 37–48.

O'Brien, A.D., Metcalf, E.S. and Rosenstreich, D.L. (1982) Defect in macrophage effector function confers *Salmonella typhimurium* susceptibility on C3H/HeJ mice. *Cellular Immunology* 67, 325–333.

O'Brien, A.D., Rosenstreich, D.L., Scher, I., Campbell, G.H., MacDermott, R.P. and Formal, S.B. (1980) Genetic control of susceptibility to *Salmonella typhimurium* infection in mice: role of the *Lps* gene. *Journal of Immunology* 124, 20–24.

Ochman, H., Soncini, F.C., Solomon, F. and Groisman, E.A. (1996). Identification of a pathogenicity island for *Salmonella* survival in host cells. *Proceedings of the National Academy of Sciences USA* 93, 7800–7804.

Owen, R.L. (1977) Sequential uptake of horsradish peroxidase by lymphoid follicle epithelium of Peyer's patches in the normal unobstructed mouse intestine: an ultrastructural study. *Gastroenterology* 72, 440–451.

Pappo, J. and Owen, R.L. (1988) Absence of secretory component expression by epithelial cells overlying rabbit gut-associated lymphoid tissue. *Gastroenterology* 95, 1173–1177.

Que, J.U. and Hentges, D.J. (1985) Effect of streptomycin administration on colonization resistance to *Salmonella typhimurium* in mice. *Infection and Immunity* 48, 169–174.

al-Ramadi, B., Meissler, J.J., Huang, D. and Eisenstein, T.K. (1992) Immunosuppression induced by nitric oxide and its inhibition by interleukin-4. *European Journal of Immunology* 22, 2249–2254.

Rathman, M., Barker, L.P. and Falkow, S. (1997) The unique trafficking pattern of *Salmonella typhimurium*-containing phagosomes in murine macrophages is independent of the mechanism of bacterial entry. *Infection and Immunity* 65, 1475–1485.

Rathman, M., Sjaastad, M.D. and Falkow, S. (1996) Acidification of phagosomes containing *Salmonella typhimurium* in murine macrophages. *Infection and Immunity* 64, 2765–2773.

Rhen, M., Riikonen, P. and Taira, S. (1993) Transcriptional regulation of *Salmonella enterica* virulence plasmid genes in cultured macrophages. *Molecular Microbiology* 10, 45–56.

Richter-Dahlfors, A., Buchan, A.M.J. and Finlay, B.B. (1997) Murine salmonellosis studied by confocal microscopy: *Salmonella typhimurium* resides intracellularly inside macrophages and excerts a cytotoxic effect on phagocytes *in vivo. Journal of Experimental Medicine* 186, 569–580.

Riikonen, P., Mäkälä, P.H., Saarilahti, H., Sukupolvi, S., Taira, S. and Rhen, M. (1992) The virulence plasmid does not contribute to growth of *Salmonella* in cultured murine macrophages. *Microbial Pathogenesis* 13, 281–291.

Roudier, C., Krause, M., Fierer, J. and Guiney, D.G. (1990) Correlation between the presence of sequences homologous to the *vir* region of *Salmonella dublin* plasmid pSDL2 and the virulence of twenty-two *Salmonella* serovars in mice. *Infection and Immunity* 58, 1180–1185.

Schonwetter, B.S., Stolzenberg, E.D. and Zasloff, M.A. (1995) Epithelial antibiotics induced at sites of inflammation. *Science* 267, 1645–1648.

Schwacha, M.G. and Eisenstein, T.K. (1997) Interleukin-12 is critical for induction of nitric oxide-mediated immunosuppression following vaccination of mice with attenuated *Salmonella typhimurium*. *Infection and Immunity* 65, 4897–4903.

Schwacha, M.G., Meissler, J.J., Jr and Eisenstein, T.K. (1998) *Salmonella typhimurium* infection in mice induces nitric oxide-mediated immunosuppression through a natural killer cell-dependent pathway. *Infection and Immunity* 66, 5862–5866.

Selsted, M.E., Miller, S.I., Henschen, A.H. and Ouellette, A.J. (1992) Enteric defensins: antibiotic peptide components in intestinal host defense. *Journal of Cell Biology* 118, 929–936.

Shea, J.E., Hensel, M., Gleeson, C. and Holden, D.W. (1996) Identification of a virulence locus encoding a second type III secretion system in *Salmonella typhimurium*. *Proceedings of the National Academy of Sciences USA* 93, 2593–2597.

Soncini, F.C. and Groisman, E.A. (1996) Two-component regulatory systems can interact to process multiple envi ronmental signals. *Journal of Bacteriology* 178, 6796–6801.

Stephen, J., Wallis, T.S., Starkey, W.G., Candy, D.C.A., Osborne, M.P. and Haddon, S. (1985) Salmonellosis· in retrospect and prospect. In: *Microbial Toxins and Diarrhoeal Disease*. Pitman, London, pp. 175–192.

Thirumalai, K., Kim, K.S. and Zychlinsky, A. (1997) IpaB, a *Shigella flexneri* invasin, colocalizes with interleukin-1 beta-converting enzyme in the cytoplasm of macrophages. *Infection and Immunity* 65, 787–793.

Tsolis, R.M., Adams, L.G., Ficht, T.A. and Bäumler, A.J. (1999a) Contribution of *Salmonella typhimurium* virulence factors to diarrhoeal doseases in calves. *Infection and Immunity* 67, 4897–4885.

Tsolis, R.M., Townsend, S.M., Ficht, T.A., Adams, L.G. and Bäumler, A.J. (1999b) Identification of a putative *Salmonella typhimurium* host range factor with homology to IpaH and YopM by signature-tagged mutagenesis. *Infection and Immunity* (in press).

Vaara, M. (1992) Agents that increase the permeability of the outer membrane. *Microbiology Review* 56, 395–411.

Vescovi, E.G., Ayala, Y.M., Dicera, E. and Groisman, E.A. (1997) Characterization of the bacterial sensor protein PhoQ – evidence for distinct binding sites for Mg^{2+} and Ca^{2+}. *Journal of Biological Chemistry* 272, 1440–1443.

Vescovi, E.G., Soncini, F.C. and Groisman, E.A. (1996) Mg^{2+} as an extracellular signal: environmental regulation of *Salmonella* virulence. *Cell* 84, 165–174.

Wallis, T.S., Vaughan, A.T., Clarke, G.J., Qi, G.M., Worton, K.J., Candy, D.C., Osborne, M.P. and Stephen, J. (1990) The role of leucocytes in the induction of fluid secretion by *Salmonella typhimurium*. *Journal of Medical Microbiology* 31, 27–35.

Watson, P.R., Galyov, E.E., Paulin, S.M., Jones, P.W. and Wallis, T.S. (1998) Mutation of *invH*, but not *stn*, reduces *Salmonella*-induced enteritis in cattle. *Infection and Immunity* 66, 1432–1438.

Wood, M.W., Jones, M.A., Watson, P.R., Hedges, S., Wallis, T.S. and Galyov, E.E. (1998) Identification of a pathogenicity island required for *Salmonella* enteropathogenicity. *Molecular Microbiology* 29, 883–891.

Wood, M.W., Rosqvist, R., Mullan, P.B., Edwards, M.H. and Galyov, E.E. (1996) SopE, a secreted protein of *Salmonella dublin*, is translocated into the target eukaryotic cell via a *sip*-dependent mechanism and promotes bacterial entry. *Molecular Microbiology* 22, 327–338.

Woodward, M.J., McLaren, I. and Wray, C. (1989) Distribution of virulence plasmids within salmonellae. *Journal of General Microbiology* 135, 503–511.

Zychlinsky, A., Prevost, M.C. and Sansonetti, P.J. (1992) *Shigella flexneri* induces apoptosis in infected macrophages. *Nature* 358, 167–169.

Chapter 5

Host Susceptibility, Resistance and Immunity to *Salmonella* in Animals

Peter S. Holt

US Department of Agriculture, Agricultural Research Service, Southeast Poultry Research Laboratory, 934 College Station Road, Athens, GA 30605, USA

Introduction

Avian and mammalian food animals come in contact with a multitude of potentially pathogenic microorganisms during their life. Intensive rearing procedures swell the density of susceptible hosts within a living space, subsequently increasing the potential for the entry and rapid dissemination of pathogens through the flock or herd. Producers wage a continual battle to reduce the incidence, severity and final impact of disease in the livestock and poultry under their care. The application of good management practice, stringent biosecurity measures and vaccination helps in this endeavour.

Members of the genus *Salmonella* pose a serious threat to the domestic food-animal industry. These organisms are responsible for significant morbidity and mortality in their respective hosts (Bullis, 1977; Wray and Sojka, 1977), as well as causing substantial disease to humans consuming processed meats (Bisplinghoff, 1985; Parham, 1985; Humphrey *et al.*, 1988) and eggs (St Louis *et al.*, 1988; Humphrey, 1990) derived from the infected animals. Human salmonellosis originating from consumption of meat or poultry products is not a new problem and has been dealt with for decades (Wray and Sojka, 1977). What may be considered new is the emergence of different food vehicles, such as eggs (St Louis *et al.*, 1988; Humphrey, 1990), as the source of these infections, forcing industry to examine its management practices and incorporate new proce-

dures to reduce the incidence and severity of the problem. To assist the producer in this endeavour, the scientific community embarked on research to define the nature of the problem, delineate the host–parasite interactions occurring during the infection and outline possible remedial actions to ameliorate the situation. Likewise, studies of the development of an immune response against *Salmonella* infection in domestic animals provide some of the vital information needed by industry to deal with the *Salmonella* problem on the farm. This review attempts to glean the available information regarding the development of immunity in domestic farm animals against *Salmonella* infection. Because of the voluminous information available regarding the immune system and *Salmonella* immunity in mice and humans, data derived from studies conducted in these species will, in many cases, also be presented to provide a backdrop for what is known regarding immunity to *Salmonella* in the food-animal species.

The Immune System

The immune system consists of two broad categories: innate and specific immunity. Innate, or non-specific, immunity includes serum components, such as complement, and non-specific defence cells, such as polymorphonuclear neutrophils (heterophils in poultry), macrophages and natural killer cells (Dietert *et al.*, 1991;

Sharma and Schat, 1991; Portnoy, 1992; Kogut *et al.*, 1994b), and provides much of the early 'front-line' defence against microbial invasion. The specific immune system is itself divided into two broad components: humoral and cellular immunity. The antibody, produced by B lymphocytes, provides the active effector function for humoral immunity. Antibodies protect the host against infection by binding to the surface of infecting organisms to prevent them from attaching to and invading host cells (McGhee *et al.*, 1992; Michetti *et al.*, 1992), enhance their engulfment and killing by phagocytic cells (Johnson *et al.*, 1985) and activate complement-mediated killing (Morgan, 1990). The protective activity of antibodies therefore occurs during the extracellular phase of the bacterial infection. Cellular immunity is mediated by T lymphocytes and these cells can serve either a direct effector function (cytolytic T lymphocytes (Tctl)) or a regulatory function (helper (Th) and suppressor (Ts) T cells) by modifying the activation of B cells or other T cells. This component of the immune response is important in protecting against intracellular pathogens, such as viruses, parasites and certain bacterial species (Lillehoj, 1993; Schat, 1994) and acts through direct killing of the infected host cell or the activation of the phagocytic cell defences. Both humoral and cellular immunity appear to play a role in protection against a *Salmonella* infection (Mastroeni *et al.*, 1993a), although the importance of each in the ultimate protection of the host still remains controversial.

The T lymphocytes are classified according to the function that they perform: Th, Ts and Tctl. Along with function, each cell produces certain unique cell-surface determinants or markers, which simplifies their detection and identification in the host (Cantor *et al.*, 1975a,b). Th cells produce the marker CD4 or L3T4 (Dialynas *et al.*, 1983), while Ts and Tctl produce CD8 or Lyt2 (Swain, 1983). The T cells can be further characterized by the T-cell receptors they produce: αβ (Haskins *et al.*, 1983) and γδ (Brenner *et al.*, 1986). Homologues to these markers have been described in poultry (Chen *et al.*, 1990), cattle (Park *et al.*, 1992) and pigs (Pescovitz *et al.*, 1984).

Current concepts regarding the development of an immune response indicate that the response is controlled by Th and that different Th subsets will drive different responses. Originally described by Mosmann *et al.* in 1986, the Th concept helped to mould thinking regarding the mechanisms of how a host mounts a response against a pathogen. The Th subsets, Th1 and Th2, are classified according to the cytokines they elicit. The Th1 cells produce interferon γ (IFN-γ) and interleukin 2 (IL-2) and primarily control cell-mediated immunity (CMI). Th2 cells produce IL-4, IL-5, IL-6 and IL-10 and primarily regulate humoral immunity. The differentiation of Th subpopulations is not total, since certain cells have been shown to produce lymphokines common to both subsets and may be precursors to Th1 and Th2 cells (Mosmann and Coffman, 1989a). Further, cytokines produced by Th1 cells affect the immunoglobulin isotype produced by the B cell (Finkelman *et al.*, 1988). The decision to elicit a Th1 or Th2 response is cytokine driven, with IL-12 mediating a Th1 response (Manetti *et al.*, 1993) and IL-4 mediating a Th2 response (Swain *et al.*, 1990). Further, IFN-γ inhibits proliferation by Th2 cells (Mosmann and Coffman, 1989b) and IL-10 inhibits cytokine synthesis by Th1 cells (Fiorentino *et al.*, 1989), indicating a regulation of response by the cytokines produced. The sources of these cytokines originate from the innate immune system, with macrophages producing IL-12 and basophils and mast cells producing the IL-4. The non-specific innate immune system therefore exerts significant control over the type of specific immune response that will be generated against a particular pathogen.

Early Infection by *Salmonella*

Following ingestion, the *Salmonella* cell must resist killing or elimination by defences such as gastric acidity (Blaser and Newman, 1982), intestinal motility (Formal *et al.*, 1958; Kent *et al.*, 1966) and the autogenous microbial flora in the intestinal tract (Hentges and Freter, 1961; Corrier *et al.*, 1991). The intestinal tract rapidly becomes colonized (Fanelli *et al.*, 1971; Turnbull and Snoeyenbos, 1974; Fedorka-Cray *et al.*, 1994; Holt *et al.*, 1995; Gray *et al.*, 1996a,b) and invasion through the bowel wall occurs via the binding of *Salmonella* to specialized epithelial cells overlying intestinal lymphoid tissue and entry through them (Kohbata *et al.*, 1986). It is at this juncture that the bacteria encounter professional phagocytic cells, which provide the first cellular defence against invasion. These cells provide important early protection against invasion, as evidenced by the exacerbation of infection when phagocytes are eliminated (O'Brien *et al.*, 1979; Kogut *et al.*, 1993) and, conversely, increased killing of *Salmonella* when professional phagocytes are activated (McGruder *et al.*, 1993, 1995; Kogut *et al.*, 1995a,b). Extraintestinal dissemination to multiple organ sites occurs in many *Salmonella* infections (Timoney *et al.*, 1989; Gast and Beard, 1990a; Gray *et al.*, 1996a), possibly via mononuclear phagocytes (Popiel and Turnbull, 1985). *Salmonella* replication occurs within the reticuloendothelial system (Hormaeche *et al.*, 1993) and their capacity to survive and to replicate within mononuclear phagocytes (Lissner *et al.*, 1983; Fields *et al.*, 1986) prompted the classification of most *Salmonella* as intracellular parasites (Hormaeche *et al.*, 1993), although this categorization is not universally accepted (Hsu, 1989).

Serum Humoral Immunity to *Salmonella*

Once an animal becomes infected with *Salmonella*, a rapid humoral immune response ensues. This response can be affected by many factors, including the dose of the challenge organism (Humphrey *et al.*, 1991a; Hassan *et al.*, 1993; Gray *et al.*, 1996b), virulence of the organism (Hassan and Curtiss, 1994a; Gray *et al.*, 1995), route of administration (Chart *et al.*, 1992) and age (Humphrey *et al.*, 1991a; Hassan and Curtiss, 1994a; Thorns *et al.*, 1996). Regarding this latter category, very young individuals respond poorly to infection (Williams and Whittemore, 1975) or respond selectively to different determinants (Thorns *et al.*, 1996). Williams and Whittemore (1975) described a phenomenon of immunological paralysis in chicks infected with *S. typhimurium* at 1–2 weeks of age and a similar poor response was observed in other laboratories (P.S. Holt and R.K. Gast, 1997, unpublished data). Such hyporesponsiveness may be a reflection of the immaturity of the immune system at this age (Jeurissen *et al.*, 1989) or the decimation of the lymphoid tissue in very young individuals

infected with virulent *Salmonella* organisms (Hassan and Curtiss, 1994a). Older birds may also experience difficulty in responding serologically to a *Salmonella* challenge (Humphrey *et al.*, 1991a).

By 1 week post-infection an antibody response can be detected in the sera of chickens (Lee *et al.*, 1981; Chart *et al.*, 1990; Gast and Beard, 1990a; Hassan *et al.*, 1991; Humphrey *et al.*, 1991b; Kim *et al.*, 1991; Baay and Huis in't Veld, 1993; Corkish *et al.*, 1994), pigs (Gray *et al.*, 1996a), cattle (Lindberg and Robertsson, 1983) and sheep (Brennan *et al.*, 1994) and this response can persist for 10 weeks or more. Serum immunoglobulin M (IgM) anti-*Salmonella* responses appear first, followed by IgG and IgA (Hassan *et al.*, 1991). The IgM and IgA levels gradually decline, while IgG levels can persist for extended periods (Hassan *et al.*, 1991; Chart *et al.*, 1992). Reinfection results in a rapid, enhanced antibody response to the challenge organism (Hassan *et al.*, 1991). Serum immunoglobulins can be passed vertically into the egg yolk (Gast and Beard, 1991; McLeod and Barrow, 1991; Nicholas and Cullen, 1991; Thorns *et al.*, 1996; Gast *et al.*, 1997) and colostrum (Royal *et al.*, 1968; Jones *et al.*, 1988). The presence of these immunoglobulins can inhibit the growth of the organism in egg contents (Holt *et al.*, 1996) and possibly protect the offspring (Thain *et al.*, 1984; Jones *et al.*, 1988). Serum antibodies will, in many cases, be detectable after individuals are no longer culture-positive for the organism (van Zijderveld *et al.*, 1993; Corkish *et al.*, 1994; Gast *et al.*, 1997).

A variety of bacterial determinants exist on or within the *Salmonella* cell, many of which will be recognized as foreign by the host, following infection or vaccination. The intensity of antibody production to individual antigens can vary during the course of the response, with certain antigens being immunodominant early on, while others are delayed in the initiation of a response. Baay and Huis in't Veld (1993) noted that serum anti-flagellum titres in chickens infected with *S. enteritidis* were highest early in the infection, while lipopolysaccharide (LPS) titres predominated in later stages of the infection. Antibody responses to flagella and LPS were weak in *Salmonella*-infected young birds but robust in adult hens, while serum responses to the fimbrial antigen SEF14 was strong in the chicks but modest in adult birds (Thorns *et al.*, 1996). Strong serum antibody responses could also be detected against outer membrane proteins (OMP) (Hassan *et al.*, 1991; Kim *et al.*, 1991) and cell-wall components (Hassan *et al.*, 1991; Gray *et al.*, 1995).

Serum antibody responses to these determinants may or may not offer protection for the host. Brownell *et al.* (1970) showed that bursectomized chickens orally infected with S. *typhimurium* shed higher levels of the organism from the intestinal tract than their non-bursectomized counterparts. Calves could be protected from lethal *Salmonella* challenge when fed colostrum from *Salmonella*-immune cows (Royal *et al.*, 1968; Jones *et al.*, 1988), although this protection could not be correlated with serum antibodies to flagellar or somatic antigens (Jones *et al.*, 1988). Similarly, serum anti-LPS antibodies were not found to offer solid protection in calves (Segall and Lindberg, 1993). Serum antibody responses to OMP (Bouzoubaa *et al.*, 1987, 1989; Charles *et al.*, 1994) following vaccination offered good protection against invasive *Salmonella* infections and individuals possessing robust agglutinating antibody responses to whole bacterial cells exhibited decreased intestinal shedding (Barbour *et al.*, 1993; Gast *et al.*, 1993), reduced extraintestinal dissemination to different organ sites (Timms *et al.*, 1990, 1994; Gast *et al.*, 1992; Charles *et al.*, 1994) and protection against lethal challenge (Aitken *et al.*, 1982).

The specificity of the antibody response differs as well. Serum antibody responses in individuals infected with different *Salmonella* species show marked serological cross-reactivity to LPS (Nicholas and Cullen, 1991; Barrow, 1992; Barrow *et al.*, 1992), in many cases due to the presence of antigen 12 in the somatic antigens of many *Salmonella* serovars (Chart *et al.*, 1990). Cross-reactivity is less pronounced in responses against flagellum antigens, although some degree of cross-reaction does occur (Baay and Huis in't Veld, 1993; Thorns *et al.*, 1996). Certain *Salmonella* species produce antigenic determinants that are relatively unique to that species (Thorns *et al.*, 1990; Thorns, 1995) and antibodies generated against them are fairly specific (Thorns *et al.*, 1996). The incorporation of modern molecular genetic techniques will enable scientists to define even more exactly the immune response to a particular antigen by examining responses to cloned fragments of that antigen (Kwang and Littledike, 1995).

Mucosal Humoral Immunity to *Salmonella*

The initiation of infection by *Salmonella* occurs primarily at mucosal surfaces and a significant humoral immune response generally occurs within this region. The predominant immunoglobulin of mucosal immunity is IgA (Bienenstock *et al.*, 1973; Schat and Myers, 1991), although IgG and/or IgM responses can also be observed (Hassan *et al.*, 1991; Segall *et al.*, 1994; Gray *et al.*, 1995, 1996a). The IgA can be found in both bile and mucosal secretions and can exist in both monomeric and multimeric forms (Bienenstock *et al.*, 1973; Lim and Maheswaran, 1977; Schat and Myers, 1991). In mice and humans, IgA has been shown to: mediate protection against infection through antibody-dependent cellular cytotoxicity (Tagliabue *et al.*, 1983), potentiate bactericidal action by iron-sequestering compounds (Funakoshi *et al.*, 1982), serve as a possible opsonin for mucosal phagocytes (Kilian *et al.*, 1988), inhibit bacterial adherence (McGhee *et al.*, 1992; Michetti *et al.*, 1992) and neutralize toxin moieties (Lange and Holmgren, 1978). Secretory IgA therefore exhibits exceptional diversity in its ability to mediate protection at mucosal surfaces.

The local immune response to *Salmonella* infection in domestic farm animals follows post-challenge kinetics similar to that of systemic immunity. In chickens, strong mucosal immune responses can be observed within 2–3 weeks post-challenge against flagella, LPS or OMP in the intestine (Hassan *et al.*, 1991; Holt and Porter, 1993) or bile (Lee *et al.*, 1981; Hassan *et al.*, 1991). Similar kinetics of infection could be observed in challenged pigs, although intestinal IgA responses were either very low (Gray *et al.*, 1996a) or non-existent (Gray *et al.*, 1995) compared with IgG and IgM. Calves orally infected with a live S. *dublin* vaccine strain exhibited detectable mucosal IgA and IgM anti-LPS responses by 3 days post-infection, which remained elevated for 30 days (Segall *et al.*, 1994). In chickens, reinfection of individuals resulted in rapid increases in IgA anti-*Salmonella* levels in the gut and bile (Hassan *et al.*, 1991), indicating that, at least in avian species, mucosal immunity is capable of an anamnestic response.

Cellular Immunity to *Salmonella*

Along with the elicitation of a humoral immune response, CMI can also be detected in individuals infected with *Salmonella*. Detectable delayed-type hypersensitivity (DTH) skin reactions were observed at 2–5 weeks post-challenge in chickens infected with *S. typhimurium* and these responses paralleled, to a certain degree, the serum humoral response (Hassan *et al.*, 1991). Skin responses could be observed against sonicated whole organism, flagella, LPS and OMP. The sonicated preparations and the OMP elicited the strongest reaction and LPS the poorest. In studies performed by Lee *et al.* (1981), organ levels of an *S. typhimurium* challenge remained high in the face of a strong serum and biliary antibody response, but these levels decreased following the appearance of cellular immunity, indicating an important role for CMI in protection. Cattle injected intradermally with *Salmonella* antigens after previous systemic infection with *S. dublin* or *S. typhimurium* (Aitken *et al.*, 1978; Robertsson *et al.*, 1982a) exhibited marked DTH skin responses which were detectable a year post-challenge (Aitken *et al.*, 1978).

In many cases, there is not a good correlation between serum antibody titres and the cellular immune responses. Merritt *et al.* (1984) found that calves vaccinated with a modified live *S. typhimurium* displayed a strong DTH response to injected antigens and generally survived the lethal challenge with the parent strain of the *S. typhimurium*. Serum antibody titres were modest or undetectable. Similarly, calves vaccinated with a live *S. typhimurium* vaccine strain exhibited strong cell-mediated immune responses and modest antibody responses, while calves immunized with a killed *S. typhimurium* preparation responded poorly in the cellular assays but strongly in the antibody assays (Lindberg and Robertsson, 1983). Protection against lethal challenge was poor in the bacterin-vaccinated group but excellent in the live-vaccine group, providing evidence for the important role of CMI in protection, although the role of mucosal immunity was not delineated. Cross-reactivity for different antigens could also be observed for the

DTH response in *Salmonella*-infected cattle. Porins elicited a strong cross-reactivity between *Salmonella* serovars, while LPS mediated a more modest effect (Robertsson *et al.*, 1982b). In pigs orally inoculated with an attenuated *S. choleraesuis* strain, a detectable serum or mucosal humoral response to the *Salmonella* antigen was not found, but the animals did mount a strong DTH response (Stabel *et al.*, 1993). Lymphocytes purified from pigs infected with *S. typhimurium* responded strongly *in vitro* to LPS or heat-extracted cell wall at 7–63 days post-infection (Gray *et al.*, 1995, 1996a,b). As was observed with humoral immunity, dose affected the response, with cells from animals receiving 10^6 organisms responding the most vigorously to the antigenic stimulation, while those animals receiving a higher dose (10^9 organisms) or a lower dose (10^3 organisms) responded more modestly (Gray *et al.*, 1996b).

Cytokines

Cytokines play a critical role in the protection mediated by the immune system. As was alluded to previously, cytokines regulate whether a predominantly humoral or cell-mediated response will be mounted. Also, the type of response is characterized by the cytokines elicited. Besides modulating the type and intensity of an immune response, these protein signals can also alter the activity of the effector cells. In mice, IFN-γ was shown to activate macrophages (Murray *et al.*, 1985; Black *et al.*, 1987), resulting in increased killing capacity of the cells for a variety of microbial pathogens (Murray *et al.*, 1985; Kagaya *et al.*, 1989). Incubation of macrophages with recombinant IFN-γ enhanced their bactericidal activity against *S. typhimurium* (Kagaya *et al.*, 1989), and mice administered exogenous IFN-γ exhibited reduced disseminated infection by that organism (Muotiala and Makela, 1990). Similar protection against *Salmonella* infection was observed in mice receiving tumour necrosis factor alpha (TNF-α), an inflammatory cytokine (Nakano *et al.*, 1990), or IL-12 (Kincy Cain *et al.*, 1996). Conversely, treatment of mice with antibodies to IFN-γ, TNF-α or IL-12 during a *Salmonella* infection exacerbated the severity of the disease in these animals (Nauciel and Espinasse-Maes, 1992; Kincy Cain *et al.*, 1996) and reduced the protection offered by vaccination with an avirulent *Salmonella* vaccine strain (Mastroeni *et al.*, 1993b). During *Salmonella* infection, elevated levels of IFN-γ (Ramarathinam *et al.*, 1991; Klimpel *et al.*, 1995), TNF-α (Arnold *et al.*, 1993), IL-2 (Klimpel *et al.*, 1995) and IL-12 (Kincy Cain *et al.*, 1996) can be detected. The prominent involvement of IFN-γ and IL-12 in protecting against a *Salmonella* infection indicates the probable role of the Th1 cell subset and subsequent involvement of cell-mediated immunity in providing specific *Salmonella* immunity.

Knowledge of the role of cytokines in animal health continues to grow. A number of cytokines from food animals have been cloned and sequenced (Blecha, 1991; Sekellick *et al.*, 1994; Suresh *et al.*, 1995). These compounds have been used therapeutically to increase resistance to both bacterial (Sordillo and Babiuk, 1991; Reddy *et al.*, 1992) and viral infections (Babiuk *et al.*, 1987; Reddy *et al.*, 1989), to enhance the killing capacity of phagocytes (Roth and Frank, 1989) and to increase vaccine efficacy (Hughes *et al.*, 1992; Nash *et al.*, 1993; Reddy *et al.*, 1993). The possible role of these agents in the manifestation of disease problems or the protection and resolution of different disease states appears to mirror those for mice and humans (Lillehoj *et al.*, 1992; Bielefeldt-Ohman, 1995; Wannemuehler, 1995).

While substantial research has been conducted on the host–parasite interactions between domestic animals and *Salmonella*, only a relatively small amount of information has been forthcoming with regard to cytokine activity during *Salmonella* infection in these animals. Calves septicaemic with *S. typhimurium* did not exhibit elevated serum levels of TNF-α (Peel *et al.*, 1990b), nor did pigs challenged orally with the organism (Stabel *et al.*, 1995). However, serum levels of this cytokine were elevated in pigs receiving the challenge intranasally and the levels remained high for several weeks, indicating that the route of infection can influence the severity of inflammation. Chickens stressed through feed withdrawal exhibit increased intestinal inflammation during an infection with *S. enteritidis* (Holt and Porter, 1992; Porter and Holt, 1993; Holt *et al.*, 1994), resulting in elevated levels of TNF in the alimentary tract of these birds (Arnold and Holt, 1996). The thera-

peutic use of cytokine reagents to ameliorate *Salmonella* infection has also been undertaken. Peel *et al.* (1990a) showed that calves administered recombinant bovine IFN-γ exhibited reduced septicaemia and diminished symptoms, compared with controls, following infection with *S. typhimurium*. Chicks and turkey poults administered supernatant fluids from immune lymphocytes were more resistant to challenge with *S. enteritidis* (McGruder *et al.*, 1993, 1995; Kogut *et al.*, 1995a; Tellez *et al.*, 1993). The mechanism of action of this supernatant preparation appeared to be through the induction of increased levels of heterophils in circulation (McGruder *et al.*, 1995) and into the site of infection (Kogut *et al.*, 1995a), as well as enhanced phagocytic capabilities of these cells (Kogut *et al.*, 1995b). The activity was observed from supernatants derived from immune lymphocytes and not from macrophages, indicating that the soluble factor is of T-cell and not macrophage origin (McGruder *et al.*, 1993). This factor has not been purified and physically characterized but the pharmacological activity of the preparation indicates that a colony-stimulating factor may be involved (Kogut *et al.*, 1994a).

Susceptibility and Genetics

Genetics plays a very important role in the relationship between *Salmonella* and its potential host animal. Some *Salmonella* serovars display a very narrow host specificity (Bullis, 1977; O'Brien, 1982), while many of the remaining members of this genus express a more wide-ranging host infectivity. Furthermore, members of a particular *Salmonella* serovar express differential capabilities for infecting a particular host (Shivaprasad *et al.*, 1990; Gast and Beard, 1992; Gast and Benson, 1995). Conversely, genetics plays an important role in enabling the host to resist infection by a *Salmonella* pathogen. Several gene loci control the innate *S. typhimurium* resistance of mice (O'Brien, 1986). The gene *Ity* regulates how well *S. typhimurium* replication is controlled in the animal (Plant and Glynn, 1976) and is expressed *in vivo* within 24 h (Swanson and O'Brien, 1983). Susceptibility in mice to *S. typhimurium* appears to be due to a macrophage defect in microbicidal capability (Lissner *et al.*, 1983). A second gene locus, *Lps*, controls the response of mice to LPS. The *Lps*-deficient mice are hyporesponsive to LPS stimulation and therefore fail to produce LPS-inducible cytokines (O'Brien, 1986). Fewer macrophages are recruited during infection and the cells exhibit a reduced ability to restrict *Salmonella* growth (Weinstein *et al.*, 1986). Like *Ity*-susceptible mice, *Lps*-defective mice are extremely susceptible to S. *typhimurium* infection (Weinstein *et al.*, 1986).

The inherent resistance (or susceptibility) to *Salmonella* in domestic food animals is not nearly as well defined, although our understanding is increasing in certain species. Pevzner *et al.* (1981) showed that response to S. *pullorum* vaccination in chickens could be selected for genetically and that responsiveness was linked, at least partially, to the B complex within the major histocompatibility complex. Bumstead and Barrow (1988) found substantial differences in lines of chicks to lethal challenge with S. *typhimurium* and later showed that bird lines that were susceptible or resistant to infection with S. *typhimurium* exhibited similar susceptibility or resistance to several other *Salmonella* serovars (Bumstead and Barrow, 1993). Gast and Benson (1995) and Guillot *et al.* (1995) showed that various lines of day-old chicks differed in their susceptibility to infection with S. *enteritidis* and that table-egg producing birds tended to be more susceptible to the challenge. Chicken-line differences were also observed with respect to dissemination, egg contamination and faecal shedding (Lindell *et al.*, 1994; Protais *et al.*, 1996). In pigs, Lacey *et al.* (1989) found that peripheral blood monocytes from certain breeds exhibited higher phagocytic capacity and microbicidal activity against S. *typhimurium*. Breed differences were also observed regarding humoral and cell-mediated responses to a vaccine strain of S. *typhimurium*, similar to those observed previously for monocyte phagocytosis (Lumsden *et al.*, 1993).

Because the immune system, both innate and specific, is of such importance in protecting the animal against *Salmonella*, any factor or occurrence that might adversely affect these defences could increase the incidence and/or severity of a *Salmonella* infection. Very young animals possess an immature immune system and are acutely susceptible to *Salmonella* infections (Gast and Beard, 1989). Pre-existing disease can also affect a *Salmonella* infection. Viruses, such as

the agent for infectious bursal disease (IBD) in poultry, infect B cells and therefore depress humoral immunity. Infections by S. *enteritidis* in birds concurrently infected with the IBD agent tended to have higher mortality and produced more eggs contaminated with S. *enteritidis* (Phillips and Opitz, 1995). Chickens infected with Coccidia may have more severe *Salmonella* infections (Qin *et al.*, 1995), while under other circumstances they tend to be protected from infection (Tellez *et al.*, 1994). Stressors, such as weaning, transport or feed removal, can exert a suppressive effect on immunity (Blecha *et al.*, 1984, 1985; Holt, 1992a,b), which could subsequently affect a *Salmonella* infection. Increased *Salmonella* shedding was observed in cattle (Corrier *et al.*, 1990), chickens (Rigby and Pettit, 1980) and pigs (Williams and Newell, 1968) subjected to shipping stress. S. *enteritidis* infections in chickens undergoing an induced moult through feed withdrawal were found to be more severe than in unmoulted hens. Moulted hens exhibited more intestinal pathology (Holt and Porter, 1992; Porter and Holt, 1993; Holt *et al.*, 1994), increased intestinal shedding (Holt and Porter, 1992; Holt *et al.*, 1994, 1995) and enhanced extraintestinal dissemination (Holt *et al.*, 1995). Similarly, inanition experienced by sheep during long-term transit resulted in increased *Salmonella* infection and death of a portion of the animals (Norris *et al.*, 1989).

Vaccination

The use of vaccination as a means of protecting a flock or herd has received substantial attention over the years. The choice of using a live organism versus a killed preparation is weighed according to the positive and negative factors of each. Auxotrophic and rough mutants of a number of *Salmonella* have provided substantial protection against infection in animals (Lindberg and Robertsson, 1983; Robertsson *et al.*, 1983; Merritt *et al.*, 1984; Barrow *et al.*, 1990, 1991; Cooper *et al.*, 1993, 1994; Segall *et al.*, 1994) and also cross-protection against different *Salmonella* serovars (Hassan and Curtiss, 1994b).

However, mass introduction of such a vaccine must be weighed against the potential introduction, at least in the case of the paratyphoid vaccines, of a potential human pathogen into animals eventually used for human consumption. The use of bacterins eliminates this potential problem but protection is generally modest and certainly not complete. The use of subunit preparations in special vaccine formulations (Allaoui-Attarki et al., 1997) or as part of a vectored vaccine system (Nazerian and Yanagida, 1995; Nazerian et al., 1996) also offers promise. However, the choice must be made as to what components should be included in such a vaccine and a clear-cut candidate has not been found. Because of the immunomodulatory nature of cytokines, their use in modulating immune responsiveness in food animals is being investigated (Hughes et al., 1992; Nash et al., 1993; Reddy et al., 1993). It may be possible in the future to incorporate genes for a particular cytokine into a vectored vaccine to direct the elicitation of an enhanced mucosal response (Ramsey and Kohonen Corish, 1993; Leong et al., 1994) or cellular immunity (Leong et al., 1994) against a particular antigenic determinant. Such a multi-component vaccine could provide a more complete and long-lasting protection for the host and ultimately the consumer. The future research promises to bring about, at least from the Salmonella standpoint, a healthier food animal, as well as a safer food product.

References

Aitken, M.M., Hall, G.A. and Jones, P.W. (1978) Investigation of a cutaneous delayed hypersensitivity response as a means of detecting Salmonella dublin infection in cattle. Research in Veterinary Science 24, 370–374.

Aitken, M.M., Jones, P.W. and Brown, G.T. (1982) Protection of cattle against experimentally induced salmonellosis by intradermal injection of heat-killed Salmonella dublin. Research in Veterinary Science 32, 368–373.

Allaoui-Attarki, K., Pecquet, S., Fattal, E., Trolle, S., Chachaty, E., Couvreur, P., and Andremont, A. (1997) Protective immunity against Salmonella typhimurium elicited in mice by oral vaccination with phosphoryl-choline encapsulated in poly (DL-lactide-co-glycolide) microspheres. Infection and Immunity 65, 853–857.

Arnold, J.W. and Holt, P.S. (1996) Cytotoxicity in chicken alimentary secretions as measured by a derivative of the tumour necrosis factor assay. Poultry Science 75, 329–334.

Arnold, J.W., Niesel, D.W., Annable, C.R., Hess, C.B., Asuncion, M., Cho, Y.J., Peterson, J.W. and Klimpel, G.R. (1993) Tumor necrosis factor-alpha mediates the early pathology in Salmonella infection of the gastrointestinal tract. Microbial Pathogenesis 14, 217–227.

Baay, M. F. D. and Huis in't Veld, J. H. J. (1993) Alternative antigens reduce cross-reactions in an ELISA for the detection of Salmonella enteritidis in poultry. Journal of Applied Bacteriology 74, 243–247.

Babiuk, L.A., Lawman, M.J. and Gifford, G.A. (1987) Use of recombinant bovine alpha 1 interferon in reducing respiratory disease induced by bovine herpesvirus type 1. Antimicrobial Agents and Chemotherapy 31, 752–757.

Barbour, E.K., Frerichs, W.M., Nabbut, N.H., Poss, P.E. and Brinton, M.K. (1993) Evaluation of bacterins containing three predominant phage types of Salmonella enteritidis for prevention of infection in egg-laying chickens. American Journal of Veterinary Research 54, 1306–1309.

Barrow, P.A. (1992) Further observations on the serological response to experimental Salmonella typhimurium in chickens measured by ELISA. Epidemiology and Infection 108, 231–241.

Barrow, P.A., Hassan, J.O., Lovell, M.A. and Berchieri, A. (1990) Vaccination of chickens with aroA and other mutants of Salmonella typhimurium and S. enteritidis. Research in Microbiology 141, 851–853.

Barrow, P.A., Lovell, M.A. and Berchieri, A. (1991) The use of two live attenuated vaccines to immunize egg-laying hens against Salmonella enteritidis phage type 4. Avian Pathology 20, 681–692.

Barrow, P.A., Berchieri, A.J. and al Haddad, O. (1992) Serological response of chickens to infection with Salmonella gallinarum–S. pullorum detected by enzyme-linked immunosorbent assay. Avian Diseases 36, 227–236.

Bielefeldt-Ohman, H. (1995) Role of cytokines in the pathogenesis and treatment of respiratory disease. In: Meyers, M.J. and Murtaugh, P.P. (eds) Cytokines in Animal Health and Disease. Marcel Dekker, New York, pp. 291–332.

Bienenstock, J., Perey, D.Y., Gauldie, J. and Underdown, B.J. (1973) Chicken γ A: physicochemical and immuno-chemical characteristics. Journal of Immunology 110, 524–533.

Bisplinghoff, F.D. (1985) Bacterial contamination of animal proteins. In: Snoeyenbos, G.H. (ed.) Proceedings of the International Symposium on Salmonella. American Association of Avian Pathologists, Kennett Square, Pennsylvania, pp. 232–238.

Black, C.M., Catterall, J.R. and Remington, J.S. (1987) *In vivo* and *in vitro* activation of alveolar macrophages by recombinant interferon-gamma. *Journal of Immunology* 138, 491–495.

Blaser, M.J. and Newman, L.S. (1982) A review of human salmonellosis: I. Infective dose. *Reviews in Infectious Diseases* 4, 1096–1106.

Blecha, F. (1991) Cytokines: applications in domestic food animals. *Journal of Dairy Science* 74, 328–339.

Blecha, F., Boyles. S.L. and Riley, J.G. (1984) Shipping suppresses lymphocyte blastogenic responses in Angus and Brahman 3 Angus feeder calves. *Journal of Animal Science* 59, 576–583.

Blecha, F., Pollmann, S. and Nichols, D.A. (1985) Immunologic reactions of pigs regrouped at or near weaning. *American Journal of Veterinary Research* 46, 1934–1937.

Bouzoubaa, K., Nagaraja, K.V., Newman, J.A. and Pomeroy, B.S. (1987) Use of membrane proteins from *Salmonella gallinarum* for prevention of fowl typhoid infection in chickens. *Avian Diseases* 31, 699–704.

Bouzoubaa, K., Nagaraja, K.V., Kabbaj, F.Z., Newman, J.A. and Pomeroy, B.S. (1989) Feasibility of using proteins from *Salmonella gallinarum* vs. 9R live vaccine for the prevention of fowl typhoid in chickens. *Avian Diseases* 33, 385–391.

Brennan, F.R., Oliver, J.J. and Baird, G.D. (1994) Differences in the immune responses of mice and sheep to an aromatic-dependent mutant of *Salmonella typhimurium*. *Journal of Medical Microbiology* 41, 20–28.

Brenner, M.B., McLean, J., Dialynas, D.P., Strominer, J.L., Smith, J.A., Owen, F.L., Seidman, J.G., Ip, S., Rosen, F. and Krangel, M.S. (1986) Identification of a putative second T-cell receptor. *Nature* 322, 145–149.

Brownell, J.R., Sadler, W.W. and Fanelli, M.J. (1970) Role of bursa of Fabricius in chicken resistance to *Salmonella typhimurium*. *Avian Diseases* 14, 142–152.

Bullis, K.L. (1977) The history of avian medicine in the U.S. II. Pullorum disease and fowl typhoid. *Avian Diseases* 21, 422–429.

Bumstead, N. and Barrow, P.A. (1988) Genetics of resistance to *Salmonella typhimurium* in newly hatched chicks. *British Poultry Science* 29, 521–529.

Bumstead, N. and Barrow, P. (1993) Resistance to *Salmonella gallinarum*, *S. pullorum*, and *S. enteritidis* in inbred lines of chickens. *Avian Diseases* 37, 189–193.

Cantor, H. and Boyse, E.A. (1975a) Functional subclasses of T-lymphocytes bearing different Ly antigens. I. The generation of functionally distinct T-cell subclasses is a differentiative process independent of antigen. *Journal of Experimental Medicine* 141, 1376–1389.

Cantor, H. and Boyse, E.A. (1975b) Functional subclasses of T lymphocytes bearing different Ly antigens. II. Cooperation between subclasses of Ly+ cells in the generation of killer activity. *Journal of Experimental Medicine* 141, 1390–1399.

Charles, S.D., Hussain, I., Choi, C.U., Nagaraja, K.V. and Sivanandan, V. (1994) Adjuvanted subunit vaccines for the control of *Salmonella enteritidis* infection in turkeys. *American Journal of Veterinary Research* 55, 636–642.

Chart, H., Rowe, B., Baskerville, A., and Humphrey, T.J. (1990). Serological response of chickens to *Salmonella enteritidis* infection. *Epidemiology and Infection* 104, 63–71.

Chart, H., Baskerville, A., Humphrey, T.J. and Rowe, B. (1992) Serological responses of chickens experimentally infected with *Salmonella enteritidis* PT4 by different routes. *Epidemiololgy and Infection* 109, 297–302.

Chen, C.L., Bucy, R.P. and Cooper, M.D. (1990) T cell differentiation in birds. *Seminars in Immunology* 2, 79–86.

Cooper, G.L., Venables, L.M., Nicholas, R.A.J., Cullen, G.A. and Hormaeche, C.E. (1993) Further studies of the application of live *Salmonella enteritidis* aroA vaccines in chickens. *Veterinary Record* 133, 31–36.

Cooper, G.L., Venables, L.M., Woodward, M.J. and Hormaeche, C.E. (1994) Vaccination of chickens with strain CVL30, a genetically defined *Salmonella enteritidis* aroA live oral vaccine candidate. *Infection and Immunity* 62, 4747–4754.

Corkish, J.D., Davies, R.H., Wray, C. and Nicholas, R.A. (1994) Observations on a broiler breeder flock naturally infected with *Salmonella enteritidis* phage type 4. *Veterinary Record* 134, 591–594.

Corrier, D.E., Purdy, C.W. and DeLoach, J.R. (1990) Effects of marketing stress on faecal excretion of *Salmonella* spp. in feeder calves. *American Journal of Veterinary Research* 51, 866–869.

Corrier, D.E., Hargis, B., Hinton, A.J., Lindsey, D., Caldwell, D., Manning, J. and DeLoach, J. (1991) Effect of anaerobic cecal microflora and dietary lactose on colonization resistance of layer chicks to invasive *Salmonella enteritidis*. *Avian Diseases* 35, 337–343.

Dialynas, D.P., Wilde, D.B., Marrack, P., Pierres, A., Wall, K.A., Havran, W., Otten, G., Loken, M.R., Pierres, M., Kappler, J. and Fitch, F.W. (1983) Characterization of the murine antigenic determinant, designated L3T4a, recognized by monoclonal antibody GK1.5: expression of L3T4a by functional T cell clones appears to correlate primarily with class II MHC antigen-reactivity. *Immunological Reviews* 74, 29–56.

Dietert, R.R., Golemboski, K.A., Bloom, S.E. and Qureshi, M.A. (1991) The avian macrophage in cellular immunity. In: Sharma, J.M. (ed.) *Avian Cellular Immunology*. CRC Press, Boca Raton, Florida, pp. 71–95.

Fanelli, M.J., Sadler, W.W., Franti, C.E. and Brownell, J.R. (1971) Localization of salmonellae within the intestinal tract of chickens. *Avian Diseases* 15, 366–375.

Fedorka-Cray, P.J., Whipp, S.C., Isaacson, R.E., Nord, N. and Lager, K. (1994) Transmission of *Salmonella typhimurium* to swine. *Veterinary Microbiology* 41, 333–344.

Fields, P.I., Swanson, R.V., Haidaris, C.G. and Heffron, F. (1986) Mutants of *Salmonella typhimurium* that cannot survive within the macrophage are avirulent. *Proceedings of the National Academy of Sciences USA* 83, 5189–5193.

Finkelman, F.D., Katona, I.M., Mosmann, T.R. and Coffman, R.L. (1988) IFN-gamma regulates the isotypes of Ig secreted during *in vivo* humoral immune responses. *Journal of Immunology* 140, 1022–1027.

Fiorentino, D.F., Bond, M.W. and Mosmann, T.R. (1989) Two types of mouse T helper cell. IV. Th2 clones secrete a factor that inhibits cytokine production by Th1 clones. *Journal of Experimental Medicine* 170, 2081–2095.

Formal, S.B., Dammin, G.J., LaBrec, E.H. and Schneider, H. (1958) Experimental *Shigella* infections: characteristics of a fatal infection produced in guinea pigs. *Journal of Bacteriology* 75, 604–610.

Funakoshi, S., Doi, T., Nakajima, T., Suyama, T. and Tokuda, M. (1982). Antimicrobial effect of human serum IgA. *Microbiology and Immunology* 26, 227–239.

Gast, R.K. and Beard, C.W. (1989) Age-related changes in the persistence and pathogenicity of *Salmonella typhimurium* in chicks. *Poultry Science* 68, 1454–1460.

Gast, R.K. and Beard, C.W. (1990a) Serological detection of experimental *Salmonella enteritidis* infections in laying hens. *Avian Diseases* 34, 721–728.

Gast, R.K. and Beard, C.W. (1990b) Isolation of *Salmonella enteritidis* from internal organs of experimentally infected hens. *Avian Diseases* 34, 991–993.

Gast, R.K. and Beard, C.W. (1991) Research note: detection of *Salmonella* serogroup D-specific antibodies in the yolks of eggs laid by hens infected with *Salmonella enteritidis*. *Poultry Science* 70, 1273–1276.

Gast, R.K. and Beard, C.W. (1992) Evaluation of a chick mortality model for predicting the consequences of *Salmonella enteritidis* infections in laying hens. *Poultry Science* 71, 281–287.

Gast, R.K. and Benson, S.T. (1995) The comparative virulence for chicks of *Salmonella enteritidis* phage type 4 isolates and isolates of phage types commonly found in poultry in the United States. *Avian Diseases* 39, 567–574.

Gast, R.K., Stone, H.D., Holt, P.S. and Beard, C.W. (1992) Evaluation of the efficacy of an oil-emulsion bacterin for protecting chickens against *Salmonella enteritidis*. *Avian Diseases* 36, 992–999.

Gast, R.K., Stone, H.D. and Holt, P.S. (1993) Evaluation of the efficacy of oil-emulsion bacterins for reducing faecal shedding of *Salmonella enteritidis* by laying hens. *Avian Diseases* 37, 1085–1091.

Gast, R.K., Porter, R.E., Jr and Holt, P.S. (1997) Applying tests for specific yolk antibodies to predict contamination by *Salmonella enteritidis* in eggs from experimentally infected laying hens. *Avian Diseases* 41, 195–202.

Gray, J.T., Fedorka-Cray, P.J., Stabel, T.J. and Ackermann, M.R. (1995) Influence of inoculation route on the carrier state of *Salmonella choleraesuis* in swine. *Veterinary Microbiology* 47, 43–59.

Gray, J.T., Fedorka-Cray, P.J., Stabel, T.J. and Kramer, T.T. (1996a) Natural transmission of *Salmonella choleraesuis* in swine. *Applied and Environmental Microbiology* 62, 141–146.

Gray, J.T., Stabel, T.J. and Fedorka-Cray, P.J. (1996b) Effect of dose on the immune response and persistence of *Salmonella choleraesuis* infection in swine. *American Journal of Veterinary Research* 57, 313–319.

Guillot, J.F., Beaumont, C., Bellatif, F., Mouline, C., Lantier, F., Colin, P. and Protais, J. (1995) Comparison of resistance of various poultry lines to infection by *Salmonella enteritidis*. *Veterinary Research* 26, 81–86.

Haskins, K., Kubo, R., White, J., Pigeon, M., Kappler, J. and Marrack, P. (1983) The major histocompatibility complex-restricted antigen receptor on T cells. I. Isolation with a monoclonal antibody. *Journal of Experimental Medicine* 157, 1149–1169.

Hassan, J.O., Mockett, A.P.A., Catty, D. and Barrow, P.A. (1991) Infection and reinfection of chickens with *Salmonella typhimurium*: bacteriology and immune responses. *Avian Diseases* 35, 809–819.

Hassan, J.O., Porter, S. B. and Curtiss, R., III (1993) Effect of infective dose on humoral immune responses and colonization in chickens experimentally infected with *Salmonella typhimurium*. *Avian Diseases* 37, 19–26.

Hassan, J.O. and Curtiss, R., III (1994a) Virulent *Salmonella typhimurium*-induced lymphocyte depletion and immunosuppression in chickens. *Infection and Immunity* 62, 2027–2036.

Hassan, J.O. and Curtiss, R., III (1994b) Development and evaluation of an experimental vaccination program using a live avirulent *Salmonella typhimurium* strain to protect immunized chickens against challenge with homologous and heterologous *Salmonella* serovars. *Infection and Immunity* 62, 5519–5527.

Hentges, D.J. and Freter, R. (1961) *In vivo* and *in vitro* antagonism of intestinal bacteria against *Shigella flexneri*. 1. Correlation between various tests. *Journal of Infectious Diseases* 37, 30–37.

Holt, P.S. (1992a) Effect of induced molting on B cell and CT4 and CT8 T cell numbers in spleens and peripheral blood of White Leghorn hens. *Poultry Science* 71, 2027–2034.

Holt, P.S. (1992b) Effects of induced moulting on immune responses of hens. *British Poultry Science* 33, 165–175.

Holt, P.S. and Porter, R.E., Jr (1992) Microbiological and histopathological effects of an induced-molt fasting procedure on a *Salmonella enteritidis* infection in chickens. *Avian Diseases* 36, 610–618.

Holt, P.S. and Porter, R.E., Jr (1993) Effect of induced molting on the recurrence of a previous *Salmonella enteritidis* infection. *Poultry Science* 72, 2069–2078.

Holt, P.S., Buhr, R.J., Cunningham, D.L. and Porter, R.E., Jr (1994) Effect of two different molting procedures on a *Salmonella enteritidis* infection. *Poultry Science* 73, 1267–1275.

Holt, P.S., Macri, N.P. and Porter, R. E., Jr (1995) Microbiological analysis of the early *Salmonella enteritidis* infection in molted and unmolted hens. *Avian Diseases* 39, 55–63.

Holt, P.S., Stone, H.D., Gast, R.K. and Porter, R. E., Jr (1996) Growth of *Salmonella enteritidis* (SE) in egg contents from hens vaccinated with an SE bacterin. *Food Microbiology* 13, 417–426.

Holt, P.S., Gast, R.K., Porter, R. E., Jr and Stone, H.D. (1999) Hyporesponsiveness of the systemic and mucosal immune systems in chickens infected with *Salmonella enteritidis* serovar enteriditis at one day of age. *Poultry Science* (in press).

Hormaeche, C.E., Villarreal, B., Dougan, G. and Chatfield, S.N. (1993) Immunity mechanisms in experimental salmonellosis. In: Cabello, F., Hormaeche, C., Mastroeni, P. and Bonina, L. (eds) *Biology of* Salmonella. Plenum Press, New York, pp. 223–235.

Hsu, H.S. (1989) Pathogenesis and immunity in murine salmonellosis. *Microbiological Reviews* 53, 390–409.

Hughes, H.P., Campos, M., van Drunen Littel-van den Hurk, S., Zamb, T., Sordillo, L.M., Godson, D. and Babiuk, L. A. (1992) Multiple administration with interleukin-2 potentiates antigen-specific responses to subunit vaccination with bovine herpesvirus-1 glycoprotein IV. *Vaccine* 10, 226–230.

Humphrey, T.J. (1990) Public health implications of the infection of egg-laying hens with *Salmonella enteritidis* phage type 4. *World's Poultry Science Journal* 46, 5–13.

Humphrey, T.J., Mead, G.C. and Rowe, B. (1988) Poultry meat as a source of human salmonellosis in England and Wales: epidemiological overview. *Epidemiology and Infection* 100, 175–184.

Humphrey, T.J., Baskerville, A., Chart, H., Rowe, B. and Whitehead, A. (1991a) *Salmonella enteritidis* PT4 infection in specific pathogen free hens: influence of infecting dose. *Veterinary Record* 129, 482–485.

Humphrey, T.J., Chart, H., Baskerville, A. and Rowe, B. (1991b) The influence of age on the response of SPF hens to infection with *Salmonella enteritidis* PT4. *Epidemiology and Infections* 106, 33–43.

Jeurissen, S.H.M., Janse, E.M., Koch, G. and DeBoer, G.F. (1989) Post-natal development of mucosa-associated lymphoid tissues in chickens. *Cell and Tissue Research* 258, 119–124.

Johnson, E.H., Hietala, S. and Smith, B.P. (1985) Chemiluminescence of bovine alveolar macrophages as in indicator of developing immunity in calves vaccinated with aromatic-dependent salmonella. *Veterinary Microbiology* 10, 451–464.

Jones, P.W., Collins, P. and Aitken, M.M. (1988) Passive protection of calves against experimental infection with *Salmonella typhimurium*. *Veterinary Record* 123, 536–541.

Kagaya, K., Watanabe, K. and Fukazawa, Y. (1989) Capacity of recombinant gamma interferon to activate macrophages for *Salmonella*-killing activity. *Infection and Immunity* 57, 609–615.

Kent, T.H., Formal, S.B. and LaBrec, E.H. (1966) Acute enteritis due to *Salmonella typhimurium* in opium-treated guinea pigs. *Archives of Pathology* 81, 501–508.

Kilian, M., Mestecky, J. and Russell, M.W. (1988) Defence mechanisms involving Fc-dependent functions of immunoglobulin A and their subversion by bacterial immunoglobulin A proteases. *Microbiological Reviews* 52, 296–303.

Kim, C.J., Nagaraja, K.V. and Pomeroy, B.S. (1991) Enzyme-linked immunosorbent assay for the detection of *Salmonella enteritidis* infection in chickens. *American Journal of Veterinary Research* 52, 1069–1074.

Kincy Cain, T., Clements, J.D. and Bost, K.L. (1996) Endogenous and exogenous interleukin-12 augment the protective immune response in mice orally challenged with *Salmonella dublin*. *Infection and Immunity* 64, 1437–1440.

Klimpel, G.R., Asuncion, M., Haithcoat, J. and Niesel, D.W. (1995) Cholera toxin and *Salmonella typhimurium* induce different cytokine profiles in the gastrointestinal tract. *Infection and Immunity* 63, 1134–1137.

Kogut, M.H., Tellez, G., Hargis, B.M., Corrier, D.E. and DeLoach, J.R. (1993) The effect of 5-fluorouracil treatment of chicks: a cell depletion model for the study of avian polymorphonuclear leucocytes and natural host defences. *Poultry Science* 72, 1873–1880.

Kogut, M.H., McGruder, E.D., Hargis, B.M., Corrier, D.E. and DeLoach, J.R. (1994a) Dynamics of avian inflammatory response to *Salmonella*-immune lymphokines: changes in avian blood leucocyte populations. *Inflammation* 18, 373–388.

Kogut, M.H., Tellez, G.I., McGruder, E.D., Hargis, B.M., Williams, J.D., Corrier, D.E. and DeLoach, J.R. (1994b)
    Heterophils are decisive components in the early responses of chickens to *Salmonella enteritidis* infections.
    Microbial Pathogenesis 16, 141–151.
Kogut, M.H., McGruder, E.D., Hargis, B.M., Corrier, D.E. and DeLoach, J.R. (1995a) Characterization of the
    pattern of inflammatory cell influx in chicks following the intraperitoneal administration of live *Salmonella
    enteritidis* and *Salmonella enteritidis*-immune lymphokines. *Poultry Science* 74, 8–17.
Kogut, M.H., McGruder, E.D., Hargis, B.M., Corrier, D.E. and DeLoach, J.R. (1995b) *In vivo* activation of
    heterophil function in chickens following injection with *Salmonella enteritidis*-immune lymphokines. *Journal
    of Leukocyte Biology* 57, 56–62.
Kohbata, S., Yokoyama, H. and Yabuuchi, E. (1986) Cytopathogenic effect of *Salmonella typhi* GIFU 10007 on M
    cells of murine ileal Peyer's patches in ligated ileal loops: an ultrastructural study. *Microbiology and Immunology*
    30, 1225–1237.
Kwang, J. and Littledike, E.T. (1995) Production and identification of recombinant proteins of *Salmonella
    typhimurium* and their use in detection of antibodies in experimentally challenged animals. *Microbiology
    Letters* 130, 25–30.
Lacey, C., Wilkie, B.N., Kennedy, B.W. and Mallard, B.A. (1989) Genetic and other effects on bacterial phago-
    cytosis and killing by cultured peripheral blood monocytes of SLA-defined miniature pigs. *Animal Genetics* 20,
    371–381.
Lange, S. and Holmgren, J. (1978) Protective antitoxic cholera *Vibrio cholerae* immunity in mice: influence of route
    and number of immunizations and mode of action of protective antibodies. *Acta Pathologica Microbiologica
    Scandinavica* 86, 145–152.
Lee, G.M., Jackson, G.D. and Cooper, G.N. (1981) The role of serum and biliary antibodies and cell-mediated
    immunity in the clearance of *S. typhimurium* from chickens. *Veterinary Immunology and Immunopathology* 2,
    233–252.
Leong, K.H., Ramsay, A.J., Boyle, D.B. and Ramshaw, I.A. (1994) Selective induction of immune responses by
    cytokines coexpressed in recombinant fowlpox virus. *Journal of Virology* 68, 8125–8130.
Lillehoj, H.S. (1993) Avian gut-associated immune system: implication in coccidial vaccine development. *Poultry
    Science* 72, 1306–1311.
Lillehoj, H.S., Kaspers, B., Jenkins, M.C. and Lillehoj, E.P. (1992) Avian interferon and interleukin-2: a review by
    comparison with mammalian homologues. *Poultry Science Reviews* 4, 67–85.
Lim, O.J. and Maheswaran, S.K. (1977) Purification and identification of turkey immunoglobulin-A. *Avian
    Diseases* 21, 675–696.
Lindberg, A.A. and Robertsson, J.A. (1983) *Salmonella typhimurium* infection in calves: cell-mediated and humoral
    immune reactions before and after challenge with live virulent bacteria in calves given live or inactivated
    vaccines. *Infection and Immunity* 41, 751–757.
Lindell, K.A., Saeed, A.M. and McCabe, G.P. (1994) Evaluation of resistance of four strains of commercial laying
    hens to experimental infection with *Salmonella enteritidis* phage type eight. *Poultry Science* 73, 757–762.
Lissner, C.R., Swanson, R.N. and O'Brien, A.D. (1983) Genetic control of the innate resistance of mice to
    Salmonella typhimurium: expression of the *Ity* gene in peritoneal and splenic macrophages isolated *in vitro*.
    Journal of Immunology 131, 3006–3013.
Lumsden, J.S., Kenndey, B.W., Mallard, B.A. and Wilkie, B.N. (1993) The influence of the swine major histo-
    compatibility genes on antibody and cell-mediated immune responses to immunization with an aromatic-
    dependent mutant of *Salmonella typhimurium*. *Canadian Journal of Veterinary Research* 57, 14–18.
McGhee, J.R., Mestecky, J., Dertzbaugh, M.T., Eldridge, J.H., Hirasawa, M. and Kiyono, H. (1992) The mucosal
    immune system: from fundamental concepts to vaccine development. *Vaccine* 10, 75–88.
McGruder, E.D., Ray, P.M., Tellez, G.I., Kogut, M.H., Corrier, D.E., DeLoach, J.R. and Hargis, B.M. (1993)
    Salmonella enteritidis immune leucocyte-stimulated soluble factors: effects on increased resistance to *Salmonella*
    organ invasion in day-old Leghorn chicks. *Poultry Science* 72, 2264–2271.
McGruder, E.D., Kogut, M.H., Corrier, D.E., DeLoach, J.R. and Hargis, B.M. (1995) Comparison of prophylactic
    and therapeutic efficacy of *Salmonella enteritidis*-immune lymphokines against *Salmonella enteritidis* organ
    invasion in neonatal Leghorn chicks. *Avian Diseases* 39, 21–27.
McLeod, S. and Barrow, P.A. (1991) Lipopolysaccharide-specific IgG in egg yolk from two chicken flocks infected
    with *Salmonella enteritidis*. *Letters in Applied Microbiology* 13, 294–297.
Manetti, R., Parronchi, P., Giudizi, M.G., Piccinni, M.P., Maggi, E., Trinchieri, G. and Romagnani, S. (1993)
    Natural killer cell stimulatory factor (interleukin 12 [IL-12]) induces T helper type 1 (Th1)-specific immune
    responses and inhibits the development of IL-4-producing Th cells. *Journal of Experimental Medicine* 177,
    1199–1204.

Mastroeni, P., Villarreal Ramos, B. and Hormaeche, C.E. (1993a) Adoptive transfer of immunity to oral challenge with virulent salmonellae in innately susceptible BALB/c mice requires both immune serum and T cells. *Infection and Immunity* 61, 3981–3984.

Mastroeni, P., Villarreal Ramos, B. and Hormaeche, C.E. (1993b) Effect of late administration of anti-TNF alpha antibodies on a *Salmonella* infection in the mouse model. *Microbial Pathogenesis* 14, 473–480.

Merritt, F.F., Smith, B.P., Reina Guerra, M., Habasha, F. and Johnson, E. (1984) Relationship of cutaneous delayed hypersensitivity to protection from challenge exposure with *Salmonella typhimurium* in calves. *American Journal of Veterinary Research* 45, 1081–1085.

Michetti, P., Mahan, M.J., Slauch, J.M., Mekalanos, J.J. and Neutra, M.R. (1992) Monoclonal secretory immunoglobulin A protects mice against oral challenge with the invasive pathogen *Salmonella typhimurium*. *Infection and Immunity* 60, 1786–1792.

Morgan, P.B. (1990) Complement and infectious diseases. In: Morgan, P.B. (ed.) *Complement: Clinical Aspects and Relevance to Disease*. Harcourt Brace Jovanovich, London, pp. 97–111.

Mosmann, T.R. and Coffman, R.L. (1989a) Heterogeneity of cytokine secretion patterns and functions of helper T cells. *Advances in Immunology* 46, 111–147.

Mosmann, T.R. and Coffman, R.L. (1989b) TH1 and TH2 cells: different patterns of lymphokine secretion lead to different functional properties. *Annual Review of Immunology* 7, 145–173.

Mosmann, T.R., Cherwinski, H., Bond, M.W., Giedlin, M.A. and Coffman, R.L. (1986) Two types of murine helper T cell clone. I. Definition according to profiles of lymphokine activities and secreted proteins. *Journal of Immunology* 136, 2348–2357.

Muotiala, A. and Makela, P.H. (1990) The role of IFN-gamma in murine *Salmonella typhimurium* infection. *Microbial Pathogenesis* 8, 135–141.

Murray, H.W., Spitalny, G.L. and Nathan, C.F. (1985) Activation of mouse peritoneal macrophages *in vitro* and *in vivo* by interferon-gamma. *Journal of Immunology* 134, 1619–1622.

Nakano, Y., Onozuka, K., Terada, Y., Shinomiya, H. and Nakano, M. (1990) Protective effect of recombinant tumour necrosis factor-alpha in murine salmonellosis. *Journal of Immunology* 144, 1935–1941.

Nash, A.D., Lofthouse, S.A., Barcham, G.J., Jacobs, H.J., Ashman, K., Meeusen, E.N., Brandon, M.R. and Andrews, A.E. (1993) Recombinant cytokines as immunological adjuvants. *Immunology and Cell Biology* 71, 367–379.

Nauciel, C. and Espinasse-Maes, F. (1992) Role of gamma interferon and tumour necrosis factor alpha in resistance to *Salmonella typhimurium* infection. *Infection and Immunity* 60, 450–454.

Nazerian, K. and Yanagida, N. (1995) A recombinant fowlpox virus expressing the envelope antigen of subgroup A avian leukosis/sarcoma virus. *Avian Diseases* 39, 514–520.

Nazerian, K., Witter, R.L., Lee, L.F. and Yanagida, N. (1996) Protection and synergism by recombinant fowl pox vaccines expressing genes from Marek's disease virus. *Avian Diseases* 40, 368–376.

Nicholas, R.A. and Cullen, G.A. (1991) Development and application of an ELISA for detecting antibodies to *Salmonella enteritidis* in chicken flocks. *Veterinary Record* 128, 74–76.

Norris, R.T., Richards, R.B. and Dunlop R.H. (1989) Pre-embarkation risk factors for sheep deaths during export by sea from Western Australia. *Australian Veterinary Journal* 66, 309–314.

O'Brien, A.D. (1982) Innate resistance of mice to *Salmonella typhi* infection. *Infection and Immunity* 38, 948–952.

O'Brien, A.D. (1986) Influence of host genes on resistance of inbred mice to lethal infection with *Salmonella typhimurium*. *Current Topics in Microbiology and Immunology* 124, 37–48.

O'Brien, A.D., Scher, I. and Formal, S.B. (1979) Effect of silica on the innate resistance of inbred mice to *Salmonella typhimurium* infection. *Infection and Immunity* 25, 513–520.

Parham, G.L. (1985) Salmonellae in cooked beef products. In: Snoeyenbos, G.H. (ed.) *Proceedings of the International Symposium on* Salmonella. American Association of Avian Pathologists, Kennett Square, Pennsylvania, pp. 275–280.

Park, Y.H., Fox, L.K., Hamilton, M.J. and Davis, W.C. (1992) Bovine mononuclear leucocyte subpopulations in peripheral blood and mammary gland secretions during lactation. *Journal of Dairy Science* 75, 998–1006.

Peel, J.E., Kolly, C., Siegenthaler, B. and Martinod, S.R. (1990a) Prophylactic effects of recombinant bovine interferon-alpha I1 on acute *Salmonella typhimurium* infection in calves. *American Journal of Veterinary Research* 51, 1095–1099.

Peel, J.E., Voirol, M.J., Kolly, C., Gobet, D. and Martinod, S. (1990b) Induction of circulating tumour necrosis factor cannot be demonstrated during septicemic salmonellosis in calves. *Infection and Immunity* 58, 439–442.

Pescovitz, M.D., Lunney, J.K. and Sachs, D.H. (1984) Preparation and characterization of monoclonal antibodies reactive with porcine PBL. *Journal of Immunology* 133, 368–375.

Pevzner, I.Y., Stone, H.A. and Nordskog, A.W. (1981) Immune response and disease resistance in chickens. I. Selection for high and low titre to *Salmonella pullorum* antigen. *Poultry Science* 60, 920–926.

Phillips, R.A. and Opitz, H.M. (1995) Pathogenicity and persistence of *Salmonella enteritidis* and egg contamination in normal and infectious bursal disease virus-infected Leghorn chicks. *Avian Diseases* 39, 778–787.

Plant, J. and Glynn, A.A. (1976) Genetics of resistance to infection with *Salmonella typhimurium* in mice. *Journal of Infectious Diseases* 133, 72–78.

Popiel, I. and Turnbull, P.C. (1985) Passage of *Salmonella enteritidis* and *Salmonella thompson* through chick ileocecal mucosa. *Infection and Immunity* 47, 786–792.

Porter, R. E., Jr and Holt, P.S. (1993) Effect of induced molting on the severity of intestinal lesions caused by *Salmonella enteritidis* infection in chickens. *Avian Diseases* 37, 1009–1016.

Portnoy, D.A. (1992) Innate immunity to a facultative intracellular bacterial pathogen. *Current Opinions in Immunology* 4, 20–24.

Protais, J., Colin, P., Beaumont, C., Guillot, J.F., Lantier, F., Pardon, P. and Bennejean, G. (1996) Line differences in resistance to *Salmonella enteritidis* PT4 infection. *British Poultry Science* 37, 329–339.

Qin, Z.R., Fukata, T., Baba, E. and Arakawa, A. (1995) Effect of *Eimeria tenella* infection on *Salmonella enteritidis* infection in chickens. *Poultry Science* 74, 1–7.

Ramarathinam, L., Shaban, R.A., Niesel, D.W. and Klimpel, G.R. (1991) Interferon gamma (IFN-gamma) production by gut-associated lymphoid tissue and spleen following oral *Salmonella typhimurium* challenge. *Microbial Pathogenesis* 11, 347–356.

Ramsay, A.J. and Kohonen Corish, M. (1993) Interleukin-5 expressed by a recombinant virus vector enhances specific mucosal IgA responses *in vivo*. *European Journal of Immunology* 23, 3141–3145.

Reddy, D.N., Reddy, P.G., Xue, W., Minocha, H.C., Daley, M. J. and Blecha, F. (1993) Immunopotentiation of bovine respiratory disease virus vaccines by interleukin-1 beta and interleukin-2. *Veterinary Immunology and Immunopathology* 37, 25–38.

Reddy, P.G., Blecha, F., Minocha, H.C., Anderson, G.A., Morrill, J.L., Fedorka-Cray, P.J. and Baker, P.E. (1989) Bovine recombinant interleukin-2 augments immunity and resistance to bovine herpesvirus infection. *Veterinary Immunology and Immunopathology* 23, 61–74.

Reddy, P.G., Reddy, D.N., Pruiett, S.E., Daley, M.J., Shirley, J.E., Chengappa, M.M. and Blecha, F. (1992) Interleukin 2 treatment of *Staphylococcus aureus* mastitis. *Cytokine* 4, 227–231.

Rigby, C.E. and Pettit, J.R. (1980) Changes in the *Salmonella* status of broiler chickens subjected to simulated shipping conditions. *Canadian Journal of Comparative Medicine* 44, 374–381.

Robertsson, J.A., Svenson, S.B. and Renstrom, L.H. (1982a) Delayed hypersensitivity skin test for detection of immune responses against *Salmonella* in cattle. *Research in Veterinary Science* 32, 225–230.

Robertsson, J.A., Svenson, S.B., Remstrom, L.H.M. and Lindberg, A.A. (1982b) Defined *Salmonella* antigens for detection of cellular and humoral immune responses in *Salmonella* infected calves: *Salmonella typhimurium*, *Salmonella dublin*. *Research in Veterinary Science* 33, 221–227.

Robertsson, J.A., Lindberg, A.A., Hoiseth, S. and Stocker, B.A.D. (1983) *Salmonella typhimurium* infection in calves: protection and survival of virulent challenge bacteria after immunization with live or inactivated vaccines. *Infection and Immunity* 41, 742–750.

Roth, J.A. and Frank, D.E. (1989) Recombinant bovine interferon-gamma as an immunomodulator in dexamethasone-treated and nontreated cattle. *Journal of Interferon Research* 9, 143–151.

Royal, W.A., Robinson, R.A. and Duganzich, D.M. (1968) Colostral immunity against *Salmonella* infection in calves. *New Zealand Veterinary Journal* 16, 141–145.

St Louis, M.E., Morse, D.L., Potter, M.E., DeMelfi, T.M., Guzewich, J.J., Tauxe, R.V. and Blake, P.A. (1988) The emergence of grade A eggs as a major source of *Salmonella enteritidis* infections: new implications for the control of salmonellosis. *Journal of the American Medical Association* 259, 2103–2107.

Schat, K.A. (1994) Cell-mediated immune effector functions in chickens. *Poultry Science* 73, 1077–1081.

Schat, K.A. and Myers, T.J. (1991) Avian intestinal immunity. *Critical Reviews in Poultry Biology* 3, 19–34.

Segall, T. and Lindberg, A.A. (1993) Oral vaccination of calves with an aromatic-dependent *Salmonella dublin* (O9,12) hybrid expressing O4,12 protects against *S. dublin* (O9,12) but not against *Salmonella typhimurium* (O4,5,12). *Infection and Immunity* 61, 1222–1231.

Segall, T., Jacobsson, S.O., Karlsson, K. and Lindberg, A.A. (1994) Mucosal immune responses in calves orally vaccinated with a live auxotrophic aroA *Salmonella dublin* strain. *Zentralblatt für Veterinarmedizin Reihe B* 41, 305–312.

Sekellick, M.J., Ferrandino, A.F., Hopkins, D.A. and Marcus, P.I. (1994) Chicken interferon gene: cloning, expression, and analysis. *Journal of Interferon Research* 14, 71–79.

Sharma, J.M. and Schat, K.A. (1991) Natural immune functions. In: Sharma, J.M. (ed.) *Avian Cellular Immunology*. CRC Press, Boca Raton, Florida, pp. 51–70.

Shivaprasad, H.L., Timoney, J.F., Morales, S., Lucio, B. and Baker, R.C. (1990) Pathogenesis of *Salmonella enteritidis* infection in laying chickens. I. Studies on egg transmission, clinical signs, faecal shedding, and serological responses. *Avian Diseases* 34, 548–557.

Sordillo, L.M. and Babiuk, L.A. (1991) Controlling acute *Escherichia coli* mastitis during the periparturient period with recombinant bovine interferon gamma. *Veterinary Microbiology* 28, 189–198.

Stabel, T.J., Mayfield, J.E., Morfitt, D.C. and Wannemuehler, M.J. (1993) Oral immunization of mice and swine with an attenuated *Salmonella choleraesuis* [delta cya-12 delta(crp-cdt)19] mutant containing a recombinant plasmid. *Infection and Immunity* 61, 610–618.

Stabel, T.J., Fedorka-Cray, P.J. and Gray, J.T. (1995) Tumor necrosis factor-alpha production in swine after oral or respiratory challenge exposure with live *Salmonella typhimurium* or *Salmonella choleraesuis*. *American Journal of Veterinary Research* 56, 1012–1018.

Suresh, M., Karaca, K., Foster, D. and Sharma, J.M. (1995) Molecular and functional characterization of turkey interferon. *Journal of Virology* 69, 8159–8163.

Swain, S. L. (1983) T cell subsets and the recognition of MHC class. *Immunological Reviews* 74, 129–142.

Swain, S.L., Weinberg, A.D., English, M. and Huston, G. (1990) IL-4 directs the development of Th2-like helper effectors. *Journal of Immunology* 145, 3796–3806.

Swanson, R. N. and O'Brien, A.D. (1983) Genetic control of the innate resistance of mice to *Salmonella typhimurium*: *Ity* gene is expressed *in vivo* by 24 hours after infection. *Journal of Immunology* 131, 3014–3020.

Tagliabue, A., Nencioni, L., Villa, L., Keren, D.F., Lowell, G.H. and Boraschi, D. (1983) Antibody-dependent cell-mediated antibacterial activity of intestinal lymphocytes with secretory IgA. *Nature* 306, 184–186.

Tellez, G.I., Kogut, M.H. and Hargis, B.M. (1993) Immunoprophylaxis of *Salmonella enteritidis* infection by lymphokines in Leghorn chicks. *Avian Diseases* 37, 1062–1070.

Tellez, G.I., Kogut, M.H. and Hargis, B.M. (1994) *Eimeria tenella* or *Eimeria adenoeides*: induction of morphological changes and increased resistance to *Salmonella enteritidis* infection in Leghorn chicks. *Poultry Science* 73, 396–401.

Thain, J.A., Baxter Jones, C., Wilding, G.P. and Cullen, G.A. (1984) Serological response of turkey hens to vaccination with *Salmonella hadar* and its effect on their subsequently challenged embryos and poults. *Research in Veterinary Science* 36, 320–325.

Thorns, C.J. (1995) *Salmonella* fimbriae: novel antigens in the detection and control of *Salmonella* infections. *British Veterinary Journal* 151, 643–658.

Thorns, C.J., Sojka, M.G. and Chasey, D. (1990) Detection of a novel fimbrial structure on the surface of *Salmonella enteritidis* by using a monoclonal antibody. *Journal of Clinical Microbiology* 28, 2409–2414.

Thorns, C.J., Bell, M.M., Sojka, M.G. and Nicholas, R.A. (1996) Development and application of enzyme-linked immunosorbent assay for specific detection of *Salmonella enteritidis* infections in chickens based on antibodies to SEF14 fimbrial antigen. *Journal of Clinical Microbiology* 34, 792–797.

Timms, L.M., Marshall, R.N. and Breslin, M.F. (1990) Laboratory assessment of protection given by an experimental *Salmonella enteritidis* PT4 inactivated, adjuvant vaccine. *Veterinary Record* 127, 611–614.

Timms, L.M., Marshall, R.N. and Breslin, M.F. (1994) Laboratory and field trial assessment of protection given by a *Salmonella enteritidis* PT4 inactivated, adjuvant vaccine. *British Veterinary Journal* 150, 93–102.

Timoney, J.F., Shivaprasad, H.L., Baker, R.C. and Rowe, B. (1989) Egg transmission after infection of hens with *Salmonella enteritidis* phage type 4. *Veterinary Record* 125, 600–601.

Turnbull, P.C. and Snoeyenbos, G.H. (1974) Experimental salmonellosis in the chicken. 1. Fate and host response in alimentary canal, liver, and spleen. *Avian Diseases* 18, 153–177.

Wannemuehler, M.J. (1995) Role of cytokines in intestinal health and disease. In: Myers, M.J. and Murtaugh, M.P. (eds) *Cytokines in Animal Health and Disease*. Marcel Dekker, New York, pp. 333–356.

Weinstein, D.L., Lissner, C.R., Swanson, R.N. and O'Brien, A.D. (1986) Macrophage defect and inflammatory cell recruitment dysfunction in *Salmonella* susceptible C3H/HeJ mice. *Cellular Immunology* 102, 68–77.

Williams, J.E. and Whittemore, A.D. (1975) Influence of age on the serological response of chickens to *Salmonella typhimurium* infection. *Avian Diseases* 19, 745–760.

Williams, L.P.J. and Newell, K.W. (1968). Sources of salmonellas in market swine. *Journal of Hygiene* 66, 281–293.

Wray, C. and Sojka, W.J. (1977) Reviews of the progress of dairy science: bovine salmonellosis. *Journal of Dairy Research* 44, 383–425.

van Zijderveld, F.G., van Zijderveld van Bemmel, A.M., Brouwers, R.A., de Vries, T.S., Landman, W.J. and de Jong, W.A. (1993) Serological detection of chicken flocks naturally infected with *Salmonella enteritidis*, using an enzyme-linked immunosorbent assay based on monoclonal antibodies against the flagellar antigen. *Veterinary Quarterly* 15, 135–137.

Chapter 6
Antibiotic Resistance in *Salmonella*

Reiner Helmuth

*Federal Institute for Health Protection of Consumers and Veterinary Medicine BgVV,
Laboratory for Molecular Biology and National Salmonella Reference Laboratory,
Diedersdorfer Weg 1, 12277 Berlin, Germany*

General Aspects of Antibiotic Resistance

Resistance of bacteria to antimicrobial agents is, in principle, not a phenomenon of our century or industrialized countries only. Its biological roots are the result of an old ecological phenomenon, in which microorganisms compete for nutritional niches. The production of compounds that kill competitors has a long history in the evolution of life, as has the production of antibiotics by Streptomycetes, the major producers of natural products with antibiotic activity (Krügel, 1997). Consequently, defence mechanisms have evolved in the competing species. However, since the start of the widespread use of antimicrobial agents by humans in the late 1940s, resistance phenomena have been observed in almost all bacterial species and against all drugs available. Antimicrobial resistance increases the morbidity, mortality and costs associated with disease. Today, antimicrobial resistance is one of the major health problems in human and veterinary medicine. It has tremendous social and economic consequences and leads to strong scientific and public-health efforts to improve the situation. Antimicrobial resistance has been recognized by the World Health Organization (WHO) and national authorities as a major emerging problem of public health (John E. Fogarty International Center, 1987; Centers for Disease Control and Prevention, 1994; Cassell, 1995; World Health Organization, 1997; Rosdahl and Pedersen,

1998). The major problems of concern today are shown in Table 6.1.

Among this list are the *Enterobacteriaceae* and other zoonotic families or species that can cause food poisoning. For the veterinarian, two areas are of importance in respect to antibiotic resistance. The first problem is therapy failure in the animal patient itself due to a resistant microorganism. If this is a pathogen that is confined to one or only a few animal species, it is a pure veterinary problem and it should be kept in mind that the animal has an ethical right to receive efficient therapy like humans. Consequently, the development of resistance in animal pathogens needs to be prevented so that the situation does not worsen. Even more important is the development of resistance in zoonotic bacterial pathogens. These can spread to humans and are consequently a matter of public health and human medicine as well. Furthermore, resistance traits selected in animal pathogens or commensal bacteria can spread by the exchange of DNA and finally reach zoonotic and pure human pathogens.

Use of Antibiotics in Animal Husbandry and Selective Pressures

Two circumstances have to come together for a resistance problem to develop. First of all, a resistance trait has to be present within the population. As pointed out above, this is quite natural,

Table 6.1. The most serious antibiotic-resistant bacteria (after Davies, 1996). Information from World Health Organization and Centers for Disease Control.

Organism	Diseases	Resistance to
Enterobacteriaceae	Diarrhoea, bacteraemia, pneumonia, urinary-tract and surgical-wound infections	Aminoglycosides, beta-lactams, trimethoprim, chloramphenicol
Haemophilus influenzae	Pneumonia, sinusitis, epiglottitis, meningitis, ear infections	Beta-lactams, tetracycline, chloramphenicol, ampicillin, trimethoprim and sulphonamide
Mycobacterium spp.	Tuberculosis	Aminoglycosides, isoniazid, ethambutol, pyrazinamide, rifampicin
Neisseria gonorrhoeae	Gonorrhoea	Beta-lactams, penicillins, spectinomycin, tetracycline, ciprofloxacin
Shigella dysenteriae	Severe diarrhoea	Ampicillin, chloramphenicol, trimethoprim + sulphonamide, tetracycline
Pseudomonas aeruginosa	Bacteraemia, pneumonia, urinary-tract infections	Aminoglycosides, beta-lactams, tetracycline, chloramphenicol, ciprofloxacin, sulphonamides
Staphylococcus aureus	Bacteraemia, pneumonia, surgical-wound infections	Chloramphenicol, methicillin, rifampicin, ciprofloxacin, clindamycin, erythromycin, beta-lactams, tetracycline, trimethoprim
Streptococcus pneumoniae	Meningitis, pneumonia	Chloramphenicol, penicillins, erythromycin
Bacteroides spp.	Anaerobic infections, septicaemia	Penicillins, clindamycin
Enterococcus spp.	Catheter infections, bacteraemia	Penicillins, aminoglycosides, vancomycin, erythromycin, tetracycline

because evolutionary processes also offer genetic and phenotypic variability for resistance genes. Consequently, there is a certain background level of resistance permanently present in all habitats. However, efficient proliferation of resistant microorganisms will only occur when selective pressures exist. Without selective pressure, resistance levels are very low and so are the chances for therapy failures. Levy (1997) has put this principle into the drug resistance equation, which states that the resistance problem is the result of the selective pressure (amount and time of antimicrobial use in a special area) and the prevalence of resistance traits against this drug.

In animal husbandry, antimicrobial agents are used for three purposes: therapy, prophylaxis and growth enhancement. In addition, the antibiotic residues are released into the environment. In all instances, a selective pressure is imposed on bacterial populations and antibiotic resistances are selected. The major concern of the public is that the pool of resistance genes is increased and that resistance genes residing on bacterial plasmids or other mobile genetic elements are spread. Since 1987, the nutritional use of antimicrobials in Europe has been regulated by the 'Council directive fixing guidelines for the assessment of additives in animal nutrition' (Anon., 1987). It demands microbiological testing of all antimicrobials used and special investigations in their potential to increase drug resistance. However, compounds used for nutritional purposes that were licensed before the guidelines have led to resistance problems, e.g. avoparcin and virginiamycin, although not in *Salmonella*, because the primary action of most growth promoters is only on Gram-positive organisms, e.g. Enterococci. In other parts of the world, the situation is different, because therapeutic antimicrobials, such as tetracyclines, are still used as feed additives.

The major selective pressures on *Salmonella* stem from the overuse of antimicrobials for prophylaxis and therapy. In particular, the wide use of mixtures of antimicrobials, antimicrobials mixed into animal feeds and therapy without diagnosis leads to a long-lasting, strong selective pressure in animal husbandry and on *Salmonella* in intensive production units. The importance of selective pressure for high resistance levels in *Salmonella* is reflected by the low resistance level in areas with low antibiotic use, such as Sweden

(Björnerot *et al.*, 1996; Franklin, 1997; Wierup, 1997). In contrast, the increasing selection of resistant organisms is well documented in monitoring programmes after the licensing of veterinary antimicrobials such as apramycin or quinolones (Threlfall *et al.*, 1986; Wray *et al.*, 1986; Piddock *et al.*, 1990b; Helmuth and Protz, 1997).

Genetic and Molecular Background of Resistance

The antimicrobial agents used today are targeted against the major biochemical and physiological processes in prokaryotic cells. Among them are cell-wall synthesis, protein synthesis, DNA and RNA duplication and transcription processes and specific biochemical pathways, such as folic acid metabolism. It was, perhaps, one of the most exiting and admirable successes of natural science that numerous scientists in the last decades could overcome resistance by developing new antimicrobial agents directed against resistance traits and new targets in resistant microorganisms. Keeping in mind that the development of a new antibiotic costs up to $US300 million and takes about 10 years, industry and other research institutions have spent large financial resources in this respect (Labischinki and Johannsen, 1997). Unfortunately our increasing knowledge about the physiology of bacterial cells has now come close to an end-point, where this approach can not be continued. Since the introduction of the fluoroquinolones directed against DNA topoisomerases, no new target sites for antimicrobial action have been found, nor are they in sight in the near future, and it will be very difficult to introduce completely new substances, although new concepts have been proposed. These focus on four strategies: modification of existing antibiotic classes, inactivation of bacterial resistance mechanisms, new target sites and the specific destruction of resistance gene mRNA (Service, 1995; Guerrier-Takada *et al.*, 1997; Labischinki and Johannsen, 1997). On the other hand, resistance is widespread and there are several basic molecular mechanisms for resistance. Table 6.2 gives an overview of the main resistance mechanisms in *Enterobacteriaceae* against important antimicrobial agents used today.

In general, three major mechanisms exist.

Table 6.2. Resistance mechanisms in *Enterobacteriaceae* against important therapeutic antibiotics.

Class of antibiotic	Mode of action	Main resistance mechanism
β-Lactams	Cell-wall synthesis	β-Lactamases, altered penicillin-binding proteins, reduction in permeability
Aminoglycosides	Protein synthesis, ribosomes 30S unit	Modifying enzymes, reduction in uptake
Tetracyclines	Protein synthesis, ribosomes 30S unit	Efflux
Quinolones	DNA gyrase	Altered target
Chloramphenicol	Protein synthesis, ribosomes 50S unit	Acetyltransferase
Folate inhibitors	Inhibition of folic acid metabolism	Altered target

First of all, an antibiotic can be destroyed or modified so that it is no longer active against microorganisms. This approach is especially prominent in resistance against β-lactam antibiotics. They can be classified into 11 distinct enzyme types, including the penicillinases, cephalosporinases, broad- and extended-spectrum and other enzymes. Today, more than 80 β-lactamases have been described, of which the TEM-type and PSE-type are frequently found in *Salmonella* (Medeiros, 1997; Massova and Mobashery, 1998). Secondly, the access of the antibiotic into the cell or to the target site can be prevented. In addition, active efflux mechanisms exist, which have been well documented in the case of tetracycline resistance (Roberts, 1996) and multidrug resistance pumps (MDRs) (Speer *et al.*, 1992; Paulsen *et al.*, 1996). Finally, the target site on its own can be modified and rendered insensitive, of which good examples are trimethoprim (Amyes and Smith, 1974) and fluoroquinolone resistance (Hooper and Wolfson, 1993; Wiedemann and Heisig, 1994).

Under therapeutic conditions, three types of resistance can lead to therapy failures. The first is the so called intrinsic or natural resistance. It is based on the lack of the target site or the presence of intrinsic, sometimes low-level, resistance mechanisms (Smith and Lewin, 1993; Paulsen *et al.*, 1996). However, more public health concern is raised about the so-called acquired resistance. It can be based on mutations in the chromosomal genes for the target site. Sulphonamide resistance in pathogenic Staphylococci was one of the earli-

est types recognized, being first reported in the 1940s (Cohen, 1992). This type of resistance is still of importance today; however, even more threatening is the frequent possession of specific resistance genes, which destroy or inactivate the antimicrobial compound by exhibiting one of the resistance mechanisms shown in Table 6.2. It is specifically these resistance genes that are of concern, because in many cases they can be spread among bacterial populations, thus increasing the overall resistance gene pool.

Spread of Resistance Genes

It is well recognized today that resistance genes can be, and have been, exchanged among bacterial populations and even passed to mammalian cells (Courvalin *et al.*, 1995; Courvalin, 1996; Davies, 1998). Two distinct kinds of spread have to be considered: first, the spread of individual resistance genes and, secondly, the spread of more complex genetic entities. Individual resistance genes can be spread by transformation and transduction, which require DNA homology and recombination. Transformation is a process in which naked DNA is taken up by cells. It is most efficient when the cells are in a special physiological stage called competence. Transformation also has an important role in nature, especially in Gram-positive organisms. Its importance for Gram-negative species is, however, still a matter of ongoing research. Transduction is the delivery of a gene by a bacteriophage into a new host cell.

In *Salmonella,* it might be of importance, because many of the strains carry prophages or inhabit biotopes where transducing phages are found (Schicklmaier and Schmieger, 1995; Schicklmaier *et al.,* 1998).

Furthermore, the importance of conjugation in the spread of resistance genes in *Salmonella* is well established. Conjugation is a parasexual, horizontal, gene-transfer process, in which self-replicating extra-chromosomal DNA molecules, called plasmids, are exchanged by a donor and recipient. This is achieved by the formation of mating pairs or mating aggregates. During this process, the plasmid is nicked at a special origin of transfer (*ori*T), leading to replication and transfer from the donor to the recipient cell. In this way, resistance genes can spread efficiently among bacterial populations. Plasmids carrying resistance genes are called R factors and are composed of several genetic units. Self-transferable R factors carry the resistance transfer factor (RTF) and resistance (R) determinants. The latter provide the resistance phenotype and the RTF the fertility. R factors have been grouped into the so-called incompatibility groups (Inc groups). An Inc group defines a set of plasmids that are not compatible, e.g. cannot coexist in one cell. Today, more than 20 incompatibility groups are known and some of them share additional genetic characteristics (Kado, 1998; del Solar *et al.,* 1998). In general, DNA transfer is most efficient at elevated temperatures. For *Salmonella,* in contrast, the *Inc*H group of plasmids are of special importance, because their transfer is temperature-dependent (Smith *et al.,* 1978). *Inc*H plasmids do not transfer at the elevated temperatures frequently encountered in the *Salmonella* host strain. Their transfer system works most efficiently at ambient temperatures. This reflects the two habitats *Salmonella* might encounter during their life-cycle, one at high temperatures in the infected host and the other at lower temperatures in the environment. In this way, the exchange of R factors under both circumstances is ensured.

R factors and other DNA molecules might carry other genetic structures, called transposons or integrons (Hall, 1997; Mahillon and Chandler, 1998). These are DNA elements that are capable of translocation from one DNA molecule to another by integrating into non-homologous regions. Transposons carry genes promoting their own transposition (transposases) and additional resistance, virulence, or catabolic genes can be located on transposons. Transposition is one of the most efficient modes for the exchange of DNA and sometimes leads to DNA rearrangements, such as replicon fusions, deletions or inversions, in the recipient molecules. Consequently, it is one of the driving forces for evolution. During transposition, target sequences that flank the inserted transposon are duplicated. In general, composite and non-composite transposons can be distinguished. Composite transposons carry insertion (IS) elements at their ends. IS elements are small (0.2–2 kb) transposons encoding only transposition functions (Stanley and Saunders, 1996; Mahillon and Chandler, 1998). Non-composite transposons lack terminal IS elements. In respect of antibiotic resistance, Tn10, encoding tetracycline resistance (Kleckner *et al.,* 1995), Tn5, encoding kanamycin, bleomycin and streptomycin resistance (Reznikoff, 1993), and Tn3, encoding ampicillin resistance, are good examples of well-defined transposons that carry resistance genes (Sherratt, 1998).

Recently, mobile gene cassettes and integrons have received large amount of scientific attention (Hall and Collins, 1995; Hall, 1997). Gene cassettes are genes that are generally integrated into the host-cell DNA. In contrast, prior to integration or after excision they can exist freely in a circularized form within the cell. Such a cassette can become part of an integron by site-specific integration into the *attI* site of an integron via the cassette-associated recombination site, called the 59-base element. Three classes of integrons have been described. Class I integrons contain two conserved segments (CS), called 5'- and 3'-CS, which flank a variable region (Hall and Collins, 1995). 5'-CS has a promoter that allows expression of the cassette genes and the genes responsible for site-specific integration, the integrase *int*I and the integration site *attI*. In class I integrons, 3'-CS carries genes for resistance to disinfectants (*qacE*Δ1) and sulphonamide. Class II integrons are characterized by a defective integrase gene *int*I2. They are related to Tn7 and its close relatives (Flores *et al.,* 1990). Still another integrase *int*I3 has been detected in class 3 integrons (Osano *et al.,* 1992).

Resistance in *Salmonella*

Resistance levels in individual regions and countries

The overall picture of resistance levels in all *Salmonella* isolates sent to an individual reference laboratory might be quite low, but it is not a suitable figure for estimating the seriousness of the resistance problem. As in other microorganisms, a resistance problem in *Salmonella* can be confined to a specific geographical region, a specific serovar or even a single production site. However, the data available for long-lasting monitoring programmes and the resulting molecular studies, which form the basis for this review, show five general trends for resistance development in *Salmonella*, which are of scientific and public-health importance.

1. Isolates without anthropogenic selective pressures are mainly sensitive against antimicrobial agents.
2. There is a general increase in the resistance levels, especially in some serovars, but not all serovars, such as *S. typhimurium*, which are linked to animal husbandry.
3. There is a general increase in the degree of multiple resistance, e.g. strains having more than one resistance.
4. There is a general increase in the number of resistance determinants that are integrated into the *Salmonella* chromosome.
5. The use of a newly licensed drug in animal husbandry often leads to increased resistance levels against the drug.

Long-standing monitoring programmes in many countries have documented the development and spread of individual resistance genes and resistant clones. Investigations on strains that were isolated in the 1920s (Datta and Hughes, 1983) or from wild or rural animals not receiving antimicrobial treatment in Africa (Levy, 1983; Helmuth and Minga, 1998) indicated low-level resistance, although self-transmissible plasmids are present in these strains. In addition, a survey of the properties of enteric isolates conserved in ice in the Arctic confirms low levels of resistance (Dancer *et al.*, 1997).

This low level of resistance is in contrast to recent reports originating from countries in Europe, the USA or Japan, in which intensive farming is quite common. At present, the overall prevalence of resistant isolates among all *Salmonella* sent to the investigating laboratories is between 10% and 30% in these countries (Kidd, 1996; Bager, 1997; Centers for Disease Control and Prevention, 1997). However, these figures are sometimes biased and not very conclusive, because of the inclusion of relatively rare serovars of subspecies I and even other subspecies that are mainly found in cold-blooded animals. Consequently, the real magnitude of the problem is more obvious when one concentrates on those food-producing animals that are held under strong antibiotic selective pressures and are of importance to human health as well, e.g. calf fattening and the predominant serovar *S. typhimurium*. In such cases, the resistance levels can be very high, lying in the range of 60–91% (Kidd, 1996; Bager, 1997; Brisabois *et al.*, 1997; Centers for Disease Control and Prevention, 1997).

History of resistance development

The first reports on resistant *Salmonella* go back to the early 1960s and describe mainly cases with monoresistant strains (Bulling *et al.*, 1973; Sojka and Hudson, 1976; Van Leeuwen *et al.*, 1979; Yoshimura *et al.*, 1980). In a retrospective study, Van Leeuwen *et al.* (1979) described the increase of tetracycline resistance from 1959 to 1974 in The Netherlands and its subsequent decline after the ban of this drug as a growth promoter. Comparable investigations are described by Bulling *et al.* (1973) for Germany. They compared resistance rates among 1000 isolates between 1961 and 1970. In 1961, the authors detected a resistance level of 2.5%, which increased to 9.5% in 1970. In addition, the predominance of transferable tetracycline monoresistance in *Salmonella* of animal origin in Germany was described (Bulling *et al.*, 1973). Similar investigations were performed by Japanese (Terakado *et al.*, 1980; Yoshimura *et al.*, 1980) and American authors (Cohen and Tauxe, 1986; Pocurull *et al.*, 1971). The early American data were reviewed by Cherubin in 1981. He concluded that resistance to antibiotics was not noted in specimens from 1948, that resistance to tetracycline first appeared in isolates obtained from both animals and humans in 1956–1957 and that it increased equally in subsequent samples. In addition, it was observed that S. *typhimurium* displayed a higher frequency of resistance than other serovars.

This observation was confirmed by Sojka and Hudson (1976) in the mid-1970s, when they detected high resistance levels in S. *typhimurium* isolated in Great Britain. The overall resistance rate was more than 90% and especially high rates of 97% were detected in S. *dublin* and S. *typhimurium*. At that time, tetracycline was still used as a growth promoter in the UK.

The increasing observations of drug resistance led to an increasing interest in its genetic and molecular bases. Datta (1962) described transmissible drug resistance in an epidemic strain of S. *typhimurium*. These observations led to more genetic studies on resistance transfer in *Salmonella*. In the mid-1960s, Anderson published a more intensive but purely genetic study on isolates of S. *typhimurium* phage type (PT) 29 (Anderson and Lewis, 1965). This type had become predominant in late 1963 in bovine infections in Britain, and had caused many human infections which were bovine in origin (Anderson, 1968). PT29 isolates were one of the first pentaresistant phage types. Starting with streptomycin and sulphonamide resistance early in 1963, it acquired resistance to tetracyclines, ampicillin, neomycin, kanamycin and furazolidone. Apparently, it had built up its drug resistance in the order streptomycin and sulphonamides, which seemed to be linked, and tetracycline and ampicillin. The investigations showed that PT29 isolates transferred their high-level resistance independently, in contrast to the genetic properties of the preceding PT1a isolates.

However, it took another 10 years for DNA isolation procedures to allow more detailed descriptions of the genetic elements involved. A good example of well-investigated strains from this period are those of PT505 in The Netherlands. This phage type predominated in the 1970s and about 90% were tetracycline-resistant. In one of the first molecular investigations of resistance in *Salmonella*, van Embden *et al.* (1976) showed that this phage type had a unique plasmid pattern. The tetracycline resistance determinant was located on a 5.8 MDa non-conjugative resistance plasmid which could be mobilized by a 58 MDa conjugative plasmid.

Similar observations stem from molecular investigations on the multiply resistant PT201, which predominated in central Europe in the 1970s. Its incidence peaked in 1974, when more than 70% of the calf isolates in The Netherlands were of resistant PT201. The isolates showed resistance against tetracycline, ampicillin, chloramphenicol and kanamycin (Helmuth *et al.*, 1981) and the molecular and genetic investigations showed two important phenomena: first, the temperature-dependent transfer of the resistance markers, which is quite common in *Salmonella* possessing Inc H1 group incompatibility plasmids (Smith *et al.*, 1978); and secondly, the formation of a plasmid cointegrate, which, upon selective pressure, transferred the complete multiresistance. The cointegrate is a large 230 MDa structure, composed of two independent plasmids providing the resistance and transfer genes. Some of the resistance determinants were obviously located on transposons which could integrate into other self-transferable plasmids.

The phenomena described above, namely the localization of resistance genes on conjugative plasmids, cointegrate formation and the localization of resistance genes on transposons, led to the efficient spread of multiresistance in *Salmonella*. It was provoked by the selective pressures applied in animal husbandry, especially in calf rearing units, and the subsequent spread to humans. A good example of this is the incidence of apramycin and gentamicin resistance in *Salmonella*. The aminoglycoside apramycin was licensed in the UK in 1980 for veterinary use. The initiated monitoring programme showed an increase in apramycin resistance from 0.1% in 1982 to 1.4% in 1984. Almost all isolates produced the enzyme aminoglycoside 3-*N*-acetyl transferase IV (ACC(3)IV) and carried various conjugative plasmids (Wray *et al.*, 1986). In subsequent years, isolates of *Salmonella* and other *Enterobacteriaceae* were isolated from human sources as well and provided evidence of the transmission of the resistance gene ACC(3)IV from bacteria of animal origin to human isolates (Threlfall *et al.*, 1985, 1986; Johnson *et al.*, 1994; Wall *et al.*, 1995).

So, from today's viewpoint, it was not a surprise that such selective pressures led to integration of resistance genes into the bacterial chromosome, making drug resistance one of the important perquisites for the survival of

Salmonella. Chromosomal integration of resistance genes was first described by Seiler and Helmuth (1986) in multiply resistant *S. dublin* isolates in Germany. Their incidence peaked between 1975 and 1980, when 75% of all isolates from cattle belonged to multiply resistant *S. dublin*. In their study, the authors could show that the first isolates carried the resistance genes on plasmids and transposons. In contrast, strains isolated in the second half of the 1970s showed plasmid profiles indistinguishable from those of sensitive strains. DNA hybridization and genetic methods elucidated the fact that the resistance genes for tetracycline, ampicillin, chloramphenicol and kanamycin were located on the chromosome, ensuring stable inheritance. Today, such chromosomal integration is widespread and will be discussed in detail in respect of the predominating clone of *S. typhimurium*, DT104.

Another feature of resistance development in *Salmonella* that led to the current situation is the increase in multiply resistant isolates. In 1992, Threlfall reviewed the use of antibiotics and the selection of food-borne pathogens. In principle, he compared the resistance levels of animal and human isolates obtained in 1980 and 1990. In this decade, the resistance rate for *S. typhimurium* increased from 36 to 54% and from 16 to 76% for *S. virchow*. Together with this overall increase, multiple resistance increased significantly during this time, namely, from 6 to 18% for *S. typhimurium* and from less than 1 to 11% for *S. virchow*. Threlfall defines multiple resistance as strains carrying resistance to four or more antimicrobial agents. This increase could be confirmed in the following years. For 1994–1996, Threlfall *et al.* (1997) showed that for *S. typhimurium*, multiple resistance had continued to increase, reaching a level of 81%. Similar data from our own laboratory confirm the observations made by Threlfall and other authors. In 1992, the majority (11%) of all German isolates carried just one resistance determinant. Four years later, in 1996, the number of monoresistant strains had dropped and the whole distribution had shifted towards isolates with two, three and more resistance determinants. The peak of the distribution was at 9% of all the strains carrying four resistance genes. This increase, observed in many countries of the world, can be attributed to the clonal spread of multiply resistant strains.

Taking a historical view, it can be concluded, that, in the 1960s, reports on resistant isolates increased and molecular studies revealed the presence of R factors with self-transfer and mobilization capabilities. However, at that time, monoresistance prevailed. In the 1970s, multiple resistance built up, due to transposition and cointegrate formation which gave rise to new plasmids conferring multiple resistance. Furthermore, efficient transfer systems at 37°C and 22°C could be detected. At the beginning of the 1980s, studies on multiply resistant *S. dublin* showed chromosomal integration of resistance genes for the first time. More recently, the 1980s and 1990s have been dominated by the appearance and spread of multiply resistant *S. typhimurium* clones, such as DT204, DT204c and, currently, DT104. In the following sections, a more detailed description of the overall phenomena is given.

Properties and spread of multiply resistant *Salmonella* clones

In the numerous investigations performed during the last decades, the prevalence of multiply resistant isolates, of the same serovar, resistance profile, phage type and molecular markers, has repeatedly been described. As mentioned above, *S. typhimurium* PT29 was one of the early examples. In the following years, three multiply resistant clones have received special attention, because of their epidemiological importance and well-documented molecular characteristics. They are described in more detail.

The *S.* wien *epidemic in the Middle East and Mediterranean countries*

The first is the *S. wien* epidemic in the Middle East and France. Before 1969, *S. wien* was only sporadically encountered and did not receive any special attention. This changed when large and serious epidemics occurred in paediatric units of Algerian hospitals (World Health Organization, 1950). In general, a high mortality – 30% of the affected children – was observed in Algeria. The epidemic was difficult to control because, in contrast with earlier sporadic isolates which were antibiotic-sensitive, the outbreak strains were multiply resistant (Domart *et al.*, 1974). Their resistance profiles covered resistances to ampi-

cillin, chloramphenicol, streptomycin, sulphonamides and, in some cases, to tetracycline and
kanamycin as well. In 1970, this particular clone
spread across the Mediterranean Sea and led to
serious nosocomial outbreaks in the major
French cities of Grenoble, Lyons and Marseilles.
It was the second commonest isolate in France in
1971 (Domart *et al.*, 1974; Adam *et al.*, 1993)
and became the source of major outbreaks in
Austria, Iraq, Italy (Marranzano *et al.*, 1976;
Schisa *et al.*, 1978; Maimone *et al.*, 1979), and
Yugoslavia (World Health Organization, 1950,
1977). Clonal identity and the genetic elements
involved were described by Maimone *et al.* in
1979. They found that most of the resistance
determinants were on an *Inc*FIme plasmid
(ACKSSPSuT) which occurred in conjugative
and non-conjugative forms and had varying sizes
between 90 and 100 MDa (McConnell *et al.*,
1979; Maimone *et al.*, 1979). Other resistances
(ASSu) resided on a smaller non-conjugative
plasmid (McConnell *et al.*, 1979). *Inc*FIme plasmids are incompatible with *Inc*FI plasmids and
the serovar-specific plasmid of *S. typhimurium*
(Willshaw *et al.*, 1978).

The *S. wien* epidemic is special in the sense
that no animal reservoir was ever described;
therefore, the many outbreaks in paediatric units
in many countries and the well-traced movement
of the epidemic can best be explained by human-
to-human spread.

Multiply resistant S. typhimurium *of phage types*
DT193, DT204 and DT204c

In contrast to *S. wien* the spread of resistant *S.
typhimurium* DT193, DT204 and its related phage
type DT204c is certainly zoonotic, and cattle and
calves seemed to be the major reservoir. The start
of the increasing incidence of DT204 is well documented in the paper of Threlfall *et al.* (1978),
when its prevalence in bovine isolates increased
from 2.4% in 1974 to 34% in 1977. In addition
to its prevalence, the degree of multiple resistance also increased. Between 1974 and 1977,
the most common resistance pattern was just
against sulphonamides and tetracyclines. None of
the resistances was transferable. However, this
situation changed dramatically when an *Inc*H2
plasmid, which carried resistances against chloramphenicol, streptomycin, sulphonamides and
tetracyclines, was established in the clone. This
was first observed during a small local outbreak in

Leicestershire. But, due to intensive calf trading,
it became widespread in calf units in Britain.
Finally, it reached the food-chain and caused
human infections as well (Rowe *et al.*, 1979). In
addition, it was not only limited to the British
Isles. Based on molecular studies, Rowe *et al.*
(1979) describe how such clones are spread by
international trade to Belgium and The
Netherlands, due to linked chains of food production. Furthermore, they showed that the
Belgian isolates were trimethoprim-resistant,
reflecting the increasing use of this antimicrobial
agent in calf rearing. The genetic and molecular
properties of these phage types were elucidated
by Willshaw *et al.* (1980). They showed that
both phage types were related and were derived
from DT49.

In the years after 1979, a new phage type,
DT204c, became most prominent. The first isolates originated from calves in Somerset,
England. In the following years, its incidence in
isolates of bovine origin increased rapidly reaching peak values of 64% in cattle and 70% in
calves. Because of its prevalence and multiple
resistance, DT204c is probably one of the best-
investigated phage types at the epidemiological,
genetic and molecular level (Threlfall *et al.*,
1980, 1986; Willshaw *et al.*, 1980; Graeber *et al.*,
1995). Isolates of this particular phage type
showed high genetic flexibility and variability,
through the aquisition of R factors, transposons
and prophages (Wray *et al.*, 1987, 1998). Like
DT204 and DT193, these multiply resistant isolates of DT204c were not only limited to the UK.
Initially, resistance to chloramphenicol, streptomycin, sulphonamides, tetracycline and
trimethoprim predominated, but in later years
almost all of the isolates of this particular phage
type had a core resistance against, ampicillin,
chloramphenicol, kanamycin and tetracycline.
Special attention was paid to and investigations
were made into resistance against aminoglycosides (Threlfall *et al.*, 1985, 1986) and high-level
resistance against the fluoroquinolones in
Germany (Graeber *et al.*, 1995; Helmuth and
Protz, 1997).

Gentamicin resistance in DT204c first
appeared in 1983 and the molecular characterization determined the existence of three lines of
isolates (Threlfall *et al.*, 1985, 1986). Two were
confined to cattle and one occurred in cattle and
humans. However, it was suggested that the use

of apramycin in animal husbandry was responsible for the appearance of gentamicin resistance in DT204c isolates. It seems that, at this time, the use of antimicrobial agents in calf-rearing created such a strong selective pressure that a certain host adaptation had occurred in some of the clones. Another example of this kind of adaptation was the increasing incidence of fluoroquinolone-resistant DT204c isolates in Germany. The use of fluoroquinolones for the treatment of calves in Germany was licensed in 1989. After that, a steep increase in multiply resistant S. typhimurium O5-negative isolates that had high minimal inhibitory concentrations (MICs) against fluoroquinolones were observed. MICs were in the range of 32–128 µg ml^{-1} for enrofloxacin and other fluoroquinolones. Peak incidence was reached in 1990, with 15% total resistance and 50% fluoroquinolone resistance in all S. typhimurium of bovine origin (Graeber et al., 1995). Fortunately, these isolates did not reach humans frequently and only one therapy failure in a young girl was described (Hof et al., 1991).

Multiply resistant S. typhimurium *DT104*

A much more widespread distribution among humans and all food-producing and other animals is attributed to the currently predominating S. typhimurium phage type, DT104. This phage type is remarkable for several reasons. It is multiply resistant, is geographically widespread and can be isolated from humans, a large variety of animal species, including all food-producing animals, and its resistance genes are located on the chromosome. As pointed out above, this contrasts with the location of resistance genes in other major clones of multiply resistant *Salmonella*, which were in general located on extrachromosomal elements, which left the possibility that resistance could be lost under conditions of low selective pressure. As stated above, chromosomal integration had so far only been described by Seiler and Helmuth (1986) in multiply resistant S. *dublin* isolates from Germany, which disappeared from Germany almost completely in the 1990s.

EPIDEMIOLOGY OF DT104. The majority of DT104 isolates are characterized by a pentaresistance against ampicillin, chloramphenicol, strepto-

mycin, sulphonamides and tetracyclines (R-type ACSSuT). Today, resistance against gentamycin, trimethoprim and fluoroquinolones has been described as well. The origin of the first pentaresistant DT104 isolates seem to be exotic birds, such as parrots and parakeets. The first isolate recorded at the Veterinary Laboratories Agency in the UK was from a seagull (C. Wray, Weybridge, personal communication); subsequently the infection was detected in cattle, which are the major reservoir of DT104 today.

In humans the first isolates were obtained in England and Wales as early as 1984. However, before 1988, fewer than 50 isolates of this particular phage R type were detected (Threlfall et al., 1994). In the following years, the incidence of DT104 increased dramatically in England and Wales, from 259 isolates in 1990 to 3837 isolates in 1995 (Threlfall et al., 1996), and in 1997, of the 2956 isolates, 95% were pentaresistant (Threlfall et al., 1998b). This epidemic is not confined to the British Isles. Reports from Germany show a prevalence of DT104 of 30% in humans (Liesegang et al., 1997), 75% in cattle, 50% in pigs, 30% in poultry and 39% in other animals in 1997 (Rabsch et al., 1997). Increasing incidences were described for Austria (Thiel, 1997), several other European countries (Anon., 1997; Rabsch et al., 1997); north-east USA (Benson et al., 1997) and Canada (Besser et al., 1997). For the USA, the incidence of DT104 or related pentaresistant isolates increased from 0.6% in 1979–1980 to 34% in 1996 and it is estimated that about 9% of human *Salmonella* infections have been caused by DT104 isolates (Glynn et al., 1998). On the basis of these data, it remains highly speculative, although scientifically challenging, to trace the real source from which DT104 emerged. Nevertheless, due to its widespread dissemination, it has become a global problem. Just the fact that it was first detected in the UK is not evidence that it emerged there.

First reports concentrated on the prevalence of DT104 in calves and adult cattle (Low et al., 1996a); however, as early as 1996, several reports pointed to isolations of DT104 from sheep (SAC Veterinary Services, 1996) and companion animals, such as cats (Low et al., 1996b; Wall et al., 1996). Today DT104 has been reported in a large variety of livestock and in horses, goats, dogs, elk, mice, coyotes, squirrels, racoons, chipmunks, pigeons and the environment as well. In cattle, DT104 can persist in symptomless carriers for more than 6 months and recurrent infections have been observed (Low *et al.*, 1996a). This poses a higher risk of occupationally acquired infections, e.g. for veterinarians, food handlers and farmers (Fone and Barker, 1994).

The importance of DT104 for humans was investigated in several outbreak and case–control studies. The case–control study performed by Wall *et al.* (1994) showed an association with the consumption of several food items and contact with animals, particularly sick farm animals. Surprisingly, the association was primarily with pork sausage, chicken and meat paste, but not with beef or beef products. However, Davies *et al.* (1996) described an outbreak of DT104 associated with beef.

How cattle acquired DT104 was the topic of a case–control study of infections in cattle performed by Evans and Davies in 1996. It turned out that major factors of herd management and hygiene were risk factors. Among them were the housing conditions, purchase of new cattle, contaminated buildings, lack of isolation facilities and vectors such as wild birds and cats.

Some reports have described a higher virulence for DT104 isolates in comparison with other *Salmonella* serovars. In their case–control study, Wall *et al.* (1994) described a higher degree of hospitalization, reaching 36% of the infected humans. In 1997, a study by Miller pointed in the same direction. It was found that 13% of pentaresistant DT104 isolates were obtained from blood cultures, compared with only 4% of other R types (Miller, 1997). However, Threlfall *et al.* (1996b) contradicted these findings when they compared the prevalence of certain serovars in almost 70,000 faecal and blood isolates. It turned out that, with 1.3%, DT104 blood isolates were not significantly more frequent than other serovars.

LOCATION OF RESISTANCE GENES IN DT104. The first report describing the chromosomal location of the resistance genes for DT104 was given by Threlfall *et al.* (1994), when the authors showed that pentaresistant isolates carried the *S. typhimurium* virulence plasmid (Helmuth *et al.*, 1985; Brown *et al.*, 1986). The genetic elimination of the virulence plasmid did not eliminate

the pentaresistance, which was neither transferable nor mobilizable. Since 1992, trimethoprim resistance could also be detected among the isolates. The resistance resided on a 4.6 MDa, nontransferable but mobilizable plasmid (Threlfall *et al.*, 1996). However, this study did not give the exact location of the resistance genes, nor did it describe the resistance genes involved. Consequently, Sandvang *et al.* (1997) performed a detailed molecular study on eight Danish DT104 isolates originating from five pig herds. By using the polymerase chain reaction (PCR) for integron detection and sequencing for integron identification, the presence of two integrons among five of their eight strains could be demonstrated. The first integron specified resistance against the aminoglycosides streptomycin and spectinomycin by the *ant* (3″)-Ia gene cassette. The β-lactamase gene cassette *PSE-1* was located on the second integron. These Danish data were confirmed absolutely by Ridley and Threlfall (1998) in a larger study performed on 45 human and 21 animal isolates from various European countries, the USA, Trinidad and South Africa. Interestingly, all isolates of DT104 contained the same inserted gene cassettes, one a 1 kb product containing streptomycin and spectinomycin resistance and the other a 1.2 kb product of the *PSE-1* gene. Both integrons were encoded on a single 10 kb *Xba*I DNA fragment, which brings them in close vicinity and makes a tandem arrangement likely. These findings are persuading evidence that all the different isolates are clonal in nature. Drug-sensitive isolates did not generate any PCR products, indicating that integrons without resistance gene cassettes are absent from DT104.

The papers described above did not establish the location of the tetracycline and chloramphenicol genes in pentaresistant DT104 isolates. Therefore, Briggs and Fratamico (1998) investigated the arrangement for all five resistance genes by synthesizing a long (13 kb) PCR product. The sequence of this product and its comparison with gene-bank sequences showed the presence of all resistance genes responsible for the pentaresistance phenotype. However, some striking observations were made: the *cmlA* gene, responsible for chloramphenicol resistance, appeared to be homologous to the *cmlA* exporter gene of *Pseudomonas aeruginosa*. The two type I integrons are separated by an R-plasmid sequence of *Pasteurella piscicida* and the 5'-CS is part of a larger IncG plasmid, pCG4, of *Corynebacterium glutamicum*.

Altogether, the data presented above shed light on important questions about antibiotic resistance in *Salmonella*. If the resistance genes are part of a larger transposon or related to pathogenicity islands, antibiotic use would select not only for resistance but also for increased virulence. However, further research is needed on this aspect.

Fluoroquinolone resistance in *Salmonella* spp.

In addition to the increasingly prevalent multiply resistant clones, the resistance against fluoroquinolones has received special attention among the public and the scientific community. Fluoroquinolones belong to a class of the newer antimicrobial agents, with a broad spectrum of activity, high efficiency and widespread application in human and veterinary medicine (World Health Organization, 1998).

Under treatment conditions, the emergence of fluoroquinolone resistance in *Salmonella* had already been observed at the end of the 1980s (Howard *et al.*, 1990; Piddock *et al.*, 1990a). However, first reports on an increasing incidence of fluoroquinolone resistance in *Salmonella* from animals stem from British and German observations (Piddock *et al.*, 1990b; Helmuth, 1992). In Great Britain, the isolates exhibited high nalidixic acid resistance and a decreased susceptibility to ciprofloxacin, a fluoroquinolone commonly used for human therapy. High fluoroquinolone resistance was observed after the licensing of fluoroquinolones for veterinary use in calves in Germany in 1989. Subsequently, the number of fluoroquinolone-resistant *Salmonella* isolates increased dramatically (Helmuth and Protz, 1997). The prevalence peaked in 1990, with a level of 15% of all *Salmonella* isolates and 50% of all isolates of bovine origin. Interestingly, almost all of these strains originated from the area close to the Dutch border, where the German calf-fattening industry is concentrated. The MIC levels for enrofloxacin were in the range 28–128 µg ml^{-1} and complete cross-resistance to the other fluoroquinolones was shown. Heisig (1993) showed that these isolates exhibited their high-level resistance due to

alterations in the *gyr*A and *gyr*B subunit, and Hof *et al.* (1991) reported a therapy failure in a patient infected with this highly resistant DT204c clone.

Other reports on veterinary isolates from various countries concentrate on strains that show high MICs against nalidixic acid, but lower MICs against the fluoroquinolones. In the Dutch study, 8.5% of *Salmonella* isolates from three broiler flocks showed resistance to nalidixic acid and flumequin, but no increased resistance to cipro- or enrofloxacin was observed (Jacobs-Reitsma *et al.*, 1994). This was in contrast to the study of Griggs *et al.* (1994), which showed that nalidixic-acid-resistant veterinary isolates had increased MICs against fluoroquinolones of 0.12–2 µg ml^{-1}. The underlying mutations could be complemented by plasmids carrying *gyr*A. This increased fluoroquinolone resistance in nalidixic-acid-resistant *Salmonella* isolates has been observed by others as well (Ruiz *et al.*, 1997).

An increase of ciprofloxacin resistance in *Salmonella* in England and Wales during 1991–1994 was described by Frost *et al.* (1996). In general, there was an increase from 0.3% to 2.1%; however, for *S. hadar* the increase was most pronounced, ranging from 2% to 39.6%. These observations led to a number of studies with the aim of investigating the nature of the resistance phenotype and if animal and human isolates shared the same properties (Reyna *et al.*, 1995; Griggs *et al.*, 1996; Piddock *et al.*, 1998). Both studies showed that resistance was the result of point mutations which lead to amino acid substitutions in the gyrase A subunit, and that there was no difference between the animal and human isolates.

Because of their multiple resistance, DT104 infections are difficult to treat and consequently special attention has been drawn to fluoroquinolone resistance in *S. typhimurium* DT104 isolates. In 1997, Threlfall *et al.* described an increasing incidence of resistance to trimethoprim and ciprofloxacin in epidemic *S. typhimurium* DT104 in England and Wales. On the basis of a cut-off of ≥ 0.125 µg ml^{-1}, the authors describe an exponential increase for ciprofloxacin resistance from almost zero in 1994 to 14% in 1996. As in Germany, this increase paralleled the licensing of enrofloxacin for veterinary use in the UK in 1993 (Threlfall *et al.*,

1998a). In the WHO meeting on 'Use of quinolones in food animals and potential impact on human health', these isolates were referred to as non-resistant but strains with reduced sensitivity (World Health Organization, 1998). This led to a discussion on current breakpoints and the importance of low-level resistance for therapy (Watson *et al.*, 1998). In 1998 this discussion was intensified after a patient died in Denmark from a food-borne DT104 infection with a strain exhibiting reduced sensitivity to ciprofloxacin (Anon., 1998). In total, 22 people were infected with a DT104 strain that exhibited an unusual resistance profile, including resistance to nalidixic acid. Identical strains were found in a slaughterhouse, two municipal food laboratories and a farm from which animals had been delivered to the abattoir. Among the infected people, seven persons were admitted into a hospital and six were treated with ciprofloxacin. None of the patients responded to therapy and one, a 62-year-old woman, died of an intestinal rupture that led to septicaemia.

Fluoroquinolones are recommended therapeutics for life-threatening *Salmonella* infections. Consequently, the level of resistance in *Salmonella* should be kept low and the MICs against this class of antimicrobial agents should be low too. The Danish case has shown that resistance levels based on a single mutation, and in the area just above or below some standard breakpoints, can lead to fatal infections and that some of our ideas about the threat of antimicrobial-resistant microorganisms might need revision.

What needs to be done to keep resistance levels low?

As shown above, many *Salmonella* serovars show high resistance levels against therapeutically useful drugs. However, it should be kept in mind that uncomplicated gastrointestinal infections, which are the majority of the cases, do not require treatment with antimicrobial agents, either in humans or animals. In contrast, generalized infections, such as bacteraemia, meningitis or infections of other extra-intestinal organs, especially in the very young or old or otherwise compromised patient, require chemotherapy. Furthermore, it has been observed that salmonel-

losis caused by resistant organisms could complicate antimicrobial therapy for other infections (Cohen and Tauxe, 1986). Most of the studies cited above were performed for epidemiological purposes and show the acquisition of resistance genes due to the selective pressure applied. In this respect, *Salmonella* is a good indicator of what might happen in other bacterial species, and resistant *Salmonella* might be an efficient source for the dissemination of resistance genes to other species. For these reasons, the monitoring of resistance development and levels in *Salmonella* is of great importance to public health.

Recently, European and international conferences called for a prudent use of antimicrobial agents (World Health Organization, 1997, 1998; Rosdahl and Pedersen, 1998). Their conclusions and recommendations focus on the prescription of antimicrobial agents by a veterinarian for purely therapeutic purposes after microbiological identification of the causative agent. The use of any antimicrobial agent should never substitute for poor hygiene and, wherever possible, alternative management methods should be used. If possible, therapy should be performed on individual animals with the narrowest-spectrum agent available and the dose and duration of therapy should limit the selection of resistant microorganisms.

Personal Remarks

I am aware that it is impossible to cover all the literature on many aspects that have been published by so many distinguished scientists on this highly controversial topic during the last decades. It was just impossible to cite them all, especially with the limited space available. In addition, some of the data seemed inconclusive and others are highly contradictory. Consequently, I decided to concentrate on some of those publications that seemed, at least for me, milestones in this field. I tried to put them in a context that describes reality and our current knowledge, on purely scientific bases and grounds. I hope that all the work put into this area, no matter if cited or not, and this review will help to improve the situation on antimicrobial resistance and lead to the more prudent use of antibiotics.

References

Adam, D., Görtz, G., Helwig, H., Knothe, H., Lode, H., Naber, G., Petersen, E.E., Stille, W., Tauchnitz, C., Ullmann, U., Vogel, F. and Wiedemann, B. (1993) Rationaler Einsatz oraler Antibiotika in der Praxis. *Münchner Medizinische Wochenschrift* 135, 591–599.

Amyes, S.G.B. and Smith, J.T. (1974) R-factor trimethoprim resistance mechanism: an insusceptible target site. *Biochemical and Biophysical Research Communications* 58, 412–418.

Anderson, E.S. (1968) Drug resistance in *Salmonella typhimurium* and its implications. *British Medical Journal* 3, 333–339.

Anderson, E.S. and Lewis, M.J. (1965) Characterization of a transfer factor associated with drug resistance in *Salmonella typhimurium*. *Nature* 208, 843–849.

Anon. (1987) Council directive fixing guidelines for the assessment of additives in animal nutrition. *Official Journal of the European Communities* L 64/19.

Anon. (1997) *Multi-drug Resistant* Salmonella. WHO Fact Sheet 139, Geneva.

Anon. (1998) Outbreak of quinolone-resistant, multiple resistant *Salmonella typhimurium* DT104, Denmark. *Weekly Epidemiologic Record* 42, 327.

Bager, F. (1997) Consumption of antimicrobial agents and occurrence of antimicrobial resistance in bacteria from food animals, food and humans in Denmark. *DANMAP* 97, 1–59.

Benson, C.E., Munro, D.S. and Rankin, S. (1997) *Salmonella typhimurium* DT104 in the north-east USA. *Veterinary Record* 141, 503–504.

Besser, T.E., Gay, C.C., Gay, J.M., Hancock, D.D., Rice, D.H., Pritchett, L.C. and Erickson, E.D. (1997) DT104 in Canada. *Veterinary Record* 140, 75.

Björnerot, L., Franklin, A. and Tysen, E. (1996) Usage of antibacterial and antiparasitic drugs in animals in Sweden between 1988 and 1993. *Veterinary Record* 21, 282–286.

Briggs, C.E. and Fratamico, P.M. (1998) Molecular characterization of an antibiotic resistance gene cluster of *Salmonella typhimurium* DT104. *Antimicrobial Agents and Chemotherapy* 43, 846–849.

Brisabois, A., Fremy, S., Moury, F., Casin, I., Breuil, J. and Collatz, E. (1997) Epidemiological survey on antibiotic susceptibility of *Salmonella typhimurium* strains from bovine origin in France. Proceedings of: *Salmonella* and *Salmonellosis*. Zoopole, Ploufragan, France, pp. 337–341.

Brown, D.J., Munro, D.S. and Platt, D.J. (1986) Recognition of the cryptic plasmid, pSLT, by restriction finger-printing and a study of its incidence in Scottish *Salmonella* isolates. *Journal of Hygiene, Cambridge* 97, 193–197.

Bulling, E., Stephan, R. and Sebek, V. (1973) The development of antibiotic resistance among *Salmonella* bacteria of animal origin in the Federal Republic of Germany and West Berlin: 1st communication: a comparison between the years of 1961 and 1970–71. *Zentralblatt für Bakteriologie, Mikrobiologie und Hygiene, 1. Abteilung Originale A* 225, 245–256.

Cassell, G.H. (1995) ASM task force urges broad program on antimicrobial resistance. *ASM News* 61, 116–120.

Centers for Disease Control and Prevention (1997) *National Antimicrobial Resistance Monitoring System. 1997 Annual Report*. CDC, Atlanta, Georgia, USA.

Centers for Disease Control and Prevention (1994) *Addressing Emerging Infectious Disease Threats: a Prevention Strategy for the United States*. CDC, Atlanta, Georgia, USA.

Cherubin, C.E. (1981) Antibiotic resistance of *Salmonella* in Europe and the United States. *Reviews of Infectious Diseases* 3, 1105–1125.

Cohen, M.L. (1992) Epidemiology of drug resistance: implications for a post-antimicrobial era. *Science* 257, 1050–1055.

Cohen, M.L. and Tauxe, R.V. (1986) Drug resistant *Salmonella* in the United States: an epidemiological perspective. *Science* 234, 964–969.

Courvalin, P. (1996) The Garrod lecture. evasion of antibiotic action by bacteria. *Journal of Antimicrobial Chemotherapy* 37, 855–869.

Courvalin, P., Goussard, S. and Grillot-Courvalin, C. (1995) Gene transfer from bacteria to mammalian cells. *Comptes Rendus de l'Académie des Sciences – Serie Iii, Sciences de la Vie* 318, 1207–1212.

Dancer, S.J., Shears, P. and Platt, D.J. (1997) Isolation and characterization of coliforms from glacial ice and water in Canada's high Arctic. *Journal of Applied Microbiology* 82, 597–609.

Datta, N. (1962) Transmissible drug resistance in an epidemic strain of *Salmonella typhimurium*. *Journal of Hygiene, Cambridge* 60, 301–310.

Datta, N. and Hughes, V.M. (1983) Conjugative plasmids in bacteria of the pre-antibiotic era. *Nature* 302, 725–726.

Davies, A., Neill, P.O., Towers, L. and Cooke, M. (1996) An outbreak of *Salmonella typhimurium* DT104 food poisoning associated with eating beef. *Communicable Disease Report* 6, R159–R162.

Davies, J.E. (1996) Bacteria on the rampage. *Nature* 383, 219–220.

Davies, J.E. (1998) Origins, acquisition and dissemination of antibiotic resistance determinants. In: *Antibiotic Resistance: Origin, Evolution, Selection and Spread*: Ciba Foundation Symposium 207, Wiley, Chichester, pp. 15–27.

Domart, A., Robineau, M., Stroh, A., Dubertret, L.M., J.F. and Modai, J. (1974) Septicémie à *Salmonella wien*: problèmes diagnostiques, thérapeutiques et épidémiologiques. *Annales de Médecine Interne* 125, 915–918.

Evans, S. and Davies, R. (1996) Case control study of multiple-resistant *Salmonella typhimurium* DT104 infection of cattle in Great Britain. *Veterinary Record* 139, 557–558.

Flores, C.M., Quadri, M.I. and Lichtenstein, C. (1990) DNA sequence analysis of five genes: *tns*A, B, C, D, and E, required for Tn7 transposition. *Nucleic Acid Research* 18, 901–911.

Fone, D.L. and Barker, R.M. (1994) Associations between human and farm animal infections with *Salmonella typhimurium* DT104 in Herefordshire. *Communicable Disease Report* 4, R136–R140.

Franklin, A. (1997) Current status of antibiotic resistance in animal production in Sweden. In: *Report of a WHO meeting*. WHO/EMC/ZOO/97.4, Geneva, pp. 229–235.

Frost, J.A., Kelleher, A. and Rowe, B. (1996) Increasing ciprofloxacin resistance in Salmonellas in England and Wales 1991–1994. *Journal of Antimicrobial Chemotherapy* 37, 85–91.

Glynn, M.K., Bopp, C., Dewitt, W., Dabney, P., Mokhtar, M. and Angulo, F.J. (1998) Emergence of multidrug-resistant *Salmonella enterica* serovar typhimurium DT104 infections in the United States. *New England Journal of Medicine* 338, 1333–1338.

Graeber, I., Montenegro, M.A., Bunge, C., Boettcher, U., Tobias, H., Heinemeyer, E.A. and Helmuth, R. (1995) Molecular marker analysis of *Salmonella typhimurium* from surface waters, humans, and animals. *European Journal of Epidemiology* 11, 325–331.

Griggs, D.J., Gensberg, K. and Piddock, L.J.V. (1996) Mutations in gyrA gene of quinolone-resistant *Salmonella* serovars isolated from humans and animals. *Antimicrobial Agents and Chemotherapy* 40, 1009–1013.

Griggs, D.J., Hall, M.C., Jin, Y.F. and Piddock, L.J.V. (1994) Quinolone resistance in veterinary isolates of *Salmonella*. *Journal of Antimicrobial Chemotherapy* 33, 1173–1189.

Guerrier-Takada, C., Salavati, R. and Altman, S. (1997) Phenotypic conversion of drug-resistant bacteria to drug sensitivity. *Proceedings of the National Academy of Sciences USA* 94, 8468–8472.

Hall, R.M. (1997) Mobile gene cassettes and integrons: moving antibiotic resistance genes in gram-negative bacteria. In: *Antibiotic Resistance: Origin, Evolution, Selection and Spread*. Ciba Foundation Symposium 207, Wiley, Chichester, pp. 192–202.

Hall, R.M. and Collins, C.M. (1995) Mobile gene cassettes and integrons: capture and spread of genes by site-specific recombination. *Molecular Microbiology* 14, 593–600.

Heisig, P. (1993) High-level fluoroquinolone resistance in a *Salmonella typhimurium* isolate due to alterations in both gyrA and gyrB genes. *Journal of Antimicrobial Chemotherapy* 32, 367–377.

Helmuth, R. (1992) Increase of drug resistance after prophylactic and therapeutic use of antimicrobial agents in livestock. In: Hinton, M.H. and Mulder, R.W.A.W. (eds) *Prevention and Control of Potentially Pathogenic Microorganisms in Poultry and Poultry Meat Processing*. FLAIR No. 6, Beekbergen, The Netherlands, 7, pp. 77–78.

Helmuth, R. and Minga, U. (1998) Molecular characterization of *Salmonella* from wild and domestic animals in Tanzania. (Unpublished.)

Helmuth, R. and Protz, D. (1997) How to modify conditions limiting resistance in bacteria in animals and other reservoirs. *Clinical Infectious Diseases* 24, 8136–8138.

Helmuth, R., Stephan, R., Bulling, E., Van Leeuwen, W.J., van Embden, J.D.A., Guinee, P.A.M., Portnoy, D.A. and Falkow, S. (1981) R-factor cointegrate formation in *Salmonella typhimurium* bacteriophage type 201 strains. *Journal of Bacteriology* 146, 444–452.

Helmuth, R., Stephan, R., Bunge, C., Hoog, B., Steinbeck, A. and Bulling, E. (1985) Epidemiology of virulence-associated plasmids and outer membrane protein patterns within seven common *Salmonella* serovars. *Infection and Immunity* 48, 175–182.

Hof, H., Ehrhard, I. and Tschäpe, H. (1991) Presence of quinolone resistance in a strain of *Salmonella typhimurium*. *European Journal of Clinical Microbiology and Infectious Diseases* 10, 747–749.

Hooper, D.C. and Wolfson, J.S. (1993) Mechanisms of bacterial resistance to quinolones. In: Hooper, D.C. and Wolfson, J.S. (eds) *Quinolone Antimicrobial Agents*. American Society for Microbiology, Washington, DC, pp. 97–118.

Howard, A.J., Joseph, T.D., Bloodworth, L.L., Frost, J.A., Chart, H. and Rowe, B. (1990) The emergence of ciprofloxacin resistance in *Salmonella typhimurium*. *Journal of Antimicrobial Chemotherapy* 26, 296–298.

Jacobs-Reitsma, W.F., Koenraad, P.M.F.J., Bolder, N.M. and Mulder, R.W.A.W. (1994) *In vitro* susceptibility of *Campylobacter* and *Salmonella* isolates from broilers to quinolones, ampicillin, tetracycline, and erythromycin. *Veterinary Quarterly* 16, 206–208.

John E. Fogarty International Center (1987) Antibiotic use and antibiotic resistance worldwide. *Reviews of Infectious Diseases* 9, S231–S316.

Johnson, A.P., Burns, L., Woodford, N., Threlfall, E.J., Naidoo, J., Cooke, C.E. and George, R.C. (1994) Gentamicin resistance in clinical isolates of *Escherichia coli* encoded by genes of veterinary origin. *Journal of Medical Microbiology* 40, 221–226.

Kado, C.I. (1998) Origin and evolution of plasmids. *Antonie Van Leeuwenhoek* 73, 117–126.

Kidd, S. (1996) Salmonella *in Livestock Production 1996*. Veterinary Laboratories Agency, CVL, Addlestone, UK, 100 pp.

Kleckner, N., Chalmers, R., Kwon, D., Sakai, J. and Bolland, S. (1995) Tn10 and IS10 transposition and chromosome rearrangements: mechanism and regulation *in vivo* and *in vitro*. *Current Topics in Microbiology and Immunology* 204, 49–82.

Krügel, H. (1997) How microorganisms produce and survive antibiotics. *Biospektrum Sonderausgabe*, 38–41.

Labischinki, H. and Johannsen, L. (1997) New antibiotics with novel mode of action. *Biospektrum Sonderausgabe*, 59–62.

Levy, S.B. (1983) Antibiotic resistant bacteria in food of man and animals. In: Woodbine, M. (ed.) *Antibiotics and Agriculture*. Butterworth, Sevenoaks, UK, pp. 525–532.

Levy, S.B. (1997) Antibiotic resistance: an ecological imbalance. In: *Antibiotic Resistance: Origin, Evolution, Selection and Spread*. Ciba Foundation Symposium 207, Wiley, Chichester, pp. 1–14.

Liesegang, A., Prager, R., Streckel, W., Rabsch, W., Gericke, B., Seltmann, G., Helmuth, R. and Tschaepe, H. (1997) Wird der *Salmonella*-enterica-Stamm DT104 des Serovars Typhimurium der neue führende Epidemietyp in Deutschland? *Infektionsepidemiologische Forschung* 1, 6–10.

Low, J.C., Hopkins, G., King, T. and Munro, D. (1996a) Antibiotic resistant *Salmonella typhimurium* DT104 in cattle. *Veterinary Record* 138, 650–651.

Low, J.C., Tennant, B. and Munro, D. (1996b) Multiple-resistant *Salmonella typhimurium* DT104 in cats. *Lancet* 348, 1391.

Mahillon, J. and Chandler, M. (1998) Insertion sequences. *Microbiology and Molecular Biology Reviews* 62, 725–774.

Maimone, F., Colonna, B., Bazzicalupo, P., Oliva, B., Nicoletti, M. and Casalino, M. (1979) Plasmids and transposable elements in *Salmonella wien*. *Journal of Bacteriology* 139, 369–375.

Marranzano, M., Nastasi, A., Scarlata, G., Salvo, S. and Falcidia, A. (1976) Resistenza agli antibiotici e fattori di resistenza in stipiti di *Salmonella wien* isolati in Sicilia. *Bollettino dell'Instituto Sieroterapico Milanese* 55, 187–190.

Massova, I. and Mobashery, S. (1998) Kinship and diversification of bacterial penicillin-binding proteins and β-lactamases. *Antimicrobial Agents and Chemotherapy* 42, 1–17.

McConnell, M.M., Smith, H.R., Leonardopoulos, J. and Anderson, E.S. (1979) The value of plasmid studies in the epidemiology of infections due to drug resistant *Salmonella wien*. *Journal of Infectious Diseases* 139, 178–190.

Medeiros, A.A. (1997) Evolution and dissemination of beta-lactamases accelerated by generations of beta-lactam antibiotics. *Clinical Infectious Diseases*. 24, S19–S45.

Miller, M.A. (1997) Widespread emergence in the United States of multiple-drug resistant type *Salmonella typhimurium*. *FDA Veterinarian* 12, 5–6.

Osano, E., Arakawa, R., Wacharotayankun, R., Ohta, M., Horii, T., Ito, H., Yoshimura, N. and Kato, N. (1992) Molecular characterization of an enterobacterial metallo-β-lactamase found in a clinical isolate of *Serratia marcescens* that shows imipenem resistance. *Antimicrobial Agents and Chemotherapy* 38, 71–78.

Paulsen, I.T., Brown, M.H. and Skurray, R.A. (1996) Proton-dependent multidrug efflux systems. *Microbiological Reviews* 60, 575–608.

Piddock, L.J.V., Ricci, V., McLaren, I. and Griggs, D.J. (1998) Role of mutation in the *gyrA* and *parC* genes of nalidixic-acid-resistant *Salmonella* serovars isolated from animals in the United Kingdom. *Journal of Antimicrobial Chemotherapy* 41, 635–641.

Piddock, L.J.V., Whale, K. and Wise, R. (1990a) Quinolone resistance in *Salmonella*: clinical experience. *Lancet* 335, 1459.

Piddock, L.J.V., Wray, C., McLaren, I. and Wise, R., (1990b) Quinolone resistance in *Salmonella* species: veterinary pointers. *Lancet* 336, 125.

Pocurull, D.W., Gaines, S.A. and Mercer, H.D. (1971) Survey of infectious multiple drug resistance among *Salmonella* isolated from animals in the United States. *Applied Microbiology* 21, 358–362.

Rabsch, W., Schroeter, A. and Helmuth, R. (1997) Prevalence of *Salmonella enterica* subsp. *enterica* serovar Typhimurium DT104 in Germany. *Newsletter Community Reference Laboratory for* Salmonella 3, 10–11.

Reyna, F., Huesca, M., González, V. and Fuchs, L.Y. (1995) *Salmonella typhimurium gyrA* mutations associated with fluoroquinolone resistance. *Antimicrobial Agents and Chemotherapy* 39, 1621–1623.

Reznikoff, W.S. (1993) The Tn5 transposon. *Annual Review Microbiology* 47, 945–963.

Ridley, A.M. and Threlfall, E.J. (1998) Molecular epidemiology of antibiotic resistance genes in multiple resistant epidemic *Salmonella typhimurium* DT 104. *Microbial Drug Resistance* 4, 113–118.

Roberts, M.C. (1996) Tetracycline resistance determinants: mechanisms of action, regulation of expression, genetic mobility, and distribution. *FEMS Microbiology Reviews* 19, 1–24.

Rosdahl, V.T. and Pedersen, K.B. (1998) In: Rosdahl, V.T. and Pedersen, K.B. (eds) *The Copenhagen Recommendations*. Danish Veterinary Laboratory, Copenhagen, pp. 1–52.

Rowe, B., Threlfall, E.J., Ward, L.R. and Ashley, A.S. (1979) International spread of multiresistent strains of *Salmonella typhimurium* phage types 204 and 193 from Britain to Europe. *Veterinary Record* 105, 468–469.

Ruiz, J., Castro, D., Goñi, P., Santamaria, J.A., Borrego, J.J. and Vila, J. (1997) Analysis of the mechanism of quinolone resistance in nalidixic acid-resistant clinical isolates of *Salmonella* serovar Typhimurium. *Journal of Medical Microbiology* 46, 623–628.

SAC Veterinary Services. (1996) *Salmonella typhimurium* DT104 causes problems in both cattle and sheep. *Veterinary Record* 138, 607–609.

Sandvang, D., Aarestrup, F.M. and Jensen, L.B. (1997) Characterisation of integrons and antibiotic resistance genes in Danish multiple resistant *Salmonella enterica* Typhimurium DT104. *FEMS Microbiology Letters* 157, 177–181.

Schicklmaier, P., Moser, E., Wieland, T., Rabsch, W. and Schmieger, H. (1998) A comparative study on the frequency of prophages among natural isolates of *Salmonella* and *Escherichia coli* with emphasis on generalized transducers. *Antonie Van Leeuwenhoek* 73, 49–54.

Schicklmaier, P. and Schmieger, H. (1995) Frequency of generalized transducing phages in natural isolates of the *Salmonella typhimurium* complex. *Applied and Environmental Microbiology* 61, 1637–1640.

Schisa, C., Biondi, M., Fiorentino, F., Sirgiovanni, M.C., Mancini, A. and Manzillo, G. (1978) Presenza di *Salmonella wien* in Campania nel 1974–1975. *Minerva Medica* 69, 2125–2128.

Seiler, A. and Helmuth, R. (1986) Epidemiology and chromosomal location of genes encoding multiple resistance in *Salmonella dublin*. *Journal of Antimicrobial Chemotherapy* 18, 179–181.

Service, R.F. (1995) Antibiotics that resist resistance. *Science* 270, 724–727.

Sherratt, D. (1998) Tn3 and related transposable elements: site specific recombination and transposition. In: Berg, D.E. and Howe, M.M. (eds) *Mobile DNA*. American Society for Microbiology, Washington DC.

Smith, H.W., Parsell, Z. and Green, P. (1978) Thermosensitive antibiotic resistance plasmids in enterobacteria. *Journal of General Microbiology* 109, 37–47.

Smith, J.T. and Lewin, C.S. (1993) Mechanisms of antimicrobial resistance and implications for epidemiology. *Veterinary Microbiology* 35, 233–242.

Sojka, W.J. and Hudson, E.B. (1976) A survey of drug resistance in *Salmonella* isolated from animals in England and Wales during 1972. *British Veterinary Journal* 132, 95–104.

del Solar, G., Giraldo, R., Ruiz-Echevarria, M.J., Espinosa, M. and Diaz-Orejas, R. (1998) Replication and control of circular bacterial plasmids. *Microbiology and Molecular Biology Reviews* 62, 434–464.

Speer, B.S., Shoemaker, N.B. and Salyers, A.A. (1992) Bacterial resistance to tetracycline: mechanisms, transfer, and clinical significance. *Clinical Microbiology Reviews* 5, 387–399.

Stanley, J. and Saunders, N. (1996) DNA insertion sequences and the molecular epidemiology of *Salmonella* and *Mycobacterium*. *Journal of Medical Microbiology* 45, 236–251.

Terakado, N., Ohya, T. and Veda, H. (1980) A survey on drug resistance and R plasmids in *Salmonella* isolated from domestic animals in Japan. *Japanese Journal of Veterinary Science* 42, 543–550.

Thiel, W. (1997) Preliminary data about isolations of *S. typhimurium* DT104 in Austria. *Newsletter Community Reference Laboratory for* Salmonella 3, 4–5.

Threlfall, E.J. (1992) Antibiotics and the selection of food-borne pathogens. *Journal of Applied Bacteriology* 73, 96S–102S.

Threlfall, E.J. and Rowe, B. (1997) Increasing incidence of resistance to trimethoprim and ciprofloxacin in epidemic *Salmonella typhimurium* DT104 in England and Wales. *European Surveillance* 2, 81–84.

Threlfall, E.J., Ward, L.R. and Rowe, B. (1978) Epidemic spread of a chloramphenicol-resistant strain of *Salmonella typhimurium* phage type 204 in bovine animals in Britain. *Veterinary Record* 103, 438–440.

Threlfall, E.J., Ward, L.R., Ashley, A.S. and Rowe, B. (1980) Plasmid-encoded trimethoprim resistance in multiple resistant epidemic *Salmonella typhimurium* phage types 204 and 193 in Britain. *British Medical Journal* 280, 1210–1211.

Threlfall, E.J., Rowe, B., Ferguson, J.L. and Ward, L.R. (1985) Increasing incidence of resistance to gentamicin and related aminoglycosides in *Salmonella typhimurium* phage type 204c in England, Wales and Scotland. *Veterinary Record* 117, 355–357.

Threlfall, E.J., Rowe, B., Ferguson, J.L. and Ward, L.R. (1986) Characterization of plasmids conferring resistance to gentamicin and apramycin in strains of *Salmonella typhimurium* phage type 204c isolated in Britain. *Journal of Hygiene, Cambridge* 97, 419–426.

Threlfall, E.J., Frost, J.A., Ward, L.R. and Rowe, B. (1994) Epidemic in cattle and humans of *Salmonella typhimurium* DT 104 with chromosomally integrated multiple drug resistance. *Veterinary Record* 134, 577.

Threlfall, E.J., Frost, J.A., Ward, L.R. and Rowe, B. (1996) Increasing spectrum of resistance in multiple resistant *Salmonella typhimurium*. *Lancet* 347, 1053–1054.

Threlfall, E.J., Ward, L.R., Skinner, J.A. and Rowe, B. (1997) Increase in multiple antibiotic resistance in non-typhoidal Salmonellas from humans in England and Wales: a comparison of data for 1994 and 1996. *Microbial Drug Resistance* 3, 263–266.

Threlfall, E.J., Angulo, F.J. and Wall, P.G. (1998a) Ciprofloxacin-resistant *Salmonella typhimurium* DT104. *Veterinary Record* 142, 255.

Threlfall, E.J., Ward, L.R. and Rowe, B. (1998b) Multiple resistant *Salmonella typhimurium* DT 104 and *Salmonella* bacteraemia. *Lancet* 352, 287–288.

van Embden, J.D.A., Van Leeuwen, W.J. and Guinee, P.A.M. (1976) Interference with propagation of typing bacteriophages by extrachromosomal elements in *Salmonella typhimurium* bacteriophage type 505. *Journal of Bacteriology* 127, 1414–1426.

Van Leeuwen, W.J., van Embden, J.D.A., Guinee, P.A.M., Kampelmacher, E.H., Manten, A., Van Schothorst, M. and Voogd, C.E. (1979) Decrease of drug resistance in *Salmonella* in the Netherlands. *Antimicrobial Agents and Chemotherapy* 16, 237–239.

Wall, P.G., Morgan, D., Lamden, K., Ryan, M., Griffin, M., Threlfall, E.J., Ward, L.R. and Rowe, B. (1994) A case control study of infection with an epidemic strain of multiple resistant *Salmonella typhimurium* DT104 in England and Wales. *Communicable Disease Report* 4, R130–R135.

Wall, P.G., Morgan, D., Lamden, K., Griffin, M., Threlfall, E.J., Ward, L.R. and Rowe, B. (1995) Transmission of multi-resistant strains of *Salmonella typhimurium* from cattle to man. *Veterinary Record* 136, 591–592.

Wall, P.G., Threlfall, E.J., Ward, L.R. and Rowe, B. (1996) Multiple resistant *Salmonella typhimurium* DT104 in cats: a public health risk. *Lancet* 348, 471.

Watson, P.M., Bell, G.D., Webster, C.M.M. and Fitzgerald, R.A. (1998) Fluoroquinolone susceptibility of *S. typhimurium* DT104. *Veterinary Record* 142, 374.

Wiedemann, B. and Heisig, P. (1994) Mechanisms of quinolone resistance. *Infection* 22, 73–79.

Wierup, M. (1997) Ten years without antibiotic growth promoters – results from Sweden with special reference to production results, alternative disease preventive methods and the usage of antibacterial drugs. In: *Report of a WHO Meeting.* WHO/EMC/ZOO/97.4, Geneva, pp. 229–235.

Willshaw, G.A., Smith, H.R. and Anderson, E.S. (1978) Molecular studies of FIme resistance plasmids particularly in epidemic *Salmonella typhimurium*. *Molecular and General Genetics* 159, 111–116.

Willshaw, G.A., Threlfall, E.J., Ward, L.R., Ashley, A.S. and Rowe, B. (1980) Plasmid studies of drug-resistant epidemic strains of *Salmonella typhimurium* belonging to phage types 204 and 193. *Journal of Antimicrobial Chemotherapy* 6, 763–773.

World Health Organization (1950) *Salmonella* surveillance 1973. *Weekly Epidemiologic Record* 50, 437–444.

World Health Organization (1977) *Salmonella* surveillance in 1974. *Weekly Epidemiologic Record* 52, 53–61.

World Health Organization (1997) The medical impact of the use of antimicrobials in food animals. In: *Report of a WHO Meeting.* WHO/EMC/ZOO/97.4, Geneva, pp. 1–24.

World Health Organization (1998) Use of quinolones in food animals and potential impact on human health. In: *Report of a WHO Meeting.* WHO/EMC/ZDI/98.10, Geneva, pp. 1–26.

Wray, C., Hedges, R.W., Shannon, K.P. and Bradley, D.E. (1986) Apramycin and gentamicin resistance in *Escherichia coli* and Salmonellas isolated from farm animals. *Journal of Hygiene, Cambridge* 97, 445–456.

Wray, C., McLaren, I., Parkinson, N.M. and Beedell, Y.E. (1987) Differentiantion of *Salmonella typhimurium* DT204c by plasmid profile and biotyping. *Veterinary Record* 121, 514–516.

Wray, C., McLaren, I.M. and Jones, Y.E. (1998) The epidemiology of *Salmonella* typhimurium in cattle: plasmid profile analysis of definitive phage type (DT) 204c. *Journal of Medical Microbiology* 47, 483–487.

Yoshimura, H., Nakamura, M., Koeda, T. and Sato, S. (1980) Antibiotic sensitivity of *Salmonellae* isolated from animal feed ingredients. *Japanese Journal of Veterinary Science* 42, 595–597.

Chapter 7

Salmonella Infections in the Domestic Fowl

Cornelius Poppe

Health Canada, 110 Stone Road West, Guelph, Ontario N1G 3W4, Canada

Historical Perspective

At the turn of the century, pullorum disease, caused by S. *pullorum*, was one of the most important diseases in commercial poultry. The disease was first described by Rettger in 1900 as a 'septicaemia of young chicks' (Rettger, 1900); later he called it a 'fatal septicaemia in young chicken, or white diarrhoea' and named its cause *Bacterium pullorum* (Rettger, 1909). During ensuing years, pullorum disease became recognized as a common, worldwide, egg-transmitted disease of chickens. A macroscopic tube-agglutination test to detect carriers of pullorum disease was developed by Jones (1913); later, it was largely replaced by a simple whole-blood test, in which stained antigen is used (Schaffer *et al.*, 1931).

Fowl typhoid in poultry, caused by S. *gallinarum*, is a septicaemic disease affecting primarily chickens and turkeys. The first described outbreak of the disease occurred in England in 1888, when a chicken breeder lost 400 chickens (Klein, 1889). Fowl typhoid has a worldwide distribution and has been found in almost all poultry-producing areas of the world (Pomeroy, 1984).

During the first five to six decades of the 1900s, the main issues with respect to salmonellosis were the occurrence of S. *typhi*, causing typhoid fever, a host-specific disease in humans (Mandal, 1979), which declined dramatically during that period in Europe and North America (Ranta and Dolman, 1947; Edwards, 1958; Bynoe

and Yurack, 1964; Nicolle, 1964; Seeliger and Maya, 1964; Sommers, 1980), and the widespread prevalence of pullorum disease and fowl typhoid in chickens and turkeys, which caused high mortality in flocks worldwide and prevented the establishment and growth of the poultry industry until the development and widespread application of testing and control measures (Rettger, 1909; Hewitt, 1928; Schaffer *et al.*, 1931; Hinshaw and McNeil, 1940; Moore, 1946; Chase, 1947; McDermott, 1947; Bullis, 1977a; Pomeroy, 1984; Snoeyenbos, 1984). A voluntary National Poultry Improvement Program (NPIP), aimed at preventing disease transmission to progeny by testing of breeder flocks, became operative in the USA in 1935 (Snoeyenbos, 1984). The pullorum and fowl-typhoid testing and control measures resulted in a much lower prevalence of these diseases in most of the developed countries in the 1950s and 1960s (Sojka *et al.*, 1975; Pomeroy, 1984; Snoeyenbos, 1984). More recently, Canada, the USA and several European countries reported complete absence or a low prevalence of pullorum disease and fowl typhoid (Thain and Blandford, 1981; Johnson *et al.*, 1992; Salem *et al.*, 1992; Erbeck *et al.*, 1993; Waltman and Horne, 1993) but these diseases still occur more or less frequently in countries in Eastern Europe, Central and South America (Silva *et al.*, 1981), Africa (Bouzoubaa and Nagaraja, 1984) and Asia (Chishti *et al.*, 1985; Kaushik *et al.*, 1986).

Since the 1940s, there has been a rapid

increase in the isolation of the non-host specific *Salmonella* serovars from humans and animals (Galton *et al.*, 1964; Guthrie, 1992). This was particularly the case with *S. typhimurium*, which until more recently has been the most prevalent serovar isolated from humans and animals (especially cattle) in many countries (Kelterborn, 1967; McCoy, 1975; Bullis, 1977b; Wray, 1985; Lior, 1989; Ferris and Miller, 1990; Rodrigue *et al.*, 1990; Kühn *et al.*, 1993; Hargrett-Bean and Potter, 1995). Poultry and poultry products have been the main sources of non-host-specific *Salmonella* infecting humans (Schaaf, 1936; Edwards, 1958; Galton *et al.*, 1964; Seeliger and Maya, 1964; McCoy, 1975; Laszlo *et al.*, 1985; Humphrey *et al.*, 1988; St Louis *et al.*, 1988), but dairy products (D'Aoust, 1985; Ryan *et al.*, 1987), beef (Bryan, 1981; Parham, 1985; Bean and Griffin, 1990) and pork (Maguire *et al.*, 1993) have also been associated with large outbreaks of salmonellosis.

Epidemiology

Prevalent serovars

Most of the information regarding the prevalence of *Salmonella* has been based on passive laboratory-based *Salmonella* surveillance (Faddoul and Fellows, 1966; Sojka *et al.*, 1975). Often no distinction has been made between symptomatic and asymptomatic infection or chronic carriage. Such surveillance systems have inherent biases (Bynoe and Yurack, 1964; Galton *et al.*, 1964). Many factors, including intensity of surveillance (Galton *et al.*, 1964), submission for serotyping (Sojka *et al.*, 1975), severity of illness and association with a recognized outbreak in the human population (Philbrook *et al.*, 1960; Telzak *et al.*, 1990; van de Giessen *et al.*, 1992; Altekruse *et al.*, 1993) and lack of a systematic method of reporting (Edwards, 1958), affect whether an infection will be reported. Reporting of animal and human salmonellosis has been substantially underestimated. However, the surveillance data allow broad comparisons and identify trends, reservoirs and routes of transmission of *Salmonella* serovars.

There has been considerable variation in the occurrence of the most common *Salmonella* serovars in domestic fowl in different countries and at different times. Certain serovars have

been known to become widespread in a country or geographical area for a given period and then decrease in incidence to a point of little importance (Faddoul and Fellows, 1966; Borland, 1975). *S. typhimurium* has been among the most common serovars isolated from poultry in many countries, especially during the period from about 1950 until the late 1970s (Faddoul and Fellows, 1966; Williams, 1984). In 1948, a list of the commonest serovars in chickens and turkeys in the USA ranked in declining order, included *S. typhimurium*, *S. anatum*, *S. derby*, *S. bareilly*, *S. meleagridis*, *S. oranienburg*, *S. give*, *S. bredeney*, *S. newport* and *S. montevideo* (Bullis, 1977b). In 1976, the commonest serovars isolated from chickens and turkeys in the USA ranked in declining order, were *S. typhimurium*, *S. heidelberg*, *S. saintpaul*, *S. infantis*, *S. thompson*, *S. montevideo*, *S. worthington*, *S. johannesburg*, *S. enteritidis* and *S. anatum* (Bullis, 1977b). *S. typhimurium* was the commonest *Salmonella* serovar isolated from poultry and other birds in the UK during the period 1968–1973; it accounted for 41.1% of all isolates and was followed, in decreasing frequency of isolation, by *S. enteritidis* (6.2%), *S. pullorum* (3.9%) and *S. gallinarum* (2.8%) (Sojka *et al.*, 1975).

During the last 10–15 years, *S. enteritidis* has replaced *S. typhimurium* as the commonest serovar in poultry in many countries worldwide. A comprehensive monograph on all aspects of *S. enteritidis* has been published (Saeed, 1999). In England and Wales, the percentage of *S. enteritidis* isolates from poultry rose from 3.3% of all *Salmonella* serovars in 1985, to 6.9% in 1986, to 22.3% in 1987, to 47.8% in 1988 and to 48.3% in 1989. The most frequently reported *S. enteritidis* phage type (PT) was PT4, which accounted for 71% of the isolates from poultry in 1988 (McIlroy and McCracken, 1990). However, more recently, there has been a reduction in the isolation rates of *S. enteritidis* in the UK: the number of isolations of *S. enteritidis* from chickens declined from 823 in 1993 to 407 in 1994 and to 245 in 1995. During the same period (1993–1995), *S. senftenberg* and *S. mbandaka* were the second and third most commonly isolated *Salmonella* serovars from chickens in the UK, but their number of isolations was much smaller than that of *S. enteritidis*. A study in The Netherlands in 1989 to determine the presence of *Salmonella* in 49 and 52 randomly selected layer and broiler flocks showed that *S. infantis* and *S. virchow* were each isolated from 30.9% of the flocks, while *S. typhimurium*, *S. enteritidis* and *S. hadar* were isolated from 25%, 20.6%, and 17.6%, respectively, of the flocks. *Salmonella* were isolated from the faecal samples of 47% of the layer and 94% of the broiler flocks. *S. enteritidis* was isolated from nine of the 49 (18.4%) layer flocks and from six of the 52 (11.5%) broiler flocks (van de Giessen *et al.*, 1991).

Nation-wide studies were carried out in Canada in 1989–1990 to determine the prevalence of *Salmonella* in randomly selected layer, broiler and turkey flocks. Faecal samples and/or scrapings from egg belts from 52.9% of 295 layer flocks were contaminated with *Salmonella*. Thirty-five different *Salmonella* serovars were isolated. The most prevalent serovars were *S. heidelberg*, *S. infantis*, *S. hadar* and *S. schwarzengrund*; they were isolated from samples of 20%, 6.1%, 5.8% and 5.1% of the flocks, respectively. *S. enteritidis* was isolated from 2.7% of the layer flocks; *S. enteritidis* PT8 was isolated from five flocks, PT13a from two flocks and PT13 from one flock (Poppe *et al.*, 1991a). A similar study of 294 broiler flocks showed that environmental samples (litter and/or water) from 76.9% of the flocks were contaminated by *Salmonella*. Fifty different serovars were isolated. The most prevalent serovars were *S. hadar*, *S. infantis* and *S. schwarzengrund*; they were isolated from samples of 33.3%, 8.8% and 7.1% of the flocks, respectively. *S. enteritidis* was isolated from 3.1% of the flocks; *S. enteritidis* PT8 was isolated from seven flocks and PT13a from two flocks (Poppe *et al.*, 1991b).

In the USA, in 1990 a survey of spent laying hens was conducted over a period of 3 months to estimate the prevalence and distribution of *S. enteritidis* in commercial egg-production flocks. It showed that any *Salmonella* serovar and *S. enteritidis* were isolated from 24% and 3%, respectively, of 23,431 pooled caecal samples collected from 406 layer houses. Regionally, the estimated prevalence of *S. enteritidis*-positive houses (i.e. at least one positive sample found in a house) for the northern, south-eastern and central/western regions was 45%, 3% and 17%, respectively. Overall, the prevalence of *Salmonella*-positive houses was 86% (Ebel *et al.*, 1992).

Sources of infection and routes of transmission

The sources of *Salmonella* infection for domestic fowl are numerous. Poultry and many other animals are often unapparent carriers, latently infected or, less frequently, clinically ill, and they may excrete *Salmonella* in their faeces and form a large reservoir and source of contamination for other animals, humans and the environment. Poultry often become infected via horizontal transmission by litter, faeces, feed, water, fluff, dust, shavings, straw, insects, equipment and other fomites contaminated with *Salmonella* and by contact with other chicks or poults, rodents, pets, wild birds, other domestic and wild animals and personnel contaminated with *Salmonella*. Vertical transmission occurs when follicles in the ovary are infected or the developing eggs become infected in the oviduct. Management practices may have a significant influence on the degree of transmission of *Salmonella*. Many of the factors that influence horizontal and vertical transmission are interrelated.

Horizontal transmission

AT HATCHING. Horizontal spread of *Salmonella* occurs during the hatching of chicks. This was shown when contaminated and *Salmonella*-free eggs were incubated together. Unincubated, fertile hatching eggs were inoculated by immersion for 15 min in a 16°C physiological saline solution containing 10^8 *S. typhimurium*. After 17–18 days incubation, the eggs were transferred to hatchers and control eggs at the same stage of incubation were added to the same tray and to those trays above and below. Fertile inoculated eggs hatched at a rate of 86%, despite the high level of contamination, indicating that chicks in eggs contaminated with *Salmonella* are likely to hatch and may contaminate other chicks in the same hatcher cabinet. Air samples showed a sharp increase in contamination in the hatcher at 20 days of incubation. About 90% of chick rinses were *Salmonella*-positive in samples from both inoculated and control eggs. In samples from inoculated eggs, *Salmonella* was detected in the digestive tract of 8% of embryos at transfer from incubator to hatcher and in 55% of chicks at hatch, whereas, in samples from control eggs, 44% of digestive tracts of hatched chicks were positive (Cason *et al.*, 1994).

AIR. Poultry may become infected by aerosols containing *Salmonella*. Laying hens exposed to aerosols containing 10^3 or 10^5 but not those given 10^2 *S. enteritidis* PT4 bacteria developed diarrhoea and some lost weight or died (Baskerville *et al.*, 1992).

LITTER. Chicks or poults in barns often contaminate the litter with faeces containing *Salmonella*, and contaminated litter is an important source and means of transmission of *Salmonella*. Poultry often ingest large numbers of *Salmonella* by picking at faecal and caecal droppings of littermates. *Salmonella* bacteria belonging to the most common serovars were found to spread rapidly from infected day-old chicks to penmates reared on litter (Snoeyenbos *et al.*, 1969; Rigby and Pettit, 1979). Infection of contact chicks reached about 100% within 7 days of contact (Snoeyenbos *et al.*, 1969). *Salmonella* were isolated from 30% of litter samples in 55% of broiler chicken houses (Long *et al.*, 1980). Another study of *Salmonella* contamination of the environment of broilers showed that *Salmonella* were commonly isolated from litter: 47.4% of 3534 litter samples contained *Salmonella* and one or more litter samples of 223 of 294 (75.9%) broiler flocks were positive for *Salmonella* (Poppe *et al.*, 1991b). Chickens on litter transmit *Salmonella* more readily than when in wire cages (Brownell *et al.*, 1969; Rigby and Pettit, 1979). Placing chicks infected orally with *S. typhimurium* in wire cages hastened the age-related decline in faecal excretion of *S. typhimurium* (Rigby and Pettit, 1979).

The infectivity and mortality rate among chicks were higher on fresh litter than on built-up litter (Botts *et al.*, 1952). Several studies have shown that *Salmonella* bacteria, such as *S. pullorum*, *S. gallinarum*, *S. infantis* and *S. typhimurium*, persist longer in fresh litter than in built-up litter (Botts *et al.*, 1952; Tucker, 1967; Fanelli *et al.*, 1970). *S. pullorum* and *S. gallinarum* persisted for 11 weeks in new litter, but for only 3 weeks in built-up litter. Similarly, *S. thompson* survived for 8–20 weeks in new litter, but for only 4–5 weeks in old litter. When the infected pens were left unoccupied, the survival time of *Salmonella* bacteria in both types of litter increased to more than 30 weeks (Tucker, 1967). Cycling of *Salmonella* bacteria between litter and the intestinal tract appeared to be significant in maintaining intestinal infection. This cycling was more evident in unchanged new litter than in built-up litter or in fresh litter changed periodically (Fanelli *et al.*, 1970).

A positive relationship between moisture content of litter and survival of *Salmonella* bacteria has been reported (Tucker, 1967). Water activity was positively correlated with *Salmonella*-positive drag swabs of litter in broiler chicken houses (Opara *et al.*, 1992). Increased moisture and a high pH as a result of dissolved ammonia were shown to be the cause of the higher bactericidal activity of old or built-up litter compared with new litter (Turnbull and Snoeyenbos, 1973).

WATER. Contaminated drinking-water may facilitate the spread of *Salmonella* among domestic fowl (Gauger and Greaves, 1946). In a study of *Salmonella* contamination in broiler flocks, *Salmonella* were isolated from 108 of 875 (12.3%) water samples and one or more water samples from 63 of 292 (21.6%) flocks were positive for *Salmonella* (Poppe *et al.*, 1991b). Chicks and poults often defecate in the drinkers and contaminate the drinking-water with their beaks and by stepping or walking in the drinkers. This happens particularly with young chicks or poults before the drinkers are raised from the floor of the barn. It was found that contamination of drinking-water also depends on the kind of drinkers used in the barns. There was a significantly greater risk of contamination of drinking-water with *Salmonella* from trough drinkers and plastic bell drinkers than from nipple drinkers (Renwick *et al.*, 1992). Laying hens inoculated with 10^5 *S. enteritidis* PT4 spread the infection via drinking-water to other uninoculated hens in 1–5 days (Nakamura *et al.*, 1994).

FEED. The feed may contain *Salmonella* and form a source of infection for poultry (see Chapter 17). Serovars of *Salmonella* found in broilers have been traced to feeds, water and the breeder flocks from which the broilers originated (Morris *et al.*, 1969). In a study of nine broiler flocks, the frequency of contamination of feed was lower than that of litter, water and dust (Higgins *et al.*, 1982). Likewise, *Salmonella* contamination of the environment of layer flocks' feed samples were less often contaminated with *Salmonella* than egg-belt scrapings: *Salmonella* were isolated from 7.2% of feed samples and 25.7% of egg belt-samples (Poppe *et al.*, 1991a). Examination of environmental samples from broiler flocks showed that 13.4% of feed samples, 12.3% of water samples and 47.4% of litter samples were contaminated with *Salmonella* (Poppe *et al.*, 1991b).

RODENTS. Rodents play an important role in the epidemiology of *Salmonella* infection in poultry (Henzler and Opitz, 1992; Anon., 1995b). A microbiological survey of ten mice-infected poultry farms showed that, on five farms where no *S. enteritidis* was isolated, 29.5% of 696 environmental samples and 6% of 232 mice were culture-positive for other *Salmonella* serovars. On another five farms on which *S. enteritidis* was isolated, *Salmonella* of any serovar were isolated from 41.4% of 1407 environmental samples and from 31.8% of 483 mice, and *S. enteritidis* from 7.5% of 1407 environmental samples and from 24% of the 483 mice. A bacterial count from the faeces of a mouse yielded more than 10^5 *S. enteritidis* bacteria per faecal pellet (Henzler and Opitz, 1992).

STRESS. The stress of induced moulting in ageing hens – a management practice to bring about a second egg-laying cycle by temporary removal of feed and water and altering the photoperiod – was shown to affect the shedding of *Salmonella*. When White Leghorn hens aged 69–84 weeks were deprived of feed to induce a moult and orally infected on day 4 of the fast with 5×10^6 *S. enteritidis* bacteria, significantly more moulted hens shed *S. enteritidis* in their faeces than unmoulted infected hens on day 14 and day 21 post-infection (p.i.). Intestinal levels of *S. enteritidis* were increased 100–1000-fold in the moulted compared with unmoulted hens on day 7 and day 14 p.i. (Holt and Porter, 1992). *S. enteritidis* was transmitted more rapidly to unchallenged hens in cages adjacent to moulted infected hens than to those in cages adjacent to unmoulted infected hens (Holt, 1995). Short-term exposure to environmental stress, such as the introduction of young chickens in the same rearing room and the removal of feed and water for 2 days, resulted in an increase in the shedding rate of *S. enteritidis* by laying hens (Nakamura *et al.*, 1994).

SEWAGE. Chickens may become infected by the transmission of *Salmonella* from sewage via feral and domestic animals. Following the diagnosis of *S. enteritidis* PT4 infection in a commercial layer flock in southern California, effluent from a nearby sewage treatment plant was shown to contain the same phage type. *S. enteritidis* PT4 was isolated from the sewage effluent, from hens, from mice trapped in the hen-houses, and from cats and skunks on the premises (Kinde *et al.*, 1996a,b).

HOUSING. The manner in which chickens are kept may influence the percentage of infected hens within a flock. Investigation of an outbreak of *S. enteritidis* PT4 among 176,000 laying hens showed a lower prevalence among caged than among free-range hens (1.7 vs. 50%). The prevalence in culled hens kept in dirt-floor houses ranged from 14 to 42% (Kinde *et al.*, 1996a).

PENETRATION OF THE EGGSHELL. Contamination of the egg contents may occur by contamination of the eggshell with faeces from hens excreting *Salmonella* (Schaaf, 1936; Stokes *et al.*, 1956; Williams *et al.*, 1968; Borland, 1975; Timoney *et al.*, 1989). Hens may have only an enteric infection and the *Salmonella* in the faeces may penetrate the eggshell pores as the egg cools and before the establishment on its surface of the proteinaceous cuticular barrier, which prevents bacterial invasion of the egg (Stokes *et al.*, 1956; Board, 1966; Forsythe *et al.*, 1967; Williams *et al.*, 1968). Alternatively, faecal matter adherent to the shell may contaminate the egg contents when the eggs have cracks or when the eggs are broken open for the preparation of food products (Borland, 1975). There appears to be a link between the shedding of high numbers of *S. enteritidis* bacteria and contamination of eggs. When cloacal tissues were heavily contaminated with *S. enteritidis*, the eggs were culture-positive, whereas, if these tissues had a low rate of infection, the eggs were culture-negative (Keller *et al.*, 1995).

Vertical transmission of Salmonella *through eggs*
Transmission of *Salmonella* via hatching eggs may occur as a result of infection of the ovary and oviduct. For an oophoritis to occur, the bird must have experienced a systemic infection. The poultry-specific serovars *S. pullorum* and *S. gallinarum*, causing pullorum disease and fowl typhoid,

respectively, are the main serovars transmitted vertically. Other serovars that may cause a transovarian infection include *S. typhimurium*, *S. enteritidis*, *S. heidelberg* and *S. menston* (Schaaf, 1936; Gordon and Tucker, 1965; Snoeyenbos *et al.*, 1969; Hopper and Mawer, 1988; Cooper *et al.*, 1989; Humphrey *et al.*, 1989b; McIlroy *et al.*, 1989; Timoney *et al.*, 1989; Gast and Beard, 1990; Barnhart *et al.*, 1991; Hoop and Popischil, 1993; Corkish *et al.*, 1994; Keller *et al.*, 1995).

S. enteritidis strains not uncommonly colonize the reproductive tract, which results in eggs containing *Salmonella*. Examination of 1119 eggs of two small flocks of 35 egg-laying hens showed that the contents of 11 eggs contained *Salmonella* bacteria. The production of infected eggs was clustered, though intermittent. The positive eggs were produced by ten of the 35 hens (Humphrey *et al.*, 1989b). Inoculation of approximately 10^6 *S. enteritidis* into the crop of adult hens was followed by a bacteraemia with infection of many body sites, including peritoneum, ovules and oviduct, in the majority of the hens (Timoney *et al.*, 1989). The organisms were present in the yolk or albumen of eggs of about 10% of the hens shortly after infection and again 10 days later, which is evidence for egg or transovarian transmission of the infection. The finding that the albumen, but not yolk samples, of some of the eggs were positive suggests that some eggs became infected in the oviduct (Timoney *et al.*, 1989).

Examination of 37 laying hens from three small flocks with a naturally acquired infection showed that in six hens the ovaries and in ten hens the oviduct were colonized by *S. enteritidis* PT4. By immunohistochemical labelling, *S. enteritidis* was demonstrated in seven of eight culture-positive hens on the surface of and within the epithelial cells in the lumen and in the tuberine glands of the oviduct and intracellularly in the follicular epithelium of the ovary (Hoop and Pospischil, 1993). When adult laying hens were inoculated orally with 10^8 *S. enteritidis*, the microorganisms were isolated 2 days p.i. from the spleen, liver, heart, gall-bladder and intestinal tissues and from various sections of the ovary and oviduct (Keller *et al.*, 1995). Detection of microorganisms by immunohistochemical staining was rare for most tissues, despite their culture-positive status, but they could be detected in oviduct tissues associated with forming eggs, indicating a heavier colonization in the egg during its development. Forming eggs taken from the oviduct were culture-positive at a rate of c. 30%, while freshly laid eggs in the same experiment were positive at a rate of less than 0.6%, suggesting that forming eggs are colonized in the reproductive tract but that factors within the eggs significantly control the pathogen before the eggs are laid. When the cloacal tissues were heavily contaminated with *S. enteritidis* the eggs were culture-positive, whereas if these tissues had a low rate of infection the eggs were culture-negative. It was concluded that, prior to eggshell deposition, forming eggs are subject to descending infections from colonized ovarian tissue, lateral infection from colonized upper oviduct tissues and ascending infections from colonized vaginal and cloacal tissues (Keller *et al.*, 1995).

Examination of the attachment and invasion of chicken ovarian granulosa cells by *S. enteritidis* PT8 showed that the organism can invade and multiply in these cells. The bacteria were found, with or without a surrounding membrane, in the cytoplasm of granulosa cells. It was suggested that the granulosa cell layer of the pre-ovulatory follicles may be a preferred site for the colonization of the chicken ovaries by invasive strains of *S. enteritidis* (Thiagarajan *et al.*, 1994, 1996).

A study was conducted in which ten hens were taken from each of three flocks naturally infected with *S. enteritidis* and from one flock naturally infected with *S. typhimurium* (Cooper *et al.*, 1989). In the first flock, *Salmonella* could not be isolated by cloacal swab or by the culture of 78 eggs, whilst *S. enteritidis* PT4 was isolated on post-mortem examination from three birds, from the caeca, the oviduct and, or, the ovary. In the second flock, *S. enteritidis* was isolated by cloacal swab on one occasion and from a pooled sample of two eggs on one occasion from a total of 39 eggs, but it was not isolated post-mortem from any organ. In the third flock, *Salmonella* spp. could not be isolated by cloacal swab or by the culture of 47 eggs; however, three unshelled yolks taken from one bird yielded *S. enteritidis* PT1, four hens had congested ovaries, shrunken follicles and evidence of inspissation and six birds yielded *S. enteritidis* PT1 from the reproductive organs. In the fourth flock, infected with *S. typhimurium*, the organism could not be isolated from 32 eggs and no gross pathological changes were observed on post-mortem in any of the hens (Cooper *et al.*, 1989). Likewise, others observed ovarian infection, congestion of ovules, misshapen ovules and egg peritonitis in hens from a laying flock infected with *S. enteritidis* PT4 (Hopper and Mawer, 1988), and a diffuse yellow fibrinous peritonitis, internal laying of soft-shelled eggs, clots of inspissated yolk in the peritoneal cavity and/or the oviduct and shrunken thick-walled congested follicles containing coagulated yolk in hens of another layer flock infected with *S. enteritidis* PT4 (Read *et al.*, 1994).

Environmental Considerations

The faecal excretion of *Salmonella* by poultry, the transportation and disposal of slurry and manure from poultry houses and barns, the transportation of slaughter offal to rendering plants, the cross-contamination of rendered meat meal and other poultry and animal by-products by dust and contaminated conveyor belts in rendering departments of slaughtering plants and in feed mills, hoppers, bins, and trucks transporting feed to poultry barns all contribute to spreading *Salmonella* in the environment (further details will be found in Chapter 16). Pigeons, sparrows, other birds, rodents, cats, dogs and insects may be contaminated by contact with or the ingestion of spilled meat meal, feather meal and other animal by-products outside rendering departments at slaughtering plants and at poultry houses from conveyor belts, hoppers and open trucks. This may lead to contamination of effluents, surface waters, creeks, rivers, lakes, pastures and soil, to the colonization of birds, cattle, pigs, sheep, horses, rodents and other animals or to contamination of animal feeds or may contribute directly to the recolonization of farm animals (Tannock and Smith, 1971; Borland, 1975; Diesch, 1978; Morse, 1978; Oosterom, 1991; Kirkwood *et al.*, 1994).

Animal Infection

Infections caused by *S. gallinarum* and *S. pullorum*

S. pullorum and *S. gallinarum* are host-specific *Salmonella*, which cause pullorum disease and

fowl typhoid, respectively, diseases specific to chickens, turkeys and several other avian species (Klein, 1889; Rettger, 1900).

Fowl typhoid

The first described outbreak of fowl typhoid was characterized by high mortality, especially during the first 2 months of the outbreak. The chicken appeared normal until about 24–36 h before death. The disease symptoms began with diarrhoea and yellow-to-green droppings. The birds were very quiet but no somnolence occurred and they died a day after developing diarrhoea (Klein, 1889). Symptoms described by others include huddling, laboured breathing and gasping, a greenish diarrhoea and mortality commencing in day-old to 5-day-old chicks and seen also in 21- and 28-day-old chicks (Kaushik *et al.*, 1986).

On post-mortem examination pathologically, the liver and spleen were enlarged, the serosa and mucosa of the intestines were engorged with blood and the rectum contained yellow, fluid faeces (Klein, 1889). Others noted an enlarged liver with a greenish-bronze colour and whitish necrotic foci distributed uniformly on its surface; a much enlarged spleen with embedded whitish necrotic foci; catarrhal enteritis; an enlarged heart with small pinhead-sized white foci on the myocardium; haemorrhages in the pericardial fat, the endocardium and the proventriculus; severely congested lungs; and enlarged and congested kidneys (Chishti *et al.*, 1985; Kaushik *et al.*, 1986). Histologically, the liver showed congestion, fatty degeneration and diffuse parenchymatous hepatitis, with focal areas of necrosis and mononuclear infiltration in the periportal areas; severe hyperplasia of the reticuloendothelial cells in the spleen, occasionally accompanied by areas of necrosis; disintegration of muscle fibres and infiltration of mononuclear cells in the myocardium; congestion and infiltration by mononuclear cells subepithelially in the mesobronchus of the lungs; and congestion and perivascular infiltration of mononuclear cells in the kidneys (Kaushik *et al.*, 1986).

The bacteria isolated from the liver grew slowly on nutrient agar and were non-motile. Subcutaneous inoculation of blood or spleen tissue of sick and dead chickens caused death in chickens but not in pigeons and rabbits (Klein, 1889).

Outbreaks of fowl typhoid occurred increasingly in chickens and turkeys in the USA during the 1939–1945 period and were most prevalent during the summer and early autumn. The species affected included chickens, turkeys and guinea-fowl. Birds of all ages and breeds were affected. Fowl typhoid and pullorum disease often tended not to occur simultaneously on the same farm. The disease appeared to persist on farms from year to year (Moore, 1946). Possible reasons for the persistence of *S. gallinarum* in chicken barns may be its longevity in environmental sources, such as fluff, chicken mash and drinking water and in frozen carcasses of birds that had died but had not been destroyed (Orr and Moore, 1953).

Pullorum disease

The first occurrence of the disease was described as 'a peculiar epidemic which occurred among young chickens two to three weeks old'. Fourteen of 17 chicks purchased became sick and died within 14 days. The first signs of disease were 'loss of appetite, the feathers then became rough, and diarrhoea prevailed'. The chicks remained standing in one place, refused to eat and later were unable to stand and died. The remaining chicks recovered from the disease, but remained stunted in growth for 2 or more months (Rettger, 1900). Chicks were sometimes found dead within a day or two after they had hatched. It was not unusual for poultry raisers to lose from 50 to 70% of the chicks under 4 weeks to 'fatal septicaemia' or 'white diarrhoea' (Rettger, 1909). Less characteristics lesions, such as runting, swollen joints and lameness, were found in an outbreak of pullorum disease in 4-week-old roasters (Salem *et al.*, 1992). A recent outbreak of atypical pullorum disease in two backyard chicken flocks was characterized by the sudden death of many adult birds in one flock and chronic wasting with high mortality in a second flock (Erbeck *et al.*, 1993). Pullorum disease tends to occur in younger chickens than fowl typhoid. In a study of 200 chickens with grossly enlarged livers and spleens, the isolation rates of *S. pullorum* were 38.9%, 2.3%, 7.9% and 13.0% in chickens below 4 weeks of age, of 4–8 weeks of age, of 9–24 weeks of age and above 24 weeks of age, respectively, whereas *S. gallinarum* isolations were made from 1.1%, 31.8%, 10.5% and 4.3% of chickens of these age-groups (Chishti *et al.*, 1985).

Pullorum disease in laying hens caused reduced egg production, fertility and hatchability (Bullis, 1977a). *S. pullorum* was frequently carried in the ovaries of hens, and, while a large percentage of the ova from infected ovaries never matured, such birds laid some fully developed eggs capable of hatching, which contained the organism, from which infected chicks developed. These, in turn, infected other chicks by their droppings. In the majority of cases the presence of *S. pullorum* in adult hens produced no noticeable symptoms; occasionally, the organism produced a fatal septicaemia in a few birds. The disease was most prevalent during the 4 months from March to June (Hewitt, 1928). Chilling of chicks during transport or crowding, overheating or chilling during the critical first week of brooding aggravated pullorum disease in young chicks. Some breeds of chickens, especially the White Leghorn, were more resistant to pullorum disease than others (Hutt and Scholes, 1941). The genetic resistance to *S. pullorum* appeared to be related to a superior control of the thermo-regulatory mechanism in the White Leghorn chicks (Scholes and Hutt, 1942).

The whole-blood test, employing killed *S. pullorum* bacteria stained with crystal violet as the antigen (Schaffer *et al.*, 1931), was used widely and successfully in chickens (see Chapter 24 for details of this and other tests). Examination of the ability of the standard, rapid, whole-blood, plate agglutination test and serum tube agglutination test antigens to detect infection in Leghorn hens inoculated with large oral doses of six recent *S. pullorum* isolates (three standard, two intermediate and one variant strain) showed that the traditional NPIP serological strategy (relying principally on the whole-blood plate test as a screening tool) remains an effective and economical method of detecting infection of chickens with current field strains of *S. pullorum* (Gast, 1997).

In recent years, most of the pullorum disease outbreaks in the USA have been reported in small or backyard poultry flocks, which may serve as potential reservoirs for disease transmission to commercial flocks (Erbeck *et al.*, 1993). A large outbreak of pullorum disease occurred among Delaware roasters. The outbreak involved 19 breeder flocks and more than 261 grow-out premises in five states in the USA (Johnson *et al.*, 1992; Salem *et al.*, 1992). Twenty-two parent (multiplier) breeder flocks became infected. The transmission occurred vertically through the egg and horizontally by contact in the hatcheries and by placement of chicks on contaminated litter (Johnson *et al.*, 1992).

In addition to chickens and turkeys, many other avian species are susceptible to pullorum disease, although often to a lesser degree. Oral and intravenous infection of northern bobwhite quail and mallard ducks revealed that the mortality in the former ranged from 65 to 100%, whereas none of the latter died or exhibited any signs of morbidity (Buchholz and Fairbrother, 1992).

At post-mortem examination, chickens affected by pullorum disease were emaciated and anaemic, the livers pale, the crops empty, the intestines pale and almost empty, the caeca nearly empty or filled with a semi-solid or rather firm cheesy matter and the yolk-sacs unabsorbed (Rettger, 1900, 1909). Severe articular and peri-articular swelling, especially of the hock joints and also the wing joints, was observed in an outbreak of pullorum disease in 4-week-old roasters (Salem *et al.*, 1992). The livers were swollen and showed multiple small white foci and petechial haemorrhages. Hydropericardium and turbid pericardial fluid, and white nodular lesions that looked like tumours were found in the heart and gizzard. The spleens were mottled or pale and brown nodules were found in the lungs of some of the birds. A white discoloration of the gizzard muscle was observed.

Histologically, separation of muscle fibres and infiltrations of macrophages, lymphocytes, plasma cells and heterophils were seen in the heart and the gizzard muscles. Lesions in the liver consisted of hepatocellular coagulation necrosis, small foci of mixed heterophil–mononuclear cells, enlarged portal lymphoid nodules and foci of extramedullary granulocytopoiesis. An acute, severe lymphocytic depletion was seen in the spleen, and acute cortical lymphocyte loss and medullary expansion were observed in multiple thymic lobes. The joint lesions consisted of acute synovitis and extensive infiltration of the synovial membrane by plasma cells. A broncho-pneumonia, with consolidation and filling of tertiary bronchi with mononuclear cells and heterophils, was observed in some of the roasters (Salem *et al.*, 1992).

Pure growth of a bacillus was obtained most often from hepatic blood. The organism was an 'actively motile, aerobic and facultative anaerobic bacillus'. Subcutaneous inoculation of chicks with the organism caused a septicaemic disease characterized by a pronounced diarrhoea (Rettger, 1900). Although Rettger (1900) described the organism as an 'actively motile' bacillus, *S. pullorum* has long been considered to lack flagella and found to be non-motile. However, recently it has been shown that *S. pullorum* reacted with *Salmonella* anti-flagellar antisera (Ibrahim *et al.*, 1986), genes encoding flagella in *S. pullorum* have been identified (Kilger and Grimont, 1993) and motility could be induced and observed in 39 of 44 *S. pullorum* strains examined (Holt and Chaubal, 1997).

Infections caused by other *Salmonella* serovars

Poultry are commonly infected with a wide variety of *Salmonella* serovars. The infection is mostly confined to the gastrointestinal tract and the birds often excrete *Salmonella* in their faeces.

Pathogenesis

Studies on the pathogenesis of experimental *S. typhimurium* infection in chickens showed that, after invasion, the bacteria multiplied in liver and spleen and spread to other organs, producing a systemic infection. The cause of death was probably a combination of anorexia and dehydration resulting from general malaise and diarrhoea. Invasiveness was the virulence determinant of overriding importance. There was no correlation between route of infection and virulence (Barrow *et al.*, 1987). Following oral inoculation of *S. enteritidis* in 1-day-old chicks, the organisms spread rapidly from the caeca and crop to internal tissues. The incidence of invasion decreased rapidly with age. Intraluminal phagocytosis of *Salmonella* bacteria was observed occasionally. A few organisms were seen in close association with the epithelial brush border in the lower ileal lumina. In 1-day-old birds, the mode of entry appeared to be translocation and an early mucosal heterophil response was noted in the intestines but not in the crop. The invading organisms were usually single and, on entering the tissues, were often surrounded by a halo. Phagocytosis was not observed in the epithelium but was seen occasionally in the lamina propria of the ileocaecal junction and of the caeca. Intratissue *Salmonella* bacteria were seen occasionally in the caeca of 2-week-old birds but were not detected in adult birds (Turnbull and Snoeyenbos, 1973). The occurrence of salmonellosis is influenced by factors including the age of the bird, the infectious dose, the route of infection, the invasiveness of the strain or serovar and the breed of chicken. Many factors influence the occurrence of salmonellosis in the domestic fowl. These include the following.

AGE AND DOSE. Young-age chicks and poults are highly susceptible to infection by *Salmonella* and many shed the bacteria with the faeces. A definite correlation exists between the age of the chicken and the number of organisms required to induce infection detectable by shedding. Experimental oral infection of chickens aged 2 days and 1, 2, 4 and 8 weeks with either 10^2, 10^4, 10^6, 10^8 or 10^{10} *S. typhimurium* bacteria showed that 10^2 *S. typhimurium* bacteria induced infection in all the 2-day-old chicks, two-thirds of the 1- and 2-week-old birds and none of the 4- and 8-week-old birds. Similarly, 10^4 bacteria induced infection in all 2-day-old and 1-week-old chicks but in only about 25% of chicks 2, 4, and 8 weeks old. Even 10^6 bacteria were insufficient to induce infection in half of the 8-week-old chickens and a quarter of the 4-week-olds. Mortality occurred only between 2 and 12 days p.i. in chicks infected when 2 days old (Sadler *et al.*, 1969).

Others observed that oral inoculation of newly hatched chicks with fewer than ten organisms of *S. enteritidis* was followed by multiplication in the caecal contents. *S. enteritidis* strains multiplied only *in vitro* in samples taken from the ileum and duodenum, irrespective of age, but multiplied in caecal samples from newly hatched chicks. Invasion from the intestines by *S. enteritidis* strains was both age- and dose-dependent. The numbers of *S. enteritidis* organisms recovered from the spleen and liver were similar when chicks were dosed orally at 1 day old and at 3 days old. However, when chicks were dosed at 5 or 6 days old, the invasion and colonization of livers and spleens were dose-dependent, with an inoculum of at least 10^9 colony-forming units (cfu) being required to ensure invasion (Cooper *et al.*, 1994a). The age of the hens at the time of infection with *S. enteritidis* influenced intestinal colonization of the hens, egg production, isolation rates of *S. enteritidis* from eggshells and isolation rates of *S. enteritidis* from the egg albumen. Hens of 62 weeks of age inoculated orally with 10^9 *S. enteritidis* of a PT13a strain were more often colonized intestinally, produced fewer eggs and produced a higher percentage of eggs with contamination of the eggshells and the egg albumen than hens of 37 and 27 weeks of age (Gast and Beard, 1990). Hens at 20 weeks of age were less susceptible to *S. enteritidis* infection than mature birds at 55 weeks of age. Hens at 20 weeks of age showed no clinical signs, while hens at 60 weeks experienced an acute septicaemia and associated mortality (Humphrey *et al.*, 1991a). The period during which specific-pathogen-free (SPF) hens infected by direct inoculation into their crop with either 10^3, 10^6 or 10^8 cells of *S. enteritidis* PT4 excreted the organism in faeces was closely related to the size of the inoculum, with the birds excreting for mean periods of 3, 16 and 37 days, respectively (Humphrey *et al.*, 1991b). After oral inoculation of food-poisoning *Salmonella* serovars, the number of chickens excreting *Salmonella* bacteria in their faeces gradually declined over a period of at least 4 weeks. In contrast, serovars adapted to other animal hosts, such as *S. choleraesuis* and *S. abortusovis*, were excreted for no longer than a few days (Barrow *et al.*, 1988).

ROUTE OF INFECTION. Chicks are more susceptible to *Salmonella* infection by inhalation and parenteral routes than via the oral route (Barrow *et al.*, 1987; Poppe and Gyles, 1987; Baskerville *et al.*, 1992; Poppe *et al.*, 1993b; Cooper *et al.*, 1994b).

SEROVAR AND STRAIN. Colonization of the alimentary tract of chickens depends on the *Salmonella* serovar. Following experimental challenge, *S. typhimurium* (strain F98) and *S. menston* were excreted in the faeces by chickens for weeks, whereas *S. choleraesuis*, a serovar host-specific for pigs, was eliminated in a short time (Smith and Tucker, 1980). After inoculation of 1-day-old chicks, the numbers and distribution of *S. typhimurium* F98, *S. menston* and *S. choleraesuis* were similar for up to 5 days. The highest counts were found in the caeca, cloaca and distal small intestine, with somewhat smaller numbers in the crop. By 7 days p.i., *S. choleraesuis* was not recovered from the crop and, by 14 days p.i., it had been completely cleared from the alimentary tract, whereas at this time the other two serovars were still present in considerable numbers in the lower part of the alimentary tract. The three serovars behaved differently from each other in 3-week-old chickens. *S. choleraesuis* was isolated once from the caeca at 1 day p.i. but not thereafter. Both *S. menston* and *S. typhimurium* had almost ceased to be excreted by 14 days p.i. and only small numbers of *S. typhimurium* were present in the caeca at 21 days p.i. *S. menston* but not *S. typhimurium* was isolated from the crop for up to 7 days p.i. (Barrow *et al.*, 1988). When chicks 1 and 7 days old were given feed contaminated artificially with 30–200 organisms g^{-1} of either *S. typhimurium* or *S. kedougou*, *S. typhimurium* was recovered from caeca, lungs, liver, spleen and kidneys but *S. kedougou* only from the caeca. *S. kedougou* proved a more effective intestinal colonizer than *S. typhimurium* in the young chicks, while the reverse was true in the older birds (Xu *et al.*, 1988).

INVASIVENESS OF *SALMONELLA* STRAINS. Strains of *S. typhimurium* differed markedly in their invasive abilities (Barrow *et al.*, 1987). The invasive ability of *S. enteritidis* appeared to be not PT- but strain-related (Timoney *et al.*, 1989; Poppe *et al.*, 1993b; Gast and Benson, 1996) and major differences in invasiveness for adult hens were seen among *S. enteritidis* PT8 strains. One PT8 strain behaved like an invasive PT4 strain isolated originally from the pericardial fluid of a chicken, whereas other strains were shed in the faeces but did not cause invasive bacteraemia (Timoney *et al.*, 1989). One strain of *S. enteritidis* PT4 was more virulent than another PT4 strain for orally inoculated 1-day-old chicks and laying hens (Poppe *et al.*, 1993b). Similarly, inoculation of groups of 5-day-old Leghorn chicks with a range of oral doses of three PT4 and three other PTs showed that, whilst some significant differences were observed between individual *S. enteritidis* PT4 strains in the frequencies at which they colonized the intestinal tract and invaded to reach the spleen, no consistent overall pattern differentiated PT4 isolates from isolates of other PTs (Gast and Benson, 1996).

Other studies have shown that colonies of *S. enteritidis* PT13a with an unusual wrinkled morphology were the result of the production of lipopolysaccharide (LPS) with a high-molecular-weight O antigen and an elevated O-antigen-to-core (O/C) ratio, and that such colonies were more virulent for intraperitoneally (i.p.) inoculated 5-day-old chicks, in that they yielded higher numbers g^{-1} of spleen 3 days p.i and caused a higher percentage of contaminated eggs after intravenous (i.v.) inoculation of laying hens than non-wrinkled colonies with a lower O/C ratio (Petter, 1993; Guard-Petter *et al.*, 1995). An *S. enteritidis* PT4 strain producing colonies with a wrinkled appearance (Cox, 1996) was more invasive for the reproductive and other tissues of laying hens and more tolerant to heat, acid and hydrogen peroxide than non-wrinkled colonies (Humphrey *et al.*, 1996). When day-old Leghorn chicks were inoculated orally with 10^6 of each of two *S. enteritidis* PT13 strains – one strain derived from the ovary of a hen from a laying flock with evidence of environmental contamination with *S. enteritidis* infection and the other isolated from the heart blood of a septicaemic broiler – the ovarian isolate could not be reisolated from the liver, spleen and caecal contents of the chicks, but the blood isolate was recultured from 4/10 livers, 4/10 spleens and 10/10 caecal contents of the asymptomatic chicks upon euthanasia and necropsy of the chicks at 14 days p.i. (C. Poppe, unpublished data). The two strains were indistinguishable: they were of the same PT, possessed a 36 MDa plasmid, had outer-membrane proteins (OMPs) of 45, 42, 40 and 37 kDa and both were sensitive to antimicrobial agents (Poppe *et al.*, 1993a). However, the ovarian isolate formed an entire and smooth colony, whereas the blood isolate developed a corrugated colony appearance after 2 days of growth at room temperature on Luria-Bertani broth (LB) agar (C. Poppe, unpublished data).

Others, however, have reported that strains of *S. enteritidis* PT4 were more invasive for young chicks than strains of PT7, 8 and 13a, and suggested that the increased invasiveness of PT4 may be one of the factors that contributed to the establishment of *S. enteritidis* PT4 in the UK (Hinton *et al.*, 1990a). The same authors also found that more recent isolates of *S. enteritidis* PT4 were more invasive than strains isolated in previous years and suggested that recent isolates of PT4 may have an enhanced virulence for chickens (Hinton *et al.*, 1990b).

BREED OF CHICKEN. A survey of inbred and partially inbred lines of chickens showed pronounced difference in mortality following challenge of newly hatched chicks with S. typhimurium (Barrow et al., 1987; Bumstead and Barrow, 1988). Two lines of chickens (Light Sussex and White Leghorn line W) were highly resistant to challenge, two lines (Rhode Island Red and Brown Leghorn) were moderately susceptible and one line (White Leghorn line C) was highly susceptible (Bumstead and Barrow, 1988). This difference in susceptibility was observed with a range of five strains of S. typhimurium of differing degrees of virulence, following both oral and intramuscular challenge. The inheritance of resistance was studied in detail by examining a series of crosses between a susceptible White Leghorn line C and a resistant White Leghorn line W. The pattern of mortality in crosses and back-crosses between these lines indicated that resistance is dominant and that it was consistent with the inheritance of a dominant autosomal resistance gene. There was no evidence of association with the major histocompatability complex. Later, it was shown that lines that had previously been shown to be resistant to S. typhimurium were also resistant to S. gallinarum, S. pullorum and S. enteritidis, and lines susceptible to S. typhimurium were also more susceptible to the other serovars (Bumstead and Barrow, 1993).

Diagnosis

SYMPTOMS. Salmonellosis in broilers due to S. typhimurium infection has been characterized by growth retardation, blindness, twisted necks, lameness and mortality and cull rates that varied between 1.7% to 10.6% in flocks during the first 2 weeks of age (Padron, 1990). Mortality rates in poultry infected with S. enteritidis PT4 were 2% in broilers during the first 48 h of life, with a cumulative mortality and morbidity rate of 6% and 20%, respectively, at 5 days of age (McIlroy et al., 1989). Affected young chicks may exhibit symptoms including anorexia, adipsia, depression, ruffled feathers, huddling together in groups, reluctance to move, drowsiness, somnolence, dehydration, white diarrhoea and stained or pasted vents (Schaaf, 1936; McIlroy et al., 1989; Baskerville et al., 1992). During the second week of life, chicks infected with S. enteritidis PT4 failed to grow and had a stunted appearance

(O'Brien, 1988). Laying flocks are often clinically normal, despite the isolation of S. enteritidis from caecal droppings, dust and litter (Hinton et al., 1989; McIlroy et al., 1989). However, clinical signs are sometimes found in laying hens. Increased mortality (1.6% month^{-1}) and decreased egg production (8% over a period of 6 weeks) were found in a large commercial laying flock infected with S. enteritidis PT4 (Kinde et al., 1996a).

SEROLOGY. Serological tests do not reliably detect birds with intestinal infection with Salmonella. Agglutination tests with S. typhimurium and S. pullorum antigens did not dependably detect infection in 4-week-old White Leghorn chickens infected orally with 10^6 S. typhimurium (Olesiuk et al., 1969). Similarly, 6-month-old SPF Brown Leghorn cockerels inoculated orally with 10^{10} S. typhimurium were colonized 1 day p.i., but a rising titre against the H antigen of a monophasic S. typhimurium strain was only noted during 3–6 days p.i. and reached a peak at 9 days p.i., after which it persisted at moderate level until the end of the experiment 45 days p.i., at which time none of the birds were still colonized with Salmonella. Development of a titre depends on the occurrence of a systemic infection (Brown et al., 1975). A similar study of cockerels infected with S. infantis showed that the serological response was lower, fell more quickly and frequently became negative 3–6 weeks p.i. (Brown et al., 1976).

While examining chickens from flocks naturally infected with S. enteritidis, serological results could not be related to the isolation of Salmonella. In individual birds, the rapid slide test, using S. pullorum-derived O antigen, and the tube agglutination tests specific for somatic group D antigen and S. enteritidis PT4-derived flagellar antigen could not be relied upon to detect infection. The microantiglobulin test, employing intravitally stained S. enteritidis PT4 antigen, and the enzyme-linked immunosorbent assay (ELISA) method, using LPS antigen from S. enteritidis, were more sensitive than the slide and tube agglutination tests, detected more infected birds that were negative by the rapid slide and tube agglutination tests and showed high titres in some birds from which Salmonella could not be isolated post-mortem. Sera from two flocks that had a history of natural *S. enteritidis* infection were evaluated by all the tests; evidence of infection was found with the microantiglobulin and ELISA tests but not with the other tests (Cooper *et al.*, 1989). Examination of blood samples from 21 *Salmonella*-free hens and ten hens naturally infected with *S. enteritidis* showed that none of the 21 *Salmonella*-free hens tested gave a positive result but three of the ten (30%) infected birds tested were positive, which suggested that the pullorum test, which uses a common O antigen with *S. enteritidis*, appears to detect only a low proportion of infected birds (Hopper and Mawer, 1988). The ability to raise an immune response is influenced by the age of the hens. Hens of 20 weeks infected orally with *S. enteritidis* developed high titres of antibodies of the immunoglobulin M (IgM) class, while those that were 1 year old at infection developed relatively little antibody (Humphrey *et al.*, 1991a). The production of antibodies has been linked to the size of the infecting dose. SPF hens infected by introduction into the crop of either 10^3, 10^6 or 10^8 cells of *S. enteritidis* PT4 produced the lowest level of IgG and IgM after receiving 10^3 organisms, intermediate levels after receiving 10^6 organisms and the highest levels of antibodies after receiving 10^8 bacteria (Humphrey *et al.*, 1991b). Further information on serology will be found in Chapter 24.

CULTURE. Numerous methods for the isolation of *Salmonella* have been described. Cultural methods have recently been reviewed (Barrow, 1995; Tietjen and Fung, 1995) and Chapter 21 provides a comprehensive review of the different media and isolation techniques.

Pathological findings

Post-mortem lesions seen in salmonellosis may consist of dehydration, emaciation, an unresorbed or poorly resorbed yolk-sac, infection of the yolk-sac and necrotic debris in the yolk-sac, air-sacculitis, splenomegaly, hepatomegaly, necrotic foci and petechiae in the liver and spleen, pseudomembranous perihepatitis and peritonitis, involving the serosal surfaces of the intestines, ovary and oviduct, enteritis, characterized by a watery intestinal contents and reddened areas of the mucosal surface of the duodenum, ileum and colon, typhlitis, with or without bloodstained or inspissated caecal cores,

diarrhoea, mild to severe mucopurulent or fibrinous pericarditis and dilatation of the pericardial sac by a thin bloodstained or a thick purulent fluid (Schaaf, 1936; Turnbull and Snoeyenbos, 1973; Brown *et al.*, 1975; O'Brien, 1988; McIllroy *et al.*, 1989; Rampling *et al.*, 1989; Baskerville *et al.*, 1992; Gorham *et al.*, 1994). Characteristic pathological findings in 2-day-old chicks inoculated orally with 10^2, 10^4, 10^6, 10^8 or 10^{10} *S. typhimurium* included omphalitis, necrotic foci in the liver, swollen kidneys and fibrinous peritonitis. *Salmonella* bacteria were isolated from the livers, yolk-sacs, kidneys and pericardial fluids, as well as intestinal contents (Sadler *et al.*, 1969). Other pathological findings in 1- to 2-week-old broilers naturally infected with *S. typhimurium* were hypopyon, panophthalmitis and a purulent arthritis (Padron, 1990). Histopathological findings in hens infected with *S. enteritidis* PT4 were foci of necrosis and infiltration by heterophils and lymphocytes in the liver parenchyma and around the portal tracts, increased lymphoid tissue in the lamina propria and submucosa of the duodenum, ileum, caeca and colon, fusion and distortion of the intestinal villi and lymphocytic infiltration in the oviduct and in the tubules and around the larger collecting ducts in the kidneys (Baskerville *et al.*, 1992).

The caeca are the site within the intestinal tract of infected chickens that most commonly contain *Salmonella* (Fanelli *et al.*, 1971; Brown *et al.*, 1975, 1976; Barrow *et al.*, 1988; Xu *et al.*, 1988). Examination of 12 different sites of the intestinal tract of chickens inoculated orally at 4 weeks of age with 10^8 *S. typhimurium* and *S. infantis* bacteria and examined at 0, 1, 2, 3, 6, 9, 13 and 17 days p.i. showed that 39.6% were positive by cloacal swab taken just prior to necropsy, 46.7% by caecal tonsil culture, 58.2% by cloacal-content culture and 85.2% by caecal-content culture. There were no significant differences in localization between the two serovars of *Salmonella* (Fanelli *et al.*, 1971). *Salmonella* were more frequently isolated from the caeca than from any other organs, tissues or other parts of the gastrointestinal tract of 6-month-old SPF Brown Leghorn cockerels infected orally with 10^{10} *S. typhimurium*. The other organs more frequently positive in the birds were the spleen, liver and kidneys, followed by less frequent isolation from lungs and heart (Brown *et al.*, 1975).

Treatment

Many antibiotics have been used either singly, in combination with other antimicrobial agents or in conjunction with the administration of caecal competitive exclusion (CE) flora from adult birds to prevent *Salmonella* infection or to treat chickens with clinical salmonellosis (Seuna *et al.*, 1985). However, administration of antimicrobials to chickens is often followed by the development of resistance and prolonged excretion of *Salmonella* (Barrow, 1989; Manning *et al.*, 1994). Gentamicin, an aminoglycoside, and enrofloxacin, a fluoroquinolone, have been used to eliminate *Salmonella* from layer-type hatching eggs artificially infected with *S. enteritidis*. Eggs, prewarmed at 37°C, were immersed for 5 min in a cold (4–8°C) broth culture containing 10^2 cfu ml^{-1} of *S. enteritidis* PT4. The artificially infected hatching eggs were treated with the pressure-differential-dipping (PDD) method, using 1500 p.p.m. gentamicin sulphate or 600 p.p.m. of enrofloxacin for 10 min and 5 min at reduced pressure. The effect of the treatment of the hatching eggs was measured by examining the eggs and newly hatched chicks for infection with *S. enteritidis* PT4. Treatment for 10 min in the gentamicin solution eliminated *S. enteritidis* from the eggs and the hatched chicks, whereas 1% were infected if the treatment lasted for only 5 min. When the eggs were exposed for 10 min to enrofloxacin, 1% of the eggs and hatching chicks were infected with *S. enteritidis* PT4, whereas treatment with enrofloxacin for 5 min resulted in an infection rate of 4.1% of the eggs and chicks (Hafez *et al.*, 1992). When groups of chicks and turkey poults were infected orally with various doses of *S. typhimurium* var. *copenhagen* and were treated for 5–10 days *per os* with water containing enrofloxacin (dosage 50 p.p.m. of the aqueous solution), the *Salmonella* were eliminated from the faeces for a short period of time, after which they were again excreted, albeit at lower numbers (Guillot and Millemann, 1992). These results are reminiscent of those in humans, where treatment with ciprofloxacin failed to completely curtail excretion of *Salmonella* (Neill *et al.*, 1991).

In The Netherlands, an *S. enteritidis* eradication programme in poultry breeder flocks was carried out with elimination by slaughter of élite, grandparent and other breeder flocks that were positive for *S. enteritidis* during the period 1989–1992. This was followed by a control programme, which consisted of treatment with enrofloxacin, followed by two treatments with CE flora (Nurmi and Rantala, 1973). Four *S. enteritidis* PT4-positive flocks (in total approximately 100,000 birds) became serologically negative after treatment (Edel, 1994).

Prevention

COMPETITIVE EXCLUSION. Nurmi and Rantala (1973) demonstrated that oral administration of caecal bacterial flora from adult birds to newly hatched chickens increased the chick's resistance to *Salmonella* infection. Evaluations under field conditions were carried out in several countries; CE treatment of 88 broiler units in Germany showed that 42% of the 38 treated flocks were colonized with *Salmonella* bacteria, whereas 48% of the 50 untreated flocks were colonized (Hüttner *et al.*, 1981). Trials in Sweden, involving 2.86 million broilers, showed that 0.65% of the treated broilers were colonized vs. 1.5% of the untreated broilers (Wierup *et al.*, 1987), whereas trials in The Netherlands with 284 flocks consisting of 8 million broilers showed that 14.7% of the treated flocks and 0.9% of the treated broilers were colonized with *Salmonella* bacteria vs. 24.1% of untreated flocks and 3.5% of untreated broilers (Goren *et al.*, 1988). However, examination of faeces collected from the broiler transport crates and skin samples from the broilers after slaughter showed that the differences between treated and untreated flocks and broilers were minimal (Goren *et al.*, 1988). Studies on the control of *S. enteritidis* colonization in broiler and Leghorn chicks with dietary lactose and a defined caecal flora maintained as a continuous-flow (CF) culture showed that treatment with defined cultures of caecal and dietary lactose had an increased protective effect when compared with treatment with either caecal culture or dietary lactose alone (Corrier *et al.*, 1992). Further information will be found in Chapter 18.

VACCINES. Live, attenuated *S. enteritidis*, *S. gallinarum* and *S. typhimurium* strains and bacterins have been used with the aim of reducing the colonization of liver, spleen, ovaries and caeca in chickens and lowering the shedding of *S. enteritidis* in faeces and eggs. Barrow *et al.* (1990a, 1991) vaccinated laying hens of 24 weeks of age twice, separated by a 2-week interval, with spectinomycin-resistant mutants of either the *S. gallinarum* 9R mutant or a rough *aroA* insertion mutant of an *S. enteritidis* PT4 strain. Two weeks after the second vaccination, the immunized and a control group of non-immunized birds were challenged with a nalidixic acid-resistant mutant of a fully virulent *S. enteritidis* PT4 strain. Immunization with the 9R strain produced a marked reduction in the number of isolations of the challenge strain from a number of organs, including the ovaries. In contrast, immunization with the rough, *aroA* strain produced little change in the isolation rate from the ovaries and small reductions in isolation rates from the livers and spleens. Both vaccine strains produced a reduction in the number of isolations from laid eggs. The 9R strain was isolated from the ovaries throughout the period of examination, whereas the *aroA* strain was not. The 9R strain was isolated from one of 473 eggs before challenge, whereas the *aroA* strain was not isolated before challenge from any of the 492 eggs examined. Sera taken from the birds immediately prior to challenge were examined by slide agglutination, using live *S. enteritidis* cells. Sera from birds vaccinated with the 9R strain contained agglutinins, whereas sera from most of the chickens immunized with the *aroA* mutant did not (Barrow *et al.*, 1991). Oral vaccination of newly hatched chickens with 10^5 or 10^9 cfu of a genetically defined *S. enteritidis aroA* strain and i.v. challenge with 10^8 cfu of a Nal^R *S. enteritidis* when the chickens were 8 weeks old showed a reduction of colonization of spleens, livers and caeca of vaccinated chickens compared with unvaccinated controls. In another experiment, two groups of newly hatched female chicks were vaccinated orally with 10^9 cfu of the *aroA* mutant, and one group was revaccinated intramuscularly (i.m.) with 10^9 cfu at 16 weeks of age. When challenged i.v. with 10^9 cfu of a Nal^R *S. enteritidis* at 23 weeks of age, there was a reduction in the colonization of spleens, livers, ovaries and caeca of the vaccinated chickens compared with unvaccinated controls. Inclusion of the i.m. booster gave increased protection to the ovary, although the vaccine strain was isolated on one occasion from a batch of eggs laid at 20 weeks old (Cooper *et al.*, 1994b).

Another approach was the use of a stable, live, avirulent, genetically modified Δ*cya* Δ*crp S.*

typhimurium vaccine strain, χ3985, in chickens for the control of *Salmonella* infection. Oral vaccination of chickens at 1 and at 14 days of age with 10^8 cfu of χ3985 protected against invasion of spleen, ovary and bursa of Fabricius and colonization of the ileum and caeca in chickens challenged with 10^6 cfu of virulent homologous *Salmonella* strains from the serogroup B (to which *S. typhimurium* belongs). Chickens challenged with heterologous *Salmonella* strains from the serogroups C, D and E were protected against visceral invasions of spleen and ovary, while invasion of the bursa of Fabricius and colonization of ileum and caeca were reduced in vaccinated chickens. Oral vaccination at 2 and at 4 weeks of age induced an excellent protection against challenge with virulent serogroup B *Salmonella* serovars and very good protection against challenge with serogroup D or E *Salmonella* serovars, while protection against challenge with group C *Salmonella* serovars was marginal but significant. Vaccination at 2 and 4 weeks of age also protected vaccinated chickens against challenge with 10^8 cfu of highly invasive *S. typhimurium* or *S. enteritidis* strains. Chickens from vaccinated hens had significantly higher antibody responses than did the progeny of non-vaccinated hens after oral infection with *Salmonella* strains (Hassan and Curtiss, 1994, 1996). Killed *Salmonella* vaccines have not produced convincing levels of protection against wild-type *Salmonella* challenge (Truscott, 1981; Barrow *et al.*, 1990b; Gast *et al.*, 1992, 1993), although others have seen protection when chickens were challenged via the i.v. or i.m. route (Timms *et al.*, 1990). Further information on vaccines will be found in Chapter 19.

Public Health Aspects

Sources of infection

Many of the human *S. enteritidis* infections have been traced to contaminated eggs and to the laying hens at the farm that supplied the eggs (Telzak *et al.*, 1990; CDC, 1992; van de Giessen *et al.*, 1992; Altekruse *et al.*, 1993; Henzler *et al.*, 1994). Shell eggs, scrambled eggs, soft-boiled eggs, lightly cooked eggs, lightly cooked omelettes, food products containing raw or partly cooked eggs, including mayonnaise, sauce tartare, egg-nog, milk shakes, mousses, egg sandwiches, dishes containing raw egg-white, ice-cream containing uncooked eggs and poultry meat have all been implicated in outbreaks of S. enteritidis infection (Anon., 1988; Coyle et al., 1988; Humphrey et al., 1988; Paul and Batchelor, 1988; Perales and Audicana, 1988; Cowden et al., 1989; Mawer et al., 1989; Stevens et al., 1989; Hennessy et al., 1996).

The largest nosocomial infection in the USA occurred in 1987 in a New York City acute and long-term care hospital, where 404 of 965 patients (42%) were affected and nine patients died. Pooled batches of eggs from the hospital kitchen tested positive for S. enteritidis PT8, as did 383 of 555 (69%) of ovaries from hens at the farm from which the eggs were obtained. The eggs had been used to prepare mayonnaise, which was used in turn to prepare a tuna–macaroni salad (Telzak et al., 1990). Large outbreaks of S. enteritidis infection have occurred rarely in Canada. A notable exception was an outbreak of 95 cases of infection with S. enteritidis PT13, which occurred among patients and staff of a regional hospital in Owen Sound, Ontario. A mixer used to blend raw shelled eggs, minced ham and sandwich fillings was the most likely vehicle of transmission (Anon., 1992).

S. enteritidis infections in humans have also been associated with the consumption of meat (Humphrey et al., 1988; Reilly et al., 1988). People may become infected with S. enteritidis as a result of infected broiler breeder flocks and broiler rearing flocks and contamination of broilers at slaughter (McIlroy et al., 1989; Corkish et al., 1994). Broilers may exhibit a pericarditis due to S. enteritidis infection, and pure cultures of S. enteritidis have been obtained from such infections (O'Brien, 1988; Rampling et al., 1989). Broiler carcasses may become contaminated by crop contents during the slaughtering process (Hargis et al., 1995). In the UK, S. enteritidis PT4 has been isolated from 16–21% of chilled and frozen chicken (Roberts, 1991). In another study in the UK, S. enteritidis was isolated from 51% of raw chicken and from 23% of giblet samples (Plummer et al., 1995).

Other serovars have also been identified as a cause of food-borne infections associated with the consumption of insufficiently cooked or raw eggs, egg-white or yolk or food products containing such ingredients. During an epidemiological investigation of recurring outbreaks of S. typhimurium infections in an institution for mental disease in Massachusetts, which involved 104 cases and six fatalities, egg-nog was established as the vehicle of transmission. S. typhimurium infection was found in the flock of birds that supplied the eggs and was recovered from some of the eggs. The S. typhimurium isolates from the eggs were of the same PT as those isolated from the patients (Philbrook et al., 1960). Raw eggs used to make the icing of a birthday cake, vanilla slices made by mixing ingredients in a mixing bowl also used for mixing the contents of raw shelled eggs, and raw eggs used to make savoury quiche, banana pie, fruit flan and meringues were all associated with outbreaks of S. typhimurium PT141 (Chapman et al., 1988). Foods containing mayonnaise contaminated with S. typhimurium PT49 were the cause of gastrointestinal illness in 120 of 700 people. The same PT was isolated from the mayonnaise and from samples taken from the chicken house of the main egg supplier whose eggs were used to prepare the mayonnaise (Mitchell et al., 1989). In another egg-associated outbreak, S. heidelberg affected 91 of a total of about 1000 persons who attended a convention in New Mexico and consumed fried eggs; the eggs appeared to be 'runny' and insufficiently cooked (Weisse et al., 1986).

Prevention

Outbreaks of S. enteritidis infections due to consumption of contaminated eggs or foods containing such eggs have led to several recommendations to eliminate or curtail salmonellosis in humans. The tracing of outbreak-associated eggs to farms where laying hens were infected with S. enteritidis led the US Department of Agriculture to implement a national programme to control the spread of S. enteritidis in commercial layer flocks (Hedberg et al., 1993). After a large nosocomial infection in New York City (Telzak et al., 1990), all health care facilities in New York State were directed to eliminate raw or undercooked eggs from the diets of persons who are institutionalized, elderly and/or immunocompromised. People immunocompromised by the human immunodeficiency virus were shown to be particularly vulnerable to recurrent septicaemia caused by S. enteritidis, S. typhimurium and

S. dublin, but not by *S. heidelberg* (Levine *et al.*, 1991). It has been recommended to pasteurize eggs for use in nursing homes, in other institutional settings and in commercial foods that may not be adequately cooked before eating (Telzak *et al.*, 1990; Hedberg *et al.*, 1993).

Consumers have been advised to avoid eating raw or undercooked eggs and to avoid eating foods that contain raw eggs, e.g. home-made products, such as Caesar salad, egg-nog, mayonnaise and ice-cream (Steinert *et al.*, 1990; Buckner *et al.*, 1994). Strains of *S. enteritidis*, *S. typhimurium* and *S. senftenberg* inoculated into the yolk of shell eggs were found to survive forms of cooking where some of the yolk remained liquid (Humphrey *et al.*, 1989b). Cooking the eggs at a sufficiently high temperature and for a sufficient period of time has been shown to prevent infections. During a study of sporadic cases of *S. enteritidis* and *S. typhimurium* infections in adults in Minnesota, it was observed that the extent to which eggs were cooked was inversely related to illness (Hedberg *et al.*, 1993). Boiling the eggs for 7–8 min was sufficient to destroy *S. typhimurium* (Schaaf, 1936; Baker *et al.*, 1983; Humphrey *et al.*, 1989a). The cooking-time–temperature relationship for complete kill of *S. typhimurium* depended on the cooking method with fried eggs: 3 min on each side at 64°C for turned-over eggs, 4 min at 70°C for covered eggs and 1 min at 74°C for scrambled eggs. In contrast, cooking for 7.5 min at 64°C for sunny-side-up eggs was not sufficient for destruction of the test organism (Baker *et al.*, 1983).

Consumers have been advised to refrigerate eggs (Humphrey, 1990). It has been found that several *S. enteritidis* phage types, *S. typhimurium* and other *Salmonella* serovars did not grow in the egg or egg yolk when stored below 10°C (Kim *et al.*, 1989; Humphrey, 1990; Saeed and Koons, 1993). *Salmonella* remained viable but did not multiply in the egg-white at 20°C or 30°C and many died out at 4°C (Bradshaw *et al.*, 1990; Lock and Board, 1992). Other recommendations are not to microwave dishes containing raw eggs but to cook them (Evans *et al.*, 1995), in order to prevent cross-contamination from raw eggs to other foods, to wash hands, cooking utensils and food-preparation surfaces with soap and water after contact with raw eggs, not to sample food products containing raw egg, such as biscuit batter, and to promptly refrigerate foods containing eggs (Humphrey *et al.*, 1994; CDC, 1996).

Regulations to Control Transmission

Regulatory authorities in many countries have adopted orders and directives to prevent the occurrence and spread of *Salmonella*. The Council of European Communities issued a directive requiring member countries to specify measures to avoid the introduction of *Salmonella* on the farm and to control *Salmonella* in flocks of layers. Faecal samples must be taken and pooled from breeder flocks consisting of more than 250 birds. When *S. enteritidis* or *S. typhimurium* is detected, the authorities must be notified. The flock is then sampled under supervision of the veterinary authority, the birds are grouped in batches of five and samples of liver, ovaries and intestines are taken. If positive, the birds must be slaughtered or slaughtered and destroyed, the poultry house cleansed and disinfected and the manure and litter safely disposed of. Eggs at the hatchery derived from a positive flock must be destroyed or treated as high-risk material. Sampling is also to be carried out of compound feeding stuffs used to feed poultry. The Poultry Breeding Flocks and Hatcheries Order 1993 adopted in the UK specifies the sampling of chick-box liners and dead breeder chicks upon arrival at the farm and the taking of faecal samples of breeder hens at 4 weeks of age and 2 weeks before lay. It further specifies that grandparent and élite breeding flocks supplying eggs to the hatchery must be sampled each week and a composite sample of meconium taken from chicks or the carcasses of all dead-in-shell and cull chicks and that parent breeder flocks must be sampled every 2 weeks.

The NPIP contains a US *S. enteritidis* monitored programme intended to reduce the incidence of *Salmonella* organisms in hatching eggs and chicks through a sanitation programme at the breeder farm and in the hatchery. It lists requirements to sample meconium from chick boxes and chicks that died within 7 days hatching and for the samples to be sent to an authorized laboratory. It also states requirements regarding the pelletizing and heating of feed and its components and the environmental samples to be taken to verify if they contain group D *Salmonella*, regulates the use of a federally licensed *S. enteritidis* bacterin in multiplier breeding flocks provided that they were examined serologically and bacteriologically after having reached the age of 4 months, states that hatching eggs may be incubated only at hatcheries meeting the NPIP provisions, stipulates sampling plans for flocks from which S. enteritidis has been isolated to determine if they are eligible under the programme, and stipulates the conditions for hatcheries to meet the requirements of the programme.

Programmes to control and eradicate S. enteritidis, S. typhimurium and/or all Salmonella serovars from poultry breeder flocks have been established in many countries. The Swedish programme is intended to control all Salmonella serovars. Imported grandparent chicks are kept for 15 weeks in quarantine and tested four times for Salmonella. A mandatory programme exists to test all layer and broiler flocks bacteriologically for Salmonella before slaughter. If S. enteritidis is isolated, the flock is destroyed (Wierup et al., 1995). A programme to prevent and control Salmonella in grandparent breeder flocks in New Zealand was described by Bates and Granshaw (1995). In Canada, eggs from layer flocks infected or environmentally contaminated with S. enteritidis have been sent to egg breaking sta-

tions for pasteurization. When human infection was traced to a layer flock, the flock has been destroyed and compensation has been paid.

Practices to control Salmonella have been instituted and promoted by the poultry industry in several countries. A 'Canadian Egg Industry Code of Practice' and 'Best Management Practices for Turkey Production' to prevent and control Salmonella have been produced by the Canadian Egg Marketing Agency and the Canadian Turkey Marketing Agency, respectively. Management practices to reduce the risk of Salmonella infection in broilers and turkeys in the USA were described by Holder (1993) and Nagaraya and Halvorson (1993). After depopulation of the flock infected with Salmonella, a cleansing and disinfection procedure must be carried out before the premises may be repopulated. Procedures to disinfect and to sample poultry houses after cleansing and disinfection and before placing chicks or hens have been described (Davies and Wray, 1996a; Davison et al., 1996) but complete elimination of Salmonella and rodents carrying the pathogen has been difficult to achieve (Davies and Wray, 1996b).

References

Altekruse, S., Koehler, J., Hickman-Brenner, R., Tauxe, R.V. and Ferris, K. (1993) A comparison of Salmonella enteritidis phage types from egg-associated outbreaks and implicated laying flocks. Epidemiology and Infection 110, 17–22.

Anon. (1988) Salmonella enteritidis phage type 4: chicken and egg. Lancet ii, 720–722.

Anon. (1992) Hospital outbreak of S. enteritidis infection: Ontario. Canadian Communicable Diseases Report 18, 57–60.

Anon. (1995a) Salmonella in Livestock Production. Ministry of Agriculture, Fisheries and Food, Welsh Office, Agriculture Department, Scottish Office, Agriculture and Fisheries Department, and Veterinary Laboratories Agency, pp. 43–55.

Anon. (1995b) Salmonella Enteritidis Pilot Project Progress Report. A cooperative effort of the Pennsylvania Poultry Producers, Pennsylvania Poultry Federation, Egg Association of America, Pennsylvania Department of Agriculture, Pennsylvania State University, and the United States Department of Agriculture, Washington, DC, pp. 48–49.

Baker, R.C., Hogarty, S., Poon, W. and Vadehra, D.V. (1983) Survival of Salmonella typhimurium and Staphylococcus aureus in eggs cooked by different methods. Poultry Science 62, 1211–1216.

Barnhart, H.M., Dreesen, D.W., Bastien, R., and Pancorbo, O.C. (1991) Prevalence of Salmonella enteritidis and other serovars in ovaries of layer hens at time of slaughter. Journal of Food Protection 54, 488–491.

Barrow, P.A. (1989) Further observations on the effect of feeding diets containing avoparcin on the excretion of salmonellas by experimentally infected chickens. Epidemiology and Infection 102, 239–252.

Barrow, P.A. (1995) Recent Progress in the Diagnosis and Control of Salmonella Infections in Poultry. Document 63 SG/10A, Office International des Épizooties, Paris, 12 pp.

Barrow, P.A., Huggins, M.B., Lovell, M.A. and Simpson, J.M. (1987) Observations on the pathogenesis of experimental Salmonella typhimurium infection in chickens. Research in Veterinary Science 42, 194–199.

Barrow, P.A., Simpson, J.M. and Lovell, M.A. (1988) Intestinal colonisation in the chicken by food-poisoning Salmonella serovars; microbial characteristics associated with faecal excretion. Avian Pathology 17, 571–588.

Barrow, P.A., Lovell, M.A. and Berchieri, A. (1990a) Immunisation of laying hens against *Salmonella enteritidis* with live attenuated vaccines. *Veterinary Record* 126, 241–242.

Barrow, P.A., Hassan, J.O. and Berchieri, A. (1990b) Reduction in faecal excretion of *Salmonella typhimurium* strain F98 in chickens vaccinated with live and killed *S. typhimurium* organisms. *Epidemiology and Infection* 104, 413–426.

Barrow, P.A., Lovell, M.A. and Berchieri, A. (1991) The use of two live attenuated vaccines to immunize egg-laying hens against *Salmonella enteritidis* phage type 4. *Avian Pathology* 20, 681–692.

Baskerville, A., Humphrey, T.J., Fitzgeorge, R.B., Cook, R.W., Chart, H., Rowe, B. and Whitehead, A. (1992) Airborne infection of laying hens with *Salmonella enteritidis* phage type 4. *Veterinary Record* 130, 395–398.

Bates, C. and Granshaw, D. (1995) *Salmonella* control – a working example. In: *Proceedings of the Forty-fourth Western Poultry Disease Conference*, Sacramento, California, pp. 69–73.

Bean, N.H. and Griffin, P.M. (1990) Food borne disease outbreaks in the United States, 1973–1987: pathogens, vehicles, and trends. *Journal of Food Protection* 53, 804–817.

Board, R.G. (1966) The course of microbial infection of the hen's egg. *Journal of Applied Bacteriology* 29, 319–341.

Borland, E.D. (1975) *Salmonella* infection in poultry. *Veterinary Record* 97, 406–408.

Botts, C.W., Ferguson, L.C., Birkeland, J.M. and Winter, A.R. (1952) The influence of litter on the control of *Salmonella* infections in chicks. *American Journal of Veterinary Research* 13, 562–565.

Bouzoubaa, K. and Nagaraja, K.V. (1984) Epidemiological studies on the incidence of salmonellosis in chicken breeder/hatchery operations in Morocco. In: Snoeyenbos, G.H. (ed.) *Proceedings of the International Symposium on Salmonella*. American Association of Avian Pathologists, Kennett Square, Pennsylvania, p. 337.

Bradshaw, J.G., Shah, D.B., Forney, E. and Madden, J.M. (1990) Growth of *Salmonella enteritidis* in yolk of shell eggs from normal and seropositive hens. *Journal of Food Protection* 53, 1033–1036.

Brown, D.D., Ross, J.G. and Smith, A.F.G. (1975) Experimental infection of cockerels with *Salmonella typhimurium*. *Research in Veterinary Science* 18, 165–170.

Brown, D.D., Ross, J.G. and Smith, A.F.G. (1976) Experimental infection of poultry with *Salmonella infantis*. *Research in Veterinary Science* 20, 237–243.

Brownell, J.R., Sadler, W.W. and Fanelli, M.J. (1969) Factors influencing the intestinal infection of chickens with *Salmonella typhimurium*. *Avian Diseases* 13, 804–816.

Bryan, F.L. (1981) Current trends in food borne salmonellosis in the United States and Canada. *Journal of Food Protection* 44, 394–402.

Buchholz, P.S. and Fairbrother, A. (1992) Pathogenicity of *Salmonella pullorum* in Northern Bobwhite quail and Mallard ducks. *Avian Diseases* 36, 304–312.

Buckner, P., Ferguson, D., Anzalone, F., Anzalone, D., Taylor, J., Hlady, W.G. and Hopkins, R.S. (1994) Outbreak of *Salmonella enteritidis* associated with homemade ice-cream – Florida, 1993. *Morbidity and Mortality Weekly Report* 43, 669–671.

Bullis, K.L. (1977a) The history of avian medicine in the US. II. Pullorum disease and fowl typhoid. *Avian Diseases* 21, 422–429.

Bullis, K.L. (1977b) The history of avian medicine in the US. III. Salmonellosis. *Avian Diseases* 21, 430–435.

Bumstead, N. and Barrow, P.A. (1988) Genetics of resistance to *Salmonella typhimurium* in newly hatched chicks. *British Poultry Science* 29, 521–529.

Bumstead, N. and Barrow, P.A. (1993) Resistance to *Salmonella gallinarum*, *S. pullorum*, and *S. enteritidis* in inbred lines of chickens. *Avian Diseases* 37, 189–193.

Bynoe, E.T. and Yurack, J.A. (1964) Salmonellosis in Canada. In: van Oye, E. (ed.) *The World Problem of Salmonellosis*. W. Junk Publishers, The Hague, pp. 397–420.

Cason, J.A., Cox, N.A. and Bailey, J.S. (1994) Transmission of *Salmonella typhimurium* during hatching of broiler chicks. *Avian Diseases* 38, 583–588.

CDC (1992) Outbreak of *Salmonella enteritidis* infection associated with consumption of raw shell eggs, 1991. *Morbidity and Mortality Weekly Report* 41, 369–372.

CDC (1996) Outbreaks of *Salmonella* serovar Enteritidis infection associated with consumption of raw shell egg – United States, 1994–1995. *Morbidity and Mortality Weekly Report* 45, 737–742.

Chapman, P.A., Rhodes, P. and Rylands, W. (1988) *Salmonella typhimurium* phage type 141 infections in Sheffield during 1984 and 1985: association with hens' eggs. *Epidemiology and Infection* 101, 75–82.

Chase, F.E. (1947) *Salmonella* studies in fowl. *Canadian Journal of Public Health* 38, 82–83.

Chishti, M.A., Khan, M.Z. and Siddique, M. (1985) Incidence of salmonellosis in chicken in and around Faisalabad. *Pakistan Veterinary Journal* 5, 79–82.

Cooper, G.L., Nicholas, R.A. and Bracewell, C.D. (1989) Serological and bacteriological investigations of chickens from flocks naturally infected with *Salmonella enteritidis*. *Veterinary Record* 125, 567–572.

Cooper, G.L., Venables, L.M., Woodward, M.J. and Hormaeche, C.E. (1994a) Invasiveness and persistence of *Salmonella enteritidis, Salmonella typhimurium*, and a genetically defined *S. enteritidis aroA* strain in young chicks. *Infection and Immunity* 62, 4739–4746.

Cooper, G.L., Venables, L.M., Woodward, M.J. and Hormaeche, C.E. (1994b) Vaccination of chickens with strain CVL30, a genetically defined *Salmonella enteritidis aroA* live oral vaccine candidate. *Infection and Immunity* 62, 4747–4754.

Corkish, J.D., Davies, R.H., Wray, C. and Nicholas, R.A.J. (1994) Observations on a broiler breeder flock naturally infected with *Salmonella enteritidis* phage type 4. *Veterinary Record* 134, 591–594.

Corrier, D., Nisbet, D., Scanlan, C., Hargis, B. and Deloach, J. (1992) Control of *Salmonella* colonization in broiler and Leghorn chicks with dietary lactose and defined caecal flora maintained in continuous-flow cultures. In: *Reports and Communications, International Symposium on* Salmonella *and Salmonellosis*. Ploufragan/Saint-Brieuc, France, pp. 406–412.

Cowden, J.D., Chisholm, D., O'Mahony, M., Lynch, D., Mawer, S.L., Spain, G.E., Ward, L. and Rowe, B. (1989) Two outbreaks of *Salmonella enteritidis* phage type 4 infection associated with the consumption of fresh shell-egg products. *Epidemiology and Infection* 103, 47–52.

Coyle, E.F., Palmer, S.R., Ribeiro, C.D., Jones, H.I., Howard, A.J., Ward, L. and Rowe, B. (1988) *Salmonella enteritidis* phage type 4 infection: association with hens' eggs. *Lancet* ii, 1295–1297.

Cox, J.M. (1996) What makes *Salmonella enteritidis* stick in chickens? *World Poultry-Misset*, May, 22–23.

D'Aoust, J.-Y. (1985) Infective dose of *Salmonella typhimurium* in Cheddar cheese. *American Journal of Epidemiology* 122, 717–720.

Davies, R.H. and Wray, C. (1996a) Determination of an effective sampling regime to detect *Salmonella enteritidis* in the environment of poultry units. *Veterinary Microbiology* 50, 117–127.

Davies, R.H. and Wray, C. (1996b) Studies of contamination of three broiler breeder houses with *Salmonella enteritidis* before and after cleansing and disinfection. *Avian Diseases* 40, 626–633.

Davison, S., Benson, C.E. and Eckroade, R.J. (1996) Evaluation of disinfectants against *Salmonella enteritidis*. *Avian Diseases* 40, 272–277.

Diesch, S.L. (1978) Environmental aspects of salmonellosis. In: *Proceedings National Salmonellosis Seminar*. United States Animal Health Association and co-sponsors, Washington, DC, 11 pp.

Ebel, E.D., David, M.J. and Mason, J. (1992) Occurrence of *Salmonella enteritidis* in the U.S. commercial egg industry: report on a national spent hen survey. *Avian Diseases* 36, 646–654.

Edel, W. (1994) *Salmonella enteritidis* eradication programme in poultry breeder flocks in The Netherlands. *International Journal of Food Microbiology* 21, 171–178.

Edwards, P.R. (1958) Salmonellosis: observations on incidence and control. *Annals of the New York Academy of Sciences* 70, 598–613.

Erbeck, D.H., McLaughlin, B.G. and Singh, S.N. (1993) Pullorum disease with unusual signs in two backyard chicken flocks. *Avian Diseases* 37, 895–897.

Evans, M.R., Parry, S.M. and Ribeiro, C.D. (1995) *Salmonella* outbreak from microwave cooked food. *Epidemiology and Infection* 115, 227–230.

Faddoul, G.P. and Fellows, G.W. (1966) A five-year survey of the incidence of salmonellae in avian species. *Avian Diseases* 10, 296–304.

Fanelli, M.J., Sadler, W.W. and Brownell, J.R. (1970) Preliminary studies on persistence of salmonellae in poultry litter. *Avian Diseases* 14, 131–141.

Fanelli, M.J., Sadler, W.W., Franti, C.E. and Brownell, J.R. (1971) Localization of salmonellae within the intestinal tract of chickens. *Avian Diseases* 15, 366–375.

Ferris, K.E. and Miller, D.A. (1990) *Salmonella* serovars from animals and related sources reported during July 1988–June 1989. In: *Proceedings of the 93th Annual Meeting of the United States Animal Health Association*. Carter Printing, Richmond, Virginia, pp. 521–538.

Forsythe, R.H., Ross, W.J. and Ayres, J.C. (1967) *Salmonella* recovery following gastrointestinal and ovarian inoculation in the domestic fowl. *Poultry Science* 46, 849–855.

Galton, M.M., Steele, J.H. and Newell, K.W. (1964) Epidemiology of salmonellosis in the United States. In: van Oye, E. (ed.) *The World Problem of Salmonellosis*. W. Junk Publishers, The Hague, pp. 421–444.

Gast, R.K. (1997) Detecting infections of chickens with recent *Salmonella pullorum* isolates using standard serological methods. *Poultry Science* 76, 17–23.

Gast, R.K. and Beard, C.W. (1990) Production of *Salmonella enteritidis*-contaminated eggs by experimentally infected hens. *Avian Diseases* 34, 438–446.

Gast, R.K. and Benson, S.T. (1996) Intestinal colonization and organ invasion in chicks experimentally infected with *Salmonella enteritidis* phage type 4 and other phage types isolated from poultry in the United States. *Avian Diseases* 40, 853–857.

Gast, R.K., Stone, H.D., Holt, P.S. and Beard, C.W. (1992) Evaluation of the efficacy of oil-emulsion bacterins for protecting chickens against *Salmonella enteritidis. Avian Diseases* 37, 1085–1091.

Gast, R.K., Stone, H.D. and Holt, P.S. (1993) Evaluation of the efficacy of oil-emulsion bacterins for reducing faecal shedding of *Salmonella enteritidis* by laying hens. *Avian Diseases* 36, 992–999.

Gauger, H.C. and Greaves, R.E. (1946) Isolations of *Salmonella typhimurium* from drinking water in an infected environment. *Poultry Science* 25, 476–478.

Gordon, R.F. and Tucker, J.F. (1965) The epizootiology of *Salmonella menston* infection of fowls and the effect of feeding poultry food artificially infected with *Salmonella. British Poultry Science* 6, 251–264.

Goren, E., de Jong, W.A., Doornenbal, P., Bolder, N.M., Mulder, R.W.A.W. and Jansen, A. (1988) Reduction of *Salmonella* infection of broilers by spray application of intestinal microflora: a longitudinal study. *Veterinary Quarterly* 10, 249–255.

Gorham, S.L., Kadavil, K., Vaughan, E., Lambert, H., Abel, J. and Pert, B. (1994) Gross and microscopic lesions in young chickens experimentally infected with *Salmonella enteritidis. Avian Diseases* 38, 816–821.

Guard-Petter, J., Lakshmi, B., Carlson, R. and Ingram, K. (1995) Characterization of lipopolysaccharide heterogeneity in *Salmonella enteritidis* by an improved gel electrophoresis method. *Applied and Environmental Microbiology* 61, 2845–2851.

Guillot, J.F. and Millemann, Y. (1992) Intestinal colonization of chickens and turkeys by *Salmonella* and antibiotic decontamination. In: *Reports and Communications, International Symposium on* Salmonella *and Salmonellosis*. Ploufragan/Saint-Brieuc, France, pp. 413–420.

Guthrie, R.K. (1992) Salmonella. CRC Press, Boca Raton, Florida, pp. 1–20.

Hafez, H.M, Jodas, S., Kösters, J. and Schmidt, H. (1992) Treatment of *Salmonella enteritidis* artificially contaminated eggs with pressure-differential-dipping (PDD) using antibiotics. In: *Reports and Communications, International Symposium on* Salmonella *and Salmonellosis*. Ploufragan/Saint-Brieuc, France, pp. 421–427.

Hargis, B.M., Caldwell, D.J., Brewer, R.L., Corrier, D.E. and Deloach, J.R. (1995) Evaluation of the chicken crop as a source of *Salmonella* contamination for broiler carcasses. *Poultry Science* 74, 1548–1552.

Hargrett-Bean, N.A. and Potter, M.E. (1995) *Salmonella* serovars from human sources, January 1992 through December 1992. In: *Proceedings of the 98th Annual Meeting of the United States Animal Health Association*. Promiter Communications and Spectrum Press, Richmond, Virginia, pp. 439–442.

Hassan, J.O. and Curtiss, R., III (1994) Development and evaluation of an experimental vaccination program using a live avirulent *Salmonella typhimurium* strain to protect immunized chickens against challenge with homologous and heterologous *Salmonella* serovars. *Infection and Immunity* 62, 5519–5527.

Hassan, J.O. and Curtiss, R., III (1996) Effect of vaccination of hens with an avirulent strain of *Salmonella typhimurium* on immunity of progeny challenged with wild-type *Salmonella* strains. *Infection and Immunity* 64, 938–944.

Hedberg, C.W., David, M.J., White, K.E., MacDonald, K.L. and Osterholm, M.T. (1993) Role of egg consumption in sporadic *Salmonella enteritidis* and *Salmonella typhimurium* infections in Minnesota. *Journal of Infectious Diseases* 167, 107–111.

Hennessy, T.W., Hedberg, C.W., Slutsker, L., White, K.E., Besser-Wiek, J.M., Moen, M.E., Feldman, J., Coleman, W.W., Edmonson, L.M., MacDonald, K.L., Osterholm, M.T. and the investigation team (1996) A national outbreak of *Salmonella enteritidis* infections from ice-cream. *New England Journal of Medicine* 334, 1281–1286.

Henzler, D.J. and Opitz, H.M. (1992) The role of mice in the epizootiology of *Salmonella enteritidis* infection on chicken layer farms. *Avian Diseases* 36, 625–631.

Henzler, D.J., Ebel, E., Sanders, J., Kradel, D. and Mason, J. (1994) *Salmonella enteritidis* in eggs from commercial chicken layer flocks implicated in human outbreaks. *Avian Diseases* 38, 37–43.

Hewitt, E.A. (1928) Bacillary white diarrhoea in baby turkeys. *Cornell Veterinarian* 18, 272–276.

Higgins, R., Malo, R., René-Roberge, E. and Gauthier, R. (1982) Studies on the dissemination of *Salmonella* in nine broiler-chicken flocks. *Avian Diseases* 26, 26–33.

Hinshaw, W.R. and McNeil, E. (1940) Eradication of pullorum disease from turkey poults. In: *Proceedings of the 44th Annual Meeting of the United States Livestock Sanitary Association*, Chicago, Illinois, pp. 178–194.

Hinton, M., Pearson, G.R., Threlfall, E.J., Rowe, B., Woodward, M. and Wray, C. (1989) Experimental *Salmonella enteritidis* infection in chicks. *Veterinary Record* 124, 223.

Hinton, M., Threlfall, E.J. and Rowe, B. (1990a) The invasive potential of *Salmonella enteritidis* phage types for young chickens. *Letters in Applied Microbiology* 10, 237–239.

Hinton, M., Threlfall, E.J. and Rowe, B. (1990b) The invasiveness of different strains of *Salmonella enteritidis* phage type 4 for young chickens. *FEMS Microbiology Letters* 70, 193–196.

Holder, T. (1993) Best management practices for *Salmonella* risk reduction in broilers. In: *Proceedings of the 97th Annual Meeting of the United States Animal Health Association*. Carter Printing, Richmond, Virginia, pp. 486–504.

Holt, P.S. (1995) Horizontal transmission of *Salmonella enteritidis* in molted and unmolted laying chickens. *Avian Diseases* 39, 239–249.

Holt, P.S. and Chaubal, L.H. (1997) Detection of motility and putative synthesis of flagellar proteins in *Salmonella pullorum* cultures. *Journal of Clinical Microbiology* 35, 1016–1020.

Holt, P.S. and Porter, R.E. (1992) Microbiological and histopathological effects of an induced-molt fasting procedure on a *Salmonella enteritidis* infection in chickens. *Avian Diseases* 36, 610–618.

Hoop, R.K. and Pospischil, A. (1993) Bacteriological, serological, histological and immunohistochemical findings in laying hens with naturally acquired *Salmonella enteritidis* phage type 4 infection. *Veterinary Record* 133, 391–393.

Hopper, S.A. and Mawer, S. (1988) *Salmonella enteritidis* in a commercial layer flock. *Veterinary Record* 123, 351.

Humphrey, T.J. (1990) Growth of salmonellas in intact shell eggs: influence of storage temperature. *Veterinary Record* 126, 292.

Humphrey, T.J., Mead, G.C. and Rowe, B. (1988) Poultry meat as a source of human salmonellosis in England and Wales. *Epidemiology and Infection* 100, 175–184.

Humphrey, T.J., Greenwood, M., Gilbert, R.J., Rowe, B. and Chapman, P.A. (1989a) The survival of salmonellas in shell eggs cooked under simulated domestic conditions. *Epidemiology and Infection* 103, 35–45.

Humphrey, T.J., Baskerville, A., Mawer, S., Rowe, B. and Hopper, S. (1989b) *Salmonella enteritidis* phage type 4 from the contents of intact eggs: a study involving naturally infected hens. *Epidemiology and Infection* 103, 415–523.

Humphrey, T.J., Chart, H., Baskerville, A. and Rowe, B. (1991a) The influence of age on the response of SPF hens to infection with *Salmonella enteritidis* PT4. *Epidemiology and Infection* 106, 33–43.

Humphrey, T.J., Baskerville, A., Chart, H., Rowe, B. and Whitehead, A. (1991b) *Salmonella enteritidis* PT4 infection in specific pathogen free hens: influence of infecting dose. *Veterinary Record* 129, 482–485.

Humphrey, T.J., Martin, K.W. and Whitehead, A. (1994) Contamination of hands and work surfaces with *Salmonella enteritidis* PT4 during the preparation of egg dishes. *Epidemiology and Infection* 113, 403–409.

Humphrey, T.J., Williams, A., McAlpine, K., Lever, M.S., Guard-Petter, J. and Cox, J.M. (1996) Isolates of *Salmonella enterica* Enteritidis PT4 with enhanced heat and acid tolerance are more virulent in mice and more invasive in chickens. *Epidemiology and Infection* 117, 79–88.

Hutt, F.B. and Scholes, R.D. (1941) Genetics of the fowl. XIII. Breed differences in susceptibility to *Salmonella pullorum*. *Poultry Science* 20, 324–352.

Hüttner, B., Landgraf, H. and Vielitz, E. (1981) Kontrolle der Salmonelleninfektionen in Mastelterntier-Bestanden durch Verabreichung von SPR-Darmflora on Eintagsküken. *Deutsche Tierärtzlichen Wochenschrift* 88, 527–532.

Ibrahim, G.F., Lyons, M.J., Walker, R.A. and Fleet, G.H. (1986) Rapid detection of salmonellae in foods using immunoassay systems. *Journal of Food Protection* 49, 92–98.

Johnson, D.C., David, M. and Goldsmith, S. (1992) Epizootiological investigation of an outbreak of pullorum disease in an integrated broiler operation. *Avian Diseases* 36, 770–775.

Jones, F.S. (1913) The value of the macroscopic agglutination test in detecting fowls that are harboring *Bacterium pullorum*. *Journal of Medical Research* 27, 481–495.

Kaushik, R.K., Singh, J., Kumar, S. and Kulshreshtha, R.C. (1986) Fowl typhoid in a few poultry farms of Haryana state. *Indian Journal of Animal Sciences* 56, 511–514.

Keller, L.H., Benson, C.E., Krotec, K., and Eckroade, R.J. (1995) *Salmonella enteritidis* colonization of the reproductive tract and forming and freshly laid eggs of chickens. *Infection and Immunity* 63, 2443–2449.

Kelterborn, E. (1967) Salmonella *Species: First Isolations, Names, and Occurrence*. W. Junk, The Hague, The Netherlands, pp. 20, 140–141, 377–379.

Kilger, G. and Grimont, P.A.D. (1993) Differentiation of *Salmonella* phase flagellar antigen types by restriction of the amplified *fliC* gene. *Journal of Clinical Microbiology* 31, 1108–1110.

Kim, C.J., Emery, D.A., Rinke, H., Nagaraja, K.V. and Halvorson, D.A. (1989) Effect of time and temperature on growth of *Salmonella enteritidis* in experimentally inoculated eggs. *Avian Diseases* 33, 735–742.

Kinde, H., Read, D.H., Chin, R.P., Bickford, A.A., Walker, R.L., Ardans, A., Breitmeyer, R.E., Willoughby, D., Little, H.E., Kerr, D. and Gardner, I.A. (1996a) *Salmonella enteritidis*, phage type 4 infection in a commercial layer flock in Southern California: bacteriologic and epidemiological findings. *Avian Diseases* 40, 665–671.

Kinde, H., Read, D.H., Ardans, A., Breitmeyer, R.E., Willoughby, D., Little, H.E., Kerr, D., Gireesh, R. and Nagaraja, K.V. (1996b) Sewage effluent: likely source of *Salmonella enteritidis*, phage type 4 infection in a commercial chicken layer flock in Southern California. *Avian Diseases* 40, 672–676.

Kirkwood, J.K., Cunningham, A.A., Macgregor, S.K., Thornton, S.M. and Duff, J.P. (1994) *Salmonella enteritidis* excretion by carnivorous animals fed on day-old chicks. *Veterinary Record* 134, 683.

Klein, E. (1889) Ueber eine epidemische Krankheit der Hühner, verursacht durch einen Bacillus – *Bacillus galli-narum*. *Zentralblatt für Bakteriologie und Parasitenkunde* 5, 689–693.

Kühn, H., Rabsch, W., Gericke, B. and Reissbrodt, R. (1993) Infektionsepidemiologische Analysen von Salmonellosen, Shigellosen und anderen Enterobacteriaceae-Infektionen. *Bundesgesundheitsblatt* 36, 324–333.

Laszlo, V.G., Csorian, E.S. and Paszti, J. (1985) Phage types and epidemiologicalal significance of *Salmonella enteritidis* strains in Hungary between 1976 and 1983. *Acta Microbiologica Hungarica* 32, 321–340.

Levine W.C., Buehler, J.W., Bean, N.H. and Tauxe, R.V. (1991) Epidemiology of nontyphoidal *Salmonella* bacteraemia during the human immunodeficiency virus epidemic. *Journal of Infectious Diseases* 164, 81–87.

Lior, H. (1989) Isolations of enteric pathogens from people in Canada. *Safety Watch* 14, 3.

Lock J.L. and Board, R.G. (1992) Persistence of contamination of hens' egg albumen *in vitro* with *Salmonella* serovars. *Epidemiology and Infection* 108, 389–396.

Long, J.R., DeWitt, W.F. and Ruet, J.L. (1980) Studies on *Salmonella* from floor litter of 60 broiler chicken houses in Nova Scotia. *Canadian Veterinary Journal* 21, 91–94.

McCoy, J.H. (1975) Trends in salmonella food poisoning in England and Wales 1941–72. *Journal of Hygiene (Cambridge)* 74, 271–282.

McDermott, L.A. (1947) The K formula stained antigen in the detection of standard and Younie types of *Salmonella pullorum* infection. *Canadian Journal of Public Health* 38, 83–84.

McIlroy, S.G. and McCracken, R.M. (1990) The current status of the *Salmonella enteritidis* control programme in the United Kingdom. In: *Proceedings of the 94th Annual Meeting of the United States Animal Health Association.* Carter Printing, Richmond, Virginia, pp. 450–462.

McIlroy, S.G., McCracken, R.M., Neill, S.D. and O'Brien, J.J. (1989) Control, prevention and eradication of *Salmonella enteritidis* infection in broiler and broiler breeder flocks. *Veterinary Record* 125, 545–548.

Maguire, H.C.F., Codd, A.A., Mackay, V.E., Rowe, B. and Mitchell, E. (1993) A large outbreak of human salmonellosis traced to a local pig farm. *Epidemiology and Infection* 110, 239–246.

Mandal, B.K. (1979) Typhoid and paratyphoid fever. *Clinics in Gastroenterology* 8, 715–735.

Manning, J.G., Hargis, B.M., Hinton, A., Corrier, D.E., DeLoach, J.R. and Creger, C.R. (1994) Effect of selected antibiotics and anticoccidials on *Salmonella enteritidis* caecal colonization and organ invasion in Leghorn chicks. *Avian Diseases* 38, 256–261.

Mawer, S.L., Spain, G.E. and Rowe, B. (1989) *Salmonella enteritidis* phage type 4 and hens' eggs. *Lancet* i, 280–281.

Mitchell, E., O'Mahony, M., Lynch, D., Ward, L.R., Rowe, B., Uttley, A., Rogers, T., Cunningham, D.G. and Watson, R. (1989) Large outbreak of food poisoning caused by *Salmonella typhimurium* definitive type 49 in mayonnaise. *British Medical Journal* 298, 99–101.

Moore, E.N. (1946) *The Occurrence of Fowl Typhoid.* Circular No. 19, University of Delaware, Newark, Delaware, pp. 1–20.

Morris, G.K., McMurray, B.L., Galton, M.M. and Wells, J.G. (1969) A study of the dissemination of salmonellosis in a commercial broiler chicken operation. *American Journal of Veterinary Research* 30, 1413–1421.

Morse, E.V. (1978) Salmonellosis and pet animals. In: *Proceedings of the National Salmonellosis Seminar.* United States Animal Health Association and other co-sponsors, Washington, DC, 6 pp.

Nagaraja, K.V. and Halvorson, D.A. (1993) Best management practices for *Salmonella* risk reduction in turkeys. In: *Proceedings of the 97th Annual Meeting of the United States Animal Health Association.* Carter Printing, Richmond, Virginia, pp. 505–523.

Nakamura, M., Nagamine, N., Takahashi, T., Suzuki, S., Kijima, M., Tamura, Y. and Sato, S. (1994) Horizontal transmission of *Salmonella enteritidis* and effect of stress on shedding in laying hens. *Avian Diseases* 38, 282–288.

Neill, MA., Opal, S.M., Heelan, J., Giusti, R., Cassidy, J.E., White, R. and Mayer, K.H. (1991) Failure of ciprofloxacin to eradicate convalescent faecal excretion after acute salmonellosis: experience during an outbreak in health care workers. *Annals of Internal Medicine* 114, 195–199.

Nicolle, P. (1964) La lysotypie de *Salmonella typhi*: son principe, sa technique, son application à l'épidémiologie de la fièvre typhoïde. In: van Oye, E. (ed.) *The World Problem of Salmonellosis.* W. Junk Publishers, The Hague, pp. 67–88.

Nurmi, E. and Rantala, M. (1973) New aspects of *Salmonella* infection in broiler production. *Nature* 214, 210–211.

O'Brien, J.D.P. (1988) *Salmonella enteritidis* infection in broiler chickens. *Veterinary Record* 122, 214.

Olesiuk, O.M., Carlson, V.L., Snoeyenbos, G.H. and Smyser, C.F. (1969) Experimental *Salmonella typhimurium* infection in two chicken flocks. *Avian Diseases* 13, 500–508.

Oosterom, J. (1991) Epidemiological studies and proposed preventive measures in the fight against human salmonellosis. *International Journal of Food Microbiology* 12, 41–52.

Opara, O.O., Carr, L.E., Rusek-Cohen, E., Tate, C.R., Mallinson, E.T., Miller, R.G., Stewart, L.E., Johnston, R.W.

and Joseph, S.W. (1992) Correlation of water activity and other environmental conditions with repeated detection of *Salmonella* contamination on poultry farms. *Avian Diseases* 36, 664–671.

Orr, B.B. and Moore, E.N. (1953) Longevity of *Salmonella gallinarum*. *Poultry Science* 32, 800–805.

Padron, M.N. (1990) *Salmonella typhimurium* outbreak in broiler chicken flocks in Mexico. *Avian Diseases* 34, 221–223.

Parham, G.L. (1985) Salmonellae in cooked beef products. In: Snoeyenbos, G.H. (ed.) *Proceedings of the International Symposium on Salmonella*. American Association of Avian Pathologists, New Orleans, Louisiana, pp. 275–280.

Paul, J. and Batchelor, B. (1988). *Salmonella enteritidis* phage type 4 and hens' eggs. *Lancet* ii, 1421.

Perales, I. and Audicana, A. (1988) *Salmonella enteritidis* and eggs. *Lancet* ii, 1133.

Petter, J.G. (1993) Detection of two smooth colony phenotypes in a *Salmonella enteritidis* isolate which vary in their ability to contaminate eggs. *Applied and Environmental Microbiology* 59, 2884–2890.

Philbrook, F.R., MacCready, R.A., Van Roekel, H., Anderson, E.S., Smyser, C.F., Sanen, F.J. and Groton, W.M. (1960) Salmonellosis spread by a dietary supplement of avian source. *New England Journal of Medicine* 263, 713–718.

Plummer, R.A.S., Blissett, S.J. and Dodd, C.E.R. (1995) *Salmonella* contamination of retail chicken products sold in the UK. *Journal of Food Protection* 58, 843–846.

Pomeroy, B.S. (1984) Fowl typhoid. In: Hofstad, M.S., Barnes, H.J., Calnek, B.W., Reid, W.M. and Yoder, H.W. (eds) *Diseases of Poultry*, 8th edn, Iowa State University Press, Ames, Iowa, pp. 79–91.

Poppe, C. and Gyles, C. (1987) Relation of plasmids to virulence and other properties of salmonellae from avian sources. *Avian Diseases* 31, 844–854.

Poppe, C., Irwin, R.J., Forsberg, C.M., Clarke, R.C. and Oggel, J. (1991a) The prevalence of *Salmonella enteritidis* and other *Salmonella* spp. among Canadian registered commercial layer flocks. *Epidemiology and Infection* 106, 259–270.

Poppe, C., Irwin, R.J., Messier, S., Finley, G.G. and Oggel, J. (1991b) The prevalence of *Salmonella enteritidis* and other *Salmonella* spp. among Canadian registered commercial chicken broiler flocks. *Epidemiology and Infection* 107, 201–211.

Poppe, C., McFadden, K.A., Brouwer, A.M. and Demczuk, W. (1993a) Characterization of *Salmonella enteritidis* strains. *Canadian Journal of Veterinary Research* 57, 176–184.

Poppe, C., Demczuk, W., McFadden, K. and Johnson, R.P. (1993b) Virulence of *Salmonella enteritidis* phage types 4, 8, and 13 and other *Salmonella* spp. for day-old chicks, hens and mice. *Canadian Journal of Veterinary Research* 57, 281–287.

Rampling, A., Anderson, J.R., Upson, R., Peters, E., Ward, L.R. and Rowe, B. (1989) *Salmonella enteritidis* phage type 4 infection of broiler chickens: a hazard to public health. *Lancet* ii, 436–438.

Ranta, L.E. and Dolman, C.E. (1947) Experience with *Salmonella* typing in Canada. *Canadian Journal of Public Health* 38, 286–294.

Read, D.H., Kinde, H. and Daft, B.M. (1994) Pathology of *Salmonella enteritidis* phage type 4 infection in commercial layer chickens in southern California. In: *Proceedings of the 44th Meeting of the Western Poultry Disease Conference, Sacramento, California, 5–7 March 1995.* pp. 76–78.

Reilly, W.J., Forbes, G.I., Sharp, J.C.M., Oboegbulem, S.I., Collier, P.W. and Paterson, G.M. (1988) Poultry-borne salmonellosis in Scotland. *Epidemiology and Infection* 101, 115–122.

Renwick, S.A., Irwin, R.J., Clarke, R.C., McNab, W.B., Poppe, C. and McEwen, S.A. (1992) Epidemiological associations between characteristics of registered broiler chicken flocks in Canada and the *Salmonella* culture status of floor litter and drinking water. *Canadian Veterinary Journal* 33, 449–458.

Rettger, L.F. (1900) Fatal septicaemia among young chicks. *New York Medical Journal* 71, 803–805.

Rettger, L.F. (1909) Further studies on fatal septicaemia in young chickens, or 'white diarrhoea'. *Journal of Medical Research* 21, 115–123.

Rigby, C.E. and Pettit, J.R. (1979) Some factors affecting *Salmonella typhimurium* infection and shedding in chickens raised on litter. *Avian Diseases* 23, 442–455.

Roberts, D. (1991) *Salmonella* in chilled and frozen chicken. *Lancet* 337, 984–985.

Rodrigue, D.C., Tauxe, R.V. and Rowe, B. (1990) International increase in *Salmonella enteritidis*: a new pandemic? *Epidemiology and Infection* 105, 21–27.

Ryan, C.A., Nickels, M.K., Hargrett-Bean, N.T., Potter, M.E., Endo, T., Mayer, L., Langkop, C.W., Gibson, C., McDonald, R.C., Kenney, R.T., Puhr, N.D., McDonnell, P.J., Martin, R.J., Cohen, M.L. and Blake, P.A. (1987) Massive outbreak of antimicrobial-resistant salmonellosis traced to pasteurized milk. *Journal of the American Medical Association* 258, 3269–3274.

Sadler, W.W., Brownell, J.R. and Fanelli, M.J. (1969) Influence of age and inoculum level on shed pattern of *Salmonella typhimurium* in chickens. *Avian Diseases* 13, 793–803.

Saeed, A.M. (1999) Salmonella enterica *Serovar* enteritidis *in Humans and Animals: Epidemiology, Pathogenesis and Control.* Iowa State University Press, Ames, Iowa.

Saeed, A.M. and Koons, C.W. (1993) Growth and heat resistance of *Salmonella enteritidis* in refrigerated and abused eggs. *Journal of Food Protection* 56, 927–931.

St Louis, M.E., Morse, D.L., Potter, M.E., DeMelfi, T.M., Guzewich, J.J., Tauxe, R.V., Blake, P.A. and the *Salmonella enteritidis* Working Group (1988) The emergence of grade A eggs as a major source of *Salmonella enteritidis* infections. *Journal of the American Medical Association* 259, 2103–2107.

Salem, M., Odor, E.M. and Pope, C. (1992) Pullorum disease in Delaware roasters. *Avian Diseases* 36, 1076–1080.

Schaaf, J. (1936) Die Salmonellose (infektiöse Enteritis, Paratyphose) des Geflügels, ihre Bedeutung und Bekämpfung. *Zeitschrift für Infektionskrankheiten, Parasitäre Krankheiten und Hygiene der Haustiere* 49, 322–332.

Schaffer, J.M., MacDonald, A.D., Hall, W.J. and Bunyea, H. (1931) A stained antigen for the rapid whole blood test for pullorum disease. *Journal of the American Veterinary Medical Association* 79, 236–240.

Scholes, J.C. and Hutt, F.B. (1942) *The Relationship between Body Temperature and Genetic Resistance to* Salmonella pullorum *in the Fowl.* Cornell University Agriculture Experimental Station Memoir 244, Ithaca, New York.

Seeliger, H.P.R. and Maya, A.E. (1964) Epidemiologie der Salmonellosen in Europa 1950–1960. In: van Oye, E. (ed.) *The World Problem of Salmonellosis.* W. Junk Publishers, The Hague, pp. 245–294.

Seuna, E., Nagaraja, K.V. and Pomeroy, B.S. (1985) Gentamicin and bacteria culture (Nurmi culture) treatments either alone or in combination against experimental *Salmonella hadar* infection in turkey poults. *Avian Diseases* 29, 617–629.

Silva, E.M., Snoeyenbos, G.H., Wienack, O.M. and Smyser, C.F. (1981) Studies on the use of 9R strain of *Salmonella gallinarum* as a vaccine in chickens. *Avian Diseases* 25, 38–52.

Smith, H.W. and Tucker, J.F. (1980) The virulence of *Salmonella* strains for chickens; their excretion by infected chicken. *Journal of Hygiene, Cambridge* 84, 479–488.

Snoeyenbos, G.H. (1984) Pullorum disease. In: Hofstad, M.S., Barnes, H.J., Calnek, B.W., Reid, W.M. and Yoder, H.W. (eds) *Diseases of Poultry,* 8th edn. Iowa State University Press, Ames, Iowa, pp. 66–79.

Snoeyenbos, G.H., Carlson, V.L., Smyser, C.F. and Olesiuk, O.M. (1969) Dynamics of *Salmonella* infection in chicks reared on litter. *Avian Diseases* 13, 72–83.

Sojka, W.J., Wray, C., Hudson, E.B. and Benson, J.A. (1975) Incidence of *Salmonella* infection in animals in England and Wales, 1968–73. *Veterinary Record* 96, 280–287.

Sommers, H.M. (1980) Infectious diarrhoea. In: Youmans, G.P., Paterson, P.Y. and Sommers, H.M. (eds) *The Biologic and Clinical Basis of Infectious Diseases.* W.B. Saunders, Philadelphia, Pennsylvania, pp. 525–553.

Steinert, L., Virgil, D., Bellemore, E., Williamson, B., Dinda, E., Harris, D., Scheider, D., Fanella, L., Bogacki, V., Liska, F., Birkhead, G.S., Guzewich, J.J., Fudala, J.K. Kondracki, S.F., Shayegani, M., Morse, D.L., Dennis, D.T., Healey, B., Tavris, D.R., Duffy, M. and Drinnen, K. (1990) Update: *Salmonella enteritidis* infections and grade A shell eggs – United States, 1989. *Morbidity and Mortality Weekly Report* 38, 877–880.

Stevens, A., Joseph, C., Bruce, H., Fenton, D., O'Mahony, M., Cunningham, D., O'Connor, B. and Rowe, B. (1989) A large outbreak of *Salmonella enteritidis* phage type 4 associated with eggs from overseas. *Epidemiology and Infection* 103, 425–433.

Stokes, J.L., Osborne, W.W. and Bayne, H.G. (1956) Penetration and growth of *Salmonella* in shell eggs. *Food Research* 21, 510–518.

Tannock, G.W. and Smith, J.M.B. (1971) Studies on the survival of *Salmonella typhimurium* and *Salmonella bovismorbificans* on pasture and in water. *Australian Veterinary Journal* 47, 557–559.

Telzak, E.E., Budnick, L.D., Zweig Greenberg, M.S., Blum, S., Shayegani, M., Benson, C.E. and Schultz, S. (1990) A nosocomial outbreak of *Salmonella enteritidis* infection due to the consumption of raw eggs. *New England Journal of Medicine* 323, 394–397.

Thain, J.A. and Blandford, T.B. (1981) A long-term serological study of a flock of chickens naturally infected with *Salmonella pullorum. Veterinary Record* 109, 136–138.

Thiagarajan, D., Saeed, A.M. and Asem, E.K. (1994) Mechanism of transovarian transmission of *Salmonella enteritidis* in laying hens. *Poultry Science* 73, 89–98.

Thiagarajan, D., Saeed, M., Turek, J. and Asem, E. (1996) *In vitro* attachment and invasion of chicken ovarian granulosa cells by *Salmonella enteritidis* phage type 8. *Infection and Immunity* 64, 5015–5021.

Tietjen, M. and Fung, D.Y.C. (1995) Salmonellae and food safety. *Critical Reviews in Microbiology* 21, 53–83.

Timms, L.M., Marshall, R.N. and Breslin, M.F. (1990) Laboratory assessment of protection given by an experimental *Salmonella enteritidis* PT4 inactivated, adjuvant vaccine. *Veterinary Record* 127, 611–614.

Timoney, J.F., Shivaprasad, H.L., Baker, R.C. and Rowe, B. (1989) Egg transmission after infection of hens with *Salmonella enteritidis* phage type 4. *Veterinary Record* 125, 600–601.

Truscott, R.B. (1981) Oral *Salmonella* antigens for the control of *Salmonella* in chicks. *Avian Diseases* 25, 810–820.

Tucker, J.F. (1967) Survival of salmonellae in built-up litter for housing of rearing and laying fowls. *British Veterinary Journal* 123, 92–103.

Turnbull, P.C.B. and Snoeyenbos, G.H. (1973) Experimental salmonellosis in the chicken. I. Fate and host response in alimentary canal, liver, and spleen. *Avian Diseases* 18, 153–177.

van de Giessen, A.W., Peters, R., Berkers, P.A.T.A., Jansen, W.H. and Notermans, S.H.W. (1991) *Salmonella* contamination of poultry flocks in The Netherlands. *Veterinary Quarterly* 13, 41–46.

van de Giessen, A.W., Dufrenne, J.B., Ritmeester, W.S., Berkers, P.A.T.A., van Leeuwen, W.J. and Notermans, S.H.W. (1992) The identification of *Salmonella enteritidis*-infected poultry flocks associated with an outbreak of human salmonellosis. *Epidemiology and Infection* 109, 405–411.

Waltman, W.D. and Horne, A.M. (1993) Isolation of *Salmonella* from chickens reacting in the pullorum-typhoid agglutination test. *Avian Diseases* 37, 805–810.

Weisse, P., Libbey, E., Nims, L., Gutierrez, P., Madrid, T., Weber, N., Voorhees, C., Crocco, V., Hules, C., Hill, S., Ray, T.M., Gurule, R., Ortiz, F., Eidson, M., Sewell, C.M., Castle, S., Hayes, P. and Hull, H.F. (1986) *Salmonella heidelberg* outbreak at a convention – New Mexico. *Morbidity and Mortality Weekly Report* 35, 91.

Wierup, M., Wold-Troell, M., Nurmi, E. and Hakkinen, M. (1987) Epidemiological evaluation of the *Salmonella* controlling effect of a nationwide use of a 'competitive exclusion' culture in poultry. In: *Proceedings of an International Workshop on Competitive Exclusion of Salmonellas from Poultry*. Bristol Laboratory, Langford, UK, p. 7.

Wierup, M., Engström, B., Engvall, A. and Wahlström, H. (1995) Control of *Salmonella enteritidis* in Sweden. *International Journal of Food Microbiology* 25, 219–226.

Williams, J.E. (1984) Avian salmonellosis. In: Hofstad, M.S., Barnes, H.J., Calnek, B.W., Reid, W.M. and Yoder, H.W. (eds) *Diseases of Poultry*, 8th edn. Iowa State University Press, Ames, Iowa, pp. 65–66.

Williams, J.E., Dillard, L.H. and Hall, G.O. (1968) The penetration patterns of *Salmonella typhimurium* through the outer structures of chicken eggs. *Avian Diseases* 12, 445–466.

Wray, C. (1985) Is salmonellosis still a serious problem in veterinary practice? *Veterinary Record* 116, 485–489.

Xu, Y.M., Pearson, G.R. and Hinton, M. (1988) The colonization of the alimentary tract and visceral organs of chicks with salmonellas following challenge via the feed: bacteriological findings. *British Veterinary Journal* 144, 403–410.

Chapter 8
Salmonella Infections in Turkeys

Hafez M. Hafez[1] and Silvia Jodas[2]

[1]*Free University of Berlin, Faculty of Veterinary Medicine, Institue of Poultry Disease, Koserstrasse 21, 14195 Berlin, Germany; [2]Poultry Health Service, Azenbergerstr. 16, 70174 Stuttgart, Germany*

Introduction

The worldwide incidence of *Salmonella* food poisoning has increased dramatically during the last few years. Although the proportion of food-poisoning outbreaks and cases in which the source of infection can be positively identified is small, poultry and poultry products are repeatedly implicated. *Salmonella* from poultry currently enter the human food-chain mainly as a result of carcass contamination from infected faecal material or from eggs.

Salmonellosis and *Salmonella* infections in turkeys are distributed worldwide and result in severe economic losses when no effort is made to control them. The large economic losses are caused by high poult mortality during the first 4 weeks of age, high medication costs, reductions in egg production in breeder flocks, poor poult quality and high costs for eradication and control measures. The most important aspect, however, is the continuing effect of *Salmonella*-contaminated turkey meat and meat products on public health.

Various efforts are made at the breeder flock hatchery level to control mortality and morbidity losses from *Salmonella* infections. It is estimated that these efforts cost the turkey industry in the USA approximately $US10 million yearly (Pomeroy *et al.*, 1989). When the expense of investigations, control measures, disposal of contaminated material, lost leisure-time spending and other costs are added and the total multi-

plied worldwide, the losses become massive (Davies and Wray, 1994). The economic costs of salmonellosis in humans are enormous and $US1 billion was lost because of absence from work and medical treatment in the USA in 1987 (Roberts, 1988). The failure of the human population to apply hygienically acceptable food-handling and cooking practice and the fact that the processing plants are not able to reduce the level of pathogenic bacteria in poultry products mean that every effort must be made to reduce the *Salmonella* contamination of the live birds before despatch to processing plants.

In turkeys, as well as in chickens, a distinction is usually made between infections caused by the two non-motile host-adapted serovars of *S. pullorum* (pullorum disease) and *S. gallinarum* (fowl typhoid) and the remainder of the motile *Salmonella* (paratyphoid (PT) infection), including the *S. arizonae* subgenera (arizonosis).

Historical Perspective

The first case of PT infection in domestic poultry was reported by Moore (1895), who described an outbreak of infectious enteritis in pigeons due to a bacillus of the hog-cholera group. The first occurrence of PT infections in turkey poults was reported by Pfaff (1921) in the USA. Pomeroy and Fenstermacher (1939) observed the infection in turkeys in Minnesota in 1932. PT infections cause major losses, predominantly in young

© CAB *International 2000. Salmonella in Domestic Animals*
(eds C. Wray and A. Wray)

poults, and Hinshaw and McNeil (1940) found that *S. typhimurium* accounted for approximately 50% of the PT outbreaks in turkeys that they investigated.

The causative agent of pullorum disease was first isolated by Rettger in 1899. In 1909, he named it *Bacterium pullorum* and later changed it to *S. pullorum*. Pullorum disease in turkeys was first described in 1928 by Hewitt. According to Hinshaw and McNeil (1940), the infection appeared to be introduced into turkeys by contact with infected chickens in hatcheries and/or by brooding chicks and poults together. Likewise, contact with chickens or yards used by chickens is an important factor in the spread of fowl typhoid to turkeys. Pfeiler and Roepke (1917) and other authors reported the disease in turkeys reared on farms where it was also prevalent in chickens.

The first description of an organism now classified as *S. arizonae* was by Caldwell and Ryerson (1939), in which attention was drawn to a bacterium isolated from diseased chuckwallas, horned lizards and gila monsters. The first reports of *S. arizonae* in turkeys were by Peluffo *et al.* (1942) and Edwards *et al.* (1943). Later, Hinshaw and McNeil (1946) isolated Arizona serovar 7:1,7,8, now classified as O18:z_4:z_{32} and O18:z_4:z_{23}, from a number of outbreaks in poults and showed that all infections were traceable to eggs produced in a defined area in California.

Aetiology

Salmonella belong to the family *Enterobacteriaceae* and all members are Gram-negative, non-sporing rods that do not have capsules. According to the latest nomenclature, which reflects recent advances in *Salmonella* taxonomy, the genus *Salmonella* consists of two species, *S. bongori* and *S. enterica* (Le Minor and Popoff, 1987), and their taxonomy is dealt with in Chapter 1. During the last few years, many modern molecular biological tests for the sub-typing of *Salmonella* have been developed (Brunner *et al.*, 1983; Eisenstein, 1990). The genus *Salmonella* of the family *Enterobacteriaceae* can be roughly classified into three categories or groups.

Group 1: highly host-adapted and invasive serovars

This group includes species-restricted and invasive *Salmonella*, such as *S. pullorum* and *S. gallinarum* in poultry and *S. typhi* in humans.

Group 2: non-host-adapted and invasive serovars

This group consists of approximately ten to 20 serovars that are able to cause an invasive infection in poultry and may be capable of infecting humans. Currently, the most important serovars are *S. typhimurium*, *S. hadar*, *S. arizonae* and *S. enteritidis*.

Group 3: non-host-adapted and non-invasive serovars

Most serovars of the genus *Salmonella* belong to this group and may cause disease in humans and other animals.

Epidemiology

Many different *Salmonella* serovars have been isolated from turkeys; their exact number, however, is difficult to estimate. Some serovars may be predominant for a number of years in a region or certain countries and then disappear to be replaced by another serovar. Table 8.1 lists *Salmonella* serovars isolated from turkeys, based on the available literature.

Salmonella surveillance in commercial turkey flocks was carried out between 1993 and 1995 by Hafez *et al.* (1997), whose results showed that the most frequently isolated serovars were *S. newport* (34.6%) and *S. reading* (30.3%), followed by *S. bredeney* (10.6%); *S. enteritidis* phage type 8 was detected for only a short period (5 weeks) in one flock. *Salmonella* shedding was of intermittent duration and varied between 1 and 20 weeks. In eight of 24 monitored meat turkey flocks, *Salmonella* could not be detected during the entire rearing period. Seven flocks (29.2%) appeared to be infected with only one serovar and in another three flocks (12.5%) two different serovars were isolated during the rearing period,

Table 8.1. *Salmonella* serovars isolated from turkeys.

Serovar	Author	Serovar	Author
S. abortusequi (B)*	2	*S. heidelberg* (B)	1; 2; 8; 7, 11;
S. agona (B)	13; 14; 16; 21; 22		15; 18; 21; 22; 23; 24; 25
S. albany (C$_3$)	25	*S. illinois* (E$_3$)	2; 25
S. aluchua (O)	6; 25	*S. indiana* (B$_1$)	22; 24; 25
S. amager (E$_1$)	25	*S. infantis* (C$_1$)	2; 6; 8; 11; 13; 14; 18;
S. amherstiana (C$_3$)	2; 25		21; 22; 23; 24; 25;
S. anatum (E$_1$)	1; 2; 5; 6; 14; 17; 18; 22,	*S. irumu* (C$_1$)	25
	23; 24; 25	*S. javiana* (D$_1$)	2; 25
S. arizonae	8; 15; 20; 22; 24	*S. johannesburg* (R)	22
S. banana (B)	25	*S. kaapstad* (B)	7; 25
S. bareilly (C$_1$)	2; 25	*S. kentucky* (C$_3$)	2; 5; 8; 25
S. berkeley (U)	2; 4; 25	*S. kingston* (B)	2; 25
S. berta (D$_1$)	2; 15; 25	*S. lexington* (E$_1$)	2; 25
S. binza (E$_2$)	1; 5; 6; 25	*S. litchfield* (C$_2$)	2; 22; 25
S. blockley (C$_2$)	6; 23; 25	*S. livingstone* (C$_1$)	22
S. bovismorbificans (C$_2$)	2; 10; 25	*S. london* (E$_1$)	13; 24; 25
S. braenderup (C$_1$)	7, 10; 25	*S. madelia* (H)	2; 25
S. brancaster (B)	2; 25	*S. manchester* (C$_2$)	1
S. bredeney (B)	1; 2; 5; 7, 9; 11; 22;	*S. manhattan* (C$_2$)	2; 6; 9; 10; 14; 22; 24; 25
	23; 25	*S. manila* (E$_2$)	6
S. budapest (B)	25	*S. mbandaka* (C$_1$)	7; 24
S. california (B)	2; 9; 12; 25	*S. meleagridis* (E$_1$)	2; 17; 22; 25
S. cambridge (E$_2$)	25	*S. menston* (C$_1$)	2; 25
S. canoga (E$_3$)	2; 25	*S. minneapolis* (E$_3$)	25
S. cerro (K)	2; 25	*S. minnesota* (L)	2; 25
S. chester (B)	1; 2; 6; 17; 25	*S. mission* (C$_1$)	25
S. choleraesuis (C$_1$)	2; 25	*S. montevideo* (C$_1$)	1; 2; 5; 6; 7 11; 21; 24; 25
S. concord (C$_1$)	2; 25	*S. muenchen* (C$_2$)	1; 2; 6; 25
S. corvallis (C$_3$)	2; 25	*S. muenster* (E$_1$)	25
S. cubana (G$_2$)	2; 5; 6; 25	*S. newbrunswick* (E$_2$)	2; 5; 25
S. derby (B)	2; 6; 10; 24; 25	*S. newington* (E$_2$)	1; 2; 6; 22; 25
S. djugu (C$_1$)	7	*S. newport* (C$_2$)	1; 2; 6; 7, 9; 10; 11; 22; 25
S. drypool (E$_2$)	8; 11; 22	*S. ohio* (C$_1$)	23
S. dublin (D$_1$)	2; 25	*S. onderstepoort* (H)	2; 25
S. duesseldorf (C$_2$)	25	*S. oranienburg* (C$_1$)	2; 25
S. duisburg (B)	24	*S. oregon* (C$_2$)	5
S. eastbourne (D$_1$)	2; 25	*S. orion* (E$_1$)	25
S. edinburg (C$_1$)	25	*S. panama* (D$_1$)	2; 10; 25
S. eimsbuettel (C$_1$)	8	*S. paratyphi B* (B)	2; 25
S. emek (C$_3$)	23	*S. pomona* (M)	2; 25
S. enteritidis (D$_1$)	2; 7; 8, 11; 14; 21; 24; 25	*S. poona* (G$_1$)	25
S. florida (H)	25	*S. pullorum* (D$_1$)	25
S. fresno (D$_2$)	25	*S. reading* (B)	2; 6; 7, 9; 10; 13; 18; 22;
S. gallinarum (D$_1$)	2; 25		25
S. gaminara (I)	2; 25	*S. rubislaw* (F)	2; 25
S. give (E$_1$)	2; 6; 10; 25	*S. rutgers* (E$_1$)	25
S. grumpensis (G$_2$)	25	*S. saintpaul* (B)	1; 2; 5; 7, 8; 9; 10; 11; 13;
S. hadar (C$_2$)	7, 15; 21; 23; 24		15; 19; 22; 23; 24; 25
S. haifa (B)	7	*S. sandiego* (B)	1; 2; 6; 8; 9; 10; 13; 22; 25
S. halmstad (E$_2$)	1	*S. schwarzengrund* (B)	2; 6; 10; 21; 25
S. hamilton (E$_2$)	25	*S. senftenberg* (E$_4$)	1; 2; 6; 7; 8; 9; 10, 22; 24;
S. harrisonburg (E$_3$)	2; 3; 25		25
S. havana (G$_2$)	24		

continued on next page

Table 8.1. *Continued*

Serovar	Author	Serovar	Author
S. shomron (K)	2; 11	*S. typhimurium* (B)	1; 2; 6; 8; 9; 11; 14; 17; 19; 21; 22; 23; 24; 25
S. siegburg (K)	25		
S. simsbury (E$_4$)	2; 25	*S. typhimurium* var. Copenhagen (B)	7, 11; 18; 22; 24; 25
S. sofia (B)	23		
S. stanley (B)	25	*S. uganda* (E$_1$)	25
S. takoradi (C$_2$)	25	*S. urbana* (N)	2; 25
S. taksony (E$_4$)	5; 25	*S. vejle* (E$_1$)	25
S. telaviv (M)	2; 25	*S. westhampton* (E$_1$)	7; 25
S. tennessee (C$_1$)	2; 18; 25	*S. wichita* (G$_1$)	2; 25
S. thomasville (E$_3$)	5; 25	*S. worchester* (G$_2$)	25
S. thompson (C$_1$)	2; 6; 13; 14; 17; 25	*S. worthington* (G$_2$)	1; 2; 5; 7, 18; 25

* Indicates antigenic group in the Kauffmann–White schema to which each serovar belongs.
Authors: 1, Bryan *et al.* (1968); 2, Dräger (1971); 3, Edwards and McWorther (1953a); 4, Edwards and McWorther (1953b); 5, Boyer *et al.* (1962); 6, Faddoul and Fellows (1966); 7, Hafez *et al.* (1997); 8, Kumar *et al.* (1971); 9, Kumar *et al.* (1972); 10, Nivas *et al.* (1973); 11, Willinger *et al.* (1986); 12, Hugh-Jones *et al.* (1975); 13, McBride *et al.* (1978); 14, Hirschmann and Seidel (1992); 15, Opengart *et al.* (1991); 16, Shreeve and Hall (1971); 17, Shahata *et al.* (1984); 18, Zecha *et al.* (1976); 19, Schellner (1985); 20, Pollan and Vasicek (1991); 21, Pomeroy (1991); 22, Pomeroy *et al.* (1984); 23, Samberg and Klinger (1984); 24, Pietzsch (1982); 25, Hinshaw (1959).

in some cases concurrently. More than two serovars could be detected in the remaining six flocks (25.0%). In the UK, *S. newport* was the commonest serovar during the period (Anon., 1995).

Transmission

Transmission and spread of *Salmonella* occurs by vertical and/or horizontal routes. Primary vertical transmission occurs by true ovarian transmission, by passage through the oviduct or by contact with infected peritoneum or air sac. Secondary vertical transmission happens by contamination of the egg content as a result of faecal contamination of the eggshell from cloaca and/or contaminated nests, floor or incubators, with subsequent penetration into the eggs.

Vertical transmission is the most important route of infection in turkey for *S. gallinarum*, *S. pullorum*, *S. arizonae*, *S. senftenberg*, *S. typhimurium* and *S. hadar*. Williams and Dillard (1968) found that unpigmented turkey eggs were more frequently penetrated by *S. typhimurium* than normal pigmented eggs. Such unpigmented eggs have thinner shells and more gross pores.

Hatcheries are one of the major sources of

horizontal transmission and *Salmonella* can survive for long periods in eggshells, meconium, dust and litter. Organisms can also spread by air throughout the hatchery, resulting in rapid transmission.

On the farm, infection is transmitted horizontally (laterally) by direct contact between infected and uninfected turkeys, and by indirect contact with contaminated environments through ingestion or inhalation of *Salmonella* organisms. Subsequently, there are many possibilities for lateral spread of the organisms through live and dead vectors. Transmission frequently occurs via faecal contamination of feed, water, equipment, environment and dust, in which *Salmonella* can survive for long periods. Failure to clean and disinfect properly after an infected flock has left the site can result in infection of the next batch of birds. *Salmonella* have been reported to survive in turkey litter for up to 9 months after removal of an infected flock.

Significant reservoirs for *Salmonella* are humans, farm animals, pigeons, waterfowl and wild birds. Rodents, pets and insects are also potential reservoirs and transmit the infection to birds and between houses. The organisms are often localized in the gut of these carriers, which shed *Salmonella* intermittently in their faeces, thus contaminating the poultry environment. Rodents constitute a persistent reservoir of *Salmonella* infection, from which poultry houses and stored feeding stuffs must be protected as far as possible (Kumar *et al.*, 1971; Baxter-Jones and Wilding, 1981; McCapes *et al.*, 1991).

Probably one of the commonest sources for lateral spread of the organisms is feed. Nearly every ingredient ever used in the manufacture of poultry feedstuffs has been shown at one time or another to contain *Salmonella*. The organism occurs most frequently in protein from animal products, such as meat and bone-meal, blood meal, poultry offal, feather meal and fish-meal. Proteins of vegetable origin have also been shown to be contaminated with *Salmonella*. Turkey feed samples have been investigated by Bryan *et al.* (1968), Willinger *et al.* (1986) and Hafez *et al.* (1997), who detected *Salmonella* contamination rates of between 3 and 9%.

Factors Influencing the Course of Infection

The course of the infection and the prevalence of salmonellosis in turkeys depend on different factors, such as the *Salmonella* serovar, age of birds, infectious dose and route of infection. Further stress-producing circumstances, such as bad management, poor ventilation, high stocking density or concurrent diseases, may also contribute to the development of a systemic infection, with possible heavy losses among young poults. After recovery, birds continue to excrete *Salmonella* in their faeces, and such birds must be considered as a potential vector of the microorganisms.

Paratyphoid Infection

Infections of avian species with motile *Salmonella*, with the exception of *S. arizonae*, are designated as paratyphoid infections and, generally, such infections are more prevalent in turkey than in any other avian species. Moran (1959) reported that the *S. typhimurium* serovar was encountered four times more frequently in turkeys than in chickens.

This group of infections is one of the most important bacterial diseases in the turkey breeder sector and results in high losses among turkey

poults during the first month after hatching, with maximum losses occurring in the first 10 days.

Among valuable breeder flocks, infections are generally accompanied by severe economic losses, because of their chronic nature and the difficulty of eradication. In many cases, the infection seriously impairs fertility, hatchability and egg production.

Clinical signs

Incubation periods range from 2 to 5 days. Mortality in young poults varies from negligible to 10–20% and, in severe outbreaks, may reach 80% or more. The severity of an outbreak in young poults depends on the serovar involved, its virulence, the degree of exposure, the age of the birds (Bierer, 1960), environmental conditions and the presence of concurrent infections. The age at which the disease is first observed in poults will depend on whether the poults are infected in the incubator or after being placed in the brooder. In poults infected orally at 1 day of age with 10^8 colony-forming units (cfu) per bird with either *S. typhimurium*, *S. anatum*, *S. thompson*, *S. meleagridis* or *S. chester*, losses started 2 days and stopped 8 days post-exposure. The mortality rate ranged between 70 and 90% (Shahata *et al.*, 1984). Similar results were obtained by Pomeroy (1944), Mitrovic (1956) and Bierer (1960). On the other hand, neither mortality nor clinical signs were observed after experimental oral infection of 3-day-old poults with 10^6 cfu per bird of *S. enteritidis* phage type 8 or 4 (Hafez and Stadler, 1997).

Poults may be infected from a few days after hatching to maturity. Clinical signs may be absent even if infection occurs in the incubator or a few days after hatching. If infection was egg-transmitted or occurred in the incubator, there are many unpipped eggs and pipped eggs with dead embryos.

Symptoms usually seen in young poults are somnolence, weakness, drooping wings, ruffled feathers and huddling together near heat sources. Many poults that survive for several days will become emaciated and the feathers around the vent will be matted with faecal material. However, in young poults, diarrhoea is not a constant symptom. Lameness, caused by arthritis, may also be present. Adult birds usually show little or no evidence of the infection and serve mostly as intestinal or internal-organ carriers over long periods.

Experimental PT infections have resulted in an acute disease of short duration, when the birds showed inappetence, increased water consumption, diarrhoea, dehydration and general listlessness. Also, Chaplin and Hamilton (1957) reported synovitis in turkeys infected intravenously with *S. thompson*. Leg weakness in mature birds is not uncommon. Higgins *et al.* (1944) encountered a flock of 24-week-old turkeys infected with PT in which 10% of the birds were so severely affected with an arthritic condition that they were unsuitable for marketing.

Lesions

Lesions may be entirely absent in extremely severe outbreaks. Birds that die in an acute phase of the disease show a persistent yolk-sac, catarrhal and haemorrhagic enteritis and necrotic foci in liver, spleen and heart muscle. Furthermore, congestion of the liver, kidney, gall-bladder and heart muscle are the most constant post-mortem findings. The pericardial sac is often filled with a straw-coloured fluid. Another common finding is a caseous caecal core, which is sometimes filled with blood. Lung and heart lesions are rare but air-sac involvement is common (Hinshaw, 1959). In adult turkeys, marked inflammation of the intestine, with occasional necrotic ulcers, has been observed and in these cases the liver and spleen are usually swollen and congested.

Pullorum Disease

Pullorum disease is an acute systemic disease of young poults, characterized by sudden death and high mortality. The disease is mostly egg-transmitted. In adult turkeys, the disease is often localized and can cause lifelong latent infections and lead to a decrease in egg production and fertility, as well as hatchability.

Clinical symptoms

The incubation period is 3–5 days. Morbidity and mortality are highly variable, from less than 10% to as high as 100%. The first indication of disease by vertical transmission is usually an increase in the number of dead-in-shell poults. Infected hatched poults appear moribund and sudden death without clinical symptoms may occur. Mortality begins to increase around the 4 or 5 days of age, with a peak mostly between the second and third week.

The symptoms are not characteristic and are similar to those of PT infection. Laboured breathing due to pneumonia is commonly observed. Some poults show white diarrhoea, with a pasting of feathers round the vent ('pasty vent'). Conjunctivitis, swelling of joints and synovial sheaths and lameness may also be seen. In rare instances, nervous symptoms have been observed. Survivors are often irregular in size, stunted or poorly feathered, and many remain carriers and disseminate the causative agent. Adult turkeys usually show no clinical signs, though some appear unthrifty. Variable degrees of decreased egg production, fertility and hatchability have been observed.

Lesions

S. pullorum may cause severe systemic lesions, although their severity is highly variable. Lesions are limited in young poults that die suddenly in the early stages. The liver may be enlarged, congested or discoloured and may be streaked with haemorrhages. Changes are commonly accompanied by small, white, focal necrosis. Necrotic foci or greyish-white nodules are also seen in heart, lungs, gizzard muscle and caeca. The intestine usually lacks tone and contains an excessive mucous discharge. The caeca may contain a caseous core and are sometimes filled with blood. Peritonitis is frequently manifested and pericarditis may be observed. The yolk-sac and its contents reveal slight or no alteration. In more protracted cases, the absorption of the yolk-sac may be poor and the contents may be of creamy or pasty consistency. The spleen may be enlarged, the kidneys congested and the ureters distended with urates. In septicaemia forms, hyperaemia may also be found in other organs.

In adult birds, the lesions most frequently found in the chronic carrier hen are misshapen, pedunculated, discoloured cystic ova, which usually contain oily and caseous material enclosed in a thickened capsule. The ovary may be haemor-rhaged, with atrophic discoloured follicles. Ovarian and oviduct dysfunction may lead to abdominal ovulation or impassable oviduct, which in turn brings about extensive peritonitis and adhesions of the abdominal viscera. In male birds, the testes may be atrophied, with thickening of the tunica albuginea and multiple abscesses. Occasional myocarditis with pericarditis and ascites may also occur, in both sexes.

Fowl Typhoid

Fowl typhoid is an septicaemic disease caused by S. *gallinarum;* the mortality and course of the disease are variable, depending on the virulence of the strain involved.

Clinical signs

The incubation period is about 5 days and the losses may extend over 2–3 weeks with a tendency for recurrence. The initial outbreak is mostly accompanied by high mortality rates of up to 26.5% (Hinshaw, 1930), followed by intermittent recurrence with less severe losses.

The clinical symptoms of fowl typhoid in turkeys are similar to those described for pullorum disease and are not characteristic. However, some infected poults show green to greenish-yellow diarrhoea, with pasting of feathers around the vent, increased thirst, anorexia, somnolence, retarded growth and respiratory distress.

In growing and mature turkeys, increased thirst, listlessness and a tendency to separate from healthy birds have been observed. Body temperatures increase several degrees to as high as 44–45°C just before death.

Lesions

In young poults, the lesions resemble those observed in pullorum disease. In peracute cases, post-mortem lesions may be absent and, in acute cases, subcutaneous blood-vessels are infected. The skeletal muscles are congested and dark in colour and often appear as if partially cooked. The heart is swollen and small, greyish, necrotic foci are seen in the myocardium. The liver is enlarged and friable, with necrotic foci and a bronze coloration. The spleen and kidneys are enlarged. The lungs are congested and have small grey areas of focal necrosis. Haemorrhagic enteritis, especially in the duodenum, and ulceration of the intestine are more or less consistent lesions in turkeys. Usually an increased percentage of large retained yolks is present. In adult carriers, there is a predilection for the reproductive organs, and lesions are similar to those described in pullorum disease.

Arizonosis

Arizonosis, caused by serovars O18:z_4;z_{32} and O18:z_4;z_{23} (formerly 7:1,7,8), is a serious problem in the turkey industry, due to high mortality and reduced production and hatchability. The turkey appears to be the primary poultry host for these particular serovars, although the host range for other serovars of S. *arizonae* is unlimited. The infection is egg-transmitted and most of the outbreaks occur in young turkeys during the first 3 weeks of life. The infection can be masked in day-old poults by antibiotic treatment.

Clinical signs

The incubation period ranges between 5 and 10 days. Mortality varies greatly: 3.5–15% is common, although losses up to 90% have been reported. Mortality is generally highest during the first 3 weeks after hatching and may continue until 5 weeks after hatch (Greenfield *et al.,* 1971). In turkeys, S. *arizonae* infection is indistinguishable from other forms of salmonellosis. Young poults appear in poor condition, listless, shivering, huddling near heat sources and sitting on their hocks. In addition, diarrhoea, with pasting in the vent area and uni- or bilateral blindness and nervous disorders, such as uncoordinated gait, convulsions, twisted neck and torticollis, have been observed in birds with brain lesions (Kowalski and Stephans, 1968). Poor growth and moderate to marked uneven growth in the flock are often seen after the clinical disease has ended.

Clinical signs are rarely found in mature turkeys and they seldom die from the infection (Sato and Adler, 1966). However, they remain latent carriers and shed the organisms. Infected parent stocks may show decreased egg production, fertility and hatchability.

Lesions

In poults that die with septicaemia, lesions may be absent. Frequently, however, there is an enlarged, congested, mottled, yellow liver, with pinpoint necrotic foci, distended gall-bladder and retained yolk-sac. Marked congestion and erosion of the gastrointestinal tract and caseous casts filling the caecal lumen are common findings. Other lesions include accumulation of caseous exudate in air sacs and the abdominal or thoracic cavities.

A small but significant number of poults have eye lesions. The eye usually has a normal cornea but an opaque lens and a cheesy exudate covers the retina. This lesion is not pathognomonic, because it also occurs with paratyphoid, aspergillosis and colibacillosis, although it occurs more frequently with arizonosis. Purulent exudate in the meninges, lateral ventricles of the brain or the middle and inner ear is seen in birds with central nervous system signs. In adult birds, small caseous mesenteric lesions and cystic ovules have been described by Hinshaw and McNeil (1946).

Diagnosis of *Salmonella* Infections in Turkeys

Clinical signs and lesions are of little value in diagnosis. Accurate diagnosis must be substantiated by isolation and identification of the causative bacteria and/or detection of antibodies using serological examination.

Isolation and identification

Usually the organism can be detected in hatching eggs, dead-in-shell embryos, heart blood, liver, spleen, kidney, crop, intestinal content, unabsorbed yolk, faeces and environmental samples, such as drag swabs, floor litter, nest litter and dust.

Yamamota *et al.* (1961) found *S. typhimurium* in higher number in caecal faeces than in intestinal faeces after experimental inoculation of bacteria into the crop of adult turkey. Faddoul and Fellows (1966) found that, in 70% of the positive turkey consignments, *Salmonella* could be isolated only from intestinal tracts, where they have a predilection to establish a chronic infection in the caeca. Hafez and Stadler (1997) inoculated 3-day-old poults orally with 10^6 cfu per bird of *S. enteritidis* (one group with phage type 8 and the other with phage type 4). Samples of heart, liver, lung, spleen, crop, proventriculus, duodenum, caeca, bursa of Fabricius and bone marrow were collected at 21 days of age and cultured separately for the presence of *Salmonella*. The highest detection rates were obtained by culturing the spleen and caeca. Culturing caseous material covering the retina is useful for the detection of *S. arizonae* in infected birds (Kowalski and Stephans, 1968).

Method of isolation

A large number of studies on *Salmonella* isolation from poultry flocks and poultry products have been published (see Chapter 21). Because the experimental designs concerning the kind of specimens, the *Salmonella* serovars involved and culture procedures have varied widely, it is difficult to compare the results. In addition, there is no single, ideal scheme for isolation of all serovars. According to Fricker (1987):

> It is a foolish person who suggests the use of a single procedure for the isolation of all salmonellas from all types of samples. One must decide upon which medium or media to use in the light of knowledge of the type of sample being studied and the types of *Salmonella* likely to be present.

To choose a method of isolation, the following aspects must be considered: sensitivity, applicability, duration and cost. In every step of isolation, different factors, such as type of sample, amount of inoculum, temperature and time of incubation, may negatively influence the results (Hafez *et al.*, 1993a). There is no general recommendation for the media types. In most publications, pre-enrichment, followed by selective enrichment and streaking on selective agar, is accepted as providing the most satisfactory results.

The World Health Organization (WHO) provided guidelines on the detection and monitoring of *Salmonella*-infected poultry flocks, with particular reference to *S. enteritidis* (Wray and

Table 8.2. Comparison of different enrichment media for detection of *Salmonella* in turkey faeces (Hafez *et al.*, 1993a). No. of samples tested = 110.

Results	Enrichment media and temperature used (°C)			
	Diasalm/41.5	RV/41.5	TT/41.5	TT/37
Positive samples	38	30	19	20
	34.5%	27.3%	17.3%	18.2%
Negative samples	72	80	91	90
	65.5%	72.7%	82.7%	81.8%

RV, Rappaport–Vassiliadis broth; TT, tetrathionate broth.

Davies, 1994). They recommend the use of buffered peptone water broth as a non-selective pre-enrichment, followed by selective enrichment in Rappaport–Vassiliadis (RV) medium and streaking on xylose lysine desoxycholate (XLD) and Brilliant green agar (BGA). In addition, bismuth sulphite (BS) agar is an excellent plating medium for the isolation of *S. arizonae* (Mallinson and Snoeyenbos, 1989).

Hafez *et al.* (1993b) carried out a study in turkey flocks to determine the best technique for isolation of *Salmonella* from turkey faeces, using different selective enrichment media. The results revealed that using Diasalm agar (Van Netten *et al.*, 1991) as selective enrichment gives the best results (Table 8.2).

Conventional cultural procedures for hatching eggs using egg yolk, albumen or shell samples have generally resulted in low isolation rates. Other sample and culture techniques, such as the 'egg moulding method' described by Baxter-Jones and Wilding (1982), have shown increased isolation rates of *Salmonella* in turkey hatching eggs in comparison with conventional culture methods. Hafez *et al.* (1986) described a further modification using enrichment media in empty eggshells. The reisolation rate of *S. senftenberg* using this method was always higher than examination of yolk and/or albumen alone from artificially contaminated broiler chicken, turkey and quail hatching eggs. Recent investigation on isolation of *S. enteritidis* from experimentally contaminated chicken hatching eggs (layer type) using pre-enrichment of empty eggshell samples led to significantly higher detection rates, in comparison with the same samples cultured without pre-enrichment after contamination with 10^2 cfu ml^{-1} *S. enteritidis* (Hafez and Jodas, 1992).

Serological examination

Invasive host-adapted *Salmonella* are generally able to stimulate the production of circulating antibodies, and different serological techniques may be used to detect the infection. The advantage of serological tests over bacteriological examination is that antibodies in serum of infected birds persist for a longer time and the bacteria shedding in faeces of infected birds is intermittent. On the other hand, some poultry with a positive serological response may not be infected with *Salmonella* organisms and poultry in the early stages of infection may be serologically negative. Antigens and macroserological tests for the detection of antibodies against *Salmonella* in poultry have been described and discussed elsewhere (Anon., 1989; Wray and Davies, 1994; see also Chapter 24).

Several serological tests have been developed for the diagnosis of *Salmonella* infections in turkeys. The rapid whole-blood test (WBT), first used in the 1920s by Runnells *et al.* (1927), progressed into a stained antigen slide agglutination test (Schaffer *et al.*, 1931). The WBT was shown to be undependable for the detection of either pullorum or fowl typhoid in turkeys (Hinshaw and McNeil, 1940; Winter *et al.*, 1952). However, the serum agglutination test (SAT) has been shown to be effective for testing turkey flocks for antibodies against *S. typhimurium* and *S. arizonae* infections (Timms, 1971; Kumar *et al.*, 1974). The standard tube agglutination test is used to test turkey serum for pullorum–typhoid, *S. typhimurium* and *S. arizonae* infection (DeLay *et al.*, 1954; Mallinson and Snoeyenbos, 1989). Further methods, such as microagglutination/ microantiglobulin, have been used, with varying degrees of success (Williams and Whittemore, 1971; Kumar *et al.*, 1977). Recently, the enzyme-linked immunosorbent assay (ELISA) using a variety of antigens, including somatic lipopolysaccharide, flagella, SEF14 fimbriae, outer-membrane proteins (OMP) and crude antigen preparations, has been used, especially for detection of *S. enteritidis* and *S. typhimurium* carriers in chicken, but there is no available literature on the using of these methods in turkey flocks. Nagaraja *et al.* (1984, 1986) successfully used the OMP as ELISA antigen for detection of antibodies against *S. arizonae* infection in turkey breeder flocks.

Differential Diagnosis

Young birds with generalized salmonellosis and arizonosis infections may show signs and lesions identical to any acute septicaemia caused by a wide variety of bacteria, including *Escherichia coli*. In severe outbreaks of salmonellosis, liver and heart lesions may be very similar to those seen with secondary invaders in avian mycoplasmosis. Nervous symptoms may resemble those of Newcastle disease, aspergillosis (Jodas and Hafez, 1997) or other diseases affecting the central nervous system. The heavy yellowish-white cheesy exudate covering the retina is, as indicated previously, not pathognomic of turkey poults with arizonosis.

Treatment

Treatment of salmonellosis in turkeys with antimicrobial drugs is highly effective in reducing mortality and clinical signs, if the treatment is administered sufficiently early in the outbreak. In turkeys, administration of tetracyclines, neomycin, amoxicillin, trimethoprim/sulphonamide, polymixin B, chloramphenicol, nitrofurans, fluoroquinolones and gentamicin in drinking water, in feed or by injection has been shown to be very effective. However, the majority of survivors become carriers. So far, no drug or combination of drugs has been found that is able to eliminate the infection and symptomless carriers from treated flocks. Hafez *et al.* (1997) found that the treatment of turkey flocks with

different antimicrobials to combat respiratory disease conditions or coccidiosis did not reduce the shedding of *Salmonella* in the faeces of naturally infected flocks. On the other hand, Guillot and Milleman (1990) reported that the use of enrofloxacin appears to be very effective in reducing the intestinal carriage of *S. typhimurium* after experimental infection.

Generally, the use of antimicrobials will tend to promote the development of antibiotic resistance in bacteria, either by mutational resistance or by the acquisition of antibiotic resistance (R) plasmids, which can be transferred to *Salmonella* from intestinal *E. coli*. Resistance to furazolidone and the quinolones has so far been unable to spread to different bacterial species, because it is mediated by chromosomal genes (Barrow, 1992a).

Although there are some gradual increases in the resistance to some antibiotics (Wray *et al.*, 1992), in Germany most *Salmonella* isolates from turkeys tested *in vitro* were sensitive to all antimicrobials used (Stadler, 1995). On the other hand, Hirsh *et al.* (1983) were able to isolate R plasmid-mediated gentamicin-resistant *S. arizonae* and *S. thompson* from turkey poults, hatching eggs and litter. Likewise, Wray *et al.* (1998) reported in 1996 that 41% of *Salmonella* isolates from turkeys in the UK were resistant to three or more antimicrobials, primarily because of the presence of multiply resistant *S. typhimurium* DT104. In addition, 75% of the DT104 isolated showed reduced susceptibility to fluoroquinolones. Ekperigin *et al.* (1983) controlled severe outbreaks of disease in flocks of poults caused by a gentamicin-resistant *S. arizonae* strains by the use of oral and parenteral administration of tetracyclines. In addition, antimicrobials can promote the development of L forms.

Sensitivity testing of the involved *Salmonella* isolates should be done concurrently with the commencement of medication, because the results of resistance tests in different regions are not comparable and the sensitivity of *Salmonella* varies from time to time.

The European Commission has ruled that chloramphenicol and furazolidone cannot be used in food-production animals, including poultry, since maximum residue limits cannot be established (EEC 2377/90, Annex IV).

Control of *Salmonella* Infections in Turkey

In countries with intensive poultry production it has been determined that, under current conditions, it would be very difficult to eliminate *Salmonella* contamination in poultry production. However, the possibility of eliminating host-specific serovars and reducing non-host-specific invasive serovars (paratyphoid) is realistic. In 1992, the European Union adopted a directive to monitor and control *Salmonella* infections (Zoonoses Directive 92/117/EEC) in breeding flocks of domestic fowl. However, the directive does not include measures for control of *Salmonella* in turkey breeding flocks. Studies in Canada have shown that *Salmonella* are widespread in the environment of turkey farms; Irwin *et al.* (1994) recovered *Salmonella* from 9.8% of feed samples, 79.6% of litter samples and 80.4% of dust samples. Many countries, however, have introduced control strategies which are based on different approaches, with the goal of reducing and eliminating *Salmonella* infections in turkeys.

In general, the major strategy for the control of *Salmonella* should include the following.

1. Cleaning the production chain from the top.
2. Feed hygiene.
3. Feed additives.
4. Competitive exclusion (CE).
5. Vaccination.
6. Education programmes.
7. Hygienic measures.
8. Eradication.

Cleaning the production chain from the top

The success or failure of any *Salmonella* reduction programme begins at the primary breeder level, the suppliers of seed stock to most of the world's turkey industry. The primary breeders must be committed to reducing and eliminating *Salmonella* in their stock and also in grandparent stocks (Pomeroy, 1991).

The major strategy for the control of *Salmonella* in poultry should now be directed to cleaning the production chain from the top in order to prevent the vertical transmission of *Salmonella*. Control measures to prevent the introduction and spread of *Salmonella* infection in breeder flocks should concentrate on high standards of animal management, with bacteriological and serological monitoring of breeding birds. These measures must be coupled with meticulous attention to all stages of hatching egg production.

All eggs should be collected not less than three times daily and the shells should be disinfected soon after collection on the farm, since the penetration of the shell by microorganisms is particularly rapid. If the bacteria penetrate the shell before the egg reaches the hatchery, it is difficult to find an effective method to counteract such contamination (Clayton *et al.*, 1985). Generally, two methods are used to disinfect turkey hatching eggs under field conditions, namely, fumigation and dipping in a solution of detergent or disinfectant.

Fumigation is best done with formaldehyde gas for at least 20 min, with a concentration of 35 ml formalin mixed with 17.5 g potassium permanganate and 20 ml water m^{-3} space. Temperature during fumigation must be maintained at a minimum of 20–24°C and relative humidity at 70%. The eggs should be placed in trays that will permit the fumigant to contact as much of the shell surface as possible. After fumigation, hygienic measures should be followed to preclude recontamination. Because of the unpleasant nature of formaldehyde gas and its possible health hazards to the operator, some owners elect to use wet treatments. Different sanitizing solutions are used and most of them are based on chlorine, glutaraldehyde or quaternary ammonium compounds. Weand and Horsting (1978) reported that chlorine, quaternary ammonium compounds and formalin are the most effective and practical disinfectants for sanitizing hatching eggs.

Dipping eggs in detergents or disinfectants is highly effective in greatly reducing or eliminating the bacteria from the shell, when performed correctly. However, there is little or no effect on those bacteria that have already penetrated the shell (Baxter-Jones and Wilding, 1981). The manufacturer's instructions for the chemicals used should be followed, particularly those concerning the number of eggs that may be dipped per litre of solution and how often fresh solution has to be provided. Attention must also be directed to the temperature of the detergent, which must be higher than the egg temperature.

Eggs sent to the hatchery should not be dry-cleaned, as damage to the cuticle increases the risk of subsequent microbial penetration.

Hatcheries must be designed to permit only a one-way flow of traffic from the egg entry room through egg trays, incubation, hatching and holding rooms to the van-loading area. The ventilation system must prevent recirculation of contaminated air. Trays used in the hatchery should be thoroughly cleaned and disinfected before eggs are placed on them. Fumigation and disinfection programmes should not be used to replace cleanliness but to support it. All eggs should be sanitized on arrival at the hatchery (presetting treatment), using fumigation. Additionally fumigation can be carried out after setting. This provides a final disinfection following handling, transport and various environmental contaminations, during storage of hatching eggs. Further fumigations are mostly carried out immediately after the transfer of hatching eggs from the setter to the hatcher.

Dipping turkey hatching eggs in disinfectant and/or antibiotics, using temperature-differential dipping (TDD) or pressure-differential dipping (PDD), to control egg-transmitted *Salmonella* and other bacterial pathogens has been widely investigated and is of great value. Baxter-Jones and Wilding (1981) pointed out that egg treatment methods to reduce egg-borne *Salmonella* should: (i) not be antibiotics, in order to avoid resistance problems; (ii) be effective for at least 48 h; (iii) affect neither hatchability nor poult quality; and (iv) be inexpensive, easy to use and safe for the staff. Use of disinfectants in TDD on turkey hatching eggs artificially infected with *S. typhimurium* resulted in elimination of the infection. Similar results have been obtained by treatment of broiler chicken hatching eggs experimentally infected with *S. senftenberg* (Mandel *et al.*, 1987). However, application of the TDD and PDD methods, using different disinfectants, on hatching eggs from layer birds artificially infected with *S. enteritidis* did not significantly reduce the *S. enteritidis* reisolation rate of newly hatched chicks (Jodas, 1992). Lucas *et al.* (1970) dipped infected turkey hatching eggs in kanamycin, neomycin and spectinomycin. The first two were able to significantly reduce *S. saintpaul* but not *S. typhimurium*. Investigation by Saif *et al.* (1971) revealed that *S. arizonae* could be eliminated from artificially infected turkey hatching eggs using the TDD-method with gentamicin sulphate. In addition, Saif and Shelly (1973) were able to reduce the reisolation rate significantly, using the TDD method and 1000 p.p.m. gentamicin sulphate on turkey hatching eggs artificially infected with 25 different *Salmonella* strains. To control the egg transmission of *S. arizonae*, Ghazikhanian *et al.* (1984) dipped preheated turkey hatching eggs for 2–3 min into 400 p.p.m. gentamicin sulphate and 300 p.p.m. quaternary ammonium compound. The eggs were then injected via the small end with 0.6 mg of gentamicin per egg. Results indicated that the proper application of the dual hatching-egg treatment with effective antibiotic would totally remove *S. arizonae* contamination distributed in different segments of hatching eggs.

Techniques and equipment for dipping turkey hatching eggs, using PDD in solutions of antibiotics are now commercially available. These methods are largely used for eradication of *Mycoplasma* infection in turkeys but are still used to combat other egg-borne infections, including *Salmonella*.

Some turkey hatcheries in Germany wash the hatching eggs after delivery with disinfectants and water and then dry them with hot air, followed by dipping in 1000 p.p.m. gentamicin and/or 500 p.p.m. enrofloxacin under a reduced pressure of 500 mbar for 5 min. The partial vacuum is then released and the eggs are allowed to soak in the antibiotics at atmospheric pressure for a further 10 min. After removal from the dip, eggs are allowed to drain and dry before setting in the incubator.

In a 2-year investigation of parent turkey flocks and a hatchery, different *Salmonella* serovars (*S. montevideo*, *S. mbandaka*, *S. braenderup* and *S. hadar*) were isolated from hatching eggs delivered to the hatchery from four out of six examined parent flocks. Bacteriological examinations of 485 samples collected from the hatchery (dead-in-shell, hatchery debris, meconium, 1-day old chicks, transport cartons) on 18 different hatching days failed to isolate *Salmonella*. The possibility of hatching eggshell contamination with isolated *Salmonella* serovars and vertical transmission to hatched poults could not be demonstrated, since all hatching eggs were sanitized by fumigation on the farm and PDD, using enrofloxacin, at the hatchery (Hafez *et al.*, 1997).

Precautions should be followed, since dipping solutions can become excessively contaminated with resistant microorganisms such as pseudomonads, and organic material. To prevent bacterial contamination of the solution, filtering with subsequent cool storage and/or addition of disinfectants is the most effective method. Thorough and continuous bacteriological monitoring of dip solution is also required. The concentration of the antibiotics must be examined regularly and renewed routinely. By using enrofloxacin, the pH value of the dipping solution can be corrected during storage. According to Froyman (1994), the use of egg dipping in antimicrobials should be critically evaluated, because of the irregular uptake of dip solution, uneven distribution of active substance in the egg compartments and lack of standardization in dipping technique.

Additionally, it is known that different disinfectants used for washing can negatively influence the antibiotic uptake of hatching eggs. Therefore it is recommended that the compatibility of different disinfectants used for egg washing and/or used in dipping solution is examined before application (Bickford et al., 1973; Horrox, 1987). Because the uptake of active agent by the hatching egg can be very irregular during dipping, individual egg injection with accurate delivery of the proper dose is preferred in breeding stock of élite and grandparent turkeys. Automated systems for in ovo drug disposition before hatch are being developed (Froyman, 1996).

Feed hygiene

Contaminated feed has long been recognized as the commonest source of new *Salmonella* serovars for poultry flocks. Investigations in different countries have shown that many poultry feed ingredients are contaminated with *Salmonella*. The level of contamination frequently varies between the time of feed manufacture and its delivery, which indicates the highly important role of transport in the recontamination of both raw materials and finished feed (McIllroy, 1996). Different approaches have been utilized to reduce the *Salmonella* contamination of feed ingredients, as well as finished feed. The subject is reviewed in Chapter 17, and the reader is also referred to

Häggblom (1993), who reviewed the Swedish control methods. Renggli (1996) summarized the different methods used and the level of decontamination achieved.

Feed additives

Short-chain organic acids

Chemical methods have been established to prevent the recontamination of finished feed. Short-chain organic acids (formic acid, propionic acid) have been used recently as feed additives. There are two fundamentally different applications: first, for the decontamination, very high acidification is needed (6% propionic acid) and, secondly, for the prevention of (re)contamination, lower doses (0.5–0.7%) are required and have been shown to reduce *Salmonella* colonization in birds consuming the treated feed (Hinton and Mead, 1991a). The antimicrobial effect of the organic acids decontaminates infected feed on contact and, as a further effect, its residual activity prevents subsequent reinfection (Hinton, 1996). The acids do not eliminate *Salmonella* from dry feed, but they kill the organisms in the crop of the birds when the feed has been moistened after consumption, by the combined effects of the acids, higher water activity and temperature. Recently, Berchieri and Barrow (1996) showed that formic and propionic acid (Bio-Add) in feed was able to reduce significantly the transmission of *S. gallinarum* strain 9R between birds after experimental infection. Also, the morbidity and mortality in 1-week-old birds were significantly lower (33.3%) in a group receiving acid-treated feed, compared with 75.6% in a group receiving untreated feed. The results suggest that some protection could be possible against other invasive *Salmonella*, such as *S. pullorum*, *S. enteritidis* and *S. arizonae*, and that the treatment might be used to reduce the extent of vertical transmission by these pathogens. Treatment of feed with organic acids must be considered as an important support to good hygiene and husbandry on all production chains.

Carbohydrates

Other feed additives, such as carbohydrates (lactose, mannose, galactose, saccharose), which are able to influence the caecal environment by increasing the amount of acid produced by bacterial fermentation, thus decreasing the pH, have been found to reduce *Salmonella* colonization (Oyofo *et al.*, 1989; DeLoach *et al.*, 1990; Hinton and Mead, 1991b). Corrier *et al.* (1991) found that the addition of lactose to the diet of turkey poults decreased the caecal pH and, when it was combined with volatile fatty acid (VFA)-producing anaerobic caecal microflora, markedly increased the concentration of undissociated VFA, thus preventing *Salmonella* colonization. Other sugars within more complex carbohydrates have been successful (Bailey *et al.*, 1991). Further information is to be found in Chapter 18.

Antimicrobials

Certain antimicrobials can be used as feed additives to prevent, but not to treat, clinical salmonellosis in poultry. Few antimicrobials are effective in reducing faecal shedding and they may prolong excretion. The antibiotics not only affect *Salmonella* but also several types of gut flora that are inhibitory for *Salmonella*.

Probiotics

Probiotics are products that are able to proliferate in the intestinal tract and beneficially affect the host animal by improving its intestinal microbial balance and consequently enhance the growth, production and health of farm animals. The word probiotic is derived from the Greek meaning 'for life' (Fuller, 1992). They are mainly composed of lactobacilli, streptococci, bifidobacteria, bacilli and yeasts. These microorganisms are able to inhibit the growth of potentially pathogenic microorganisms by lowering the pH through production of lactate, lactic acid and VFA (Mulder, 1996). The use of probiotics in chickens was reviewed by Barrow (1992b).

In turkeys, the *in ovo* inoculation at the time of transfer of embryonated eggs from the setter to the hatcher or spray application of *Lactobacillus reuteri* has beneficial effects in promoting greater viability, associated with enhanced colonization of the caeca with *L. reuteri* and more rapid shedding of *Salmonella* when the birds were challenged with either *S. typhimurium* or *S. senftenberg* at hatch or on the day after hatch (Edens *et al.*, 1991; Edens and Casas, 1992).

Casas *et al.* (1993) reported the use of *L. reuteri* in turkey under field conditions. Three applications using a spray formulation in the

hatchery, the first at 15% pipping (26 days of incubation), the second at 40–60% pipping and the third at 12 h before taking the poults out of the hatcher, resulted in good protection against in-hatcher contamination with *S. typhimurium*. Feed application up to 5–6 weeks of age with meat turkeys resulted in improved viability, body weight and feed conversion. Similar results have been found by Damron *et al.* (1981). Improvement of body weight and feed intake was reported in male turkeys at 20 weeks of age (Jirophocaked *et al.*, 1990) after using dried *Bacillus subtilis* culture.

However, in a number of trials in chickens, in which birds have been given the probiotic product and then challenged with small numbers of *Salmonella*, in no instance did administration of the probiotic reduce the *Salmonella* population of the caeca (Hinton and Mead, 1991a; Stavric *et al.*, 1992; Bolder *et al.*, 1993). Hinton and Mead (1991b) concluded that it is unlikely that the present formulations of probiotics will have a part to play in the control of *Salmonella* infections in poultry.

Competitive exclusion

CE, also named the 'Nurmi concept' or exclusion flora (EF), in combination with conventional hygienic measures, has been shown to be very effective as a preventive measure against *Salmonella* infection in poultry, and further details will be found in Chapter 18. Schneitz and Nurmi (1996) also presented an excellent review on CE, with special reference to history, development, safety, administration in chickens and turkeys, benefits and mechanism of function. CE is a culture of an undefined mixture of microorganisms from the crop and intestinal-tract contents of adult birds (Nurmi and Rantala, 1973) or defined cultures (Impey *et al.*, 1984; Stavric, 1992). Commercial CE products are accepted and used in several countries today (see Chapter 18). However, reluctance of authorities to grant licences for undefined products is one reason for the slow market penetration (Nurmi *et al.*, 1995). This may also result in attempts to introduce characterized CE cultures as defined ones.

CE should be applied to newly hatched chicks or turkey poults as soon as possible at the hatchery or farm. Administration can be carried

out by spray in the hatchery, using either a garden sprayer or an automatic spray cabinet, similar
to that used for Newcastle disease or infectious
bronchitis vaccine. The droplet size should be at
least 1 mm in diameter (Pivnick and Nurmi,
1982; Goren *et al.*, 1989; Mead, 1994).
According to Mulder (1996), however, the
administration of the flora in the hatchery causes
some opposition, because it is illogical to spray a
large number of microorganisms in an environment which has, by all possible means, to be kept
more or less sterile to prevent contamination. An
alternative is to apply the CE preparation via
drinking-water when chicks or poults are
delivered to the farm (Schneitz and Nuotio,
1992; Wierup *et al.*, 1992; Mead, 1994). Also,
spray application in the hatchery followed by
drinking-water administration on the farm has
been found to be effective in controlling
Salmonella in commercial broiler chickens
(Blankenship *et al.*, 1993). Further applications
in older birds, after antibiotic treatment to regenerate the intestinal microflora, have shown satisfactory results. According to Fowler and Mead
(1990), the combination of antibiotic therapy
and CE was able to reduce the *Salmonella* infection rate in broiler breeder flocks and the vertical
transmission to the end-product. Inoculation of
turkeys with intestinal extracts has also been
demonstrated to increase resistance to *Salmonella*
colonization (Lloyd *et al.*, 1977). Turkey poults
are fully protected with CE prepared from chickens or turkeys (Impey *et al.*, 1984; Reid and
Barnum, 1984; Seuna *et al.*, 1984; Schneitz and
Nuotio, 1992).

The results of Corrier *et al.* (1991) indicate
that intracloacal inoculation of newly hatched
poults with intestinal flora from adult chickens or
addition of lactose to the feed both significantly
decreased caecal colonization by *S. senftenberg*.
Furthermore, combined treatment with intestinal
flora and provision of dietary lactose resulted in
generally lower levels of *Salmonella* colonization
than did either of the two treatments alone.

Many authors reported that treatment must
precede exposure to *Salmonella* to be protective
(Anderson *et al.*, 1984). However, evidence
exists that CE treatment given after *Salmonella*
challenge reduces the number of *Salmonella* in
the chicken caeca and the number of infected
birds in a flock (Schneitz and Nurmi, 1996). For
the control of *Salmonella* in turkeys, use of caecal

cultures may also be limited by the interference
of antibiotics (Anderson *et al.*, 1984).

Vaccination

Vaccines may be useful in controlling clinical salmonellosis caused by host-adapted or invasive
serovars. However, the application of vaccines to
control colonization with non-invasive
Salmonella is unlikely to be effective since
humoral or cell-mediated immunity will have little influence on events in the lumen of the gastrointestinal tract (Barrow, 1990). In addition,
there is evidence that highly invasive strains are
likely to stimulate a stronger immune response
and to be eliminated earlier than would occur
with less invasive strains (Barrow *et al.*, 1988).

Live and inactivated (killed) vaccines are
used to control *Salmonella* in poultry, though
there appear to be no reports of the use of live
vaccines in turkeys. McCapes *et al.* (1967) studied the use of a *Salmonella* bacterin in turkey
breeders, particularly to determine whether any
parental resistance would be passed to their
poults. Poults originating from *S. typhimurium*-
vaccinated hens exhibited resistance to yolk-sac
challenge with both the B group *S. typhimurium*
and *S. schwarzengrund*, but not with the E group
S. anatum. According to Thain *et al.* (1984),
administration of an inactivated *S. hadar* vaccine
to turkey breeding stock may be of value in limiting the spread of this serovar in young poults.
They found that the use of this vaccine produced
high levels of immunoglobulin G (IgG) antibodies, which were passed on through the egg to the
poults. Also, successful vaccination of turkey
breeders for the control of *Salmonella* with autogenous mineral-oil adjuvant vaccines, prepared
from the serovars *S. sandiego* and *S. arizonae*, was
applied by Nagaraja *et al.* (1988). Their results
suggested that OMP of the organism give better
protection than formalin-killed whole-cell bacterin. Ghazikhanian *et al.* (1984) showed that
using autogenous oil-emulsion *S. arizonae* bacterin resulted in a significant reduction of the
overall egg transmission rate in vaccinated challenged hens compared with non-vaccinated challenged turkeys.

The use of subunit vaccines prepared from
OMP from *S. heidelberg*, incorporating them into
lipid-conjugated immunostimulating complexes

(ISCOMs), for protection against homologous and heterologous *Salmonella* challenge in turkeys has been studied by Charles *et al.* (1993). The reisolation rate of *Salmonella* from internal organs after challenge with *S. heidelberg, S. reading* or *S. enteritidis* in turkeys was completely negative for the homologous serovar and significantly lower for the heterologous serovar in vaccinated turkeys. The results also indicate that fewer turkeys shed *Salmonella* after vaccination with ISCOM preparations than after vaccination with OMP alone.

Education programmes

Since the success of any disease control programme depends on the farm and personal sanitation, it is essential to deliver education programmes about microorganisms, modes of transmission as well as awareness of the reasons behind such control programmes to people involved in poultry production. In addition, effective education programmes must be implemented to increase public awareness of the necessary measures to be taken for protection against *Salmonella* in food products from turkeys. Finally, research must continue to find additional control and preventive means.

Hygienic measures

Turkey houses should be kept locked and visitors not allowed to enter without permission. Further precautions related to staff should be taken, including regular bacteriological examination to identify the carriers and to prevent transmission and cross-contamination on the farm. The all-in, all-out principle should be adopted wherever possible. Multiple ages of birds on a farm constitute a serious disease risk, in particular if multiple-age birds are closely associated. When a turkey house is depopulated, all droppings and litter should be removed from the house prior to cleaning. Cleaning, disinfection and vector control must be integrated in a comprehensive *Salmonella* control programme. The procedure should be tailored to meet the particular needs. The cleaning and disinfection programme should include the time schedule, type of disinfectant and concentration and desired level, as well as check and

microbiological monitoring of the procedures. The procedures should be established not only for cleaning and disinfecting the house and surfaces but also for cleaning and disinfection the equipment that is itself used for cleaning. Rodents, especially rats and mice, are particularly important sources of *Salmonella* contamination of poultry houses. Intensive and sustained rodent control is essential and needs to be well planned and routinely performed, and its effectiveness should be monitored. Household pets also constitute a serious *Salmonella* hazard. Buildings should therefore be animal-proof. Guidelines on cleaning, disinfection and vector control in *Salmonella*-infected poultry flocks have been given by WHO (Anon., 1994).

Additional hygiene measures are leaving a house empty for 2–4 weeks before a new flock is placed and restocking with poults from a known *Salmonella*-free source, if possible. The rearing of several successive flocks on the same litter should be avoided.

Eradication

Pullorum disease and fowl typhoid have been nearly or totally eradicated from countries with a modern turkey industry through consequent testing and eradication of the infected breeding flocks. A similar development appears to have been achieved with *S. arizonae* and, to a lesser extent, with *S. typhimurium* in turkey breeding stocks. The methods of testing for detection of carriers and the measures adopted are described elsewhere.

Public Health Aspects

In spite of significant improvements in technology and hygienic practice at all stages of poultry production, accompanied by advanced improvement in public sanitation, salmonellosis remains a persistent threat to human and animal health. In many countries, the high incidence of salmonellosis in humans appears to be caused by infection derived from contaminated poultry meat, including turkeys. According to Dubbert (1988), even with its apparently high contamination rate, meat poultry has not been the leading cause of food poisoning. Statistics gathered by the Centers for

Disease Control in the USA from 1973 to 1983 show that turkey was implicated in 9% of salmonellosis outbreaks in which the vehicle was known, while beef or veal accounted for 19% of the outbreaks. In addition, the poultry industry does far more extensive monitoring than other meat producers, and the number of annual isolations from poultry compared with that from other meats reflects the level of testing, rather than the level of contamination (Murray, 1992). However, with the high prevalence and contamination rate of turkey meat and products with *Salmonella*, it is not surprising that turkey products are an important source of human *Salmonella* infection (Gensheimer, 1984; Smith *et al.*, 1985; Todd, 1987; Yule *et al.*, 1988; De Boer and Van der Zee, 1992; Hafez *et al.*, 1997). Bentley (1984) reported that 70% of poultry-associated food-borne salmonellosis in Canada derived from turkey meat. In the turkey industry, the practice of further processing turkey carcasses for sale as portion, sausage or burgers, etc. has increased worldwide. The greater proportion of turkey meat is still used, however, in the catering industry. The contaminated products cause disease as a result of inadequate cooking or cross-contamination of working surfaces in the kitchen environment. Large outbreaks are mostly associated with the food prepared in food-service establishments, such as hotels, restaurants and institutions, and by catered foods.

In the UK until 1980, the most significant serovar isolated from turkeys has been *S. hadar*, which was first identified in 1969 in poultry offal meal imported from Israel. In 1973, turkey breeding flocks were infected from feed, with rapid spread in both turkeys and chickens, and by 1976 this serovar had become the fourth most frequently isolated *Salmonella* from humans in England and Wales (Watson and Kirby, 1984). In an analysis of the factors contributing to the more than 1000 outbreaks of food poisoning in England and Wales between 1970 and 1979, Roberts (1983) concludes that the food most frequently responsible was poultry. In 168 human outbreaks, turkey meat was implicated.

New approaches to the problem of contamination must be adopted and the discussion on the decontamination of the end-product must be re-evaluated carefully and without emotion. In general, the basic conception of the problem and what needs to be done to solve it have remained the same.

References

Anderson, W.R., Mitchell, W.R., Barnum, D.A. and Julian, R.J. (1984) Practical aspects of competitive exclusion for the control of *Salmonella* in turkeys. *Avian Diseases* 28, 1071–1078.

Anon. (1989) *National Poultry Improvement Plan and Auxiliary Provisions*. Animal and Plant Health Inspection Service, United States Department of Agriculture, Hyattsville, Maryland.

Anon. (1994) *World Health Organization Guidelines on Cleaning, Disinfection and Vector Control in* Salmonella *Infected Poultry Flocks*. WHO/Zoon 94.172, WHO, Geneva.

Anon. (1995) Salmonella *in Livestock Production*. Veterinary Laboratories Agency, MAFF, New Haw, Addlestone, pp. 43–55.

Bailey, J.S., Blankenship, L.C. and Cox, N.A. (1991) Effect of fructo-oligosaccharide on *Salmonella* colonisation of the chicken intestine. *Poultry Science* 70, 2433–2438.

Barrow, P.A. (1990) Immunological control of *Salmonella* in poultry. In: *Proceedings of the International Symposium on Control of Human Bacterial Enteropathogens in Poultry*. Academic Press, Atlanta, Georgia, pp. 199–217.

Barrow, P.A. (1992a) Chemotherapeutic and growth-promoting antibiotics and salmonellae in poultry. In: Hinton, M.H. and Mulder, R.W.A.W. (eds) *Proceedings of Prevention and Control of Potentially Pathogenic Microorganisms in Poultry and Poultry Meat Processing 7. The Role of Antibiotics in the Control of Foodborne Pathogens*. Het Spelderholt Agriculture Research Department DLO-NL, Beekbergen, The Netherlands, pp. 111–115.

Barrow, P.A. (1992b) Probiotics for chickens. In: Fuller, R. (ed.) *Probiotics*. Chapman and Hall, London, pp. 1–8.

Barrow, P.A., Simpson, J.M. and Lovell, M.A. (1988) Intestinal colonization in the chicken by food-poisoning *Salmonella* serovars: microbial characteristics associated with faecal excretion. *Avian Pathology* 17, 571–588.

Baxter-Jones, C. and Wilding, G.P. (1981) *Salmonella* contamination of turkeys: recent progress towards improved control. *Hohenheimer Arbeiten* 116, 21–30.

Baxter-Jones, C. and Wilding, G.P. (1982) Controlling egg-borne pathogens. *Poultry International* 6, 12–22.

Bentley, A.H. (1984) The *Salmonella* situation in Canada. In: Snoeyenbos, G.H. (ed.) *Proceedings of the International Symposium on Salmonella*. American Association of Avian Pathologists, Kennett Square, Pennsylvania, pp. 54–61.

Berchieri, A. and Barrow, P.A. (1996) Reduction in incidence of experimental fowl typhoid by ncorporation of a commercial formic-acid preparation (Bio-Add™) into poultry feed. *Poultry Science* 75, 339–341.

Bickford, S.M., Soy, J.I. and Barnes, L.E. (1973) Gentamycin concentrations in turkey eggs and in tissues of progeny following egg dipping. *Avian Disease* 17, 301–307.

Bierer, W.B. (1960) Effect of age factor on mortality in *Salmonella typhimurium* infection in turkey poults. *Journal of American Veterinary Medicine Association* 137, 657–658.

Blankenship, L.C., Bailey, J.S., Cox, N.A., Stern, N.J., Brewer, R. and Williams, O. (1993) Two-step mucosal competitive exclusion flora treatment to diminish salmonellae in commercial broiler chickens. *Poultry Science* 72, 1667–1672.

Bolder, N.M., van Lith, L.A.J.T., Putirulan, F.F. and Mulder, R.W.A.W. (1993) In: Jenson, J.F., Hinton, M.H. and Mulder, R.W.A.W. (eds) *Proceedings of Prevention and Control of Potentially Pathogenic Microorganisms in Poultry and Poultry Meat Processing 12. Probiotics and Pathogenicity*. Het Spelderholt Agriculture Research Department DLO-NL, Beekbergen, The Netherlands, pp. 63–72.

Boyer, C.I., Narotsky, S., Bruner, S. and Brown, D.W. (1962) Salmonellosis in turkeys and chickens associated with contaminated feed. *Avian Disease* 6, 43–50.

Brunner, F., Margadant, A., Peduzzi, R. and Piffaretti, J.C. (1983) The plasmid pattern as an epidemiological tool for *Salmonella typhimurium* epidemics: comparison with the lysotype. *Journal of Infectious Diseases* 148, 7–11.

Bryan, F.T., Ayres, J.C. and Kraft, A.A. (1968) Contributory sources of salmonellae on turkey products. *American Journal of Epidemiology* 87, 578–591.

Bulls, K.L. (1977) The history of avian medicine in the US. II. Pullorum disease and fowl typhoid. *Avian Diseases* 21, 422–429.

Caldwell, M.E. and Ryerson, D.L. (1939) Salmonellosis in certain reptiles. *Journal of Infectious Disease* 65, 242–245. (Cited in Greenfield *et al.*, 1971.)

Casas, I.A., Edens, F.W., Dobrogosz, W.J. and Parkhurst, C.R. (1993) Performance of GAIAfeed and GAIAspray: *Lactobacillus reuteri*-based probiotic for poultry. In: Jenson, J.F., Hinton, M.H. and Mulder, R.W.A.W. (eds) *Proceedings of Prevention and Control of Potentially Pathogenic Microorganisms in Poultry and Poultry Meat Processing 12. Probiotics and Pathogenicity*. Het Spelderholt Agriculture Research Department DLO-NL, Beekbergen, The Netherlands, pp. 63–72.

Chaplin, W.C. and Hamilton, C.M. (1957) A synovitis in turkeys produced by *Salmonella thompson*. *Poultry Science* 36, 1380–1381.

Charles, S.D., Nagaraja, K.V. and Sivanandan V. (1993) A lipid-conjugated immunostimulating complex subunit vaccine against *Salmonella* infection in turkeys. *Avian Diseases* 37, 477–484.

Clayton, G.A., Lade, R.E., Nixey, C., Jones, D.R., Charles, D.R., Hopkins, J.R. Binstead, J.A. and Prickett, R. (1985) *Turkey Production: Breeding and Husbandry*. MAFF, HMSO, London.

Corrier, D.E., Hinton, A., Kubena, L.F., Zirpin, R.L. and DeLoach, J.R. (1991) Decreased *Salmonella* colonisation in turkey poults inoculated with anaerobic caecal microflora and provided dietary lactose. *Poultry Science* 70, 1345–1350.

Damron, B.L., Wilson, H.R., Voitle, R.A. and Harms, R.H. (1981) A mixed *Lactobacillus* culture in the diet of Broad Breasted Large White turkey hens. *Poultry Science* 60, 1350–1351.

Davies, R.H. and Wray, C. (1994) *Salmonella* pollution in poultry units and associated enterprises. In: Dewi, A., Axford, R., Marai, F.M. (eds) *Pollution in Livestock Production System*. CAB International, Wallingford, UK, pp. 137–165.

De Boer, E. and Van der Zee, H. (1992) *Salmonella* in food animal origin in the Netherlands. In: *Proceedings of Salmonella and Salmonellosis*. Zoopôle Developpment Ploufragan/Saint-Brieuc, France, pp. 265–272.

DeLay, P.D., Jackson, T.W., Jones, E.E. and Stover, D.E. (1954) A testing service for the control of *Salmonella typhimurium* infection in turkeys. *American Journal of Veterinary Research* 15, 122–129.

DeLoach, J.R., Oyofo, B.A., Corrier, D.E., Kubena, R.L., Zirpin, R.L. and Mollenhauer, H.H. (1990) Reduction of *Salmonella typhimurium* concentration in broiler chickens by milk or whey. *Avian Diseases* 34, 389–392.

Dräger, H. (1971) Salmonellosen – Ihre Entstehung und Verhuetung. Akademie Verlag, Berlin, 313 pp.

Dubbert, W. (1988) Assessment of *Salmonella* contamination in poultry – past, present, and future. *Poultry Science* 67, 944–949.

Edens, F.W. and Casas, I.A. (1992) GAIAfeed™ improves commercial turkey performance. In: *Proceedings of the 19th World Poultry Congress* Vol. 3. Ponsen and Looijen, Wageningen, Amsterdam, The Netherlands, pp. 91–92.

Edens, F.W., Parkhurst, C.R. and Casas, I.A. (1991) *Lactobacillus reuteri* and whey reduce *Salmonella* colonization in the ceca of turkey poults. *Poultry Science* 70 (Suppl. 1).

Edwards, P.R. and McWorther, A.C. (1953a) Two new *Salmonella* types: *Salmonella harrisonburg* and *Salmonella westhampton*. *Cornell Veterinarian* 43, 110–111.

Edwards, P.R. and McWorther, A.C. (1953b) A new *Salmonella* type: *Salmonella berkeley*. *Cornell Veterinarian* 43, 572–573.

Edwards, P.R., Cherry, W.B. and Bruner, D.W. (1943) Further studies on coliform bacteria serologically related to the genus *Salmonella*. *Journal of Infectious Disease* 73, 229–238.

Eisenstein, B.I. (1990) New molecular techniques for microbial epidemiology and the diagnosis of infectious diseases. *Journal of Infectious Disease* 161, 595–602.

Ekperigin, H.E., Jang, S. and McCapas, R.H. (1983). Effective control of a gentamicin-resistant-*Salmonella arizonae* infection in turkey poults. *Avian Diseases* 27, 822–829.

Faddoul, G.P. and Fellows, G.W. (1966) A five-year survey of the incidence of salmonellae in avian species. *Avian Diseases* 10, 296–304.

Fowler, N.G. and Mead, G.C. (1990) Competitive exclusion and *Salmonella enteritidis*. *Veterinary Record* 126, 489.

Fricker, C.R. (1987) The isolation of salmonellas and campylobacters. *Journal of Applied Bacteriology* 63, 99–116.

Froyman, R. (1994) The potential use of enrofloxacin (Baytril™), either alone or combined with competitive exclusion flora, in *Salmonella* sanitation programs for poultry. Unpublished report of the WHO-FEDESA-FEP Workshop on Competitive Exclusion, Vaccination and Antibiotics in *Salmonella* Control in Poultry, WHS/CDS/VOPH/94.134 WHO, Geneva.

Froyman, R. (1996) Antimicrobial medication in domestic poultry. In: Jordan, F.T.W. and Pattison, M. (eds) *Poultry Disease*, 4th edn. Cambridge University Press, Cambridge, pp. 484–494.

Fuller, R. (1992) History and development of probiotics. In: Fuller, R. (ed.) *Probiotics*. Chapman and Hall, London, pp. 1–8.

Gensheimer, K.F. (1984) Poultry giblet-associated salmonellosis. Maine. *Morbidity and Mortality Weekly Report* 33, 630–631. Georgia. *Morbidity and Mortality Weekly Report* 34, 707–708.

Ghazikhanian, G.Y., Kelly, B.J. and Dungan, W.M. (1984) *Salmonella arizonae* control program. In: Snoeyenbos, G.H. (ed.) *Proceedings of the International Symposium on Salmonella*. American Association of Avian Pathologists, Kennett Square, Pennsylvania, pp. 142–149.

Goren, E., deJong, W.A., Doornenbal, P., Bolder, N.M., Mulder, R.W.A.W. and Jansen, A. (1989) Protection of broilers against *Salmonella* infection by spray application of intestinal microflora: laboratory and field studies. *Proceedings of the 38th Western Poultry Disease Conference*, Tempe, Arizona, pp. 212–218.

Greenfield, J., Bigland, C.H. and Dukes, T.W. (1971) The genus *Arizona* with special reference to *Arizona* disease in turkeys. *Veterinary Bulletin* 41, 605–612.

Guillot, J.F. and Milleman, Y. (1990) Intestinal colonization of chickens and turkeys by *Salmonella* and antibiotic decontamination. In: Mulder, R.W.A.W. (ed.) *Proceedings of Prevention and Control of Potentially Pathogenic Microorganisms in Poultry and Poultry Meat Processing. 1. Colonisation Control*. Het Spelderholt Agricultural Research Service (DLO-NL), Beekbergen, The Netherlands, pp. 31–44.

Hafez, H.M. and Jodas, S. (1992) Effect of sample selection from hatching eggs on *Salmonella enteritidis* detection rate. *Deutsche Tieraerztliche Wochenschrift* 99, 489–490.

Hafez, H.M. and Stadler, A. (1997) *Salmonella enteritidis* colonization in turkey poults. *Deutsche Tieraerztliche Wochenschrift* 104, 118–120.

Hafez, H.M., Mandl, J. and Woernle, H. (1986) Einfache und empfindliche Methode zum Nachweis von Salmonellen in Eiern. 2. Mitteilung: Dekontamination mit Sysovet-PA und Schlupfergebnisse. *Deutsche Tieraerztliche Wochenschrift* 93, 37–38.

Hafez, H.M., Stadler, A. and Jodas, S. (1993) *Salmonella* isolation in poultry and factors influencing: review. In: Fanchini, A. and Mulder, R.W.A.W. (eds) *Proceedings of Prevention and Control of Potentially Pathogenic Microorganisms in Poultry and Poultry Meat Processing. 13. Consequences of the Zoonosis Order: Monitoring, Methodology and Data Registration*. Het Spelderholt Agricultural Research Service (DLO-NL), Beekbergen, The Netherlands, pp. 65–72.

Hafez, H.M., Stadler, A. and Kösters, J. (1997) Surveillance on *Salmonella* in turkey flocks and processing plants. *Deutsche Tieraerztliche Wochenschrift* 104, 33–35.

Häggblom, P. (1993) Monitoring and control of *Salmonella* in animal feed. In: Bengtson, S.O. (ed.) *Proceedings of International Course on Salmonella Control in Animal Production and Products. A Presentation of the Swedish Salmonella Programme*. WHO, Malmö, pp. 127–137.

Hewitt, E.A. (1928) Bacillary white diarrhea in baby turkeys. *Cornell Veterinarian* 18, 272–276. (Cited in Hinshaw, 1959.)

Higgins, W.A., Christiansen, J.B. and Schroeder, C.H. (1944) A *Salmonella enteritidis* infection associated with leg deformity in turkeys. *Poultry Science* 23, 340–341.

Hinshaw, W.R. (1930) Fowl typhoid of turkeys. *Veterinary Medicine* 25, 514–517.

Hinshaw, W.R. (1959) Diseases of the turkeys. In: Biester, H.E. and Schwarte, L.H. (eds) *Diseases of Poultry*, 4th edn. Iowa State University Press, Ames, Iowa, pp. 974–1076.

Hinshaw, W.R. and McNeil, E. (1940) Eradication of pullorum disease from turkey flocks. In: *Proceeding of 44th Annual Meeting United States Livestock Sanitation Association*, pp.178–194. (Cited in Snoeyenbos, 1991.)

Hinshaw, W.R. and McNeil, E. (1946) The occurrence of type 10 paracolon in turkeys. *Journal of Bacteriology* 53, 281–280.

Hinton, M. (1996) Organic acids for control of *Salmonella*. *World Poultry Misset* Salmonella Special, 33–34.

Hinton, M. and Mead, G.C. (1991a) Control of *Salmonella* infections in broiler chickens. In: Mulder, R.W.A.W. and Lan, C.A. (eds) *Proceeding of Prevention and Control of Potentially Pathogenic Microorganisms in Poultry and Poultry Meat Processing. 5. Zoonoses Control in Europe – Post 1993*. Het Spelderholt Agricultural Research Service (DLO-NL), Beekbergen, The Netherlands, pp. 25–27.

Hinton, M. and Mead, G.C. (1991b) Opinion. *Salmonella* control in poultry: the need for the satisfactory evaluation of probiotics for this purpose. *Letters in Applied Microbiology* 13, 49–50.

Hirschmann, R.-U. and Seidel, A. (1992) Die Entwicklung des *Salmonella*-Geschehens bei der Tierart Rind in Mecklenburg-Vorpommern von 1980 bis 1989. *Tieraerztliche Umschau* 47, 249–257.

Hirsh, D.C., Ikeda, J.S., Martin, L.D., Kelly, B.F. and Ghazikhanian, G.Y. (1983) R plasmid-mediated gentamicin resistance in salmonellae isolated from turkeys and their environment. *Avian Disease* 27, 766–772.

Horrox, N.E. (1987) *Mycoplasma* – to dip or not to dip? *International Hatchery Practice* 2, 13–17.

Hugh-Jones, M.E., Harvey, R.W.S. and McCoy, J.H. (1975) A *Salmonella california* contamination of a turkey feed concentrate. *British Veterinary Journal* 131, 673–680.

Impey, C.S., Mead, G.C. and George, S.M. (1984). Evaluation of treatment with defined and undefined mixtures of gut microorganisms for preventing *Salmonella* colonization in chicks and turkey poults. *Food Microbiology* 1, 143–147.

Irwin, R.J., Poppe, C., Menier, S., Finley, G.G. and Oggel, J. (1994). A national survey to estimate the prevalence of *Salmonella* species among Canadian registered commercial turkey flocks. *Canadian Journal of Veterinary Research* 58, 263–267.

Jirophocaked, S., Sullivan T.W. and Shahani, K.M. (1990) Influence of a dried *Bacillus subtilis* culture and antibiotics on performance and intestinal microflora in turkeys. *Poultry Science* 69, 1966–1973.

Jodas, S. (1992) Behandlungsverfahren zur Bekämpfung von Salmonella enteritidis bei künstlich infizierten Bruteiern (Legehybriden). Doctorate of Veterinary Medecine thesis, Munich University, Germany.

Jodas, S. and Hafez, H.M. (1997) Mykosen. In: Hafez, H.M. and Jodas, S. (eds) *Putenkrankheiten*. Ferdinand Enke Verlag, Stuttgart, pp. 130–134.

Kowalski, L.M. and Stephans, J.F. (1968) Arizona 7:1,7,8 infection in young turkeys. *Avian Diseases* 12, 317–329.

Kumar, M.C., York, M.D., McDowell, J.R. and Pomeroy, B.S. (1971) Dynamics of *Salmonella* infection in fryer roaster turkeys. *Avian Diseases* 15, 221–232.

Kumar, M.C., Olson, H.R. and Ausherman, L.T. (1972) Evaluation of monitoring programs for *Salmonella* infection in turkey breeding flocks. *Avian Diseases* 16, 644–648.

Kumar, M.C., Nivas, S.C., Bahl, A.K., York, M.D. and Pomeroy, B.S. (1974) Studies on natural infection and egg transmission of *Arizona hinshawii* 7:1,7,8 in turkeys. *Avian Diseases* 18, 416–426.

Kumar, M.C., York, M.D. and Pomeroy, B.S. (1977) Development of microagglutination test for detecting *Arizona hinshawii* 7:1,7,8 in turkeys. *American Journal of Veterinary Research* 38, 255–257.

Le Minor, L. and Popoff, M.Y. (1987) Request for an opinion. Designation of *Salmonella enterica* sp. nov. norm. rev. as the type and only species of the genus *Salmonella*. *International Journal of Systematic Bacteriology* 37, 465–468.

Lloyd, A.B., Cumming, R.B. and Kent, R.D. (1977) Prevention of *Salmonella typhimurium* infection in poultry by pretreatment of chickens and poults with intestinal extracts. *Australian Veterinary Journal* 53, 82–87.

Lucas, T.E., Kumar, M.C., Kleven, S.H. and Pomeroy, B.S. (1970) Antibiotic treatment of turkey hatching eggs preinfected with salmonellae. *Avian Diseases* 14, 455–462.

McBride, G.B., Brown, B. and Skura, B.J. (1978) Effect of bird type, growers and season on the incidence of salmonellae in turkeys. *Journal of Food Science* 43, 323–326.

McCapes, R.H., Coffland, R.T. and Christie, L.E. (1967) Challenge of turkey poults originating from hens vaccinated with *Salmonella typhimurium* bacteria. *Avian Diseases* 11, 15–24.

McCapes, R.H., Osburn, B.I. and Riemann, H. (1991) Safety of foods of animal origin: model for elimination of *Salmonella* contamination of turkey meat. *Journal of American Veterinary Medicine Association* 199, 875–880.

McIlroy, S.G. (1996) How did birds become infected by a *Salmonella* serovar. *World Poultry Misset* Salmonella *Special*, May, 15–17.

Mallinson, E.T. and Snoeyenbos, G.H. (1989) Salmonellosis. In: Purchase, H.G., Arp, L.H., Domermuth, C.H. and Pearson, J.E. (eds) *Isolation and Identification of Avian Pathogens*, 3rd edn. Kendall/Hunt, Dubuque, Iowa, pp. 3–11.

Mandel, J., Hafez, H.M., Woernle, H. and Kösters, J. (1987) Wirksamkeit unterschiedlicher Methoden zur Desinfektion Salmonella-kontaminierter Huehnerbruteier (Mastrasse) 2. Mitteilung: Dekontamination mit Lysovet-PA und Schlufpergebnisse. *Archiv für Gefluegelkunde* 51, 141–144.

Mead, G.C. (1994) Use of exclusion flora (EF) to control food-poisoning salmonellas in poultry in the United Kingdom. Unpublished report of the WHO-FEDESA-FEP Workshop on Competitive Exclusion, Vaccination and Antibiotics in *Salmonella* Control in Poultry, WHS/CDS/VOPH/94.134, WHO, Geneva.

Mitrovic, M. (1956) First report of paratyphoid infection in turkey poults due to *Salmonella reading*. *Poultry Science* 35, 171–174.

Moore, V.A. (1895) On a pathogenic bacillus of the hog-cholera group associated with a fatal disease in pigeons. *USDA BAI Bulletin* 8, 71–76. (Cited in Nagaraja *et al.*, 1991a.)

Moran, A.B. (1959) *Salmonella* in animals: a report for 1957. *Avian Diseases* 3, 85–88.

Mulder, R.W.A.W. (1996) Probiotics and competitive exclusion microflora against *Salmonella. World Poultry Misset* Salmonella *Special*, May, 30–32.

Murray, C. (1992) Zoonotic origin of human salmonellosis in Australia. In: *Proceedings of* Salmonella *and Salmonellosis*. Zoopôle Developpement Ploufragan/Saint-Brieuc, France, pp. 319–325.

Nagaraja K.V., Emery, D.A., Newman, J.A. and Pomeroy, B.S. (1984) Detection of *Salmonella arizonae* in turkey flocks by ELISA. In: *Proceeding of the 27th Annual Meeting of American Association of Veterinary Laboratory Diagnosis*, pp. 185–204. (Cited in Nagaraja *et al.*, 1991b.)

Nagaraja, K.V., Ausherman, L.T., Emery, D.A. and Pomeroy, B.S. (1986) Update on enzyme linked immunosorbent assay for its field application in the detection of *Salmonella arizonae* infection in breeder flocks of turkeys. In: *Proceedings of the 29th Annual Meeting of American Association of Veterinary Laboratory Diagnosis*, pp. 347–356.

Nagaraja K.V., Kim, C.J. and Pomeroy, B.S. (1988) Prophylactic vaccines for the control and reduction of *Salmonella* in turkeys. In: *Proceedings of 92nd Annual Meeting US Animal Health Association*, pp. 347–348.

Nagaraja K.V., Pomeroy, B.S. and Williams, J.E. (1991a) Paratyphoid infections. In: Calnek, B.W., Barnes, H.J., Reid, W.M. and Yoder, H.W., Jr (eds) *Diseases of Poultry*, 9th edn. Iowa State University Press, Ames, Iowa, pp. 99–130.

Nagaraja K.V., Pomeroy, B.S. and Williams, J.E. (1991b) Arizonosis. In: Calnek, B.W., Barnes, H.J., Reid, W.M. and Yoder, H.W., Jr (eds) *Diseases of Poultry*, 9th edn. Iowa State University Press, Ames, Iowa, pp. 130–137.

Nivas, S.C., Kumar, M.C., York, M.D. and Pomeroy, B.S. (1973) *Salmonella* recovery from three turkey-processing plants in Minnesota. *Avian Diseases* 17, 605–616.

Nurmi, E., and Rantala, M. (1973) New aspects of *Salmonella* infections in broiler production. *Nature, London* 241, 210–211.

Nurmi, E., Hakkinen, M. and Nuotio, L. (1995) *Salmonella* situation, control programme and means for control, especially by use of the competitive exclusion method in Finland. Presentation at the VI Symposium on the WPSA, Portugese Division (SPAMCA), Sanatrem, Portugal. (Cited in Schneitz and Nurmi, 1996.)

Opengart, K.N., Tate, C.R., Miller, R.G. and Mallinson, E.T. (1991) The use of the drag-swab technique and improved selective plating media in the recovery of *Salmonella arizona* (7:1,7,8) from turkey breeder hens. *Avian Diseases* 35, 228–230.

Oyofo, B.A., De Loach, J.R., Corrier, D.E., Norman, J.O., Zirpin, R.L. and Mollenhauer, H.H. (1989) Effect of carbohydrates on *Salmonella typhimurium* colonization in broiler chickens. *Avian Diseases* 33, 531–534.

Peluffo, C.A., Edwards, P.R. and Brunner, D.W. (1942) *Journal of Infectious Disease* 70, 185–192. (Cited in Geenfield *et al.*, 1971.)

Pfaff, F. (1921) Eine Truthühnerseuche mit Paratyphus-Befund. *Zentralblatt für Infektionskrankheite der Haustiere* 22, 285–292.

Pfeiler, W. and Roepke, W. (1917) Zweite Mitteilung ueber das Auftreten des Huehnertyphus und die Eigenschaften seines Erregers. *Zentralblatt für Bakteriologie I. Original* 79, 125. (Cited in Hinshaw, 1959.)

Pietzsch, O. (1982) *Salmonellose Überwachung in der Bundesrepublik Deutschland. ZVS-Jahresbericht 1980 der Zentralstelle für veterinaermedizinische Salmonellose-Forschung des Bundesgesundheitsamtes. Veterinärmedizinsche* Berichte Institut für Veterinärmedizin des Bundesgesundheitsamtes 2/1982, Robert-von-Ostertag-Institut, Dietrich Reimer Verlag, Berlin.

Pivnick, H. and Nurmi, E. (1982) The Nurmi concept and its role in the control of *Salmonella* in poultry. In: Davies, R. (ed.) *Developments in Food Microbiology – 1*. Applied Science, Barking, Essex, UK, pp. 41–70.

Pollan, B. and Vasicek, L. (1991) Vorkommen von Salmonella arizona bei Puten in Österreich. *Wiener Tieraerztliche Monatsschrift* 78, 81–83.

Pomeroy, B.S. (1944) Salmonellosis of turkeys. Ph. D. dissertation, University of Minnesota. (Cited in Nagaraja *et al.*, 1991a.)

Pomeroy, B.S. (1991) Reducing *Salmonella*: the industry has a choice. *Turkey World* 4, 24–28.

Pomeroy, B.S. and Fenstermacher, R. (1939) Paratyphoid infections in turkeys. *Journal of American Veterinary Medicine Association* 94, 90–97.

Pomeroy, B.S., Nagaraja, K.V., Olson, H., Ausherman, L.T., Nivas, S.C. and Kumar, M.C. (1984) Control of *Salmonella* infection in turkeys in Minnesota. In: *Proceedings of the International Symposium on* Salmonella. New Orleans, Lousiana, pp.115–123.

Pomeroy, B.S., Nagaraja, K.V., Ausherman, L.T., Peterson, I.L. and Friendshuh, K.A. (1989) Studies of feasibility of producing salmonella free turkeys. *Avian Diseases* 33, 1–7.

Reid, C.R. and Barnum, D.A. (1984) The effects of treatments of cecal contents on their protective properties against *Salmonella* in poults. *Avian Diseases* 29, 1–11.

Renggli, F. (1996) *Salmonella* contamination of poultry products is an important problem, and could be prevented though the routine practice of feed decontamination and proper feed handling: different methods of decontamination exist and some of them can even be combined to enhance efficiency. *World Poultry Misset* Salmonella *Special*, May, 34–35.

Rettger, L.F. (1909) Further studies on fatal septicaemia among young chickens or white diarrhoea. *Journal of Medical Research* 21, 115

Roberts, D. (1983) Factors contributing to outbreaks of food poisoning in England and Wales 1970–1979. *Journal of Hygiene, Cambridge* 89, 491.

Roberts, T. (1988) Salmonellosis control: estimated economic costs. *Poultry Science* 67, 936–943.

Runnells, R.A., Coon, C.J. , Farley, H. and Thorp, F. (1927) An application of the rapid-method agglutination test to the diagnosis of bacillary white diarrhea infection. *Journal of American Veterinary Medicine Association* 70, 660–662.

Saif, Y.M., and Shelly, S.M. (1973) Effect of gentamycin sulfate dip on *Salmonella* organisms in experimentally infected turkey eggs. *Avian Diseases* 17, 574–581.

Saif, Y.M., Ferguson, L.C. and Nestor, K.E. (1971) Treatment of turkey hatching eggs for control of Arizona infection. *Avian Diseases* 15, 446–461.

Samberg, Y. and Klinger, I. (1984) Some epidemiological aspects of *Salmonella* contamination in poultry. In: *Proceedings of the International Symposium on* Salmonella. New Orleans, Lousiana, pp. 48–53.

Sato, G. and Adler, H.E. (1966) Experimental infection of adult turkeys with *arizona* group organisms. *Avian Diseases* 10, 329–336.

Schaffer, J.M., McDonald, A.D., Hall, W.J. and Bunyea H. (1931) A stained antigen for the rapid whole blood test for pullorum disease. *Journal of American Veterinary Medicine Association* 79, 236–240.

Schellner, H.-P. (1985) Ergebnisse aus mehrjährigen Salmonellen-Untersuchungen (1977–1983). *Berliner und Muenchner Tieraerztliche Wochenschrift* 98, 10–14.

Schneitz, C. and Nuotio, L. (1992) Efficacy of different microbial preparations for controlling *Salmonella* colonisation in chicks and turkey poults by competitive exclusion. *British Poultry Science* 33, 207–211.

Schneitz, C. and Nurmi, E. (1996) Competitive exclusion in poultry production. In: American Association of Avian Pathologists (ed.) *Proceeding of Enteric Disease Control, 39th Annual Meeting.* Louisville, Kentucky, pp. 44–55.

Seuna, E., Nagaraja, K.V. and Pomeroy, B.S. (1984) Gentamicin and bacterial culture (Nurmi culture) treatments either alone of in combination against experimental *Salmonella hadar* infection in turkey poults. *Avian Diseases* 29, 617–629.

Shahata, M.A., El-Timawy, A.A.M. and El-Dimerdash, M.Z. (1984) Some studies on salmonellosis of turkeys. *Assuit Veterinary Medicine Journal* 11, 203–208.

Shreeve, B.J. and Hall, B. (1971) Excretion of *Salmonella agona* in turkeys. *Veterinary Record* 88, 480–481.

Smith, M., Fancher, W., Blumberg, R., Bohan, G., Smith, D., McKinley, T. and Sikes, R.K. (1985) Turkey-associated salmonellosis at an elementary school. *Morbidity and Mortality Weekly Report, no. 34.*

Snoeyenbos, G.H. (1991) Pullorum disease. In: Calnek, B.W., Barnes, H.J., Reid, W.M. and Yoder, H.W., Jr (eds) *Diseases of Poultry*, 9th edn. Iowa State University Press, Ames, Iowa, pp.73–86.

Stadler, A. (1995) Untersuchungen über das Vorkommen und die Verbreitung von Salmonellen bei Mastputen. Doctor in Veterinary Medicine thesis, Munich University, Germany.

Stavric, S. (1992) Defined cultures and prospects. *International Journal of Food Microbiology* 15, 245–263.

Stavric, S., Gleeson, T.M., Buchanan, B. and Blanchard, B. (1992) Experience of the use of probiotics for *Salmonella* control in poultry. *Letters in Applied Microbiology* 14, 69–71.

Thain, J.A., Baxter-Jones, C., Wilding, G.P. and Cullen, G.A. (1984) Serological response of turkey hens to vaccination with *Salmonella hadar* and its effect on their subsequently challenged embryos and poults. *Research in Veterinary Science* 36, 320–325.

Timms, L. (1971) Arizona infection in turkeys in Great Britain. *Journal of Medical Laboratory Technology* 28, 150–156.

Todd, E.C.D. (1987) Foodborne and water-borne disease in Canada – 1980 annual summary. *Journal of Food Protection* 50, 420–428.

Van Netten, P. Van Der Zee, H. and Van De Moosdijk, A. (1991) The use of a diagnostic semi-solid medium for the isolation of *Salmonella enteritidis* from poultry. In: Mulder, R.W.A.W. and De Vries, A.W. (eds) *Proceedings of the Combined Session of the 10th Symposium on the Quality of Poultry Meat and the 4th Symposium on the Quality of Eggs and Egg Products*, Vol. III, *Safety and Marketing Aspects*. Doorwerth, The Netherlands, pp. 59–66.

Watson, W.A. and Kirby, F.D. (1984) The *Salmonella* problem and its control in Great Britain. In: Snoeyenbos, G.H. (ed.) *Proceedings of the International Symposium on* Salmonella. American Association of Avian Pathologists, Kennett Square, Pennsylvania, pp. 35–47.

Weand, D.C. and Horsting, A.G. (1978) Sanitation of hatching eggs. In: *Proceedings of 16th World Poultry Congress*, pp. 797–808. (Cited in Nagaraja *et al.*, 1991.)

Wierup, M., Wahlstroem, H. and Engstroem, B. (1992) Experience of a 10-year use of competitive exclusion treatment as part of the *Salmonella* control programme in Sweden. *International Journal of Food Microbiology* 15, 287–291.

Williams, J.E. and Dillard, L.H. (1968) *Salmonella* penetration of fertile and infertile chicken eggs at progressive stages of incubation. *Avian Diseases* 12, 629–635.

Williams, J.E. and Whittemore, A.D. (1971) Serological diagnosis of pullorum disease with the microagglutination system. *Applied Microbiology* 21, 394–399.

Willinger, H., Flatscher, J., Dreier, F. and Wildner, T. (1986) Epidemiologische Untersuchungen zum Vorkommen von Salmonellen in Gefluegelhaltungen. *Wiener Tieraerztliche Monatsschrift* 73, 141–148.

Winter, A.R., Burkhart, B. and Widley, H. (1952) Further studies on the whole blood and tube methods of testing turkeys for *Salmonella pullorum* infection. *Poultry Science* 31, 399–404.

Wray, C. and Davies, R.H. (1994) *Guidelines on Detection and Monitoring of Salmonella Infected Poultry Flocks with Particular Reference to* Salmonella enteritidis. Veterinary Public Health Unit. WHO/Zoon 94.173, WHO, Geneva.

Wray, C., McLaren, I.M. and Beedell, Y.E. (1992) In: J.F., Hinton, M.H. and Mulder, R.W.A.W. (eds) Antimicrobial resistance in salmonella isolated from poultry: 20 years of monitoring. In: *Proceeding of Prevention and Control of Potentially Pathogenic Microorganisms in Poultry and Poultry Meat Processing 7. The Role of Antibiotics in the Control of Foodborne Pathogens*. Het Spelderholt Agriculture Research Department DLO-NL, Beekbergen, The Netherlands, pp. 25–30.

Wray, C., Jones, Y.E., McLaren, I.M. and Davies, R.H. (1998) Antibiotic resistance in *Salmonella* from turkeys. In: *Proceedings of First International Symposium on Turkey Diseases, Berlin, 19–21 February, 1998*. German Veterinary Medicine Society, Giessen, pp. 304–306.

Yamamota, R., Adler, H.E., Sadler, W.W. and Stewart, G.F. (1961) A study of *Salmonella typhimurium* infection in market-age turkeys. *American Journal of Veterinary Research* 22, 382–387.

Yule, B.F., Macleod, A.F., Sharp, J.C.M. and Forbes, G.I. (1988) Costing of a hospital-based outbreak of poultry-borne salmonellosis. *Epidemiology and Infection* 100, 35–42.

Zecha, B.C., McCapes, R.H., Dungan, W.M., Holte, R.J., Worchester, W.W. and Williams, J.E. (1976) The Dillon Beach project – a five-year epidemiological study of naturally occurring *Salmonella* infection in turkeys and their environment. *Avian Diseases* 21, 141–159.

Chapter 9
Salmonella Infection in Ducks

Robert R. Henry

Cherry Valley Farms, North Kelsey Moor, Lincoln LN7 6HH, UK

Historical Perspectives

Duck farming for the production of meat and eggs has been practised for several thousand years. The association between ducks and *Salmonella* has been known for probably about as long, if not in exactly those terms, at least as a 'cause-and-effect' situation, in that the consumption of duck eggs was associated with a high probability of 'stomach upset'. That this was probably due largely to the presence of *S. typhimurium* transmitted to the egg by a clinically perfectly healthy duck was a discovery that came very much later. The concept of '*Salmonella* food poisoning' is of relatively recent origin and the present general interest even more recent, being a side-effect of urbanization and intensive farming, combined with mass catering.

Although food poisoning as a result of eating duck eggs has been recognized for centuries, so far as can be determined (on the basis of lack of reported incidents) *Salmonella* food poisoning from duck meat is very rare. This is probably due largely to the culinary methods and eating habits of the relatively small proportion of the population that eats duck meat, rather than to any great difference between duck farming and other poultry farming methods.

In spite of the long history of the association between *Salmonella* and duck, the literature on the subject is scanty and is largely confined to reports of serovars isolated on routine monitoring of post-mortem specimens either targeting poultry in general or as part of a more general *Salmonella* prevalence survey. This lack of information may simply reflect the view that the source of infection, epidemiology, diagnosis and treatments reported for the other avian species apply equally to the duck. This is, of course, largely true, but there are some differences, which may be relevant when considering monitoring procedures or treatment regimes.

The following commentary reflects mainly experiences in the UK, with references where applicable to work from other countries. All references relate to the Pekin/mallard type of duck (*Anas platyrhynchos*), unless otherwise stated.

Pathogenesis

Although the prevalence of *Salmonella* in the duck population (both farmed and feral) is well recognized internationally, the degree to which the duck population itself suffers as a consequence appears to be minimal. It may be that the gregarious habits of the species in its natural habitat and the fact that it tends to swim, feed and drink in what can only be regarded as its own effluent has, over the millennia, produced a genetic strain tolerant to the potentially pathogenic organisms residing in the gut to which this behaviour pattern exposes it. Whatever the reason, the duck is not uncommonly infected with *Salmonella* but is only very rarely diseased, so far as can be determined by clinical observation.

Septicaemia due to *Salmonella* may occasionally arise but can usually be traced to some form of environmental or managemental stress. Such infections are most common in very young ducklings that have been exposed to chilling stress or where lack of good hatchery hygiene has permitted massive challenge.

Clinical Disease

The usual signs of clinical disease following infection are more common in the very young duckling, and generally commence in the immediate post-hatch period. The recent experience with *S. enteritidis* phage type (PT) 4 was that, although immediate post-hatch infections were seen, in birds of about 3 weeks of age or more the clinical condition also occurred under those circumstances that otherwise commonly give rise to *Escherichia coli* septicaemia.

Where clinically significant *Salmonella* infections do arise, the signs are those common to almost any bacterial infection: general listlessness, hunched appearance, diarrhoea, dehydration (as evidenced by semi-closed eyes with possibly some 'gumming' of the eyelids), reduced food and water intake, possibly some signs of nervous incoordination and eventual coma and death. The progress through the various stages may take only a few hours (hence the American term 'keel disease', because 'the birds just keel over and die'). Perhaps more commonly, the clinical condition runs a course of 4–5 days. Recovered birds may show some degree of unthriftiness, which, in some cases, can be associated with a hock or foot arthritis or synovitis. When adult birds become infected there are neither gross clinical signs nor even a drop in egg production, so far as can be determined from field cases and experimental infections. It has, however, been reported that, following experimental challenge, there is a change in eggshell structure, which is consistent with the presence of stress within the flock (S.E. Solomon, Glasgow, personal communication, 1997). If changes in egg fertility or hatchability follow the change from negative to positive *Salmonella* status of a flock, they are too slight to be apparent, even when true egg transmission results. Whether this lack of obvious effect results from the sometimes slow spread of infection through an adult population

or whether it reflects the tolerance of the duck is not clear.

Post-mortem Findings

On post-mortem examination of birds that have died or are showing clinical signs due to *Salmonella* infection, the pathological findings cover the full range of reactions that may be noted in the fowl, with, to varying degrees, carcass congestion, pericarditis, perihepatitis, hepatic microabscesses, typhlitis and the emaciation, dehydration and nephrosis/nephritis, that probably reflect more the general malaise of the bird than any specific effect caused by the *Salmonella*. In this context, it should be noted that nephritis, particularly with accompanying visceral gout, is, in the duck, pathognomonic for lack of water intake, irrespective of the cause. Although microabscesses (demonstrable as small pale spots on the liver) may result from *Salmonella* infection, in birds of between 1 and 2 weeks of age, streptococcal infections may be found to have a similar effect.

Zhakov and Prudnikov (1987) reported changes in the bursa of Fabricius and thymus in ducks immunized with live *S. typhimurium* vaccine and challenged subsequently. Bursal follicular depletion was observed, with enlargement of the medulla of the thymus and reduction in the size of the cortex. These findings may have significance with regard to the immune mechanisms being activated, but bursal changes, at least, are not specifically associated with salmonellosis and bursal regression in ducks has also been associated with reovirus infection (Smyth and McNulty, 1994).

Diagnosis

There are, in general, two situations in which 'diagnosis' of *Salmonella* infections is called for; the first of these is the technically trivial situation in which clinical signs of disease are present, with, and occasionally without, gross pathological changes to be found in dead or sacrificed birds, while the second is the rather more challenging requirement for accurate but economically sustainable monitoring of layer/breeder flocks for *Salmonella* infection and transmission of *Salmonella* to the egg.

For the purpose of diagnosis of clinical disease, it is usual to culture affected organs on blood agar and MacConkey agar media, with subsequent aerobic incubation at 37°C for 18–24 h. In cases of acute salmonellosis, an almost pure growth of *Salmonella* can be expected if the specimens are reasonably fresh. Identification from initial culture by serological or biochemical means is usually possible. Although multiple-serovar carrier status is not uncommon, clinical salmonellosis would appear to be due to single serovars.

It is also recommended practice to take cultures from the brains of ducks that have died or have shown either nervous signs or general malaise, because *Salmonella* can be recovered with reasonable certainty. There is an advantage that contaminants seem to be less prevalent in this tissue than, for instance, in liver or spleen. Recovery of *Salmonella* from the brain of birds that merely show presumptive *Salmonella*-associated malaise is less common, possibly because the level of malaise at which culls may be taken varies considerably, or possibly because the assumed association is false, despite the fact that the bird may be a *Salmonella* carrier.

Monitoring Programmes

The situation with regard to the diagnosis of carrier status/egg-transmitter status is more complex, mainly because of the limitations imposed by costs and practicalities. The wide range of programmes designed for monitoring flocks for the presence of *Salmonella* demonstrates the unsuitability of any one of them for all situations.

In considering the suitability of a programme for any given set of circumstances the following factors should be considered.

1. Egg transmission can occur without clinical signs in either parent or offspring.
2. Infection results in faecal shedding of the organism for a greater or lesser length of time, irrespective of when the first infection occurred.
3. On a flock basis, egg transmission of one *Salmonella* serovar may take place with faecal shedding of a different serovar. It has not been established whether this implies dual infections in some cases.
4. As far as can be determined, seroconversion

in the parent always accompanies egg transmission.

Sampling strategy

The strategy for monitoring parent flocks for *Salmonella* status starts with the assumption that any infection that may be present has, at some time or other, entered via the oropharynx and will have, shortly thereafter, appeared at the cloaca, where it will be present for a variable length of time. As far as is known, wound infection/insect bites in the duck are not a significant factor in the spread of salmonellosis, although injection of *S. virchow* has been reported to result in faecal excretion and vertical transmission (Gosh *et al.*, 1990).

The optimum strategy with respect to monitoring parent flocks for potential to spread *Salmonella* to their offspring involves sufficiently frequent faecal sampling from 1 day old onwards, together with possibly serological investigations.

Sampling tactics

In the UK, the Poultry Breeding Flocks and Hatcheries Order (HMSO, 1993) does not apply to duck breeders, but conformation with its requirements appears to be a sound basis for *Salmonella* monitoring. This Order lays down requirements, *inter alia*, to take box liners and 'dead on arrival' from newly placed parent flocks, followed by all birds that die or are culled in the first 4 days. These examinations are useful for monitoring infection of parent or hatchery origin, provided that an adequate number of samples are taken. Although the prevalence of infection emanating from a contaminated hatchery may be high, this is not necessarily so, and equally the tendency for *Salmonella* infection to spread amongst day-old ducklings is high, but this also does not necessarily occur. Personal observation (R.R. Henry) shows the possibility of nine faecal excretors of *S. enteritidis* in a pen of 400 birds, part of a total flock size of 2500 up to an age of 5 weeks (when the infected birds were identified and removed).

There are further requirements in this Order to take composite faeces samples at 4 weeks of age and 2 weeks before 'entering the laying phase'. This frequency of monitoring may be inadequate to identify an infected flock early enough for some remedial action to be taken.

Experience has shown that the use of cloacal swabs to determine faecal excretion of *Salmonella* is less efficacious than the use of litter samples. The belief that a high percentage of the birds may be excreting at any one time is not well founded. Examination of litter samples is capable of revealing levels of excretion that may be missed by cloacal swabbing unless a very high percentage of the birds is examined. Perhaps because of the nature of duck litter management, it appears that examination of litter is more effective and more practical than the examination of drag swabs. Examination of water from drinking troughs is also a possible sample source for monitoring for *Salmonella*, but the number of recovered organisms (per gram of material) is not as high as from litter. Samples taken at fortnightly or monthly intervals appear to be an appropriate compromise between cost/work load and effective surveillance. There are situations in which examination of litter samples or other sources of faecal material are not appropriate – for instance, during medication for other conditions.

Depending on the past history and age of the birds, reliance may have to be placed on serology or, if appropriate, bacteriological examination of egg material. For examination of egg material to be an effective monitoring procedure, selection of an appropriate number of eggs is essential. If the *Salmonella* status of a potential breeder flock is in doubt due to lack of testing, from a management point of view it is optimal to screen those eggs produced at the onset of lay that are not considered suitable for the production of commercially viable offspring. Almost by definition, under these circumstances, the egg numbers will be low and infertility high, but, since examination of live eggs is not optimal for *Salmonella* recovery, this is not a problem. In the early stages of lay, it may be necessary to examine all available eggs to obtain reliable data. These eggs should be incubated for at least 10 days prior to bacteriological examination. The examination of fresh eggs is not an effective monitoring procedure.

For eggs being taken on to hatch, the eggs discarded at the candling of eggs after 10 days of incubation appear to be the most practical and productive material for *Salmonella* detection. Early dead embryos are preferred to 'clears' (i.e. infertile eggs), although *Salmonella* may not be the cause of early embryo death. However, culture of such eggs, if it produces *Salmonella*, usually produces it as a monoculture, with a bacterial count of the order of 10^6 colony-forming units (cfu) g^{-1}.

Apart from identifying positive parent flocks with complete accuracy (and reasonable sensitivity), the technique has the advantage that it leaves time for the eggs to be hatched in isolation or diverted to another hatchery, should such options be available, thus preventing spread to other hatchlings. Given a moderate degree of care, it is possible and, in some parts of the world commonplace to move duck eggs at about 23 days of incubation over considerable distances, for the final stages of hatching.

The examination of meconium or hatch debris/dead-in-shell, as required by the Poultry Health Scheme in the UK, is clearly too late for much useful action to be taken and, unless the hatcher contains the offspring of only one parent flock, it is considerably less accurate with regard to the identity of the parent source of any isolate(s).

Antibiotic Sensitivity

Salmonella isolates in the UK show good *in vitro* sensitivity to amoxycillin, apramycin, chloramphenicol, colistin, cotrimoxazole, enrofloxacin, framycetin, neomycin, spectinomycin, streptomycin, the tetracyclines and trimethoprim. Wasniewski and Galazka (1991) reported from Poland that, of 41 *S. typhimurium* and 74 *S. enteritidis* isolates, 95% and 92%, respectively, were enrofloxacin-sensitive and 'much fewer' were sensitive to streptomycin, oxytetracycline, neomycin and chloramphenicol.

Treatment

Treatment of ducks for clinical salmonellosis is only rarely necessary. Since the disease usually follows some form of stress – frequently chilling – the primary action is to remedy the predisposing cause if this is still relevant. Should veterinary advice indicate that antibiotic therapy is required, the only antibiotic licensed in UK for water administration is amoxycillin. For in-feed medication, chlortetracycline is suitably licensed. The evidence that either of these is likely to be of clinical benefit for the control of salmonellosis is scanty. The clinically affected birds tend not to eat or drink and hence do not benefit from the medication provided. Antibiotics may help to reduce the weight of challenge to the in-contact birds but neither of those mentioned above appears to prevent production of the carrier state. For both the above licensed medications, the withdrawal time is 7 days.

Treatment of faecal excretors of *Salmonella* to limit environmental contamination is possible using neomycin in the feed at a standard rate of 320 g activity t^{-1}. The feed supply rate should be taken into account when calculating the appropriate medication rate. If birds are not fed *ad libitum*, there may be considerable intake of litter, with resulting medicament dilution. The use of neomycin in the food prior to moving birds from rearing to laying premises has also coincided with the disappearance of faecal shedding of some *Salmonella* serovars.

Treatment of layers to eliminate S. *enteritidis* (PT4) has been shown to be possible by the use of enrofloxacin medication, following the manufacturer's recommended regime. The treatment of flocks to prevent egg transmission of other *Salmonella* is presumably possible by the same means, although faecal transmission may well cease spontaneously without treatment and without the appearance of egg transmission. However, the successful eradication of S. *typhimurium* from a positive laying flock has not been achieved so far. Egg transmission of the organism has been very significantly reduced but, in a flock of only 100 layers, a low level of transmission eventually reappeared following enrofloxacin treatment. Tenk *et al.* (1994) report treatment of a flock of geese with a 31% excretion rate using enrofloxacin. *Salmonella* excretion was suspended for a period of 42 days, at which time a 4% excretion rate was observed. That this followed plucking stress is probably significant.

Given a suitably large selection of small flocks, suitable hatchery and rearing arrangements and an intensive monitoring programme, a medication regime would permit *Salmonella*-negative parent flocks to be established.

Prevention

Although the use of anaerobic cultures of adult bowel flora to colonize the intestinal tract of the fowl has proved a useful technique in limiting *Salmonella* carrier status (Nurmi and Rantala, 1973), the technique has not, as yet, proved useful for the duck. Ducklings infected with *Salmonella* at 1 day old or shortly after, whether given a strictly anaerobically grown culture of adult bowel flora or not, will become carriers for a variable length of time, which may last until point of lay. This is, perhaps, all the more surprising because attempts to evaluate the preventive effects of adult bowel flora on carrier status with artificially challenged day-old ducklings have foundered over the failure to establish infection and transmission in the negative controls. The factors that differ between the field and experimental situation are not yet clear. That there is possibly a requirement for further investigation in this respect is suggested by the fact that *Salmonella* carrier status may spontaneously disappear from a flock coming into lay and that the spread of even S. *typhimurium* between adjacent houses does not necessarily take place. Even pen-to-pen spread of infection is not inevitable, at least in the short term, in spite of movement of stockman foot traffic between pens.

Vaccination of day-old ducklings to prevent either S. *enteritidis* or S. *typhimurium* infections has not been attempted. Immunization of older birds with either Bovivac (Hoescht Rousel) for its S. *typhimurium* content or the S. *enteritidis* vaccine (Salenvac™) failed to prevent egg transmission of these *Salmonella*. There is reason to believe that the prevalence of egg transmission was reduced and these vaccines (as appropriate to the problem) may prove a useful adjunct to medication in an eradication scheme.

The use of live vaccines has not been considered because none are commercially available in the UK but, in view of the immune mechanism of the duck, these might be the only ones likely to be fully effective.

Sources of Infection

Salmonella infections can be spread both horizontally and vertically.

Horizontal transmission

Horizontal transmission would appear, from a comparison of raw feed-material isolates and duck isolates, to account for the changes in prevalent serovar that occur on farms from time to time, and possibly in a significant proportion of the ongoing incidences. The role of litter-material contamination, residual environmental contamination and feral vectors in the latter context should not, however, be underestimated. The latter two may account for the recrudescence of infections considered to have been eliminated from the farm.

Feed

Although heat treatment of duck feed and the incorporation of organic acids into the feed have significantly reduced the prevalence of feed-borne *Salmonella*, the extent of treatment is not yet total. Thus it will not prevent the adventitious contamination of feed during the later stages of manufacture, storage and transport. The situation in this respect is currently improving but until such time as raw materials are free of *Salmonella* contamination it is likely to be a major source of infection.

Environment and feral vectors

The horizontal transmission of *Salmonella* from contaminated environments and from wild birds and vermin is universally acknowledged. The role of fomites and possibly even human vectors tends to be forgotten, but lapses in biosecurity from these sources are equally probable. In flocks that conform to the requirements of the UK Poultry Health Scheme, horizontal transmission from these sources is under good control. In the UK, a dramatic change has taken place over the last 10 years and *Salmonella*-negative breeder flocks are becoming more common and are probably as common as *Salmonella*-positive ones.

The *Salmonella*-negative status of grower flocks is less common, mainly because of economic factors, which give rise to the requirement for multi-age sites, where satisfactory inter-crop clean-down may be compromised by the proximity of other livestock and a safe haven for vermin may be readily available. There is a significant move towards resolving the situation.

Litter

One of the main differences between the management practices of duck growing and broiler or turkey growing is the need to replenish the duck litter material on a daily basis. In the UK, the litter can be either of wood by-product origin (shavings or sawdust) or straw (usually barley or wheat straw).

The concept that either litter source will be *Salmonella*-free is far from certain. Contamination of straw can be expected to occur both in the field and, more probably, in the stack, which, unless preventive steps are taken, can become home to rats/mice and wild birds, and home/hunting-ground to cats, all of which are potential carriers and disseminators of *Salmonella* contamination. The situation with regard to wood shavings or sawdust is somewhat similar. Although the faecal material that might carry the contamination is (relatively) easier to identify in wood shavings than in straw, there is little practical that can be done on farm to remove any such contamination. The behaviour pattern of ducks is such that they avidly sift through the fresh litter provided, for extraneous and possibly tasty/nutritious material. Straw, shavings, litter and almost anything else of ingestible size are likely to be swallowed. Proper storage of the bulk material at source before delivery to the distributor and proper packaging of the final product prior to farm delivery are essential if *Salmonella* contamination from these sources is to be avoided.

Vertical transmission

True egg transmission occurs when the egg is infected before oviposition. The alternative route of vertical transmission, egg contamination, may occur in conjunction with or independently of egg transmission. The duck egg usually has a very thick cuticle, which may help to prevent invasion of the egg by *Salmonella* via eggshell pores, but it is observed that cuticle cover can vary considerably. Additionally, cuticle removal may commonly be seen as a result of the surface of the egg being scraped by the parent or another duck while the cuticle is still soft after oviposition. Ducks have a habit of defecating in the nesting-boxes provided and duck faeces are of a fluid consistency. Baker *et al.* (1985) observed that good egg hygiene is helpful in minimizing hatchery contamination. The occurrence of egg transmission of S. *typhimurium* and the faecal excretion of a different *Salmonella* serovar by birds in a flock is not unknown, although whether individual birds were dual-infected has not been ascertained.

Under normal circumstances, true egg transmission of *Salmonella* in the duck is very largely confined to transmission of S. *typhimurium*. Whether the egg transmission of S. *enteritidis* by the duck was true egg transmission, or fell into the category more properly referred to as egg contamination, does not appear to have been investigated fully for the duck. However, the apparently successful use of antibiotic to control faecal excretion and the continued appearance of S. *enteritidis*-infected day-old ducklings tends to suggest the former.

Vertical transmission has been identified in mature layers as soon as 8 days after challenge following artificial infection through the drinking water with S. *typhimurium*, supplied at the rate of 10^3 cfu bird^{-1} in their water.

The use of neomycin in the food at up to 640 g t^{-1} failed to prevent egg transmission of S. *typhimurium*, although extensive bacteriological examination indicated the absence of faecal excretion over the period of lay (R.R. Henry, personal observation).

The natural egg transmission of *Salmonella* serovars other than S. *typhimurium* has not been demonstrated in the author's experience, although vertical transmission of S. *livingstone* and S. *nagoya* have been noted. Egg transmission of S. *hadar* has been noted in the Muscovy (Barbary) duck (*Cairina moschata*). This type of bird is, in many ways, more goose-like than duck-like in many of its attributes. Gosh *et al.* (1990) reported the vertical transmission of S. *virchow* following both oral and intraperitoneal challenge of adult birds.

Hatchery transmission

Whether Salmonella egg infection takes place by true egg transmission, shell penetration or external shell contamination, the end effect is contamination of, first, the hatchlings and hatcher in which the eggs are hatched, secondly, ducklings of the same and other parentage in the same hatcher and, thirdly, the general hatchery environment, from which further duckling infection may take place at a later date.

Salmonella Serovars

The *Salmonella* serovars that may infect ducks and give rise to general infection and/or gut carrier status appear to differ little from those that would be expected to behave similarly in the fowl or turkey, except that the fowl-adapted S. *pullorum/gallinarum* does not appear to be a particular duck pathogen (Buchholz and Fairbrother, 1992), although its presence has been recorded on occasion.

A survey in the USA by Price *et al.* (1962) on 7029 post-mortem occasions carried out over 10 years isolated 491 *Salmonella*. Of these, 457 (93%) isolates were S. *typhimurium* and there were fewer than ten each of the other serovars (Table 9.1). These organisms were isolated from ducklings considered to be clinically affected with 'keel disease', but in retrospect the role of some of the isolates as the primary cause of disease may be doubtful. A similar survey in Slovakia by Simko (1988) reported on 679 *Salmonella* isolations, of which 61% were S. *typhimurium*, 22% S. *anatum*, 4% S. *meleagridis* and other serovars (Table 9.1) were less than 1%. He also reported 706 isolations from goose farms, of which 41% were S. *enteritidis*, 17% S. *typhimurium*, 16% S. *anatum* and 11% S. *newport* and other serovars listed in Table 9.1.

Because of changes in both UK legislation and public attitude, considerable attention has been paid to the presence of *Salmonella* in the gut of many species, including ducks. Routine surveillance has shown carrier status of the serovars listed in Table 9.1, both in apparently healthy ducks and, rarely, associated with clinical infections of ducklings in the first few days of life. In recent years (1989–1996) in the UK, S. *typhimurium* and S. *enteritidis* have been associated with septicaemia in young duckling, but as a primary cause of disease they did not and do not figure strongly.

The statistic of apparent prevalence of *Salmonella* in general or any one serovar in particular seems to depend as much on the level of interest shown in the subject as on the true incidence. Multiple isolations – for instance, in the process of ascertaining the longevity of carrier status – tend to bias the statistic as to the actual numbers of birds infected and the overall potential health impact (on birds or humans), unless

Table 9.1. *Salmonella* serovars isolated from ducks.

Serovar	Author	Serovar	Author	Serovar	Author
S. agona	2, 2*, 3	*S. hadar*	2, 2*, 3	*S. nagoya*	3
S. anatum	1, 2*, 3	*S. heidelberg*	2, 2*, 3	*S. newport*	1, 2*, 3
S. bareilly	2, 2*	*S. indiana*	3	*S. oregon*	1
S. binza	3	*S. infantis*	3	*S. orion*	3
S. braenderup	3	*S. kedougou*	3	*S. panama*	1, 2*
S. bredeny	2*, 3	*S. lille*	2*	*S. regent*	3
S. choleraesuis	2*	*S. livingstone*	3	*S. saintpaul*	1
S. derby	3	*S. london*	2	*S. seftenberg*	1, 3
S. enteritidis	1, 2*, 3	*S. manchester*	1	*S. thomasville*	3
S. frenso	3	*S. manhatten*	1	*S. typhimurium*	1, 2, 2*, 3
S. gallinarium/ pullorum	2	*S. mbandaka*	3	*S. virchow*	2
S. give	1, 3	*S. montevideo*	2	*S. wangata*	3

Authors: 1, Price *et al.*, 1962; 2, Simko, 1988: ducks; 2*, Simko, 1988: geese; 3, MAFF (1997).

very careful correlation of the reported isolations is carried out.

These UK isolates from ducks have usually been seen as transient gut infections and occasionally as short-term hatchery or farm residents. *S. typhimurium* is almost certainly the predominant serovar in the UK, as elsewhere, probably because, in addition to horizontal transmission, it is the subject of vertical transmission (q.v.) by all routes.

In general, the lists of isolated serovars seem to reflect mainly feed-material contamination and, as such, may be expected to be broadly similar around the world and across species at risk, following the common economic pressures for cheap feed ingredient raw material.

Isolation of *Salmonella*

The culture of *Salmonella* from material of duck origin may be carried out with almost any of the recognized pre-enrichment, enrichment and selective media in common use (see also Chapter 21). Although the growth characteristics of *Salmonella* on selective media are well documented, the different media are usually designed to be optimal under specific conditions, which may not include unforeseen types of background flora such as occurs in ducks. In the case of under-investigated species, such as ducks, *ad hoc* investigations of new media are recommended, in view of the findings of Mann and McNabb (1984), who investigated the prevalence of

Salmonella in geese. To investigate the use of different media for isolation of *Salmonella* from duck material, a trial was carried out by R.R. Henry (unpublished observations). Samples of 25 g of neck-flap or sliced giblets and 10 g amounts of meconium, all known to be contaminated with *Salmonella*, were diluted with buffered peptone water at ratios of 25 g : 225 ml and 10 g : 90 ml, respectively, and incubated overnight at 37°C. Appropriate subcultures were made in the following enrichment media: tetrathionate (Tet.), Rappaport–Vassiliadis broth (RV), RV soya peptone broth (RVS), selenite (Sel.) and modified semi-solid RV (MSRV). After incubation, subcultures were made on to xylose lysine desoxycholate (XLD) agar, XLD agar + novobiocin (XN), brilliant green agar (BG) and xylose–lysine + tergitol 4 agar (XLT4).

No media combination was totally successful and slight differences between different combinations of enrichment broths and selective agars were apparent in relation to the different *Salmonella* serovars present, e.g. MSRV favoured the isolation of *S. enteritidis* but missed *S. mbandaka*, whereas RV favoured *S. typhimurium* and *S. indiana*. When two or more different serovars were present, none of the combinations detected the maximum of three different serovars. Differences were also apparent in the ease of media preparation, e.g. XLT4 equated in use to XN, but required the addition of four components, in contrast to one for XN. The selection of the optimal media combination must largely remain a matter of individual choice, but this study indicated that the RVS with BG and XN, with Sel with BG and XN as a back-up, appears to be the medium of choice.

The procedure for the examination of egg material may follow any of the procedures outlined above, but the findings (over 4000 accessions involving 15,000 eggs) suggest that, if eggshells are suitably disinfected, the pre-enrichment incubation of whole, macerated 10-day-incubated dead-embryo or clear eggs at 37°C overnight, followed by direct plating to BG and XN, is a more cost-effective procedure for detecting contaminated eggs with a view to categorizing a parent flock as *Salmonella*-positive or -negative. This method saves on media, time and incubator space. Should results, for some reason, prove equivocal, the pre-enrichment broth can still be subjected to selective enrichment at 48 h post-inoculation.

It has been noted that XLD (with or without 10 p.p.m. novobiocin) has the unexplained and unmatched property of showing duck-associated *S. typhimurium* as a noticeably smaller colony than other *Salmonella* or even *S. typhimurium* from other sources. This property is retained even after the *Salmonella* has been subcultured on non-selective media four times. The colony size, which is also independent of the H_2S production characteristics of the isolate, is a useful marker in the laboratory, enhancing the ability to 'pick off' multiple isolates and saving on time spent on identification of the selected small colony isolate.

Serology

Ducks that are challenged by a disease entity will, like any other species, mount a combination of cell-mediated and humoral responses. The humoral response produced by the duck differs from that of the other farmed avian species, both in the detail of the immunoglobulin G (IgG) produced and in the amount and nature of the IgM. These differences in the humoral response have been summarized by Higgins and Warr (1993).

From a practical serological point of view, the result of these differences is that serological tests that depend on some secondary effect (e.g. agglutination, complement fixation, passive haemagglutination, agar gel precipitation, etc.)

tend to be insensitive. Experience has shown that, when using the rapid slide agglutination test against 'pullorum antigen', a 10% false-positive reaction may be expected, although the reason for this has never been investigated. The foregoing is not to say that ducks will not give a response to some of these, tests but the significance of the results has to be considered in relation to the sensitivity of the test (and possibly also its specificity).

Enzyme-linked immunosorbent assay (ELISA) tests seem at present to be the most useful for examining the humoral response of the duck to a wide variety of antigens, including the flagellar and lipopolysacchride (LPS) antigens of *Salmonella*. Unfortunately, there is no common agreed protocol. Workers in this field either use their own standards or utilize commercial kits, where these are available. Details of ELISAs will be found in Chapter 24.

There is no laid-down standard methodology for ELISAs with regard to initial serum dilution, how many dilutions should be looked at or the time/temperature of incubation. Each test system relates the optical density (OD) to a predetermined level on any of a statistically valid number of samples. In the author's laboratory, a single 1/100 serum dilution in normal saline is used, with an incubation period of *c.* 30 min at a temperature of 37°C. ODs are read when the (duplicated) positive control shows an OD of *c.* 0.8. The ODs of the test sera are quoted as a percentage of the positive control. The advantages of such a system are that it is: (i) robust; (ii) sensitive; (iii) reproducible; and (iv) relatively simple.

The disadvantages are that: (i) the test depends on the continuing presence of a good-quality standard serum; (ii) the ODs obtained are diagnostically only useful 'in house' and therefore comparison of results with workers in other laboratories is difficult and depends on comparing interpreted results, rather than actual titre values, or exchanging control sera; (iii) The relatively high serum concentration (1/500, 1/1000 or even greater dilutions are more usual) may render the test 'oversensitive'. It is arguable that, for a monitoring system, oversensitivity is only a problem if it leads to false conclusions or a waste of time in retesting. So far, this has not been the case.

Serology interpretation

The humoral response is only a very minor part of the duck's immunological defence mechanism and, until a serological test is developed that relates titre to either the actual carriage of a particular *Salmonella* serovar or the bird's resistance to *Salmonella* infections, there appears to be little point in regarding ELISA results as other than an indication of *Salmonella* infection in a duck or within a group of ducks.

Examination of sera from the Muscovy duck can be carried out using the same methodology. Within limited experience, it would appear that this type of bird reacts, serologically, rather more vigorously than the Pekin bird, as judged by the ODs obtained under similar conditions of husbandry and challenge and in spite of using the second antibody designed for the Pekin duck.

To judge that, given a set of 'positive' reactions, the birds have encountered the particular *Salmonella* from which the antigen was prepared (either LPS or flagellar) is an overoptimistic view of the test system. There is a tendency to regard serological reactions as a response to the organism providing the test antigen (e.g. pullorum test-positive or *S. enteritidis*-positive), whereas the only valid inference from the test is that the serum is showing a reaction to the antigen in use.

In the above example, birds would be reacting to the *Salmonella*'s O9 LPS antigen, which may or may not have come from either of these organisms and may equally be indicating contact with another group D *Salmonella*, e.g. *S. dublin*.

The use of flagellar antigens is beset with the same difficulties. The use of the flagellar H gm antigen of *S. enteritidis* gives strong positive reactions to sera from ducks vaccinated with *S. dublin* (H gp). It is suggested that the common poultry *Salmonella* sharing the flagellar antigenic type with *S. enteritidis* would be unlikely to stimulate antibody (Timoney *et al.*, 1990) and that therefore flagellar antigens would provide a suitable test for *S. enteritidis* infection. The truth of this observation is probably 'period-dependent' rather than absolute. Some 20 years ago, *S. montevideo* or *S. agona* would probably have produced more serological reactions in poultry in general than the *S. enteritidis* circulating at the time.

It is possible to produce a *Salmonella* H 'm' antigen, using various serovars of *Salmonella*, with a considerable increase in specificity but, in duck sera at least, an apparent drop in sensitivity that precluded the use of the test (at that time) for certification purposes. Possibly further refinement of the test would be worth consideration, should the subject matter remain of general interest.

References

Baker, R.C., Quershi, R.A., Sandhu, T.S. and Timoney, J.F. (1985) The frequency of *Salmonella* on duck eggs. *Poultry Science* 644, 646–652.

Buchholz, P.S. and Fairbrother, A. (1992) Pathogenicity of *Salmonella* pullorum in northern bobwhite quail and mallard ducks. *Avian Diseases* 36, 304–312.

Gosh, S.S., Roy, A.K.B. and Nanda, S.K. (1990) Transmission of *S. virchow* in fowls and Khaki Campbell ducks. *Indian Veterinary Journal* 671, 84–85.

Higgins, D.A. and Warr, G.W. (1993) Duck immunoglobulins: structure, functions and molecular genetics. *Avian Pathology* 22, 211–236.

HMSO (1993) *The Poultry Breeding Flocks and Hatcheries Order 1993*. Statutory Instrument 1994 No. 1898.

MAFF (1997) Salmonella *in Livestock Production*. Veterinary Laboratories Agency, New Haw, Addlestone.

Mann, E. and McNabb, C.D. (1984) Prevalence of *Salmonella* contamination in market-ready geese in Manitoba. *Avian Disease* 284, 978–983.

Nurmi, E. and Rantala, M. (1973) New aspects of *Salmonella* infection in broiler production. *Nature* 241, 210–211.

Price, J.I., Dougherty, E., 3rd and Bruner, D.J.W. (1962) *Salmonella* infections in white pekin duck: a short summary of the years 1950–60. *Avian Diseases* 6, 145–147.

Smyth, J.A. and McNulty, M.S. (1994) A transmissible disease of the bursa of Fabricius of the duck. *Avian Pathology* 23, 447–460.

Simko, S. (1988) Salmonellae in ducks and geese on farms with latent infections and in centres of salmonellosis. *Imunoprofylaxia* 1–2, 92–101.

Tenk, I., Kovacs, Z. and Matray, D. (1994) Effect of enrofloxacin on the *Salmonella* excretion of plucked geese. *Magyar Allatorvosok Lapja* 498, 501–503.

Timoney, J.F., Sikora, N., Shivaprasad, H.L. and Opitz, M. (1990) Detection of antibody to *Salmonella enteritidis* by a gm flagellin-based ELISA. *Veterinary Record* 127, 168–169.

Wasniewsky, A. and Galazka, V. (1991) Baytril Bayer – a preparation for treatment for salmonellosis in poultry. *Zeszty Naukowe Akademii Rolniczej we Wrocawiu, Weterynaria*, 49, 109–114.

Zhakov, M. and Prudnikov, V.S. (1987) Morphological changes in the central immunoreactive organs of ducks after oral immunisation against salmonellosis. *Veterinarnaya Nauka Porizvostvu* 25, 56–58.

Chapter 10
Salmonella Infections in Cattle

Clifford Wray and Robert H. Davies

Veterinary Laboratories Agency (Weybridge), New Haw, Addlestone, Surrey KT15 3NB, UK

Salmonella infections are an important cause of mortality and morbidity in cattle and subclinically infected cattle are frequently found. Cattle thus constitute an important reservoir for human infections. There have been a number of reviews over the years (see Buxton, 1957; Gibson, 1965; Wray and Sojka, 1977) and this chapter brings these previous accounts up to date.

Historical Perspective

The early history of the disease is confused because at that time the genus *Salmonella* was inadequately characterized. Few of the serovars were clearly defined and they were commonly confused with members of other genera.

Calf paratyphoid was first recorded in Europe in the middle of the nineteenth century, when a form of diarrhoea or *Kälberruhr* attracted considerable attention in The Netherlands, Germany and Denmark. The disease was recognized in certain localities where it tended to occur year after year and to cause heavy losses. Obich (1865) was the first to consider the disease to be caused by an infectious agent, but it was not studied bacteriologically until Jensen (1891) investigated an outbreak in Denmark and isolated a coliform bacillus from the viscera of affected calves; which he named *Bacillus paracoli*. Thomassen (1897) isolated an organism which he named 'pseudotyphoid bacillus', thus indicating a relationship with the recently isolated 'typhoid bacillus'. Further *Salmonella* isolations from calves were made by a number of workers; some of these appeared to be similar to *Bacillus enteritidis*, which was isolated by Gärtner from an outbreak of food poisoning, and others resembled *S. paratyphi* B. Further confusion was caused by the isolation of *Bacterium typhimurium* from mice by Löffler in 1889, and Kaensche's description of an organism isolated during an outbreak of food poisoning in Breslau in 1893. This outbreak occurred when condemned meat from an emergency-slaughtered cow was stolen and marketed. The cultured organism was generally known as the Breslau bacillus, although Kaensche used the name *Bacillus kaensche*.

Salmonella infection was subsequently reported in adult cattle by Mohler and Buckley (1902), who described a disease outbreak in the USA caused by an organism resembling *S. enteritidis*. Many of these earlier identifications of *S. enteritidis* were subsequently shown by serology to be erroneous and are now recognized as *S. dublin*, which has been recognized subsequently in many countries. Similar confusion existed between *S. paratyphi* B and *S. typhimurium* and it was not accepted until the 1920s that these two serovars were distinct. When the genus *Salmonella* was more clearly defined, it was apparent that the serovar most commonly associated with bovine salmonellosis in those parts of Europe where the disease was endemic was *S. dublin*, and *S. typhimurium* was next in order of frequency.

Epidemiology

Incidence of salmonellosis in cattle

S. typhimurium and S. dublin appear to be the commonest serovars isolated from cattle, although the distribution of these two serovars may differ between countries, and S. dublin is thought not to be present in some countries.

Prior to the introduction in 1975 of the Zoonoses Order (HMSO, 1989) in the UK, estimates of the prevalence of salmonellosis were based on the number of cases diagnosed by the Ministry of Agriculture's laboratories. During the 1960s and early 1970s, S. dublin was the predominant serovar, and the number of cases increased from 115 in 1958 to a peak of 4012 cases in 1969 (Wray and Sojka, 1977). Its prevalence has subsequently declined, so that S. typhimurium is now the more frequent (MAFF, 1998). With the introduction of the Zoonoses Order, isolations of Salmonella from certain species of food animals became reportable. For practical purposes, an 'incident' is defined as the isolation of Salmonella sp. from an individual animal, from identifiable groups of animals or from their products or surroundings.

In the UK, S. dublin currently causes some 400–500 incidents annually, infection being equally common in adult cattle and calves. Infection of cattle with S. dublin is more frequent than with S. typhimurium in Denmark and Sweden, while the converse is the case in France. In The Netherlands, the prevalence of S. dublin appears to be increasing (Visser et al., 1993). Although S. dublin has been isolated from humans in New Zealand, it has not been detected in farm animals (Sanson and Thornton, 1997). S. dublin infection was formerly more frequent in cattle in Australia than S. typhimurium, but since 1990 the latter is more common. Since 1980, S. dublin has spread eastward from California to other states and northwards into Canada, where it had not been detected previously (Robinson et al., 1984). Its prevalence now appears to be declining, although it was detected in 26 states during 1995/96 (Ferris and Miller, 1996).

A study of allelic variation in S. dublin by Selander et al. (1992) identified three electrophoretic types, marking three closely related clones, one of which (Du1) was globally distrib-uted. Non-motile variants are common in the USA, comprising 24% of the S. dublin isolates (Ferris and Miller, 1996). The non-motile variants recovered from the USA belonged to Du1. Strains possessing the Vi antigen were confined to clone Du3, which is apparently limited to France and Great Britain (although in our experience British isolates do not express this antigen). Studies by Ferris et al. (1992) of 100 clinical isolates of S. dublin identified 26 groups by phage typing, but 52% belonged to one group. Seven plasmid profile groups were identified, of which 90 isolates belonged to three groups.

Many different phage types of S. typhimurium have been isolated from cattle and in the UK one particular type appears to become dominant until it is replaced by another phage type. During the 1960s, multiply resistant S. typhimurium phage type 29 became widespread in calves (Anderson, 1968). This then declined and, in 1977, multiply resistant phage types 204 and 193 became frequent, and the variant DT204c was the predominant type in calves (Wray et al., 1987b).

Although the prevalence of DT204c decreased during the early 1990s, a subsequent increase in the prevalence of multiply resistant DT104 in both adult cattle and calves has occurred (Wray and Davies, 1996; MAFF, 1996). In the UK, one or two phage types usually predominate, although some 20–30 different phage types are detected annually at a much lower incidence. Surveillance in The Netherlands has shown a similar progression of multiply resistant phage types. In 1972, PT201 (Dutch typing system) was detected in calves; this was replaced by PT193 in 1977, in 1979 PT202 (corresponding to DT204c) was detected and currently PT206DT104) is occurring (van Leeuwen et al., 1984; W.J. van Leeuwen, RIVM, Bilthoven, The Netherlands, personal communication, 1998).

Many Salmonella serovars other than S. dublin and S. typhimurium have been isolated from cattle (Kelterborn, 1967). During the period 1968–1974, Sojka et al. (1977) recorded the isolation of 101 different Salmonella serovars from cattle in England and Wales. Currently in the UK, 40–50 different Salmonella serovars, usually at a low prevalence, are detected annually in cattle (MAFF, 1996). In the USA, 48% of the 730 Salmonella, other than S. dublin and S. typhimurium, isolated from cattle were represented by seven serovars (Ferris and Miller, 1996).

Antimicrobial resistance

More detailed information will be found in Chapter 6. Antimicrobial resistance has been monitored in *Salmonella* isolated from cattle in England and Wales since 1971. During this period, *S. dublin* has remained fully susceptible to the antimicrobials used for sensitivity testing (Wray, 1997). A similar situation occurs in Scandinavia, some other European countries and Australia. However, in Germany (Helmuth and Seiler, 1986), the USA (Blackburn *et al.*, 1984) and The Netherlands (van Leeuwen *et al.*, 1984), multiply resistant *S. dublin* strains have been isolated. Likewise, in France, multiply resistant *S. dublin* has been isolated from calves but not from adult cattle (Martel and Coudert, 1993).

Multiple antibiotic resistance in the UK is usually associated with a small number of phage types of *S. typhimurium*, e.g. PT29 (Anderson, 1968), DT204c (Wray *et al.*, 1987b) and more recently DT104 (Evans and Davies, 1996; Wray and Davies, 1996). In The Netherlands, multiply resistant phage types were 201, 193, 202 and, more recently, 206 (van Leeuwen *et al.*, 1984; W.J. van Leeuwen, personal communication). However, in the USA, most *S. typhimurium* from cattle are multiply resistant (B.P. Smith, Davis, California, personal communication). Other serovars may also acquire antimicrobial resistance (Martel *et al.*, 1995; Wray, 1997) and an account of the epidemiology of multiply resistant *S. newport*, was given by Holmberg *et al.* (1984).

Route of infection

Infection is usually by the mouth and numerous experiments have shown that oral doses ranging from 10^6 to 10^{11} *S. dublin* and 10^4 to 10^{11} *S. typhimurium* are necessary to cause disease in healthy cattle (see Wray and Sojka, 1977). With the higher doses, the results of experimental infections are more consistent and the symptoms more acute. These doses are likely to be higher than those encountered under natural conditions, where concurrent disease and stress con-

tribute. Nazer and Osborne (1977) found that intraduodenal inoculation of 10^4 *S. dublin* produced disease.

It has long been suggested that aerosol transmission may be a means by which *Salmonella* are transmitted, although De Jong and Ekdahl (1965) and Nazer and Osborne (1977) found that large doses (10^4–10^{11} *S. dublin*) injected supraconjunctivally were required to produce disease. More recently, experimental infection of calves by aerosol has been reported (Wathes *et al.*, 1988).

Transmission of infection

Most infection is introduced into *Salmonella*-free herds by the purchase of infected cattle, either as calves for intensive rearing or adult cattle for replacements. Purchased animals may have acquired infection on their home-farm premises, in transit or on dealers' premises (Wray *et al.*, 1990, 1991). In the USA, however, up to 75% of large western dairy herds are infected with *Salmonella*, but it is difficult to determine the source of serovars, other than *S. dublin*, because of the widespread environmental contamination.

Adult cattle

A case–control study of *S. typhimurium* DT104 infection in cattle showed that the introduction of newly purchased cattle to a farm increased the risk of disease and that the period of highest risk was the first 4 weeks after purchase. Purchase through dealers was associated with a fourfold increase in risk, compared with purchasing cattle directly from other farms (Evans and Davies, 1996). As indicated previously, *S. dublin* has become widespread in some countries, after being localized in an area for many years. Clonal spread within a country is also a common feature of the epidemiology of *S. typhimurium* in cattle herds (Aarestrup *et al.*, 1997).

Some adult cattle which recover from *Salmonella* infection, especially in the case of *S. dublin*, may become active carriers and excrete the organism continuously or intermittently in their faeces for years, if not life. Sojka *et al.* (1974) investigated two active *S. dublin* carriers and found that continuous excretion of the organism occurred, the number of organisms ranging from 10^2 to $10^4\ \mathrm{g}^{-1}$ faeces. In contrast, animals recovering from S. typhimurium infection may continue to excrete the organism in their faeces, but this period is usually limited to a few weeks or months after recovery (Gibson, 1965). However, Evans and Davies (1996) in their investigation of S. typhimurium DT104 infection commented that, while the disease outbreak was of short duration, subclinical infection in the herd could persist for up to 18 months and that recurrence of infection may occur in some herds 2–3 years after the original infection (Davies, 1997). During a study of disease caused by serovars other than S. dublin and S. typhimurium, Richardson (1975a) found that some cows excreted Salmonella for up to 11 months after the disease outbreak.

Ingestion of Salmonella does not necessarily lead to infection or disease, and Richardson (1975b) considered that S. dublin organisms, and presumably other serovars, may be ingested by adult cattle and pass through the alimentary tract, with little or no invasion of lymph nodes. Grazing contacts of active carriers often yield positive rectal swabs, but when removed from pasture and stalled they cease shedding Salmonella in about 2 weeks. Likewise, during an outbreak of S. newport, Clegg et al. (1983) frequently isolated the organism from rectal swabs from a number of animals, but when the two longest excreters were housed in clean premises, the organism was isolated neither from rectal swabs nor at post-mortem examination.

Some animals in a herd may harbour localized infection in lymph nodes or tonsils, especially following S. dublin infection, without excreting the organism in their faeces. Such animals may be termed latent carriers, which are of importance in the epidemiology of S. dublin infection and the persistence of the organism on farms. Watson et al. (1971) pointed out that no investigation had been made into cattle that had negative faeces in herds in which S. dublin infection was present. During their investigation, they found that 14 of 59 cows with negative rectal swabs were infected at post-mortem examination. Latent S. dublin carriers may become active carriers or even clinical cases during stress, especially during pregnancy, when abortion may occur and the organism may be excreted in the genital discharges, urine and milk.

Smith et al. (1994) found that some cattle in California had chronic Salmonella infection of the udder, and some of these cattle shed the organism in both faeces and milk. Experimental studies (Spier et al., 1991) induced infection by inoculation of low numbers of organisms (c. 5000) into the teat canal, and the resulting milk contained 10^2–10^5 colony-forming units (cfu) ml^{-1} milk. Such milk may be a source of infection for calves. House et al. (1993) found that three heifers that were infected with S. dublin as calves had infected supramammary lymph nodes when slaughtered at 13 months of age. It is likely that S. dublin septicaemia may result in lymph-node infections and consequent shedding during lactation. In contrast with California, where chronic S. dublin infection of the udder is not uncommon, European studies indicate that mammary-gland infection with Salmonella is uncommon, although epidemiologically important. Giles et al. (1989) described an outbreak of S. typhimurium in a dairy herd in which a cow excreted the organism intermittently in its milk for 2.5 years. More recently, Sharp and Rawson (1992) described possible mammary-gland infection with S. typhimurium DT104 and Wood et al. (1991) the persistence of S. enteritidis in the udder.

A number of studies have shown that Salmonella infection may be present on farms in the absence of clinical disease. Thus, Heard et al. (1972) studied patterns of Salmonella infection on four farms and found that, on a well-run farm, although the incidence of Salmonella infection was high, there was little disease. A 3-year study of a farm on which S. dublin was present detected the organism in cattle and their environment in the absence of clinical disease (Wray et al., 1989). Morisse et al. (1992) isolated Salmonella from 6.9% of 145 cows in 17 herds without a history of Salmonella. Smith et al. (1994) found a 75% prevalence of Salmonella-infected dairy herds by serological testing. Thus, failure to recognize that the introduction of infection may sometimes precede the development of clinical disease by several months or years may lead investigators to attribute infection to somewhat unlikely sources. A sudden change in herd resistance, due to concurrent disease, nutritional stress or severe weather, can result in clinical disease.

Calves

The mixing of young susceptible calves and their subsequent transport is an effective means for the rapid dissemination of *Salmonella*, especially *S. typhimurium*. Gronstol *et al.* (1974) studied the effect of transport experimentally and found that cross-infection occurred and the stress activated latent infection in calves whose faeces had been negative for 5 weeks. Indeed, it is likely that DT204c was introduced into The Netherlands by the import of British calves (Rowe *et al.*, 1979). A number of studies have shown that few calves are infected on arrival at the rearing farm (Rankin *et al.*, 1969; Osborne *et al.*, 1974; Wray *et al.*, 1987a), but infection then spreads rapidly, often in the absence of clinical disease. Our studies (Wray *et al.*, 1987a) showed that the infection rate for *S. typhimurium* increases during the first week to reach a peak at 14–21 days, although the spread of *S. dublin* appears to be slower, with a peak at 4–5 weeks. On all of the farms studied, there was little clinical disease. It was also found that the *Salmonella* infection rate and duration of excretion was higher in single-penned calves than in those that were group-housed. During an analysis of the spatial and temporal patterns of *Salmonella* excretion by calves penned individually, the results showed that non-contagious or indirect routes were more important than direct contagious routes in disease spread (Hardman *et al.*, 1991). These authors suggested that the avoidance of aerosol production and the effective cleaning and disinfection of utensils between feeding and of buildings between batches were likely to be of great importance in the control of salmonellosis in calves.

Both active and latent carriers of *S. dublin* may produce congenitally infected calves and, though most of these calves are often stillborn, some may survive to infect in-contact animals or to become a carrier. Calves may also be infected by drinking milk from cows with *Salmonella* infection in the udder or associated lymph nodes (Smith *et al.*, 1989).

Additional factors in the aetiology of disease

Many factors, conveniently labelled as 'stress', may either exacerbate the disease or increase the susceptibility of cattle to *Salmonella* infections. Frik (1969) suggested that the persistent carrier state in adult cattle follows concurrent *S. dublin* and *Fasciola hepatica* infection, and Richardson

and Watson (1971) reported that salmonellosis in adult cattle was more prevalent on fluke-infested farms, although Taylor and Kilpatrick (1975) suggested that the two infections were independent of each other, but were influenced by similar external conditions. However, in experimental studies, Aitken *et al.* (1976) found that fascioliasis increased the susceptibility to *S. dublin* infection.

Similarly, in calves, combined *Salmonella* and bovine viral diarrhoea virus (BVDV) infection was more severe than *Salmonella* infection alone (Wray and Roeder, 1987), and severe disease was observed in a group of pregnant dairy heifers that had natural BVDV and *S. typhimurium* DT104 infection (Penny *et al.*, 1996). Morisse and Cotte (1994), on the other hand, could find no association between BVDV and *F. hepatica* and *Salmonella* infections, because both agents had an identical prevalence in infected and control herds. They suggested, however, that metabolic and hepatic changes observed during the peripartum period could result in disturbances in the intestinal ecosystem and in the emergence of *Salmonella* populations.

In a case–control study of *S. typhimurium* DT104 in cattle, Evans (1996) found that the feeding of purchased maize or root crops was associated with a reduced risk of disease. It is known that the normal flora of the alimentary tract is inhibitory to colonization by *Salmonella* (Chambers and Lysons, 1979). The inhibitory effect is probably most dependent on the volatile fatty acid content and pH of rumen fluid, and a change in feed or an inadequate level of feeding can alter the balance of the intestinal flora to favour the multiplication of *Salmonella* (Frost *et al.*, 1988; Morisse and Cotte, 1994). Thus, it is possible that the susceptibility of an animal to *Salmonella* may be modulated by feeding changes that affect the intestinal flora balance.

Environmental Aspects

Persistence in buildings

An important aspect of the epidemiology of *Salmonella* infection in cattle is the persistence of the organism in animal accommodation after depopulation. *Salmonella* were isolated from the environment of six of nine calf units investigated after cleansing and disinfection (Wray et al., 1987b). Similar results were found in markets (Wray et al., 1991) and on calf dealers' premises, when the organism was detected in the environment of ten of the 12 premises investigated (Wray et al., 1990). The cleaning and disinfection routines were often ineffective and Salmonella were isolated from 7.6% and 5.3% of the wall and floor samples, respectively, before disinfection and 6.8% and 7.6% afterwards. Eight different Salmonella serovars were detected, of which the commonest was S. typhimurium. On one of the premises, the Salmonella isolation rates increased after cleansing and disinfection, and it is possible that the use of pressure hoses may spread Salmonella by aerosols to other parts of the building or that vigorous cleansing may have washed Salmonella out from previously untouched areas. S. typhimurium DT204c has been shown by plasmid-profile analysis to persist in calf units for periods ranging from 4 months to 2 years (mean 14 months) (McLaren and Wray, 1991). A case–control study of S. typhimurium DT104 indicated an increased risk of disease when cattle were housed, possibly indicating that contaminated buildings may be a source of infection or that close confinement or stress when housed increases the risk of clinical disease (Evans and Davies, 1996). In laboratory experiments, S. dublin survived for almost 6 years in faeces on different building materials (Plym-Forshell and Eskesbo, 1996).

Animal wastes

Modern, intensive, cattle production systems produce large amounts of slurry, which has highlighted the risk of pasture contamination because of the disposal problems. The subject was reviewed comprehensively by Jones (1992), although many of the studies were performed in the laboratory rather than on the farm. Jones and Matthews (1975) isolated small numbers of Salmonella, fewer than 100 g^{-1} from 20 of 187 different samples of cattle slurry. In Sweden, Thunegard (1975) found that 35% of cattle slurry samples contained Salmonella, compared with only 6% of solid manure samples. Multiplication of Salmonella has been reported during storage of slurry (Blum, 1968; Findlay, 1973), but, in general, their numbers are reduced

by storage. Salmonella have been shown to survive for up to 286 days in slurry, but this is a function of the initial number of organisms, storage temperature and Salmonella serovar. Survival is greatest at temperatures below 10°C and in slurries containing more than 5% solids. Although Salmonella die rapidly in slurry during storage, there are occasions when disinfection may be necessary, and various chemicals, such as lime, formalin or chlorine, have been recommended (Jones, 1992). Aeration of slurry is also an effective method of reducing Salmonella (Munch et al., 1987), who found a 90% reduction within 2 days.

Pasture contamination

Salmonella may survive for long periods in infected faeces, where their survival is dependent on a number of factors, especially climatic conditions. Field (1948) reported that S. dublin survived for at least 72 days in faeces on pasture in winter and 119 days in summer. In moist, unheaped faeces, survival up to 3–4 months has been reported, although properly composted faeces heat up rapidly and Salmonella reduction is much quicker. Findlay (1972) found that S. dublin survived for 13–24 weeks when spread in slurry on pasture, although Taylor (1973), using a smaller inoculum, did not isolate S. dublin from soil or pasture for more than 76 days. Taylor and Burrows (1971) found that S. dublin applied in slurry could be recovered from the bottom 3 inches of grass for 19 days, although it could not be recovered from the upper levels for more than 10 days. When the grass was cropped, no survival was reported beyond 7 days. From their experiments, they concluded that the disease risk was not great after 7 days and that to infect calves it was necessary to use slurry containing 10^6 S. dublin. A similar experiment by Kelly and Collins (1978), in which cattle were allowed to graze pasture 10 days after it had been treated with slurry from a dairy herd, did not result in any animal becoming infected. However, outbreaks of disease in cattle have been described following the grazing of cattle on slurry-contaminated pasture (Rasch and Richter, 1956; Jack and Hepper, 1969), when unstored slurry has been spread or waste allowed to enter watercourses.

Salmonella survival has been estimated to be

from less than 30 days to 1 year in soil, from 200 to 259 days in soil contaminated with animal faeces, from 57 to 300 days in soil contaminated with cattle slurry and from 11 days to 9 months in soil containing sewage sludge (Jones, 1992). Davies and Wray (1996) found that, when calf carcasses contaminated with *S. typhimurium* were placed in either a decomposition pit or a deep burial pit, *Salmonella* were isolated from the soil around the pit for 27 weeks and for 15 weeks around the burial site. *Salmonella* reappeared in soil samples during the cold weather after an apparent 9-week absence. Spread to a nearby drainage ditch occurred and wild birds became contaminated from eating *Salmonella*-infected maggots.

Sewage sludge

Sewage sludge is used as a fertilizer, and examination of 882 samples of settled sewage, sewage sludges and final effluent from eight sewage-treatment works found that 68% of the samples were positive for *Salmonella* (Jones et al., 1980). An investigation of 26 outbreaks of salmonellosis in animals found that the attributed sources of infection were sewage effluent (ten incidents), septic-tank effluent (eight), sewage sludge (three), seagulls (three) and abattoir effluent (two) (Reilly et al., 1981). Because sewage sludge is used as a fertilizer, guidelines on the agricultural use of sewage sludge have been produced to minimize possible risks (HMSO, 1977); these have subsequently been incorporated into UK legislation through the Sludge (Use in Agriculture) Regulation 1989, which implements the EC Directive 86/278. In the UK, there is little evidence of risk to animals when known contaminated sludge is spread on pasture, provided that the guidelines are followed (Linklater et al., 1985).

In contrast, on the continent, there have been a number of reports of outbreaks that were associated with grazing cattle on pasture contaminated with sludge (Jones, 1992). Consequently, in Germany, the agricultural disposal of sewage sludge is not allowed if it is not considered hygienically safe.

Water-borne infection

There are many reports on the isolation of *Salmonella* from rivers and streams and, once a water-supply is contaminated, rapid spread of infection may occur. In Wales, a number of cases of *S. dublin* infection have been associated with watercourses contaminated by grazing animals and farm effluents (Williams, 1975). Polluted water may also contaminate pasture whenever flooding occurs, and evidence indicates that many clinical outbreaks in cattle arise from grazing recently flooded pasture.

Biological treatment of sewage removes most of the suspended and dissolved organic matter but it does not necessarily remove pathogenic bacteria. Consequently, surface watercourses may be heavily contaminated from this source (Harbourne et al., 1978). In The Netherlands, many rivers, canals and lakes are contaminated, although the likely numbers of *Salmonella* are small (< 10 salmonella l^{-1}).

In the USA, Gay and Hunsaker (1993) found that recycled water that had been used to flush out the accommodation and was then being pumped into the stalls was contaminated with *Salmonella* and responsible for the continued presence of *Salmonella* on the farm.

The role of wild animals

Infection may be introduced on to farms by free-living animals and birds. Rats and mice may acquire *S. dublin* infection (Tablante and Lane, 1989), but available evidence suggests that they do not play a major role in the spread of infection (Gibson, 1965). Infected mice and rats may prolong the persistence of *Salmonella* on farms, but it is probably of limited duration. Evans and Davies (1996) found that wild birds and cats were possible vectors of *S. typhimurium* DT104, particularly if they had access to feed stores. A high population density of cats around the farm and evidence of access to feed stores by wild birds were both associated with an increased risk of disease. Cats have also been shown to be infected with *S. typhimurium* DT104 (Wall et al., 1995, 1996). The presence of infection was also detected in mice, rats and, to a lesser extent, dogs, foxes and badgers, and contamination of feed, grain stores and bedding by their faeces was common (Davies, 1997).

Feedstuffs

The role of animal feedstuffs in the epidemiology of *Salmonella* is the subject of Chapter 17 and this section serves only to emphasize the more important aspects. The importation into Europe of contaminated products has often been followed by the appearance of *Salmonella* serovars previously unknown in Europe. Although *S. dublin* and *S. typhimurium* are the commonest serovars, the increasing prevalence in the UK of other serovars in cattle during the 1970s was associated with imported animal protein, e.g. meat and bone-meal and fish-meal. Finished feeds have also been found to be contaminated with *Salmonella*, as have vegetable proteins such as soya, rape-seed meal and cottonseed. Likewise, waste from the food industry, such as biscuit-meal, is often contaminated.

The introduction of contaminated ingredients into compound-feed mills that prepare mixed animal feedstuffs may lead to contamination of cooling systems, storage bins and other equipment, from which, in turn, other products may be contaminated. Although many of the protein ingredients are treated with heat at temperatures high enough to ensure the destruction of *Salmonella*, faulty processing and contamination during and after processing may often lead to heavy contamination of supposedly sterilized products. Various methods have been used to reduce *Salmonella* contamination of animal feeds, such as direct injection of steam, expansion–extrusion, pelletization and treatment with organic acids, such as propionic and formic acid. In recent years, many farmers in the UK have been mixing their own feeds, and as Evans and Davies (1996) pointed out, poor storage conditions on the farm may allow *Salmonella* contamination by animals and birds.

Clinical Findings

Adult cattle

Adult cattle of all ages may be affected; acute and subacute forms of the disease are recognized. There are no significant differences between infections caused by the different *Salmonella* serovars.

In acute salmonellosis, the onset is sudden, with fever, dullness, loss of appetite and depressed milk yield; Kahrs *et al.* (1972) recorded that milk production dropped from 5000 to 1000 lb day^{-1} during an outbreak of *S. typhimurium* infection in a herd of 111 cows. The fever often subsides precipitously with the onset of severe diarrhoea, which may vary from watery green-brown to fetid watery faeces containing blood, mucus and shreds or casts of necrotic bowel lining. Pregnant animals may abort. The fever tends to persist for a few days and then declines before death, which occurs most frequently 4–7 days from the onset of clinical signs. In untreated cases, the case fatality rate may reach 75%, although this may be reduced to 10% by treatment. In all cases, the animals show signs of toxaemia, dehydration and associated weight loss, with the faeces remaining liquid for 10–14 days, and complete recovery may take up to 2 months.

Subacute salmonellosis is less dramatic. Fever varies or is absent and the other symptoms are less severe. The prognosis, even without treatment, is quite good, although complete recovery takes about 2 months. In some instances, variations in the clinical disease arise from the fact that illness is not due to recently acquired infection, but to activation by some other disease or the stress of a latent infection or infection cycling through a herd (Gibson, 1958). Such cases may show a wide gradation in the severity of clinical signs, and it may be difficult to determine whether disease is caused primarily by *Salmonella* or whether *Salmonella* are playing a secondary role (Hughes *et al.*, 1971).

As mentioned earlier, sick, pregnant cows may abort. However, it has become apparent that cows may abort from *S. dublin* infection without showing clinical signs. In the UK, *S. dublin* is the second most commonly diagnosed bacterial cause of abortion. Hinton (1971, 1974) studied the clinical aspects of *S. dublin* abortion and, in 86 of 111 cases, abortion was the only clinical sign. There was no evidence of congenital infection of the calf when abortion cases that were not faecal excreters calved again. On the other hand, when constant faecal excreters were examined, congenital infection of the calf and/or transient vaginal excretion could be demonstrated. Field observations (Henner and Lugmayr, 1972) and experimental studies (Hall and Jones, 1977) have suggested that abortion is preceded by a period of pyrexia, when the organism is multiplying in the placenta.

Calves

In calves, the clinical disease is more common after the first week of life and clinical signs are often seen at 2–6 weeks of age. The clinical picture in calves is subject to wide variation, including pulmonary infections, but the enteric form of the disease predominates. The typical clinical case is characterized by fever, dullness and loss of appetite, followed by a brown scour, with fluid, offensive faeces, which often contain blood or mucus. Affected calves quickly lose condition and become dehydrated, weak and emaciated. With *S. dublin*, bacteraemia and respiratory signs often predominate; the calves often tend to be older than those affected with other serovars (Wray *et al.*, 1987a).

Some calves may suffer septicaemia and collapse with no diarrhoea. Affected animals show profound depression, dullness, prostration, high fever (40.5–42°C) and death within 24–48 h. Because the infection also commonly causes pneumonia, evidence of this may be present. Jaundice is a feature in some cases and nervous signs of encephalomeningitis may be seen in other cases. Polyarthritis and osteitis have also been described, and a sequel to some cases of enteric salmonellosis is the development of dry gangrene of the extremities, including ear tips, tail tip and the limbs from the fetlock down (O'Connor *et al.*, 1972). Calves from dams infected with *S. dublin* may be stillborn, non-viable or sickly from birth.

The severity and duration of clinical signs are often related to the standard of husbandry and hygiene. On poorly run farms, 80% or more calves may develop clinical signs, with mortality rates of 10–50%. Contributory factors are important in determining the occurrence and severity of clinical disease. Concurrent viral diseases, such as BVDV infection, have been shown to be of importance.

Haematological and Biochemical Aspects

The haematological changes during bovine salmonellosis have not been studied in depth and their interpretation is rendered difficult by such factors as the age of the animal, state of hydration and possible effects of endotoxaemia.

In calves that died from salmonellosis, Smith *et al.* (1979) and Endresen (1970) found that a leucopenia and assoicated neutropenia preceded death, whereas in experimentally infected calves, Wray (1980) observed that a terminal leucocytosis followed the leucopenia. In calves that survive salmonellosis, a leucocytosis, with an increase in immature neutrophils, occurs (Smith *et al.*, 1979; Wray, 1980; Rings, 1985).

Elevated erythrocyte counts, haemoglobin content and packed-cell volumes have been observed in cattle during the terminal phase of salmonellosis (Hall *et al.*, 1980; Wray 1980), and in some an anaemia may occur, which is masked by the haemoconcentration. In calves that survive, erythrocyte counts remain within normal values.

Alterations in the blood biochemical levels occurred terminally in some of the calves that died, blood urea levels increased markedly and alterations in sodium and potassium levels indicated the severity of dehydration (Fisher, 1971; Wray, 1980). In diarrhoeic calves that died of salmonellosis, Fisher and Martinez (1975) found that faecal solids increased, and this, coupled with a reduction in fluid intake, resulted in considerable extravascular loss of water.

Blood urea analysis, clinical chemistry and leucocyte examination may be of value in the prognosis of *Salmonella* infections of cattle and provide guidance for therapy, although care must be exercised in their interpretation because of the variability.

The Pathogenesis of *Salmonella* Infection

Calves

As indicated previously, the intestine does not appear to be the sole portal of entry for *Salmonella*, and infection of oesophagectomized calves with *S. typhimurium* resulted in haematogenous spread via the tonsil to the intestine (De Jong and Ekdahl, 1965). Most experimental studies have been made with oral challenge, where the organism gains entry to the tissues by an invasive process, mainly in the lower small intestine. Here the bacteria are able to invade the intestinal mucosa via both M cells, which overlie the lymphoid follicles, and the enterocytes. The bacterial infection results in distinct pathological changes: the villi become oedematous and shortened and there is an abnormal extrusion of enterocytes. Using specific staining techniques, Segall and Lindberg (1991) showed that S. dublin had a special affinity for the columnar enterocytes of the terminal jejunum and ileum, the follicle-associated epithelium over the Peyer's patches and glandular tissues in the duodenum, tonsilar area and lungs. Using ligated bovine intestinal loops, Watson et al. (1995) found no difference in the ability of S. typhimurium and S. dublin to invade the intestinal mucosa. When isogenic mutants were used, mutations in the slyA (a gene postulated to regulate the expression of virulence factors) had only a small effect on invasiveness, whereas mutations in invH (a gene involved in secretion of proteins from Salmonella) caused a significant reduction in epithelial invasion.

The mechanisms involved in the induction of diarrhoea are poorly understood; the involvement of bacterial toxins has long been speculated upon, but, to date, there is little evidence to support a role for toxins. There are numerous reports of cytotonic, cytotoxic and enterotoxic activities in vitro, but these reports are often contradictory. (Lax et al., 1994). A putative Salmonella enterotoxin gene (Stn) has been cloned (Chopra et al., 1987). The effect of mutations in the invH, slyA and Stn genes on the enteropathogenicity of S. typhimurium and S. dublin was studied in bovine ligated intestinal loops by Watson et al. (1997). They found that the invH mutation markedly affected the secretory and inflammatory response, implicating secreted effector molecules and enteropathogenicity. Surprisingly, the slyA mutation caused only a small effect, whereas the Stn mutation had no effect on enteropathogenicity. There appeared to be a direct correlation between the inflammatory response and the secretory response, although it could not be concluded that one occurred as a consequence of the other or vice versa. A role for secreted effector molecules in Salmonella-induced fluid secretion has recently been confirmed. Disruption of Sop B (a 60 kDa secreted protein of S. dublin) does not affect intestinal invasion but abolishes secretory and inflammatory responses in bovine ileal loops (Galyov et al., 1997).

Many Salmonella serovars contain large plasmids that are associated with virulence (viru-lence plasmids), and a common nomenclature (Salmonella plasmid virulence (SPV)) has been agreed for the operon. The S. dublin virulence plasmid mediates systemic infection but not the enteric phase of the disease in cattle (Wallis et al., 1995), although Libby et al. (1997) in their studies observed diarrhoea and enhanced intracellular proliferation in intestinal tissue associated with the presence of the spv genes. The S. dublin virulence plasmid also appears to be involved in the lysis of macrophages, although the spv genes do not appear to be involved (Guilloteau et al., 1996). However, slyA and invH mutations have also been shown to affect macrophage lysis (Watson et al., 1997), again implicating secreted effector molecules in this process.

Many of the systemic pathological effects of salmonellosis leading to shock are believed to be caused by the release of endotoxins (cell-wall polysaccharides). Peel et al. (1990) did not detect increased blood levels of tumour necrosis factor in experimentally infected calves and consequently postulated that local release of endotoxin may be responsible for the pathological changes in the intestine.

Adult cattle

While there have been numerous studies of experimental Salmonella infections in calves, these have been very much less frequent in adult cattle. In cows that had been starved and refed, Brownlie and Grau (1967) observed multiplication of S. dublin in the rumen. However, Hall et al. (1978) failed to detect multiplication of S. dublin after intraruminal inoculation of animals that had not been starved. Similarly, Chambers and Lysons (1979) found that rumen fluid was inhibitory to S. typhimurium. When nine pregnant heifers were dosed orally with 10^{10}–10^{11} S. dublin, the response was variable: two became severely ill, three developed mild illness and four remained healthy (Hall and Jones, 1979). One of the mildly ill animals aborted. In earlier experiments (Hall and Jones, 1977), using intravenous challenge, they found that S. dublin multiplied rapidly in the connective tissue of the cotyledons just before abortion, which resulted in placental destruction and hormonal changes, which initiated the abortion.

Spier *et al.* (1991) produced intramammary infection in five post-parturient cows with low numbers of *S. dublin* and, while excretion of *S. dublin* was intermittent, none of the cows showed clinical signs, even though persistently infected. However, following the administration of dexamethasone, one cow became acutely ill and chronic mastitis was demonstrated in the others.

Post-mortem Findings

Adult cattle

The post-mortem findings of a typical case reveal an acute muco/necrotic enteritis, especially of the ileum and large intestine. The wall is thickened and covered with yellow-grey necrotic material overlying a red, granular surface. The mesenteric lymph nodes and spleen may be enlarged. These lesions, however, are insufficient to form a firm diagnosis, and bacteriological confirmation is necessary.

Calves

Animals dying of acute septicaemia show extensive submucosal and subserosal petechial haemorrhages. More prolonged cases are characterized by poor bodily condition, usually with evidence of fetid diarrhoea. The small intestine typically shows a diffuse mucoid or mucohaemorrhagic enteritis and the mesenteric lymph nodes are oedematous, congested and greatly enlarged. A severe haemorrhagic and diphtheritic enteritis has been seen in some cases of *S. typhimurium* infection. The liver commonly shows jaundice, with thick turbid bile, and many reports refer to the presence of necrotic foci in the liver and kidneys. Gibson (1965) commented that, although such liver lesions were commonly reported in mainland Europe, they were uncommon in his experience. Sharply defined areas of pneumonia may be present in the anterior lobes of the lungs. When joints are affected, the joint cavities and adjacent tendon sheaths contain a gelatinous or serofibrinous fluid. Skeletal lesions of epiphyseal separation, osteoperiostitis and rarefying osteomyelitis of the distal limb bones were described in calves suffering from chronic *S. dublin* infection (Gitter *et al.*, 1978).

Laboratory Diagnosis of *Salmonella* Infections

Bacteriological examination

In the live animal, confirmation of clinical salmonellosis is performed by culture of rectal swabs or, preferably, freshly voided faeces. In adult cattle with clinical salmonellosis, the organism is excreted continuously and in large numbers and isolation of the organism presents no difficulties. Faecal culture, however, may give negative results in the early stages of the disease before the onset of acute diarrhoea. In the febrile stages of the disease, especially *S. dublin* infection, the organism may be isolated by blood culture and, sometimes, from the milk.

In the case of active carriers, usually *S. dublin* but occasionally other serovars, faecal cultures carried out on three occasions at 7–14 day intervals are recommended to confirm diagnosis. Latent carriers of *S. dublin* may best be detected at parturition, when vaginal swabs, faeces and milk may be positive (Richardson, 1973). During an investigation of *S. dublin* abortion, Hinton (1974) found that excretion of the organism in milk and the vaginal mucus did not persist beyond 4–5 weeks and faecal excretion was usually transient. When abortions occur, it is usually possible to isolate *Salmonella* from fetal stomach contents or placenta. In some cases, these materials may be culture-negative, because abortion was caused by endotoxaemia, release of endogenous prostaglandin $F_{2\alpha}$ ($PGF_{2\alpha}$) and lysis of the corpus luteum.

Faeces or rectal swabs may be used for calves, but swabs from up to 50% of the infected calves may be negative on culture, because of intermittent excretion, and samples should therefore be taken from a number of calves. The salivary excretion of *S. dublin* has been described (Richardson and Fawcett, 1973), but a number of workers found that faeces were more reliable than mouth swabs.

At necropsy, the isolation of *Salmonella* from tissues and intestinal contents usually presents few problems, but care must be taken in the interpretation of the findings, especially when small numbers of the organism are isolated from a few tissues or in cases where other pathogens are present. The isolation of *Salmonella* should be correlated with clinical signs and pathological lesions in order to determine the significance of the isolation.

To identify infected herds or to monitor persistence of infection, slurry or environmental samples collected with swabs may be cultured. Likewise, milk filters, when available, may also be used to monitor dairy herds.

Culture media

This is dealt with in Chapter 21, and the following is provided for guidance. Although many different media have been described, it should be remembered that *S. dublin* is inhibited by brilliant green and care should be exercised in the use of media containing brilliant green. Likewise, *S. dublin* may be inhibited by the incubation of enrichment broths at 43°C, e.g. Mueller–Kauffman tetrathionate (Gitter *et al.*, 1978) or Rappaport (Peterz *et al.*, 1989).

For cattle faeces and rectal swabs, enrichment in selenite broth and plating on brilliant green agars or deoxycholate–citrate agar are usually used in the UK. In the USA, many use xylose–lysine–Tergitol 4 (XLT4) after enrichment in selenite and tetrathionate. Our recent studies obtained better results by using pre-enrichment in buffered peptone water, subculture (1 : 100 dilution) into semi-solid Rappaport–Vassiliadis and plating on Rambach agar. The last method is also suitable for slurry and milk.

Serological diagnosis

The different tests and principles behind their use are more fully dealt with in Chapter 24.

Serum samples from cattle may be tested by the serum agglutination test (SAT) (Wray *et al.*, 1977), when flagellar titres of 1 : 320 or higher and somatic titres of 1 : 40 or higher are considered indicative of infection. In the cases of calves less than 4 months of age, a good flagellar response is produced, but there is little or no production of somatic agglutinins. When a calf is 3–4 months old, it is capable of producing high-titre agglutinins. The SAT has been used mostly for *S. dublin* but, when Lawson *et al.* (1974) evaluated the test in a known infected herd, many serologically positive animals were found not to be infected at slaughter. Other serological tests

that have been used include the complement fixation test, indirect haemagglutination test and anti-globulin test (see Wray and Sojka, 1977), but the results have been no better than the SAT.

More recently, indirect enzyme-linked immunoabsorbent assays (ELISAs) have been developed and Robertsson (1984) studied *S. dublin*-infected herds. Smith *et al.* (1989) used the ELISA to predict the carrier status of cows in a herd where animals had mammary and intestinal infections with *S. dublin*. They found that, with eight ELISA-positive cows, only 3.4% of 985 faeces and 2.5% of 756 milk samples were culture-positive (Smith *et al.*, 1992). In calves, 17.3% of 643 faeces were culture-positive. They considered that ELISAs were a better guide than culture for identified carrier status. Nielsen and Vestergaard (1992) found a sensitivity of 92% and specificity of 91% when an ELISA, read at an end-point titre of 800, was used to detect *S. dublin* carriers. A comparison of persisting anti-lipopolysaccharide (LPS) antibodies and post-mortem culture of animals from *Salmonella*-infected herds showed that only half (14/31) *S. dublin*-seropositive adult cattle were culture-positive, although 6/13 calves were positive. However, none of the cattle seropositive for *S. typhimurium* was culture-positive at postmortem. They concluded that serology was useful to identify infected herds but insufficient for the identification of persistently infected animals. Because of cross-agglutination between the *Salmonella* groups B- and D-LPS in the ELISA, Konrad *et al.* (1994) oxidized the LPS with sodium periodate to produce pure O9 antigen and so eliminate these cross-reactions.

Hoorfar *et al.* (1996) evaluated a fimbrial (SEF14) antigen to differentiate *S. dublin*- and *S. typhimurium*-infected herds and, while there was not complete agreement in an ELISA using a somatic antigen, they considered that the high specificity of the SEF14 antigen may increase the predictive value of the test in areas with a low prevalence of *Salmonella*.

The use of milk assays for *Salmonella* antibodies

A number of earlier assays based on the milk-ring test were not reliable and false-positive results were not uncommon (see Wray and Sojka,

1977). Although Hinton (1973) considered the whey agglutination test to be nearly comparable to the SAT, it was only practical to examine whey for flagellar agglutinins. However, ELISAs have been used to test milk for *Salmonella* antibodies with promising results. The ELISA was used by Smith *et al.* (1989) to detect *S. dublin* antibodies in milk; they found a good correlation between serum and milk titres and concluded that ELISAs were a useful test for detection of cattle with chronic *S. dublin* infection of the intestine and mammary gland. All of the cattle shed *S. dublin* in the milk and faeces; 46% of 1733 milk samples and 4% of 1733 faeces were culture-positive. Antibody titres remained elevated for long periods and there was a significant difference in milk ELISA titres between the infected group and the uninfected control group.

Likewise, Hoorfar *et al.* (1994) and Hoorfar and Wedderkopp (1995) used the ELISA to screen milk samples for *S. dublin* and *S. typhimurium* antibodies; they found a good correlation between serum and milk antibody titres, but considered that further modifications were necessary before the assay could be used for bulk milk samples.

Treatment

There are differences of opinion amongst veterinarians about the rationale and wisdom of treating cases of salmonellosis with antimicrobials (Whitlock, 1984), because of their efficacy and the likelihood of producing carriers. The general view of most practitioners is that prompt treatment with broad-spectrum antibiotics is beneficial. Aggressive treatment early in the course of the disease, especially in calves, where infection may become septicaemic, is required.

A number of experiments have shown that some antimicrobials may prolong *Salmonella* shedding, but the importance under farm conditions is uncertain, especially as studies of calves on farms have shown that, during outbreaks of salmonellosis, there is concomitantly a high prevalence of subclinical infection in untreated animals.

The choice of drugs to be used depends on tests to determine the antibacterial sensitivity of the isolate, especially as some *Salmonella* isolates may be multiply resistant, though early treatment

based on a likely successful regimen must be used before these results are available. Parenteral fluid therapy will increase the survival rate when given intravenously. Oral fluids will also help animals to survive the period of acute dehydration and toxaemia. Intravenous (i.v.) hypertonic saline (4 mg kg^{-1} of 7% NaCl), combined with oral rehydration with water or hypotonic sodium-containing fluid via a stomach tube, has proved particularly useful.

Vaccination

Prevention of salmonellosis by immunization has been practised for many years and both live and inactivated vaccines have been used. Although many publications attest to the efficacy of live vaccines in experimental trials, few are commercially available. Likewise, the relative importance of humoral and cell-mediated immunity has been the subject of debate, but it is now generally accepted that both play a part in producing protection. In the UK, only a formalin-inactivated preparation of *S. dublin* and *S. typhimurium* is now available for protection of cattle against *Salmonella* infections. In seven dairy herds that were infected with *S. typhimurium*, the use of the vaccine resulted in a rapid cessation of *Salmonella* excretion, which contrasted with the prolonged *Salmonella* excretion observed in five non-vaccinated herds (Evans and Davies, 1996). The vaccine has been the subject of recent field and experimental trials and has been shown to induce antibodies in adults, which are transferred to calves in colostrum (Jones *et al.*, 1988). Since calves may become infected within the first few weeks of life, passive protection by colostral antibodies is essential, because the development of active immunity depends to some extent on the age of the calf. Although attempts at passive protection have met with conflicting results, Jones *et al.* (1988) immunized cows with formalin-inactivated *S. typhimurium* approximately 7 and 2 weeks pre-partum. Calves were allowed to suck for 48 h and were then fed cold, stored colostrum from their own dam for a further 8 days. The calves were protected against a lethal challenge given at 5 days and they only excreted *Salmonella* in their faeces for a short period. Mortality was also reduced in calves that sucked from a vaccinated dam and were then fed on normal colostrum and, in calves born to non-immunized cows and later fed on immune colostrum, the degree of protection was correlated with the colostral antibody titre.

Most recent attempts at developing improved vaccines for calves have utilized live vaccines, which are considered to produce better cell-mediated immunity. These have either been naturally occurring avirulent strains, laboratory-derived or mutated strains or, more recently, genetically engineered strains. A live *S. dublin* vaccine, consisting of a part-rough strain, was shown during use in the field to give good protection when administered parenterally to calves, but it is no longer available commercially in the UK.

In Germany, live *S. typhimurium* and *S. dublin* vaccines have been available commercially for many years; these consist of adenine and histidine auxotrophs and adenine and thiamine auxotrophs, respectively. These are administered in the calves' milk at 1 day of age, and good immunity has been shown. The vaccine strains can be differentiated from wild-type strains by their inability to grow in minimal media or by genetic techniques (Schwarz and Liebisch, 1994; Liebisch and Schwarz, 1996).

Control of Bovine Salmonellosis

In general, measures for the control of bovine salmonellosis are equally applicable to all serovars, despite the differences in some aspects of their epidemiology.

Herd disease security measures

If possible, a closed-herd policy should be maintained. If stock are purchased or returned to the farm from markets, etc., they should be kept in isolation for 4 weeks, as many disease outbreaks occur within 2–3 weeks of arrival and the quarantine period allows the infection to be contained and reduces spread to the resident stock. The quarantine buildings should be located as far away as possible from the resident herd, and good hygiene (separate boots, overalls and equipment) and disinfection procedures observed. Animals should be inspected daily for signs of diarrhoea or other illness and, after about 3 weeks, faeces may

be cultured for the presence of *Salmonella*. In the case of adults, a positive serological test for *S. dublin* or *S. typhimurium* will indicate that the animal should be kept in isolation and its faeces cultured.

As mentioned previously, waterways and pasture may be contaminated during the application of slurry and sewage sludge; therefore, surface water should be fenced off and pastures rested for 4–5 weeks after the application of slurry and sludge. Effective rodent and bird control should be carried out, because these animals have been shown to contaminate stored feed.

Visitors who have been on to other farms and in contact with animals should be provided with clean protective clothing and disinfected boots. Foot-baths with active disinfectants should be provided at the entry to all livestock farms.

Calf-rearing units

Control measures on farms rearing purchased calves should be planned on the probability that *Salmonella* infection will be introduced sooner or later. Disease is not, however, inevitable and its prevention depends on two main factors:

1. Minimizing the number of *Salmonella* to which the calves are exposed;
2. Maintaining maximum resistance by optimal nutrition and management practices.

An all-in, all-out system should be adopted and buildings thoroughly cleaned and disinfected, allowing at least 2 weeks between batches to minimize carry-over of infection. Ideally, calves should be bought in age-matched groups directly from local farms, using the purchaser's own transport; if this is not possible, calves should be bought from as few sources as possible, using reputable dealers or reliable buying groups. Markets should be avoided if possible. Body weight, condition and adequate colostral antibody status are useful guides. Journey times should be as short as possible, using suitable vehicles and complying with appropriate regulations. Housing, i.e. the general layout of the site and construction of the calf house, should take account of drainage to minimize cross-infection. All fittings, utensils and surfaces should lend themselves to effective cleansing and disinfection. A strict hygiene routine for buckets and teats should be maintained. Each calf should have its own bucket and automatic feeders should be dismantled, cleaned and disinfected daily. Concurrent infection, especially respiratory diseases or BVDV, may precipitate clinical disease and ill calves should be nursed in isolation facilities. The use of vaccines has been questioned on the grounds that calves may be infected on arrival, but their use is preferable to the prophylactic use of antibiotics.

Control measures during outbreaks

Once salmonellosis has occurred, the priority is to stop it spreading to other animals and people. Clinically affected animals should be kept in isolation or segregated, to reduce the large weight of environmental contamination. Animals that have recovered may still excrete the organism and should remain in isolation until *Salmonella* excretion ceases (usually 2–12 weeks). In many *Salmonella* outbreaks, infection will be found to be widespread, with healthy animals excreting *Salmonella* in their faeces. Stressful events, such as parturition, nutritional stress, anorexia, other disease, etc., may trigger clinical disease, and, because newly calved cows are susceptible, disease often occurs in calving boxes or stalls. These should not be used for other animals until they have been thoroughly cleaned and disinfected.

If the problem reaches outbreak proportions and it is not possible to isolate all the affected animals, attempts should be made to segregate groups of different-aged animals as far as possible. *Salmonella* spreads around the farm environment on boots, tractors, other equipment, surface water, effluent from animal accommodation, birds, rodents, domestic animals, etc., and every effort should be made to attain a very high standard of cleanliness and discipline at all times. All manure and effluent should be contained for disposal or treated in such a way as to minimize environmental contamination. Slurry should be stored for at least 3 months and bedding from the isolation areas is best burned. Manure and slurry are best spread on arable land.

Vaccination may have a part to play in the control of disease outbreaks, and vaccination of cows will help to protect the calves and reduce *Salmonella* contamination.

Calves

Sick calves or calves from known infected cows should not be introduced into an already occupied calf house or unit. One infected calf can give rise to widespread infection in a calf house, irrespective of whether the calves are single-penned or group-housed. This is especially the case with purchased calves, which may come from a large number of sources.

The calf attendant should follow a disciplined procedure of disinfection on entering and leaving the house and, if possible, should not deal with healthy calves in another house, or should care for healthy calves first.

Public Health Aspects

Although *Salmonella* are an important cause of food poisoning, it should not be forgotten that humans may be infected by direct contact with infected cattle. Skin lesions caused by S. *dublin* and other serovars have been described in veterinarians following obstetrical manipulations (Williams, 1980; Visser, 1991). It was reported by Evans and Davies (1996) that, on 20% of the cattle farms where S. *typhimurium* DT104 infection was present, possible or confirmed associated human illness occurred in farm workers or their families. A study of human cases of salmonellosis found that 30% were associated with contact with infected cattle (Fone and Barker, 1994) and, in a case–control study, 10% of the cases originated on farms (Wall *et al.*, 1994, 1995b).

Milk-borne *Salmonella* infection

Contamination of milk may occur by a variety of routes. Cattle may occasionally, during the febrile stages of salmonellosis, excrete the organism in their milk (Gibson, 1965; Maclachlan, 1974) or, more commonly, infected faeces from either a clinical case or a healthy carrier may contaminate milk during the milking process (McEwen *et al.*, 1988). In Western Europe, udder infection with *Salmonella* appears infrequent, although the excretion of S. *typhimurium* has been reported (Ogilvie, 1986; Giles and King, 1987). In herds that had experienced acute outbreaks of salmonellosis, Morisse *et al.* (1984)

detected *Salmonella* in 10–60% of aseptically drawn milk samples. During a herd investigation by Giles *et al.* (1989), a cow was detected which excreted *S. typhimurium* intermittently from the udder for 2.5 years. In California, udder infections with *S. dublin* do not appear to be uncommon (Smith *et al.*, 1989; Spier *et al.*, 1991), and *S. dublin* infection of humans from drinking raw milk has been documented (Werner *et al.*, 1979; 1984). Milk may also be contaminated by the use of polluted water (Gay and Hunsaker, 1993) or dirty equipment. Indirect contamination may also occur when the udders, flanks, etc. of cattle become contaminated when they wade into polluted streams and the *Salmonella* enter the milk during milking (George *et al.*, 1972).

Many outbreaks of human salmonellosis have been associated with the drinking of raw milk (Sharpe *et al.*, 1980; Potter *et al.*, 1984) and, in Scotland, an outbreak of human illness caused by *S. dublin* involved at least 700 persons (Small and Sharp, 1979). During the period 1992–1996, Djuretic *et al.* (1997) described 11 general outbreaks associated with milk and dairy products, such as cheese.

Some of these outbreaks involved pasteurized milk and, although the heat treatment used during pasteurization is adequate to destroy *Salmonella* in milk, pasteurization-plant failure and post-pasteurization contamination can occur. Ryan *et al.* (1987) reported one such outbreak when over 16,000 culture-confirmed cases of human *S. typhimurium* infection occurred, after drinking pasteurized milk from a plant where post-processing contamination occurred.

Milk products, such as cheese (Rampling, 1996), ice-cream (Djuretic *et al.*, 1997), dried milk powder (Rowe *et al.*, 1987) and others (Becker and Terplan, 1986), have frequently been associated with human salmonellosis.

Meat and meat products

The most serious hazard to public health is that arising from the sick or casualty animal; in such animals, latent *Salmonella* infection may become generalized and so infect other animals in the lairage or during transit. Although there have been many surveys of the incidence of *Salmonella* in healthy cattle at abattoirs, there has been no uniformity of the material examined or the sampling techniques, and consequently the results are not comparable. Surveys in various countries have shown that in adult cattle the incidence of *Salmonella* varied from 0.3 to 11.6% and in calves from 4.3 to 14.3% (see Wray and Sojka, 1977).

The transfer of *Salmonella* from animals carrying the organisms both externally and internally can be limited only by cleanliness of the lairage and strict adherence to hygienic procedures during slaughter and dressing. The presence of even small numbers of *Salmonella* in carcass meat and edible offal may lead to heavy contamination of minced meat and sausage. In 1971, Edel and Kampelmacher found 33.7% minced meat samples, 20% tartar, 16.7% beefburgers and 16.9% raw sausages to be contaminated with *Salmonella*. Roberts *et al.* (1975) found the prevalence of *Salmonella* in sausages and sausage meat from several different producers to range from about 2 to 60%. A more recent survey found 17% of 786 samples of raw sausage to be contaminated with *Salmonella*, which included *S. typhimurium* DT104 (Nichols and de Louvois, 1995). Improperly cooked and stored beef was associated with an outbreak of *S. typhimurium* DT104 (Davies *et al.*, 1996) and a study of food in Manchester found 0.5% of cooked meat and 3.2% of tripe and udder to be contaminated with *Salmonella* (Barrell, 1987). Such contaminated meat and meat products may either cause direct infection or indirect contamination through cross-contamination in the kitchen.

References

Aarestrup, F.M., Jensen, N.E. and Baggesen, D.L. (1997) Clonal spread of tetracycline resistant *Salmonella typhimurium* in Danish dairy herds. *Veterinary Record* 140, 313–314.

Aitken, M.M., Jones, P.W., Hall, G.A. and Hughes, D.L. (1976) The effect of fascioliasis on susceptibility of cattle to *Salmonella dublin*. *British Veterinary Journal* 132, 119–120.

Anderson, E.S. (1968) Drug resistance in *Salmonella typhimurium* and its implications. *British Medical Journal* 3, 333–339.

Barrell, R.A. (1987) Isolations of salmonellas from humans and foods in the Manchester area: 1981–1985. *Epidemiology and Infection* 98, 277–284.

Becker, H. and Terplan, G. (1986) *Salmonella* in milk and milk products. *Zentralblatt für Veterinärmedizin B* 33, 1–25.

Blackburn, R.D., Schlater, L.K. and Swanson, M.R. (1984) Antibiotic resistance of members of the genus *Salmonella* isolated from chickens, turkeys, cattle and swine in the United States, October 1981–September 1982. *American Journal of Veterinary Research* 45, 1245–1249.

Blum, J. (1968) Studies on the occurrence, tenacity, growth and disinfection of salmonellas in waste waters of agricultural enterprises. *Schweierz Archiv Tierheilkunde* 110, 243–261.

Brownlie, L.E. and Grau, F.H. (1967) The effect of food intake on growth and survival of salmonellas and *E. coli* in the bovine rumen. *Journal of General Microbiology* 46, 125–134.

Buxton, A. (1957) *Salmonellosis in Animals*. Commonwealth Agricultural Bureaux, Farnham Royal, pp. 1–209.

Chambers, P.G. and Lysons, R.J. (1979) The inhibitory effect of bovine rumen fluid on *Salmonella typhimurium*. *Research in Veterinary Science* 26, 273–276.

Chopra, A.K., Houston, C.W., Peterson, J.W., Prasad, R. and Mekalanos, J.J. (1987) Cloning and expression of the *Salmonella* enterotoxin gene. *Journal of Bacteriology* 169, 5095–5100.

Clegg, F.G., Chiejina, S.N., Duncan, A.L., Kay, R.N. and Wray, C. (1983) Outbreaks of *Salmonella newport* infection in dairy herds and their relationship to management and contamination of the environment. *Veterinary Record* 112, 580–584.

Davies, A., O'Neil, P., Towers, L. and Cooke, M. (1996) An outbreak of *Salmonella typhimurium* DT104 food poisoning associated with eating beef. *PHLS Communicable Disease Report* 6, R159–R162.

Davies, R.H. (1997) A two year study of *Salmonella typhimurium* DT104 infection and contamination on cattle farms. *Cattle Practice* 5, 189–194.

Davies, R.H. and Wray, C. (1996) Seasonal variations in the isolation of *Salmonella typhimurium*, *Salmonella enteritidis*, *Bacillus cereus* and *Clostridium perfringens* from environmental samples. *Journal of Veterinary Medicine B* 43, 119–127.

De Jong, H. and Ekdahl, M.O. (1965) Salmonellosis in calves – the effect of dose rate and other factors on transmission. *New Zealand Veterinary Journal* 13, 59–64.

Djuretic, T., Wall, P.G. and Nichols, G. (1997) General outbreaks of infectious intestinal disease associated with milk and dairy products in England and Wales: 1992–1996. *PHLS Communicable Disease Report* 7, R41–R45.

Edel, W. and Kampelmacher, E.H. (1971) *Salmonella* infection in fattening calves at the farm. *Zentralblatt für Veterinärmedizin (B)* 18, 617–621.

Endresen, H.A. (1970) Evaluation of leucopenia in cattle. *Journal of the American Veterinary Medicine Association* 156, 858–866.

Evans, S.J. (1996) *A Case Control Study of Multiple-resistant S. typhimurium DT104 Infection of Cattle in Great Britain*. Report AHVG, MAFF, Tolworth, Surrey.

Evans, S.J. and Davies, R.H. (1996) Case control of multiple-resistant *Salmonella typhimurium* DT 104 infection in cattle in Great Britain. *Veterinary Record* 139, 557–558.

Ferris, K.E. and Miller, D.A. (1996) *Salmonella* serovars from animals and related sources reported during July 1995–June 1996. *Proceedings of the US Animal Health Association* 100, 505–526.

Ferris, K.E., Andrews, R.E., Thoen, C.O. and Blackburn, B.O. (1992) Plasmid profile analysis, phage typing and antibiotic sensitivity of *Salmonella dublin* from clinical isolates in the United States. *Veterinary Microbiology* 32, 51–62.

Field, H.I. (1948) A survey of bovine salmonellosis in Mid and West Wales. *Veterinary Journal* 104, 294–302.

Findlay, C.R. (1972) The persistence of *Salmonella dublin* in slurry in tanks and on pasture. *Veterinary Record* 91, 233–235.

Findlay, C.R (1973) Salmonellae in sewage sludge: multiplication. *Veterinary Record* 93, 102–103.

Fisher, E.W. (1971) Hydrogen ion and electrolyte disturbances in neonatal calf diarrhoea. *Annals of the New York Academy of Sciences* 176, 223–230.

Fisher, E.W. and Martinez, A.A. (1975) Studies of neonatal calf diarrhoea III. Water balance studies in neonatal salmonellosis. *British Veterinary Journal* 131, 643–651.

Fone, D.L. and Barker, R.M. (1994) Association between human and farm animal infection with *Salmonella typhimurium* DT104 in Herefordshire. *Communicable Disease Report* 4, 136–140.

Frik, J.F. (1969) *Salmonella dublin* infecties bij runderen in Nederland. Thesis, University of Utrecht.

Frost, A.J., O'Boyle, D. and Samuel, J.L. (1988) The isolation of *Salmonella* spp. from feed lot cattle managed under different conditions before slaughter. *Australian Veterinary Journal* 65, 224–225.

Galyov, E.E., Wood, M.W., Rosqvist, R., Mullan, P.B., Watson, P.R., Hedges, S. and Wallis, T.S. (1997) A secreted effector protein of *Salmonella dublin* is translocated into eukaryotic cells and mediates inflammation and fluid secretion in infected ileal mucosa. *Molecular Microbiology* 25, 903–912.

Gay, J.M. and Hunsaker, M.E. (1993) Isolation of multiple *Salmonella* serovars from a dairy two years after a clinical salmonellosis outbreak. *Journal of the American Veterinary Medicine Association* 203, 1314–1320.

George, J.T.A., Wallace, J.G., Morrison, H.R. and Harbourne, J.F. (1972) Paratyphoid in man and cattle. *British Medical Journal* iii, 208–211.

Gibson, EA. (1958) Studies on the epidemiology of *Salmonella* infection in cattle. Thesis, University of London.

Gibson, E.A. (1965) Reviews of the progress of dairy science: *Salmonella* infection in cattle. *Journal of Dairy Research* 32, 97–134.

Giles, N. and King, S.C. (1987) Excretion of *S. typhimurium* from a cow's udder. *Veterinary Record* 120, 23.

Giles, N., Hopper, S.A. and Wray, C. (1989) Persistence of *S. typhimurium* in a large dairy herd. *Epidemiology and Infection* 103, 235–241.

Gitter, M., Wray, C., Richardson, C. and Pepper, R.T. (1978) Chronic *Salmonella dublin* infection in calves. *British Veterinary Journal* 134, 113–121.

Gronstol, H., Osbourne, A.D. and Pethiyagoda, S. (1974) Experimental *Salmonella* infection in calves. (1) The effect of stress factors on the carrier state. *Journal of Hygiene, Cambridge* 72, 155–162.

Guilloteau, L.A., Wallis, T.S., Gautier, A.V., MacIntyre, S., Platt, D.J. and Lax, A.J. (1996) The *Salmonella* virulence plasmid enhances *Salmonella*-induced lysis of macrophages and influences inflammatory responses. *Infection and Immunity* 64, 3385–3393.

Hall, G.A. and Jones, P.W. (1977) A study of the pathogenesis of experimental *Salmonella dublin* abortion in cattle. *Journal of Comparative Pathology* 87, 53–65.

Hall, G.A. and Jones, P.W. (1979) Experimental oral infection of pregnant heifers with *Salmonella dublin*. *British Veterinary Journal* 135, 75–82.

Hall, G.A., Jones, P.W. and Aitken, M.M. (1978) The pathogenesis of experimental intra-ruminal infection of cows with *Salmonella dublin*. *Journal of Comparative Pathology* 88, 409–417.

Hall, G.A., Jones, P.W., Parsons, K.R., Young, E.R. and Aitken, M. (1980) The haematology of experimental *Salmonella dublin* infections in pregnant heifers. *British Veterinary Journal* 136, 182–189.

Harbourne, J.F., Thomas, G.W. and Luery, K.W. (1978) *Salmonella* in effluent waters and river sites in North Yorkshire. *British Veterinary Journal* 134, 350–357.

Hardman, P.M., Wathes, C.M. and Wray, C. (1991) Transmission of salmonellae among calves penned individually. *Veterinary Record* 129, 327–329.

Heard, T.W., Jennet, N.E. and Linton, A.H. (1972) Changing patterns of *Salmonella* excretion in various cattle populations. *Veterinary Record* 90, 359–364.

Helmuth, R. and Seiler, A. (1986) Epidemiology and chromosonal location of genes encoding multiresistance in *Salmonella dublin*. *Journal of Antimicrobial Chemotherapy* 18 (Suppl. C), 179–181.

Henner, S. and Lugmayr, D. (1972) Enzootisches Verwerfen bei rindern verursacht durch Salmonellen. [Enzootic abortion in cattle caused by salmonellae.] *Tierärztliche Umschau* 27, 271–272.

Hinton, M.H. (1971) *Salmonella* abortion in cattle. *Veterinary Bulletin* 41, 973–980.

Hinton, M.H. (1973) *Salmonella dublin* abortion in cattle. ii. Observations on the whey agglutination test and the milk ring test. *Journal of Hygiene, Cambridge* 71, 471–479

Hinton, M.H. (1974) *Salmonella dublin* abortion in cattle: studies on the clinical aspects of the condition. *British Veterinary Journal* 130, 556–563.

HMSO (1977) *Report of the Working Party on the Disposal of Sewage Sludge to Land*. Report No. 5, Department of the Environment Standing Technical Committee, London.

HMSO (1989) *The Zoonoses Order*. Statutory Instrument No. 285, HMSO, London.

Holmberg, S.D., Osterholm, M.T., Senger, K.A. and Cohen, M.L. (1984) Drug-resistant *Salmonella* from animals fed antimicrobials. *New England Journal of Medicine* 311, 617–622.

Hoorfar, J. and Wedderkopp, A. (1995) Enzyme-linked immunosorbent assay for screening of milk samples for *Salmonella typhimurium* in dairy herds. *American Journal of Veterinary Research* 56, 1549–1554.

Hoorfar, J., Lind, P. and Bitsch, V. (1994) Evaluation of an O-antigen enzyme-linked immunosorbent assay for screening of milk samples for *Salmonella dublin* infection in dairy herds. *Canadian Journal of Veterinary Research* 59, 142–148.

Hoorfar, J., Lind, P., Bell, M.M. and Thorns, C.J. (1996) Seroreactivity of *Salmonella*-infected cattle herds against a fimbrial antigen in comparison with lipopolysaccharide antigens. *Journal of Veterinary Medicine B* 43, 461–467.

House, J.K., Smith, B.P., Dilling, G.W. and Roden, LD. (1993) Enzyme-linked immunosorbent assay for serological detection of *Salmonella dublin* carriers on a large dairy. *American Journal of Veterinary Research* 54, 1391–1399.

Hughes, L.E., Gibson, E.A., Roberts, H.E., Davies, E.T., Davies, G. and Sojka, W.J. (1971) Bovine salmonellosis in England and Wales. *British Veterinary Journal* 127, 225–238.

Jack, E.J. and Hepper, P.T. (1969) An outbreak of *Salmonella typhimurium* infection associated with the spreading of slurry. *Veterinary Record* 84, 196–198.

Jensen, C.O. (1891) Om den infektiose Kalvediarrhoe og dens Aarsag. *Maanedsskr Dyrlaeger* 4, 140–162.

Jones, P.W. (1992) Salmonellas in animal wastes and hazards for other animals and humans from handling animal wastes. In: Salmonella *and Salmonellosis, 15–17 September, 1992.* Ploufragan/St Brieuc, France, pp. 280–294. (Obtainable from ISPAIA, BPT, 22440 Ploufragan, France. 450 FF + postage.)

Jones, P.W. and Matthews, P.R.J. (1975) Examination of slurry from cattle for pathogenic bacteria. *Journal of Hygiene, Cambridge* 74, 57–64.

Jones, P.W., Rennison, L.M., Lewin, V.H. and Redhead, D.L. (1980) The occurrence and significance to animal health of salmonellas in sewage and sewage sludge. *Journal of Hygiene, Cambridge* 84, 47–62.

Jones, P.W., Collins, P. and Aitken, M.M. (1988) Passive protection of calves against experimental infection with *Salmonella typhimurium*. *Veterinary Record* 123, 536–541.

Kaensche, C. (1896) Zur Kenntnis der Krankheit-serreger bei Fleischvergiftungen. *Zeitschrift für Hygiene* 22, 53–67.

Kahrs, R.F., Bentinck-Smith, J., Bjorck, G.R., Bruner, D.W., King, J.M. and Lewis, N.F. (1972) Epidemiologic investigation of an outbreak of fatal enteritis and abortion associated with dietary change and *Salmonella typhimurium* infection in a dairy herd. a case report. *Cornell Veterinarian* 62, 175–191.

Kelly, W.R. and Collins, J.D. (1978) The potential health significance of antibiotic-resistant *E. coli* and other infectious agents present in farm effluent. In: *Animals and Human Health.* CEC Eur. 6009 EN, pp. 172–188.

Kelly, W.R. and Collins, J.D. (1978) Animals and human health hazards associated with the utilisation of animal effluents. In: Kelly, W.R. (ed.) CEC Eur. 6009 FW.

Kelterborn, E. (1967) Salmonella *Species: First Isolation, Names and Occurrence.* W. Junk, The Hague.

Konrad, H., Smith, B.P., Dilling, G.W. and House, J.K. (1994) Production of *Salmonella* serogroup D (O9)-specific enzyme-linked immunosorbent assay antigen. *American Journal of Veterinary Research* 55, 1647–1651.

Lawson, G.H.K., McPherson, E.A. and Wooding, P. (1974) The epidemiology of *Salmonella dublin* infection in a dairy herd. II. Serology. *Journal of Hygiene, Cambridge* 72, 329–337.

Lax, A.J., Barrow, P.A., Jones, P.W. and Wallis, T.S. (1994) Current perspectives in salmonellosis. *British Veterinary Journal* 151, 351–377.

Libby, S.J., Adams, L.G., Ficht, T.A., Allen, C., Whitford, H.A., Buchmeier, N.A., Bossie, S. and Guiney, D.G. (1997) The *spv* genes on the *Salmonella dublin* virulence plasmid are required for severe enteritis and systemic infection in the natural host. *Infection and Immunity* 65, 1768–1792.

Liebisch, B. and Schwarz, S. (1996) Evaluation and comparison of molecular techniques for epidemiological typing of *Salmonella enterica* subsp. *enterica* serovar *dublin. Journal of Clinical Microbiology* 34, 641–646.

Linklater, K.A., Graham, M.M. and Sharp, J.C.M. (1985) Salmonellae in sewage sludge and abattoir effluent in south-east Scotland. *Journal of Hygiene, Cambridge* 94, 301–307.

Löffler, F. (1892) Über Epidemien unter den im hygienischen Institute zu Greifswald gehaltenen Mäusen und über Bekampfung der Feldmaus plage. *Zentralblatt für Bakteriologie* 11, 129–141.

MAFF (1998) Salmonella *in Livestock Production.* Veterinary Laboratories Agency (Weybridge), Addlestone, UK.

McEwen, S.A., Martin, W.S., Clarke, R.C., Tamblyn, S.E. and McDermott, J.J. (1988) The prevalence, incidence, geographical distribution, antimicrobial sensitivity patterns and plasmid profiles of milk filter *Salmonella* isolates from Ontario dairy farms. *Canadian Journal of Veterinary Research* 52, 18–22.

Maclachlan, J. (1974) Salmonellosis in Midlothian and Peebleshire. *Public Health, London* 88, 79–87.

McLaren, I.M. and Wray, C. (1991) Epidemiology of *Salmonella typhimurium* infection in calves: persistence of salmonellae in calf units. *Veterinary Record* 129, 461–462.

Martel, J.L. and Coudert, M. (1993) Bacterial resistance monitoring in animals: the French national experience of surveillance schemes. *Veterinary Microbiology* 35, 321–338.

Martel, J.L., Chaslus-Dancla, E., Coudert, M., Poumarat, F. and Lafont, J.P. (1995) Survey of antimicrobial resistance in bacterial isolates from diseased cattle in France. *Microbial Drug Resistance* 1, 273–283.

Mohler, J.R. and Buckley, J.S. (1902) *19th Report US Bureau Animal Industry*, p. 297.

Morisse, J.P. and Cotte, J.P. (1994) Evaluation of some risk factors in bovine salmonellosis. *Veterinary Research* 25, 185–191.

Morisse, J.P., Cotte, J.P. and Huonnic, D. (1984) Dissemination of salmonellae by chronically infected dairy cows. *Point Vétérinaire* 15, 55–59.

Morisse, J,P., Cotte, J,P., Argente, G. and Daniel, L. (1992) Approche épidémiologique de l'excrétion de salmonelles dans un réseau de 50 exploitations bovine laitiers avec et sans antécédents cliniques. *Annales de Médicine Vétérinaire* 136, 403–409.

Munch, B., Errebo Larsen, H. and Aalbaek, B. (1987) ii. Experimental studies on the survival of pathogenic and indicator bacteria in aerated and non-aerated cattle and pig slurry. *Biological Wastes* 22, 49–65.

Nazer, A.H.K. and Osborne, A.D. (1977) Experimental *Salmonella dublin* infection in calves. *British Veterinary Journal* 133, 388–398.

Nichols, G.L. and de Louvois, J. (1995) The microbiological quality of raw sausages sold in the UK. *PHLS Microbiology Digest* 12, 236–242.

Nielsen, B.B. and Vestergaard, E.M. (1992) Use of ELISA in the eradication of *Salmonella dublin* infection. In: *Proceedings International* Salmonella *and Salmonellosis*. Ploufragen, France, pp. 220–224. (Obtainable from ISPAIA, BPT, 22440 Ploufragan, France. 450 FF + postage.)

Obich (1865) *Woschenschrift für Thierheilkunde und Viehzucht.* Cited by Jensen, C.O. (1893) *Ueber die Kälberruhr und deren Aetiologie. Monatschefte für Praktishe ThierheilkundeI*, Band 597–124. Reprinted in Jensen, C.O. (1948) *Selected Papers 1886–1908*, vol. 1. Einar Munksgaard, Copenhagen, p. 180.

O'Connor, P.J., Rogers, P.A.M., Collins, J.D. and McErlean, B.A. (1972) On the association between salmonellosis and the occurrence of osteomyelitis and terminal dry gangrene in calves. *Veterinary Record* 90, 459–460.

Ogilvie, T.H. (1986) The persistent isolation of *Salmonella typhimurium* from the mammary gland of a dairy cow. *Canadian Veterinary Journal* 27, 329–331.

Osborne, A.D., Linton, A.H. and Pethiyagoda, S. (1974) Epidemiology of *Salmonella* infection in calves. (2) Detailed study in a large beef rearing unit. *Veterinary Record* 94, 604–610.

Peel, J.M., Voirol, M.J., Kolly, C., Gobet, D. and Martinod, S. (1990) Induction of circulating tumour necrosis factor cannot be demonstrated during septicemic salmonellosis in calves. *Infection and Immunity* 58, 439–442.

Penny, C.D., Low, J.C., Nettleton, P.F., Scott, P.R., Sargison, N.D., Strachan, W.D. and Honeyman, P.C. (1996) Concurrent bovine viral diarrhoea virus and *Salmonella typhimurium* DT104 infection in a group of pregnant dairy heifers. *Veterinary Record* 138, 485–489.

Peterz, M., Wiberg, C. and Norberg, P. (1989) The effect of incubation temperature and magnesium chloride concentration on growth of *Salmonella* in home-made and in commercially available dehydrated Rappaport–Vassiliadis broths. *Journal of Applied Bacteriology* 66, 523–528.

Plym-Forshell, L. and Ekesbo, I. (1996) Survival of salmonellas in urine and dry-faeces from cattle – an experimental study. *Acta Veterinaria Scandinavica* 37, 127–131.

Potter, M.E., Kaufmann, A.F., Blake. P.A. and Feldman, R.A. (1984) Unpasteurised milk: the hazards of a health fetish. *Journal of the American Medical Association* 252, 2048–2052.

Rampling, A. (1996) Raw milk, cheeses and *Salmonella*. *British Medical Journal* 312, 67–68.

Rankin, J.D., Taylor, R.J. and Burrows, M.R. (1969) *Salmonella* infection in young calves. *Veterinary Record* 85, 582.

Rasch, K. and Richter, J. (1956) Endemiologisches um einen bovinen Dauerausscheider von *Salmonella heidelberg*. *Berliner Müchener Tierarztliches Wochenschrift* 69, 211–214.

Reilly, W.J., Forbes, G.I., Paterson, G.M. and Sharp, J.C.M. (1981) Human and animal salmonellosis in Scotland associated with environmental contamination. *Veterinary Record* 108, 553–555.

Richardson, A. (1973) The transmission of *Salmonella dublin* to calves from adult carrier cows. *Veterinary Record* 92, 112–115.

Richardson, A. (1975a) Outbreaks of a bovine salmonellosis caused by serovars other than *S. dublin* and *S. typhimurium*. *Journal of Hygiene, Cambridge* 74, 195–203.

Richardson, A. (1975b) Salmonellosis in cattle [review]. *Veterinary Record* 96, 329–331.

Richardson, A. and Fawcett, A.R. (1973) *Salmonella dublin* infection in calves: the value of rectal swabs in diagnosis and epidemiologicalal studies. *British Veterinary Journal* 129, 151–156.

Richardson, A. and Watson, W.A. (1971) A contribution to the epidemiology of *Salmonella dublin* infection in cattle. *British Veterinary Journal* 127, 172–183.

Rings, D.M. (1985) Salmonellosis in calves. *Veterinary Clinics North America Food Animal Practitioners* 1, 529–530.

Roberts, D., Boag, K., Hall, M.L. and Shipp, C.R. (1975) The isolation of salmonellas from British pork sausages and sausagemeat. *Journal of Hygiene, Cambridge* 75, 173–184.

Robertsson, J. (1984) Humoral antibody response to experimental and spontaneous *Salmonella* infections in cattle measured by ELISA. *Zentralblatt für Veterinarmedizin* B 31, 367–380.

Robinson, R.A., Blackburn, B.O., Murphy, C.D., Morse, E.V. and Potter, M.E. (1984) Epidemiology of *Salmonella dublin* in the USA. In: *Proceedings of the International Symposium on* Salmonella, *New Orleans 1984*. Avian Association of Avian Pathologists, New Bolton Center, Kennet Square, Pennsylvania, USA, pp. 182–193.

Rowe, B., Begg, N.T., Hutchinson, D.N., Dawkins, H.C., Gilbert, R.J., Jacob, M. Hales, B.H, Rae, F.A. and Jepson, M. (1987) *Salmonella ealing* infections associated with consumption of infant dried milk. *Lancet* ii, 900–903.

Rowe, B., Threlfall, E.J., Ward, L.R. and Ashley, A.S. (1979) International spread of multi resistant strains of *Salmonella typhimurium* phage types 204 and 193 from Britain to Europe. *Veterinary Record* 105, 468–469.

Ryan, C.A., Nickels, M.K., Hargett-Bean, N.T., Potter, M.E., Endo, T., Mayer, L., Langkop, C.W., Gibson, C., McDonald, R.C., Kenney, R.T., Buhr, N.D., McDonnell, P.J., Martin, R.J., Cohen, M.L. and Blake, P.A. (1987) Massive outbreak of antimicrobial-resistant salmonellosis traced to pasteurised milk. *Journal of American Medical Association* 258, 3269–3274.

Sanson, R.L. and Thornton, R.N. (1997) A modelling approach to the quantification of the benefits of a national surveillance programme. *Preventive Veterinary Medicine* 30, 37–47.

Schwarz, S. and Liebisch, B. (1994) Pulsed-field gel electrophoretic identification of *Salmonella enterica* serovar *typhimurium* live vaccine strain Zoosaloral H. *Letters in Applied Microbiology* 19, 469–472.

Segall, T. and Lindberg, A.A. (1991) Experimental oral *Salmonella dublin* infection in calves: a bacteriological and pathological study. *Journal of Veterinary Medicine* B 38, 169–185.

Selander, R.K., Smith, N.H., Li, J., Beltran, P., Ferris, K.E., Kopecko, D.J. and Rubin, F.A. (1992) Molecular evolutionary genetics of the cattle-adapted serovar *Salmonella dublin*. *Journal of Bacteriology* 174, 3587–3592.

Sharp, J.C.M., Paterson, G.M. and Forbes, G.I. (1980) Milk-borne salmonellosis in Scotland. *Journal of Infection* 2, 333–340.

Sharp, M.W. and Rawson, B.C. (1992) Persistent *Salmonella typhimurium* DT 104 in a dairy herd. *Veterinary Record* 131, 375–376.

Small, R.G. and Sharp, J.C.M. (1979) A milk-borne outbreak due to *Salmonella dublin*. *Journal of Hygiene, Cambridge* 82, 95–100.

Smith, B.P., Habasha, F., Reina-Guerra, M. and Hardy, A.J. (1979) Bovine salmonellosis: experimental production and characterisation of the disease in calves, using oral challenge with *Salmonella typhimurium*. *American Journal of Veterinary Research* 40, 1511–1513.

Smith, B.P., Oliver, D.G. and Singh, P. (1989) Detection of *Salmonella dublin* mammary gland infection in carrier cows using enzyme linked immunosorbent assay for antibody in milk and serum. *American Journal of Veterinary Research* 50, 1352–1360.

Smith, B.P., House, J.K., Dilling, G.W., Roden, L.D. and Spier, S.J. (1992) Identification of *Salmonella dublin* infected cattle. In: *Proceedings of Salmonella and Salmonellosis*. Ploufragan, France, pp. 225–230. (Obtainable from ISPAIA, BPT, 22440 Ploufragan, France. 450 FF + postage.)

Smith, B.P., Roden, L.D., Tharmond, M.C., Dilling, G.W., Konrad, H., Pelton, J.A. and Picanso, J.P. (1994) Prevalence of salmonellae in cattle and in the environment on California dairies. *Journal of American Veterinary Medicine Association* 205, 467–471.

Sojka, W.J., Thomson, P.D. and Hudson, E.B. (1974) Excretion of *Salmonella dublin* by adult bovine carriers. *British Veterinary Journal* 130, 482–488.

Sojka, W.J., Wray, C., Shreeve J and Benson, J.A. (1977) Incidence of *Salmonella* infections in animals in England and Wales, 1968–74. *Journal of Hygiene, Cambridge* 78, 43–56.

Spier, S.J., Smith, B.P., Cullor, J.S., Olander, H.J., Roden, L.D. and Dilling, G.W. (1991) Persistent experimental *Salmonella dublin* intramammary infection in dairy cows. *Journal of Veterinary Internal Medicine* 5, 341–350.

Tablante, N. and Lane, V.M. (1989) Wild mice as potential reservoirs of *Salmonella dublin* in a closed dairy herd. *Canadian Veterinary Journal* 30, 590–592.

Taylor, R.J. and Burrows, M.R. (1971) The survival of *Escherichia coli* and *Salmonella dublin* in slurry on pastures and the infectivity of *S. dublin* for grazing calves. *British Veterinary Journal* 127, 536–543.

Taylor, R.J. (1973) ii. A further assessment of the potential hazard in calves allowed to graze pasture contaminated with *Salmonella dublin* in slurry. *British Veterinary Journal* 129, 354–358.

Taylor, S.M. and Kilpatrick, D. (1975) The relationship between concurrent live fluke infection and salmonellosis in cattle. *Veterinary Record* 96, 342–343.

Thomassen, M. (1897) Une nouvelle septicémie des baux avec néphrite et urocystite (bactériurie) consécutives. *Annales de l'Institut Pasteur* 11, 523–540.

Thunegard, E. (1975) On the persistence of bacteria in manure: a field and experimental study with special reference to *Salmonella* in liquid manure. *Acta Veterinaria Scandinavica* 16 (Suppl. 56), 86 pp.

van Leeuwen, W.J., Voogd, C.E. and Guinee, P.A.M. (1984) Antibiotic resistance in *Salmonella* in the Netherlands. In: Pohl and Leunen, J. (eds) *Resistance and Pathogenic Plasmids*. CEC Seminar, Brussels, p.1.

Visser, I.J.R. (1991) Cutaneous salmonellosis in veterinarians. *Veterinary Record* 129, 364.

Visser, I.J.R., Veen, M., vander Giessen, J.W.B., Peterse, D.J. and Wouda, W. (1993) *Salmonella dublin* in dairy herds in north Netherlands. *Tijdschrift voor Diergeneeskunde* 118, 84–87.

Wall, P.G., Morgan, D., Lamden, K., Ryan, M., Giffin, M., Threlfall, E.J., Ward, L.R. and Rowe, B. (1994) A case control study of infection with an epidemic strain of multiresistant *Salmonella typhimurium* DT104 in England and Wales. *Communicable Disease Review* 4, R130.

Wall, P.G., Davis, S., Threlfall, E.J., Ward, L.R. and Ewbank, A.J. (1995a) Chronic carriage of multidrug resistant *Salmonella typhimurium* in a cat. *Journal of Small Animal Practice* 36, 279–281.

Wall, P.G., Morgan, D., Lamden, K., Griffin, M., Threlfall, E.J., Ward, L.T. and Rowe, B. (1995b) Transmission of multi-resistant strains of *Salmonella typhimurium* from cattle to man. *Veterinary Record* 136, 591–592.

Wall, P.G., Threlfall, E.J., Ward, L.R. and Rowe, B. (1996) Multiresistant Salmonella typhimurium DT104 in cats: a public health risk. *Lancet* 348, 471.

Wallis, T.S., Paulin, S.M., Plested, J.S., Watson, P.R. and Jones, P.W. (1995) The *Salmonella dublin* virulence plasmid mediates systemic but not enteric phases of salmonellosis in cattle. *Infection and Immunity* 63, 2755–2761.

Wathes, C.M., Zaidan, W.A.R., Pearson, G.R., Hinton, M. and Todd, N. (1988) Aerosol infection of calves and mice with *Salmonella typhimurium*. *Veterinary Record* 123, 590–594.

Watson, P.R., Paulin, S.M., Bland, A.P., Jones, P.W. and Wallis, T.S. (1995) Characterisation of intestinal invasion by *Salmonella typhimurium* and *Salmonella dublin* and effect of a mutation in the *inv* H gene. *Infection and Immunity* 63, 2743–2754.

Watson, P.R., Paulin, S.M., Jones, P.W. and Wallis, T.S. (1997) Role of *stn*, *sly* A and *inv* H in *Salmonella* induced enteritis in cattle. In: *Proceedings* Salmonella *and Salmonellosis*. ISPAIA, Ploufragan, pp. 153–156.

Watson, W.A., Wood, B. and Richardson, A. (1971) *Salmonella dublin* infection in a beef herd. *British Veterinary Journal* 127, 294–298.

Werner, S.B., Humphrey, G.L. and Kamei, I. (1979) Association between raw milk and human *Salmonella dublin* infection. *British Medical Journal* 2, 238–241.

Werner, S.B., Morrison, F.R. and Humphrey, G.L. (1984) *Salmonella dublin* and raw milk consumption. *Journal of the American Medical Association* 251, 2195–2199.

Whitlock, R.H. (1984) Therapeutic strategies involving antimicrobial treatment of the gastrointestinal tract in large animals. *Journal of the American Veterinary Medicine Association* 185, 1210–1213.

Williams, B.M. (1975) Environmental considerations in salmonellosis. *Veterinary Record* 96, 318–321.

Williams, E. (1980) Veterinary surgeons as vectors of *Salmonella dublin*. *British Medical Journal*, 22 March, 815–818.

Wood, J.D., Chalmers, G.A., Fenton, R.A., Pritchard, J., Schoonderwoerd, M. and Lichtenberger, W.L. (1991) The persistent shedding of *Salmonella enteritidis* from the udder of a cow. *Canadian Veterinary Journal* 32, 738–741.

Wray, C. (1980) Some haematological and blood biochemical findings during experimental *Salmonella typhimurium* infection in calves. *Zentralblätt für Veterinarmedizin* B 27, 365–373.

Wray, C. (1997) Development of antibiotic resistance: a vet's tale. *Journal of Medical Microbiology* 46, 26–28.

Wray, C. and Davies, R.H. (1996) *Salmonella* in animals – a veterinary viewpoint. *PHLS Microbiology Digest* 13, 44–48.

Wray, C. and Roeder, P.L. (1987) Effect of bovine virus-diarrhoea-mucosal disease infection on *Salmonella* infections in calves. *Research in Veterinary Science* 42, 213–218.

Wray, C. and Sojka, W.J. (1977) Reviews of the progress of dairy science: bovine salmonellosis. *Journal of Dairy Research* 44, 383–425.

Wray, C., Sojka, W.J. and Callow, R.J. (1977) The serological response in cattle to *Salmonella* infection. *British Veterinary Journal* 133, 25–32.

Wray, C., Todd, N. and Hinton, M.H. (1987a) The epidemiology of *Salmonella typhimurium* infection in calves: excretion of *S. typhimurium* in faeces of calves in different management systems. *Veterinary Record* 121, 293–296.

Wray, C., McLaren, I., Parkinson, N.M. and Beedell, Y. (1987b) Differentiation of *Salmonella typhimurium* DT204c by plasmid profile and biotyping. *Veterinary Record* 121, 514–516.

Wray, C., Wadsworth, Q.C., Richards, D.W. and Morgan, J.H. (1989) A three-year study of *Salmonella dublin* in a closed dairy herd. *Veterinary Record* 124, 532–535.

Wray, C., Todd, N., McLaren, I.M., Beedall, Y.E. and Rowe, B. (1990) The epidemiology of *Salmonella* infection in calves: the role of dealers. *Epidemiology and Infection* 105, 295–305.

Wray, C., Todd, N., McLaren, I.M. and Beedell, Y.E. (1991) The epidemiology of Salmonella in calves: the role of markets and vehicles. *Epidemiology and Infection* 107, 521–525.

Chapter 11

Salmonella Infections in Pigs

Paula J. Fedorka-Cray,[1] Jeffrey T. Gray[2] and Clifford Wray[3]

[1]*Richard Russell Research Center, 950 College Station Road, Athens, GA, 30605-2720, USA;*
[2]*University of Nebraska-Lincoln, Veterinary Diagnostic Center, Lincoln, Nebraska, USA;*
[3]*Central Veterinary Laboratories Agency (Weybridge), New Haw,*
Addlestone, Surrey KT15 3NB, UK

Historical Perspectives

The organism now known as S. *choleraesuis* was first isolated from pigs by Salmon and Smith (1886), when they considered it to be the cause of swine fever (hog cholera). The importance of the organism as a cause of disease in pigs was neglected when the viral aetiology of swine fever was discovered, and a number of years elapsed before S. *choleraesuis* was recognized as a primary pathogen that was capable of causing several different disease syndromes.

Pigs are also susceptible to S. *typhimurium*, which may be the more important pathogen in a number of countries, not only from the animal-disease aspects but also from the public-health point of view (Buxton, 1957). A wide variety of other serovars have been isolated from pigs and, although they may occasionally cause disease, in general, infected pigs remain healthy carriers and, as a consequence, are of public-health importance. In addition to the economic impact of salmonellosis on the human population, it is also a major cause of economic loss in pig production, resulting in millions of dollars in lost income to the pork industry (Schwartz, 1990).

Epidemiology

General introduction

Members of the genus *Salmonella* are extremely ubiquitous in nature, recovered from nearly all vertebrates, as well as insects, and are often referred to as universal pathogens (Taylor and McCoy, 1969; Falkow and Mekalanos, 1990). It is convenient to consider porcine *Salmonella* as consisting of two groups. The first group would consist of the host-adapted serovar S. *choleraesuis*. The second group would consist of all other serovars, which have a much broader host range and include S. *typhimurium*, which is most often associated with gastroenteritis worldwide (Falkow and Mekalanos, 1990). However, some host-adapted serovars can cause severe disease in humans, which may result in high mortality, as observed following infection by S. *choleraesuis* (Cherubin, 1980).

Prevalent serovars and phage types

In the UK, S. *choleraesuis* was the predominant serovar in pigs during the 1950s and 1960s and, in 1958 and 1968, it constituted 90% and 74.2%, respectively, of all *Salmonella* isolates from pigs (Sojka *et al.*, 1977). Subsequently its prevalence has declined in the UK and only one incident of S. *choleraesuis* was reported in pigs in 1997. It is now regarded as an infrequent isolate. This seems to be the current situation in a number of other European countries (Laval *et al.*, 1992; Baggesen and Christensen, 1997a; Helmuth *et al.*, 1997) and also in Australia, where the organism was last isolated in 1987 (C. Murray, Adelaide, personal communication, 1998). In contrast, in

North America, S. *choleraesuis* remains a major problem for the pig industry (Wilcock and Schwartz, 1992). The reasons for the differences between North America and Europe are not known but may be related to husbandry practices.

In Europe, the predominant *Salmonella* serovars in a number of different counties are S. *typhimurium* and S. *derby*. In the UK in 1997, of the 338 *Salmonella* incidents reported in pigs, 62% were caused by S. *typhimurium* and 12% by S. *derby* (Report, 1997a). Regional differences are also apparent and, in Germany, the second commonest serovar in 1998 was S. *agona*, while in Denmark it was S. *infantis*.

In the USA, S. *typhimurium* (including var. Copenhagen), S. *derby* and S. *choleraesuis* (var. Kunzendorf) are the three commonest serovars recovered from clinical submissions to the National Veterinary Services Laboratory (Ferris and Miller, 1998). The top five serovars recovered during a national prevalence study conducted in 1995 were S. *derby* (33.5%), S. *agona* (13.0%), S. *typhimurium* (including var. Copenhagen) (14.7%), S. *brandenberg* (8.0%), S. *mbandaka* (7.7%) (E. Bush, personal communication). However, the recovery of a serovar varies by source, as well as geographical location (Currier et al., 1986; Davies et al., 1997).

The number of reports of some other serovars has increased during the last decade, but it is not known whether this is the result of better monitoring or whether it indicates an increased disease or environmental prevalence.

Of the different phage types of S. *typhimurium*, the most frequent in pigs in the UK are DT104, DT193 and DT208, all of which are resistant to a number of antibiotics (Wray et al., 1997). In 1997, only 15.8% of the *Salmonella* isolates from predominantly diseased pigs were fully susceptible to the antibiotics used for monitoring; tetracycline resistance was the most frequent. In Denmark in 1996, 80% of the *Salmonella* isolates from pigs were fully susceptible; although the multiply resistant S. *typhimurium* phage type DT104 was present, its frequency was low and control measures had been instituted (Report, 1997b). Phage typing of veterinary isolates of S. *typhimurium* in the USA was initiated in 1996 and S. *typhimurium* DT104 has been isolated from a number of species, including pigs. In 1997, 19% of the diagnostic and 44.2% of the slaughter isolates were fully sus-

ceptible to all antimicrobials. Multiple resistance to five or more antimicrobials was observed in 23.3% of diagnostic and 16.8% of slaughter isolates (NARMS-EB, 1997).

Transmission

Historically, transmission of *Salmonella* between hosts is thought to occur via the faecal–oral route of exposure. Since *Salmonella* are often shed in large numbers in the faeces, it is not improbable to expect that this is a major route for transmission of the organism. A number of studies have reproduced experimental infection by the oral route and, during acute disease, pigs will shed up to 10^6 S. *choleraesuis* (Smith and Jones, 1967) or 10^7 S. *typhimurium* (Gutzmann et al., 1976) g^{-1} faeces. Generally, high doses have to be used and disease is frequently difficult to reproduce. Dawe and Troutt (1976) produced moderate disease following oral inoculation of 10^6 cells, but most authors report successful disease reproduction with doses of 10^8–10^{11} *Salmonella* (Gray et al., 1995, 1996a,b). During a study on two farms, Williams et al. (1981) were unable to reach any firm conclusions as to the natural route of transmission of S. *choleraesuis* and they suggested that cannibalism of piglets may be of importance in the spread of the organism.

However, as early as 1965, De Jong and Ekdahl, after oesophagectomizing calves and giving an oral challenge, proposed that both haematogenous and lymphatogenous routes of infection are important in the dissemination of *Salmonella*, particularly S. *typhimurium*. Hardman et al. (1991) further demonstrated that calves penned individually were susceptible to infection with *Salmonella*.

Aerosol experiments in chickens and mice have shown that infections with *Salmonella* can be achieved more regularly via the lungs than by oral inoculation (Clemmer et al., 1960; Darlow et al., 1961). This gave support to the role that aerosols may also play in the transmission and dissemination of *Salmonella*.

Further studies in pigs using oesophagotomy indicated that the upper respiratory tract may be equally important in transmission (Fedorka-Cray et al., 1995) and that the tonsils and lungs may be important sites for the invasion and dissemination of *Salmonella*. Pneumonia associated with

S. *choleraesuis* infection has been previously described (Baskerville and Dow, 1973) and a recent increase in S. *choleraesuis*-associated pneumonia has been reported (Turk *et al.*, 1992). It is unclear whether this predilection for the lung is due solely to the pathogen, poor ventilation in large confinement buildings or some combination of these and other factors. Experimental infection models have not provided good answers, because positive lung samples have been regarded as an artefact of intranasal or *per os* inoculation. However, in addition to the oesophagotomy work by Fedorka-Cray *et al.* (1995), Gray *et al.* (1996b) also demonstrated that the lung is colonized in pigs that are naturally exposed to pigs infected with S. *choleraesuis*, indicating that lung colonization is not an artefact of experimental inoculation. Collectively, these studies indicate that the traditional paradigm of faecal–oral transmission is no longer valid and that other possibilities, such as aerosol or dust-borne transmission, need to be considered.

Environmental considerations

Although transmission of *Salmonella* typically occurs through faecal–oral or aerogenous transmission, other vectors must be considered when discussing dissemination of the organism. In pigs, observed sources of contamination include rodents, insects, humans and contaminated feed and feedstuffs (Clarke and Gyles, 1993; McChesney *et al.*, 1995). Mouse faecal pellets have been shown to contain up to 10^5 colonyforming units (cfu) *Salmonella* (Henzler and Opitz, 1992). During an investigation of *Salmonella* contamination, which involved 23 pig farms, Davies and Wray (1997) found a wide range of wild animals, including rats, mice, cats and birds, to be infected. Cats and birds were associated with contamination of feed and grain stores, and rodents were involved in the perpetuation of infection in specific buildings on the farm. Infected foxes were most common on outdoor breeding farms. Flies and dust can also act as mechanical vectors that spread *Salmonella* throughout the facility or environment (Edel *et al.*, 1967, 1970).

It is well known that animal feeds frequently contain *Salmonella* and that animals fed contaminated feed often become infected (Linton and Jennet, 1970). The rate of contamination of animal protein delivered to a large feed-mill in the south-eastern part of the USA was reported over a 10-month period (Williams *et al.*, 1969). Of 311 samples, 68% contained one or more of the 68 *Salmonella* serovars identified in the study. Eighty-six per cent of the meat-meal and 18% of the fish-meal sampled were found to be contaminated. In another study to determine the prevalence of *Salmonella* in pig feeds, 2.8% of the feeds and feed ingredients taken from farm environments were positive for *Salmonella* (Harris *et al.*, 1997). Feed trucks have also been implicated as a source for feed and feedstuff contamination (Fedorka-Cray *et al.*, 1997a).

Salmonella may persist in the environment for long periods and Linton *et al.* (1970) considered such persistence an important risk factor. Berends *et al.* (1996) suggested that contamination of endemic flora in finishing sites was the predominant source of infection of finished pigs, rather than infection originating from breeding farms or other sources, though the latter is more important when pigs from various sources are mixed on finishing farms. Baggesen *et al.* (1997) isolated *Salmonella* from faeces, pens, dust, equipment, ventilation equipment and slurry during their studies on pig farms. Gray and Fedorka-Cray (1995) showed that S. *choleraesuis* survives in dry faeces for at least 13 months post-shedding, demonstrating the importance of cleaning organic matter from the environment. Davies and Wray (1997) found a high level of *Salmonella* persisting in pig pens after disinfection, and the organism persisted in soil for at least 6 months on outdoor farms (R.H. Davies, VLA, Weybridge, personal communication, 1999).

Virulence Factors/Pathogenesis

Many potential virulence factors have been identified for *Salmonella*, but few have been tested critically for their contribution to virulence and the subject is reviewed in Chapter 4. It has been estimated that *Salmonella* possess over 200 virulence factors, of which only a fraction has been characterized (R. Curtiss III, personal communication, 1999). Many studies have relied on *in vitro* data to draw their conclusions. This makes it difficult to develop meaningful extrapolations for human and animal disease. In addition, many studies utilize mice as a model for disease and these results often cannot be repeated in other hosts and few studies have been done in pigs.

Pospischil *et al.* (1990) used immunolabelling techniques (peroxidase and immunogold) to detect and locate *Salmonella* in the tissues of experimentally infected pigs. *S. typhimurium* had a low tendency to invade the enteric mucosa and did not reveal any predilection for a specific intestinal location. In contrast, *S. choleraesuis* was located preferentially in the colon and on the luminal surface of the ileal M cells of the Peyer's patches, from where it had a tendency to invade. Studies *in vitro* suggest that *S. typhimurium* may adhere by means of mannose-resistant adhesins (Jones *et al.*, 1981), and invasion of Madin Darby canine kidney cells by *S. choleraesuis* has been shown to be an active process, requiring bacterial RNA and protein synthesis (Finlay *et al.*, 1988).

Several serovars have been shown to produce enterotoxins, specifically cholera-like toxin (Prasad *et al.*, 1990, 1992). Very little is known about this toxin as it relates to the pathogenesis of *Salmonella* infection. A common feature of *Salmonella*-induced enteritis is severe damage to intestinal epithelial cells, which is likely to be the result of a cytotoxin. At least three cytotoxins have been identified. A wide variety of serovars possess a heat-labile cytotoxin, described by Ashkenazi *et al.* (1988). Another cytotoxin is a low-molecular-weight membrane-associated toxin, which has not been characterized (Reitmeyer *et al.*, 1986). A third toxin, described by Libby *et al.* (1990), appears to be present in nearly all *Salmonella*, *Shigella* and enteroinvasive *Escherichia coli*. This cloned protein is a 26 kDa cell-associated haemolysin and its role in virulence is under study.

Roof and Kramer (1989) showed that virulent *S. choleraesuis* were able to survive within porcine neutrophils by inhibiting superoxide anion production and resisting the bactericidal activity of the cells. Heat-shock proteins have been shown to be produced by *S. typhimurium* inside murine macrophages. Mutants that are defective in the ability to produce these proteins are less virulent in mice and do not survive well in macrophages (Falkow and Mekalanos, 1990). The phagocytosis of *S. choleraesuis* in the lungs of pigs was studied by Baskerville *et al.* (1972) during the period 6 h–14 days after intranasal infection. All bacteria were phagocytosed soon after arrival by polymorphonuclear leucocytes and macrophages; many were destroyed but some survived and multiplied within the cells. Between days 5 and 7, the *Salmonella* caused necrosis of the phagocytes, were liberated in large numbers and caused damage of the lung tissues. Later studies showed that the free bacteria did not attach to or penetrate pulmonary cells and they suggested the damage was caused by toxin (Baskerville *et al.*, 1973).

The lipopolysaccharide (LPS) of *Salmonella* is a major determinant of host specificity and virulence (see Chapter 2). The intact LPS affords resistance to phagocytosis and killing by macrophages and complement-mediated killing (Saxen *et al.*, 1987; Robbins *et al.*, 1992). In addition, it has been shown that LPS is a major contributor to survival of *Salmonella* in the intestinal tract (Nnalue and Lindberg, 1990). The LPS component of *Salmonella* also contributes to vascular damage and thrombosis. The endotoxic properties of *Salmonella* cause fever, disseminated intravascular coagulation, circulatory collapse and shock (Takeuchi and Sprinz, 1967; Clarke, 1985).

Motility provided by flagella appears to be important for invasion for some, but not all, serovars of *Salmonella*. Regardless of the other contributions the flagella may make, their presence increases the probability that the organism will come in contact with an epithelial cell. It has been shown that strains with polar rather than peritrichous flagella have an increased ability to come in contact with, and potentially invade, epithelial cells (Jones *et al.*, 1992).

A siderophore called enterobactin has been identified in *S. typhimurium* (Benjamin *et al.*, 1985). This protein does not appear to be necessary for full virulence and its importance may be relative to the amount of extracellular growth, that occurs. Interestingly, pigs infected with *S. choleraesuis* have a reduction in serum iron, total iron-binding capacity and transferrin. The intracellular environment is low in iron and it has been suggested that *S. choleraesuis* has a non-siderophore mechanism for scavenging iron (Clarke and Gyles, 1993).

Disease in Pigs

Associated serovars

Clinical porcine salmonellosis can be separated into two syndromes; one involves S. *typhimurium*, which is associated with enterocolitis, while the other involves S. *choleraesuis* and is usually associated with septicaemia. In the USA, clinical porcine salmonellosis is almost solely due to infection with S. *typhimurium* or S. *choleraesuis*. Clinical disease has also been associated with S. *typhisuis*. This serovar is difficult to isolate and, because of this difficulty, may be responsible for more outbreaks than it is directly associated with by culture (Wilcock and Schwartz, 1992). In addition, there have been reports of both S. *dublin* (Lawson and Dow, 1966) and S. *enteritidis* (Reynolds et al., 1967) causing disease in pigs. In contrast, clinical disease in other countries has been associated with many other serovars and S. *choleraesuis* may or may not be one of them (B. Nielsen, Copenhagen, personal communication, 1995).

The vast majority of S. *choleraesuis* outbreaks in pigs are due to the H_2S-producing variant Kunzendorf (Wilcock and Schwartz, 1992). However, the non-H_2S producing S. *choleraesuis* has been as high as number 2 in the top 10 commonest *Salmonella* isolates from pigs in a given year (Ferris and Thomas, 1993).

Populations affected

Intensively reared, weaned pigs are most often affected by *Salmonella* infection. In general, S. *typhimurium* tends to cause disease in young pigs from 6 to 12 weeks of age. Disease from this serovar is rare in adult animals, but infection is frequent. S. *choleraesuis* causes disease among a wider range of ages. Mortality tends to be higher in younger than in older pigs, while morbidity is often equal regardless of age. Disease from S. *choleraesuis* in the adult is not a common occurrence. However, if a susceptible population is exposed, the animals will be significantly affected (Wilcock and Schwartz, 1992). It is not known how common subclinical infection is in the adult. Normally, only moribund suspect cases are cultured for S. *choleraesuis*. The occurrence of salmonellosis in suckling pigs is rare, presumably

because of lactogenic immunity, but infection is not uncommon (Gooch and Haddock, 1969; Wilcock et al., 1976). However, neonatal pigs are susceptible to oral challenge with *Salmonella* and develop a disease similar to that observed in weaned pigs (Wilcock and Olander, 1978).

Septicaemia

The septicaemic form of porcine salmonellosis is usually caused by S. *choleraesuis*. Affected pigs are inappetent, lethargic and febrile, with temperatures of up to 107°F (41.7°C). Respiratory signs may consist of a shallow, moist cough and diaphragmatic breathing. Clinical signs first appear after 24–36 h of infection (Reed et al., 1986). Often, producers will find the first evidence of disease as dead pigs with cyanotic extremities and abdomens. In most outbreaks, mortality is high and morbidity is variable but generally less than 10% (Reed et al., 1986; Wilcock and Schwartz, 1992). Diarrhoea is normally not a feature of S. *choleraesuis* infection until at least the fourth or fifth day of infection. It may last from 5 to 7 days after onset if chronic reinfection is not occurring.

Gross lesions include colitis, infarction of gastric mucosa, swollen mesenteric lymph nodes, splenomegaly, hepatomegaly and lung congestion. Random white foci of necrosis are often observed on the liver (Reed et al., 1986; Wilcock and Schwartz, 1992).

The microscopic lesion that is most often associated with S. *choleraesuis* in pigs is the paratyphoid nodule. This lesion can be viewed in the liver as clusters of histiocytes amid foci of acute coagulative hepatocellular necrosis and corresponds to the white foci seen grossly (Lawson and Dow, 1966). Other lesions may include fibrinoid thrombi in venules of gastric mucosa and cyanotic skin and glomerular capillaries. Swelling of histiocytes and epithelial cells, typical of Gram-negative sepsis, as well as hyperplasia of reticular cells of the spleen and lymph nodes, is often observed (Wilcock et al., 1976).

Enterocolitis

Enterocolitis in pigs is typically associated with S. *typhimurium* infection and occasionally with S. *choleraesuis* infection. In contrast to the septicaemic disease, the initial sign of infection is often watery, yellow diarrhoea. Infected pigs are inappetent, febrile and lethargic. Mortality is usually low. However, morbidity can be high within a few days of infection (Wilcock and Schwartz, 1992).

The major gross lesion at necropsy is focal or diffuse necrotic colitis and typhlitis. Mesenteric lymph nodes are greatly enlarged. Intestinal lesions develop as red, rough mucosal surfaces, which may also have grey-yellow debris. Colon and caecal contents are bile-stained and scant, often with black or sand-like gritty material on the surface. Intestinal necrosis may be seen as sharply delineated button ulcers, often associated with resolving lesions (Wilcock and Olander, 1978; Wilcock and Schwartz, 1992; Wood and Rose, 1992). In cases of S. typhimurium enterocolitis, the liver and spleen are not enlarged except by terminal congestion (Wilcock and Schwartz, 1992).

Histopathological examination reveals necrosis of cryptic and surface enterocytes, which may be local or diffuse. The lamina propria and submucosa contain macrophages and lymphocytes, with neutrophils observed only in the very early stages of disease. It is not uncommon to see lymphoid atrophy or regenerative hyperplasia associated with this disease (Wilcock et al., 1976; Jubb et al., 1985; Reed et al., 1986).

Treatment

Various antibiotics have been used to treat severe Salmonella infections in pigs, but actual controlled trials to judge their efficacy are few. Wilcock and Schwartz (1992) mention a number of trials, but the general conclusion was that therapy was equivocal or of little merit. Antimicrobials have also been used to reduce, but not eliminate, the shedding of Salmonella by sick or recovered pigs (Holcomb and Fedorka-Cray, 1997). However, anecdotal information from practitioners suggests that severe Salmonella infections will respond to appropriate antimicrobial therapy. Ancillary therapy, such as the use of fluids to replace lost electrolytes and to prevent dehydration, will assist recovery.

Detection of Salmonella

Culture

A great interest has developed in the animal-production and food-processing industries to create and evaluate new methods to detect, either directly or indirectly, the presence of Salmonella. Traditional culture methods may take 3–5 days to complete and much effort has been directed towards finding more rapid methods. However, the culture of Salmonella is the standard by which all other methods are measured. Recovery of the organism is the only means by which definitive serotyping can be achieved. In addition, the isolation of the organism serves as an invaluable source of epidemiological data, which cannot be overlooked.

Most of the original methods were developed for the diagnosis of clinical salmonellosis in humans and other animals. In pigs, clinical salmonellosis is, with the exception of S. choleraesuis infection, uncommon. The sensitivity of the culture method may also be affected by the phase of the infection. In acute salmonellosis, large numbers of Salmonella are shed in the faeces, whereas a chronically infected pig or a carrier may excrete only low numbers of Salmonella intermittently. Thus, for clinical samples, direct culture may suffice, whereas samples from chronically infected pigs or from the environment will almost certainly require pre-enrichment and selective enrichment.

Many different culture media and methods have been developed and used for Salmonella detection and these are reviewed in Chapter 21. For the isolation of S. choleraesuis, Smith (1952) found it absolutely necessary to utilize media other than tetrathionate broth and selenite F broth, both of which have been reported to be toxic for the organism. It has been suggested that this may explain the infrequent isolation of S. choleraesuis in pigs during epidemiological surveys (Ewing, 1986). During a study on the prevalence of Salmonella in finishing pigs, Davies et al. (1997) failed to isolate S. choleraesuis on xylose–lysine–Tergitol 4 (XLT4) medium, though the organism grew on modified brilliant green agar. When possible, a combination of enrichment media should be employed and may include Gram-negative (GN)–Hajna broth and tetrathionate broth for the isolation of host-adapted serovars, as well as broad-host-range *Salmonella* (Ewing, 1986; Fedorka-Cray *et al.*, 1996).

Many plating media have been devised for the isolation and differentiation of *Salmonella* (see Chapter 21) and the choice of the media will be governed by the operator's experience and requirements. A recent study compared Hektoenenteric (HE) agar, Rambach agar, *Salmonella* identification (SM-ID) medium, XLT4, novobiocin–brilliant green–glycerol–lactose agar (NBGL) and modified semi-solid Rappaport–Vassiliadis medium (MSRV) for the isolation of *Salmonella*. Tests of these relatively new media found MSRV to be the most sensitive and specific, but it was also the most difficult to use. The XLT4 plates were found to be as sensitive as HE, with improved specificity. The other media did not perform as well (Dusch and Altwegg, 1995). It should also be remembered that the classical *S. choleraesuis* does not produce H_2S and may be missed on media, such as xylose lysine desoxycholate (XLD), which incorporate an H_2S indicator.

In all cases, pooled faecal samples are preferred over rectal swabs for the detection of *Salmonella*-carrier pigs (McCall *et al.*, 1966).

Enzyme-linked immunosorbent assays

Enzyme-linked immunosorbent assays (ELISA) can be used to detect either the organism or a humoral immune response to the organism. Antigen-capture ELISA to detect microorganisms in food and feedstuffs are gaining widespread use in the industry. Whereas culture may take 3–7 days to identify the organism, ELISA can detect the organism in a much shorter period of time, usually 1 day or less. However the reliability of some of these assays is questionable. In general, the cleaner the sample, the better the assay will perform. Usually faeces or faeces-contaminated samples do not test as well as food and feedstuffs. Feng (1992) listed and described several commercial rapid-screening assays. Several antigen-capture immunoassays have been utilized to detect *Salmonella* in pig faeces (Araj and Chugh, 1987; Lambiri *et al.*, 1990; van Poucke, 1990). They have the same disadvantage of many ELISA tests in that they require 10^4–10^5 cfu

Salmonella ml^{-1} to detect the organism (Dziezak, 1987). In order to achieve these numbers, a time-consuming and expensive concentration protocol or a lengthy pre-enrichment must be employed. Some investigators have had success in utilizing rapid-enrichment protocols to detect *Salmonella* in pig faeces (Cherrington and Huis in't Veld, 1993).

The second use of ELISA is to detect animals that have been, or are currently, infected with *Salmonella* (see Chapter 24). The detection of antibodies to the O antigen of *Salmonella* has been used successfully in pigs (Nielsen *et al.*, 1994). The mixed ELISA utilizes LPS produced from both *S. typhimurium* and *S. choleraesuis*. The majority of pig-related *Salmonella* serovars produce high titres to the O antigens that are present. Although the test can be utilized at the herd level, it is not suitable as an individual pig test (Nielsen *et al.*, 1995). The mixed ELISA has been used for routine screening of breeding, multiplying and slaughter pig herds in Denmark since 1993. The screening of breeding and multiplying herds is performed on serum samples, whereas meat juice is used for slaughter pigs. The meat juice is obtained by freezing a 10 g sample of muscle tissue at −20°C overnight and then allowing it to thaw, thereby releasing antibody-containing tissue fluid. On the basis of the ELISA results, further farm investigations may be undertaken, using culture methods. Unfortunately, experimentally and naturally infected pigs have been shown to have a titre to LPS for at least 12 weeks after exposure to *S. choleraesuis*, even after clearing the bacteria (Gray *et al.*, 1996b). This may result in a number of ELISA-positive pigs that are no longer infected. It is unclear what effect vaccination has on the outcome of this assay. However, data indicate that pigs vaccinated with a commercially available, modified, live, plasmid-free *S. choleraesuis* vaccine do not initiate a humoral immune response to *S. choleraesuis* antigens. This would suggest that pigs vaccinated with this strain would appear as non-infected pigs in a diagnostic test.

Another ELISA has been used to detect antibodies in *Salmonella*-carrier pigs, employing a heat-extracted antigen (Kramer *et al.*, 1994). The results from this study indicate that most pigs infected with *S. typhimurium* or *S. choleraesuis* have an antibody response to this antigen.

This assay shows a correlation between prevalence and severity of infection and the magnitude of the antibody response. A second ELISA, utilizing a different type of heat-extract antigen in a mixed-ELISA format, has also been developed and shows good sensitivity and specificity (Gray and Fedorka-Cray, 1999).

Polymerase chain reaction

The extraordinary ability of the polymerase chain reaction (PCR) to exponentially replicate a target DNA sequence has made it a very powerful tool in the armamentarium of the diagnostician, epidemiologist and molecular biologist. This assay is based on the ability of target (organism)-specific primers, through complementary DNA base-pairing, to anneal only to the target sequence. Thermostable DNA polymerase recognizes the template primer complex as a substrate, which results in the simultaneous copying of both strands of the segment of DNA between the two annealed primers. The denaturation annealing and elongation steps take place in a cyclical fashion, relying on the thermostability of the *Taq* polymerase, until the target sequence is amplified to detectable amounts (Ehrlich and Sirko, 1994).

The PCR assay has been used to identify *Salmonella* in food and clinical samples (Araj and Chugh, 1987; Rahn *et al.*, 1992; Cohen *et al.*, 1993). However, obstacles in the detection of organisms include the presence of substances inhibitory to PCR (Wilde *et al.*, 1990; Rossen *et al.*, 1992) and the inability to detect $<10^3$ cfu g^{-1} of sample without pre-enrichment (Ehrlich and Sirko, 1994). Investigators have improved detection methods in PCR assays by combining it with immunomagnetic separation (Widjojoatmodjo *et al.*, 1991, 1992) or by enrichment culture (Stone *et al.*, 1994).

Vaccination

It is generally accepted that live, attenuated, orally administered *Salmonella* vaccines provide the best protection against *Salmonella* infection. The superior protection achieved in comparison with killed *Salmonella* bacterins and subunit vaccines is generally attributed to the ability of live vaccines to stimulate a more effective cell-mediated immune response. Oral administration allows the attenuated mutant to utilize natural routes of infection, which facilitates the crucial step of antigen presentation to lymphocytes in the gut-associated lymphoid tissue. These events induce the production of secretory immunoglobulin A (IgA) on mucosal surfaces (Clarke and Gyles, 1993).

In many countries, however, inactivated vaccines are the only products available and their efficacy is equivocal. Linton *et al.* (1970) considered that immunization with a killed vaccine conferred only a weak protection against *Salmonella* infection in general. Davies and Wray (1997) showed that vaccination of breeding stock on a farm with an inactivated *S. typhimurium/S. dublin* vaccine was associated with a reduction of *Salmonella* from 67% to 12% in weaned pigs and from 52% to 5% in the adults. In Denmark, Dahl *et al.* (1997) demonstrated that the use of killed vaccines reduced the clinical impact of *S. typhimurium* infection in pigs but did not reduce subclinical infection.

Recently, the development of specific non-reverting mutations to construct both homologous and heterologous vaccine vehicles, with multiple attenuating mutations, has been achieved (Chatfield *et al.*, 1992).

A mutation in the *galE* region in *S. typhi* results in a deficiency in UDP-glucose-4-epimerase, the enzyme that converts UDP-glucose to UDP-galactose, an essential component of *Salmonella* smooth LPS (Levine *et al.*, 1989). In several large trials on humans, this mutant has appeared to be very efficacious. Because of this success, this mutation has been employed for many *Salmonella* serovars, including *S. typhimurium* (Nnalue and Lindberg, 1990). However, a *galE* mutation in *S. choleraesuis* does not reduce virulence in pigs. The somatic antigens of *Salmonella* serogroups are the main component of host specificity and facilitate survival in the gastrointestinal tract and entry into deeper tissues (Nnalue and Lindberg, 1990). This is due to the fact that galactose is missing from the oligosaccharide repeating unit of the O-antigen side-chain of *S. choleraesuis* (Nnalue and Stocker, 1986).

Another common attenuation involves the creation of auxotrophic mutants that require metabolites not available in animal tissues. Aromatic mutants, which have a complete block in the aromatic biosynthetic pathway, have a requirement for aromatic metabolites, such as *p*-aminobenzoate and 2,3-dihydroxybenzoate. Oral vaccination with *aro*A, *aro*D mutants in mice and calves has been effective in reducing disease and has been shown to be safe (Robertsson *et al.*, 1983; Smith *et al.*, 1984; Hook, 1990). Experiments using an *aro*A mutant of *S. typhimurium* indicated that vaccinated pigs shed *Salmonella* significantly less frequently than non-vaccinated pigs (Lumsden *et al.*, 1991).

Mutations in global regulatory pathways have also been a popular means of attenuation. Several studies have utilized strains with deletions in the genes for adenylate cyclase (*cya*) and for cAMP-receptor protein (*crp*). Cyclic AMP and cAMP-receptor protein regulate at least 200 genes, many of which are required for the breakdown of catabolites. *Salmonella* with deletion mutations in the *cya*, *crp* genes have been shown to be safe and effective in eliciting protective immunity in mice, chickens and pigs (Curtiss and Kelly 1987; Stabel *et al.*, 1990, 1991; Coe and Wood 1992). A large study evaluating the safety and efficacy of a battery of *S. choleraesuis* Δ*cya*, Δ*crp* isogenic mutants in mice indicates that several of these strains are protective and safe (Kelly *et al.*, 1992).

An *S. choleraesuis* strain that has been cured of the 50 kb virulence plasmid has been shown to be safe and efficacious in pigs (Kramer *et al.*, 1992). The non-specific mutation was obtained by repeated passage through porcine neutrophils. The plasmid-free variant lacks the ability to invade Vero cell monolayers and porcine neutrophils, as well as having increased resistance to killing by hydrogen peroxide (H_2O_2) and phagocytic killing by porcine neutrophils (Roof *et al.*, 1992).

In Germany, an adenine-deficient strain of *S. choleraesuis* is used for immunization of pigs and is commercially available (Meyer *et al.*, 1993).

The use of vaccines should be considered as part of an overall strategy to control *Salmonella* on the farm and their use should be in conjunction with other measures.

Competitive Exclusion

To prevent the *Salmonella* carrier state in pigs, new intervention strategies need to be investigated. Competitive exclusion is one approach and has been used successfully with poultry

(Bailey, 1987; Bailey *et al.*, 1992, Blankenship *et al.*, 1993) and further information on the subject will be found in Chapter 18. A mucosal competitive-exclusion culture from swine (MCES) was developed by Fedorka-Cray *et al.* (1999). Following application in suckling pigs and subsequent challenge with *S. choleraesuis*, 28% of the gut tissues from the MCES-treated pigs were poitive versus 79% positive tissues from the control pigs. A 2–5 $\log_{10}$ reduction of *Salmonella* in the caecal contents or ileocolic junction was observed in the MCES-treated pigs when compared with the controls. These data indicate that use of MCES may be a useful approach for control of *Salmonella* in pigs.

Control Measures

During the last decade, the structure of the pig industry has changed markedly, with the introduction of large integrated systems and breeding pyramids, akin to those of the poultry industry. Control measures are of increasing importance, because of consumer concerns about food-borne zoonoses. Epidemiological studies in recent years suggest that *Salmonella* infection of suckling piglets is much lower than that of older animals, because of lactogenic immunity, and that, by the application of integrated quality-control systems agreed upon by all staff, the prevalence of *Salmonella* on pig farms can be reduced. Application of these systems requires some knowledge of the *Salmonella* prevalence on an individual farm and this can be monitored, as indicated earlier, by either serology or culture. The main components of an integrated quality-control system are as follows.

Biosecurity

As indicated previously there are many routes by which *Salmonella* can be introduced on to a farm and the organism is often disseminated widely on farms. Control measures include changes of clothing and boots for visitors, bird and rodent control, foot-baths containing active disinfectant outside houses, limiting access to the site by visitors and lorries, etc. Farm size, stocking densities and pig density within a region all have a negative effect on the *Salmonella* status of a farm, perhaps by predisposing to *Salmonella* spread within and between farms.

All-in, all-out systems

Effective cleaning and disinfection are important aspects of disease control. It is generally accepted that farms should operate an all-in, all-out policy, with adequate cleaning and disinfection after the pen is empty. Linton *et al.* (1970) found that uninfected animals, which remained in disinfected pens usually stayed free of *Salmonella*, but, as the number of pigs per pen increased, a higher prevalence of infection was found. Tielen *et al.* (1997) found that *Salmonella*-negative piglets placed in clean accommodation remained free, despite serological evidence of *Salmonella* in the sows. Fedorka-Cray *et al.* (1997b) weaned pigs at 14–21 days and removed them to clean accommodation, where the piglets remained free of *Salmonella*. Improved disinfection of weaner and grower pens on several farms produced a significant reduction in the incidence of positive batches (from 80% to 11% on one farm) (Davies and Wray, 1997).

A word of caution should be introduced, however, as other investigators have shown no reduction in the prevalence of *Salmonella* using all-in, all-out management of finishing pigs, compared with conventional farrow-to-finish systems, in North Carolina, USA (Davies *et al.*, 1997).

Feeding systems

Many batches of animal feed are contaminated with *Salmonella* (see Chapter 17). In The Netherlands, it was found that the prevalence of *Salmonella* infection was ten times lower in pigs fed a porridge than those fed dry feed (Tielen *et al.*, 1997). A study of 40 fattening farms in the Netherlands isolated *Salmonella* from 19.4% of the samples from farms using whey, compared with 64.4% of farms using water (Van Schie and Overgoor, 1987). Field studies in Denmark found a lower prevalence of *Salmonella* on farms mixing their own feed and feeding liquid feed (Bager, 1994). It is possible that the size of the feed particles can influence the intestinal flora or the distribution of organic acids, which have been shown to be inhibitory (Prohaszka *et al.*, 1990). Naturally fermented feed is now being recommended for reduction of *Salmonella* infection in Denmark. Likewise, the use of acidified feed may improve the response to treatment and reduce the spread of disease.

Vaccination and strategic medication will also assist in *Salmonella* reduction on a farm, although the latter should be discouraged because of the risk of producing resistant microorganisms.

Salmonella control presents a challenge to all involved in the pig industry. It is important that control is based on a detailed knowledge of the epidemiology of infection and a specific control programme for each enterprise. Many strategies have been tried and tested, e.g. vaccination, use of antibacterials, etc. None have been successful on its own, but improved hygiene and disease security, combined with vaccination, are methods that have been used successfully in other sectors of the livestock industry. Implementation of the hazard-analysis critical control point (HACCP) principles on the farm, which are currently being studied in the USA and elsewhere, may provide a model for future production practices.

Public-health Aspects

Pigs are an important reservoir of *Salmonella* for humans and there have been many reports on the isolation of the organism from pork and pork products (Buxton, 1957; Bryan, 1980). In the USA, 12.4% of fresh pork sausages were positive for *Salmonella* (Johnston *et al.*, 1982). More recent reports indicate that a wide range of *Salmonella* may be present in fresh pork (Bozzano *et al.*, 1993; Fernandez-Escartin *et al.*, 1995). In Greece, 28% of pork pig carcasses were found to be contaminated with *Salmonella* (Epling *et al.*, 1993). Carcass contamination was 17.5% in Canada (Lammerding *et al.*, 1998), 21% in The Netherlands (Oosterom *et al.*, 1985), 6.5% in the USA (FSIS, 1999) and 27% in Belgium (Korsak *et al.*, 1998). Pig meat products, such as hot vacuum-packaged pork (Van Laack *et al.*, 1993) especially those incorporating low-grade material, such as mechanically recovered meat (Banks and Board, 1983), are frequently contaminated with *Salmonella*.

In many countries, the relationship between human illness and *Salmonella* contamination of pig meat is unclear, with the possible exception of Denmark, where approximately 15–20% of human cases are considered to be related to pork and pork products (Wegener *et al.*, 1994). In the UK, there have been reports of recent outbreaks of food poisoning associated with pork products (Maguire *et al.*, 1993; Cornell and Neal, 1998).

The prevalence of *Salmonella* in the intestine of individual pigs from different sources is extremely variable (Gray *et al.*, 1995, 1996a,b). Individual animals may remain as carriers for up to 36 weeks (Wood and Rose, 1992). In Denmark, 6.2% of caecal samples were found positive, usually with one serovar or phage type predominating from each farm source (Baggesen *et al.*, 1996). Sampling methods are important in these surveys, because it has been shown that rectal swabs provide an underestimate of the level of infection, as may carcass swabs. Davies *et al.* (1999) investigated *Salmonella* contamination at a large abattoir and isolated *Salmonella* from 7.0% of 2211 carcass swabs and 11.6% of 2205 samples of large-intestinal contents. Many of the isolates were resistant to antibiotics. It is commonly supposed that lairage of pigs will increase the chance of cross-infection and contamination of pigs (Morgan *et al.*, 1987). *Salmonella* shedding may be increased by any stress factor, including transport (Scheepens *et al.*, 1994; Berends *et al.*, 1996), but stress levels, and possibly *Salmonella* excretion, are reduced by overnight lairage (Warris *et al.*, 1988). There appears to be little difference in the magnitude of this effect in relation to the distance travelled (Rajkowski *et al.*, 1998). Davies *et al.* (1999) found a reduced rate of intestinal carriage of *Salmonella* and carcass contamination in pigs that had been held overnight in lairage, as compared with pigs slaughtered within 2–3 h of arrival.

The slaughter process in a well-run pig abattoir is capable of reducing the level of surface contamination of carcasses, because of the scalding and singeing stages, but any *Salmonella* that survive these stages can be spread between carcasses by the dehairing equipment (Gill and Bryant, 1993). Most of the contamination, however, results from escape of intestinal contents during evisceration (Saide-Albornoz *et al.*, 1995). The studies of Davies *et al.* (1999) indicated that there was little increase in carcass contamination after evisceration, although further increases caused by trimming and meat inspection have been described (Mousing *et al.*, 1997). HACCP-style procedures have been widely adopted in abattoirs, but the level of microbial monitoring to verify the critical control points and the correct application of procedures is often insufficient (von Langer, 1995). Employment of HACCP principles in the abattoir, however, appear to have decreased the recovery of *Salmonella* from pig carcasses (FSIS, 1999). It is clear that biosecurity and hygiene precautions to control *Salmonella* should be taken throughout the pig meat production and distribution chain, from nucleus breeders to hot-dog stalls (Simonsen *et al.*, 1987). The use of isolated weaning techniques may increase the likelihood that pigs can be raised free of *Salmonella* (Fedorka-Cray *et al.*, 1997b).

References

Araj, G. and Chugh, T.D. (1987) Detection of *Salmonella* spp. in clinical specimens by capture enzyme-linked immunosorbent assay. *Journal of Clinical Microbiology* 25, 2150–2153.

Ashkenazi, S., Cleary, T.G., Murray, B.C., Wagner, A. and Pickering, L.K. (1988) Quantitative analysis and partial characterisation of cytotoxin production by *Salmonella* strains. *Infection and Immunity* 56, 3089–3094.

Bager, F. (1994) *Salmonella* in Danish pigherds: risk factors and sources of infection. In: *Proceedings of the XVII Nordic Veterinary Congress.* Reykjavik, pp. 79–82.

Baggesen, D.L. and Christensen, J. (1997) Distribution of *Salmonella enterica* serovars and phage types in Danish pig herds. In: *Proceedings of the 2nd International Symposium on Epidemiology and Control of* Salmonella *in Pork, Copenhagen, Denmark, 20–22 August, 1997*, pp. 107–109. (Obtainable from Federation of Danish Pig Producers and Slaughterhouses, Axelborg, Axeltorv 3, 1609 Copenhagen V, Denmark. Price DK 350 + postage.)

Baggesen, D.L., Wegener, H.C., Bager, F., Stege, H. and Christensen, J. (1996) Herd prevalence of *Salmonella enterica* infections in Danish slaughter pigs determined by microbiological testing. *Preventive Veterinary Medicine* 26, 201–213.

Baggesen, D.L., Dahl, J., Wingstrand, A. and Nielsen, B. (1997) Detection of *Salmonella enterica* in different materials from the environment of pig farms In: *Proceedings of the 2nd International Symposium on Epidemiology and Control of* Salmonella *in Pork, Copenhagen, Denmark, 20–22 August, 1997*, pp. 173–175.

Bailey, J.S. (1987) Factors affecting microbial competitive exclusion in poultry. *Food Technology* 47, 88–92.

Bailey, J.S., Cox, N.A., Blankenship, L.C. and Stern, N.J. (1992) Effect of competitive exclusion microflora on the distribution of *Salmonella* serovars in an integrated poultry operation. *Poultry Science* 71 (Suppl. 1).

Banks, J.G. and Board, R.G. (1983) Incidence and level of contamination of British pork sausages and ingredients with salmonellas. *Journal of Hygiene, Cambridge* 90, 213–224

Baskerville, A. and Dow, C. (1973) Pathology of experimental pneumonia in pigs produced by *Salmonella choleraesuis*. *Journal of Comparative Pathology* 83, 207–215.

Baskerville, A., Dow, C., Curran, W.L. and Hanna, J. (1972) Ultrastructure of phagocytosis of *Salmonella choleraesuis* by pulmonary macrophages *in vivo*. *British Journal of Experimental Pathology* 53, 641–647.

Baskerville, A., Dow, C., Curran, W.L. and Hanna, J. (1973) Further studies on experimental bacterial pneumonia: ultrastructure changes produced in the lungs by *Salmonella choleraesuis*. *British Journal of Experimental Pathology* 54, 90–98.

Benjamin, W.H., Turnbough, C.L., Posey, B.S. and Briles, D.E. (1985) The ability of *Salmonella typhimurium* to produce the siderophore enterobactin is not a virulence factor in mouse typhoid. *Infection and Immunity* 50, 392–397.

Berends, B.R., Urlings, H.A.P., Snijders, J.M.A. and Van Knapen, F. (1996) Identification and quantification of risk factors in animal management and transport regarding *Salmonella* spp. in pigs. *International Journal of Food Microbiology* 30, 37–53.

Blankenship, L.C., Bailey, J.S., Cox, N.A., Stern, N.J., Brewer, R. and Williams, O. (1993) Two-step mucosal competitive exclusion flora treatment to diminish *Salmonella* in commercial broiler chickens. *Poultry Science* 72, 1667–1672.

Bozzano, A.I., Diguerdos, G., Saccares, S., Bilet, S., Zottola, T., Brozzi, A.M., Panfilli, G. and Fontanelli, G. (1993) Occurrence of *Salmonella* in porkmeat and processed pork products in the Latium region, central Italy from 1980–1989. *Italian Journal of Food Science* 5, 167–172.

Bryan, F.L. (1980) Foodborne diseases in the United States associated with meat and poultry. *Journal of Food Protection* 43, 140–150.

Buxton, A. (1957) Public health aspects of salmonellosis in animals. *Veterinary Record* 69, 105–109.

Chatfield, S., Li, L.J.L., Sydenham, M., Douce, G. and Dougan, G. (1992) *Salmonella* genetics and vaccine development. In: Hoermache, C.E., Penn, C.W. and Smyth, C.J. (eds) *Molecular Biology of Bacterial Infections: Current Status and Future Perspectives*. Cambridge University Press, Cambridge, pp. 299–312.

Cherrington, C.A. and Huis in't Veld, J.H.J. (1993) Development of a 24 h screen to detect viable *Salmonellas* in faeces. *Journal of Applied Bacteriology* 75, 58–64.

Cherubin, C.E. (1980) Epidemiologic assessment of antibiotic resistance in *Salmonella*. In: *CRC Handbook – Zoonosis*. CRC Press, Boca Raton, Florida, pp. 173–200.

Clarke, R.C. (1985) Virulence of wild and mutant strains of *Salmonella typhimurium* in calves. PhD dissertation, University of Guelph, Ontario, Canada.

Clarke, R.C. and Gyles, C.L. (1993) *Salmonella*. In: Gyles, C.L. and Thoen, C.O. (eds) *Pathogenesis of Bacterial Infections in Animals*, 2nd edn. Iowa State University Press, Ames, Iowa, pp. 133–153.

Clemmer, D.I., Hickey, J.L.S., Bridges, J.F., Schliessmann, D.J. and Shaffer, M.F. (1960) Bacteriologic studies of experimental air-borne salmonellosis in chicks. *Journal of Infectious Disease* 106, 197–210.

Coe, N. and Wood, R.L. (1991) The effect of exposure to a *cya/crp* mutant *of Salmonella typhimurium* on the subsequent colonization of swine by the wild-type parent strain. *Veterinary Microbiology* 31, 207–220.

Cohen, N.D., Neibergs, H.L., McGruder, E.D., Whitford, H.W., Behle, R.W., Ray, P.M. and Hargis, B.M. (1993) Genus-specific detection of salmonellae using the polymerase chain reaction (PCR). *Journal of Veterinary Diagnostic Investigation* 5, 368–371.

Cornell, J. and Neal, K.R. (1998) Protracted outbreak of *Salmonella typhimurium* definitive type 170 food poisoning related to tripe, pig bag and chitterlings. *Communicable Disease and Public Health* 1, 28–30.

Currier, M., Singleton, M., Lee, J. and Lee, D.R. (1986) *Salmonella* in swine at slaughter: incidence and serovar distribution at different seasons. *Journal of Food Protection* 49, 366–368.

Curtiss, R., III and Kelly, S.M. (1987) *Salmonella typhimurium* deletion mutants lacking adenylate cyclase and cyclic AMP receptor protein are avirulent and immunogenic. *Infection and Immunity* 55, 3035–3043.

Dahl, J., Wingstrand, A., Baggesen, D.L. and Nielsen, B. (1997) Strategies for elimination of *Salmonella typhimurium*. In: *Proceedings of the 2nd International Symposium on Epidemiology and Control of* Salmonella *in Pork, Copenhagen, Denmark, 20–22 August*, pp. 153–156.

Darlow, H.M., Bale, W.R. and Carter, G.B. (1961) Infection of mice by the respiratory route with *Salmonella typhimurium*. *Journal of Hygiene, Cambridge* 59, 303–308.

Davies, P.R., Morrow, W.E.M., Jones, F.T., Deen, J., Fedorka-Cray, P.J. and Harris, I.T. (1997) Prevalence of salmonella in finishing swine raised in different production systems in North Carolina, USA. *Epidemiology and Infection* 119, 237–244.

Davies, R.H. and Wray, C. (1997) Distribution of *Salmonella* on 23 pig farms in the UK. In: *Proceedings of the 2nd International Symposium on Epidemiology and Control of* Salmonella *in Pork, Copenhagen, Denmark, 20–22 August*, pp. 137–141.

Davies, R.H., McLaren, I.M. and Bedford, S. (1999) Observations on the distribution of *Salmonella* in a pig abattoir. *Veterinary Record* 145, 655–661.

Dawe, D.L. and Troutt, H.F. (1976) Treatment of experimentally induced salmonellosis in weanling pigs with trimethoprim and sulfadiazine. In: *Proceedings of the 4th International Pig Veterinary Association, Ames*, p. M4.

De Jong, H. and Ekdahl, M.O. (1965) Salmonellosis in calves – the effect of dose rate and other factors on transmission. *New Zealand Veterinary Journal* 13, 59–64.

Dusch, H. and Altwegg, M. (1995) Evaluation of five new plating media for isolation of *Salmonella* species. *Journal of Clinical Microbiology* 33, 802–804.

Dziezak, J.D. (1987) Rapid methods for microbiological analysis of food. *Food Technology* 41, 56–73.

Edel, W., Guinee, P.A.M., van Schothorst, M. and Kampelmacher, E.H. (1967) *Salmonella* infection in pigs fattened with pellets and unpelleted meal. *Zentralblatt für Veterinarmedizin* 14, 393–401.

Edel, W., van Schothorst, M., Guinee, P.A.M. and Kampelmacher, E.H. (1970) Effect of feeding pellets on the prevention and sanitation of *Salmonella* infections in fattening pigs. *Zentralblatt für Veterinarmedizin* 17, 730–738.

Ehrlich, G.D. and Sirko, D.A. (1994) PCR and its role in clinical diagnostics. In: Ehrlich, G.D. and Greenberg, S.J. (eds) *PCR-based Diagnostics in Infectious Disease*. Blackwell Scientific Publications, Boston, Massachssetts, pp. 1.1–1.3.

Epling, L.K., Carpenter, J.A. and Blankenship, L.C. (1993) Prevalence of *Campylobacter* and *Salmonella* spp. on pork carcasses and the reduction effected by spraying with lactic acid. *Journal of Food Protection* 53, 536–537.

Ewing, W.H. (1986) Isolation and preliminary identification. In: Edwards, P.R. and Ewing, W.H. (eds) *Identification of Enterobacteriaceae*, 4th edn. Elsevier Science Publishing, pp. 27–45.

Falkow, S. and Mekalanos, J. (1990) The enteric bacilli and vibrios. In: Davis, B., Dulbecco, R., Eisen, H. and Ginsberg, H. (eds) *Microbiology*, 4th edn. J.B. Lippincott, Philadelphia, pp. 576–579.

Fedorka-Cray, P.J., Kelley, L.C., Stabel, T.J., Gray, J.T. and Laufer, J.A. (1995) Alternate routes of invasion may affect pathogenesis of *Salmonella typhimurium* in swine. *Infection and Immunity* 63, 2658–2664.

Fedorka-Cray, P.J., Gray, J.T. and Thomas, L.A. (1995) Comparison of culture media for the isolation of *Salmonella* species. In: *Symposium Diagnosis of* Salmonella *Infections, Proceedings of the US Animal Health Association*, Reno, Nevada, pp. 116–123.

Fedorka-Cray, P.J., Hogg, A., Gray, J.T., Lorenzen, K., Velasquez, J. and von Behren, P. (1997a) Feed and feed trucks as sources of *Salmonella* contamination in swine. *Swine Health and Production* 5, 189–193.

Fedorka-Cray, P.J., Harris, D.L. and Whipp, S.C. (1997b) Using isolated weaning to raise *Salmonella*-free swine. *Veterinary Medicine*, April, 375–382.

Fedorka-Cray, P.J., Bailey, J.S., Stern, N. and Cox, N.A. (2000) Mucosal competitive exclusion to reduce *Salmonella* in swine. *Journal of Food Protection* 62, 1376–1380.

Feng, P. (1992) Rapid methods for detecting food-borne pathogens. In: *FDA Bacteriological Analytical Manual* 7th edn. Appendix 1.

Fernandez-Escartin, F., Saldana Lozano, J., Rodriguez, D., Martinez-Gonzales, N. and Torres, J.A. (1995) Incidence and level of *Salmonella* serovars in raw pork obtained from Mexican butcher shops. *Food Microbiology* 12, 435–439.

Ferris, K. and Thomas, L.A. (1993) *Salmonella* serovars from animals and related sources during July 1992–June 1993. In: *Proceedings 97th Annual Meeting US Animal Health Association*. Cummings Corporation and Carter Printing, Richmond, Virginia, p. 538.

Ferris, K.E. and Miller, D.A. (1998) *Salmonella* serovars from animals and related sources reported during July 1997–June 1998. In: *Proceedings Annual Meeting US Animal Health Association*. Minneapolis, Minnesota, pp. 584–605.

Finlay, B.B., Starnbach, M.N., Francis, C.L., Stocker, B.A.D., Chatfield, S., Dougan, G. and Falkow, S. (1988) Identification and characterization of Tn *phoA* mutants of *Salmonella* that are unable to pass through a polarized MDCK epithelial cell monolayer. *Molecular Microbiology* 6, 757–766.

FSIS (1999) HACCP Implementation: First Year: Salmonella. Http://www.fsis.usda.gov/OPHS/salmdata.htm.

Gill, C.O. and Bryant, J. (1993) The presence of *Escherichia coli, Salmonella* and *Campylobacter* in pig carcasses dehairing equipment. *Food Microbiology* 10, 337–344.

Gooch, J.M. and Haddock, R.L. (1969) Swine salmonellosis in a Hawaiian piggery. *Journal of the American Veterinary Medical Association* 154, 1051–1054.

Gray, J.T. and Fedorka-Cray, P.J. (1995) Survival and infectivity of *Salmonella choleraesuis* in swine feces. In: *76th Conference on Research Workers in Animal Disease*, P94.

Gray, J.T. and Fedorka-Cray, P.J. (1999) Detection of swine exposed to *Salmonella* spp. *American Association of Veterinary Laboratory Diagnosticians* (in press).

Gray, J.T., Fedorka-Cray, P.J., Stabel, T.J. and Ackermann, M.R. (1995) Influence of inoculation route on the carrier state of *Salmonella choleraesuis* in swine. *Veterinary Microbiology* 47, 43–59.

Gray, J.T., Stabel, T.J. and Fedorka-Cray, P.J. (1996a) Effect of dose on the immune response and persistence of *Salmonella choleraesuis* infection in swine. *American Journal of Veterinary Research* 57, 313–319.

Gray, J.T., Fedorka-Cray, P.J., Stabel, T.J. and Kramer, T.T. (1996b) Natural transmission of *Salmonella choleraesuis* in swine. *Applied and Environmental Microbiology* 62, 141–146.

Gutzmann, F., Layton, H., Simiins, K. and Jarolman, H. (1976) Influence of antibiotic-supplemented feed on the occurrence and persistence of *S. typhimurium* in experimentally infected swine. *American Journal of Veterinary Research* 37, 649–655.

Hardman, P.M., Wathes, C.M. and Wray, C. (1991) Transmission of salmonellae among calves penned individually. *Veterinary Record* 129, 327–329.

Harris, I.T., Fedorka-Cray, P.J., Gray, J.T., Thomas, L.A. and Ferris, K. (1997) Prevalence of *Salmonella* organisms in swine feed. *Journal of the American Veterinary Medicinal Association* 210, 382–385.

Helmuth, R., Kasbohrer, A., Geue, L., Rabsch, W. and Protz, D. (1997) Pilot study on the prevalence of *Salmonella* in slaughter pigs in Germany. III Detection of *Salmonella* by PCR, serovar distribution and population analysis of isolates. In: *Proceedings of the 2nd International Symposium on Epidemiology and Control of Salmonella in Pork*, Copenhagen, Denmark, 20–22 August, 1997, pp. 103–106. (Obtainable from Federation of Danish Pig Producers and Slaughterhouses, Axelborg, Axeltorv 3, 1609 Copenhagen V, Denmark. Price DK 350 + postage.)

Henzler, D.J. and Opitz, H.M. (1992) The role of mice in the epizootiology of *Salmonella enteritidis* infection on chicken layer farms. *Avian Diseases* 336, 625–631.

Holcomb, H.L. and Fedorka-Cray, P.J. (1997) Effect of Naxel and Baytril on persistence of *Salmonella heidelberg* in swine. In: *Abstracts of the Annual Meeting of the American Society for Microbiology*. ASM, Washington, DC, p. 445.

Hook, E.W. (1990) *Salmonella* species (including typhoid fever). In: Mandell, G.L., Douglas, R.G. and Bennet, J.E. (eds) *Principles and Practices of Infectious Disease*. Churchill Livingstone, New York, pp. 1700–1722.

Johnston, R.W., Green, S.S., Chiu, J., Proatt, M. and Rivera, J. (1982) Incidence of *Salmonella* in fresh pork sausages in 1979 compared with 1969. *Journal of Food Science* 47, 1369–1371.

Jones, B.D., Lee, C.A. and Falkow, S. (1992) Invasion of *Salmonella typhimurium* is affected by the direction of flagellar rotation. *Infection and Immunity* 60, 2475–2480.

Jones, G.W., Richardson, L.A. and Uhlman, D. (1981) The invasion of HeLa cells by *Salmonella typhimurium*: reversible and irreversible bacterial attachment and the role of bacterial motility. *Journal of General Microbiology* 127, 351–360.

Jubb, K.V., Kennedy, P.C. and Palmer, N.C. (1985) Infectious and parasitic diseases. In: Jubb, K.V., Kennedy, P.C. and Palmer, N. (eds) *Pathology of Domestic Animals*, 3rd edn. Academic Press, New York, pp. 135–143.

Kelly, S.M., Bosecker, B.A. and Curtiss, R., III (1992) Characterization and protective properties of attenuated mutants of *Salmonella choleraesuis*. *Infection and Immunity* 60, 4881–4890.

Korsak, N., Daube, G., Ghafir, Y, Chahed, A., Jolly, S. and Vindevogel, H. (1998) An efficient sampling technique used to detect for food-borne pathogens on pork and beef carcasses in nine Belgian abattoirs. *Journal of Food Protection* 61, 535–541.

Kramer, T.T., Roof, M.B. and Matheson, R.R. (1992) Safety and efficacy of an attenuated strain of *Salmonella choleraesuis* for vaccination of swine. *American Journal of Veterinary Research* 53, 444–448.

Kramer, T.T., Rhiner, J., Gray, J.T. and Fedorka-Cray, P.J. (1994) Comparison of IgG ELISA and bacteriologic diagnosis in experimental swine salmonellosis. In: *Proceedings of the 75th Annual Conference Research Workers in Animal Disease*.

Lambiri, M., Mavridou, A., Richardson, S.C. and Papadakis, J.A. (1990) Comparison of the TECRA *Salmonella* immunoassay with the conventional culture method. *Letters in Applied Microbiology* 11, 182–184.

Lammerding, A.M., Garcia, M.M., Mann, E.D., Robinson, Y., Dorward, W.J., Truscott, R. and Tittinger, F. (1998) Prevalence of *Salmonella* and thermophyllic *Campylobacter* in fresh pork. *Journal of Food Protection* 51, 47–52.

Laval, A., Morvan, H., Despez, G. and Corbion, B. (1992) Salmonellosis in swine. In: *Reports and Communications, Salmonella and Salmonellosis, 15–17 September, 1993.* Ploufragan/St Brieuc, France, pp. 164–175. (Obtainable from ISPAIA, BPT, 22440 Ploufragan, France. 450 FF + postage.)

Lawson, G. and Dow, C. (1966) Porcine salmonellosis. *Journal of Comparative Pathology* 76, 363–371.

Levine, M.M., Ferreccio, C., Black, R.E., Tacket, C.O. and Germanier, R. (1989) Progress in vaccines against typhoid fever. *Review of Infectious Disease* 11, S552–S567.

Libby, S.J., Goebel, W., Muir, S., Songer, G. and Heffron, F. (1990) Cloning and characterization of a cytotoxin gene from *Salmonella typhimurium. Research in Microbiology* 141, 775–783.

Linton, A.H. and Jennet, N.E. (1970) Multiplication of *Salmonella* in liquid feed and its influence on the duration of excretion in pigs. *Research in Veterinary Science* 11, 452–457.

Linton, A.H., Heard, T.W., Grimshaw, J.J. and Pollaed, P. (1970) Computer-based analysis of epidemiological data arising from salmonellosis in pigs. *Research in Veterinary Science* 11, 523–532.

Lumsden, J.S., Wilkie, B.N. and Clarke, R.C. (1991) Resistance to faecal shedding in pigs and chickens vaccinated with an aromatic dependent mutant of *Salmonella typhimurium. American Journal of Veterinary Research* 52, 1784–1787.

McCall, C.E., Martin, W.T. and Boring, J.R. (1966) Efficiency of cultures of rectal swabs and faecal specimens in detecting salmonellae carriers: correlation with numbers of salmonellas excreted. *Journal of Hygiene, Cambridge* 64, 261–269.

McChesney, D.G., Kaplan, G. and Gardner, P. (1995) FDA survey determines *Salmonella* contamination. *Feedstuffs* 7, 20–23.

Maguire, H.C.F., Codd, A.A., MacKay, V.F., Rowe, B. and Mitchell, E. (1993) A large outbreak of human salmonellosis traced to a local pig farm. *Epidemiology and Infection* 110, 239–246.

Meyer, H., Koch, H., Methner, U. and Steinbach, G. (1993) Vaccines in salmonellosis control in animals. *Zentralblatt für Bakteriologie* 278, 407–415.

Morgan, I.R., Krautil, F.L. and Craven, J.A. (1987) Effect of lairage on caecal and *Salmonella* carcass contamination of slaughter pigs. *Epidemiology and Infection* 98, 323–330.

Mousing, J., Kyrrval, J., Jensen, T.K., Aalbaek, B., Buttenschon, J., Svensmark, B. and Willeberg, P. (1997) Meat safety consequences of implementing visual post-mortem meat inspection procedures in Danish slaughter pigs. *Veterinary Record* 140, 472–477.

NARMS-EB (1997) *Annual Report.* FDA, USDA-ARS, USDA-APHIS, USDA-FSIS, Athens, Georgia.

Nielsen, B., Baggesen, D., Bager, F. and Lind, P. (1994) Serological diagnosis of *Salmonella* infections in swine by ELISA. In: *Proceedings of the 13th International Pig Veterinary Science Congress,* Bangkok, Thailand, 26–30 June.

Nielsen, B., Baggesen, D., Bager, F., Haugegaard, J. and Lind, P. (1995) The serological response to *Salmonella* serovars typhimurium and infantis in experimentally infected pigs. The time course followed with an indirect anti-LPS ELISA and bacteriological examinations. *Veterinary Microbiology* 47, 205–218.

Nnalue, N.A. and Lindberg, A.A. (1990) *Salmonella choleraesuis* strains deficient in O antigen remain fully virulent for mice by parenteral inoculation but are avirulent by oral administration. *Infection and Immunity* 58, 2493–2501.

Nnalue, N.A. and Stocker, B.A.D. (1986) Some *gal*E mutants of *Salmonella choleraesuis* retain virulence. *Infection and Immunity* 54, 635–640.

Oosterom, J., Dekker, R., Dewilde, G.J.A., Van Kempende, T.F. and Engels, G.B. (1985) Prevalence of *Campylobacter jejuni* and *Salmonella* during pig slaughtering. *Veterinary Quarterly* 7, 31–34.

Pospischil, A., Wood, R.L. and Anderson, T.D. (1990) Peroxidase–antiperoxidase and immunogold labeling of *Salmonella typhimurium* and *Salmonella choleraesuis* var *kunzendorf* in tissues of experimentally infected swine. *American Journal of Veterinary Research* 51, 619–624.

Prasad, R., Chopra, A.K., Peterson, J.W. and Pericas, R. (1990) Biological and immunological characterization of a cloned cholera toxin-like enterotoxin from *Salmonella typhimurium. Microbial Pathogenesis* 9, 315–329.

Prasad, R., Chopra, A.K., Charry, P. and Peterson, J.W. (1992) Expression and characterization of the cloned *Salmonella typhimurium* enterotoxin. *Microbial Pathogenesis* 13, 109–121.

Prohaszka, L., Jayarao, B.M., Fabian, A. and Kovacs, S. (1990) The role of intestinal volatile fatty acids in the *Salmonella* shedding of pigs. *Journal of Veterinary Medicine B* 37, 570–574.

Rahn, K., De Grandis, S.A., Clarke, R.C., McEwen, S.A., Galan, J.E., Ginocchio, R., Curtiss, R., III and Gyles, C.L. (1992) Amplification of an *inv*A gene sequence of *Salmonella typhimurium* by polymerase chain reaction as a specific method of detection of *Salmonella. Molecular and Cellular Probes* 6, 271–279.

Rajkowski, K.T., Eblen, S. and Laubauch, C. (1998) Efficiency of washing and sanitising trailers used for swine transport in reduction of salmonella and *Escherichia coli. Journal of Food Protection* 61, 35–37.

Reed, W.M., Olander, H.J. and Thacker, H.L. (1986) Studies on the pathogenesis of *Salmonella typhimurium* and *Salmonella choleraesuis* var *kunzendorf* infection in weanling pigs. *American Journal of Veterinary Research* 47, 75–83.

Reitmeyer, J.C., Peterson, J.W. and Wilson, K.J. (1986) *Salmonella* cytotoxin: a component of the bacterial outer membrane. *Microbial Pathogenesis* 1, 503–510.

Report (1997a) Salmonella *in Livestock Production 1997*, Veterinary Laboratories Agency, New Haw, Addlestone, UK.

Report (1997b) *Annual Report of Zoonoses in Denmark*. Danish Zoonoses Centre, Copenhagen.

Reynolds, I.M., Miner, P.W. and Smith, R.E. (1967) *Salmonella enteritidis* from porcine meningitis: a case report. *Cornell Veterinarian* 58, 180–185.

Robbins, J.B., Chu, C. and Sneerson, R. (1992) Hypothesis for vaccine development: protective immunity to enteric diseases caused by nontyphoidal salmonellae and shigellae may be conferred by serum IgG antibodies to the O-specific polysaccharide of their lipopolysaccharides. *Clinical Infectious Diseases* 15, 346–361.

Robertsson, J.A., Lindberg, A.A., Hoiseth, S. and Stocker, B.A.D. (1983) *Salmonella typhimurium* infection in calves: protection and survival of virulent challenge bacteria after immunization with live or inactivated vaccines. *Infection and Immunity* 41, 742–750.

Roof, M.B. and Kramer, T.T. (1989) Porcine neutrophils function in the presence of virulent and avirulent *Salmonella choleraesuis*. *Veterinary Immunology and Immunopathology* 23, 365–376.

Roof, M.B., Kramer, T.T., Roth, J.A. and Minion, F.C. (1992) Characterization of a *Salmonella choleraesuis* isolate after repeated neutrophil exposure. *American Journal of Veterinary Research* 53, 1328–1332.

Rossen, L., Norskov, K., Holmstrom, K. and Rasmussen, O.F. (1992) Inhibition of PCR by components of food samples, microbial diagnostic assays and DNA-extraction solutions. *International Journal of Food Microbiology* 17, 37–45.

Saide-Albornoz, J.J., Knipe, C.L., Murano, E.A. and Beran, G.W. (1995) Contamination of carcasses during slaughter, fabrication and chilled storage. *Journal of Food Protection* 58, 993–997.

Salmon, D.E. and Smith, T. (1886) The bacterium of swine plague. *American Monthly Microbiology Journal* 7, 204.

Saxen, H., Reima, I. and Makela, P.H. (1987) Alternative complement pathway activation by *Salmonella* O polysaccharide as a virulence determinant in the mouse. *Microbial Pathogenesis* 2, 15–28.

Scheepens, C.J.M., Hessing, M.J.C., Hensen, E.J. and Hendricks, P.A.J. (1994) Effect of climatic stress on the immunological reactivity of weaned pigs. *Veterinary Quarterly* 16, 137–143.

Schwartz, K. (1990) Salmonellosis in midwestern swine. In: *Proceedings of the 94th Annual Meeting US Animal Health Association*, pp. 443–449.

Simonsen, B., Bryan, F.L., Christian, J.H.B., Roberts, T.A., Tomkins, R.B. and Silliker, J.H. (1987) Prevention and control of food-borne salmonellosis through application of hazard analysis critical control (HACCP). *International Journal of Food Microbiology* 4, 227–247.

Smith, B.P., Reina-Guerra, M., Hoiseth, S.K., Stocker, B.A.D., Habasha, F., Johnson, E. and Merritt, F. (1984) Aromatic-dependent *Salmonella typhimurium* as modified live vaccines for calves. *American Journal of Veterinary Research* 45, 59–66.

Smith, H.W. (1952) The evaluation of culture media for the isolation of *Salmonella* from faeces. *Journal of Hygiene, Cambridge* 50, 21–36.

Smith, H.W. and Jones, J.E.T. (1967) Observations on experimental oral infections with *S. dublin* in calves and *S. choleraesuis* in pigs. *Journal of Pathology and Bacteriology* 93, 141–156.

Sojka, W. J., Wray, C., Shreeve, J., Benson, A. J. (1977) Incidence of *Salmonella* infections in animals in England and Wales, 1968–1974. *Journal of Hygiene, Cambridge* 78, 43–56.

Stabel, T.J., Mayfield, J.E., Tabatabai, L.B. and Wannemuehler, M.J. (1990) Oral immunization of mice with attenuated *Salmonella typhimurium* containing a recombinant plasmid which codes for production of a 31-kilodalton protein of *Brucella abortus*. *Infection and Immunity* 58, 2048–2055.

Stabel, T.J., Mayfield, J.E., Tabatabai, L.B. and Wannemuehler, M.J. (1991) Swine immunity to an attenuated *Salmonella typhimurium* mutant containing a recombinant plasmid which codes for production of a 31-kilodalton protein of *Brucella abortus*. *Infection and Immunity* 59, 2941–2947.

Stone, G.G., Oberst, R.D., Hays, M.P., McVey, S. and Chengappa, M.M. (1994) Detection of *Salmonella* serovars from clinical samples by enrichment broth cultivation–PCR procedure. *Journal of Clinical Microbiology* 32, 1742–1749.

Takeuchi, A. and Sprinz, H. (1967) Electron microscope studies of experimental *Salmonella* infections in the preconditioned guinea pig. 2. Response of the intestinal mucosa to the invasion by *Salmonella typhimurium*. *American Journal of Pathology* 51, 137–146.

Taylor, J. and McCoy, J.H. (1969) *Salmonella* and *Arizona* infections and intoxications. In: Riemann, H. (ed.) *Foodborne Infections and Intoxications*. Academic Press, New York, pp. 3–71.

Tielen, M.J.M., van Schie, F.W., van der Wolf, P.J., Elbers, A.R.W., Koppens, J.M.C.C. and Wolbers, W.B. (1997) Risk factors and control measures for subclinical infection in pig herds (1997). In: *Proceedings of the 2nd International Symposium on Epidemiology and Control of* Salmonella in Pork, Copenhagen, Denmark, 20–22 August, pp. 32–35.

Turk, J.R., Fales, W.H., Maddox, C., Miller, M., Pace, L., Fischer, J., Kreeger, J., Johnson, G., Turnquist, S., Ramos, J.A. and Gosser, H.S. (1992) Pneumonia associated with *Salmonella choleraesuis* infection in swine: 99 cases (1987–1990). *Journal of the American Veterinary Medical Association* 201, 1615–1616.

Van Laack, R.L.J.M., Johnson, J.L., van der Palen, C.J.N.M., Smulders, F.J.M. and Snijders, J.M.A. (1993) Survival of pathogenic bacteria on pork loins as influenced by hot processing and packaging. *Journal of Food Protection* 56, 847–851.

Van Schie, F.W. and Overgoor, G.H.A. (1987) An analysis of the possible effects of different feed upon the excretion of *Salmonella* bacteria in clinically normal groups of fattening pigs. *Veterinary Quarterly* 9, 185–188.

van Poucke, L.S.G. (1990) *Salmonella*-TEK, a rapid screening method for *Salmonella* species in food. *Applied and Environmental Microbiology* 56, 924–927.

von Langer, L. (1995) The need for microbiological examination to improve the standards of fresh meat hygiene. *Tierärztiches Umshau* 50, 47–52.

Warris, P.D., Brown, S.N., Edwards, J.E. and Knowles, T.G. (1998) Effect of lairage time on levels of stress and meat quality in pigs. *Animal Science* 66, 255–261.

Wegener, H.C., Baggesen, D.L. and Gaardslev, K. (1994) *Salmonella typhimurium* phagetypes from human salmonellosis in Denmark 1988–1993. *APMIS* 102, 521–525.

Widjojoatmodjo, M.N., Fluit, A.C., Torensma, R., Keller, B.H.I. and Verhoef, J. (1991) Evaluation of the magnetic immuno PCR assay for the rapid detection of *Salmonella*. *European Journal of Clinical Microbiology and Infectious Disease* 10, 935–938.

Widjojoatmodjo, M.N., Fluit, A.C., Torensma, R., Verdonk, G.P.H.T. and Verhoef, J. (1992) The magnetic immuno-polymerase chain reaction assay for direct detection of salmonellae in fecal samples. *Journal of Clinical Microbiology* 30, 3195–3199.

Wilcock, B. and Olander, H. (1978) Influence of oral antibiotic feeding on the duration and severity of clinical disease, growth performance and pattern of shedding in swine inoculated with *Salmonella typhimurium*. *Journal of the American Veterinary Medical Association* 172, 472–477.

Wilcock, B., Armstrong, C. and Olander, H. (1976) The significance of the serovar in the clinical and pathologic features of naturally occurring porcine salmonellosis. *Canadian Journal of Comparative Medicine* 40, 80–88.

Wilcock, B.P. and Schwartz, K. (1992) Salmonellosis. In: Leman, A.D., Straw, B.E., Mengeling, W.E., D'Allaire, S. and Taylor, D.J. (eds) *Diseases of Swine*, 7th edn. Iowa State University Press, Ames, Iowa, pp. 570–583.

Wilde, J., Eiden, J. and Yolken, R. (1990) Removal of inhibitory substances from human fecal specimens for detection of group A rotavirus by reverse transcriptase and polymerase chain reaction. *Journal of Clinical Microbiology* 28, 1300–1307.

Williams, D.R., Hunter, D., Binder, J. and Hough, E. (1981) Observations on the occurrence of *Salmonella choleraesuis* and other salmonellas in two herds of feeder pigs. *Journal of Hygiene, Cambridge* 86, 369–377.

Williams, L.P., Vaughn, J.P., Scott, A. and Blanton, V. (1969) A ten-month study on Salmonella contamination in animal protein meals. *Journal of the American Veterinary Medical Association* 155, 167–174.

Wood, R.L. and Rose, R. (1992) Populations of *Salmonella typhimurium* in internal organs of experimentally infected carrier swine. *American Journal of Veterinary Research* 53, 653–657.

Wray, C., Davies, R.H., Jones, Y.E. and Chappell, S. (1997) Antibiotic resistance and *Salmonella* infections in pigs. In: *Proceedings of the 2nd International Symposium on Epidemiology and Control of* Salmonella *in Pork, Copenhagen, Denmark, 20–22 August, 1997,* pp. 247–250. (Obtainable from Federation of Danish Pig Producers and Slaughterhouses, Axelborg, Axeltorv 3, 1609 Copenhagen V, Denmark. Price DK 350 + postage.)

Chapter 12
Salmonella Infections in Sheep

Clifford Wray[1] and Karl A. Linklater[2]

[1]*Veterinary Laboratories Agency (Weybridge), New Haw, Addlestone, Surrey KT15 3NB, UK;*
[2]*SAC, West Mains Road, Edinburgh EH9 3JG, UK*

Historical

The earliest references to the occurrence of *Salmonella* causing dysentery in sheep appear to be those of Frickinger (1919) and Bruns and Gasters (1920). They described an outbreak of dysentery in 300 wethers following rail transport; of these animals, 250 were ill and 140 died. Meat from emergency-slaughtered animals was used to make sausages and 1500 people subsequently developed food poisoning. In Colorado, two outbreaks of dysentery in feedlot lambs were described by Newsome and Cross (1924, 1930), the outbreaks involving 31,369 and 1600 lambs, respectively. A *Salmonella* sp. now considered to be *Salmonella abortusovis* was first isolated from a case of abortion in Germany in 1921 (Schermer and Ehrlich, 1921) and subsequently in the UK (Bosworth and Glover, 1925). The characteristics of the British isolate were described and correlated with those of the German isolate by Lovell (1931). Subsequent isolations of the organism were made in Cyprus (Manley, 1932) and France (Lesbouryies *et al.*, 1933) and there have been many subsequent references to its occurrence in other countries.

Many other *Salmonella* serovars have subsequently been associated with deaths and abortions in sheep in all parts of the world. The number of flocks affected annually is probably small, but within these flocks the losses may be high because many animals may be affected in any one outbreak.

Epidemiology

Salmonella infections in sheep have been recorded in most countries of the world. In the UK, the number of incidents varies from 100 to 200 a year and, given the size of the UK sheep population, ovine salmonellosis is of much less economic importance than salmonellosis in other species of farm animals. In the UK, *S. abortusovis* has been isolated on two occasions since 1975 – once in 1976 and again in 1984 – although it was previously widespread in the south-west of England. The more common serovars associated with ovine salmonellosis during the period 1991–1997 were *S. typhimurium*, *S. arizonae* (O61:k:1,2,7), *S. derby* and *S. montevideo*. The last of these has regularly been associated with abortion in ewes, especially in south-east Scotland and northern England. During the period 1970–1981, a total of 67 incidents of *S. montevideo* were identified, but in 1982 a sudden upsurge occurred and infection was confirmed in 38 flocks The prevalence still remains high and some 20–30 incidents occur annually. Twenty-seven biotypes of *S. montevideo*, which belong to two biogroups, have been recognized. One biogroup (10di) has been shown to predominate in sheep in the UK, while the other biogroup (2d) is associated mainly with human, cattle and poultry infections (Reilly *et al.*, 1985). In addition to the serovars mentioned above, 20–30 other serovars are recorded annually.

In other countries for which data are available, *S. typhimurium* appears to be the commonest *Salmonella* serovar in sheep, although the predominant phage type may differ between countries, e.g. in Australia DT135 and DT9 are dominant, whereas in the UK DT104 is currently the most common. *S. abortusovis* has been described in many countries worldwide and in some it is a serious problem. *S. arizonae* (O 61:k:1,5,7), which belongs to *S. enterica* ssp. IV, was first isolated from sheep in the USA and has subsequently been isolated in a number of countries, including the UK, Canada, Norway and Germany.

Many other *Salmonella* serovars have been isolated from sheep throughout the world, but, given their large numbers, the prevalence of ovine salmonellosis is low; however, when disease outbreaks occur, the losses can have serious economic consequences.

Transmission

As *S. abortusovis* is host-specific, it is invariably introduced into a flock by an infected sheep (Mura and Contini, 1954; Ivanov *et al.*, 1966). Infection probably takes place by mouth and, in all probability, occurs a long time before the abortions are actually apparent. Continental workers have isolated *S. abortusovis* from ram's testicles (Bartmann, 1957) and from the preputial secretions of a ram serving ewes in an infected flock (Stanoev and Milev, 1967). Rams certainly become infected at the same time as ewes. Jack (1968) found that rams from 11 infected flocks all had high serological titres, but he was unable to isolate the bacterium from their genitalia. Likewise experimental studies by Sanchis and Pardon (1986) failed to detect the organism in the genitalia of 18 experimentally infected rams, and ewes made pregnant by three of the remaining rams showed no signs of infection. Jacob (1959) examined the role of rams in infected flocks. Lesions within the epididymis were present in four of the 43 rams examined, but *S. abortusovis* was isolated from only one ram with a testicular abscess and from none of the preputial washings of rams examined.

In the case of the other *Salmonella* serovars, there are many sources of infection, e.g. food, water, other animals, wild birds and humans, and infection is most probably by the oral route. It has, however, also been suggested that infection

may occur via the conjunctiva and nostrils, because of the difficulties in causing disease by experimental oral infection. Sanchis *et al.* (1991) infected pregnant ewes intraconjunctivally and eight of 18 aborted and another four gave birth to stillborn lambs. Jack (1968), on the other hand, was unable to cause abortion by the intraconjuctival route.

S. arizonae has been isolated from the nostrils of naturally (Meehan *et al.*, 1992) and experimentally infected sheep (Hannam *et al.*, 1986) and when lambs were infected intranasally; the organism was isolated throughout the 6-month study period (Brogden *et al.*, 1994). The nasal cavity appears to be a site of persistent colonization in some carrier sheep and further studies are desirable to determine its importance.

Other Factors

Other diseases may aggravate *Salmonella* infection in sheep. Vodas and Elitsina (1986) found that mixed infections with *S. abortusovis* and *Chlamydia psittaci* were more severe than *S. abortusovis* alone. Likewise, Linklater (1983) recorded the presence of *C. psittaci* during abortion outbreaks associated with *S. montevideo* and Sojka *et al.* (1983) found that *S. arizonae* (O 61:k:1,5,7) was often associated with other disease conditions. Heavy dosing with zinc oxide as a prophylaxis against facial eczema has also precipitated *Salmonella* outbreaks in young sheep (Allworth *et al.*, 1985).

Angelachov (1964) considered that environmental factors, including poor feeding, appeared to be linked with the development of abortion caused by *S. abortusovis*. Cooper (1967) has shown that starvation may activate latent infection and Baker *et al.* (1971) commented that losses caused by *S. dublin* were reduced when the ewes were provided with extra feed prior to movement.

In range sheep, the commonest occurrence of salmonellosis is during times of drought, when the sheep are concentrated on to small areas of grass, which become heavily contaminated by their faeces. Linklater (1983) found that abortions caused by *S. montevideo* reached epidemic proportions during the severe winter of 1982. This necessitated supplementary feeding and subjected the pregnant ewes to prolonged cold stress, and also led to close congregation of sheep at a high risk in early pregnancy. Radial spread probably took place by means of birds or other wild animals or by environmental contamination by watercourses. Hunter *et al.* (1976) found that *S. typhimurium* spread from infected sheep into nearby watercourses and infected animals drinking downstream.

As mentioned previously, transport can precipitate *Salmonella* infection in sheep and, in Australia, high mortality rates were associated with shipping sheep to the Middle East (Gardiner and Craig, 1970), although Higgs *et al.* (1993) concluded that inanition was probably the primary cause.

Clinical Findings

Clinical signs of salmonellosis are varied and include general systemic and enteric signs, as well as abortion. The signs vary with serovar, and with *S. abortusovis*, abortion occurs during the last 4–6 weeks of pregnancy. The fetuses are reasonably fresh and placental retention is rare, although the placenta may occasionally show chorionic thickening. Affected ewes may have a transitory pyrexia, which can go unnoticed, and are rarely seen ill unless post-parturient infection develops. The ewes do not discharge for more than a few days after abortion, and infection does not appear to cause infertility in either ewes or rams, and lambing in the flock is usually normal the following year.

Stillbirths or weak lambs may be born at term, and apparently healthy lambs may scour or die of septicaemia during their first 2–3 weeks of life.

The clinical picture is very similar with *S. montevideo* infection. Apart from aborting, affected ewes usually appear normal. Many ewes in an infected group may excrete the organism in their faeces without abortion and the infection can pass through a group of non-pregnant animals without any signs being present (Linklater, 1983). The number of ewe deaths in a flock can vary from occasional deaths up to 10–15% of the flock, and abortion losses range from 10 to 75% or more (Fincham, 1961).

With *S. typhimurium*, enteric and systemic signs may predominate (Saunders *et al.*, 1966; Hunter *et al.*, 1976). Affected animals are anorectic, have a high temperature (> 41°C) and

scour profusely. Those that do not die of septicaemia may continue to scour and die from dehydration. Pregnant animals may die of septicaemia before aborting and, if *S. typhimurium* is present in a group of in-lamb ewes, all symptoms are usually present, including apparently sudden deaths without prior signs of ill-health.

The symptoms with *S. dublin* are similar to those of *S. typhimurium*, with pyrexia, malaise and diarrhoea being common, as well as abortion (Baker *et al.*, 1971). McCaughey *et al.* (1971) found a marked haemoconcentration and decreased leucocyte count in sheep infected experimentally with *S. dublin*.

In Australia four forms of salmonellosis were recognized in sheep: septicaemia, acute, sub-acute and chronic enteric disease (Richards *et al.*, 1993).

Pathological Findings

Aborted fetuses and placentae are usually fresh and no typical macroscopic lesions are present. Histological examinations reveal severe placental damage, with desquamation and necrotic lesions of the chorionic epithelium, associated with leucocytic infiltration and bacterial emboli in the allantochorion. Examination of the caruncles of ewes slaughtered after abortion reveals areas of necrosis, associated with leucocyte demarcation and thrombosis of blood-vessels, and leucocytes within the lumen of the uterine glands.

Within the fetus, specific lesions are not present. Common findings are subcutaneous oedema and excess fluid within the abdominal and thoracic cavities, associated with fibrinous exudate from the serous membranes of the viscera, and haemorrhages within the lungs and epicardium. Histologically, hyperaemia, haemorrhages and oedema are present in many organs.

Post-mortem findings in carcasses of animals that die of salmonellosis are variable. Usually signs of septicaemia are seen, with splenic enlargement, congested organs and, often, acute enteritis (Richards *et al.*, 1993). In animals that have not eaten for several days, the gall-bladder is distended and the liver swollen and friable. In acute disease, there is severe abomasitis and enteritis, with the associated lymph nodes being grossly enlarged. The intestinal contents tend to be very fluid and inflammatory changes may also be detected in the caecum and colon.

Richards *et al.* (1993) found that, in sub-acute disease, the ileum and upper large intestine were involved and, in chronic disease, there was considerable mucosal repair in the ileum, caecum and proximal colon.

The histological changes in naturally infected sheep varied with the severity of the disease (Richards *et al.*, 1993). In acute cases, there was intense focal to diffuse haemorrhagic necrotizing inflammation of the mucosa and neutrophil infiltration. In chronic cases, the predominant lesion was focal nodular hyperplasia of the mucosa. The major visceral organs were severely congested and there was widespread evidence of disseminated intravascular coagulation, accompanied by rupture of capillaries and small vessels. Focal necrosis, with microgranuloma formation, was a common finding in the liver. Similar findings were recorded by Orr *et al.* (1977) in sheep that were infected experimentally with *S. typhimurium* and killed over a period of 21 days.

Pathogenesis

Natural infection is most probably by the oral route and the subsequent course of events is probably similar to that described for cattle (see Chapter 10).

Experimental infections with *S. abortusovis* have utilized a number of different routes and, on many occasions, disease has not been reproduced. Pardon *et al.* (1980) reported on the unreliability of intragastric infections and found that the subcutaneous route gave more consistent results. With doses varying between 2.5×10^7 and 10^{10} colony-forming units (cfu), they produced abortion in seven of 11 pregnant ewes infected between the 64th and 84th days of pregnancy.

Sanchis *et al.* (1991) infected 20 pregnant ewes intraconjunctivally; eight aborted and another four gave birth to stillborn lambs. There is an apparent effect of stage of gestation on the response of pregnant sheep to *S. abortusovis*. Sanchis and Pardon (1984a), using a subcutaneous inoculum of 10^{10} organisms, showed that ewes infected 37 days before mating did not abort in the subsequent pregnancy, one of nine infected 33 days after service aborted and eight of ten infected at 89 days' gestation aborted. The average time between infection and abortion in the last group was 19 days.

Abortion in pregnant ewes was successfully reproduced by oral dosing with *S. montevideo* (Linklater, 1985), who found that abortions occurred 12–19 days after infection. Apart from abortion, the ewes showed little signs of ill-health. Similarly, live lambs born to infected ewes and themselves excreting showed few signs of ill-health. Likewise, Linklater found that those infected at 12–14 weeks' gestation aborted, while those infected later with the same dose of organisms did not. He concluded that the syndrome produced by *S. montevideo* is therefore akin to that produced by *S. abortusovis*.

Brown *et al.* (1976) infected 6- to 9-month-old lambs with *S. typhimurium* (DT32) and slaughtered them serially; *Salmonella* were recovered from the blood between the third and eighth day after infection. Bacteriological recovery at slaughter correlated with the severity of the lesions. Systemic infection was confirmed in all lambs slaughtered up to 9 days post-infection. In two lambs killed on days 11 and 21, respectively, *Salmonella* were isolated only from the posterior mesenteric lymph node, and *Salmonella* were not isolated from two lambs killed at 3 months.

Linklater (1985) found that an oral dose of 10^9 cfu *S. typhimurium* produced an acute systemic disease, characterized by fever, anorexia, dullness and a profuse diarrhoea, similar to that described in a field outbreak (Hunter *et al.*, 1976). Death occurred 6.2 (± 1.9) days after infection and the period of faecal excretion was relatively short 13.9 (± 4.9) days and none of the animals excreted for more than a month. Therefore sheep do not appear to become carriers following recovery from oral infection. More prolonged faecal excretion has been described when other routes have been used for *S. typhimurium* infection, e.g. intravenous (Josland, 1953) and intranasal (Tannock and Smith, 1971).

Linklater (1985) found that *S. typhimurium* infections were more severe in pregnant ewes than in non-pregnant ones, and abortion was produced 5–9 days later by doses of organisms that were not lethal to non-pregnant animals. Similar finding were reported by McCaughey *et al.* (1971), who infected pregnant ewes with *S. dublin* and caused abortion 4–7 days after infection. Similar clinical responses were reported by Thomas and Harbourne (1974). Thus it would appear that *S. typhimurium* and *S. dublin* produced a similar severe febrile response in sheep, with abortion and septicaemia being common signs.

In contrast, S. *infantis* infections of sheep did not cause a bacteraemia and the organism was isolated from the intestine and mesenteric lymph nodes after challenge (Brown *et al.*, 1977).

Diagnosis

Confirmation of diagnosis depends on the isolation of the causal organisms. In the case of abortions, typical Gram-negative organisms are seen in direct smears made from stomach contents of fetuses and from placentae. These can be confirmed as *Salmonella* by direct overnight culture on selective agars, such as deoxycholate citrate agar, MacConkey agar, etc. (see Chapter 21). Typical non-lactose-fermenting colonies are present in profuse numbers and it should not be necessary to resort to enrichment media if salmonellae are the cause of abortion. However, Thomas and Harbourne (1974) commented on their failure to isolate S. *dublin* from mummified fetuses from experimentally infected sheep. Occasional problems may occur with S. *arizonae*, because the organism is a slow lactose fermenter and it belongs to the O61 group, which may not be included in the standard polyvalent 'O' typing sera. S. *abortusovis* grows more slowly than other serovars and may require up to 72 h for colonies to reach a significant size. The literature is confusing on the best media for isolation of S. *abortusovis* and Jack (1968) found selenite broth inhibitory, as were some of the selective media.

In view of the uncertainties, routine cultures of fetuses should be made on blood agar and placental culture direct on MacConkey agar. Pardon *et al.* (1990) used *Salmonella–Shigella* agar during their studies on S. *abortusovis*. *Salmonella* may also be isolated from vaginal discharges in the period immediately following abortion.

In enteric and septicaemic cases, isolation of the causal organism should be possible by direct culture of organs, faeces and intestinal lymph nodes. Enrichment broths are helpful for faeces and intestinal contents and in cases where antibiotics may have been used. In chronic enteric infections, the predilection site for infection is the posterior mesenteric lymph node, and

it may be possible to recover *Salmonella* from it after faecal excretion has ceased. Linklater (1985) isolated S. *montevideo* from the posterior mesenteric lymph node of a lamb 20 days after being born to an experimentally infected dam.

Serological Diagnosis

A serum agglutination test has been used for detecting ewes infected with S. *abortusovis*, but it is advisable to sample several sheep soon after abortion, because antibody levels usually drop within the 2–3 months of parturition. Somatic titres of 1/40 and above and flagellar titres of 1/320 and above are considered to indicate exposure. However, many apparently healthy sheep have antibodies, especially flagellar agglutinins, to common *Salmonella* serovars and Sojka *et al.* (1977) found that 17% of 882 samples gave a positive titre of 1/320 to the H 'c' antigen of S. *abortusovis*. Since many apparently healthy sheep have flagellar agglutinins to *Salmonella* at diagnostic titres, it is considered that in adult sheep interpretation of the results as positive should be based on both the somatic and the flagellar titres. Ivanov *et al.* (1972) found that the indirect haemagglutination test was more sensitive than either the serum agglutination or the complement fixation test for the diagnosis of S. *abortusovis* infection.

A microagglutination test was developed by Sanchis and Abadie (1985) and later a fast agglutination test with a rose bengal antigen (Sanchis and Abadie, 1990). They found that the rose bengal test appeared to be more sensitive and able to detect seroagglutination for a longer period than the microagglutination test. Other tests that have been developed include a milk-ring test (Kartashova, 1969) and a coloured antigen test analogous to the Castaneda Brucella test (Dhawedkar and Dimov, 1968).

Thomas and Harbourne (1974) adapted the stained pullorum test to detect antibodies in the milk of S. *dublin*-infected sheep, but found that the tests did not give clear-cut results, and many of the milk samples were positive during the first fortnight of lactation – probably reducing its value as a screening test (Arkhangel'skii and Kartashova, 1962).

Allergic tests have also been used and, 2–3 months after abortion, the number of positive skin reactions was greater than the number of positive serum results (Arkhangel'skii *et al.*, 1980).

Treatment

Usually, by the time salmonellosis is detected in a flock, the infection has spread throughout the group. Nevertheless, isolation of infected animals is worthwhile to try to limit the spread of the infection.

Animals that are septicaemic have to be treated with suitable parenteral antibiotics and those that scour may require supportive therapy with electrolyte solutions. Aborting ewes frequently develop post-parturient metritis and injections of long-acting preparations of antibiotics at the time of abortion will help to prevent this. Attempts have also been made to minimize losses by the administration of antibiotics in feed as soon as the causal *Salmonella* is identified. Results for these treatments are equivocal and evidence for their efficacy is mostly circumstantial.

Jack (1968) used furazolidone to treat ewes experimentally and naturally infected with *S. abortusovis*, but found no benefit. Likewise, Baker *et al.* (1971) and Gitter and Sojka (1970) found that antibiotic therapy before lambing failed to control *S. dublin* infection. Spence and Westwood (1978) considered antimicrobial therapy ineffective for ewes but of some benefit for the lambs during an outbreak of *S. agona*. On the other hand, Saunders *et al.* (1966) expressed their personal opinion that antibiotic therapy apparently assisted recovery, but pointed out that no group was left untreated during an outbreak of enteric *S. typhimurium* infection. On the other hand, Hunter *et al.* (1976) found antimicrobial therapy ineffective during an outbreak of *S. typhimurium* and considered that vaccination was the only apparently successful control measure.

Immunity

There does not appear to have been a systematic study of the immune response of sheep to *Salmonella*. Most of the available information results from studies on the development of vaccines against different *Salmonella* serovars.

Begg *et al.* (1990) immunized sheep with a live Aro⁻ *S. typhimurium* and found that only those that had been inoculated intramuscularly developed high levels of serum antibodies and significant delayed cutaneous hypersensitivity. In contrast, those immunized orally gave minimal responses. Most of the antibody in the intestine appears to be immunoglobulin M (IgM), with some IgG_1 and IgG_2 and an absence of IgA (Lascelles *et al.*, 1988; Brennan *et al.*, 1995; Mukkur *et al.*, 1995). On the other hand, Husband (1978, 1980), using the intraperitoneal route, detected intestinal IgA in lambs and protection against *Salmonella* infection.

Although delayed cutaneous hypersensitivity was observed following vaccination by Begg *et al.* (1990), Mukkur *et al.* (1995) and Brennan and Baird (1994), the latter (Brennan *et al.*, 1995) were unable to detect a marked T-lymphocyte response, and cultured purified cells responded very weakly to *Salmonella* antigen and failed to produce interferon-gamma (INF-γ) and interleukins 2 and 4.

Studies with *S. abortusovis* suggest that the immunological response is similar to that of *S. typhimurium*. Following intraconjunctival infection with *S. abortusovis*, the sheep produced agglutinins to both O and H antigens. Later studies, which used *in vitro* culture of lymphocytes (Berthon *et al.*, 1994), found the antibodies to be mainly IgM and IgG_1, with IgG_2 production occurring later.

Vaccination

Parenteral killed vaccines have been used for many years to protect sheep against *Salmonella* infections. Much of the early work was done in New Zealand by Josland (1954), who concluded that vaccination appeared to offer some protection against experimental challenge with *S. typhimurium*, although the number of vaccinated animals that showed resistance did not differ significantly from the controls. Field trials done by Wallace and Murch (1967) and by Beckett (1967), however, were more successful, with fewer deaths in the vaccinates than in the controls. Rudge *et al.* (1968) claimed that sheep vaccinated with a bivalent preparation that contained the antigen of both *S. typhimurium* and *S. bovismorbificans* showed significantly less bacteraemia, severe scouring and mortality rates than unvaccinated controls. These findings were confirmed by Davis (1969) after a series of field trials on eight farms.

The use of vaccines has been described since *S. abortusovis* was first isolated, but the fact that the disease subsides spontaneously in the years after an outbreak may have resulted in false claims for the use of some of these vaccines.

Subcutaneously applied live *S. abortusovis* vaccine has been used to prevent abortions in ewes in France (Sanchis and Pardon, 1984b; Pardon *et al.*, 1990). The vaccine strain RV6 represents a non-dependent reverse mutant selected from a streptomycin-dependent strain of *S. abortusovis*. One injection of RV6 protects ewes at least as much as two injections of killed adjuvant vaccine. Such subcutaneous vaccinal injections have been administered to thousands of pregnant ewes in different areas of France. Innocuity and efficacy have been tested in field conditions (Sanchis and Pardon, 1984b).

Linde *et al.* (1992) produced an attenuated live *S. typhimurium* vaccine by means of metabolic drift mutations, and found it gave significant protection against *S. abortusovis* in two large field trials. The mortality rates in the controls were 30%, in those given killed vaccine 11% and only 0.1% in those given the live vaccine.

Robinson (1970) reviewed the work on vaccines and concluded that natural infection or live vaccines gave more solid immunity than killed parenteral vaccines. Jelinek *et al.* (1982) used a live *gal*E mutant of *S. typhimurium* to protect sheep against oral challenge with 2×10^{11}

S. typhimurium. The vaccine was given by both the subcutaneous route (10^9 organisms) and orally for 4 days, with doses ranging from 2×10^{10} to 2×10^{12}. The mortality rates in the vaccinates were 0/6 and 1/15, as compared with 6/9 of the unvaccinated controls.

*Aro*A mutants have also been used experimentally in sheep (Mukkur *et al.*, 1987; Begg *et al.*, 1990; Mukkur and Walker, 1992). They found that better protection was obtained by giving two intramuscular or two subcutaneous doses. Oral dosing gave protection against mortalities but the reduction in the prevalence of severe diarrhoea was less consistent. Mukkur *et al.* (1991) also produced an *Aro*A *S. havana* vaccine, which reduced carriage of the challenge strains as compared with the controls. Mukkur and Walker (1992) found that parenteral immunization gave protection within 7 days, whereas protection after oral immunization was not evident until 3 weeks after immunization. Protection persisted for 6 months but not for 12 months.

Control and Public-health Aspects

A major concern is for people working with infected animals and there are many reports of shepherds and farmers becoming infected (e.g. Hunter *et al.*, 1976; Findlay, 1978; Spence and Westwood, 1978). It is also possible that meat, milk and milk products may be contaminated with *Salmonella*, but there have been no recent outbreaks in the UK associated with these vehicles.

References

Allworth, M.B., West, D.M. and Bruere, A.N. (1985) Salmonellosis in ram hoggets following prophylactic zinc dosing. *New Zealand Veterinary Journal* 33, 171.

Angelachov, A. (1964) Some aetiological and epidemiological features of abortion in ewes. *Veterinaria Sbirka, Bulgaria* 64, 3–6.

Arkhangel'skii, I.I. and Kartashova, V.M. (1962) Rapid methods for demonstrating salmonella in milk ring test. *Veterinariya, Moscow* 9, 74–78.

Arkhangel'skii, I.I., Sidorchuk, A.A. and Radzhabov, M.D. (1980) Allergens for the diagnosis of salmonellosis in sheep. *Veterinary Bulletin* 1948.

Baker, J.R., Faull, W.B. and Rankin, J.E.F. (1971) An outbreak of salmonellosis in sheep. *Veterinary Record* 88, 270–277.

Bartmann, E. (1957) Occurrence of *Salmonella abortusovis* in the male and female genital organs of sheep. Inaugural dissertation, Munich.

Beckett, F.W. (1967) The use of Salmonella vaccine in outbreaks of salmonellosis in sheep. *New Zealand Veterinary Journal* 15, 66–69.

Begg, A.P., Walker, K.H., Love, N.V. and Mukkur, T.K.S. (1990) Evaluation of protection against experimental salmonellosis in sheep immunised with live aromatic-dependent *Salmonella typhimurium*. *Research in Veterinary Science* 52, 147–153.

Berthon, P., Gohin, I., Lantier, I. and Oliver, M. (1994) Humoral immune response to *Salmonella abortusovis* in sheep: *in vitro* induction of an antibody synthesis from either sensitised or unprimed lymph nodes cells. *Veterinary Immunology and Immunopathology* 41, 275–294.

Bosworth, T.J. and Glover, R.E. (1925) Contagious abortion in ewes. *Veterinary Journal* 81, 319–334.

Brennan, F.R. and Baird, G.D. (1994) Differences in the immune response of mice and sheep to an aromatic-dependent mutant of *Salmonella typhimurium*. *Journal of Medical Microbiology* 41, 20–28.

Brennan, F.R., Oliver, J. and Baird, G.D. (1995) *In vitro* studies with lymphocytes from sheep orally inoculated with an aromatic-dependent mutant of *Salmonella typhimurium*. *Research in Veterinary Science* 58, 152–157.

Brogden, K.A., Meehan, J.T. and Lehmkuhl, H.D. (1994) *Salmonella arizonae* infection and colonisation of the upper respiratory tract of sheep. *Veterinary Record* 135, 410–411.

Brown, D.D., Ross, J.G. and Smith, A.F.G. (1976) Experimental infection of sheep with *Salmonella typhimurium*. *Research in Veterinary Science* 21, 335–340.

Brown, D.D., Ross, J.G. and Smith, A.F.G. (1977) Experimental infection of sheep with *Salmonella infantis*. *British Veterinary Journal* 133, 435–441.

Bruns, H. and Gasters, (1920) Paratyphus epidemie in einer hammel herde. *Zeitschrift für Hygiene und Infections Krankheiten* 90, 263–280.

Cooper, B.S. (1967) Evaluation of vaccines against salmonellosis in sheep. *New Zealand Veterinary Journal* 15, 215–216.

Davis, G.B. (1969) Field trials with an ovine *Salmonella* vaccine. *New Zealand Veterinary Journal* 17, 62–64.

Dhawedkar, R. and Dimov, I. (1968) Antigen fixation colour test for the serological diagnosis of *Salmonella abortusovis*. *Veterinary Bulletin* 1970941.

Fincham, I. (1961) Salmonellosis in sheep. *State Veterinary Journal* 16, 87–94.

Findlay, C.R. (1978) Epidemiological aspects of an outbreak of salmonellosis in sheep. *Veterinary Record* 103, 114–115.

Frickenger, H. (1919) Fleischvergiftungsepidemie in Anschluß an eine Paratyphuserkrankung beim Schaf. *Zeitschrift für Fleisch und Milch Hygiene* 29, 346–351.

Gardiner, M.R. and Craig, J. (1970) Factors affecting survival in the transportation of sheep by sea in the tropics and subtropics. *Australian Veterinary Journal* 46, 65–69.

Gitter, M. and Sojka, W.J. (1970) *Salmonella dublin* abortion in sheep. *Veterinary Record* 87, 775–778.

Hannam, D.R., Wray, C. and Harbourne, J.F. (1986) Experimental *Salmonella arizonae* infection of sheep. *British Veterinary Journal* 142, 458–466.

Higgs, A.R.B., Norris, R.T. and Richards, R.B. (1993) Epidemiology of salmonellosis in the live sheep export industry. *Australian Veterinary Journal* 70, 330–335.

Hunter, A.G., Corrigall, W., Mathieson, A.O. and Scott, J.A. (1976) An outbreak of *Salmonella typhimurium* in sheep and its consequences. *Veterinary Record* 98, 126–130.

Husband, A.J. (1978) An immunisation model for the control of infectious enteritis. *Research in Veterinary Science* 25, 173–177.

Husband, A.J. (1980) Intestinal immunity following a single intraperitoneal immunisation in lambs. *Veterinary Immunology and Immunopathology* 1, 277–286.

Ivanov, I., Hristoforou, L., Surtmadjieu, K., Stamenov, B. and Salvkov, I. (1966) Abortion in farm animals. *Academy of Agricultural Sciences, Bulgaria* 55–69.

Ivanov. I., Massalski, N. and Divoka, T. (1972) The use of serological reactions in the diagnostics of *Salmonella abortusovis* infections in sheep. *Veterinary Science: Sofia* 9, 13–15.

Jack, E.J. (1968) *Salmonella abortusovis*: an atypical *Salmonella*. *Veterinary Record* 82, 558–561.

Jacob, W.K. (1959) Breeding hygiene in sheep: examination of flocks infected with *Salmonella abortusovis*. *Berliner Munchener Tierarztliche Wochenschrift* 72, 475–479.

Jelinek, P.D., Robertson, G.M. and Millar, C. (1982) Vaccination of sheep with a live 'Gal E' mutant of *Salmonella typhimurium*. *Australian Veterinary Journal* 59, 31–32.

Josland, S.W. (1953) Observations on the aetiology of bovine and ovine salmonellosis in New Zealand. *New Zealand Veterinary Journal* 1, 131–136.

Josland, S.W. (1954) The immunogenic properties of *Salmonella typhimurium* in sheep. *New Zealand Veterinary Journal* 2, 2–7.

Kartashova, V.M. (1969) Diagnosis of salmonellosis in sheep by the ring test on milk. *Veterinariya Moscow* 9, 106–108. (In *Veterinary Bulletin* (1970), 2705.)

Lascelles, A.K., Beh, K.J., Mukkur, R.K.S. and Willis, G. (1988) Immune response of sheep to oral and subcutaneous administration of live aromatic-dependent mutant of *Salmonella typhimurium*. *Veterinary Immunology and Immunopathology* 18, 259–267.

Lesbouyries, M.M., Dadot and Berthelon (1933) Avortement paratyphique de la brebis. *Bulletin de l'Académie Vétérinaire de France* 6, 318–321.

Linde, K., Bondarenko, V. and Sviridenko, V. (1992) Prophylaxis of *Salmonella abortusovis*-induced abortion of sheep by a *Salmonella typhimurium* live vaccine. *Vaccine* 10, 337–340.

Linklater, K.A. (1983) Abortion in sheep associated with *Salmonella montevideo* infection. *Veterinary Record* 112, 372–374.

Linklater, K.A. (1985) Studies on the pathogenesis of salmonellosis in sheep. Fellowship thesis, Royal College of Veterinary Surgeons.

Lovell, R. (1931) A member of the *Salmonella* group causing abortions in sheep. *Journal of Pathology and Bacteriology* 34, 13–22.

McCaughey, W.J., Kavanagh, P.J. and McClelland, T.G. (1971) Experimental *Salmonella dublin* infection in sheep. *British Veterinary Journal* 127, 557–566.

Manley, F.H. (1932) Contagious abortion in sheep and goats in Cyprus. *Journal of Comparative Pathology and Therapeutics* 45, 293–300.

Meehan, J.T., Brogden, K.A., Courtney, C., Cutlip, R.C. and Lehmkuhl, H.D. (1992) Chronic proliferative rhinitis associated with *Salmonella arizonae* in sheep. *Veterinary Pathology* 29, 556–559.

Mukkur, T.K.S. and Walker, K.H. (1992) Development and duration of protection against salmonellosis in mice and sheep immunised with live aromatic-dependent *Salmonella typhimurium*. *Research in Veterinary Science* 52, 147–153.

Mukkur, T.K.S., MacDowell, G.H., Stocker, B.A.D. and Lascells, A.K. (1987) Protection against experimental salmonellosis in mice and sheep by immunisation with aromatic-dependent *Salmonella typhimurium*. *Journal of Medical Microbiology* 24, 11–19.

Mukkur, T.K.S., Walker, K.H. and Stocker,. B.A.D. (1991) Generation of aromatic-dependent *Salmonella havana* and evaluation of its immunogenic potential in mice and sheep. *Veterinary Microbiology* 29, 181–194.

Mukkur, T.K.S., Walker, K.H., Baker, P. and Jones, D. (1995) Systemic and mucosal intestinal antibody response to sheep immunised with aromatic-dependent live or killed *Salmonella typhimurium*. *Comparative Immunology, Microbiology and Infectious Diseases* 18, 27–39.

Mura, D. and Contini, A. (1954) Sorgenti e vie di transmissione dell' infezione abortovine e caprino da *Salmonella*: importanza del maschio reproduttore. *Veterinaria Italiana* 5, 788–802.

Newsome, I.E. and Cross, F. (1924) An outbreak of paratyphoid dysentery in lambs. *Journal of American Veterinary Medicine Association* 66, 289–300.

Newsome, I.E. and Cross, F. (1930) Paratyphoid dysentery in lambs again. *Journal of American Veterinary Medicine Association* 76, 91–92.

Orr, M.B., Hunter, A.R., Brown, D.D. and Smith, A.F.G. (1977) The histopathology of experimental infections of sheep with *Salmonella typhimurium*. *Veterinary Science Communication* 1, 191–195.

Pardon, P., Sanchis, R. and Marly, J. (1980) Experimental *Salmonella abortusovis* infection in ewes. *Veterinary Record* 106, 389–390.

Pardon, P., Sanchis, R., Marly, J., Lantier, F., Guilloteau, L., Buzoni, G.D., Oswald, I.P., Pepin, M., Kaeffer, B., Berthon, P. and Popoff, M.Y. (1990) Experimental ovine salmonellosis (*Salmonella abortusovis*): pathogenesis and vaccination. *Research in Microbiology* 141, 945–953.

Reilly, W.J., Old, D.C., Munro, D.S. and Sharp, J.C.M. (1985) An epidemiological study of *Salmonella montevideo* by biotyping. *Journal of Hygiene* 95, 23–28.

Richards, R.S., Norris, R.T. and Higgs, A.R.B. (1993) Distribution of lesions in ovine salmonellosis. *Australian Veterinary Journal* 70, 326–330.

Robinson, R.A. (1970) *Salmonella* infection diagnosis, vaccination and control. *New Zealand Veterinary Journal* 18, 259–277.

Rudge, J.M., Cooper, B.S. and Jull, D.J. (1968) Testing a bivalent vaccine against experimental infection with *Salmonella bovismorbificans* and *Salmonella typhimurium*. *New Zealand Veterinary Journal* 16, 29–30.

Sanchis, R. and Abadie, G. (1985) Sérodiagnostic de la salmonellose abortive des brebis à *Salmonella abortusovis*: microtechnique de séroagglutination. *Bulletin Laboratoire Vétérinaire I* 19/20, 45–51.

Sanchis, R. and Abadie, G. (1990) Utilisation d'un antigène coloré au rose bengale pour le sérodiagnostic de la salmonellose ovine: comparaison avec la technique de séroagglutination. *Recueil de Médecine Vétérinaire* 166, 431–436.

Sanchis, R. and Pardon, P. (1984a) Infection expérimentale de la brebis par *Salmonella abortusovis*: influence de stade de gestation. *Annales de Recherches Vétérinaires* 15, 97–103.

Sanchis, R. and Pardon, P. (1984b) Trial in contaminated environment of an attenuated living vaccine against *Salmonella abortusovis* abortion in sheep. *Annales de Recherches Vétérinaire* 15, 381–386.

Sanchis, R. and Pardon, P. (1986) Infection expérimentale du bélier par *Salmonella abortusovis*. *Annales de Recherches Vétérinaires* 17, 387–393.

Sanchis, R., Pardon, P. and Abadie, G. (1991) Abortion and serological reaction of ewes after conjunctival instillation of *S. enterica* subsp. *enterica* ser. Abortusovis. *Annales de Recherches Vétérinaires* 22, 59–64.

Saunders, C.N., Kinch, D.A. and Martin, R.D. (1966) An outbreak of *Salmonella typhimurium* infection in sheep. *Veterinary Record* 79, 554–556.

Schermer and Ehrlich, (1921) Weitere Beitrage über die Paratyphuserkrankungen der Haustiere. *Berliner Tierarztliche Wochenschift* 37, 469–473.

Sojka, W.J., Wray, C. and Brand, T.F. (1977) Agglutinins to common *Salmonellae* in the sera of apparently healthy sheep. *British Veterinary Journal* 133, 615–622.

Sojka, W.J., Wray, C., Shreeve, J.E. and Bell, J.C. (1983) The incidence of *Salmonella* infection in sheep in England and Wales, 1975–1981. *British Veterinary Journal* 139, 386–392.

Spence, J.B. and Westwood, A. (1978) *Salmonella agona* infection in sheep. *Veterinary Record* 102, 332–336.

Stanoev, S. and Milev, M. (1967) *Salmonella* carriers in sheep. *Veterinarno-meditsinski Nauki Sofia* 4, 29. (In *Veterinary Bulletin* (1967), 4069.)

Tannock, G.W. and Smith, M.M.B. (1971) A *Salmonella* carrier state of sheep following intranasal inoculation. *Research in Veterinary Science* 12, 371–373.

Thomas, G.W. and Harbourne, J.F. (1974) Experimental *Salmonella dublin* infection in housed sheep. *Veterinary Record* 94, 414–417.

Vodas, K. and Elitsina, P. (1986) On the intensity of the epizootic process in sheep with *Salmonella abortusovis* and *Chlamydia psittaci* var *ovis* infection. *Veterinarno-meditsinski Nauki Sofia* 23, 40–46.

Wallace, G.V. and Murch, O. (1967) Field trials to assess the value of a bivalent killed *Salmonella* vaccine in the control of ovine salmonellosis. *New Zealand Veterinary Journal* 15, 62–65.

Chapter 13
Salmonella in Horses

John K. House and Bradford P. Smith

Department of Medicine and Epidemiology, School of Veterinary Medicine,
University of California – Davis, Davis, CA 95616, USA

Historical Perspective

Salmonella abortusequi-induced abortion was first described by Kilborne and Smith in 1893 (Kilborne, 1893; Smith, 1893). *Salmonella typhimurium* was subsequently recognized as a cause of colitis in 1919 (Graham *et al.*, 1919). By the 1950s, *S. abortusequi* had virtually disappeared from the USA; however, there are a number of reports of *S. abortusequi*-associated abortions in horses from Italy (Condoleo *et al.*, 1983), India (Garg and Manchanda, 1986) and Albania (Kuka, 1989). Internationally, reports of the prevalence of *S. typhimurium* and other *Salmonella* serovars have progressively increased. The initial report of *S. typhimurium* by Graham described outbreaks of disease in horses undergoing cross-country rail transportation (Graham *et al.*, 1919). Many of the risk factors and characteristics of equine salmonellosis have been further elucidated by descriptions of disease on breeding farms and in hospitalized horses (Baker, 1969; Ingram and Edwards, 1980; Roberts and O'Boyle, 1981; Hird *et al.*, 1984; Ikeda *et al.*, 1986; Madigan *et al.*, 1992; van Duijkeren *et al.*, 1994). The *Salmonella* serovars isolated from clinically affected horses are also commonly associated with disease in other livestock species. Other than *S. abortusequi*, there does not appear to be an equine-adapted serovar that causes persistent infections in horses. Colitis remains the commonest clinical manifestation of salmonellosis in adult horses. Septicaemia, with or without colitis, is most commonly seen in foals.

Epidemiology

The ability of healthy horses to resist *Salmonella* infection and the proclivity for clinical disease to occur in the immunologically compromised and naïve host (foals) illustrate the opportunistic nature of the pathogen. Equine salmonellosis is most commonly observed on breeding farms and in veterinary hospitals (Morse *et al.*, 1976b; Carter *et al.*, 1979, 1986; Roberts and O'Boyle, 1981; Kikuchi *et al.*, 1982; Hird *et al.*, 1984, 1986; Faulstich, 1987; Rumschlag and Boyce, 1987; Begg *et al.*, 1988; Powell *et al.*, 1988; Castor *et al.*, 1989; Traub-Dargatz *et al.*, 1990; Walker *et al.*, 1991; Hansen *et al.*, 1993; van Duijkeren *et al.*, 1994; Pare *et al.*, 1996). In other environments, the common themes to most of the reported outbreaks of disease in adult horses are congregation of animals (high stocking density) and alteration of gastrointestinal function by feed and/or water deprivation, disease or medication (Cordy and Davis, 1946; Owen *et al.*, 1983; McClintock and Begg, 1990). Assuming there has been no acquired immunity from prior exposure, resistance to *Salmonella* is largely dependent on intact innate immunity and the defence provided by the normal intestinal bacterial flora.

The protective effect of the gut flora is in part mediated by the inhibitory effects of the volatile fatty acids (VFAs) and the low oxidation and reduction potentials that they produce in the gut (Meynell, 1963). The inhibitory effect of

VFAs is enhanced at low pH, when there is a higher proportion of undissociated VFAs present (Wolin, 1969; Chambers and Lysons, 1979). The stronger acids (lactic) are less toxic than the weak acids (acetic, propionic, butyric). Undissociated molecules of acidic substances can penetrate the cell membrane of Salmonella more readily than the corresponding ions. Strong acids have a higher pK_a and are thus more dissociated than weak acids (Chiu, 1974). At relatively low pH values, weak acids, which exist to a considerable extent in undissociated form, can enter cells and damage them by changing the internal pH, whereas a strong acid cannot (Chiu, 1974).

Anorexia, antimicrobial administration, intestinal surgery and marked changes in diet increase the susceptibility of horses to Salmonella challenge (Hird et al., 1984, 1986; Carter et al., 1986; Traub-Dargatz et al., 1990). Changes in the quality of substrate ingested have profound effects on the bacterial and protozoal populations in the large bowel (Argenzio, 1975; Clarke et al., 1990). Concentrate (grain) feeding is associated with an increase in bacterial counts and a reduction in the protozoal counts and pH in the caecum (Goodson et al., 1988). Acetic, propionic and butyric acids are the predominant end-products of carbohydrate digestion, with less lactic and succinic acid present. Increasing concentrations of lactic acid are found when there is an excess of rapidly fermentable carbohydrate (Argenzio et al., 1974; Argenzio, 1975). Reduction of motility favours the growth of anaerobes, such as Clostridia, and facultative anaerobes, such as Salmonella (Linerode and Goode, 1970). Feeding following a period of fasting is associated with an increased proportion of soluble carbohydrate delivered to the caecum, thus promoting rapid fermentation and lactate production (Breukink, 1974).

Salmonella is acquired most commonly via oral intake, involving faecal–oral contamination or contaminated feed. Many Salmonella infections are not associated with clinical manifestations of disease (McCain and Powell, 1990). Prevalence surveys based on faecal and tissue Salmonella cultures report between 0 and 70% of normal healthy horses are infected with Salmonella (Table 13.1; Smith et al., 1978; Gibbons, 1980; Roberts and O'Boyle, 1981; Begg et al., 1988; McCain and Powell, 1990).

Many of these prevalence surveys were carried out in veterinary hospitals and slaughter facilities. Interpretation of the survey results is confounded by lack of data to distinguish recently acquired infections from chronic infections. A recent survey of horses in a veterinary hospital, using a Salmonella-specific polymerase chain reaction (PCR), found evidence of Salmonella in faeces of 64.5% of hospitalized horses (Cohen et al., 1996). Lack of knowledge regarding the identity and relatedness of the serovars involved limits the interpretation of this study. Comparative surveys of hospitalized and naturally housed horses indicate a much lower prevalence of Salmonella shedding by horses housed in their normal environment (Roberts and O'Boyle, 1981; Begg et al., 1988). When all the surveys are considered, the prevalence of Salmonella in asymptomatic horses appears to reflect the level of Salmonella contamination in which the horses are placed (Table 13.1).

In experimental-challenge studies, doses of 10^6–10^{11} colony-forming units (cfu) were required to induce disease (Owen et al., 1979; Smith et al., 1979; Roberts and O'Boyle, 1982). Fasting horses for 48 h prior to challenge was necessary to produce a consistent model of disease. Clinical signs were dependent on the challenge dose, virulence of the organism and equine management. A small oral dose of S. typhimurium (10^6) caused fever and depression, a larger dose (10^9) caused fever and colitis and a still larger dose (10^{11}) caused bacteraemia and death (Smith et al., 1979; Roberts and O'Boyle, 1982). This same range of clinical signs is seen with naturally occurring disease.

A high proportion of animal and vegetable protein supplements used in feed formulations in the USA (and probably elsewhere) contain small numbers of Salmonella (Davies, 1986; McChesney et al., 1995). Other sources of Salmonella for horses include people, other horses, wild and domestic animals, water and contaminated equipment or environment. There are numerous reports suggesting that equine Salmonella carriers are a common source of infection for other horses (Baker, 1969, 1970; Begg et al., 1988; McCain and Powell, 1990), but only a few documented cases of chronic Salmonella infections have been described in the horse (Bryans et al., 1961; Binde et al., 1983). Following experimental Salmonella challenge, transportation and antimicrobial therapy exacer-

Table 13.1. Salmonella prevalence surveys of horse populations.

Author	Sample cultured	Number sampled	Positive(%)	Serovars	Location
Kuhlmann, 1964	Healthy horses	1727	0 (0)		Slaughterhouse
	Diseased horses	3832	15 (0.04)	S. dublin (15), S. typhimurium (4), S. anatum (1)	Slaughterhouse
Mann et al., 1964	Liver and mesenteric lymph node	250	5 (0.02)	S. anatum (5)	Slaughterhouse
Baker, 1970	Faeces	201 (horses)	3 (1.49)	S. rostock (1), S. typhimurium (1), S. virchow (1)	Veterinary hospital
Quevedo et al., 1973a	Faeces	100 (horses)	27 (27)	S. good (16), S. montevideo (3), S. vaertan (3), S. derby (3), S. anatum (2), S. hato (2), S. meleagridis (1)	Slaughterhouse
Quevedo et al., 1973b	Sparrows	206 (sparrows)	21 (10.2)	S. good (9), S. montevideo (8), S. newport (4), S. typhimurium (2)	Same slaughter facility as above
Smith et al., 1978	Faeces	1451	46 (3.2)*	S. agona (15), S. anatum (14), S. typhimurium (7), S. typhimurium var. Copenhagen (4), S. infantis (2), S. montevideo (1), S. meleagridis (1), S. drypool (1)	Veterinary hospital
Smith et al., 1978	Faeces	78	0 (0)		Brood-mare farm
Roberts and O'Boyle, 1981	Faeces	462 (horses)	110 (23.8)	Reported as per cent of isolates, not as number of horses; S. anatum (54%), S. ohio (11.27%), S. typhimurium (9.09%); 21 serovars isolated	Veterinary hospital
Roberts and O'Boyle, 1981	Hospital drains	143 (drains)	48 (33.56)	S. anatum (40), S. ohio (4), S. typhimurium (2); ten serovars isolated	Same veterinary hospital as above
Roberts and O'Boyle, 1981	Faeces	600	8 (1.3)	S. typhimurium (5), S. chester (2), S. saintpaul, S. anatum (3). (one from veterinary hospital); two dual infections	University farm, stud farm, training yards, spelling farm and police stables
Palmer et al., 1985a	Faeces	100 (horses)	13 (13)	S. senftenberg (9), S. typhimurium (4), S. london (2), S. agona (1)	Veterinary hospital
Begg et al., 1988	Faeces	250	7 (2.8)	S. typhimurium (4), S. anatum (2), S. tennessee (1)	Veterinary hospital
Begg et al., 1988	Faeces	75	0 (0)		Thoroughbred farm
McCain and Powell, 1990	Mesenteric lymph nodes	70	50 (71.4)	S. albany (13), S. anatum (5), S. reading (4); there were three or fewer isolates of the remaining 12 serovars	Slaughterhouse

*20 horses on admission.

bated the severity of clinical disease and increased *Salmonella* shedding (Owen *et al.*, 1983). The extrapolation of this work is the suggestion that clinical salmonellosis subsequent to antimicrobial therapy, abdominal surgery, anthelmintics and purgatives reflects recrudescence of chronic latent *Salmonella* infections (Begg *et al.*, 1988). In a 5-year study of 1931 horses hospitalized in an intensive-care facility (Mainar-Jaime *et al.*, 1998), these phenomena were not observed, suggesting that recrudescence of latent infections following severe stress is a rare event. *Salmonella* infections of compromised hospitalized horses were most frequently nosocomial (House *et al.*, 1999). Interestingly, in this study nosocomial *Salmonella* did not appear to have an impact on mortality rate of *Salmonella*-infected hospitalized horses (Mainar-Jaime *et al.*, 1998).

Although foals born to mares that are shedding *Salmonella* in their faeces are often infected and may become clinically ill shortly after birth, there are no reports of *Salmonella* shedding in the colostrum and milk of infected mares. *Salmonella* exposure may result from faecal contamination at birth or during the immediate peripartum period as the foal searches for the udder. Mares shedding *Salmonella* at and following parturition are usually asymptomatic (Kikuchi *et al.*, 1982; Binde *et al.*, 1983; Walker *et al.*, 1991). Neonates are more susceptible than adult horses, probably due to their incompletely developed immune system and relative lack of competitive gut flora.

Because veterinary hospitals are required to treat horses with salmonellosis and concurrently house high-risk patients, an effective infectious disease-control programme is essential. Drainage systems may function as reservoirs of environmental *Salmonella* contamination in stables and veterinary hospitals. The blockage of drains and reflux of effluent represent a high risk of exposure to *Salmonella*. Facility design has a great impact on sanitation efforts and should be considered prior to building veterinary facilities (Tillotson *et al.*, 1997). Horses with colic that undergo abdominal surgery suffer significant stress, e.g. the large colon or caecum may be emptied and lavaged, feed is often withheld for 24–48 h or they are treated with antimicrobials, undergo anaesthesia and experience different degrees of ileus. These events contribute to producing a patient that is extremely susceptible to develop-

ing salmonellosis when exposed to even very small numbers of virulent organisms. In particular, horses with large colon impactions appear to be at greater risk of acquiring nosocomial *Salmonella* infections in a veterinary-hospital environment (House *et al.*, 1999). The frequency of *Salmonella* outbreaks and *Salmonella* shedding by asymptomatic horses tends to increase in hot weather (Smith *et al.*, 1978; Carter *et al.*, 1986). Presumably heat stress impairs host immunity, while at the same time the reduced bacterial generation time increases the environmental challenge dose. In another study, high ambient temperatures were a risk factor for nosocomial *Salmonella* infections in hospitalized horses (House *et al.*, 1999).

Pathogenicity and Prevalence of Serovars

Certain serovars show up year after year as the most frequent isolates from ill horses in the USA. *S. typhimurium* (including var. Copenhagen) remains the most frequently identified serovar from horses having clinical signs of illness in the USA and the UK (Ferris and Miller, 1989; C. Wray, Weybridge, UK, personal communication, 1997). In 1996, *S. typhimurium* was also the most frequently reported serovar from horses in France, followed by group C *Salmonella*. *S. bovismorbificans* and *S. dublin* were each isolated from one or two horses. In Northern Ireland, *S. typhimurium* was isolated from six horses and *S. dublin* from three horses. Reports from Belgium recorded 28 equine *Salmonella* isolates in 1996, including 20 *S. typhimurium*, five *S. enteritidis*, and one each of *S. bovismorbificans*, *S. hadar* and *S. saintpaul*. Many other group B, C and E *Salmonella* serovars have been associated with equine diarrhoea. While many of the 2400-plus *Salmonella* serovars may cause or contribute to clinical illness, *S. typhimurium* (including var. Copenhagen) is generally recognized to be one of the most virulent serovars for horses and is also the most frequently isolated serovar from a wide range of animal species. Many *S. typhimurium* isolates contain a virulence plasmid, which facilitates intracellular survival (Woodward *et al.*, 1989).

Likewise, *Salmonella* isolates in veterinary hospitals contain resistance (R) plasmids, which provide resistance to antimicrobial drugs, making treatment more difficult and expensive (Sato *et al.*, 1984; Ikeda and Hirsh, 1985; Donahue, 1986; Hartmann and West, 1995). In the veterinary hospital environment, R-plasmid-mediated antimicrobial resistance may be passed between *Salmonella* serovars and between different isolates of the same serovar (Ikeda and Hirsh, 1985). This has major clinical implications because resistance to an antimicrobial drug, that has never been used in the clinic may occur. *S. typhimurium* DT104 is a phage type that has recently emerged having chromosomal resistance to five antimicrobials and marked virulence for humans and animals. *S. typhimurium* DT104 was isolated 46 times from horses in the UK in 1996, out of a total of 124 equine *Salmonella* isolates.

Clinical Signs

Foals less than 2 months of age infected with *Salmonella* are likely to become bacteraemic and to exhibit a rapid, weak pulse, and they may either die acutely from endotoxic shock or develop focal organ, bone and joint infections, in addition to experiencing dehydration and electrolyte imbalances, associated with diarrhoea. Focal infections are most common in growth plates (physes) and joints, resulting in lameness (Morgan *et al.*, 1974; Goedegebuure *et al.*, 1980; Poyade-Alvarado and Marcoux, 1993). Bacterial endocarditis, septic renal infarcts and meningoencephalitis may also occur (Stuart *et al.*, 1973). In older horses, the commonest clinical signs of *Salmonella* infection are fever, diarrhoea, anorexia and marked leucopenia. Infection with less virulent serovars or in more resistant horses may result in no clinical signs or mild signs, such as fever and anorexia. Diarrhoea associated with *Salmonella* is due to colitis and typhlitis (caecal involvement). Hind-gut fermenters, such as horses, elephants and rabbits, have a huge colonic surface area, and colitis caused by *Salmonella* can result in massive, rapid fluid and electrolyte losses. As a result, *Salmonella* colitis in horses has a higher mortality rate than salmonellosis in simple-stomached animals and ruminants (neither have a colon or caecum with a surface area approaching that of a horse colon). In the early stages of colitis, there may be cramping and fluid

distension of the bowel, resulting in mild to moderate signs of colic (often before the onset of diarrhoea). It is important for the clinician examining the horse with colic to consider impending colitis as a cause. Horses with colic that have a fever and reddened rectal mucosa should be considered suspect *Salmonella* cases and treated accordingly (Dorn *et al.*, 1975; Smith, 1979; Smith *et al.*, 1979). The diarrhoea typically associated with salmonellosis contains protein and fibrin and has a resulting foul odour. Faecal colour usually remains fairly normal, although it may appear brown or bloody and the consistency may approach liquid. Loss of mucosal integrity predisposes to systemic absorption of endotoxin, and the resulting endotoxaemia induces a depressed mental state, anorexia, discoloured mucous membranes, poor capillary refill, rapid heart rate, weak pulse quality and neutropenia. The massive fluid and electrolyte losses associated with the voluminous diarrhoea lead to dehydration, electrolyte imbalances, acid–base abnormalities and renal shut-down. The end-result may be endotoxic shock and death. Although there are numerous reports of *Salmonella* infections causing a high case-fatality rate, no effect on mortality rates was observed in another study of hospitalized horses (Mainar-Jaime *et al.*, 1998).

Clinical Pathology

The most dramatic clinical pathological changes associated with salmonellosis are neutropenia and/or a toxic, degenerative left shift, loss of plasma proteins and increases in red blood-cell packed cell volume (PCV). The neutropenia is often quite profound (< 1000 neutrophils μl^{-1}) and neutrophils may be vacuolated. In some cases, plasma fibrinogen will be elevated. When the mucosal integrity is compromised by *Salmonella*, the liver is presented via the portal blood with a large number of bacteria and large amounts of bacterial toxins that require removal via the portal bloodstream. In performing this function, hepatocytes may be irreversibly damaged and hepatocellular enzymes (sorbitol dehydrogenase (SDH), aspartate aminotransaminase (AST), etc.) may be markedly elevated in the serum. Horses dehydrated from diarrhoea may have altered electrolyte levels and blood pH and a prerenal azotaemia, indicated by elevated blood urea nitrogen and serum creatinine. When a horse is receiving fluid therapy, a poor prognosis is suggested by declining plasma-protein values, associated with an increasing PCV or a very high PCV that fails to respond to fluid therapy.

Diagnosis

Colitis with essentially identical clinical signs can be associated with a number of other agents (Staempfli *et al.*, 1992a,b; Beier *et al.*, 1994). In 50–70% of cases, an aetiological agent is not isolated; many of these cases occur as a sequel to antimicrobial therapy for an unrelated problem. In the last few years, in the Veterinary Medicine Teaching Hospital (VMTH) at the University of California, Davis, *Clostridium difficile* has been isolated more frequently from horses with severe colitis than has *Salmonella* (Magdesian *et al.*, 1997). Some of the other major differential diagnoses include Potomac horse fever (equine monocytic ehrlichiosis), necrotizing enterocolitis, other clostridial diseases, non-steroidal anti-inflammatory toxicity, cantharadin and other toxins and endotoxaemia. The definitive diagnosis of salmonellosis requires culture of the organism from faeces, blood or tissues. Standard culture methods, with and without selective enrichment, are used for faecal cultures. Culture of a rectal biopsy has been found more sensitive than faecal culture (Palmer *et al.*, 1985b). Transient bacteraemia may occur during the course of the disease and *Salmonella* may be cultured from blood collected aseptically using commercially available bottles of broth (such as trypticase soy broth). Serology using enzyme-linked immunosorbent assay (ELISA) can be used to detect an antibody response in infected animals (House *et al.*, 1996). The serological response of horses to *Salmonella* infection varies with different *Salmonella* serovars. At necropsy, *Salmonella* can usually be isolated from mesenteric lymph nodes of infected horses. Many foals and some horses that die from *Salmonella* infection are bacteraemic and, as a consequence, the organism will be isolated from any tissue cultured. The cause of death is uncertain in horses that die with colitis and have *Salmonella* culture-negative tissues, but *Salmonella* culture-positive faeces.

Treatment

The treatment of equine salmonellosis differs slightly according to the age of the patient and the manifestations of disease. The keys to successful treatment are controlling bacteraemia and focal sites of infection through judicious use of effective antimicrobial drugs, limiting inflammatory cascades to prevent septic (endotoxic) shock through use of non-steroidal anti-inflammatory drugs (NSAIDs), replacing fluid and electrolyte losses and meeting ongoing requirements, giving plasma to replace enteric protein losses and meeting the nutritional demands of the patient.

Because untreated salmonellosis has a high mortality rate in horses, early diagnosis and aggressive therapy are essential. Preliminary symptomatic treatment regimes are adjusted as clinical data regarding the patient's acid–base and electrolyte status become available and according to the antimicrobial drug susceptibilities of the organism isolated.

In neonates, early diagnosis by means of blood culture helps to select the correct antimicrobial drug. Clinical signs of *Salmonella* sepsis cannot be readily differentiated from any other cause of neonatal bacterial sepsis, and the initial antimicrobial drug selected should be effective against most Gram-negative enteric bacteria, the most common type of bacteria associated with sepsis in neonatal foals. Drugs with relatively low toxicity, such as third-generation cephalosporins or chloramphenicol, are preferred; however, drug selection should ultimately be based on antimicrobial sensitivity data. Septic neonates will also require intensive nursing and frequent feeding. NSAIDs can help mitigate the immune cascades that are initiated by endotoxin. Failure to dampen these cascades may result in endotoxic shock. The toxic effects of NSAIDS – chiefly, gastritis, colitis and renal papillary necrosis – are exacerbated by dehydration. The nephrotoxic effects of these drugs may be additive when combined with other drugs, such as aminoglycoside antibacterials, so correction of dehydration and therapeutic drug monitoring are important in the severely compromised neonate. Other recommended approaches to treating neonatal sepsis should also be followed, as outlined in textbooks of equine neonatology (Madigan, 1997).

There is generally less emphasis on antimicrobial therapy for treatment of *Salmonella*-induced colitis in adult horses unless the patient

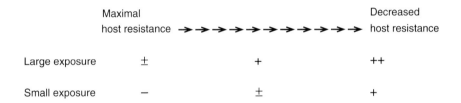

Fig. 13.1. Balance between host and pathogen is affected by resistance of host and amount of exposure to pathogen: ++, consistently severe disease; +, usually causes disease; ±, may cause illness; −, usually does not cause illness.

is severely leucopenic. Systemic antimicrobial drugs are indicated to prevent or treat bacteraemia, but they generally have little effect on the colitis *per se*. Treatment with antimicrobials does not reduce *Salmonella* shedding in the faeces even when antimicrobial sensitivity tests suggest that the drug selection is appropriate (Morse *et al.*, 1976a; van Duijkeren *et al.*, 1995). Many clinicians also feel that oral antimicrobials have little effect on the course of the colitis. NSAIDs, intravenous fluid therapy and intravenous plasma to replace serum proteins lost through the inflamed bowel are all important parts of successful therapy. Providing deep, dry bedding and support for the frog and the use of NSAIDS to block the effects of endotoxin help to prevent laminitis, a common complication that can lead to the unfortunate sequel of a surviving but chronically lame horse.

Pathological Findings

The principal gross lesions in horses that die from colitis are enlarged mesenteric lymph nodes and severe typhlitis and colitis. Inflammation of the ileum may also be present. Necrosis and ulceration of the mucosa and lamina propria, with severe oedema and congestion of the mucosal and submucosal blood-vessels, may be observed in the small colon, large colon and caecum. The most severely affected areas are usually over the submucosal lymphoid patches. The mucosa usually appears congested and a thin film of fibrin may be observed on mucosal surfaces. The mesenteric and colonic lymph nodes are

often oedematous and congested and periportal cholangiohepatitis may be observed, secondary to the necrotizing process in the gastrointestinal tract. The presence of generalized petechia and ecchymoses is compatible with septicaemia and disseminated intravascular coagulation.

In foals, bacterial localization may lead to lesions of endocarditis, septic arthritis, septic physitis, renal infarcts, meningitis and uveitis.

Prevention and Control

Salmonella are widespread in the environment and in animal populations. The three major factors that determine the outcome after exposure to *Salmonella* are host resistance, size of exposure (i.e. number of organisms), and virulence of the serovar. Pathogen exposure and host resistance are the only variables that can be manipulated to prevent clinical salmonellosis (Fig. 13.1).

Clinically affected patients rapidly amplify environmental contamination. Infected horses may shed anything from one to 10^9 *Salmonella* g^{-1} faeces, with clinically ill patients shedding the highest numbers. The magnitude of shedding is influenced by the size of the challenge dose, host immunity and pathogen virulence. Horses may shed *Salmonella* within several hours of exposure, and infected animals should be isolated as quickly as possible to minimize environmental contamination. Preventing *Salmonella* contamination of the environment and equipment by liquid faeces is exceedingly difficult. At the University of California, Davis, horses are isolated if they have clinical signs (fever, leucopenia and diarrhoea) compatible with salmonellosis. Confirmation of diagnosis by *Salmonella* isolation leads to an additional 36–48 h of environmental *Salmonella* contamination and exposure of other patients at risk. On infected premises, *Salmonella* can often be isolated from the drains, floors, dust and crevices in stall walls (Hird *et al.*, 1984; Castor *et al.*, 1989). Animal care providers may inadvertently carry the organism from one horse to another on clothing and hands. Contaminated equipment has also been implicated as a vector in veterinary hospital outbreaks (Hird *et al.*, 1986).

Protocols to minimize *Salmonella* environmental contamination and the risk of exposing susceptible animals and people to *Salmonella* include isolation of infected animals, steaming or burning of their manure and bedding to kill *Salmonella*, feed and water management to minimize *Salmonella* contamination (vermin-proof bins), chlorinated water, fly and rodent control and personnel training (Gibbons, 1980). *Salmonella* control programmes need to be tailored according to the farm or institutional resources. Monitoring environmental *Salmonella* contamination and the incidence of *Salmonella* shedding in horses at risk is important to evaluate the effectiveness of and to recognize deficiencies in *Salmonella* control strategies implemented.

The resistance of horses in veterinary hospitals is typically compromised, rendering them very susceptible to *Salmonella* infection. Many patient management decisions have an impact on the susceptibility of patients to *Salmonella* infection. Antimicrobial therapy, alterations in diet and the use of contaminated equipment have all been associated with an increased risk of salmonellosis in hospitalized horses (Owen *et al.*, 1983; Hird *et al.*, 1984, 1986; Traub-Dargatz *et al.*, 1990; Cohen and Honnas, 1996). Consideration should be given to the decisions behind the use, selection and duration of antimicrobial therapy. In one study, horses receiving parenterally administered antibiotics were 4.2 times more likely to become infected with *Salmonella* than horses not receiving antibiotics, and horses receiving both oral and parenteral antibiotics were 40.4 times as likely to be infected (Hird *et al.*, 1986). Oxytetracycline therapy has been associated with prolonged *Salmonella* shedding and precipitation of clinical disease (Owen *et al.*, 1983). In a more recent study, the only antimicrobial associated with an increased risk of nosocomial *Salmonella* infection in hospitalized horses was potassium penicillin (House *et al.*, 1999).

There are a number of reports of other species of livestock introducing pathogenic *Salmonella* serovars into veterinary hospitals (Baker, 1970; Morse *et al.*, 1976b). Hence, from the perspective of *Salmonella* control, it is desirable to hospitalize horses in facilities separate from those of other livestock. Host immunity and pathogen dose may also be affected by environmental conditions. *Salmonella* outbreaks tend to occur more commonly in the summer months. Increasing the incubation temperature of growth media from 23°C to 37°C reduces the generation time of *Salmonella* from 108 min to 18 min (Chiu, 1974). Higher environmental temperatures have an impact on the host by reducing feed intake and increasing the sensitivity to endotoxin (i.e. at higher temperatures, illness is more severe and the mortality rate is higher).

There are no controlled clinical trials that evaluate the efficacy of *Salmonella* vaccines in horses; however, a number of empirical reports suggest that the use of autogenous *Salmonella* bacterins and attenuated *S. abortusequi* vaccines help prevent disease on breeding farms (Fang *et al.*, 1978; Carter *et al.*, 1979; Dewes, 1979). The only *Salmonella* vaccine licensed for use in horses in the USA is Endovac-Equi, produced by Immvac. Endovac-Equi is an *S. typhimurium* bacterin–toxoid in aluminium hydroxide adjuvant.

References

Argenzio, R.A. (1975) Functions of the equine large intestine and their interrelationship in disease. *Cornell Veterinarian* 65, 303–330.

Argenzio, R.A., Southworth, M. and Stevens, C.E. (1974) Sites of organic acid production and absorption in the equine gastrointestinal tract. *American Journal of Physiology* 226, 1043–1050.

Baker, J.R. (1969) An outbreak of salmonellosis involving veterinary hospital patients. *Veterinary Record* 85, 8–10.

Baker, J.R. (1970) Salmonellosis in the horse. *British Veterinary Journal* 126, 100–105.

Begg, A.P., Johnston, K.G., Hutchins, D.R. and Edwards, D.J. (1988) Some aspects of the epidemiology of equine salmonellosis. *Australian Veterinary Journal* 65, 221–223.

Beier, R., Amtsberg, G. and Peters, M. (1994) Bacteriological investigation of the occurrence and significance of *Clostridium difficile* in horses. *Pferdeheilkunde* 10, 3–7.

Binde, M., Gronstol, H. and Berg, C. (1983) Outbreak of salmonellosis in horses. *Norsk Veterinaertidsskrift* 95, 627–631.

Breukink, H.J. (1974) Oral mono- and disaccharide tolerance tests in ponies. *American Journal of Veterinary Research* 35, 1523–1527.

Bryans, J.T., Fallon, H. and Shephard, B.P. (1961) Equine salmonellosis. *Cornell Veterinarian* 51, 467–477.

Carter, J.D., Hird, D.W., Farver, T.B. and Hjerpe, C.A. (1986) Salmonellosis in hospitalized horses: seasonality and case fatality rates. *Journal of the American Veterinary Medical Association* 188, 163–167.

Carter, M.E., Dewes, H.G. and Griffiths, O.V. (1979) Salmonellosis in foals. *Journal of Equine Medicine and Surgery* 3, 78–83.

Castor, M.L., Wooley, R.E., Shotts, E.B., Brown, J. and Payeur, J.B. (1989) Characteristics of *Salmonella* isolated from an outbreak of equine salmonellosis in a veterinary teaching hospital. *Equine Veterinary Science* 9, 236–241.

Chambers, P.G. and Lysons, R.J. (1979) The inhibitory effect of bovine rumen fluid on *Salmonella typhimurium*. *Research in Veterinary Science* 26, 273–276.

Chiu, M.M. (1974) Effects of pH, salt, and temperature on growth of *Salmonella typhimurium*. Masters Thesis, University of California, Davis.

Clarke, L.L., Roberts, M.C. and Argenzio, R.A. (1990) Feeding and digestive problems in horses: physiologic responses to a concentrated meal. *Veterinary Clinics of North America Equine Practice* 6, 433–450.

Cohen, N.D. and Honnas, C.M. (1996) Risk factors associated with development of diarrhoea in horses after celiotomy for colic: 190 cases (1990–1994). *Journal of the American Veterinary Medical Association* 209, 810–813.

Cohen, N.D., Martin, L.J., Simpson, R.B., Wallis, D.E. and Neibergs, H.L. (1996) Comparison of polymerase chain reaction and microbiological culture for detection of Salmonellae in equine faeces and environmental samples. *American Journal of Veterinary Research* 57, 780–786.

Condoleo, R., Roperto, F., Amaddeo, D., Guarino, G., Fontanelli, G. and Troncone, A. (1983) Abortion and perinatal mortality in horses: microbiological and virological observations and histopathological findings. *Atti della Societa Italiana delle Scienze Veterinarie* 37, 525–527.

Cordy, D.R. and Davis, R.W. (1946) an outbreak of salmonellosis in horses and mules. *Journal of the American Veterinary Medical Association* 108, 20–24.

Davies, L.E. (1986) *Salmonella* reduction program of animal protein producers industry. In: *Proceedings of the 90th USAHA, Louisville, Kentucky 1986*, pp. 368–373.

Dewes, H.F. (1979) Effects of vaccination with killed *Salmonella* bacteria (correspondence). *New Zealand Veterinary Journal* 27, 106–107.

Donahue, J.M. (1986) Emergence of antibiotic-resistant *Salmonella agona* in horses in Kentucky. *Journal of the American Veterinary Medical Association* 188, 592–594.

Dorn, C.R., Coffman, J.R., Schmidt, D.A., Garner, H.E., Addison, J.B. and McCune, E.L. (1975) Neutropenia and salmonellosis in hospitalized horses. *Journal of the American Veterinary Medical Association* 166, 65–67.

Fang, H.W., Yan, T.J., Chang, S.Y., Chang, H.C., Theng, M., Feng, W.T., Wu, P.G. and Cheng, S.S. (1978) Studies on *Salmonella abortusequi* attenuated vaccine. *Collected Papers of Veterinary Research* 5, 1–14.

Faulstich, A. (1987) Investigations on the appearance of *Salmonella* in hospitalized horses. PhD Dissertation, Fachbereich Veterinarmedizin der Freien Universitat, Berlin, German Federal Republic, 150 pp.

Ferris, K.E. and Miller, D.A. (1989) *Salmonella* serovars from animals and related sources reported during July 1988–June 1989. In: *Proceedings of the 93rd USAHA, Richmond, Virginia*.

Ferris, K.E. and Miller, D.A. (1995) *Salmonella* serovars from animals and related sources reported during July 1994–June 1995. In: *Proceedings of the 99th USAHA, Reno, Nevada*.

Garg, D.N. and Manchanda, V.P. (1986) Prevalence and aetiology of equine abortion. *Indian Journal of Animal Sciences* 56, 730–735.

Gibbons, D.F. (1980) Equine salmonellosis: a review. *Veterinary Record* 106, 356–359.

Goedegebuure, S.A., Dik, K.J., Firth, E.C. and Merkens, H.W. (1980) Polyarthritis and polyosteomyelitis in foals. *Veterinary Pathology* 17, 651.

Goodson, J., Tyznik, W.J., Cline, J.H. and Dehority, B.A. (1988) Effects of an abrupt diet change from hay to concentrate on microbial numbers and physical environment in the caecum of the pony. *Applied Environmental Microbiology* 54, 1946–1950.

Graham, R., Francois, V.C. and Reynolds, H.K. (1919) Bacteriologic studies of a peracute disease of horses and mules. *Journal of the American Veterinary Medical Association* 56, 378–393.

Hansen, L.M., Jang, S.S. and Hirsh, D.C. (1993) Use of random fragments of chromosomal DNA to highlight restriction site heterogeneity for fingerprinting isolates of *Salmonella typhimurium* from hospitalized animals. *American Journal of Veterinary Research* 54, 1648–1652.

Hartmann, F.A. and West, S.E. (1995) Antimicrobial susceptibility profiles of multidrug-resistant *Salmonella* anatum isolated from horses. *Journal Veterinary Diagnostic Investigation* 7, 159–161.

Hird, D.W., Pappaioanou, M. and Smith, B.P. (1984) Case–control study of risk factors associated with isolation of *Salmonella saintpaul* in hospitalized horses. *American Journal of Epidemiology* 120, 852–864.

Hird, D.W., Casebolt, D.B., Carter, J.D., Pappaioanou, M. and Hjerpe, C.A. (1986). Risk factors for salmonellosis in hospitalized horses. *Journal of the American Veterinary Medical Association* 188, 173–177.

House, J.K., Smith, B.P., House, A. and Fewell, M. (1996) Implications of *Salmonella* infection in horses: diagnostic and epidemiological considerations in management and prevention of disease outbreaks. In: *Proceedings of the 14th Annual Veterinary Medical Forum, American College of Veterinary Internal Medicine, San Antonio, Texas, 1996*, pp. 535–537.

House, J.K., Mainar-Jaime,R.C., Smith, B.P., House, A. and Kamiya, D.Y. (1999) Case–control study of risk factors for nosocomial *Salmonella* infection among hospitalized horses. *Journal of the American Veterinary Medical Association* 214, 1511–1516.

Ikeda, J.S. and Hirsh, D.C. (1985) Common plasmid encoding resistance to ampicillin, chloramphenicol, gentamicin, and trimethoprim-sulfadiazine in two serovars of *Salmonella* isolated during an outbreak of equine salmonellosis. *American Journal of Veterinary Research* 46, 769–773.

Ikeda, J.S., Hirsh, D.C., Jang, S.S. and Biberstein, E.L. (1986) Characteristics of *Salmonella* isolated from animals at a veterinary medical teaching hospital. *American Journal of Veterinary Research* 47, 232–235.

Ingram, P.L. and Edwards, G.B. (1980) *Salmonella* infections in horses. *Transactions of the Royal Society of Tropical Medical Hygiene* 74, 113.

Kikuchi, N., Kawakami, Y., Murase, N., Ohishi, H., Tomioka, Y., Iwata, K., Fujimura, M. and Sakazaki, R. (1982) Isolation of *Salmonella typhimurium* from foals with pyrexia and diarrhoea. *Bulletin of Equine Research Institute, Japan* 19, 43–50.

Kilborne, F.S. (1893) *Bulletin of the Bureau of Animal Industry United States Department of Agriculture* 3, 49.

Kuhlmann, W. (1964) Occurrence of *Salmonella* in slaughtered horses. *Monatshefte für Veterinarmedizin* 19, 790–792.

Kuka, A. (1989) Data on the diagnosis of *Salmonella* abortion in mares. *Buletini i Shkencave Zooteknike e Veterinare* 7, 57–63.

Linerode, P.A. and Goode, R.L. (1970) The effects of colic on the microbial activity of the equine large intestine. In: *Proceedings 16th Convention American Association of Equine Practitioners, Montreal*, pp. 321–341.

McCain, C.S. and Powell, K.C. (1990) Asymptomatic salmonellosis in healthy adult horses. *Journal of Veterinary Diagnostic Investigation* 2, 236–237.

McChesney, D.G., Kaplan, G. and Gardner, P. (1995) FDA survey determines *Salmonella* contamination. *Feedstuffs* 67, 20–23.

McClintock, S.A. and Begg, A.P. (1990) Suspected salmonellosis in seven broodmares after transportation. *Australian Veterinary Journal* 67, 265–267.

Madigan, J.E. (1997) *Manual of Equine Neonatal Medicine*, 3rd edn. Live Oak Publishing, Woodland, California.

Madigan, J.E., Walker, R.L., Hird, D.W., Case, J.T., Bogenrief, D.S., Smith, B.P. and Dyke, T.M. (1992) Equine neonatal salmonellosis: clinical observations and control measures (a case report). In: *Equine Anaesthesia, Abdominal Surgery, and Medicine of the Foal: Proceedings of Fourteenth Bain Fallon Memorial Lectures, July. Sydney, Australia*.

Magdesian, K.G., Madigan, J.E., Jang, S.S. and Hirsh, D.C. (1997) Colitis associated with *Clostridium difficile* in horses. *Journal of Veterinary Internal Medicine* 11, 110.

Mainar-Jaime, R.C., House, J.K., Smith, B.P., Hird, D.W., House, A.M. and Kamiya, D.Y. (1998) Influence of fecal shedding of *Salmonella* organisms on mortality in hospitalized horses. *Journal of the American Veterinary Medical Association* 213, 1162–1166

Mann, P.H., Cavrini, C. and Pieracci, F. (1964) *Salmonella* organisms found in healthy horses at slaughter. *Cornell Veterinarian* 54, 495.

Meynell, G.G. (1963) Antibacterial mechanisms of the mouse gut II: the role of EH and volatile fatty acids in the normal gut. *British Journal of Experimental Pathology* 44, 209–219.

Morgan, J.P., van der Watering, C. and Kersjes, A.W. (1974) *Salmonella* bone infection in colts and calves: its radiographic diagnosis. *Journal of the American Veterinary Radiology Society* 15, 66–76.

Morse, E.V., Duncan, A.M., Fessler, J.F. and Page, E.A. (1976a) The treatment of salmonellosis in Equidae. *Modern Veterinary Practice* 57, 47–51.

Morse, E.V., Duncan, M.A., Page, E.A. and Fessler, J.F. (1976b) Salmonellosis in Equidae: a study of 23 cases. *Cornell Veterinarian* 66, 198–213.

Owen, R. ap R., Fullerton, J.N., Tizard, I.R., Lumsden, J.H. and Barnum, D.A. (1979) Studies on experimental enteric salmonellosis in ponies. *Canadian Journal Comparative Medicine* 43, 247–254.

Owen, R. ap R., Fullerton, J. and Barnum, D.A. (1983) Effects of transportation, surgery, and antibiotic therapy in ponies infected with *Salmonella*. *American Journal of Veterinary Research* 44, 46–50.

Palmer, J.E., Benson, C.E. and Whitlock, R.H. (1985a) *Salmonella* shed by horses with colic. *Journal of the American Veterinary Medical Association* 187, 256–257.

Palmer, J.E., Whitlock, R.H., Benson, C.E., Becht, J.L., Morris, D.D. and Acland, H. M. (1985b) Comparison of rectal mucosal cultures and fecal cultures in detecting *Salmonella* infection in horses and cattle. *American Journal of Veterinary Research* 46, 697–698.

Pare, J., Carpenter, T.E. and Thurmond, M.C. (1996) Analysis of spatial and temporal clustering of horses with *Salmonella krefeld* in an intensive care unit of a veterinary hospital. *Journal of the American Veterinary Medical Association* 209, 626–628.

Powell, D.G., Donahue, M., Ferris, K., Osborne, M. and Dwyer, R. (1988) An epidemiologicalal investigation of equine salmonellosis in central Kentucky during 1985 and 1986. In: *Equine Infectious Diseases: Proceedings of Fifth International Conference*. University Press of Kentucky, Lexington, pp. 231–235.

Poyade-Alvarado, A. and Marcoux, M. (1993) Haematogenous septic arthritis and osteomyelitis in foals: 39 cases (1985–1989). *Pratique Vétérinaire Equine* 25, 275–280.

Quevedo, F., Dobosch, D. and Gonzalez-L., E. (1973a) Contamination of horse meat with *Salmonella*: an ecological study. I. Carrier horses. *Gaceta Veterinaria* 35, 119–123.

Quevedo, F., Lord, R.D., Dobosch, D., Granier, I. and Michanie, S.C. (1973b) Isolation of *Salmonella* from sparrows captured in horse corrals. *American Journal of Tropical Medicine and Hygiene* 22, 672–674.

Roberts, M.C. and O'Boyle, D.A. (1981) The prevalence and epizootiology of salmonellosis among groups of horses in south east Queensland. *Australian Veterinary Journal* 57, 27–35.

Roberts, M.C. and O'Boyle, D.A. (1982) Experimental *Salmonella anatum* infection in horses. *Australian Veterinary Journal* 58, 232–240.

Rumschlag, H.S. and Boyce, J.R. (1987) Plasmid profile analysis of *Salmonellae* in a large-animal hospital. *Veterinary Microbiology* 13, 301–311.

Sato, G., Nakaoka, Y., Ishiguro, N., Ohishi, H., Senba, H., Kato, H., Honma, S. and Nagase, N. (1984) Plasmid profiles of *Salmonella typhimurium* var. *copenhagen* strains isolated from horses. *Bulletin of Equine Research Institute* 21, 105–109.

Smith, B.P. (1979) Atypical salmonellosis in horses: fever and depression without diarrhoea. *Journal of the American Veterinary Medical Association* 175, 69–71.

Smith, B.P., Reina-Guerra, M. and Hardy, A.J. (1978) Prevalence and epizootiology of equine salmonellosis. *Journal of the American Veterinary Medical Association* 172, 353–356.

Smith, B.P., Reina-Guerra, M., Hardy, A.J. and Habasha, F. (1979) Equine salmonellosis: experimental production of four syndromes. *American Journal of Veterinary Research* 40, 1072–1077.

Smith, T. (1893) *Bulletin of the Bureau of Animal Industry, United States Department of Agriculture* 3, 53.

Staempfli, H.R., Prescott, J.F. and Brash, M.L. (1992a) Lincomycin-induced severe colitis in ponies: association with *Clostridium cadaveris*. *Canadian Journal of Veterinary Research* 56, 168–169.

Staempfli, H.R., Prescott, J.F., Carman, R.J. and McCutcheon, C.L. (1992b) Use of bacitracin in the prevention and treatment of experimentally-induced idiopathic colitis in horses. *Canadian Journal of Veterinary Research* 56, 233–236.

Stuart, B.P., Martin, B.R., Williams, L.P., Jr and Byern, H.V. (1973) *Salmonella*-induced meningoencephalitis in a foal. *Journal of the American Veterinary Medical Association* 162, 211–213.

Traub-Dargatz, J.L., Salman, M.D. and Jones, R.L. (1990) Epidemiologic study of *Salmonellae* shedding in the faeces of horses and potential risk factors for development of the infection in hospitalized horses. *Journal of the American Veterinary Medical Association* 196, 1617–1622.

Tillotson, K., Savage, C.J., Salman, M.D., Genty-Weeks, C.R., Rice, D., Fedorka-Cray, P.J., Hendrikson, D.A., Jones, R.L., Nelson, W. and Traub-Dargatz, J.L. (1997) Outbreak of *Salmonella infantis* infection in a large animal veterinary teaching hospital. *Journal of the American Veterinary Medical Association* 211, 1554–1557.

van Duijkeren, E., Sloet-van Oldruitenborgh-Oosterbaan, M.M., Houwers, D.J., van Leeuwen, W.J. and Kalsbeek, H.C. (1994) Equine salmonellosis in a Dutch veterinary teaching hospital. *Veterinary Record* 135, 248–250.

van Duijkeren, E., Flemming, C., van Oldruitenborgh, M.S., Kalsbeek, H.C. and van der Giessen, J.W.B. (1995) Diagnosing salmonellosis in horses: culturing of multiple versus single faecal samples. *Veterinary Quarterly* 17, 63–66.

Walker, R.L., Madigan, J.E., Hird, D.W., Case, J.T., Villanueva, M.R. and Bogenrief, D.S. (1991) An outbreak of equine neonatal salmonellosis. *Journal of Veterinary Diagnostic Investigation* 3, 223–227.

Wolin, M.J. (1969) Volatile fatty acids and the inhibition of *Escherichia coli* growth by rumen fluid. *Applied Microbiology* 17, 83–87.

Woodward, M.J., McLaren, I. and Wray, C. (1989) Distribution of virulence plasmids within Salmonellae. *Journal of General Microbiology* 135, 503–511.

Chapter 14
Salmonella Infections in Dogs and Cats

Margery E. Carter and P. Joseph Quinn

Department of Veterinary Microbiology and Parasitology, Faculty of Veterinary Medicine,
University College Dublin, Ballsbridge, Dublin 4, Ireland

Salmonellosis occurs worldwide in dogs and cats and is of clinical and public-health importance. Infections in these animals are usually asymptomatic, with intermittent excretion of *Salmonella* in faeces. Clinical syndromes, which are comparatively uncommon, are often most severe in young or debilitated animals. These syndromes include enterocolitis, septicaemia and, rarely, abortions. Conjunctivitis has been reported in cats. Outbreaks of salmonellosis have occurred in veterinary hospitals, where predisposing factors, such as immunosuppressive therapy or major surgery, can precipitate disease in a carrier animal. The significance and extent of the carrier–excreter state in dogs and cats was perhaps not fully realized until the early 1950s, when surveys of apparently healthy animals were conducted. One of the earliest comprehensive reviews of *Salmonella* infections in animals was by Buxton (1957), who documented the occurrence of the organisms in cats and dogs.

Epidemiology

Serovars

Numerous *Salmonella* serovars have been isolated from dogs. There are regional and national variations in the occurrence of serovars (Tables 14.1 and 14.2), reflecting those present in the animals' diet or general environment. S. *typhimurium* and S. *enteritidis* appear to be univer-

sally distributed, with S. *anatum* a common isolate from dogs in the USA (Morse and Duncan, 1975). No host-adapted serovars are described in dogs or cats. There are rare reports of S. *paratyphi* B isolation from dogs, usually where this serovar is also present in the human population. Some authors have noted a similarity in the distribution of serovars isolated from dogs and humans (Galton *et al.*, 1952). Comparatively few surveys have been carried out in cat populations (Table 14.3). Multiply resistant strains of S. *typhimurium* definitive types (DT) 104 and 204c have been isolated from domestic carnivores. S. *enteritidis* phage type (PT) 4, often associated with poultry and eggs (Humphrey, 1990), has also been recovered from cats.

Excreter state

Isolation rates of *Salmonella* serovars from faeces of clinically normal cats range from 0 to 14% (Center *et al.*, 1995). The prevalence of *Salmonella* in the faeces of asymptomatic dogs under various environmental and husbandry conditions is presented in Table 14.4.

Salmonella excretion rates for asymptomatic dogs ranged from 0 to 43%. Faecal samples from dogs in boarding kennels, dog pounds and veterinary hospitals occasionally yielded more than one serovar from a single sample, four serovars being isolated in one instance (McElrath *et al.*, 1952). Generally, there was a lower excretion

Table 14.1. *Salmonella* serovars isolated from dogs in various countries.

Germany[a]	Iran[b]	Ireland[c]	South Africa[d]	UK[e]
S. agona	S. adelaide	S. agona	S. berta	S. adelaide
S. anatum	S. anatum	S. brandenburg	S. enteritidis	S. alachua
S. bilthoven	S. braenderup	S. bredeney	S. gloucester	S. ardwick
S. brandenburg	S. derby	S. derby	S. haardt	S. dublin
S. bredeney	S. enteritidis	S. dublin	S. lagos	S. enteritidis (unknown PT)
S. derby	S. haifa	S. enteritidis	S. saintpaul	S. enteritidis PT1
S. dublin	S. havana	S. fyris	S. tennyson	S. goldcoast
S. enteritidis	S. heidelberg	S. kentucky	S. tsevie	S. indiana
S. give	S. hindmarsh	S. montevideo	S. typhimurium	S. infantis
S. heidelberg	S. infantis	S. ohio		S. kedougou
S. infantis	S. kiel	S. typhimurium		S. montevideo
S. java	S. kisangani			S. panama
S. manhattan	S. manhattan			S. rissen
S. newington	S. newport			S. schwarzengrund
S. oranienberg	S. reading			S. seftenberg
S. panama	S. saintpaul			S. singapore
S. saintpaul	S. II sofia			S. stanley
S. thompson	S. tallahassee			S. typhimurium (unknown PT)
S. typhimurium	S. thompson			S. typhimurium DT104
	S. typhimurium			S. typhimurium DT204c
				S. virchow
				S. worthington

[a] Förster *et al.* (1974); [b] Shimi *et al.* (1976); [c] P.T. Quinn, M.E. Carter and Y.E. Abbott, Faculty of Veterinary Medicine, Dublin, unpublished data, 1986–1996; [d] Venter (1988); [e] Veterinary Investigation Services (1988–1996) and Neil *et al.* (1981).
PT, phage type; DT, definitive type

Table 14.2. *Salmonella* serovars isolated from dogs in the USA (based on Bruner and Moran, 1949; Adler *et al.*, 1951; Ball, 1951; Gorham and Garner, 1951; Galton *et al.*, 1952; Bruner, 1973; Uhaa *et al.*, 1988).

S. anatum	S. illinois	S. newbrunswick
S. bareilly	S. infantis	S. newington
S. bonariensis	S. inverness	S. newport
S. bovismorbificans	S. javiana	S. norwich
S. bredeney	S. johannesburg	S. oranienburg
S. budapest	S. kentucky	S. oregon
S. california	S. krefeld	S. paratyphi B
S. canoga	S. litchfield	S. pomona
S. carrau	S. livingstone	S. poona
S. cerro	S. lomita	S. pullorum
S. choleraesuis	S. luciana	S. rubislaw
S. cubana	S. macallen	S. saintpaul
S. derby	S. madelia	S. sandiego
S. duval	S. manhattan	S. seftenberg
S. enteritidis	S. meleagridis	S. tallahassee
S. florida	S. memphis	S. tennessee
S. gaminara	S. miami	S. typhimurium
S. give	S. minnesota	S. urbana
S. hartford	S. montevideo	S. weslaco
S. homosassa	S. muenchen	S. worthington

Table 14.3. *Salmonella* serovars isolated from cats in Iran, UK and USA.

Iran[a]	UK[b]	USA[c]
S. adelaide	S. enteritidis (unknown PT)	S. anatum
S. anatum	S. enteritidis PT4	S. bareilly
S. blockley	S. enteritidis PT24	S. bredeney
S. braenderup	S. typhimurium (unknown PT)	S. choleraesuis
S. derby	S. typhimurium DT12	S. derby
S. gaminara	S. typhimurium DT49	S. donna
S. havana	S. typhimurium DT104	S. enteritidis
S. infantis	S. typhimurium DT193	S. javiana
S. kisangani	S. typhimurium DT204c	S. lomita
S. livingstone		S. montevideo
S. manhattan		S. pharr
S. oritamerin		S. poona
S. II sofia		S. pullorum
S. thompson		S. sandiego
S. typhimurium		S. saintjuan
S. tyresoe		S. typhimurium
		S. weslaco

[a] Shimi and Barin (1977); [b] Veterinary Investigation Services (1988–1996); [c] Ball (1951), Bruner (1973) and Bruner and Moran (1949).
PT, phage type; DT, definitive type.

rate in household pet dogs, compared with those confined to kennels. In one survey of pet dogs, a wide range of serovars was obtained and this was attributed to exposure to a variety of sources rather than to cross-infection between animals (Shimi *et al.*, 1976). The excretion rate in stray dogs was comparatively high, as these animals presumably survived by scavenging and hunting. Surveys of racing greyhounds in kennels yielded an infection rate of up to 43.5%, with an estimated exposure rate leading to more than one infection every 2 months (Stucker *et al.*, 1952). This high level of excretion was attributed to the diet of contaminated raw meat and offal. The infection rate in these greyhounds fell to lower levels (20–30%) during the racing season, indicating that infection was due to ingestion rather than through animal contact at the racing track. Working farm dogs may be exposed to high levels of *Salmonella* in the farm environment, particularly if recent outbreaks of salmonellosis have occurred in the livestock (Carter *et al.*, 1983).

Asymptomatic canine excreters appear to shed *Salmonella* in faeces intermittently for comparatively short periods. When the faeces of 49 culture-positive dogs were sampled over a 4–6 week period, only five dogs yielded the same serovar that had been isolated initially. Within this sampling period, seven dogs were found to have acquired a second, and in some instances a third, *Salmonella* serovar (Mackel *et al.*, 1952). Tanaka *et al.* (1976a) examined apparently healthy stray dogs for faecal shedding of *Salmonella*. The number of organisms fluctuated from 10^2 to 10^5 *Salmonella* 100 g^{-1} of faeces. Two dogs were found to shed small numbers of organisms (10^2 *Salmonella* 100 g^{-1}) sporadically for up to 74 days. Three other dogs excreted 10^4–10^5 *Salmonella* 100 g^{-1} of faeces initially and shedding continued for up to 115 days.

Transmission

Transmission in dogs and cats usually occurs either directly or indirectly by the faecal–oral route. Tanaka *et al.* (1976b) succeeded in experimentally producing a carrier state in dogs by oral administration of naturally contaminated faeces. Each dog received 39–92 *S. typhimurium* organisms *per os*. These experimentally infected dogs shed *Salmonella* for 18–24 days. None of three dogs dosed with 60 freshly isolated *Salmonella* organisms became shedders or produced antibodies. This apparent loss of virulence of *S.*

Table 14.4. Categories of apparently healthy dogs, from different countries, shedding *Salmonella* serovars.

Categories of dogs sampled	Geographical region/country	Dogs shedding/dogs sampled	% shedding	Reference
Household pets	Hawke's Bay/New Zealand	0/150	0.0	Timbs *et al.* (1975)
	Washington/USA	10/809	1.2	Gorham and Garner (1951)
	Tehran/Iran	21/472	4.4	Shimi *et al.* (1976)
	Florida/USA	244/1626	15.0	Galton *et al.* (1952)
Boarding kennels	Tehran/Iran	28/181	15.5	Shimi *et al.* (1976)
	Florida/USA	21/126	16.6	McElrath *et al.* (1952)
Quarantine kennels (rabies)	Honolulu/Hawaii	38/295	12.9	Adler *et al.* (1951)
	Jacksonville/USA	194/1385	14.0	Mackel *et al.* (1952)
Dog refuges/pounds	Hawke's Bay/New Zealand	5/150	3.3	Timbs *et al.* (1975)
	Florida/USA	71/895	7.9	McElrath *et al.* (1952)
	Brisbane/Australia	12/138	8.7	Frost *et al.* (1969)
Greyhound kennels	Florida/USA	930/2548	36.5	Galton *et al.* (1952)
	Jacksonville/USA	697/1602	43.5	Stucker *et al.* (1952)
Veterinary hospitals	Brisbane/Australia	7/157	4.5	Frost *et al.* (1969)
	Jacksonville/USA	13/66	19.7	Mackel *et al.* (1952)
Working dogs on sheep farms	Hawke's Bay/New Zealand	13/300	4.3	Timbs *et al.* (1975)
Working dogs at abattoir	Brisbane/Australia	2/10	20.0	Frost *et al.* (1969)
Stray dogs	Tehran/Iran	3/19	15.8	Shimi *et al.* (1976)
	Khartoum/Sudan	104/442	23.5	Khan (1970)

typhimurium for the intestinal tract following culture of the organisms was also observed in experiments with mice (Tanaka *et al.*, 1977).

Intraocular inoculation of S. *typhimurium* in cats with 9×10^5 colony-forming units (cfu) produced conjunctivitis, with faecal shedding of *Salmonella* organisms, while a lower dose (9×10^3 cfu) initiated an asymptomatic excreter state (Fox *et al.*, 1984).

Rare cases of intrauterine transmission have been reported in both bitches (Redwood and Bell, 1983) and queens (Reilly *et al.*, 1994).

Sources of infection

Domestic carnivores are exposed to diverse sources of *Salmonella* serovars (Fig. 14.1). Dogs are particularly indiscriminate in their eating habits, ingesting food irrespective of quality, freshness or source. Many cats and dogs are allowed to roam and thus have access to carrion or are able to hunt, kill and eat wildlife species. Meat, offal and meat-and-bone-meal, common ingredients in the diets of dogs and cats, are frequently contaminated with *Salmonella*. Offal sold at a wholesale meat market and judged 'fit for consumption' was found to contain 24 different *Salmonella* serovars and 56.6% of 408 samples yielded *Salmonella* organisms (Sinell *et al.*, 1984). When 112 samples of commercial raw meat, used for feeding greyhounds, were cultured for *Salmonella*, 44.6% were positive (Chengappa *et al.*, 1993). In both surveys, S. *typhimurium* was the most common serovar isolated.

Galton *et al.* (1955) isolated *Salmonella* from

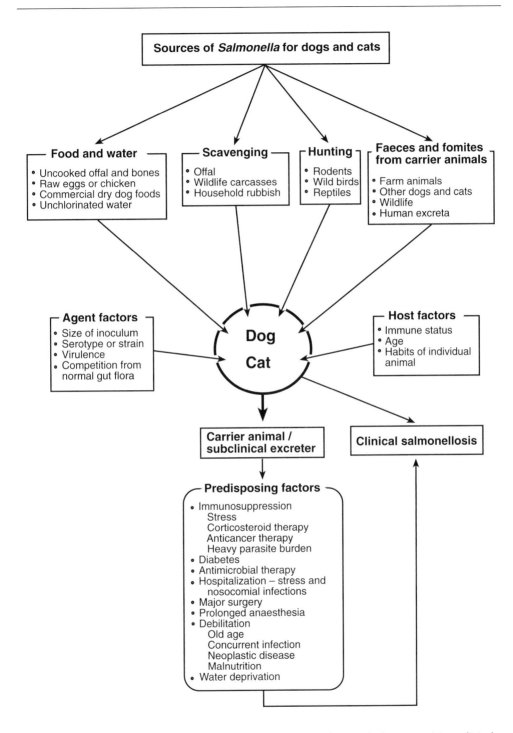

Fig. 14.1. Sources of *Salmonella* for dogs and cats and predisposing factors which may precipitate clinical salmonellosis in a previously asymptomatic carrier animal.

26 (26.5%) of 98 samples of dry dog meal, but found none in pressed foods, such as dog biscuits, or in kibbled products. Two brands of dog meal that failed to yield *Salmonella* serovars had been effectively heat-treated. Day *et al.* (1963) fed contaminated dry dog meal to young dogs and, of the nine serovars recovered from their faeces, seven were also cultured from the meal.

Raw or undercooked poultry meat or eggs can be a source of *Salmonella* for both humans and companion animals. *S. enteritidis* PT4 may be present in the yolk of intact hens' eggs (Humphrey *et al.*, 1989). Storage of eggs at temperatures of 10°C or above can allow the organisms to multiply rapidly to numbers that may permit their survival after cooking (Humphrey, 1990).

Wildlife species are a potential source of *Salmonella* for dogs and cats. Over 50 *Salmonella* serovars have been isolated from mice and rats (Weber, 1982). Wild mice can become infected with *S. enteritidis*. A naturally infected mouse excreted the organism in its faeces for 19 weeks and had 10^4 *Salmonella* g^{-1} of liver and 10^3 organisms g^{-1} of macerated intestine (Davies and Wray, 1995). Deaths from *S. typhimurium* have been recorded in garden birds (Laing, 1990; Kirkwood *et al.*, 1995). The occurrence of *Salmonella* serovars in free-living wild birds was reviewed by Wilson and MacDonald (1967), who concluded that, although salmonellosis in birds is a potential hazard to humans and domestic animals, there is a low incidence of disease in the general wild-bird population. However, rare localized outbreaks of salmonellosis do occur. Both captive and wild reptiles, including lizards, tortoises and terrapins, are notorious for *Salmonella* infections (Borland, 1975) and dogs and cats may acquire infection from these sources.

Coprophagia or ingestion of faecally contaminated food or water may result in infection by *Salmonella* serovars. The organisms are often widespread in soil on farms where outbreaks of salmonellosis have occurred and these organisms can survive for up to 9 months in shaded, moist soil (Carter *et al.*, 1979). A comparative survey of working dogs on sheep properties with and without ovine salmonellosis during the previous 12-month period indicated an excreter rate in dogs of 4.7% and 4.0%, respectively (Timbs *et al.*, 1975). *S. bovismorbificans*, the most common

serovar in sheep in the district, was isolated only from dogs on farms where recent outbreaks of salmonellosis in the sheep had occurred. *S. typhimurium* was isolated from the faeces of the two asymptomatic farm dogs where clinical disease attributed to this serovar had occurred in dairy cows (Carter *et al.*, 1983). Two recently acquired parakeets (*Melopsittacus undulatus*) that developed clinical salmonellosis were thought to have been the source of *S. typhimurium* var. Copenhagen infection for a human infant and a 4-year-old Siamese cat in the same household (Madewell and McChesney, 1975).

Carrion is a potentially rich source of *Salmonella* and replication of these organisms may occur in carcasses, depending on ambient temperatures. Fly maggots are a possible source of infection associated with carcasses. Fly eggs placed in moist food contaminated with *S. enteritidis* resulted in infected maggots, pupae and adult flies (Kintner, 1949).

Animal Infection

Pathogenesis

The numbers of *Salmonella* ingested, the serovar or strain involved and its virulence may determine the clinical outcome of infection with *Salmonella* serovars. Non-specific host factors that affect the ability of organisms to establish and produce infection include gastric acidity, peristalsis, quality of intestinal mucus, lysozyme in secretions, lactoferrin in the intestinal tract and the normal resident intestinal flora (Clarke and Gyles, 1993). The greater susceptibility of young animals to salmonellosis may be associated in part with the lack of a well-established normal flora. In addition, a naïve or incompetent immune system renders such animals vulnerable to these enteropathogens.

Salmonella are facultative intracellular pathogens and the ability of these organisms to survive and multiply inside phagocytes is critical to the outcome of infection (Salyers and Whitt, 1994). Endotoxin (lipopolysaccharide) is a major virulence factor of *Salmonella* serovars and is the probable cause of death in animals with septicaemic salmonellosis. Endotoxin also interferes with the beneficial effects of activated complement. Other toxins produced by *Salmonella* include a heat-labile enterotoxin and a cytotoxin that inhibits protein synthesis (Groisman *et al.*, 1990). Some virulence factors are associated with plasmids, such as a 50 MDa plasmid in *S. dublin* (Terakado *et al.*, 1983), a 60 MDa plasmid in *S. typhimurium* (Jones *et al.*, 1982) and a 36 MDa plasmid in *S. enteritidis* (Nakamura *et al.*, 1985).

The distribution of organisms in the asymptomatic carrier state was studied in 17 dogs naturally infected with *S. typhimurium* (Tanaka *et al.*, 1976a). *Salmonella* were isolated from the mucosae and contents of most segments of the intestinal tract, including the gastric mucosa in three cases. The mucosae of the caecum, colon and ileum were most commonly found to harbour *Salmonella* particularly at the junction of these intestinal segments. *Salmonella* were present at levels between 10^2 and 10^6 100 g^{-1} of intestinal tissue. Organisms were also recovered from the intestinal lymph nodes, with most isolates from the jejunal node. The carrier state was subsequently produced experimentally with *S. typhimurium* (Tanaka *et al.*, 1976b). *Salmonella* were isolated from the mucosa of the middle portion of the ileum, 1 day post-inoculation (PI) and within 1–2 days they had spread to the mucosae of the lower part of the ileum, the caecum and the upper part of the colon. The major site of localization and multiplication in 5–15 days PI was the mucosa adjacent to the ostium ileocaecocolium. The jejunal node yielded *Salmonella* from the second day PI. Detectable serum antibody levels were present from the seventh day PI.

A high percentage of *Salmonella* are eventually cleared by the host's defences and only a small proportion survive for comparatively short periods, leading to the carrier state (Clarke and Gyles, 1993). A range of predisposing causes can precipitate clinical salmonellosis in a previously asymptomatic carrier animal (Fig. 14.1). In enterocolitis, the organisms adhere to the intestinal cells, invade the mucosa and are taken up by mucosa-associated macrophages. They induce toxin-mediated damage to microvilli of apical enterocytes, which results in shortening of villi, degeneration and abnormal extrusion of enterocytes and increased emptying of goblet cells. There is a neutrophilic reaction in the lamina propria, accompanied by transepithelial migration of neutrophils into the lumen (Takeuchi and

Sprinz, 1967). Invasive strains of *Salmonella* are able to breach the mucosal barrier and invade underlying tissues. When these strains penetrate underlying tissues, septicaemia follows, with replication in the cells of the reticuloendothelial system. Characteristic systemic signs of fever and vascular damage occur, due to the effects of circulating endotoxin.

Factors predisposing to the development of clinical disease

Clinical salmonellosis is rare in cats and dogs. Healthy adult animals appear to have a high natural resistance to the disease. Young animals and ageing or debilitated dogs and cats are most susceptible (Buxton, 1957). This high natural resistance to clinical disease may be lowered by various immunosuppressive or stress factors (Fig 14.1). The disease is difficult to reproduce experimentally in dogs and cats. This may be due, in part, to the difficulty in experimentally mimicking the stress factors present in situations such as veterinary hospitals (Timoney, 1976). Sporadic cases of salmonellosis can occur in the general cat and dog populations, but outbreaks of the disease tend to occur in confined groups of animals, such as those in kennels, research facilities and veterinary hospitals. Animals in confinement often experience environmental stresses, such as overcrowding and changes in diet, heating, lighting and humidity. Animals in research facilities may be exposed to stress during shipping. Fifteen cats out of 142 (10.6%) in 17 shipments to a research laboratory were found to be infected with *Salmonella* serovars and two subsequently died from the septicaemic form of the disease (Fox and Beaucage, 1979). Additional stresses in veterinary hospitals may include immunosuppressive therapy, surgery, prolonged anaesthesia and antimicrobial therapy. Animals with concurrent disease or suffering from debilitating conditions, such as diabetes, are susceptible. Clinical salmonellosis developed in six dogs with multicentric lymphosarcoma that were subjected to surgery and treated with corticosteroids and anti-cancer drugs. Three of the affected animals subsequently died from salmonellosis (Calvert and Leifer, 1982).

The index case initiating an outbreak of salmonellosis in a veterinary hospital may be an animal admitted with *Salmonella*-associated diarrhoea or may result from activation of a pre-existing asymptomatic infection. Clinically affected dogs and cats usually excrete larger numbers of *Salmonella* than carrier animals. However, comparatively small numbers of *Salmonella* may be required to cause clinical salmonellosis in animals stressed by surgical or medical treatments (Timoney *et al.*, 1978).

Predisposing or risk factors have been investigated in several nosocomial outbreaks of salmonellosis. In an episode involving hospitalized dogs, Calvert (1985) listed pre-existing disease, hospitalization of 5 days or longer, major surgery, glucocorticoid therapy, cytotoxic drug administration and neoplasia as important risk factors leading to the development of clinical salmonellosis. In a nosocomial outbreak of disease in dogs due to *S. krefeld*, Uhaa *et al.* (1988) statistically analysed the risk factors involved. The use of antimicrobial agents, particularly ampicillin, was identified as an important risk factor. This has been attributed to selective elimination by the antibiotic of competing intestinal microflora that are antagonistic to *Salmonella*, thereby creating a more favourable environment for the replication of *Salmonella* serovars. A serious nosocomial outbreak of salmonellosis in cats was reported by Timoney *et al.* (1978). Most of the cats involved were less than 1 year old and had been admitted for routine surgical procedures, medical reasons or for boarding. The animals became ill 2–5 days after admission. *Salmonella* serovars were consistently isolated from the oral cavity, as well as from the faeces, of the affected cats. Organisms in the oral cavity may contaminate the cats' coats during grooming, increasing the hazard of *Salmonella* being transferred to the hands and clothing of personnel caring for the animals.

Clinical signs

The majority of *Salmonella* infections in dogs and cats are asymptomatic. The main clinical syndromes, in both animal species, are an acute enterocolitis and septicaemia, with resultant endotoxaemia. Rare syndromes include conjunctivitis in cats and *in utero* infections in bitches and queens, which result in abortions, stillbirths or the birth of weak offspring.

Acute enterocolitis usually develops within 3–5 days of exposure and is limited to mucosal invasion. It is manifested by watery or mucoid diarrhoea, containing blood in severe cases, accompanied by vomiting, fever (40–41°C), inappetence or anorexia, lethargy, abdominal pain and progressive dehydration. Some cats drool excessively during the illness and often have pale mucous membranes (Timoney, 1976; Center *et al.*, 1995). Most animals recover in 3–4 weeks, but a carrier–excreter state may persist for a further 6 weeks.

In a few instances, acute enterocolitis can develop into an overwhelming infection, with septicaemia, endotoxic shock and signs consistent with the development of disseminated intravascular coagulation. Respiratory distress is often reported in these animals. In two cases involving young pups of about 2 months of age, septicaemia associated with *S. dublin* resulted in meningoencephalitis and neurological signs, such as left-sided hemiparesis (Milstein, 1975; Nation, 1984). Septicaemia in cats due to *S. arizonae* (Krum *et al.*, 1977) and *S. enteritidis* (Fox and Beaucage, 1979) has been reported.

Abortion attributable to *Salmonella* serovars is rare. Redwood and Bell (1983) reported an *in utero* infection caused by *S. panama* in two boxer bitches in the same breeding establishment. A pure growth of *S. typhimurium* was isolated from stillborn kittens (Reilly *et al.*, 1994) and *S. choleraesuis* was cultured from the uterus of a rural cat containing three dead fetuses (Hemsley, 1956).

S. typhimurium-associated unilateral conjunctivitis has been described in a cat (Fox and Galus, 1977). The condition was reproduced experimentally using an intraocular instillation of *S. typhimurium* (Fox *et al.*, 1984). Features common to the natural infection and experimental studies included moderate to severe conjunctivitis of several days' duration, with lacrimation, blepharospasm and a prominent nictitating membrane. Scleral injection and mucopurulent ocular discharge continued for several days, with large numbers of neutrophils in conjunctival scrapings, but with no signs of systemic illness. *S. typhimurium* was recovered from the faeces.

Some cats with salmonellosis are pyrexic and have a neutropenia with a left shift and vague non-specific signs of illness but no definite intestinal signs (Dow *et al.*, 1989).

An unusual case of *Salmonella*-associated

pneumonia was reported in an 8-year-old cat previously treated for severe dental calculus. The animal appeared to have no enteric manifestation of disease (Rodriguez *et al.*, 1993). *S. typhimurium* was isolated from kidney tissue of an 8-month-old dog with unilateral pyelonephrosis. No other organ systems were involved (Crow *et al.*, 1976). The animal recovered following surgical removal of the kidney and antibiotic therapy. A 7-year-old dog with intermittent epigastric pain, extending over a period of 8 months, was found to have chronic cholecystitis (Timbs *et al.*, 1974). *S. typhimurium* was present in the liver, the bile and a mesenteric lymph node. The main post-mortem finding was severe thickening of the gall-bladder wall. Histopathological findings included numerous inflammatory foci in the liver and mucosal congestion and haemorrhage in the intestines, without evidence of diarrhoea.

Pathological findings

When salmonellosis is confined to the intestinal tract, gross pathological findings may include an acute enteritis with bloodstained luminal contents, limited to the distal ileum, caecum and colon. Histopathological findings vary from catarrhal inflammation to villous atrophy, mucosal erosion and infiltration of neutrophils and macrophages into the lamina propria.

The carcasses of cats that die or are euthanized *in extremis* tend to be emaciated and dehydrated. The mucous membranes are pale and diffuse haemorrhagic enteritis is often present. Most of the parenchymatous organs may be grossly normal but the liver frequently contains small necrotic foci, about 2 mm in diameter. Blood-tinged fluid may be present in the serous cavities (Timoney, 1976). In some animals, the main pathological findings include petechial or ecchymotic haemorrhages, involving most organs including the lungs and epicardial and endocardial surfaces. Thrombosis and haemorrhage are consistent histopathological findings (Krum *et al.*, 1977).

Pups that have died rapidly from septicaemic salmonellosis may have good fat reserves. Multifocal or local extensive areas of consolidation are present in the lungs. The spleen, liver and kidneys may be swollen, pale and friable. Petechial or ecchymotic haemorrhages are often

present in many organs. Histopathological findings can include embolic pneumonia, splenitis, myocarditis, nephritis and meningoencephalitis. *Salmonella* localize in the vascular beds of lungs, heart, spleen, liver, kidneys and brain (Nation, 1984).

Diagnosis

Salmonellosis should be suspected in hospitalized dogs or cats developing fever and acute enterocolitis, particularly in those animals that have recently undergone major surgery, those with serious medical conditions or those receiving immunosuppressive therapy. Young pups or kittens with diarrhoea should be investigated for possible infection with *Salmonella*. Bloody diarrhoea and a degenerative left shift of the leucogram are consistent with salmonellosis (Calvert, 1985).

In severe disease, abnormal biochemical changes may include hypoalbuminaemia, elevated serum alkaline phosphatase levels and hypoglycaemia. Calvert (1985) found that a combination of: (i) hypoalbuminaemia with hypoglycaemia, or (ii) hypoalbuminaemia, elevated alkaline phosphatase levels and hypoglycaemia correlated with a poor prognosis.

The definitive diagnosis of salmonellosis requires the isolation of the organisms from affected tissue, blood, cerebral–spinal fluid or transtracheal washings, specimens that do not normally yield *Salmonella* in carrier animals. Because of the comparatively common occurrence of the asymptomatic excreter state, isolation of *Salmonella* from faecal samples does not confirm a causal relationship between the organism and the clinical disease in the animal (Center *et al.*, 1995). However, high numbers of *Salmonella* are usually recovered from animals with *Salmonella*-associated diarrhoea, while asymptomatic carrier animals shed fewer organisms intermittently and usually these organisms can be isolated only after enrichment procedures (Fox, 1991). The presence of large numbers of leucocytes in faeces is indicative of an invasive infection. *Salmonella* can be isolated from conjunctival scrapings from cats with conjunctivitis (Fox *et al.*, 1984), and in abortions and stillbirths the organisms are present in fetal and placental tissues (Redwood and Bell, 1983).

Isolation procedures include the inoculation of selective indicator media, such as modified brilliant green agar, xylose lysine desoxycholate (XLD) or *Salmonella–Shigella* (SS) media, as well as the use of an enrichment broth, such as selenite F, tetrathionate and Rappaport broths, which are selective for *Salmonella* serovars. The media are incubated at 37°C aerobically for 2–3 days (see Chapter 21). Aliquots from enrichment broth are subcultured on to further plates of selective indicator media. An initial presumptive identification of *Salmonella* organisms is based on their colonial appearance. Subsequently, suspect colonies are subcultured in indicator media, such as triple-sugar iron agar and lysine decarboxylase broth. These are incubated at 37°C for 18 h and examined for the characteristic appearance given by *Salmonella* serovars. The serovar can be established using culture from the slant of the triple-sugar iron agar in a slide agglutination test with commercially available typing sera (Quinn *et al.*, 1994). Antibiotic susceptibility tests should be conducted on *Salmonella* isolates associated with the septicaemic form of salmonellosis. Antibiotic-resistant strains of *Salmonella* are not infrequently isolated from cats and dogs (Timoney, 1978). Phage typing of some serovars, such as *S. typhimurium* and *S. enteritidis*, which is usually carried out in reference laboratories, can be useful in epidemiological studies.

A polymerase chain reaction (PCR)-hybridization technique for detecting *S. typhimurium* in canine faeces has been developed. It was found to be more sensitive than standard cultural procedures (Stone *et al.*, 1995).

Standard serologicalal tests may prove useful in groups of animals but have a limited application for individual dogs and cats. Differential diagnosis should include canine parvovirus and coronavirus infections, canine enteric campylobacteriosis, canine haemorrhagic gastroenteritis and feline panleucopenia.

Treatment

Antimicrobial therapy is unnecessary and possibly undesirable in uncomplicated *Salmonella* enterocolitis. Careful management of fluid and electrolyte balances usually promotes recovery (Fox, 1991). Antimicrobial therapy, particularly by the oral route, may alter the normal flora, pro-

long the shedding period for *Salmonella* and increase the risk of the development of transferable antibiotic resistance. Ampicillin administered parenterally or orally was considered to be a significant risk factor in nosocomial outbreaks of salmonellosis in dogs (Uhaa *et al.*, 1988).

Antimicrobial therapy is required for potentially fatal septicaemic salmonellosis. The choice should be based on isolation and antibiotic susceptibility tests. However, most isolates are susceptible to the fluoroquinolone enrofloxacin, administered at 5 mg kg^{-1} *per os* every 12 h (Rutgers *et al.*, 1994). Chloramphenicol and trimethoprim–sulphonamide combinations may also be effective. Transfusion of plasma can reduce mortality in dogs with signs of severe endotoxaemia. Immunoglobulin for the treatment of endotoxaemia in dogs is available commercially in some countries.

Prevention

Prevention of both the carrier–excreter state and cases of clinical salmonellosis is desirable from the human health aspect, particularly in households with young children. Complete prevention of *Salmonella* infections is difficult, because of the numerous sources of *Salmonella* serovars available to dogs and cats, the viability of the organisms in the environment and the comparatively common occurrence of carrier animals. Because of the wide range of serovars that may infect domestic carnivores, vaccination is not currently a feasible proposition. However, thorough cooking of the animals' food, particularly meat and poultry products, good hygiene and the use of heat-processed commercial food products would eliminate major sources of *Salmonella* for dogs and cats. Uneaten, moist food should not be allowed to remain in food bowls at ambient temperatures for long periods, as *Salmonella* may be able to replicate in the food. Dry meals and biscuits must be stored carefully to avoid contamination by rodents or insects. Dogs and cats should not be allowed access to water from dubious sources or to unpasteurized milk. The hunting and scavenging activities of dogs and cats may increase their exposure to *Salmonella* serovars, but such instinctive behaviour is not easily controlled.

Preventative measures are particularly important in veterinary hospitals, where predisposing factors could initiate outbreaks of clinical salmonellosis. In addition to the measures suggested for household dogs and cats, meticulous attention should be paid to routine cleaning and to sterilization or disinfection of cages, food bowls and utensils. Cages should be steam-cleaned or disinfected with sodium hypochlorite (3%) or iodophors (2%). Endoscopic and other equipment likely to become contaminated should be sterilized after use.

Staff can spread *Salmonella* between patients on their hands, footwear or clothing. Routine hand washing, before and after handling hospitalized animals, should be observed and protective clothing must be changed and washed regularly.

If possible, boarders and animals with minor ailments should be separated from animals most at risk, particularly those undergoing major surgery or receiving immunosuppressive therapy. An ideal procedure would be to identify *Salmonella* excreters on arrival by cultural methods on 3 consecutive days, using enrichment procedures for isolation. However, practical considerations, such as emergency surgery, may render this strategy impractical. Any animal admitted with diarrhoea or developing diarrhoea while in the hospital should be isolated, a faecal sample cultured for *Salmonella* and procedures instituted to prevent fomite transmission. Animals with clinical salmonellosis usually excrete large numbers of organisms, thus posing a serious threat to those in their immediate vicinity.

Public Health Aspects

Sources of infection

Cats and dogs tend to be the most popular household pets in developed countries. It has been estimated that, by the beginning of the next millennium, the numbers of pet cats in the UK will reach 7.8 million and the dog population will have declined slightly to 6.8 million (Anon., 1996). These companion animals can spend a large proportion of their lives indoors in close contact with their owners. Some cats frequent or live on food premises as part of a rodent control programme. Because of their agility, felines can gain access to and contaminate food preparation areas. Humans can become infected, usually by the oral route, by *Salmonella* shed in faeces of either carriers or clinically affected animals. *Salmonella* can be present on hands after handling animals, on fomites or in contaminated food. Humans are considered more likely to acquire infection from cats than from dogs. Cats tend to rake soil or litter over their faeces, thus contaminating their paws. Oral swabs from infected cats have revealed the consistent presence of *Salmonella* organisms, leading to the contamination of their fur during grooming (Timoney *et al.*, 1978). This presence of the organisms over the cat's coat increases the potential for transfer of *Salmonella* to the hands and clothing of humans handling such animals. Many reported cases of pet-associated salmonellosis occur in young children, where a direct route of transmission may be of importance, as infants tend to place objects into their mouths indiscriminately (Kaufmann, 1966).

Isolation of the multiply resistant *S. typhimurium* DT104, resistance (R) type ampicillin, chloramphenicol, streptomycin, sulphonamides, tetracycline (ACSSuT) from cats and dogs is of public-health concern. This definitive type has become the second most prevalent *Salmonella*, after *S. enteritidis* PT4, in the human population in England and Wales. During the period from 1991–1995, of 110 *Salmonella* isolated from cats, 70 were *S. typhimurium* and 40 of these were DT104, R type ACSSuT. Four of the strains had additional resistance to trimethoprim (Wall *et al.*, 1996). *S. typhimurium* DT104 was shown to be excreted in the faeces of infected cats for 12 weeks after recovery from an acute episode of enterocolitis (Wall *et al.*, 1995).

Prevention

It is standard policy in many countries to exclude domestic carnivores from premises where food is prepared, stored or served. Animals should not be allowed to eat from the same food dishes as humans. Individuals handling dogs and cats must be made aware of the need to thoroughly wash their hands after contact with animals, especially before serving or consuming food. Because clinical salmonellosis is uncommon in dogs and cats, animals with diarrhoea may often be treated empirically, without submission of a faecal sample for culture. Veterinarians should consider the possibility of salmonellosis in animals with enterocolitis and advise their clients on the zoonotic potential of such infections.

Acknowledgements

The authors wish to thank Mrs Lesley Doggett for typing the manuscript and Mrs Dores Maguire for the artwork.

References

Adler, H.E., Willers, E.H. and Levine, M. (1951) Incidence of *Salmonella* in apparently healthy dogs. *Journal of the American Veterinary Medical Association* 118, 300–304.

Anon. (1996) Cats as man's best friend. *Veterinary Record* 138, 99.

Ball, M.R. (1951) Salmonellosis in dogs and cats in the Los Angeles, Honolulu and Bermuda areas. *Journal of the American Veterinary Medical Association* 118, 164–166.

Borland, E.D. (1975) *Salmonella* infection in dogs, cats, tortoises and terrapins. *Veterinary Record* 96, 401–402.

Bruner, D.W. (1973) *Salmonella* cultures typed during the years 1950–1971 for the Service Laboratories of New York State Veterinary College at Cornell University. *Cornell Veterinarian* 63, 138–143.

Bruner, D.W. and Moran, A.B. (1949) *Salmonella* infections in domestic animals. *Cornell Veterinarian* 39, 53–63.

Buxton, A. (1957) Symptoms and lesions of *Salmonella* infections in animals: dogs and cats. In: *Salmonellosis in animals: a Review*. Review Series No. 5, Commonwealth Agricultural Bureaux, Farnham Royal, p. 101.

Calvert, C.A. (1985) Salmonella infections in hospitalized dogs: epizootiology, diagnosis and prognosis. *Journal of the American Animal Hospital Association* 21, 499–503.

Calvert, C.A. and Leifer, C.E. (1982) Salmonellosis in dogs with lymphosarcoma. *Journal of the American Veterinary Medical Association* 180, 56–58.

Carter, M.E., Dewes, H.G. and Griffiths, O.V. (1979) Salmonellosis in foals. *Journal of Equine Medicine and Surgery* 3, 78–83.

Carter, M.E., Cordes, D.O. and Carman, M.G. (1983) Observations on acute salmonellosis in four Waikato dairy herds. *New Zealand Veterinary Journal* 31, 10–12.

Center, S.A., Hornbuckle, W.E. and Hoskins, J.D. (1995) The liver and pancreas. In: Hoskins, J.D. (ed.) *Veterinary Pediatrics*. W.B. Saunders, Philadelphia, pp. 212–213.

Chengappa, M.M., Staats, J., Oberst, R.D., Gabbert, N.H. and McVey, S. (1993) Prevalence of *Salmonella* in raw meat used in diets of racing greyhounds. *Journal of Veterinary Diagnostic Investigation* 5, 372–377.

Clarke, R.C. and Gyles, C.L. (1993) *Salmonella*. In: Gyles, C.L. and Thoen, C.O. (eds) *Pathogenesis of Bacterial Infections in Animals*. Iowa State University Press, Ames, pp. 133–153.

Crow, S.E., Lauerman, L.H. and Smith, K.W. (1976) Pyelonephrosis associated with *Salmonella* infection in a dog. *Journal of the American Veterinary Medical Association* 169, 1324–1326.

Davies, R.H. and Wray, C. (1995) Mice as carriers of *Salmonella enteritidis* on persistently infected poultry units. *Veterinary Record* 137, 337–341.

Day, W.H., James, E. and Heather, C.D. (1963) Salmonellosis in the dog. *American Journal of Veterinary Research* 24, 156–158.

Dow, S.W., Jones, R.L., Henik, R.A. and Husted, P.W. (1989) Clinical features of salmonellosis in cats: six cases (1981–1986). *Journal of the American Veterinary Medical Association* 10, 1464–1466.

Förster, D., Holland, U. and Tesfamariam, H. (1974) Occurrence of *Salmonella* in the dog. *Zentralblatt für Veterinärmedizin B* 21, 120–134.

Fox, J.G. (1991) *Campylobacter* infections and salmonellosis. *Seminars in Veterinary Medicine and Surgery (Small Animal)* 6, 212–218.

Fox, J.G. and Beaucage, C.M. (1979) The incidence of *Salmonella* in random-source cats purchased for use in research. *Journal of Infectious Diseases* 139, 362–365.

Fox, J.G. and Galus, C.B. (1977) *Salmonella*-associated conjunctivitis in a cat. *Journal of the American Veterinary Medical Association* 171, 845–847.

Fox, J.G., Beaucage, C.M., Murphy, J.C. and Niemi, S.M. (1984) Experimental *Salmonella*-associated conjunctivitis in cats. *Canadian Journal of Comparative Medicine* 48, 87–91.

Frost, A.J., Eaton, N.F., Gilchrist, D.J. and Moo, D. (1969) The incidence of *Salmonella* infection in the dog. *Australian Veterinary Journal* 45, 109–110.

Galton, M.M., Scatterday, J.E. and Hardy, A.V. (1952) Salmonellosis in dogs. I. Bacteriological, epidemiological and clinical considerations. *Journal of Infectious Diseases* 91, 1–5.

Galton, M.M., Harless, M. and Hardy, A.V. (1955) *Salmonella* isolations from dehydrated dog meals. *Journal of the American Veterinary Medical Association* 126, 57–58.

Gorham, J.R. and Garner, F.M. (1951) The incidence of *Salmonella* infections in dogs and cats in a non-urban area. *American Journal of Veterinary Research* 12, 35–37.

Groisman, E.A., Fields, P.I. and Heffron, F. (1990) Molecular biology of *Salmonella* pathogenesis. In: Iglewski, B.H. and Clark, V.L. (eds) *The Bacteria. XI Molecular Basis of Bacterial Pathogenesis.* Academic Press, San Diego, pp. 251–272.

Hemsley, L.A. (1956) Abortion in two cats, with the isolation of *Salmonella choleraesuis* from one case. *Veterinary Record* 68, 152.

Humphrey, T.J. (1990) Growth of salmonellas in intact shell eggs: influence of storage temperatures. *Veterinary Record* 126, 292.

Humphrey, T.J., Baskerville, A., Chart, H. and Rowe, B. (1989) Infection of egg-laying hens with *Salmonella enteritidis* PT4 by oral inoculation. *Veterinary Record* 125, 531–532.

Jones, G.W., Rabert, D.K. Svinarich, D.M. and Whitfield, H.J. (1982) Association of adhesive, invasive and virulent phenotypes of *Salmonella typhimurium* with autonomous 60-megadalton plasmids. *Infection and Immunity* 38, 476–486.

Kaufmann, A.F. (1966) Pets and *Salmonella* infection. *Journal of the American Veterinary Medical Association* 149, 1655–1661.

Khan, A.Q. (1970) *Salmonella* infections in dogs and cats in the Sudan. *British Veterinary Journal* 126, 607–611.

Kintner, L. (1949) Canine salmonellosis. *Veterinary Medicine* 44, 396–398.

Kirkwood, J.K., Holmes, J.P. and Macgregor, S. (1995) Garden bird mortalities. *Veterinary Record* 136, 372.

Krum, S.H., Stevens, D.R. and Hirsh, D.C. (1977) *Salmonella arizonae* bacteraemia in a cat. *Journal of the American Veterinary Medical Association* 170, 42–44.

Laing, P.W. (1990) *Salmonella typhimurium* in various species. *Veterinary Record* 126, 173.

McElrath, H.B., Galton, M.M. and Hardy, A.V. (1952) Salmonellosis in dogs. III Prevalence in dogs in veterinary hospitals, pounds and boarding kennels. *Journal of Infectious Diseases* 91, 12–14.

Mackel, D.C., Galton, M.M., Gray, H. and Hardy, A.V. (1952) Salmonellosis in dogs. IV Prevalence in normal dogs and their contacts. *Journal of Infectious Diseases* 91, 15–18.

Madewell, B.R. and McChesney, A.E. (1975) Salmonellosis in a human infant, a cat and two parakeets in the same household. *Journal of the American Veterinary Medical Association* 167, 1089–1090.

Milstein, M. (1975) *Salmonella dublin* septicaemia in a Scottish terrier recently imported from England. *Canadian Veterinary Journal* 16, 179–180.

Morse, E.V. and Duncan, M.A. (1975) Canine salmonellosis: prevalence, epizootiology, signs and public health significance. *Journal of the American Veterinary Medical Association* 167, 817–820.

Nakamura, M., Sato, S., Ohya, T., Suzuki, S. and Ikeda, S. (1985) Possible relationship of a 36-megadalton *Salmonella enteritidis* plasmid to virulence in mice. *Infection and Immunity* 47, 831–833.

Nation, P.N. (1984) *Salmonella dublin* septicaemia in two puppies. *Canadian Veterinary Journal* 25, 324–326.

Neill, S.D., McNulty, M.S., Bryson, D.G. and Ellis, W.A. (1981) Microbiological findings in dogs with diarrhoea. *Veterinary Record* 109, 538–539.

Quinn, P.J., Carter, M.E., Markey, B.K. and Carter, G.R. (1994) Enterobacteriaceae. In: Quinn, P.J., Carter, M.E., Markey, B.K. and Carter, G.R. (eds) *Clinical Veterinary Microbiology.* Mosby-Year Book Europe, London, pp. 209–236.

Redwood, D.W. and Bell, D.A. (1983) *Salmonella panama*: isolation from aborted and newborn canine fetuses. *Veterinary Record* 112, 362.

Reilly, G.A.C., Bailie, N.C., Morrow, W.T., McDowell, S.W.J. and Ellis, W.A. (1994) Feline stillbirths associated with mixed *Salmonella typhimurium* and leptospira infection. *Veterinary Record* 135, 608.

Rodriguez, C.O., Moon, M.L. and Leib, M.S. (1993) *Salmonella choleraesuis* pneumonia in a cat without signs of gastrointestinal tract disease. *Journal of the American Veterinary Medical Association* 202, 953–955.

Rutgers, H.C., Stepien, R.L., Elwood, C.M., Simpson, K.W. and Batt, R.M. (1994) Enrofloxacin treatment of Gram-negative infections. *Veterinary Record* 135, 357–359.

Salyers, A.A. and Whitt, D.D. (1994) *Salmonella* infections. In: Salyers, A.A. and Whitt, D.D. (eds) *Bacterial Pathogenesis – a Molecular Approach.* ASM Press, Washington, DC, pp. 229–243.

Shimi, A. and Barin, A. (1977) *Salmonella* in cats. *Journal of Comparative Pathology* 87, 315–318.

Shimi, A., Keyhani, M. and Bolurchi, M. (1976) Salmonellosis in apparently healthy dogs. *Veterinary Record* 98, 110–111.

Sinell, H.J., Klingbeil, H. and Benner, M. (1984) Microflora of edible offal with particular reference to *Salmonella. Journal of Food Protection* 47, 481–484.

Stone, G.G., Oberst, R.D., Hays, M.P., McVey, S., Galland, J.C., Curtiss, R., Kelly, S.M. and Chengappa, M.M. (1995) Detection of *Salmonella typhimurium* from rectal swabs of experimentally infected beagles by short cultivation and PCR-hybridization. *Journal of Clinical Microbiology* 33, 1292–1295.

Stucker, C.L., Galton, M.M., Cowdery, J. and Hardy, A.V. (1952) Salmonellosis in dogs. II Prevalence and distribution in greyhounds in Florida. *Journal of Infectious Diseases* 91, 6–11.

Takeuchi, A. and Sprinz, H. (1967) Electron microscope studies of experimental *Salmonella* infections in the preconditioned guinea-pig. 2. Response of the intestinal mucosa to the invasion of *Salmonella typhimurium*. *American Journal of Pathology* 51, 137–146.

Tanaka, Y., Katsube, Y. and Imaizumi, K. (1976a) Distribution of salmonellae in the digestive tract and lymph node of carrier-dogs. *Japanese Journal of Veterinary Science* 38, 215–224.

Tanaka, Y., Katsube, Y. and Imaizumi, K. (1976b) Experimental carrier in dogs produced by oral administration of *Salmonella typhimurium*. *Japanese Journal of Veterinary Science* 38, 569–578.

Tanaka, Y., Katsube, Y. and Imaizumi, K. (1977) Virulence of *Salmonella typhimurium* harboured in feces of carrier-dog for mice. *Japanese Journal of Veterinary Science* 39, 353–356.

Terakado, N., Sekizaki, T., Hashimoto, K. and Naitoh, S. (1983) Correlation between the presence of a fifty-megadalton plasmid in *Salmonella dublin* and virulence for mice. *Infection and Immunity* 41, 443–444.

Timbs, D.V., Durham, P.J.K. and Barnsley, D.G.C. (1974) Chronic cholecystitis in a dog infected with *Salmonella typhimurium*. *New Zealand Veterinary Journal* 22, 100–102.

Timbs, D.V., Davies, G.B., Carter, M.E. and Carman, M.G. (1975) The *Salmonella* excretor incidence of dogs in Hawke's Bay. *New Zealand Veterinary Journal* 23, 54–56.

Timoney, J.F. (1976) Feline salmonellosis. *Veterinary Clinics of North America: Small Animal Practice* 6(3), 395–398.

Timoney, J.F. (1978) The epidemiology and genetics of antibiotic resistance of *Salmonella typhimurium* isolated from diseased animals in New York. *Journal of Infectious Diseases* 137, 67–73.

Timoney, J.F. Neibert, H.C. and Scott, F.W. (1978) Feline salmonellosis: a nosocomial outbreak and experimental studies. *Cornell Veterinarian* 68, 211–219.

Uhaa, I.J., Hird, D.W., Hirsh, D.C. and Jang, S.S. (1988) Case–control study of risk factors associated with nosocomial *Salmonella krefeld* infection in dogs. *American Journal of Veterinary Research* 49, 1501–1505.

Venter, B.J. (1988) Epidemiology of salmonellosis in dogs – a conceptual model. *Acta Veterinaria Scandinavica* 84, 333–336.

Veterinary Investigation Services (1988–1996) Reports, published in the *Veterinary Record*.

Wall, P.G., Davis, S., Threlfall, E.J., Ward, L.R. and Ewbank, A.J. (1995) Chronic carriage of multidrug resistant *Salmonella typhimurium* in a cat. *Journal of Small Animal Practice* 36, 279–281.

Wall, P.G., Threlfall, E.J., Ward, L.R. and Rowe, B. (1996) Multiresistant *Salmonella typhimurium* DT 104 in cats: a public health risk. *Lancet* 348, 471.

Weber, W.J. (1982) Salmonellosis. In: Weber, W.J. *Diseases Transmitted by Rats and Mice*. Thomson Publications, Fresno, California, pp. 46–52.

Wilson, J.E. and MacDonald, J.W. (1967) *Salmonella* infection in wild birds. *British Veterinary Journal* 123, 212–219.

Chapter 15

Public-health Aspects of *Salmonella* Infection

Tom Humphrey

PHLS Food Microbiology Research Unit, Church Lane, Heavitree, Exeter EX2 5AD, Devon, UK

Historical Perspective

Gaffky is credited with being the first to culture the causative agent of typhoid (*S. typhi*) in 1884. At that time, it was also known that bacteria similar to *S. typhi* could cause enteric disease in humans and farm animals. This was confirmed when Salmon and Smith reported the isolation of the bacteria responsible for 'hog cholera' or 'swine fever' in 1885. The name *Salmonella* was subsequently adopted in honour of Salmon, an American veterinary surgeon. In the early to mid-20th century, there were many pioneering studies into the identification and differentiation of *Salmonella*. The schemes that were developed made use of the fact that, although *Salmonella* show considerable antigenic diversity, they elicit two principal antibody reactions in infected animals. Thus antibodies are produced against cell-surface or somatic O antigens and flagella or H antigens. In 1929, White developed a typing scheme based on this antigenic variation, which was later modified by Kauffmann. This work enabled the differentiation of *Salmonella* into serovars. In the UK, as in many other countries, it is usual to supplement this with phage typing. With common serovars or phage types (PTs), it may be necessary to take identification further and include a range of molecular or genotypic typing methods.

Space does not permit a detailed analysis of the progression of *Salmonella* epidemiology in many countries and therefore discussion on this aspect will be largely confined to England and Wales, as trends in these countries are indicative of those in many others.

Up to the Second World War, most *Salmonella* isolated from human cases belonged to 14 serovars that were largely indigenous to the UK. The epidemiological picture was dominated by *S. typhimurium*, *S. enteritidis*, *S. newport* and *S. choleraesuis*. In the early 1940s, it became obvious that there had been a marked increase in the numbers of different serovars isolated from human cases. Ten new serovars were identified in 1942 and eight more in 1943. Almost all of these were novel to the UK and had been described elsewhere, particularly in the USA. It became apparent that many of these 'new' *Salmonella* serovars had been introduced into the UK from imported spray-dried egg. The pattern of human infection has gradually become more complicated ever since, with more and more *Salmonella* serovars being identified. As this chapter will demonstrate, however, although there are over 2400 identifiably different *Salmonella* serovars, the epidemiological picture tends to be dominated by two or three serovars only. *S. enteritidis* and *S. typhimurium* continue to predominate, although, in recent years in Western Europe, the former has become markedly more common than the latter.

Recent Trends in Human Infection

Human *Salmonella* infection continues to be a major problem, in terms of both morbidity and economic cost (Barnass *et al.*, 1989).

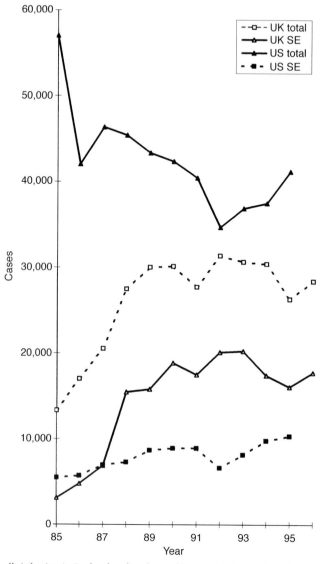

Fig. 15.1. *Salmonella* infection in England and Wales and the USA. SE, *S. enteritidis.*

In many countries, there was a marked upsurge in the number of cases of human salmonellosis between the mid-1980s and mid-1990s. Much of this has been associated with *S. enteritidis*, and the worldwide increase in human infection has been referred to as a pandemic (Rodrigue *et al.*, 1990). This bacterium is primarily chicken-associated and vehicles most commonly identified with human infection are contaminated chicken meat or eggs (see later). There are approximately 50 PTs of *S. enteritidis*. When the pandemic began

in the mid-1980s, there was a clear geographical distribution of PTs. In Eastern Europe, PT1 predominated, in Western Europe PT4 was the most important, whereas in North America most cases were caused by either PT8 or PT13a (Rodrigue *et al.*, 1990). There are no obvious reasons for the above distributions, particularly as it is thought that the *S. enteritidis* pandemic started as a single event, perhaps in one of the élite poultry breeding flocks. Since the mid-1980s, there has been a gradual change in the epidemiological pattern of

Table 15.1. The five most common *Salmonella* serovars isolated from human cases in England and Wales and the USA, 1991–1995.

	England and Wales			USA		
Year	Total	Serovar	(%)	Total	Serovar	(%)
1991	27,693	*S. enteritidis*	(63)	40,443	*S. enteritidis*	(22)
		S. typhimurium	(19)		*S. typhimurium*	(21)
		S. virchow			*S. heidelberg*	
		S. newport			*S. hadar*	
		S. agona			*S. newport*	
1992	31,355	*S. enteritidis*	(64)	34,688	*S. typhimurium*	(22)
		S. typhimurium	(17)		*S. enteritidis*	(19)
		S. virchow			*S. heidelberg*	
		S. newport			*S. hadar*	
		S. hadar			*S. newport*	
1993	30,650	*S. enteritidis*	(66)	36,917	*S. typhimurium*	(23)
		S. typhimurium	(16)		*S. enteritidis*	(22)
		S. virchow			*S. heidelberg*	
		S. newport			*S. newport*	
		S. hadar			*S. hadar*	
1994	30,559	*S. enteritidis*	(57)	37,522	*S. enteritidis*	(26)
		S. typhimurium	(18)		*S. typhimurium*	(21)
		S. virchow			*S. heidelberg*	
		S. hadar			*S. newport*	
		S. newport			*S. hadar*	
1995	28,791	*S. enteritidis*	(52)	41,222	*S. enteritidis*	(25)
		S. typhimurium	(21)		*S. typhimurium*	(22)
		S. virchow			*S. newport*	
		S. hadar			*S. heidelberg*	
		S. newport			*S. hadar*	

S. enteritidis infection. PT4 has spread to many countries other than those in Western Europe and it is becoming increasingly important in North America (Alktekruse *et al.*, 1997) and Japan (S. Kumagai, personal communication, 1997). The number of cases of human infection in England and Wales caused by *S. enteritidis* remained largely unchanged between 1991 and 1997, where PT4 predominated, although cases of *Salmonella* infection declined in 1998. In England and Wales, other PTs of *S. enteritidis*, such as PT6, are increasing in importance and cases of infection caused by these bacteria have markedly increased in the last 3 years (L. Ward, personal communication, 1998). PT4 continues to cause widespread infection throughout Western Europe (Schmidt, 1995), and it is of interest that, despite the widespread nature of *S. enteritidis*, the infection rates in different countries appear to be so variable. The increase in reported cases in England and Wales occurred in the second half of the 1980s (Fig. 15.1) and this trend is typical of much of

Western Europe. There are marked regional variations in North America and this is also seen in Europe. Thus, while cases increased markedly in Germany, from 50,633 in 1981 to 195,378 in 1992, cases in The Netherlands fell from 10,783 to 2590 in the same period (K. Stöhr, Geneva, personal communication, 1996).

The continued importance of *S. enteritidis* as a human pathogen has tended to overshadow the other *Salmonella* serovars. It should not be forgotten that the great majority of the c. 2400 serovars are capable of causing infection in humans. In any given year, many hundreds of different *Salmonella* will be isolated from human cases, food animals and foods derived from them. *S. typhimurium* is the second commonest human isolate in England and Wales (Table 15.1) and this picture is repeated throughout much of the rest of Western Europe (Schmidt, 1995). In England and Wales, *S. enteritidis* and *S. typhimurium* account for over 80% of reported human *Salmonella* infections (Table 15.1).

The ubiquity of *Salmonella* and their ability to infect or colonize food animals mean that almost any food can cause human infection, particularly foods of animal origin, and most animal-derived food groups will provide examples of vehicles of *Salmonella* infection. It is necessary to differentiate between short-term problems, which may result from a chance infection that leads to the contamination of a particular product or foodstuff, and the underlying background of human salmonellosis brought about by sustained infection or carriage in some food animals. Thus, the occasional colonization or infection of dairy cows (Gay *et al.*, 1994) may result in *Salmonella* outbreaks associated with either raw (Maguire *et al.*, 1992) or improperly pasteurized (Ryan *et al.*, 1987) milk products. A similar relationship exists between *Salmonella* infection in pigs and contamination of pork products (Cowden *et al.*, 1989a; Maguire *et al.*, 1993). In general, outbreaks of this kind show a pronounced epidemiological picture. Infection in a particular food animal will lead to contamination of a certain food product, which in turn will cause cases of human *Salmonella* infection. Two nationwide outbreaks in England and Wales, which involved S. *agona* in 1975 and S. *hadar* in the years between 1977 and 1981 (Anon., 1989), are a good illustration of this phenomenon. Both were associated with poultry infected from contaminated imported feed. The problem was identified by surveillance. Control measures were instituted and human cases showed a rapid decline. Such events are noteworthy because of their comparative rarity. In truth, human cases are dominated by only one or two *Salmonella* and/or by a defined range of food products. This will be illustrated by reference to two *Salmonella* of current international importance.

The first is S. *typhimurium* definitive type (DT) 104. Isolations from, and clinical cases in, cattle and latterly pigs, sheep and poultry, showed a marked increase in the UK in 1991 (Anon., 1995a). This was followed by an increase in human cases. The outbreak caused particular concern, as most isolates of the epidemic strains of DT104 were resistant to many commonly used antibiotics (Threlfall *et al.*, 1992, 1994). Given that S. *typhimurium* DT104 is associated with many different food animals, it is not surprising that case–control studies revealed that

a range of foods were vehicles of human infection. These included sausages, cooked chicken, a commercial brand of paté and milk products (Wall *et al.*, 1994).

A working contact with infected domestic animals has also been found to be an important risk factor (Fone and Barker, 1994). S. *typhimurium* DT104 has also shown a marked recent increase in parts of the USA (Alktekruse *et al.*, 1997), but cases in the UK peaked in 1995 and have declined by about 50% since then (Anon., 1999). Although S. *typhimurium* DT104 has many phenotypic attributes in common with S. *enteritidis* PT4 (Humphrey *et al.*, 1997a), including highly invasive behaviour in chickens (Williams *et al.*, 1998; Leach *et al.*, 1999), it may not share its longevity as a human pathogen. The continuing importance of the latter bacterium as a human and poultry pathogen requires a more detailed consideration of its behaviour in foods and food animals, and this will be discussed in the following section.

Common Vehicles of Infection

Table 15.2 shows some data on food vehicles of PT4 and DT104 infection in England and Wales. Table 15.2 is dominated by poultry meat and eggs. It cannot be emphasized too strongly, however, that the ubiquity of *Salmonella* means that any food, if not handled properly and protected from contamination, can cause infection. *Salmonella* are not overly fastidious in either their growth temperature or their nutritional requirements. They will grow at temperatures of between 7 and 46°C on a wide range of cooked or raw foods (Ingham *et al.*, 1990; Golden *et al.*, 1993) and storage at non-refrigeration temperatures has been shown to be important in many outbreaks (Roberts, 1986). For example, Luby *et al.* (1993), reported a large outbreak, in which customers became infected with either S. *agona* or S. *hadar*, because cooked turkey was held unrefrigerated in a small restaurant kitchen for several hours, rinsed with water to remove offensive odour and incompletely reheated before serving.

Outbreak analysis by the Public Health Laboratory Service (PHLS) in England and Wales has enabled the identification of important risk factors in outbreaks of food poisoning, where the

Table 15.2. Food vehicles in outbreaks of *S. enteritidis* PT4 and *S. typhimurium* DT104 infections in England and Wales. (Data from Anon., 1997.) Table only includes data from outbreaks where vehicle was identified.

Food vehicle	PT4 (1989–1996)	DT104 (1992–1995)
Egg and egg dishes	103	2
Desserts*	98	–
Poultry	75	12
Red meat/meat products	39	10
Fish/shellfish	18	1
Salad/fruit/vegetables	17	–
Sauces*	12	–
Milk/milk products	9	5
Miscellaneous foods	130	3

* Many desserts or sauces contained raw shell egg.

most commonly identified human pathogen is *Salmonella*. Most outbreaks occur as a consequence of the delay between preparation/cooking and consumption of the food, as this gives *Salmonella* time to reach high numbers, which means that infection may be more widespread and thus more easily identified and possibly more severe. The important food-associated risk factors are inadequate cooking or reheating, storage at inappropriate temperatures and cross-contamination from raw to cooked foods. Infected/carrier food handlers are relatively unimportant as the cause of *Salmonella* outbreaks and evidence, albeit indirect, suggests that people (including excreters of *Salmonella*) who have passed 'formed' stools and practise reasonable hygiene are rarely responsible for either initiating or extending an outbreak (Cruickshank and Humphrey, 1987). *Salmonella* are readily removed by hand washing (Pether and Gilbert, 1971). Catering staff are often found to have eaten the food that has caused an outbreak and are therefore victims rather than perpetrators. It should be noted, however, that food handlers suffering from diarrhoea constitute a significant risk and should be excluded from work for 48 h following recovery (Anon., 1995b).

While the direct involvement of carrier food handlers as disseminators of *Salmonella* is doubtful, their potential role in food cross-contamination is beyond dispute. *Salmonella*-contaminated raw products, principally meat and poultry, have been shown to bring about extensive conta-

mination of the kitchen environment (de Wit *et al.*, 1979; de Boer and Hahne, 1990). The homogenization of contaminated egg contents can also result in the production of contaminated droplets, which distribute *S. enteritidis* widely in the environment around mixing bowls. The bacteria were found to be capable of prolonged survival (> 24 h), whether in small egg droplets or in smears that were largely invisible to the naked eye (Humphrey *et al.*, 1994). This in itself is not necessarily important in the infection process but *Salmonella* such as PT4 and DT104, which are capable of persistence (Humphrey *et al.*, 1995; 1997a), are more likely to come into contact with foodstuffs which might permit multiplication. Recent work by Bradford *et al.* (1997) found that contact times of as little as 5 s were sufficient to transfer PT4 from dried egg droplets to either cooked beef or melon. Once on these foods, the bacterium was capable of rapid growth at 20°C. Food handlers must adopt the highest possible hygiene standards, not only in terms of personal hygiene but also in their handling and storage of foodstuffs, particularly where they are using both raw and cooked foods.

Milk and Milk Products

Clearly, the use and maintenance of proper kitchen hygiene are important in the control of salmonellosis and other types of food poisoning. There have been many outbreaks, however, that have resulted from failures in process hygiene. Milk and milk products provide good examples of outbreaks of this type. Pasteurized milk has an enviable safety record, although some outbreaks have been associated with this product. A recent outbreak in England and Wales of *S. goldcoast* infection, in which cheese made from 'pasteurized' milk was identified as the vehicle, was caused by serious faults in the pasteurization process (Anon., 1997). This is a relatively unusual occurrence, however, and contamination after heat treatment is more likely. An example of this was a large, multi-state outbreak of salmonellosis in the USA, where production faults meant that pasteurized milk was contaminated with the raw product (Ryan *et al.*, 1987).

It would seem that certain strains or serovars of *Salmonella* may also be more resistant to processes used in food production than previously thought. Such tolerance and the ability of *Salmonella* to respond to and interact with their environment present considerable challenges to the food-production industries and can lessen margins of error in food-treatment processes. All *Salmonella* studied so far can be killed by properly applied pasteurization. Many other heat treatments that are common in the production of milk products, for example, are believed to be able to kill *Salmonella*. The spray-drying process applied to milk may not necessarily fall into this category, however, and can only be applied safely if the milk used has been pasteurized and any failures in this initial process can compromise the safety of the final products. In addition, the frequent changes in temperature within the spray drier can cause cracks in the internal metal cladding. These may become contaminated with milk and milk powder and occasionally *Salmonella*. An outbreak of *S. ealing* infection in young children in England and Wales in the mid-1980s was caused by spray-dried milk powder, which was believed to have been contaminated with *Salmonella* by milk powder leaking from cracks in the metal liner in the spray drier (Rowe *et al.*, 1987).

Salmonella can survive in certain cheeses, and such products have been implicated in some large outbreaks. In Canada, 1500 people were infected after consuming Cheddar-type cheese contaminated with raw milk (D'Aoust, 1985). More recently, in the UK, an outbreak of infection with *S. dublin* was associated with soft, unpasteurized, cow's-milk cheese (Maguire *et al.*, 1992). Similarly, a large outbreak of *S. paratyphi* B (var. Java) infection in France in 1993, in which over 273 cases were identified, was caused by a contaminated goat's-milk cheese (Desenclos *et al.*, 1996).

The principal vehicle in milk-associated human salmonellosis is unpasteurized milk, and many outbreaks have been caused by its consumption. A recent PHLS survey found that raw milk on sale to the public in England and Wales was frequently of an unsatisfactory microbiological quality and sometimes *Salmonella*-positive (Anon., 1997). Prohibition of the sale of unpasteurized milk imposed in Scotland in 1983 has all but eliminated milk-borne outbreaks in that country. Sale of raw milk is also prohibited in many other European countries. In England and Wales, it is still possible for consumers to purchase unpasteurized milk, although the use of raw milk is banned in schools.

Raw milk may become *Salmonella*-positive as a consequence of systemic infection with *Salmonella*, such as *S. dublin* or *S. typhimurium* DT104, or infection of the udder. In general, however, milk becomes contaminated as a result of faecal contamination. Intestinal carriage of *Salmonella* in dairy cows can be prolonged (Giles *et al.*, 1989). An analysis of outbreaks of milk-borne *Salmonella* infection between 1951 and 1980 revealed that in 49 the milk became *Salmonella*-positive as a result of faecal contamination and in 22 by udder excretion (Galbraith *et al.*, 1982). Properly applied milking hygiene can reduce contamination levels in the raw milk but cannot be relied on as the only control measure. To ensure consumer safety, milk should be pasteurized. The example of cheese-associated outbreaks given earlier also demonstrates that acidification alone will also not control *Salmonella*, particularly as some wild-type isolates, like those of *Escherichia coli* O157:H7 (Waterman and Small, 1996), naturally show enhanced acid tolerance (Humphrey *et al.*, 1995, 1997a).

Other Foods

There are many examples of 'unusual' foods causing *Salmonella* outbreaks, often on a sufficiently large scale to cause national problems. Recent outbreaks include a savoury maize snack contaminated with *S. agona* (Killalea *et al.*, 1996), bean sprouts that contained *S. saintpaul* (O'Mahony *et al.*, 1990) and salami that was *S. typhimurium* DT124-positive (Cowden *et al.*, 1989a). Chocolate products, which normally give little cause for concern because of their low water activity, have caused a few outbreaks of salmonellosis. More than 170 cases of infection with *S. eastbourne* were reported in Canada and the USA in 1973–1974 after the consumption of chocolate products (Craven *et al.*, 1975), and in 1982 imported Italian chocolate containing *S. napoli* gave rise to many cases of disease in England and Wales (Gill *et al.*, 1983). The interesting feature in both these outbreaks was that the dose sufficient to cause illness was low, probably no more than 1000 organisms in the *S. eastbourne* outbreak, while numbers of *S. napoli* found in the Italian chocolate ranged from 2 to 23 g^{-1} (Greenwood and Hooper, 1983). These numbers are much lower than those normally associated with infection and it may be that the *Salmonella* strain involved was more virulent or that the vehicles were particularly protective. The latter is more likely and will be discussed again later (see Table 15.4).

Contamination of Meat and Meat Products

It is the view of the British Medical Association that all raw meat and raw-meat products should be regarded as being potentially contaminated with bacteria that can cause food-borne disease. Comminuted meat products, such as sausages, are often *Salmonella*-positive. A PHLS survey of sausages purchased from retail outlets found that there was a close relationship between price and presence of *Salmonella* contamination. Low-price catering or economy sausages were much more frequently contaminated than more expensive brands (Nichols and de Louvois, 1995). This is believed to be due to the use of mechanically recovered chicken meat in the cheaper sausages. The mincing of meat to produce sausages or beef-burgers will also introduce *Salmonella*, which are usually surface contaminants, into the interior of the product. *Salmonella* are able to attach to freshly exposed muscle tissues and attachment significantly increases heat tolerance (Humphrey *et al.*, 1997b), which might permit survival when cooking is not complete. Sausages were identified as vehicles in sporadic cases of *S. typhimurium* DT104 infections in England and Wales in the early 1990s (Wall *et al.*, 1994). Recent work has shown that cells of *S. typhimurium* DT104 inoculated into sausages that were then cooked from frozen were not all destroyed, even when the manufacturer's cooking instructions were followed (Anon., 1997).

Salmonella are usually introduced on to carcasses as a result of faecal contamination. The extent to which this happens is governed by the degree of carriage in the live animal and the hygiene of the slaughter process. There is much variation in reported isolation rates of *Salmonella* from red-meat animals. For example, despite the current widespread infection with DT104, *Salmonella* is isolated relatively infrequently from adult sheep and cattle in the UK (Mackey, 1989), but isolation is more common in parts of

Australia and the USA, where higher levels of intestinal carriage have been demonstrated (Samual *et al.*, 1980). The prevalence of *Salmonella*-positive red-meat animals will increase as a result of mixing at markets, where negative animals will mix with those that are colonized/infected (Wray *et al.*, 1991). The stress of transportation of all animals may lead to a recrudescence of latent infections (Williams and Spencer, 1973). *Salmonella* can also spread rapidly during transportation or in the lairage prior to slaughter. Grau and Smith (1974) in Australia demonstrated that the prevalence of *Salmonella*-positive lamb carcasses was related to the length of time the animals were held before they were killed.

Improvements in slaughterhouse hygiene will do much to reduce carcass contamination rates. The Meat Hygiene Service (MHS) in the UK has introduced a system of assessment of the fleece/hide cleanliness of animals arriving at slaughterhouses. Animals are graded from 1 to 5, with the higher scores denoting those that are unclean. Slaughterhouse operators can refuse to accept animals that are too dirty or can insist that they are cleaned before slaughter. This measure is aimed at the control of *E. coli* O157, but it should also reduce levels of contamination with *Salmonella* and other enteropathogens.

Poultry meat as a vehicle of *Salmonella* infection

Most chickens produced for the commercial market are reared using the broiler system. Birds are housed intensively and are supplied with food and water *ad libitum*. In general, chicks grow to slaughter weight in about 42 days and broiler houses can contain many thousands of birds. This concentration of potential hosts provides *Salmonella* with greater opportunities for causing infection, and spread can be rapid through infected flocks. Poultry processing is similarly intensive. Birds are transported from the farm to the processing plant in crates that can hold between 20 and 30 birds each. Crates will be stacked, which means that birds in the lower cages will become contaminated with the faeces of birds in the cages above them. Crates are also recycled during the working day, are subjected to only a cursory washing and are frequently

Salmonella-positive (J. Wilson and T. Humphrey, unpublished).

A significant proportion of live birds will arrive at the processing plant already Salmonella-positive. However, studies in a great many countries have shown that carcass contamination rates frequently exceed those of carriage in the live bird (McBride et al., 1980). This is a manifestation of the speed and intensity of poultry processing. Modern slaughter lines can process up to 200 birds min^{-1}. The slaughter operations are highly efficient and the emphasis has largely been on cost-effectiveness, because of the fierce commercial competition in the retail market and the poultry industry. Cross-contamination is frequent and can occur at almost any stage of slaughter, but is more likely at evisceration, when gut contents might be spilled, and particularly during immersion scalding, where birds are passed through a tank of 'hot' water so that feathers will be loosened prior to mechanical plucking. Faecal material adhering to the feathers will be washed into the scald water. There may also be involuntary defecation, which will introduce more faeces into the water. Most poultry-associated Salmonella are found in the intestine of chickens and will be introduced into scald-tank water from that source. Chickens destined to be sold as chilled carcasses are usually scalded using 'soft' scalding water, where water temperatures are between 50 and 52°C. At this temperature, particularly with the high loading of organic matter in scald-tank water, death rates of Salmonella can be slow and scald-tank water is frequently Salmonella-positive (Humphrey and Lanning, 1987). This can cause cross-contamination of previously Salmonella-negative birds (Mulder et al., 1978) and scalding has also been shown to facilitate the attachment of Salmonella to chicken skin, making them subsequently more difficult to remove (Notermans and Kampelmacher, 1975).

In general, Salmonella are present on carcasses, either on the outer surface or in the abdominal cavity. This poses an indirect threat to public health, because of the potential for cross-contamination in the kitchen (de Wit et al., 1979). If Salmonella are trapped in areas of the carcass where heat transfer during cooking may be slow, as in the skin between the legs and breast, they may survive incomplete cooking, particularly when cooked by people other than the consumer, and undercooked chicken meat is

internationally an important vehicle for Salmonella infection (Wall et al., 1994). The importance of undercooked chicken meat as a vehicle for infection with potentially invasive Salmonella, such as S. enteritidis PT4 (Cowden et al., 1989b) and S. typhimurium DT104 (Wall et al., 1994; Williams et al., 1998; Leach et al., 1999), could also suggest that carcass contamination may not necessarily always be confined to carcass exteriors. S. enteritidis PT4 has been isolated from muscle tissues of carcasses sampled at retail sale in England and Wales (Humphrey, 1991a; A. Rampling, Dorchester, personal communication, 1992). There are two possible routes of contamination. S. enteritidis PT4 can be highly invasive in commercial broilers, causing septicaemia, raised mortality (Lister, 1988) and pericarditis (Rampling et al., 1989). There is the possibility that, when septicaemic birds are bled, the bacterium will lodge in the small blood-vessels. Muscle contamination may also occur as a consequence of the intensity of poultry processing. Studies with Clostridium perfringens (Lillard, 1973) have shown that, when this bacterium is in scald water, it can subsequently be isolated from the edible parts of chicken carcasses. Given the frequency with which Salmonella can be isolated from scald-tank water (Humphrey and Lanning, 1987), the above route of contamination may also be important with Salmonella, such as S. enteritidis PT4. Although the significance of such contamination has not been fully evaluated, the consumption of hot, cooked, take-away chicken was found to be an important risk factor in sporadic cases of S. enteritidis PT4 infection at the beginning of the pandemic in England and Wales (Cowden et al., 1989b). Given the ability of Salmonella to attach to tissues (Firstenberg-Eden, 1981; Humphrey et al., 1997b) and the higher heat tolerance of some isolates of PT4 (Humphrey et al., 1995), it is possible that some cells of the bacterium survive in the muscle of chickens which, on cursory examination, appear cooked. The bacteria would be better protected from gastric acidity and perhaps better able to cause infection. The proper cooking of chicken is an important measure in the control of poultry-associated Salmonella infection.

The importance of chicken meat as a vehicle for human salmonellosis has led to many surveys being carried out in order to determine contamination rates. There is general agreement that chicken meat sampled from retail outlets is frequently *Salmonella*-positive. National surveys reveal that many *Salmonella* serovars can be recovered from chicken meat and that many of these will match those found in infected humans (Humphrey *et al.*, 1988; Anon., 1995c). A comparison of data from PHLS surveys since 1987 shows that carcass contamination rates have gradually fallen and, in the last survey which was undertaken in 1995, only 30% of fresh chicken carcasses yielded *Salmonella* (Anon., 1995c). This represents a substantial improvement from earlier data, where more than 80% of carcasses were contaminated. The change is a result of a sustained attempt by the UK poultry industry to lessen infection/colonization of the live bird. Control will be discussed later but poultry/meat-associated human salmonellosis continues to pose considerable challenges to the international poultry industry.

Eggs and egg products as vehicles of *Salmonella* infection

There has long been an association between the contamination of eggs and egg products with *Salmonella* and human infection. Experiences in the UK are typical of those in other developed countries. Past problems largely involved bulk, liquid, raw egg or spray-dried egg powder, which were also responsible for introducing a number of different *Salmonella* serovars into the UK during and immediately after the Second World War. The egg pasteurization legislation that was introduced in the UK in 1963 to combat this problem has proved to be a remarkably successful control measure. For about 20 years after its introduction, eggs were only infrequently implicated as vehicles for human salmonellosis. It is believed that during this long period egg contamination, if it occurred at all, was largely confined to the outer shell and was a manifestation of intestinal carriage of non-invasive *Salmonella* in commercial laying hens. *Salmonella* present on eggshells may gain access to egg contents when they are broken out and may grow in egg contents at ambient temperature, particularly when the yolk and albumen have been homogenized. Contamination of this kind can be important and is believed to have been a major contributory factor in two large outbreaks of human

Salmonella infection in the USA in 1962/63 (Thatcher and Montford, 1962; Ager *et al.*, 1967).

In the mid-1980s, a remarkable and rapid change occurred in the international epidemiology of human salmonellosis. There was a marked increase in the number of infections caused by different PTs of *S. enteritidis*. Case–control studies in the USA (St Louis *et al.*, 1988) and England and Wales (Cowden *et al.*, 1989b) revealed that the principal vehicle of infection was either undercooked or raw eggs or dishes derived from them. In Europe, contaminated poultry meat was also important (Cowden *et al.*, 1989b).

In common with other poultry-associated *Salmonella*, *S. enteritidis* can be isolated from eggshells (Humphrey *et al.*, 1989a) and, when present, *Salmonella* can migrate through the shell and associated membranes and reach the contents of intact eggs. This will occur with a higher frequency if eggshells are damaged (Vadehra *et al.*, 1969) or particularly when a freshly laid egg which is warm comes into contact with *Salmonella*-positive faeces (Sparks and Board, 1985; Padron, 1990). The new feature of the international pandemic of *S. enteritidis* infection, however, was the fact that the bacterium could be isolated with regularity from the contents of clean, dry, intact, commercially produced eggs. This new behaviour for this bacterium was and is the single most important factor for the spread and continued international success of *S. enteritidis* PTs, especially PT4. All available scientific evidence suggests that contamination of egg contents is a consequence of the infection of reproductive tissue (Hoop and Pospischil, 1993). The bacterium is almost always present in egg contents in pure culture (Humphrey, 1994) and studies on outbreak-associated eggs have shown that there is no association between the presence of *S. enteritidis* on eggshells and that of the bacterium in egg contents (Humphrey *et al.*, 1989a, 1991; Mawer *et al.*, 1989). Thus it would seem that, if *S. enteritidis* is able to pass through eggshells, it does not do so with any great frequency and is no better able to do this than other faecal organisms that will also be present on eggshells. In fact, research by Dolman and Board (1992) found that *S. enteritidis* does not compete well with other potential egg-content contaminants in either shell membranes or in albumen stored at ambient temperature. Further evidence for the relative lack of involvement of shell/faecal contamination is the finding that the bacterium can be isolated from the reproductive tissues of naturally infected hens even in the absence of intestinal carriage (Lister, 1988; Bygrave and Gallagher, 1989).

The fact that S. enteritidis can be in tissues, including reproductive tissues, in the absence of faecal carriage can create problems with regard to flock surveillance. In addition, studies with PT4 have shown that the behaviour of the bacterium in laying hens is markedly different from that seen in broiler flocks. For example, infected hens rarely show any outward clinical signs of infection, even though the bacterium is highly invasive. They continue to feed, lay and behave normally. They do, however, lay eggs with Salmonella-positive contents. The continued international importance of S. enteritidis PTs as both human and poultry pathogens has led to greater surveillance of egg contamination. Before the current pandemic, there were relatively few surveys on Salmonella contamination of egg contents. Philbrook et al. (1960) examined a sample of 1137 hens' eggs and isolated S. typhimurium from the contents of three (0.3%). Later studies by Chapman et al. (1988), in an investigation of an outbreak of S. typhimurium infection, did not find Salmonella in the contents of 1000 eggs taken from naturally infected flocks. A possible explanation for this is the recent finding (Keller et al., 1995) that S. typhimurium may survive less well in the contents of forming eggs than S. enteritidis. It should be pointed out, however, that a full understanding has yet to be gained concerning the mechanisms of egg-content contamination with S. enteritidis.

Egg contamination occurs whilst forming eggs are in the reproductive tract, but the presence of S. enteritidis in ovaries and oviducts does not necessarily mean that egg contents will become Salmonella-positive. It is a routine finding on studies of artificially infected hens that, whilst many reproductive tissue samples can be Salmonella-positive, S. enteritidis is only rarely recovered from the contents of eggs laid by hens infected orally (Keller et al., 1995; Humphrey et al., 1996). It would seem that it is rather easier to recover S. enteritidis from the contents of eggs laid by naturally infected hens, although the observed prevalence of contamination is very variable. In a study of eggs from 22 naturally

infected commercial flocks associated with outbreaks of PT4 infection, 56/8700 (0.6%) were found to be contents-positive. Rates for individual flocks ranged from 0.1 to 10% (Humphrey, 1994). A small study by Paul and Batchelor (1988) in the UK found that, of ten eggs from a batch implicated in a family outbreak of PT4 infection, five were contents-positive. In contrast, in Spain, Perales and Audicana (1989) found that only 0.1% of eggs from flocks implicated in outbreaks were contents-positive. In general, however, the examination of eggs from infected flocks yields more contaminated eggs than this.

A study where naturally infected hens were caged individually and each egg laid by each hen was examined over a 3-month period (Humphrey et al., 1989c) found that 1% of 1100 eggs were contents-positive for PT4. The study also revealed that there can be a clustering of contents-positive eggs, with a number of hens laying contaminated eggs at or around the same time. The reasons for this are not yet understood, but similar clustering has been identified in the investigation of egg-associated outbreaks.

Almost all currently available evidence on the levels of Salmonella contamination in egg contents suggests that in fresh eggs only a few cells of Salmonella are present (Humphrey et al., 1989a,b,c; 1991; Mawer et al., 1989; Timoney et al., 1989; Gast and Beard, 1990). Growth is possible, however, following storage, particularly at temperatures above 20°C. The examination of naturally contaminated eggs revealed a strong association between egg age and storage conditions and the numbers of Salmonella present in egg contents (Humphrey et al., 1991). If eggs were stored at 20°C for longer than 21 days, they were significantly more likely to be heavily contaminated than eggs that were fresher. These data were supported by a national study where eggs obtained from retail outlets in England and Wales were stored at 20–21°C for approximately 5 weeks before examination. This study revealed that 50% of contents-positive eggs contained more than 10^4 Salmonella g^{-1} of egg contents (de Louvois, 1993).

The principal site of contamination in egg contents appears to be either the yolk membrane or the albumen immediately surrounding it. Egg albumen is a strongly iron-restricted environment and the few cells of S. enteritidis that may be present are unable to multiply until they have access to the iron-rich yolk contents. *Salmonella* are unable to traverse the yolk membrane of fresh eggs but can do so when the membrane permeability has been altered as a consequence of storage. In eggs stored at a constant 20°C, yolk membranes become permeable in approximately 28 days. Membrane breakdown is accelerated by storage at high (> 25°C) ambient temperatures, where temperatures fluctuate or where humidity is high, as in the kitchen. If *S. enteritidis*, in common with spoilage organisms, is able to invade the yolk, growth may be rapid and studies demonstrated that five *Salmonella* cells can reach a population of 10^{11} cells per egg within 24 h at 20°C (Humphrey *et al.*, 1989b). The Chief Medical Officer of England and Wales advised that eggs should be stored under refrigeration following purchase (Anon., 1988) and most egg boxes in England and Wales carry this suggestion.

Control of Salmonellosis

Salmonellosis is primarily a zoonosis and intervention is possible at any stage from farm to fork. The practicality of intervention and its cost-effectiveness may depend on the *Salmonella* serovar involved and the degree of its spread. Thus, infections in the UK with *S. typhimurium* DT124 resulting from the contamination of salami sticks (Cowden *et al.*, 1989a) were prevented by a temporary withdrawal of the product and improvements in production hygiene. Milk-borne salmonellosis can be prevented by the simple expedient of proper pasteurization of milk and milk products. Where infection is widespread, such as the multinational pandemic caused by *S. enteritidis* (Rodrigue *et al.*, 1990), control may be more difficult to achieve, particularly if the vehicles of infection are foods, such as eggs or chicken, which are consumed in large numbers.

On-farm control measures

For control strategies to be successful, it is necessary to identify the routes and sources of infection of the food animal, the manifestations of disease in the infected animal and appropriate

and cost-effective control measures. In the UK, the isolation of *Salmonella* from farm animals and their environment is reportable under the Zoonoses Order (1989) and advice is offered to the farmer to assist in controlling the outbreak. The practical difficulties of control can be most easily illustrated by comparing and contrasting the epidemiologies of PT4 and *S. typhimurium* DT104. *S. enteritidis* PT4 is very strongly chicken-associated, with most infections in humans being caused by contaminated eggs or poultry meat (Table 15.2; Anon., 1997).

Animal feed has been shown to be an important source of animal infection, especially in poultry (Williams, 1981), and legislation was introduced in the UK in 1981 and 1989 to prevent *Salmonella* contamination of animal feedstuffs, i.e. the Importation of Processed Animal Feed (1981) and the Processed Animal Protein Order (1989). These orders, together with improvements in production hygiene, have been remarkably successful and have resulted in a marked decline in feed contamination rates (Anon., 1994). (Different aspects of the manufacture and processing of animal feed are covered in Chapter 17.) There is still, however, the possibility of recontamination of animal feed on the farm, and antimicrobial compounds such as organic acids have been incorporated to prevent contamination. Humphrey and Lanning (1988) found that such treatment of feed given to breeding birds reduced the vertical transmission of *Salmonella* and also limited the horizontal spread of *S. enteritidis* PT4 in some broiler flocks (Humphrey, 1991b).

Chickens can also be protected from *Salmonella* infection if they are given a mixed culture of gut bacteria derived from *Salmonella*-free adult birds. Protection is maximized if the cultures are applied as soon as possible after hatching (Seuna and Nurmi, 1979; see also Chapter 18). However, *S. enteritidis* PT4 has been isolated only rarely from finished feed and this potential route would not appear to be an important source of infection, although feed may still be an important source of other *Salmonella* serovars and continued vigilance is required.

Control of PT4 clearly required the identification of other routes of infection. It became clear that the invasive nature of PT4 in chickens, like that of some other *S. enteritidis* PTs, was resulting in vertical transmission from infected breeding flocks, which resulted in the production of infected replacement stocks. The importance of vertical transmission of S. enteritidis and possibly some S. typhimurium DTs in poultry production has been recognized by the European Union, which issued Directive 92/117 to deal with this problem (Anon., 1992). The Directive requires the regular monitoring of poultry breeding flocks and the slaughter of flocks where the presence of either S. enteritidis or S. typhimurium is confirmed. Such a policy, whilst having laudable aims, can prove to be an economic burden on the industry involved, particularly where, as with S. enteritidis in poultry, infection is widespread and often at a high prevalence.

Most poultry used for either egg or meat production are housed intensively in closed, controlled environments. This allows some control over the ingress of Salmonella into the poultry flock and measures such as proper rodent control are important in the control of S. enteritidis infection (Eckroade et al., 1991). The marked reductions in contamination rates of poultry carcasses on sale in the UK (Anon., 1996) is a tribute to the intervention measures employed by the UK poultry industry and decisive action by the UK government. Intervention has not been as successful in egg production and contamination rates in eggs in retail outlets have yet to show any significant decline (Anon., 1997), although the decision by the British Egg Industry Council to vaccinate commercial laying flocks against S. enteritidis should bring about an improvement.

Although vertical transmission of S. enteritidis may be a factor in the continued infection of egg-laying flocks it is felt that difficulties of in-house cleaning and disinfection, leading to infections of replacement flocks, may be more important (Duguid and North, 1991). The observation that isolates of PT4 better able to cause infection in chickens are also those most likely to persist on surfaces (Humphrey et al., 1995, 1996) may also be of relevance to this problem. It would seem that the ultimate control of PT4 infection requires a better understanding of the behaviour of the bacterium in the farm environment. It may also be necessary to improve the identification of infected birds. The invasive nature of S. enteritidis in chickens means that examination of faeces only may underestimate the prevalence of infection. A better approach may be to carry out environmental or serological screening, with confirmation of positive results by detailed microbiological examination of birds. Measurement of yolk antibodies against S. enteritidis is effective in the detection of positive breeder or layer flocks (van de Giessen et al., 1992) and provides a cost-effective, non-invasive technique that is worthy of wider use.

In theory, relatively host-adapted Salmonella, such as S. enteritidis PT4, offer better opportunities for control than those more widespread Salmonella, such as S. typhimurium DT104, which has been isolated from a wide range of foods and where many vehicles of infection have been identified (Anon., 1997). It has also been found in many food animals, causing severe disease in some. S. typhimurium DT104 is often found in animals, such as dairy cows (Penny et al., 1996), which have free access to the natural environment. Investigations in south-west England revealed that infection was passing from farm to farm via watercourses, cows presumably becoming infected following the consumption of contaminated water (J. Lund, Exeter, personal communication, 1995). The proper storage and handling of slurry and restricting the access of infected animals to pasture offer a degree of control, but options are clearly more limited than with housed animals, such as chickens. S. typhimurium DT104 bears a number of similarities to S. dublin and control may be best effected with vaccination.

Control at slaughter

The high prevalence of Salmonella-contaminated poultry carcasses has led to many investigations into decontamination measures during processing. In general, methods that have been shown to be successful on a pilot scale are either too expensive in a commercial situation or bring about unacceptable changes in carcass quality. One of the more effective end-product treatments is gamma-irradiation and Dempster (1985) demonstrated that this process can destroy Salmonella on chicken carcasses. There is, however, considerable consumer resistance to this procedure and it is unlikely to be adopted in the short term. With red meat, spraying with either hot water (80°C) or lactic acid has been shown to have an application as a carcass decontaminant (Smulders, 1987; Davey and Smith, 1989).

The degree of cross-contamination in the abattoir is influenced by both the number of colonized/infected animals and the slaughter hygiene. In addition to the on-farm measures outlined above and the previously mentioned need for clean outer surfaces on the live animals, the prevalence of *Salmonella*-positive animals entering the slaughterhouse can be lowered by a reduction in transport times and a limitation on the time spent by animals in the lairage. Measures such as tying of the anus to prevent the contamination of carcasses with faeces will also reduce the extent and degree of carcass contamination.

Control measures in retail outlets and during catering

Consumers rely on the expertise of food producers, retailers and caterers to ensure that the food they purchase, particularly that to be consumed with no further treatment, is *Salmonella*-free. Thus, during food manufacture, measures are put in place to ensure that foods which may be *Salmonella*-positive, such as raw milk or liquid raw egg, receive heat treatments that are lethal to *Salmonella*. Such products are protected from recontamination after heating. The scale of processed-food production, particularly in the developed world, means that some errors will probably occur and there have been many documented instances of ineffective heat treatment or post-heating contamination leading to outbreaks of salmonellosis. In general, however, such foods have an excellent safety record.

The risks to consumers are rather greater in either the home or catering establishments. The risk factors for *Salmonella* infection, as with those for other types of food-borne disease, have been well documented (Roberts, 1986). Control should be based upon the practice of good hygiene in all aspects of the handling of food. Staff should be made aware of their responsibilities and staff education should emphasize the importance of hand washing, proper storage and refrigeration of food, kitchen sanitation and prevention of cross-contamination. The risks when foods, such as shell eggs, are used raw or are lightly cooked also need to be emphasized. Staff with gastroenteric symptoms must be excluded until recovery. Such measures also need to be included in educational programmes for staff in retail outlets, where proper storage and the prevention of cross-contamination are particularly important.

There has recently been much discussion in Europe about the refrigerated storage of eggs in retail outlets as a means of limiting the risk of infection with S. *enteritidis*. The ability of this bacterium to grow in egg contents is largely controlled by the integrity of the yolk membrane. The rate of yolk membrane breakdown is accelerated by storage at high and/or fluctuating temperatures and high humidities. In the UK, consumers have been advised to store eggs under refrigeration following purchase (Anon., 1992). This has the advantage of suppressing or inhibiting the growth of *Salmonella*, rendering them more heat-sensitive (Humphrey, 1990) and preserving the integrity of the yolk membrane (Williams, 1992). In the USA, eggs are held under refrigeration in shops, whereas in most of Europe control in retail outlets is confined to limitations on shelf-life and a suggestion that storage temperatures should not exceed 20°C.

A great deal of debate has taken place over the role of the consumer in the prevention of food poisoning. In the UK, consumers were given advice by the Chief Medical Officer on the cooking of eggs for high-risk groups, such as the very young, the very old and the sick or immunocompromised (Anon., 1988). There is also much government and food-industry advice on the safe handling and storage of food. It is well recognized that the successful control of salmonellosis requires intervention at all points on the food-chain. Consumers and caterers can do much with regard to kitchen hygiene and the proper cooking of 'at-risk' foods to limit risk. The ease with which cross-contamination can occur in the kitchen (de Wit *et al.*, 1979), coupled with an increasing desire amongst consumers to eat foods that are only lightly cooked, means that control may be best applied during food and animal production.

Salmonella infection from pets

The principal focus of this chapter has been on food-borne *Salmonella* infection. It should not be forgotten, however, that person-to-person infection is common in institutions such as hospitals during outbreaks. It is also possible to become infected from exotic pets. A study in Canada between 1994 and 1996 (Woodward *et al.*, 1997) illustrates potential problems. Investigations identified that 3–5% of human cases of *Salmonella* infection were associated with exposure to exotic pets. Such pets include iguanas, turtles and hedgehogs. *Salmonella* isolates from pets and cases included *S. java*, *S. poona*, *S. jangwani* and *S. manhattan* (Woodward *et al.*, 1997). Companion animals are considered in Chapter 14.

Clinical Manifestations of Salmonellosis and Infective Dose

Salmonellosis is a potentially serious infection and in the UK there are approximately 70 *Salmonella*-associated deaths each year (Anon., 1997). As with most other enteric infections, the very young, the elderly and those who are immunocompromised or who have underlying diseases are more at risk from infection. The symptoms and outcome of infection are also likely to be more serious in these groups. In most people, however, the clinical outcome of salmonellosis is complete resolution of all symptoms, frequently with only bedrest and some supportive therapy. The common symptoms of salmonellosis are shown in Table 15.3.

The incubation period is usually within the range of 12–72 h but occasionally may extend up to a week. In some outbreaks, where large numbers of organisms are believed to have been consumed, incubation periods as short as 2.5 h have been reported (Stevens *et al.*, 1989). With the great majority of the *Salmonella* serovars found in humans, the symptoms are confined to those shown in Table 15.3. In a small proportion of cases (*c.* 1%), *Salmonella* can be invasive and this can lead to bacteraemia and Gram-negative sepsis. With certain serovars, principally *S. dublin*, *S. choleraesuis* and some strains of *S. virchow*, the incidence of extraintestinal infections is higher. In up to 25% of cases of *S. dublin*, 75% of *S. choleraesuis* and 4% of *S. virchow* infections, the organisms can be isolated from blood cultures (Threlfall *et al.*, 1992). Such cases can carry an appreciable mortality, though most will recover with antibiotic therapy. Sequelae of this invasive disease can include metastatic abscesses in organs such as bone, particularly vertebrae, the wall of

Table 15.3. Symptoms of *Salmonella* infection.

Symptom	% of cases*
Diarrhoea	87
Abdominal pain	84
Feeling feverish	75
Nausea	65
Muscle pain	64
Vomiting	24
Headache	21
Blood in stools	6

* From an egg-associated outbreak of *S. enteritidis* (Stevens *et al.*, 1989).

the aorta and the kidney. Other complications are non-specific and include reactive arthritis and ankylosing spondylitis, the pathogenesis of which is not understood.

The volunteer studies conducted by McCullough and Eisele (1951), which demonstrated that, with *Salmonella* such as *S. bareilly* and *S. newport*, a dose of at least 100,000 cells was required to cause infection, created the belief that infection was only possible if large numbers of cells were consumed. This view was supported by studies with other serovars, including *S. derby* and *S. anatum*, where it was necessary for volunteers to consume more than 10 million cells. Such numbers of *Salmonella* must be relatively rare in foods, apart from instances of gross mishandling, and epidemiological data suggest that infection can be initiated from a low infective dose, particularly where the bacteria are protected within a food matrix. Foods with a high fat content or a good buffering capacity may protect small numbers of *Salmonella* during their passage through the acid regions of the stomach, thus permitting a lower dose of organisms to initiate infection. Examples of such foods and infectious doses associated with them are shown in Table 15.4. It has also been demonstrated that people on antacids, where gastric acidity has been lowered, or antimicrobials, which would have altered the gut flora, are more susceptible to infection. The infective dose required in these people may also be lower. Minimal infective doses can also vary with age and state of health; in the young they are very low. Analysis of a large number of outbreaks demonstrated that, for non-typhoid salmonellosis, there was a relationship between infective dose and the severity of

Table 15.4. Some examples of foods associated with low infective doses in outbreaks of salmonellosis (courtesy of P.J. Stephens).

Food vehicle	Salmonella serovar	Infective dose (cells per person)
[1] Chocolate	*S. eastbourne*	< 100
[2] Chocolate	*S. napoli*	10–100
[3] Chocolate	*S. typhimurium*	≤ 10
[4] Cheddar cheese	*S. heidelberg*	100
[5] Cheddar cheese	*S. typhimurium*	1–10
[6] Maize snack	*S. agona*	2–45
[7] Hamburger	*S. newport*	10–100
[8] Potato chips	Various types	4–45

[1] Craven *et al.*, 1975; [2] Greenwood and Hooper, 1983; [3] Kapperud *et al.*, 1990; [4] Fontaine *et al.*, 1980; [5] D'Aoust, 1985; [6] Killalea *et al.*, 1996; [7] Fontaine *et al.*, 1978; [8] Lehmacher *et al.*, 1995.

illness (Glynn and Bradley, 1992; Mintz *et al.*, 1994).

The study of outbreaks can provide much interesting information concerning the pathogenicity of salmonellosis and human reaction to infection, as it permits observations on a population exposed to the same *Salmonella*, in the same foods and at the same time. Such work has revealed that there can be marked variations in attack rates, although this may be explained in part by the uneven distribution of *Salmonella* in non-liquid foods. However, early studies by Hobbs (1971) on poultry-associated outbreaks of salmonellosis found attack rates varying from 8 to 86% (mean 20%). Recent investigations from 47 *Salmonella* outbreaks, involving a number of vehicles and five different serovars, calculated the mean attack rate as 56% (Glynn and Bradley, 1992).

Salmonellosis remains an internationally important human disease and presents many challenges to the food and agriculture industries and those charged with the protection of public health. Infection rates will only be reduced if there is the closest possible working relationship between all those involved with food production and the government agencies with responsibility for food safety.

References

Ager, E.A., Kenrad, E., Nelson, M.D., Galton, M..M., Boring, J.R. and Jernigan, J.R. (1967) Two outbreaks of egg-borne salmonellosis and implications for their prevention. *Journal of the American Medical Association* 199, 372–378.

Alktekruse, S.F., Cohen, M.L. and Swerdlow, D.L. (1997) Emerging food-borne diseases. *Emerging Infectious Diseases* 3, pp. 285–293.

Anon. (1988) *Raw Shell Eggs*. EL/88/P136, Department of Health, London.

Anon. (1989) *Salmonella* in eggs: PHLS evidence to Agriculture Committee. *PHLS Microbiology Digest* 6, 1–9.

Anon. (1992) Council of the European Communities Directive 92/117/EEC, Brussels, 17 December.

Anon. (1994) Salmonella *in Animal Feeding Stuffs and Ingredients*. MAFF Animal Health (Zoonosis) Division, Tolworth, UK.

Anon. (1995a) Salmonella *in Animal and Poultry Production 1994*. MAFF, London.

Anon. (1995b) The prevention of human transmission of gastrointestinal infection and bacterial intoxication: report of a Working Party of the PHLS *Salmonella* Committee. *Communicable Diseases Review* 5, 157–172.

Anon. (1996) *Report on Poultry Meat*. Advisory Committee on the Microbiological Safety of Food, HMSO, London, 110 pp.

Anon. (1997) *PHLS Evidence to House of Commons Select Committee on Agriculture Enquiry into Food Safety*. HMSO Stationery Office, London.

Anon. (1999) The rise and fall of *Salmonella? Communicable Disease Report* 9, 32.

Barnass, S., O'Mahoney, M., Sockett, P.N., Garner, J., Franklin, J. and Tabaqchali, S. (1989) The tangible cost implications of a hospital outbreak of multiple-resistant *Salmonella*. *Epidemiology and Infection* 103, 227–234.

Bradford, M.A., Humphrey, T.J. and Lappin-Scott, H.M. (1997) The cross-contamination and survival of *Salmonella enteritidis* PT4 on sterile and non-sterile foodstuffs. *Letters in Applied Microbiology* 24, 261–264.

Bygrave, A.C. and Gallagher, J. (1989) Transmissions of *Salmonella enteritidis* in poultry. *Veterinary Record* 124, 333.

Chapman, P.A., Rhodes, P. and Rylands, W. (1988) *Salmonella typhimurium* phage type 141 infections in Sheffield during 1984 and 1985: association with hens' eggs. *Epidemiology and Infection* 101, 75–82.

Cowden, J.M., O'Mahoney, M., Bartlett, C.L.R., Rana, B., Smyth, B., Lynch, D., Tillett, H., Ward, L., Roberts, D.,

Gilbert, R.J., Baird-Parker, A.C. and Kilsby, D.C. (1989a) A national outbreak of *Salmonella typhimurium* DT124 caused by contaminated salami stick. *Epidemiology and Infection* 103, 219–225.

Cowden, J.M., Lynch, D. and Joseph, C.A. (1989b) Report of a national case control study of *Salmonella enteritidis* phage type 4 infection. *British Medical Journal* 299, 771–773.

Craven, P.C., Mackel, D.C., Baine, W.B., Barker, W.H., Gangarosa, E.J., Goldfield, M., Rosenfield, H., Altman, R., Lachapelle, G., Davies, J.W. and Swanson, R.C. (1975) International outbreak of *Salmonella eastbourne* infection traced to contaminated chocolate. *Lancet* i, 788–793.

Cruickshank, J.G. and Humphrey, T.J. (1987) The carrier food-handler and non-typhoid salmonellosis. *Epidemiology and Infection* 98, 223–230.

D'Aoust, J.Y. (1985) Infective dose of *Salmonella* typhimurium in cheese. *American Journal of Food Science and Technology* 24, 305–316.

Davey, K.R. and Smith, M.G. (1989) A laboratory evaluation of a novel hot water cabinet for the decontamination of sides of beef. *International Journal of Food Science and Technology* 24, 305–316.

de Boer, E. and Hahne, M. (1990) Cross contamination with *Campylobacter jejuni* and *Salmonella* spp from raw chicken products during food preparation. *Journal of Food Protection* 53, 1067–1068.

de Louvois, J. (1993) *Salmonella* contamination of eggs. *Lancet* ii, 366–367.

Dempster, J.F. (1985) Radiation preservation of meat and meat products: a review. *Meat Science* 52, 661–666.

Desenclos, J.-C., Bouvet, P., Benz-Lemione, E., Grimont, F., Desqueyroux, H., Rebière, I. and Grimont, P.A. (1996) Large outbreak of *Salmonella enterica* serovar *paratyphi* B infection caused by goats milk cheese, France, 1993: a case finding and epidemiological study. *British Medical Journal* 312, 91–94.

de Wit, J.C., Broekhuizen, G. and Kampelmacher, E.H. (1979) Cross-contamination during the preparation of frozen chickens in the kitchen. *Journal of Hygiene* 83, 8327–8332.

Dolman, J. and Board, R.G. (1992) The influence of temperature on the behaviour of mixed bacterial contamination of the shell membrane of hens' eggs. *Epidemiology and Infection* 108, 115–121.

Duguid, J.P. and North, R.A.E. (1991) Eggs and *Salmonella* food poisoning: an evaluation. *Journal of Medical Microbiology* 34, 65–72.

Eckroade, R.J., Davison, S. and Benson, C.E. (1991) Environmental contamination of pullet and layer houses with *Salmonella enteritidis*. In: *Proceedings of the Symposium on Diagnosis and Control of Salmonella, San Diego, California, USA, 29 October, 1991*, pp. 14–17.

Firstenberg-Eden, R. (1981) Attachment of bacteria to meat surfaces: a review. *Journal of Food Protection* 44, 6002–6007.

Fone, D.L. and Barker, R.M. (1994) Associations between human and farm animal infections with *Salmonella typhimurium* DT104 in Herefordshire. *Communicable Disease Review* 4, 11.

Fontaine, R.E., Arnon, S., Martin, W.T., Vernon, T.M., Gangarosa, E.J., Farmer, J.J., Moram, A.B., Silliliker, J.H. and Decker, D.L. (1978) Raw hamburger: an interstate common source of human salmonellosis. *American Journal of Epidemiology* 107, 36–45.

Fontaine, R.E., Cohen, M.L., Martin, W.T. and Vernon, T.M. (1980) Epidemic salmonellosis from cheddar cheese: surveillance and prevention. *American Journal of Epidemiology* 111, 247–253.

Galbraith, N.S., Forbes, P. and Clifford, C. (1982) Communicable disease associated with milk and dairy products in England and Wales 1951–1980. *British Medical Journal* l, 1761–1765.

Gast, R.K. and Beard, C.W. (1990) Production of *Salmonella enteritidis*-contaminated eggs by experimentally infected hens. *Avian Disease* 34, 438–446.

Gay, J.M., Rice, D.H. and Steiger, J.H. (1994) Prevalence of faecal *Salmonella* shedding by cull dairy cattle marketed in Washington State. *Journal of Food Protection* 57, 195–197.

Giles, N., Hopper, S.A. and Wray, C. (1989) Persistence of *Salmonella typhimurium* in a large dairy herd. *Epidemiology and Infection* 103, 235–241.

Gill, O.N., Sockett, P.N., Bartlett, C.L.R., Vaile, M.S.B., Rowe, B., Gilbert, R.J., Dulake, C., Murrell, H.C. and Salmaso, S. (1983) Outbreak of *Salmonella napoli* infection caused by contaminated chocolate bars. *Lancet* i, 574–577.

Glynn, J.R. and Bradley, D.J. (1992) The relationship between dose and severity of disease in reported outbreaks of *Salmonella* infection. *Epidemiology and Infection* 109, 371–388.

Golden, D.A., Rhodehamel, E.J. and Kautter, D.A. (1993) Growth of *Salmonella* spp. in cantaloupe, watermelon and honeydew melons. *Journal of Food Protection* 56, 194–196.

Grau, F.H. and Smith, M.G. (1974) *Salmonella* contamination of sheep and mutton carcasses related to pre-slaughter holding conditions. *Journal of Applied Bacteriology* 37, 111–116.

Greenwood, M.H. and Hooper, W.L. (1983) Chocolate bars contaminated with *Salmonella napoli*: an infective study. *British Medical Journal* 286, 1394.

Hobbs, B.C. (1971) Poultry disease and world economy. In: Gordon, R.F. and Freeman, B.M. (eds) *Poultry Disease and World Economy.* p. 65.

Hoop, R.K. and Pospischil, A. (1993) Bacteriological, serological, histological and immuno-histochemical findings in laying hens with naturally acquired *Salmonella enteritidis* phage type 4 infection. *Veterinary Record* 133, 391–393.

Humphrey, T.J. (1990) Heat resistance in *Salmonella enteritidis* phage type 4: the influence of storage temperatures before heating. *Journal of Applied Bacteriology* 69, 493–497.

Humphrey, T.J. (1991a) Food poisoning – a change in patterns? *Veterinary Annual* 31, 32–37.

Humphrey, T.J. (1991b) The influence of feed treatment with organic acids on the colonisation of broiler chickens with *Salmonella enteritidis* PT4. In: Mulder, P.W.A.W. and de Vries, A.W. (eds) *Quality of Poultry Products III. Safety Marketing Aspects. Proceedings of the Combined Sessions of the 10th Symposium on the Quality of Poultry Meat and the 4th Symposium on the Quality of Eggs and Egg Products, Doorweth, Holland, 12–17 May, 1991.* Spelderholt Centre for Poultry Research and Information Services, Beekbergen, the Netherlands, pp. 49–57.

Humphrey, T.J. (1994) Contamination of egg shell and contents with *Salmonella enteritidis*: a review. *International Journal of Food Microbiology* 21, 31–40.

Humphrey, T.J. and Lanning, D.G. (1987) *Salmonella* and *Campylobacter* contamination of broiler chickens and scald water: the influence of water pH. *Journal of Applied Bacteriology* 63, 21–25.

Humphrey, T.J. and Lanning, D.G. (1988) The vertical transmission of *Salmonella* and formic acid treatment of chicken feed: a possible strategy for control. *Epidemiology and Infection* 100, 43–49.

Humphrey, T.J., Mead, G.C. and Rowe, B. (1988) Poultry meat as a source of human salmonellosis in England and Wales. *Epidemiology and Infection* 100, 175–184.

Humphrey, T.J., Cruickshank, J.G. and Rowe, B. (1989a) *Salmonella enteritidis* phage type 4 and hens' eggs. *Lancet* I, 281.

Humphrey, T.J., Greenwood, M., Gilbert, R.J., Chapman, P.A. and Rowe, B. (1989b) The survival of *Salmonella* in shell eggs under simulated domestic conditions. *Epidemiology and Infection* 103, 33–45.

Humphrey, T.J., Baskerville, A., Mawer, S.L., Rowe, B. and Hopper, S. (1989c) *Salmonella enteritidis* PT4 from the contents of intact eggs: a study involving naturally-infected hens. *Epidemiology and Infection* 103, 415–423.

Humphrey, T.J., Chart, H., Baskerville, A. and Rowe, B. (1991) The influence of age on the response of SPF hens to infection with *Salmonella enteritidis* PT4. *Epidemiology and Infection* 106, 33–43.

Humphrey, T.J., Martin, K. and Whitehead, A. (1994) Contamination of hands and work surfaces with *Salmonella enteritidis* PT4 during the preparation of egg dishes. *Epidemiology and Infection* 113, 403–409.

Humphrey, T.J., Slater, E., McAlpine, K., Rowbury, R.J. and Gilbert, R.J. (1995) *Salmonella enteritidis* phage type 4 isolates more tolerant to heat, acid or hydrogen peroxide also survive longer on surfaces. *Applied and Environmental Microbiology* 6, 3161–3164.

Humphrey, T.J., Williams, A., McAlpine, K., Lever, S., Guard-Petter, J. and Cox, J.M. (1996) Isolates of *Salmonella enterica* Enteritidis PT4 with enhanced heat and acid tolerance are more virulent in mice and more invasive in chickens. *Epidemiology and Infection* 117, 79–88.

Humphrey, T.J., Wilde, S.J. and Williams, S. (1997a) Colony morphology and the pathogenicity of *Salmonellae*. *PHLS Microbiology Digest* 14, 8–10.

Humphrey, T.J., Wilde, S.J. and Rowbury, R.J. (1997b) Heat tolerance of *Salmonella typhimurium* DT04 isolates attached to muscle tissue. *Letters in Applied Microbiology* 25, 265–268.

Ingham, S.C., Alford, R.A. and McCowan, A.P. (1990) Comparative growth rates of *Salmonella typhimurium* and *Pseudomonas fragili* on cooked crab meat stored under air and modified atmosphere. *Journal of Food Protection* 53, 566–567.

Kapperud, G., Gustavsen, S., Hellesnes, I., Hansen, A.H., Lassen, J., Hirn, J., Jahkola, M., Montenegro, M.A. and Helmuth, R. (1990) Outbreak of *Salmonella* typhimurium infection traced to contaminated chocolate and caused by a strain lacking the 60-megadalton virulence plasmid. *Journal of Clinical Microbiology* 28, 2597–2601.

Keller, L.H., Benson, C.E., Krotec., K. and Eckroade, R. (1995) *Salmonella enteritidis* colonisation of the reproductive tract and forming and freshly laid eggs of chickens. *Infection and Immunity* 63, 2443–2449.

Killalea, D., Ward, L.R., Roberts, D., de Louvois, J., Sufi, F., Stuart, J. M., Wall, P.G., Susman, M., Schwieger, M., Sanderson, P.J., Fisher, I.S.T., Mead, P.S., Gill, O.N., Bartlett, C.L.R. and Rowe, B. (1996) International epidemiological and microbiological study of outbreak of *Salmonella agona* infection from a ready to eat savoury snack in England and Wales and the United States. *British Medical Journal* 313, 1105–1107.

Leach S.A., Williams, A., Davies, A.C., Wilson, J., Marsh, P.D. and Humphrey, T.J. (1999) Aerosol route enhances the contamination of intact eggs and muscle of experimentally infected laying hens by *Salmonella typhimumium* DT104. *FEMS Microbiology Letters* 171, 203–207.

Lehmacher, A., Bockemühl, J. and Aleksic, S. (1995) Nationwide outbreak of human salmonellosis in Germany due to contaminated paprika-powdered potato chips. *Epidemiology and Infection* 115, 501–511.

Lillard, H.S., (1973) Contamination of blood systems and edible parts of poultry with *Clostridium parfringens* during water scalding. *Journal of Food Science* 38, 131–134.

Lister, S.A. (1988) *Salmonella enteritidis* infection in broilers and broiler breeders. *Veterinary Record* 123, 50.

Luby, S.P., Jones, J.L. and Horan, J.M. (1993) A large salmonellosis outbreak associated with a frequently penalised restaurant. *Epidemiology and Infection* 110, 31–39.

McBride, G.B., Skura, B.J., Yada, R.Y. and Bowmer E.J. (1980) Relationship between incidence of *Salmonella* contamination among pre-scalded, eviscerated and post-chilled chickens in poultry processing plants. *Journal of Food Protection* 43, 538–542.

McCullough, N.B. and Eisele, C.W. (1951) Experimental human salmonellosis; pathogenicity of strains of *Salmonella meleagridis* and *Salmonella anatum* obtained from spray-dried whole egg. *Journal of Infectious Diseases* 88, 278–289.

Mackey, B.M. (1989) The incidence of food poisoning bacteria on red meat and poultry in the United Kingdom. *Food Science and Technology Today* 3, 246–249.

Maguire, H., Cowden, J., Jacob, M., Rowe, B., Roberts, D., Bruce, J. and Mitchell, E. (1992) An outbreak of *Salmonella dublin* infection in England and Wales associated with a soft unpasteurised cows' milk cheese. *Epidemiology and Infection* 109, 389–396.

Maguire, H.C.F., Codd, A.A., Mackay, V.E., Rowe, B. and Mitchell, E. (1993) A large outbreak of human salmonellosis traced to a local pig farm. *Epidemiology and Infection* 110, 239–246.

Mawer, S.L., Spain, G.E. and Rowe, B. (1989) *Salmonella enteritidis* phage type 4 and hens' eggs. *Lancet* i, 280–281.

Mintz, E.D., Carter, M.L., Hadler, J.L., Wassell, J.T., Zingeser, J.A. and Tauxe, R.V. (1994) Dose–response effects in an outbreak of *Salmonella enteritidis*. *Epidemiology and Infection* 112, 13–23.

Mulder, R.W.A.W., Dorresteijn, L.W.J. and Van Der Broek, J. (1978) Cross-contamination during the scalding and plucking of broilers. *British Poultry Science* 19, 61–70.

Nichols, G.L. and de Louvois, J. (1995) The microbiological quality of raw sausages sold in the UK. *PHLS Microbiological Digest* 12, 236.

Notermans, S. and Kampelmacher, E.H. (1975) Heat destruction of some bacterial strains attached to broiler skin. *British Poultry Science* 16, 351–361.

O'Mahoney, M., Cowden, J., Smyth, B., Lynch, D., Hall, M., Rowe, B., Teare, E.L., Tettmar, R.E., Rampling, A.M., Coles, M., Gilbert, R.J., Kingcott, E. and Bartlett, C.L.R. (1990) An outbreak of *Salmonella saint paul* infection associated with bean sprouts. *Epidemiology and Infection* 104, 229–235.

Padron, M. (1990) *Salmonella typhimurium* penetration through the eggshell of hatching eggs. *Avian Disease* 34, 463–465.

Paul, J. and Batchelor, B. (1988) *Salmonella enteritidis* phage type 4 and hens' eggs. *Lancet* ii, 1421.

Penny, C.D., Low, J.C., Nettleton, P.F., Scott, P.R., Sargison, N.D., Strachan, W.D. and Honeyman, P.C. (1996) Concurrent bovine viral diarrhoea virus and *Salmonella typhimurium* DT104 infection in a group of pregnant dairy heifers. *Veterinary Record* May, 485–489.

Perales, I. and Audicana, A. (1989) The role of hens' eggs in outbreaks of salmonellosis in north Spain. *International Journal of Food Microbiology* 8, 175–180.

Pether, J.V.S. and Gilbert, R.J. (1971) The survival of *Salmonella* on finger-tips and transfer of the organisms to food. *Journal of Hygiene* 69, 673–681.

Philbrook, F.R., MacCready, R.A. and van Rockel, H. (1960) Salmonellosis spread by a dietary supplement of avian source. *New England Journal of Medicine* 263, 713–718.

Rampling, A., Anderson, J.R., Upson, R., Peters, E., Ward, L.R. and Rowe, B. (1989) *Salmonella enteritidis* phage type 4 infection of broiler chicken: a hazard to public health. *Lancet* ii, 436–438.

Roberts, D. (1986) Factors contributing to outbreaks of food-borne infection and intoxication in England and Wales 1970–1982. In: *2nd World Congress Foodborne Infections and Intoxication, Berlin*, Vol. 1, pp. 157–159.

Rodrigue, D.C., Tauxe, R.V. and Rowe, B. (1990) International increase in *Salmonella enteritidis*: a new pandemic. *Epidemiology and Infection* 105, 21–27.

Rowe, B., Begg, N.T., Hutchinson, D.N., Dawkins, H.C., Gilbert, R.J., Jacob, M., Hales, B.H., Rae, F.A. and Jepson, M. (1987) *Salmonella ealing* infections associated with consumption of infant dried milk. *Lancet* ii, 900–903.

Ryan, C.A., Nickels, M.K., Hargett-Bean, N.T., Potter, M.E., Endo, T., Mayer, L., Langkop, C.W., Gibson, C., McDonald, R.C., Kenny, R.T., Buhr, N.D., McDonnell, P.J., Martin, R.J., Cohen, M.L. and Blake, P.A. (1987) Massive outbreak of antimicrobial-resistant salmonellosis traced to pasteurised milk. *Journal of the American Medical Association* 258, 3269–3274.

St Louis, M.E., Morse, D.L., Potter, M.E., DeMelfi, T. M., Guzewich, J.J., Tauxe, R.V. and Blake, P.A. (1988) The emergence of grade A eggs as a major source of *Salmonella enteritidis* infections: new implications for the control of salmonellosis. *Journal of American Medical Association* 259, 2103–2107.

Samual, J.L., O'Boyle, P.A., Mathers, W.J. and Frost, A.J. (1980) Distribution of *Salmonella* in the carcasses of normal cattle at slaughter. *Research in Veterinary Science* 28, 368–372.

Schmidt, K. (ed.) (1995) WHO *Surveillance Programme for Control of Foodborne Infections and Intoxications in Europe*. Federal Institute for Health Protection, Berlin.

Seuna, E. and Nurmi, E. (1979) Therapeutic trials with antimicrobial agents and a cultured caecal microflora in *Salmonella infantis* infections in chicken. *Poultry Science* 58, 1171–1174.

Smulders, F.J.M. (1987) Prospectives for microbial decontamination of meat and poultry by organic acids with special reference to lactic acids. In: Smulders, F.J.M. (ed.) *Elimination of Pathogenic Organisms from Meat and Poultry. Proceedings of an International Symposium: Prevention of Contamination, and Decontamination in the Meat Industry. Zeist, The Netherlands, 2–4 June, 1986.* Elsevier, Amsterdam, pp. 319–344.

Sparks, N.H.C. and Board, R.G. (1985) Bacterial penetration of the recently oviposited shell of hen's eggs. *Australian Veterinary Journal* 62, 169–170.

Stevens, A., Joseph, C., Bruce, J., Fenton, D., O'Mahoney, M., Cunningham, D., O'Connor, B. and Rowe, B. (1989) A large outbreak of *Salmonella enteritidis* phage type 4 associated with eggs from overseas. *Epidemiology and Infection* 103, 425–433.

Thatcher, F.S. and Montford, J. (1962) Egg products as a source of Salmonellae in processed foods. *Canadian Journal of Public Health* 53, 61–69.

Threlfall, E.J., Hall, M.L.M. and Rowe, B. (1992) *Salmonella* bacteraemia in England and Wales, 1981–1990. *Journal of Clinical Pathology* 45, 34–36.

Threlfall, E.J., Frost, J.A., Ward, L.R. and Rowe, B. (1994) Epidemic in cattle and humans of *Salmonella typhimurium* DT104 with chromosomally integrated multiple drug resistance. *Veterinary Record* 134, 577.

Timoney, J.F., Shivaprasad, H.L. and Rowe, B. (1989) Egg transmission after infection of hens with *Salmonella typhimurium* phage type 4. *Veterinary Record* 125, 600–601.

Vadehra, D.V., Baker, R.C. and Naylor, H.B. (1969) *Salmonella* infection of cracked eggs. *Poultry Science* 48, 954–957.

van de Giessen, A.W., Dufrenne, J.B., Ritmeester, W.S., Berkers, P.A.T.A., Van Leeuwen, W.J. and Notermans, S.H.W. (1992) The identification of *Salmonella enteritidis*-infected poultry flocks associated with an outbreak of human salmonellosis. *Epidemiology and Infection* 109, 405–411.

Wall, P.G., Morgan, D., Lamden, K., Ryan, M., Giffin, M., Threlfall, E.J., Ward, L.R. and Rowe, B. (1994) A case control study of infection with an epidemic strain of multiresistant *Salmonella typhimurium* DT104 in England and Wales. *Communicable Disease Review* 4, R130.

Waterman, S.P. and Small, P.L.C. (1996) Characterisation of the acid resistance phenotype and *rpos* alleles of shiga-like toxin-producing *Escherichia coli. Infection and Immunity* 64, 2808–2811.

Williams, A., Davies, A.C., Wilson, J., Marsh, P D., Leach, S. and Humphrey, T.J. (1998) Contamination of the contents of intact eggs by *Salmonella typhimurium* DT104. *Veterinary Record* 14, 562–563.

Williams, E.F. and Spencer, R. (1973) Abattoir practices and their effect on the incidence of Salmonellae in meat. In: Hobbs, B.C. and Christian, J.H.B. (eds) *The Microbiological Safety of Food.* Academic Press, London, pp. 41–46.

Williams, J.E. (1981) *Salmonella* in poultry feeds – a worldwide review. Part 1. *World's Poultry Science Journal* 37, 6–19.

Williams, K.C. (1992) Some factors affecting albumen quality with particular reference to Haugh unit score. *World's Poultry Science Journal* 48, 5–16.

Woodward, D.L., Khakhria, R. and Johnson, V.M. (1997) Human salmonellosis associated with exotic pets. *Journal of Clinical Microbiology* 35, 786–2790.

Wray, C., Todd, N., McLaren, I.M. and Beedell, Y.E. (1991) The epidemiology of *Salmonella* in calves: the role of markets and vehicles. *Epidemiology and Infection* 107, 521–525.

Chapter 16

Environmental Aspects of *Salmonella*

Christopher J. Murray

Institute of Medical and Veterinary Science, Box 14, Rundle Mall, Adelaide, SA 5000, Australia

Introduction

Salmonella are zoonotic organisms and, although potential pathogens of humans and animals, asymptomatic carriage is the more frequent occurrence. Animals form the major reservoir of *Salmonella*, but significant dissemination into the wider environment has resulted from the activities of humans and other animals. The ability of *Salmonella* to exist in many environmental niches suggests that it should be accepted as an environmental organism whose dissemination is likely to continue and increase in the future. Frequently, the source of contamination, particularly of processed foods and animals, can only be correlated with the *Salmonella* found in the local environment. This, however, is not usually the primary source of the organism, which is likely to have been introduced previously into these sites by some other vector, such as carrier animals or pollution from human or farm effluent.

Various aspects of the behaviour and distribution of *Salmonella* in the environment have been reviewed previously (Wray, 1975; Murray, 1991). It should also be remembered that other bacterial pathogens are also associated with the environment, interacting with both humans and animals (Hinton and Bale, 1991).

Biodiversity

Many studies have been reported in the literature of the incidence of *Salmonella* in animals and environmental sites, including studies of waste disposal, environmental niches and the survival rate during treatments. There has been a general worldwide trend of an increasing incidence of *Salmonella*-related disease. One explanation may be the decreasing diversity of the genetic base of some domestic livestock and changes in their management practices. Consideration must be given to the serovars involved in different animal species. Some serovars can dominate in particular animals, either transiently or long-term, potentially exposing other animals to these same serovars. Experience suggests that the capacity for a serovar or strain of a serovar to colonize different species, including humans, is variable. Some serovars, e.g. *S. dublin*, are recognized as being extremely species-specific, while the majority of *Salmonella* are commonly accepted, whether validly or not, as being potentially able to infect many species. While strains can be identified across species, e.g. *S. typhimurium* phage type 135 in Australia, other strains have not moved out of one dominant species niche, despite ample opportunity to cross the species barrier, e.g. *S. hadar* in New Zealand, *S.* II *sofia* in poultry in Australia. This species influence is also evident when reviewing serovars found in many animals, including environmental animals.

One study (Hunter and Izsák, 1990) reviewed isolations from cattle, sheep, pigs, fowl and turkeys in the UK from 1976 to 1988. They found that the diversity indices for *Salmonella* incidence have declined for fowls and other animals, except for turkeys, which were the only group showing increasing diversity indices. In fowl, there was increased frequency of S. enteritidis and a declining incidence of other serovars. These authors suggest that declining diversity may be influenced by factors such as intensification of handling practices, reduction in genetic diversity of breeding stock and increasing standardization of food types. This provides the opportunity for dissemination and persistence of a pathogenic strain, such as S. enteritidis phage type 4, which occurred in many countries.

The Kimberley region of Western Australia presents an unusual opportunity for studying Salmonella in a natural environment. The region is tropical, remote and, for much of its area, unaffected by development, having a relatively undisturbed natural fauna and a low indigenous population, with some members still living traditional lifestyles, relying on captured native fauna as a food source. This region has yielded several new and unusual Salmonella serovars. The range of serovars found in the human population in that region differs from that found in developed regions of the same state. S. typhimurium accounted for 41% of cases within the developed communities but for only 3% of the Kimberley-region cases (J.B. Iveson, Perth, personal communication, 1979). However, the serovars from humans in the Kimberleys show a similar distribution to those in native fauna, particularly lizards and snakes, which are part of the native (human) diet. Later studies in the same region showed a similar range of serovars from marsupials and other native animals (How et al., 1983) and, more recently, bush flies (I. Dadour, Perth, personal communication, 1997). Of the native animals, vegetarian etherians have a similar frequency of carriage rate and serovar distribution to carnivores. Many of the serovars isolated from animals and humans in Australian tropical regions are less common in temperate regions. This may be a reflection of fauna distribution, but may also reflect some host-species preferences of the Salmonella serovars.

Water

Salmonella are disseminated into the aquatic environment from a diverse range of sources, including effluent discharges, agricultural runoff and excretion by wild animals. Generally, Salmonella survive for a time period dependent on the nature of the water. Sediments may protect enteric bacteria from some of the stresses associated with aquatic environments and may provide nutrients that support bacterial growth. In artificial freshwater systems, Salmonella and Escherichia coli survived for at least 56 days (Fish and Pettibone, 1995); the number of cells increased initially and then stabilized in sediments, while decreasing in the water over the sediments.

Freshwater pools in a forest area contained Salmonella; however, when the pools dried up during dry periods, Salmonella could not be detected from the sites of the pools (Thomason et al., 1975), although Salmonella were detected in wet soil and vegetation associated with the pool areas. The authors suggest that consumption of contaminated water by wildlife and other animals completes one phase of the cyclic movement of these organisms in nature.

In the tropical region of Australia, freshwater pools contained Salmonella at levels up to 110 colony-forming units (cfu) 100 ml^{-1} (Townsend, 1992). Two pools were compared, one with no focal point for faecal contamination, fed from plantation and forest, and another with some households on septic systems in the catchment area. The plantation-fed pool had Salmonella in 69% of samples, while the other pool had Salmonella in 96% of samples and generally higher numbers of the organisms. Both pools presented a similar distribution of serovars, some being common in the human population of the region while other locally common human serovars were not found.

Earlier studies had reported similar results, noting that survival in water is affected by factors including presence of protozoa, antibiosis, organic material, algal toxins, dissolved nutrients, heavy metals, temperature and physiochemical properties (Burton et al., 1987). The survival of Salmonella was greater in sediments with higher clay content. In drinking-water, the survival of S. enteritidis was found to decrease as organic load increased (Pokorny, 1988). Salmonella survival was extended as cell density increased; each log increase in density resulted in a 50% extension of survival time. The survival time at 4°C was almost twice the survival time at 20°C. As the chemical oxygen demand (COD) increased from 1.3 mg l^{-1} to 15.2 mg l^{-1}, survival at 4°C decreased from 21 to 2 days, when the initial cell density was 100 cells ml^{-1}. In estuarine environments, *Salmonella* were found to survive for extended periods and would multiply when temperatures were above 18°C (Rhodes and Kator, 1988).

The occurrence in waters of viable non-culturable forms of bacteria has been reported. In fresh water, *Salmonella* appeared to die off in 2–3 days (Roszak *et al.*, 1984); however, when nutrient was added to the water, *Salmonella* began to grow. The initial cell morphology was atypical, although the cell morphology later became typical. After 21 days in the non-culturable phase, cells could no longer be resuscitated.

In marine waters, similar effects are found, sediments being important, as bacteria may attach to particles and then settle into sediments for increased survival. Open seas are important for eliminating organisms, including *Salmonella* from human activities, via discharges and rivers. Elimination of *Salmonella* in the North Sea has been demonstrated to be predominantly due to protozoa (Glaus and Heinemeyer, 1994). *Salmonella* multiply in sterile sea water; however, in natural sea water, the rate of elimination increased where higher protozoa levels were present. In areas of greater bacterial load, higher numbers of protozoa were present, which resulted in a higher rate of elimination. Seasonal influences were also noted, with higher levels of protozoa being found in warmer, summer waters, and while *Salmonella* were present in higher numbers than in colder, winter water, the rate of elimination was faster.

The inactivation of *Salmonella* is greater than for other coliforms, faecal streptococci or *Clostridium perfringens* in sea water (Morinigo *et al.*, 1989), where sunlight increases the inactivation of *Salmonella*. Sewage added to sea water aids their survival in the first 3 days, followed by a subsequent decline in numbers. In natural waters, a decline of *Salmonella* numbers occurs, although with added sewage the numbers increase before declining. In filtered fresh water and sea water, the numbers of *Salmonella* remain relatively stable, while, in unfiltered samples of the same waters, a reduction in numbers occurs. This indicates the role of other organisms in the decline of *Salmonella* in waters in the natural environment.

Sublethal injury has been demonstrated in sea water in the Antarctic at −1.8°C in 54–56 days (Smith, J.J. *et al.*, 1994). Adaptation to low temperature occurs, with *Salmonella* slowly forming colonies on solid media at −1.8°C, but losing the ability to grow at 37°C. Survival of *Salmonella* in low-nutrient sites, such as sea water, can be influenced by the production of starvation-specific proteins (Galdiereo *et al.*, 1994) and changes to the cell structure. The reactions of *Salmonella* to starvation are reviewed elsewhere (Foster and Spector, 1995).

The parameters affecting *Salmonella* also influence other microbial flora in waters. These influences can affect the total flora, with the fate of *Salmonella* being more influenced by the activity of other organisms, such as protozoa, rather than by intrinsic features of the organism itself.

Waste

Salmonella may be present in any waste from human or animal activities. The *Salmonella* content of meat waste from domestic sources going into a refuse system has been determined by Durrant and Beatson (1981). In this study, the serovars found in meat waste in the UK included *S. typhimurium* S.4,12:d:–, *S. virchow*, *S. bredeney* and *S. enteritidis*. The degree of contamination of the environment from this source may be small compared with animal waste and sewage; however, in any environmental site, the potential for proliferation exists and subsequent dissemination by birds, rodents, insects and other scavengers when conditions of temperature and moisture are favourable. Perusal of waste sites can be illuminating, as the author has personally observed the illegal dumping of large masses of chicken processing waste on to an urban refuse site and subsequent scavenging by gulls. Other studies have documented the incidence of serovars in sewage and abattoir effluents (Linklater *et al.*, 1985).

It is idealistic to expect that waste will be decontaminated before disposal and that disposals will be performed to the requirements of regulatory agencies at all times. A large lake in Spain subjected to untreated human, animal and industrial effluent and used for recreation and irrigation was found to be contaminated with *Salmonella*, of which some serovars were consistent with human and animal sources (Alcaide *et al.*, 1984).

Waste treatment

Numerous authors have investigated the occurrence of *Salmonella* in waste and the effects of different treatment processes (see Strauch, 1991).

The risks associated with sewage sludges and their disposal include: contamination of waters, leading to human and animal infections; contamination of aquatic flora and shellfish; postharvest treatments, such as smoking, failing to kill *Salmonella* in fish; risks to bathers from exposure; contamination of foods, such as vegetables, via irrigation (Danielsson, 1977). The risks are not limited to *Salmonella* and include a wide range of bacterial and viral pathogens. It may well be that *Salmonella* is far less significant than many other pathogens in these sites. Viral pathogens of humans and animals are far more species-specific than bacterial pathogens, with human waste posing a greater risk to humans than to other animals and vice versa. The effects of treatments on *Salmonella* are similar to those of a wide range of other bacteria, with treatment processes also affecting parasites within the waste. Treatment of human waste in developed countries is generally well controlled; however, constraints on the disposal of treated waste, particularly into aquatic systems, both marine and fresh waters, are increasing. Animal waste from farms does not usually receive the same attention and treatment as human waste.

Different approaches to waste treatment exist across a wide range of plants and operating authorities. *Salmonella* levels have been compared between primary treatment with removal of sludge and no primary treatment with long sludge ageing systems by Farrell *et al.* (1990), who monitored six plants in Ohio. *Salmonella* levels in sludge solids were found to be an average of 1.3 logs less than in incoming waste water (based on total solids) with no primary treatment and long sludge ageing systems, but only 0.8 logs less when primary treatment was performed. Under US legislation, the extent and type of treatment restrict reuse of the sludge, which must be treated by a 'process to significantly reduce pathogens' before any contact with crops is permitted. Where there is no treatment of the sludge after collection from the waste water treatment process, no use on land surface is permitted.

The incidence of *Salmonella* and the effects of treatments were investigated in eight treatment plants in the UK (Jones and Rennison, 1980). *Salmonella* was found to be present in 85% of settled sewage, 87% of raw sewage and 96% of anaerobically digested sludge, at levels generally less than 200 *Salmonella* cfu 100 ml^{-1}. The concentration of *Salmonella* was significantly reduced by most of the sludge treatments, including percolating filters and activated sludge processes. Aerobic digestion was more effective at reducing *Salmonella* than anaerobic digestion, which is the only process that does not reduce levels by greater than 70%. Lime and copperas conditioning, followed by dewatering, which raises pH to above 10, effectively eliminated *Salmonella*. Commenting on the risks to cattle arising from the disposal of sludges on to pasture, the authors concluded that these risks will continue to exist unless the pathogen is eliminated.

During large-scale transport of sewage sludge for disposal on to arid land, the potential exists for airborne dissemination of organisms, although a 4-month study failed to demonstrate the presence of *Salmonella* in air associated with such a system (Pillai *et al.*, 1996).

The moisture content of composted sewage sludge is critical to its bacterial stability (Russ and Yanko, 1981), and moisture content of at least 20% is required for *Salmonella* to grow. When contamination of well-composted sludge occurs, the effects are transient, with population peaks after 5 days, followed by die-off. The carbon/nitrogen (C/N) ratio is a consistent indicator of the potential for *Salmonella* growth; a C/N below 15 : 1 did not support repopulation. Over a 5-year period, dewatered, anaerobically digested sludge was found to contain an average of 10^5 *Salmonella* g^{-1} of total solids. Composting can reduce this to below detectable levels (0.2 g^{-1}). To achieve *Salmonella* control at all times, the conditions must be maintained. Should moisture increase during storage, then repopulation and regrowth can occur, re-establishing the risk of *Salmonella* contamination and dissemination from the sludge. Treatments of composted sludges must be practical for general use. Lowering pH could control *Salmonella* in composted sludge (Hussong *et al.*, 1985); however, the resultant product is undesirable for geochemical reasons.

It is common practice to use slurry as a means of removal of animal wastes. Subsequent dispersal on to land and pasture provides a means of disposal and reuse of nutrients but introduces risk of spreading contamination and cross-species infection, e.g. exposure of cattle to contaminated slurry from pigs (Jones, 1980). Although the levels of *Salmonella* in slurry are usually less than 100 ml^{-1} the survival of the organism varies with the physical conditions, e.g. the initial pH of slurry ranges from 6.2 to 8, which is not lethal for *Salmonella*. The virulence of *S. dublin* did not appear to be affected by storage in slurry (Jones, 1975), although earlier experiments found that *S. dublin* levels of 10^5 ml^{-1} in slurry were necessary to infect cattle grazing pasture spread with the contaminated slurry. Survival of *Salmonella* on pasture does occur; however, dispersal generally results in die-off with time and Josland (1951) found that the survival of *Salmonella* was less on pasture exposed to sun compared with shaded pasture.

Piggery waste treated by passage through three ponds – anaerobic then facultative then aerobic – can remain contaminated with *Salmonella* (Henry *et al.*, 1995). Water from the aerobic pond was used for wash water in the piggeries studied, because of limited water resources. Disposal of sludge from the anaerobic ponds increased the potential for spread of contamination to the wider environment; although the risks of *Salmonella* infections were assessed as minimal, the presence of these substantial reservoirs of contamination precluded complacency.

Contamination in slaughterhouse waste is similar to levels in farm wastes. Treated waste remained contaminated, with 84% of flocculated poultry sludge, 90% of flocculated pig sludge and 92% of aerobically activated pig sludge containing *Salmonella* (Fransen *et al.*, 1996).

Use of waste for biogas production as part of the digestion process has been investigated. Anaerobic thermophilic digestion may be used to treat cattle waste, with collection of gas for energy use. The control of pathogens is regarded as an important part of the evaluation process (Plym-Forshell, 1995). The studies showed that *Salmonella* are eliminated during digestion at 55°C for 24 h, but subsequent release into a manure pit was associated with recontamination of the treated waste, with *Salmonella* surviving in the pit for 35 days. Similar findings have been reported where regrowth of *Salmonella* occurred after digestion (Errebo Larsen *et al.*, 1994). Biogas generation from waste has value for energy production in conjunction with waste

disposal in developing as well as developed countries (Gadre *et al.*, 1986; Kearney *et al.*, 1993), although the pressures for such processes vary from environmental concerns to provision of affordable energy. The capacity for recontamination and regrowth of *Salmonella* and other pathogens becomes extremely significant when considering the treatment of wastes, their disposal and their subsequent behaviour in the environment.

Historically, reuse of waste water has developed from a need to prevent pollution of waterways and conserve water and nutrients, and this remains important today. Shuval (1991) considered that, although scientific and technical advances have occurred, problems associated with many pathogens still remain; however, the problem of *Salmonella* transmission to cattle and subsequently to humans as a result of wastewater irrigation of pasture land can be considered a marginal problem, although it cannot be completely ignored. Thus the pre-treatment of effluent in order to provide some measure of inactivation or removal of *Salmonella* organisms would be prudent.

As a result of expert committee assessment, World Health Organization (WHO) guidelines for reuse of waste water for irrigation to crops likely to be eaten uncooked are not more than 1000 faecal coliforms l^{-1}. For irrigation of pasture and crops, no microbiological guidelines are given; however, treatment for microbiological control should be equivalent to 8–10 days' retention in stabilization ponds (WHO, 1989).

There are increasing pressures in developed countries for reductions of biological oxygen demand (BOD) and total solids in waste water from food processing (Rajkowski *et al.*, 1996). Although reconditioning of water for agricultural and landscape use is increasing, survival of *Salmonella* and other pathogens in the treated water can occur after removal of solids. Chlorination of the treated water is the only reliable method for general decontamination.

Poultry

As commercial poultry industries have developed in countries around the world, the role of poultry as a source of *Salmonella* for human infections has been recognized. Countries have approached the problems differently; some, such as Sweden, attempted to eradicate *Salmonella* from their commercial flocks and, in others, such as Australia, the industry initiated extensive monitoring and controls. Outbreaks of *S. enteritidis* in many countries, beginning during the 1980s, increased the focus of the poultry industries and regulatory authorities on *Salmonella*. Some investigations focused on environmental aspects of the spread of *Salmonella*, some of which are discussed elsewhere in this review. Although commercial poultry and egg production favours the vertical transmission of *Salmonella* to the final progeny, environmental controls can influence the spread of infection. Except for the classical, species-specific poultry pathogens, *S. pullorum* and *S. gallinarum*, asymptomatic carriage is the dominant syndrome of salmonellosis in poultry.

Dust has been found to contain *Salmonella* in broiler rearing sheds (Morgan-Jones, 1980), while water has provided a significant source for the spread between birds. Dust samples collected from surfaces, including walls, inlets and fans, have also been found to be reliable indications of environmental and flock contamination. Likewise, Brown *et al.* (1992) isolated *Salmonella* from dust, insulation materials and wood in broiler units. Thirty-one broiler farms in Texas were surveyed with drag swabs for *Salmonella* four times over an 11-month period (Caldwell *et al.*, 1995). Not all of the farms were stocked at the time of sampling, but 25 serovars were identified, with repeated detection of the same serovar on seven farms. On five of the seven farms, the same serovar was isolated from consecutive samplings. Litter has been found to be a more effective monitor of environmental contamination in broiler and turkey units than drinking-water, and the *Salmonella* isolations from litter correlate with those in the birds (Poppe *et al.*, 1991; Irwin *et al.*, 1994). Investigation of some physical factors related to *Salmonella* contamination indicates that higher moisture content of litter correlates with increased incidence of *Salmonella*. Higher ammonia levels in wet regions are associated with increased incidence, while higher-pH litter shows decreased incidence (Opara *et al.*, 1992).

Follow-up of egg-associated outbreaks due to *S. enteritidis* in the USA include trace back to source flocks. *S. enteritidis* of the phage types associated with 18 outbreaks studied by Altekruse *et al.* (1993) were found in the internal organs of birds in 71% of source flocks and in 79% of environmental samples from the sheds. A greater range of *S. enteritidis* phage types was encountered from the environmental sampling than from either the outbreaks or the birds, suggesting that the environment may harbour other strains of potential public-health significance beside those detected as a result of outbreak tracing.

The ubiquitous presence of *Salmonella* within poultry-shed environments provides opportunities for multiple modes of transmission. Airborne exposure to *S. enteritidis* phage type 4 can produce generalized infection, even after low doses (Baskerville *et al.*, 1992). In well-ventilated chicken sheds, *S. enteritidis* phage type 4 could be found in the air with chicks becoming colonized as a result of the exposure (Lever and Williams, 1996). *S. enteritidis* phage type 4 and *S. typhimurium* can survive in aerosols up to 2 h with negligible reduction in numbers (McDermid and Lever, 1996). *S.* II *sofia* dominates the *Salmonella* flora of commercial broiler flocks in Australia (Harrington *et al.*, 1991). Evidence suggests that airborne spread may be a common route of transmission between birds (T. Grimes, Sydney, personal communication, 1996). Colonization of the respiratory tract, in addition to caecal contents, is common with this serovar.

S. enteritidis can survive for at least 1 year in empty poultry sheds, where naturally infected flocks have previously been housed (Davies and Wray, 1996a). In this study, *S. enteritidis* rapidly declined in litter, but the organisms persisted in food troughs, nesting-boxes, unbedded floor areas and dust on fans. Association with dust particles appears to aid persistence, the organisms being found in dust remaining after cleaning and disinfection. The organisms were also found outside the units, in litter and wild-bird droppings. Survival was demonstrated in feeds for at least 2 years. *S. enteritidis* could be found in sheds after cleaning and disinfection, indicating that the processes may not achieve total elimination of *Salmonella* (Davies and Wray, 1996c). Wooden nesting-boxes were the sites remaining most highly contaminated after disinfection. As with all cleaning and disinfection regimes, proper cleaning, correct use of disinfection agents and adherence to correct procedures are critical to achieve control.

In another study, Lahellec *et al.* (1986) found that, on the first day when new chicks were placed into disinfected sheds, shed walls yielded the highest frequency of *Salmonella* recovery, while, at the end of rearing, drinkers had the highest frequency. The source of contamination of walls could not be determined.

Canadian practice is to place birds into facilities that have been thoroughly emptied, cleaned and disinfected. In Australia, one large broiler producer follows these procedures, with the additions of drag swabs after cleaning to check for *Salmonella*, and units are not restocked until declared *Salmonella*-free (M. McKenzie, Bribane, personal communication). In Australia, the predominant S. II *sofia* may be providing a competitive-exclusion effect against other *Salmonella*. This organism is also frequently isolated from environmental sources associated with broilers but has not been identified as colonizing other environmental sites, animals or humans during the 17 years since it dramatically spread throughout Australian broiler flocks. Its prevalence in egg-laying flocks (3.5% of *Salmonella* isolated) is much lower than in broilers (67% of *Salmonella* isolated), with egg-laying flocks also showing a greater diversity of serovars (Murray, 1994a).

Poultry are distributed in the human food-chain. *Salmonella* from chickens are consequently spread within the food-preparation environment, also entering waste systems during preparation prior to cooking. Serovars on chickens being prepared in a hospital kitchen have been found to correspond with those in the sewers of the establishment (Reilly *et al.*, 1991). Similar serovars were found in the hospital sewers upstream from the kitchen outlet, suggesting that transmission occurred to residents, followed by asymptomatic excretion by patients. *Salmonella* are highly motile organisms; the potential also exists that *Salmonella* may spread against the normal flow in sewers, possibly within wet surface films.

S. *enteritidis* phage type 4 appeared in a commercial layer flock in southern California in 1994, this being the first flock recorded with the organism in the region (Kinde *et al.*, 1996). The farm was adjacent to a stream fed solely by effluent from an upstream sewage-treatment plant. Feral animals, including rodents and skunks, used the stream for drinking. S. *enteritidis* phage type 4 of the same plasmid type was isolated from the creek water, skunk liver, eggs, chicken liver and mouse liver. General environmental and wildlife contamination, as a consequence of the discharge of inadequately decontaminated sewage, was considered to be the source of infection in the layer flock.

Farm Environment

Farm environments may become contaminated with *Salmonella* following outbreaks of disease or colonization of animals or by general contamination. Of 60 dairy herds in California, 75% had serological evidence of recent exposure to *Salmonella* (Smith, B.P. *et al.*, 1994). In addition to direct animal-to-animal transmission, environmental and management practices were identified as contributing to increased recycling of *Salmonella*. Factors include recycling of lagoon waste water for flushing, potentially contaminated feeds, inadequately controlled rodent and wild-bird populations, contaminated rendering trucks being driven into animal areas and use of the same loader for transporting dead animals and moving feeds. An outbreak of S. *montevideo* occurred in cows on a dairy farm in California and resolved without intervention. Investigation of the environment 2 years after the outbreak revealed S. *montevideo* in 76% of environmental samples and 88% of samples of effluent (Gay and Hunsaker, 1993). S. *montevideo* was found in the recycled wash water used in the dairy and the organism had remained as part of the environmental flora of the dairy following the outbreak, although there was no evidence of clinical infection in the cows. Environmental pressures to reduce water usage by increasing recovery and reuse increase the risk of recycling *Salmonella*. Treatment of farm waste water for recycling is desirable, but practical measures to reduce organic loads to enable subsequent satisfactory decontamination may not be economic for many farms.

Two outbreaks of salmonellosis in dairy herds in the UK caused by S. *newport* were investigated by Clegg *et al.* (1983). These outbreaks involved an initial outbreak of clinical infections on one farm in May–June 1973, followed by a subsequent episode from September to February 1974. Slurry from the farm had been dispersed over land, which subsequently drained into a dam, which then fed into a stream and became a source of infection to cattle drinking from the stream. S. *newport* was found to survive for up to

6 months in soil beneath infected cow-pats, but was not isolated from grass.

Two methods of disposal of carcasses have been compared for their capacity to eliminate *Salmonella* and subsequent spread to the surrounding environment (Davies and Wray, 1996b). In this study, calf carcasses were artificially contaminated with *S. typhimurium*, *Bacillus cereus* and *C. perfringens* and then placed into either concrete-lined decomposition pits or a burial pit with no lining. It was found that the frequency of *Salmonella* isolations decreased more rapidly from the burial pit than from the decomposition pit (4 months compared with 6 months to disappear). A large population of blowfly larvae developed within 2 weeks in the decomposition pit and these were *Salmonella*-positive. *Salmonella* were also found in wild-bird droppings in the vicinity of the pit for a period of 4 weeks after loading the pit. Nearby drainage ditches were *Salmonella*-positive, initially close to the pit but, after 3 weeks, up to 12 m along the drain. Following a cold period over winter, *Salmonella* reappeared, possibly due to changes in other flora within the environment.

The role of dealers was identified in the dissemination of *Salmonella* during the supply of calves to other farms, spreading *Salmonella* from one site to another (Wray *et al.*, 1990). Australia exports live sheep, predominantly to the Middle East, on specially constructed ships, with loads in excess of 11,000 animals. Salmonellosis is one of the potential problems during the sea voyage. Different serovars may be detected in sheep during the voyage, some of which may be acquired during yarding in feedlots prior to embarkation (Higgs *et al.*, 1993), and stress in the animals in the feedlots and during the voyage accentuates the risk of salmonellosis.

During outbreaks, *Salmonella* have been found in the nasal passages of sheep, suggesting that inhalation of contaminated dust may be a factor in its spread (Tannock and Davis, 1973). The authors quote another study where isolation of *Salmonella* from nasal passages of calves in abattoir yards indicates that the upper respiratory tract can be a route of colonization, leading to infection.

S. hadar became the dominant serovar in poultry in New Zealand in 1990. Spent broiler litter is used for fertilizer on dairy pastures, being directly spread after clean-out. Some farms also graze cattle to control grass between poultry houses, with animals being potentially exposed to *S. hadar* in dust, as well as by ingestion (Christensen and Cullinane, 1993). The survival of *S. hadar* was found to be extended in pasture associated with litter compared with pasture alone. Though excretion of *S. hadar* was identified in cattle grazed on the pasture, it was concluded that the practice did not constitute a serious risk to the cattle. *S. hadar* also appeared in broilers in Australia during the same period, and experience suggests that the strain of *S. hadar* is an extremely good survivor in the environment, including waste water lagoons (M. Mackenzie, Brisbane, personal communication).

Environmental control must be included as part of the control measures for reducing infection in veterinary practice (Wray, 1985). Cleaning and disinfection routines may not always be successful in eliminating contamination and may even aggravate problems (Wray, 1995). Investigations of an outbreak of *S. ohio* in foals on a Californian farm found that control can be achieved by decontamination and management of the environment of the foals and mares (Walker *et al.*, 1991). *S. ohio* is a serovar infrequently implicated in outbreaks; however, it features commonly in summaries of *Salmonella* isolations from animal feeds in many countries. Experience in the food industry, particularly with control of *Listeria monocytogenes*, is that control is not achieved with one application of decontamination. Rather, environmental control is achieved by repeated decontamination, leading to a reduction and then elimination of the organisms, followed by adherence to rigid ongoing maintenance of control measures.

Use of treated and untreated wastes as fertilizers is likely to remain, as wastes are a valuable source of nutrients. Development of improved technologies to preserve nutrients as part of the worldwide trend for recycling wastes and reducing dumping will lead to better extraction of components of wastes as reusable nutrient sources. In the treatment, reuse or disposal of wastes, consideration must be given to the pathogen destruction during treatments, balanced with the potential for recontamination of and subsequent proliferation of pathogens in the treated product.

Animal Feeds

Feeds used in modern farming practices are dominated by prepared feeds with multiple ingredients, and the subject is dealt with in depth in Chapter 17. Many of the ingredients are of animal origin, including meat-meal, bone-meal and fish-meal, or vegetable origin, such as soya-meal, and have been through processing that should be sufficient to eliminate any *Salmonella* present in the raw materials. Contamination of the feed ingredients may also occur as post-processing contamination within the factory environment. Recontamination can occur almost immediately after materials leave a cooker (Bensink and Boland, 1979), and contamination of the plant environment may be highest at sites immediately post-cooking (Timoney, 1968). Some serovars may be better adapted for survival in factory environments and then appear in the feed ingredients.

Contaminated feed ingredients have resulted in the international spread of *Salmonella*, with the potential for introduction of serovars from other sources into animal populations. Comparisons of serovars occurring in feeds worldwide suggest that the presence of serovars in feeds does not necessarily correspond with their presence in animals. Serovars such as *S. mbandaka*, *S. tennessee* and *S. livingstone* are reported in feeds in unrelated countries (Murray, 1994b; Malmqvist *et al.*, 1995; Wegener *et al.*, 1995). Differences were demonstrated between *S. livingstone* strains of animal-feed and human origin in Sweden (Katouli *et al.*, 1994) and yet in the UK some similarities were demonstrated (Old *et al.*, 1995; Crichton *et al.*, 1996).

Apart from the obvious focus on the spread from poultry to humans, alternative uses for poultry may aid the dissemination of serovars to other animals. For instance, day-old cockerels are used as feed for carnivorous birds and animals in captivity. *S. enteritidis* infection in an owl and a lynx in London Zoo was traced to chicks used as feed (Kirkwood *et al.*, 1994). This may provide a hazard to both other animals and humans visiting the area, by increased contamination of the environment. Crocodiles are commercially farmed in Australia for meat and hides. Raw chicken meat is used as feed for both species farmed (*Crocodilus johnstoni* and *Crocodilus porosus*). While the feeding habits of the two species differ, general environmental contamination occurs, with resulting contamination of the dressed meat (Manolis *et al.*, 1991).

Aquaculture

Aquaculture in non-marine environments has been commonly practised in South-east Asia and is growing in developed countries. Extensive and intensive aquaculture ponds in South-east Asia were frequently contaminated with *Salmonella* (Reilly and Twiddy, 1992). Twenty-two per cent of mud and water samples and 16% of raw prawns from ponds investigated contained *Salmonella*. Feed may be a potential source of contamination, with chicken manure commonly used. *Salmonella* were also reported from eel ponds in Japan and farmed catfish in the USA. Additional dissemination occurs with the disposal of waste throughout the processing of these products. Increasing development of aquaculture, particularly in developing countries, where poorly treated effluent water is used, results in increasingly contaminated fish and crustaceans entering the food trade. The use of antimicrobials in aquaculture ponds can increase the potential for the development and dissemination of resistant organisms.

Rodents

Rodents are regarded as a potential risk to public health and rats are seen as an indication of unsatisfactory sanitation. Rats and mice are susceptible to *Salmonella* and, in the past, *S. enteritidis* var. Danysz was used as a control measure.

Of the 299 *Salmonella* isolated from rodents during several studies in the UK, *S. enteritidis* and *S. typhimurium* accounted for 58.5% and 28.4% respectively (Healing, 1991). Of 58 *S. enteritidis*, 55 were var. Danysz. It was concluded that rats were unlikely to be a significant source of infection for humans. In an earlier study (Healing *et al.*, 1980), *Salmonella* were not detected in 783 small rodents (excluding rats and house mice). On a broiler farm in Dorset where *S. enteritidis* was prevalent, of 141 rodents tested, including 23 house mice, none yielded *Salmonella* (Healing and Greenwood, 1991), which suggests that rodents may not be a significant source of transmission or maintenance. *Salmonella* in rodents may be a reflection of the contamination in the

area of feeding, rather than constituting a significant reservoir of Salmonella. The activities of
humans may cause rodents to act as sources of
dissemination of Salmonella.

Although S. enteritidis infections of mice on
poultry units was reported 9 years previously
(Krabisch and Dorn, 1980), their significance as
vectors of S. enteritidis on poultry units has
received attention only relatively recently
(Henzler and Opitz, 1992). The latter authors
found that the rate of S. enteritidis infections in
mice on S. enteritidis-contaminated farms ranged
from 4.8 to 71.4%, with a mean of 24%. Mice
caught in the vicinity of poultry farms were
infected with virulent strains of S. enteritidis and
were considered a source of the strains for egg-
laying flocks (Guard-Petter et al., 1997).

In the UK, wild mice were trapped, collected and fed feed pellets contaminated with S.
enteritidis phage type 4 at levels of approximately
10^5 or 10^3 cfu per mouse (Davies and Wray,
1995a). When fed 10^5 cfu, mice excreted
Salmonella for 8 months; however the excretion
decreased to a low level after 4 weeks, which was
then maintained for the next 7 months. The
mice fed 10^3 cfu excreted Salmonella for 5 weeks,
but excretion had decreased to a low level after 3
weeks. Fourteen poultry units where S. enteritidis
had been detected in poultry were monitored;
mice were found in six of the units, mice from
five yielding S. enteritidis at a prevalence ranging
from 29 to 86%. In the other unit, despite extensive contamination of the environment with S.
enteritidis, the organism could not be detected in
either mice or their droppings. Thirteen units
yielded Salmonella from environmental sampling.

Droppings from mice fed artificially infected
feed were placed into either the feed, water
drinking vessels or litter of 3-week-old chickens
(Davies and Wray, 1995a). The chickens became
infected when exposed to the faeces of mice
infected 2 months previously with 10^5 Salmonella,
while none were infected from faeces collected 5
months after initial infection of the mice. The
prevalence of infection and the numbers of
Salmonella in droppings of the mice were low,
although one dropping contained 10^2–10^3
Salmonella 6 weeks after capture. The authors
also concluded that mice need direct contact
with faeces of infected chickens to become
infected. As mice do not usually have a large ter-
ritory, they may be unlikely to spread infection

from one shed to nearby sheds. Mice can contaminate water, where survival and multiplication
of Salmonella can occur, increasing the spread of
infection. Mice need to be controlled as part of
an overall plan of management of poultry units,
in addition to efficient cleaning and disinfection.

Fifteen of 42 rats collected from farms in the
Rotarua region of New Zealand where salmonellosis in sheep occurred carried Salmonella. Three
were S. typhimurium and the remainder S. bovismorbificans (Robinson and Daniel, 1968). S.
bovismorbificans was the commonest serovar in
sheep at that time. None of the rats displayed
signs of disease when captured. In domestic environments surrounding the Veterinary Research
Institute, Ipoh, Malaysia, 71% of shrews and 25%
of rats had Salmonella (Joseph et al., 1984).
Predominant serovars included S. bareilly and S.
weltevreden, which are among the commonest
serovars infecting humans and animals in
Malaysia. It is likely that shrews and rats reflect
the general environment. The residential areas
where the animals were collected had septic-tank
disposal systems, where, as previously discussed,
mesophilic anaerobic digestion is not the most
effective process for reduction of Salmonella.

An examination of dead hedgehogs collected in Norfolk over a 10-year period (Keymer
et al., 1991) showed that, of 74 animals, ten
yielded Salmonella from the liver and other internal organs, indicating infection, while 13 animals
had Salmonella in the gut only. Of the ten
infected animals all grew S. enteritidis phage type
11, which is unusual in both humans and other
animals, although it has been isolated from
hedgehogs and other animals.

Insects

Cockroaches are found in almost any place
inhabited by humans, especially where food is
prepared. Much anecdotal evidence exists of
cockroaches being a health risk and a vehicle for
the spread of infectious organisms, including
Salmonella (Bennett, 1993). While there is little
hard evidence for such claims, there is a definite
need to keep sites clean and free of vermin, without concentrating only on cockroaches.

In French urban areas around public swim-
ming-pools, food-handling establishments and
residential flats, Salmonella were not detected in

157 cockroaches (Rivault *et al.*, 1993). In India, 4.1% of 221 cockroaches collected from urban environments were carriers of *Salmonella* and were regarded as a source for the spread of infections (Devi and Murray, 1991). Serovars included *S. typhimurium* and *S. bovismorbificans*, which had been associated with hospital outbreaks where cockroaches were implicated as the source. *Salmonella* was carried by 1.1% of cockroaches collected in the vicinity of university farms, feed stores and residences (Singh *et al.*, 1980). However, in the same study, 6.2% of rats, 10.9% of house mice, 10.5% of shrews and 6.6% of ants carried *Salmonella*. Singh *et al.* (1995) showed that when *S. paratyphi* B var. Java was fed to cockroaches at a level of 10^7 organisms g^{-1} of feed it could not be detected in faeces after 6 days, in living cockroaches after 8 days or in dead cockroaches after 10 days.

As with other insects, animals and birds, *Salmonella* in cockroaches may be a reflection of their environment, carried transiently as part of their gut flora as a result of feeding on sites contaminated with *Salmonella*, rather than a primary source of the organisms. Cockroaches are only one part of the insect flora of domestic and rural environments and may be over-emphasized in importance compared with other vehicles, because of their unpleasant appearance and association with the dark.

Lesser meal-worm beetles (*Alphitobius diaperinus*) are frequently associated with poultry sheds (Davies and Wray, 1995b). Beetles were fed for 6 days with feed contaminated with *S. enteritidis*. Over the next 6 days, *Salmonella* were not detected as surface contaminants of the beetles, nor were the *Salmonella* detected in the beetles when killed 6 days after removal from the contaminated feed. In the same investigation, beetles, mice and environmental swabs were collected from two poultry units after *S. enteritidis* infected flocks had been removed. Although 136 of 600 environmental swabs and 17 of 35 mice grew *S. enteritidis*, *Salmonella* were not isolated from 500 beetles. The authors concluded that infected mice and persistent contamination of the environment are likely to pose a greater hazard than these beetles. In an investigation of the sources for both vertical and horizontal transmission of *S. berta* in poultry in Denmark (Brown *et al.* 1992), the serovar was isolated from one of six mealworm beetles.

Flies have frequently been shown to be contaminated with *Salmonella*, and Edel *et al.* (1973) found that 1.5% of the 202 fly traps examined were contaminated with *S. typhimurium*. In flies, the organisms may establish and multiply (Greenberg *et al.*, 1970). Blowfly larvae (*Lucilia serricata*) have also been shown to be contaminated with *Salmonella* and it has been shown that maggots are a potent vehicle for *Salmonella* infection of chickens (Davies and Wray, 1994b).

Birds

Although wild birds are recognized as carriers of *Salmonella*, evidence suggests that infected birds are rarely identified. Short-term carriage of the *Salmonella* occurs, reflecting the feeding environments of the birds. The incidence of *Salmonella* carriage in wild birds appears to be low. Of 382 dead wild birds representing a diversity of UK bird types, 2.9% had *Salmonella*, while only 0.4% of trapped wild birds yielded *Salmonella* from cloacal swabs (Goodchild and Tucker, 1968). The authors noted that *Salmonella* were not recovered from predatory birds, which suggests that *Salmonella* were not common amongst the small birds and vermin on which they prey. Wild birds associated with sewage-treatment plants in the UK were found to have a low incidence; only one bird of 599 had *Salmonella*. However, gulls were not present in the 39 species trapped and tested (Plant, 1978).

While gulls may carry *Salmonella* and be a source of dissemination from contaminated sites, evidence shows that this carriage does not result in infection and lasts only a few days (Girdwood *et al.*, 1985). Of 540 wild birds trapped in Norway, 180 were gulls, which were the only birds carrying *Salmonella*, with four isolations from three species (Kupperud and Rosef, 1983). *Salmonella* were isolated from eight of 30 gulls in the Auckland region of New Zealand, associated with abattoirs and rubbish dumps; however, none of 24 gulls in the remote Ninety Mile Beach region harboured *Salmonella* (Robinson and Daniel, 1968). Black-backed gulls in the area did not disperse more than 60 miles from their nesting area, while red-billed gulls could travel more than 250 miles. Rottnest Island, 19 km off the coast from Perth in Western Australia, has a resident population of native animals and a transient human population.

Gulls on the island carried *Salmonella* serovars that were representative of those in the island's environment, rather than that of the nearby mainland (Hart *et al.*, 1987). Gulls have been identified as a source of *Salmonella* contamination for fish-processing plants, with which they are frequently associated (Berg and Anderson, 1972). Fish-meal is traded worldwide and can be a vehicle of transmission of *Salmonella* into food animals.

In 1951, a large outbreak of salmonellosis due to S. *typhimurium* occurred in sheep in South Australia (Watts and Wall, 1952). Wild birds were collected as part of investigations for a source. Of birds collected, only carnivores carried *Salmonella* and they were assessed as being a vehicle for the contamination of water troughs. This consequently aided the maintenance of the outbreak, with water being one vehicle for the spread of infection in the sheep. Several crows and magpies, which were caught alive, were found to be carriers of S. *typhimurium*. These birds remained excreters for 14 days, after which they gradually ceased excreting, the last infected faeces being obtained after 27 days. Of foxes and rabbits also captured, one of four foxes carried the *Salmonella* but all 13 rabbits were negative.

Sparrows associated with farms where calf salmonellosis occurred were infected with S. *typhimurium* (seven out of 30 birds positive), with the same phage type 101 being isolated from both calves and sparrows (Cizek *et al.*, 1994). Of 432 pigeons from *Salmonella*-free farms, two carried *Salmonella* while 1734 other wild birds (no gulls included) were negative. In black-headed gulls, 4.2% of adult birds carried *Salmonella*, while 19.2% of non-flying young in nesting colonies were positive. The authors conclude that it is unlikely that *Salmonella* would spread to individual birds in wild populations by direct contact. The *Salmonella*-contaminated environment was regarded as the main source of infections among wild birds, with the organisms being acquired during food gathering and drinking. Birds may be a secondary source of infection for humans and farm animals. Particularly in land-locked Central Europe, the black-headed gull is the only species that may have some practical role in the dissemination of *Salmonella*, because of its large numbers, spatial dispersion and range of movements.

Environmental spread can be diverse; in Antarctica, S. *enteritidis* phage type 4 was found in a penguin during a survey of animals and birds on Bird Island, South Georgia (Olsen *et al.*, 1996). Only one isolate was obtained, which, on pulse-field gel electrophoresis (PFGE), was different from strains in Europe and South America. The authors suggest that the source was likely to be related to sewage or discharges of wastes from ships or fishing vessels. It is unlikely that migratory birds or albatrosses would be a significant vehicle for transmission, as birds, as discussed elsewhere, generally carry *Salmonella* only transiently.

Wild Animals as Reservoirs and Disseminators

Animals in the natural environment may be reservoirs of *Salmonella*, irrespective of the country or region. These include wild animals, such as opossums (Runkel *et al.*, 1991), domesticated animals, including dogs, whether strays (Sugiyama *et al.*, 1993), household pets or kennel dogs (Shimi *et al.*, 1976), cats and reptiles (Borland, 1975). Even in urban environments, exotic animals kept as pets can be a source of *Salmonella*. Salmonellosis in humans has been attributed to pets, including terrapins and lizards. The serovars associated with reptiles and reptilian-associated infections are similar and are often unusual compared with the serovars within a community (Ackman *et al.*, 1995).

Many wild animals are scavengers, although it is extremely difficult to investigate their microbial flora in pristine environments, that are unaffected, either directly or indirectly, by the activities of humans. Scavengers, including rodents and opossums, in Panama had higher carriage rates of *Salmonella* than non-scavengers, most infected animals being near areas associated with human activity (Kourany *et al.*, 1976). Disturbance to the natural environment, including deforestation and inadequate sanitation, increases the potential for contamination of wild animals.

In some countries, there is field slaughter of wild animals destined for human consumption. In Australia, feral pigs and kangaroos are shot in the wild, partially eviscerated and then transported under refrigeration for processing (Bensink *et al.*, 1991). Generally, these animals are sought in areas well isolated from larger centres of human habitation. Of animals shot in the wild, *Salmonella* were isolated from 34.4% of feral pigs and 11.1% of kangaroos. The serovars from pigs were different from those in kangaroos but, in both groups, many of the serovars were those associated with humans and animals within Australia (Murray, 1994a). On Rottnest Island, as mentioned previously, the serovars found in reptiles did not correlate with those in the marsupial quokkas, even though they are sharing the same environmental niche (Hart *et al.*, 1987).

Other Sources

Bean sprouts were identified as the source of an outbreak of salmonellosis in the UK (O'Mahony *et al.*, 1990). The outbreak featured cases of *S. virchow* and *S. saintpaul*. The strains of *S. saintpaul* were of a different plasmid profile from those usually seen in the UK (C. Murray, Adelaide, unpublished data, 1988), and *S. virchow* phage type 34, while rare in humans in the UK was common in humans and poultry in Australia (B. Rowe, London, personal communication, 1988). The mung beans used for the sprouting had been imported from Australia and, at the time, the outbreak strains were relatively common in poultry in the geographical region from where the beans originated. The beans were packaged in a plant that was in the vicinity of a poultry by-product plant and the possibility exists that contaminated dust may have entered the packaging-plant environment.

Salmonella is present in feeds, foods and their ingredients. The environment is a significant source of contamination for these products and the increased trade in foods can spread *Salmonella* internationally; frequently, the *Salmonella* originate from the environment (D'Aoust, 1994). *Salmonella* are found in or on fruits, vegetables and spices. While these types of products are implicated by public-health authorities as the source of *Salmonella*, they are not the true source; rather, they are a vehicle for transmission of the *Salmonella* from an environmental source associated with the raw product.

Hazard analysis and critical control points (HACCP) are used internationally for controlling processes in the food industry, including achieving microbial control (Tietjen and Fung, 1995). In the USA, a combination of HACCP and microbiological criteria is incorporated into regulations for raw-meat and poultry processing (Dreesen, 1996; Morris, 1996). Performance standards include *Salmonella* monitoring. The principles of HACCP can be applied to a wide range of activities related to release of contaminated materials into the environment. While this may be desirable, the hazards are likely to be assessed as low compared with the economic costs of control measures.

Environmental colonization, either natural or within processing environments, is a source of contamination of prepared foods. This will continue to occur spasmodically, irrespective of the best intentions of processors. Indications suggest that the true factors involved in contamination will sometimes not be revealed, due to commercial confidentiality and public liability concerns.

S. waycross has been the major serovar associated with human infections in Guam for many years. The serovar cannot be linked to food sources; however, it is found in domestic dogs, soils, lizards and domestic vacuum-cleaner dust. While accounting for approximately 40% of human isolates, it is rarely involved in outbreaks (Haddock *et al.*, 1990). The colonization of the natural environment appears to be the source of its dissemination into the domestic environment (Haddock and Nocon, 1994).

Southern Africa has a relatively high incidence of *Salmonella* ssp. II serovars in the human population (Schrire *et al.*, 1987). These serovars are generally associated with environmental sources and cold-blooded animals. A large number of these serovars were isolated in Australia during investigations of herbal teas originating from Southern Africa (C. Murray, Adelaide, unpublished data, 1984). In Australia, a number of new serovars have been found in remote regions, e.g. *S. mowanjum*, *S. ord* and *S. kalumbaru*, undoubtedly with environmental animal reservoirs (Kelterborn, 1987; J. Iveson, Perth, personal communication, 1983).

S. abortusovis is a host-specific *Salmonella* causing abortion in sheep. This serovar was isolated from two children in outback Australia in 1994 (Murray, 1994a). These were the first recorded cases of the serovar causing infection in humans and the first recorded isolations of the serovar from any source in Australia. Subsequently, it was isolated from kangaroo meat in a processing plant, the meat originating from animals in the same geographical region as the children. No evidence of infection was found in any sheep associated with the property where the children became infected. The isolates were serologically *S. abortusovis* but did not have the virulence plasmid associated with this serovar (S. Rubino, Sassari, personal communication, 1996). It is possible that the source of this organism is an unidentified wild animal.

One arena affecting transmission to humans, which is difficult to address, is the domestic kitchen. Regulatory controls cannot be applied to this environment; *Salmonella* control, as well as control of other pathogens, is reliant on the hygiene education of the general population. Elimination of organisms on all raw foods entering homes is not likely to occur. All steps in the food, animal and waste cycles are important in the control of *Salmonella* (Oosterom, 1991).

In commercial garden fertilizers in the UK, *Salmonella* were detected in 40 of 120 samples, with mixed serovars, including *S. hadar*, isolated (Williams Smith *et al.*, 1982). While this may have little direct significance for human or animal health, it is a potential source of *Salmonella* and may be a source of contamination of vegetables, which are becoming increasingly identified as a vehicle for *Salmonella* infection (Madden, 1992). Surface decontamination can significantly reduce the level of *Salmonella* on tomatoes; however, when the organisms have penetrated the stem tissue, decontamination is difficult (Zhuang and Beuchat, 1996). In spite of the low pH of tomatoes (pH 4.0–4.3), *Salmonella* can survive and multiply when temperature conditions are favourable.

Salmonella have also been shown to form biofilms on glass and chlorinated polyvinyl chloride (CPVC) pipes (Jones and Bradshaw, 1996). This can enable the organisms to effectively colonize processing lines.

Conclusion

It is difficult to presume that a reduction in the spread of pathogens such as *Salmonella* into the environment will occur, despite the introduction and promotion of improved measures to prevent the release, spread and destruction of these organisms. Not only do practices vary in developed countries but they are subjected to even greater pressures in many developing or underdeveloped countries. Practices in one region can affect another region, particularly with increased movement of population, food and agricultural products.

Salmonella are environmental organisms and may be found in virtually any phase of natural and man-made environments. *Salmonella* should be regarded as part of the normal flora of all animals, but having the potential to cause infection within the animals. *Salmonella* incidence and distribution in the environment has been greatly increased by humans, particularly as a result of modern civilization and agriculture; this can be regarded as (unnatural) contamination of environmental sources and the food-chain. Even well-developed countries are unable to provide reliable decontamination of discharges from treatment plants and agriculture. It is difficult to imagine that all discharges will ultimately be free of *Salmonella* and other pathogens.

For practical purposes, we must accept that *Salmonella* will not be eliminated and must address our efforts to controlling its introduction and spread into the agricultural and food-chains. These controls will influence the general microbial flora, without focusing solely on *Salmonella*, and are achievable in many processes, but *Salmonella* will continue to be a feature of humans, animals and the general environment.

References

Abalaka, J.A. and Deibel, R.H. (1990) Viability of *Salmonella* in bone-meal. *Microbios* 62, 155–164.

Ackman, D.M., Drabkin, P., Birkhead, G. and Cieslack, P. (1995) Reptile-associated salmonellosis in New York State. *Pediatric Infectious Disease Journal* 14, 955–959.

Alcaide, E., Martinez, J.P. and Garay, E. (1984) Comparative study on *Salmonella* isolation from sewage-contaminated natural waters. *Journal of Applied Bacteriology* 56, 365–371.

Altekruse, S., Koehler, J., Hickman-Brenner, F., Tauxe, R.V. and Ferris, K. (1993) A comparision of *Salmonella enteritidis* phage type from egg-associated outbreaks and implicated laying flocks. *Epidemiology and Infection* 110, 17–22.

Baskerville, A., Humphrey, T.J., Fitzgeorge, R.B., Cook, R.W., Chart, H., Toew, B. and Whitehead, A. (1992) Airborne infection of laying hens with *Salmonella* enteritidis phage type 4. *Veterinary Record* 130, 395–398.

Bennett, G. (1993) Cockroaches as carriers of bacteria. *Lancet* 341, 732.

Bensink, J.C. and Boland, P.H. (1979) Possible pathways of contamination of meat and bone-meal with *Salmonella*. *Australian Veterinary Journal* 55, 521–524.

Bensink, J.C., Ekaputra, I. and Taliotis, C. (1991) The isolation of *Salmonella* from kangaroos and feral pigs processed for human consumption. *Australian Veterinary Journal* 68, 106–107.

Berg, R.W. and Anderson, A.W. (1972) *Salmonella* and *Edwardsiella tarda* in gull feces: a source of contamination in fish processing plants. *Applied Microbiology* 24, 501–503.

Borland, E.D. (1975) *Salmonella* infection in dogs, cats, tortoises and terrapins. *Veterinary Record* 96, 401–402.

Brown, D.J., Olsen, J.E. and Bisgaard, M. (1992) *Salmonella enterica*: infection, cross infection and persistence within the environment of a broiler parent stock unit in Denmark. *Zentralblatt für Bakteriologie* 277, 129–138.

Burton, G.A., Gunnison, D. and Lanza, G.R. (1987) Survival of pathogenic bacteria in various freshwater sediments. *Applied and Environmental Microbiology* 53, 633–638.

Caldwell, D.J., Hargis, B.M., Corrier, D.E., Vidal, L. and DeLoach, J.R. (1995) Evaluation of persistence and distribution of *Salmonella* serovar isolation from poultry farms using drag-swab sampling. *Avian Diseases* 36, 617–621.

Christensen, N.H. and Cullinane, L.C. (1993) Faecal excretion of *Salmonella hadar* from calves grazed on pastures fertilised with *S. hadar*-contaminated broiler litter. *New Zealand Veterinary Journal* 41, 157–160.

Cizek, A., Literak, I., Hejlicek, K., Treml, F. and Smola, J. (1994) *Salmonella* contamination of the environment and its incidence in wild birds. *Journal of Veterinary Medicine* 41, 320–327.

Clegg, F.G., Chiejina, S.N., Duncan, A.L., Kay, R.N. and Wray, C. (1983) Outbreaks of *Salmonella newport* infection in dairy herds and their relationship to management and contamination of the environment. *Veterinary Record* 112, 580–584.

Crichton, P.B., Old, D.C., Taylor, A. and Rankin, S.C. (1996) Characterisation of strains of *Salmonella* serovar livingstone by multiple typing. *Journal of Medical Microbiology* 44, 325–331.

Danielsson, M.L. (1977) *Salmonella* in sewage and sludge. *Acta Veterinaria Scandanavica (Copenhagen)* Suppl. 65, 108–114.

D'Aoust, J. (1994) *Salmonella* and the international food trade. *International Journal of Food Microbiology* 24, 11–31.

Davies, R.H. and Wray, C. (1994a) An approach to reduction of *Salmonella* infection in broiler chicken flocks through intensive sampling and identification of cross-contamination hazards in commercial hatcheries. *International Journal of Food Microbiology* 24, 147–160.

Davies, R.H. and Wray, C. (1994b). Use of larvae of *Lucilla serricata* in colonisation studies of *Salmonella enteritidis* in poultry. In: Pusztai, A., Hinton, M.H. and Mulder, R.W.A.W. (eds) *Flair No. 6. The Attachment of Bacteria to the Gut*. CVP-DLO Het, Spelderholt.

Davies, R.H. and Wray, C. (1995a) Mice as carriers of *Salmonella enteritidis* on persistently infected poultry units. *Veterinary Record* 137, 337–341.

Davies, R.H. and Wray, C. (1995b) Contribution of the lesser meal-worm beetle (*Alphitobius diaperinus*) to the carriage of *Salmonella enteritidis* in poultry. *Veterinary Record* 137, 407–408.

Davies, R.H. and Wray, C. (1996a) Persistence of *Salmonella enteritidis* in poultry units and poultry food. *British Poultry Science* 37, 589–596.

Davies, R.H. and Wray, C. (1996b) Seasonal variations in the isolation of *Salmonella typhimurium*, *Salmonella enteritidis*, *Bacillus cereus* and *Clostridium perfringens* from environmental samples. *Journal of Veterinary Medicine* 43, 119–127.

Davies, R.H. and Wray, C. (1996c) Studies of contamination of three broiler breeder houses with *Salmonella enteritidis* before and after cleansing and disinfection. *Avian Diseases* 40, 626–633.

Devi, S.J.N. and Murray, C.J. (1991) Cockroaches (*Blatta* and *Periplaneta* species) as reservoirs of drug-resistant salmonellas. *Epidemiology and Infection* 107, 357–361.

Dreesen, D.W. (1996) Effectiveness of microbiological criteria for the safety of foods of animal origin. *Journal of the American Veterinary Medical Association* 209, 2052–2054.

Durrant, D.S. and Beatson, S.H. (1981) Salmonellae isolated from domestic meat waste. *Journal of Hygiene, Cambridge* 86, 259–264.

Edel, W., van Schothorst, M., Guinée, P.A.M. and Kamplemacher, E.R. (1973). Mechanisms and prevention of *Salmonella* infections in animals. In: Hobbs, B.C. and Christian, J.H.B. (eds) *The Microbiological Safety of Food*. Academic Press, London, pp. 247–256.

Errebo Larsen, H., Munch, B. and Schlundt, J. (1994) Use of indicators for monitoring the reduction of pathogens in animal waste treated in biogas plants. *Zentralblatt für Hygiene und Umweltmedizin (Stuttgart)* 195, 544–555.

Farrell, J.B., Salotto, B.V. and Venosa, A.D. (1990) Reduction in bacterial densities of wastewater solids by three secondary treatment processes. *Research Journal of the Water Pollution Control Federation* 62, 177–184.

Fish, J.T. and Pettibone, G.W. (1995) Influence of freshwater sediment on the survival of *Escherichia coli* and *Salmonella* sp. as measured by three methods of enumeration. *Letters in Applied Microbiology* 20, 277–281.

Foster, J.W. and Spector, M.P. (1995) How *Salmonella* survive against the odds. *Annual Review of Microbiology* 49, 145–174.

Fransen, N.G., van den Elzen, A.M.G., Urlings, B.A.P. and Bijker, P.G.H. (1996) Pathogenic microorganisms in slaughterhouse sludge – a survey. *International Journal of Food Microbiology* 33, 245–256.

Gadre, R.V., Ranade, D.R. and Godbole, S.H. (1986) A note on survival of salmonellas during anaerobic digestion of cattle dung. *Journal of Applied Bacteriology* 60, 93–96.

Galdiereo, E., Donnarumma, G., de Martino, L., Marcatili, A., Cipollaro de l'Ero, G. and Merone, A. (1994) Effect of low-nutrient seawater on morphology, chemical composition, and virulence of *Salmonella typhimurium*. *Archives of Microbiology* 162, 41–47.

Gay, J.M. and Hunsaker, M.E. (1993) Isolation of multiple *Salmonella* serovars from a dairy two years after a clinical salmonellosis outbreak. *Journal of the American Veterinary Medical Association* 203, 1314–1320.

Girdwood, R.W.A., Fricker, C.R., Munro, D., Shedden, C.B. and Monaghan, P. (1985) The incidence and significance of *Salmonella* carriage by gulls (*Larus* spp.) in Scotland. *Journal of Hygiene, Cambridge* 95, 229–241.

Glaus, H. and Heinemeyer, E. (1994) The elimination of *Salmonella typhimurium* in coastal waters with varous levels of microbiogically hygienic contamination. *Zentralblatt für Hygiene und Umweltmedizin (Stuttgart)* 196, 312–326.

Goodchild, W.M. and Tucker, J.F. (1968) Salmonellae in British wild birds and their transfer to domestic fowl. *British Veterinary Journal* 124, 95–101.

Greenberg, B., Kowalski, J.A. and Klowden, M.J. (1970) Factors affecting the transmission of Salmonellae by flies. natural resistance to colonisation and bacterial interference. *Infection and Immunity* 2, 800–809.

Guard-Petter, J., Henzler, D.J., Rahman, M.M. and Carlson, R.W. (1997) On-farm monitoring of mouse-invasive *Salmonella enterica* serovar Enteritidis and a model for its association with the production of contaminated eggs. *Applied and Environmental Microbiology* 63, 1588–1593.

Haddock, R.L. and Nocon, F.A. (1994) Infant salmonellosis and vacuum cleaners. *Journal of Tropical Paediatrics* 40, 53–54.

Haddock, R.L., Nocon, F.A., Santos, E.A. and Taylor, T.G. (1990) Reservoirs and vehicles of *Salmonella* infection on Guam. *Environment International* 16, 11–16.

Harrington, C.S., Lanser, J.A., Manning, P.A. and Murray, C.J. (1991) Epidemiology of *Salmonella sofia* in Australia. *Applied and Environmental Microbiology* 57, 223–227.

Hart, R.P., Iveson, J.B. and Bradshaw, S.D. (1987) The ecology of *Salmonella* serovars in a wild marsupial (the quokka *Setonix brachyurus*) in a disturbed environment. *Australian Journal of Ecology* 12, 267–279.

Healing, T.D. (1991) *Salmonella* in rodents: a risk to man? *Communicable Diseases Review* 1, R114–R116.

Healing, T.D. and Greenwood, M.H. (1991) Frequency of isolation of *Campylobacter* spp., *Yersinia* spp. and *Salmonella* spp. from small mammals in two sites in southern Britain. *International Journal of Environmental Health Research* 1, 54–62.

Healing, T.D., Kaplan, C. and Prior, A. (1980) A note on some Enterobacteriaceae from faeces of small wild British mammals. *Journal of Hygiene, Cambridge* 85, 343–345.

Henry, D.P., Frost, A.J., O'Boyle, D. and Cameron, R.D.A. (1995) The isolation of salmonellas from piggery waste water after orthodox pondage treatment. *Australian Veterinary Journal* 72, 478–479.

Henzler, D.J. and Opitz, M.H. (1992). The role of mice in the epizootology of *Salmonella enteritidis* infection in chicken layer farms. *Avian Diseases*. 36, 625–631.

Higgs, A.R.B., Norris, R.T. and Richards, R.B. (1993) Epidemiology of salmonellosis in the live sheep export industry. *Australian Veterinary Journal* 70, 330–335.

Hinton, M. and Bale, M.J. (1991) Bacterial pathogens in domesticated animals and their environment. *Journal of Applied Bacteriology* 70, 81S–90S.

How, R.A., Bradley, A.J., Iveson, J.B., Kemper, C.M., Kitchener, D.J. and Humphreys, W.F. (1983) The natural history of salmonellae in mammals of the tropical Kimberley region, Western Australia. *Ecology of Disease* 2, 9–32.

Hunter, P.R. and Izsák, J. (1990) Diversity studies of *Salmonella* incidents in some domestic livestock and their potential relevance as indicators of niche width. *Epidemiology and Infection* 105, 501–510.

Hussong, D., Burge, W.D. and Enkiri, N.K. (1985) Occurrence, growth, and suppression of *Salmonella* in composted sewage sludge. *Applied and Environmental Microbiology* 50, 887–893.

Irwin, R.J., Poppe, C., Messier, S., Finley, G.G. and Oggel, J. (1994) A national survey to estimate the prevalence of *Salmonella* species among Canadian registered commercial turkey flocks. *Canadian Journal of Veterinary Research* 58, 263–267.

Jones, K. and Bradshaw, S.B. (1996) Biofilm formation by the Enterobacteraceae: a comparison between *Salmonella enteritidis*, *Escherichia coli* and a nitrogen-fixing strain of *Klebsiella pneumoniae*. *Journal of Applied Bacteriology* 80, 458–464.

Jones, P.W. (1975) The effect of storage in slurry on the virulence of *Salmonella dublin*. *Journal of Hygiene, Cambridge* 74, 65–70.

Jones, P.W. (1980) Animal health today – problems of large livestock units: disease hazards associated with slurry disposal. *British Veterinary Journal* 136, 529–542.

Jones, P.W. and Rennison, L.M. (1980) The occurrence and significance to animal health of *Salmonella*s in sewage and sewage sludges. *Journal of Hygiene, Cambridge* 84, 47–62.

Jones, P.W., Collins, P., Brown, G.T.H. and Aitken, M. (1982) Transmission of *Salmonella mbandaka* to cattle from contaminated feed. *Journal of Hygiene, Cambridge* 88, 255–263.

Joseph, P.G., Ham Thong Yee and Sivanandan, S.P. (1984) The occurrence of salmonellae in house shrews and rats in Ipoh, Malaysia. *Southeast Asian Journal of Tropical Medicine and Public Health* 15, 326–330.

Josland (1951) Survival of *Salmonella typhimurium* on various substances under natural conditions. *Australian Veterinary Journal* 27, 264–266.

Katouli, M., Wollin, R., Gunnarsson, A., Kühn, I. and Möllby, R. (1994) Biochemical phenotypes of *Salmonella livingstone* isolated from humans, animals and feedstuffs in Sweden. *Acta Veterinaria Scandinavica (Copenhagen)* 35, 27–36.

Kearney, T.E., Larkin, M.J., Frost, J.P. and Levett, P.N. (1993) Survival of pathogenic bacteria during mesophilic anaerobic digestion of animal waste. *Journal of Applied Bacteriology* 75, 215–219.

Kelterborn, E. (1987) *Catalogue of* Salmonella *First Isolations 1965–1984*. VEB Gustav Fisher Verlag, Jena and Martinus Nijhoff Publishers, Dordrecht, 307 pp.

Keymer, I.F., Gibson, E.A. and Reynolds, D.J. (1991) Zoonoses and other findings in hedgehogs (*Erinaceus europaeus*): a survey of mortality and review of the literature. *Veterinary Record* 128, 245–249.

Kinde, H., Read, D.H., Ardans, A., Breitmeyer, R.E., Willoughby, D., Little, H.E., Kerr, D., Gireesh, R. and Nagaraja, K.V. (1996) Sewage effluent: likely source of *Salmonella enteritidis*, phage type 4 infection in a commercial layer flock in southern California. *Avian Diseases* 40, 672–676.

Kirkwood, J.K., Cunningham, A.A., Macgregor, S.K., Thornton, S.M. and Duff, J.P. (1994) *Salmonella enteritidis* excretion by carnivorous animals fed on day old chicks. *Veterinary Record* 134, 683.

Kourany, M., Bowdre, L. and Herrer, A. (1976) Panamanian forest animals as carriers of *Salmonella*. *American Journal of Tropical Medicine and Hygiene* 25, 449–455.

Krabisch, P. and Dorn, P. (1980) Zur epidemiologischen Bedeutung von Lebendvektoren bei Verbreitung von Salmonellen in der Geflugelmast Berliner und Müncher. *Tierarzfliche Wochenschrift* 92, 232–235.

Kupperud, G. and Rosef, O. (1983) Avian wildlife reservoir of *Campylobacter fetus* subsp. *jejuni*, *Yersinia* spp., and *Salmonella* spp. in Norway. *Applied and Environmental Microbiology* 45, 375–380.

Lahellec, C., Colin, P. and Bennejean, G. (1986) Influence of resident *Salmonella* on contamination of broiler flocks. *Poultry Science* 65, 2034–2039.

Lax, A.J., Barrow, P.A., Jones, P.W. and Wallis, T.S. (1995) Current perspectives in salmonellosis. *British Veterinary Journal* 151, 351–377.

Lever, M.S. and Williams, A. (1996) Cross-infection of chicks by airborne transmission of *Salmonella* enteritidis PT4. *Letters in Applied Microbiology* 23, 347–349.

Linklater, K.A., Graham, M.M. and Sharp, J.C.M. (1985) Salmonellae in sewage sludge and abattoir effluent in south-east Scotland. *Journal of Hygiene, Cambridge* 94, 301–307.

McDermid, A.S. and Lever, M.S. (1996) Survival of *Salmonella enteritidis* PT4 and *Salm. typhimurium* Swindon in aerosols. *Letters in Applied Microbiology* 23, 107–109.

McLaren, I.M. and Wray, C. (1991) Epidemiology of *Salmonella typhimurium* infection in calves: persistence of Salmonellae on calf units. *Veterinary Record* 129, 461–462.

Madden, J.M. (1992) Microbial pathogens in fresh produce – the regulatory perspective. *Journal of Food Protection* 55, 821–823

Malmqvist, M., Jacobsson, K.-G., Hägglbom, P., Cerenius, F., Sjöland, L. and Gunnarsson, A. (1995) *Salmonella* isolated from animals and feedstuffs in Sweden during 1988–1992. *Acta Veterinaria Scandinavica (Copenhagen)* 36, 21–39.

Manolis, S.C., Webb, G.J.W., Pinch, D., Melville, L. and Hollis, G. (1991) *Salmonella* in captive crocodiles (*Crocodylus johnstoni* and *C. porosus*). *Australian Veterinary Journal* 68, 102–105.

Morgan-Jones, S.C. (1980) The occurrence of Salmonellae during rearing of broiler birds. *British Poultry Science* 21, 463–470.

Morinigo, M.A., Cornax, R., Munoz, M.A., Romero, P. and Borrego, J.J. (1989) Viability of *Salmonella* species in natural waters. *Current Microbiology* 18, 267–273.

Morris, J.G. (1996) Current trends in human diseases associated with foods of animal origin. *Journal of the American Veterinary Medical Association* 209, 2045–2047.

Murray, C.J. (1991) Salmonellae in the environment. *Revue Scientifique et Technique. Office International des Epizooties* 10, 765–785.

Murray, C.J. (1994a) *Australian Salmonella Reference Centre Annual Report*. Institute of Medical and Veterinary Science, Adelaide, 114 pp.

Murray, C.J. (1994b) *Salmonella* serovars and phage types in humans and animals in Australia 1987–1992. *Australian Veterinary Journal* 71, 78–81.

Old, D.C., McLaren, I.M. and Wray, C. (1995) A possible association between *Salmonella livingstone* strains from man and poultry in Scotland. *Veterinary Record* 137, 544.

Olsen, B., Bergström, S., McCafferty, D.J., Sellin, M. and Wiström, J. (1996) *Salmonella enteritidis* in Antarctica: zoonosis in man or humanosis in penguins. *Lancet* 348, 1319–1320.

Olsen, J.E., Brown, D.J., Skov, M.N. and Christensen, J.P. (1993) Bacterial typing methods suitable for epidemiological analysis: applications in investigations of salmonellosis in livestock. *Veterinary Quarterly* 15, 125–135.

O'Mahony, M., Cowden, J., Smyth, B., Lynch, D., Hall, M., Rowe, B., Teare, E.L., Tettmar, R.E. and Rampling, A.M. (1990) Outbreak of *Salmonella saintpaul* infection associated with beansprouts. *Epidemiology and Infection* 104, 229–235.

Oosterom, J. (1991) Epidemiological studies and proposed preventive measures in the fight against human salmonellosis. *International Journal of Food Microbiology* 12, 41–52.

Opara, O.O., Carr, L.E., Russek-Cohen, E., Tate, C.R., Mallinson, E.T., Miller, R.G., Stewart, L.E., Johnston, R.W. and Joseph, S.W. (1992) Correlation of water activity and other environmental conditions with repeated detection of *Salmonella* contamination on poultry farms. *Avian Diseases* 36, 664–671.

Pillai, S.D., Widmer, K.W., Dowd, S.E. and Ricke, S.C. (1996) Occurrence of airborne bacteria and pathogen indicators during land application of sewage sludge. *Applied and Environmental Microbiology* 62, 296–299.

Plant, C.W. (1978) Salmonellosis in wild birds feeding at sewage treatment works. *Journal of Hygiene, Cambridge* 81, 43–48.

Plym-Forshell, L. (1995) Survival of *Salmonella* and *Ascaris suum* eggs in a thermophilic biogas plant. *Acta Veterinaria Scandinavica (Copenhagen)* 36, 79–85.

Pokorny, J. (1988) Survival and virulence of *Salmonella* in water. *Journal of Hygiene, Epidemiology, Microbiology and Immunology* 32, 361–366.

Poppe, C., Irwin, R.J., Messier, S., Finley, G.G. and Oggel, J. (1991) The prevalence of *Salmonella enteritidis* and other *Salmonella* spp. among Canadian registered commercial broiler flocks. *Epidemiology and Infection* 107, 201–211.

Poppe, C., Johnson, R.P., Forsberg, C.M. and Irwin, R.J. (1992) *Salmonella enteritidis* and other *Salmonella* in laying hens and eggs from flocks with *Salmonella* in their environment. *Canadian Journal of Veterinary Research* 56, 226–232.

Rajkowski, K.T., Rice, E.W., Huynh, B. and Patsy, J. (1996) Growth of *Salmonella* spp. and *Vibrio cholerae* in reconditioned wastewater. *Journal of Food Protection* 59, 577–581

Reilly, P.J.A. and Twiddy, D.R. (1992) *Salmonella* and *Vibrio cholerae* in brackishwater cultured tropical prawns. *International Journal of Food Microbiology* 16, 293–301.

Reilly, W.J., Oboegbulem, S.I., Munro, D.S. and Forbes, G.I. (1991) The epidemiological relationship between *Salmonella* isolated from poultry meat and sewage effluents at a long-stay hospital. *Epidemiology and Infection* 106, 1–10.

Rhodes, M.W. and Kator, H. (1988) Survival of *Escherichia coli* and *Salmonella* spp. in estuarine environments. *Applied and Environmental Microbiology* 54, 2902–2907.

Rivault, C., Cloarec, A. and Le Guyader, A. (1993) Bacterial load of cockroaches in relation to urban environment. *Epidemiology and Infection* 110, 317–325.

Robinson, R.A. and Daniel, M.J. (1968) The significance of *Salmonella* isolations from wild birds and rats in New Zealand. *New Zealand Veterinary Journal* 16, 53–55.

Roszak, D.B., Grimes, D.J. and Colwell, R.R. (1984) Viable but nonrecoverable stage of *Salmonella enteritidis* in aquatic systems. *Canadian Journal of Microbiology* 30, 334–338.

Runkel, N.S., Rodriguez, L.F., Moody, F.G., LaRocco, M.T. and Blasdel, T. (1991) *Salmonella* infection of the bilary and intestinal tract of wild opossums. *Laboratory Animal Science* 41, 54–56.

Russ, C.F. and Yanko, W.A. (1981) Factors affecting Salmonellae repopulation in composted sludges. *Applied and*

Environmental Microbiology 41, 597–602.

Schrire, L., Crisp, S., Bear, N., McStay, G., Koornhof, H.J. and Le Minor, L. (1987) The prevalence of human isolates of *Salmonella* subspecies *II* in southern Africa. *Epidemiology and Infection* 98, 25–31.

Shimi, A., Keyhani, M. and Bolurchi, M. (1976) Salmonellosis in apparently healthy dogs. *Veterinary Record* 98, 110–111.

Shuval, H.I. (1991) Effects of wastewater irrigation of pastures on the health of farm animals and humans. *Revue Scientifique et Technique, Office International des Epizooties* 10, 847–866.

Singh, B.R., Khurana, S.K. and Kulshreshtha, S.B. (1995) Survival of *Salmonella paratyphi* B var. *Java* on experimentally infected cockroaches. *Indian Journal of Experimental Biology* 33, 392–393.

Singh, S.P., Sethi, M.S. and Sharma, V.D. (1980) The occurrence of Salmonellae in rodent, shrew, cockroach and ant. *International Journal of Zoonoses* 7, 58–61.

Smith, B.P., Da Roden, L., Thurmond, M.C., Dilling, G.W., Konrad, H., Pelton, J.A. and Picanso, J.P. (1994) Prevalence of Salmonellae in cattle and in the environment on California dairies. *Journal of the American Veterinary Medical Association* 205, 467–471.

Smith, J.J., Howington, J.P. and McFeters, G.A. (1994) Survival, physiological response, and recovery of enteric bacteria exposed to a polar marine environment. *Applied and Environmental Microbiology* 60, 2977–2984.

Strauch, D. (1991) Survival of pathogenic microorganisms and parasites in excreta, manure and sewage sludge. *Revue Scientifique et Technique, Office International des Epizooties* 10, 813–846.

Sugiyama, Y., Sugiyama, F. and Yagami, K. (1993) Isolation of *Salmonella* from impounded dogs introduced to a laboratory. *Experimental Animal* 42, 119–121.

Tannock, G.W. and Davis, G.B. (1973) The isolation of *Salmonella* from the nasal passages of sheep. *Research in Veterinary Science* 14, 123–124.

Thomason, B.M., Biddle, J.W. and Cherry, W.B. (1975) Detection of Salmonellae in the environment. *Applied Microbiology* 30, 764–767.

Tietjen, M. and Fung, D.Y.C. (1995) Salmonellae and food safety. *Critical Reviews in Microbiology* 21, 53–83.

Timoney, J. (1968) The sources and extent of *Salmonella* contamination in rendering plants. *Veterinary Record* 83, 541–543.

Townsend, S.A. (1992) The relationships between Salmonellas and faecal indicator concentrations in two pools in the Australian wet/dry tropics. *Journal of Applied Bacteriology* 73, 182–188.

Walker, R.L., Madigan, J.E., Hird, D.W., Case, J.T., Villanueva, M.R. and Bogenrief, D.S. (1991) An outbreak of equine neonatal salmonellosis. *Journal of Veterinary Diagnostic Investigation* 3, 223–227.

Watts, P.S. and Wall, M. (1952) The 1951 *Salmonella typhimurium* epidemic in sheep in South Australia. *Australian Veterinary Journal* 28, 165–168.

Wegener, H.C., Larsen, S.K. and Flensberg, J. (1995) *Annual Report on Zoonoses in Denmark 1995*. Danish Zoonosis Centre and Danish Veterinary Service, Copenhagen, 12 pp.

WHO (1989) *Health Guidelines for the Use of Wastewater in Agriculture and Aquaculture: Report of a WHO Scientific Group*. WHO Technical Report Series No. 778, Geneva.

Williams Smith, H., Tucker, J.F., Hall, M.L.M. and Rowe, B. (1982) *Salmonella* organisms in garden fertilizers of animal origin. *Journal of Hygiene, Cambridge* 89, 125–128.

Wray, C. (1975) Survival and spread of pathogenic bacteria of veterinary importance within the environment. *Veterinary Bulletin* 45, 543–550.

Wray, C. (1985) Is salmonellosis still a serious problem in veterinary practice? *Veterinary Record* 116, 485–489.

Wray, C. (1995) Salmonellosis: a hundred years old and still going strong. *British Veterinary Journal* 151, 339–341.

Wray, C., Todd, N., McLaren, I.M., Beedell, Y.E. and Rowe, B. (1990) The epidemiology of *Salmonella* infection in calves: the role of dealers. *Epidemiology and Infection* 105, 295–305.

Zhuang, R.-Y. and Beuchat, L.R. (1996) Effectiveness of trisodium phosphate for killing *Salmonella montevideo* on tomatoes. *Letters in Applied Microbiology* 22, 97–100.

Chapter 17
Salmonella in Animal Feed

Robert H. Davies[1] and Mike H. Hinton[2]
[1]*Veterinary Laboratories Agency (Weybridge), New Haw, Addlestone, Surrey KT15 3NB, UK;*
[2]*Home Office, 50 Queen Anne's Gate, London SW1H 9RT, UK*

Introduction

Animal feed is a recognized source of pathogenic microorganisms for farm livestock. However, the reported incidence of infectious diseases contracted from feed is, in general, extremely low, despite the fact that many million tonnes of forages and manufactured animal feed are consumed throughout the world each year. Diseases of farm animals that may be associated with feed include, *inter alia*, anthrax, botulism, bovine spongiform encephalopathy (BSE), classical swine fever, foot-and-mouth disease, listeriosis, salmonellosis, swine vesicular disease, toxoplasmosis and trichinosis (for reviews, see Hinton and Bale, 1990; Hinton and Mead, 1990; Hinton, 1993). Several of these agents are also pathogenic for humans, and animal feed may act as an indirect source of infection for people consuming foods of animal origin. Examples of illnesses in this category include salmonellosis, toxoplasmosis, trichinosis and possibly new variant Creutzfeld–Jakob disease (CJD), although, of these, it is only salmonellosis that is currently of worldwide importance.

Salmonella can be isolated regularly from animal-feed ingredients, including both animal and vegetable proteins, such as soya, rape, palm kernel, rice bran and cottonseed (e.g. Williams, 1981; Jones *et al.*, 1982; Davies, 1992) and thus feed is a major potential 'route' by which new infections may be introduced into farm livestock (Shapcott, 1984; Blackman *et al.*, 1992), particularly poultry (Advisory Committee on the Microbial Safety of Food, 1996). It has frequently been possible to trace specific serovars present in feed ingredients through the production process to the finished product (e.g. Mackenzie and Bains, 1976). Specific ingredients may also be implicated in outbreaks of salmonellosis using analytical techniques such as case–control studies (Anderson *et al.*, 1997), although, of course, their presence in feed and feed ingredients may not be associated with any clinical problems (e.g. Harris *et al.*, 1997).

Many *Salmonella* serovars that are not strictly host-adapted, such as *S. derby* in pigs and *S. montevideo* in sheep, show host preferences and may be more easily acquired from feed than the more common types found during routine feed testing (Jones and Richardson, 1996). The susceptibility of individual animal species may vary with age; for example, it is thought that some newly hatched chicks may become infected following the ingestion of a single *Salmonella* cell (Pivnic and Nurmi, 1982), compared with several thousand or million cells for adult birds (Sadler *et al.*, 1968), although the infective dose may be underestimated when *Salmonella* numbers are determined in experimental studies using selective agars.

The isolations of *Salmonella* from feedstuffs and ingredients recorded in the UK during 1995 and 1996 are summarized in Table 17.1. The feeding of mammalian meat and bone-meal (MMBM) to farmed livestock was made illegal in

Table 17.1. The prevalence of *Salmonella* in feedstuff and raw-ingredient tests in the UK (adapted from Chief Veterinary Officer, 1997, and C. Wray, personal communication).

	Year					
	1995		1996		1997	
Product	No. of tests	Proportion (%) positive	No. of tests	Proportion (%) positive	No. of tests	Proportion (%) positive
Processed animal protein at UK protein-processing premises	10,341	1.9	10,023	3.2	9,603	3.1
Linseed, rape-seed, soya-bean and sunflower meals at UK crushing premises	1,749	2.7	4,155	6.0	5,965	5.6
Non-oil-seed-meal vegetable products	12,560	1.5	14,091	1.7	14,216	1.8
Pig and poultry meals	3,826	2.8	5,712	4.4	6,039	2.2
Poultry extrusions	4,502	0.6	8,870	1.9	6,307	1.2
Pig extrusions	1,578	0.9	3,874	1.1	4,159	1.1
Ruminant concentrates	2,927	2.8	4,205	2.5	3,992	2.0
Protein concentrates	1,534	7.4	1,513	4.2	1,285	3.0
Minerals/others	951	0.4	1,037	4.2	863	0.1

the UK during 1996, as part of the measures introduced to control BSE. As a consequence, MMBM is no longer tested under the provisions of the UK's Processed Animal Protein Order 1989. There has been an increase in the use of oil-seed meals since the introduction of the ban on MMBM, and this has been associated with a twofold increase in the prevalence of *Salmonella*-positive oil-seed meal samples in 1996 compared with 1995 (Table 17.1).

The commonest *Salmonella* serovars isolated from feed are rarely the most prevalent in animals, including humans, while the two most important serovars associated with human disease, *S. enteritidis* and *S. typhimurium*, are infrequently isolated from animal feed and processed animal protein (Wilson, 1989; Bisping, 1993; Table 17.2). *Salmonella*, including these two serovars, can survive for months or years in stored animal feed (Davies and Wray, 1996a; Allen, 1997), however, and this suggests that their low prevalence is not related to their ability to survive in feed (Davies and Wray, 1996a). In addition, there has been a gradual reduction in the number of isolations made from animal feed during recent years, with no concurrent reduction in cases of

salmonellosis in humans (Gareis, 1995). This suggests that the relationship between the two situations, if any, is far from simple.

Contamination of Feed Ingredients During Storage and Processing

Vegetable ingredients

Basic cereal ingredients, such as wheat, barley, oats and maize, are considered not to be particularly prone to *Salmonella* contamination, since the porous nutritious material is contained within a protective cuticle and grains are exposed to sunlight before harvesting. Nevertheless, it is still relatively common to find evidence of contamination of grain with the same *Salmonella* serovars found in local wildlife species, for example badgers and foxes. Thus, the organisms can be found in dust taken from within combine harvesters and balers and also in spillage and dust from farm auger, grinding and milling systems. There is also a potential risk of contamination by livestock-related serovars when manure or poultry litter is stacked at the

Table 17.2. The results of testing home-produced processed animal protein under the Processed Animal Protein Order 1989 (adapted from Chief Veterinary Officer, 1997).

Year	No. of samples tested	No. (%) of positive samples		
		All serovars	*S. enteritidis*	*S. typhimurium*
1989	1647	88 (6.3)	4 (0.2)	8 (0.5)
1990	1388	40 (2.9)	0	1 (< 0.1)
1991	1360	46 (3.4)	0	2 (0.1)
1992	1318	22 (1.7)	0	3 (0.2)
1993	1274	26 (2.0)	0	1 (< 0.1)
1994	1066	27 (2.5)	0	0
1995	1041	18 (1.7)	0	0
1996	780	18 (2.3)	0	0

edge of cereal fields shortly before harvest. This material, which may harbour a wide range of *Salmonella*, is attractive to wild birds and other wildlife, which may then acquire a transient *Salmonella* infection. An outbreak of salmonellosis in a livestock unit or human population can result in large-scale contamination of surface and river water (Baver and Hormansdorfer, 1995) and this may be used for diluting agrochemical sprays, such as desiccants, which may be applied within a short time of harvest.

Work in the UK at the Veterinary Laboratories Agency has identified considerable opportunity for contamination of home-grown feed ingredients on farms on which *Salmonella* are present in cattle or pigs. When the farm grain store is within 1 km of the livestock-housing or manure-storage areas cross-contamination, mediated by wild birds, rodents and cats, which have acquired infection from the livestock areas, has been common. Thus open-topped storage bins and flat stores are vulnerable to contamination unless a high level of bird-proofing, rodent control and general disease security is practised. Other causes of contamination include the use of empty livestock buildings for short-term storage of grain before drying and trailers and bucket loaders that have not been adequately cleansed and disinfected. Grain in trailers may also become contaminated by *Salmonella* in material thrown up from the wheels passing through effluent on driveways and roads.

The conditions under which certain feed ingredients are grown and stored in Third World countries are also likely to lead to cross-contami-nation. Thus, paddy-fields used for growing rice may be flooded with water containing *Salmonella* from human and animal effluent, while low standards of sewage disposal facilitate environmental pollution and consequent infection of wild animals (e.g. Köhler, 1993). Storage conditions may also be less satisfactory in countries with low rainfall, since protection from rain is not a priority. It is common in these circumstances for bulk ingredients to be stored in containers that are not totally enclosed at the production site or during short-term storage at docks, thus allowing contamination with wild-bird and rodent droppings.

By-products of milling grains, such as bran, wheat feed or middlings, and rice bran comprise the surface layers of the grain and, as *Salmonella* may be more concentrated in them than in the original material, heat treatment is often recommended to reduce this problem. These and other by-products, such as maize gluten, sugar-beet pulp and molasses, citrus pulp and expelled or extracted meals prepared from rape-seed, palm kernel, soya, cottonseed, sunflower and linseed, may be blended with primary cereal ingredients, pulses, whole oil-seed products and mineral/vitamin supplements to produce least-cost rations for farm livestock.

Contamination of vegetable ingredients commonly occurs during cooling after cooking and oil extraction, and specific *Salmonella* serovars are frequently associated with certain processed ingredients from particular suppliers. These 'endemic' strains may persist in the plant for many years unless effective remedial measures are taken.

Mammalian and avian proteins

Salmonella infections in individual animals are frequently asymptomatic and their prevalence may increase during transport and lairage, while the rate of contamination of meat and by-products is further increased during the slaughter process and subsequent processing and transport to and storage in rendering plants (Urlings *et al.*, 1992).

The rendering of slaughterhouse waste results in products that contain high concentrations of protein and/or fat and these can be incorporated into animal feed (e.g. El Boushy *et al.*, 1990). The cooking times and temperatures used (80–145°C for 0.5–3 h) should be capable of eliminating *Salmonella*, but undercooking may occur, for example after a 'shut-down' period, when the cooking equipment is below the optimum temperature or when the material has not been sufficiently comminuted and the organisms remain protected within large fragments of tissue. The European Union standard for inactivating the agent of BSE is more than sufficient to kill *Salmonella*, namely 133°C for at least 20 min at 3 bar pressure after macerating to a maximum particle size of 50 mm.

The rendered product may also become recontaminated with *Salmonella* after cooling. For example, in some plants, cooker seals that have been damaged by bone chips allow seepage of contaminated uncooked material on to the expeller mechanism, which conveys the finished product to the later stages of processing. The leakage is less likely to be a problem with poultry protein cookers, as avian bones are broken down more easily. Once *Salmonella* establish in the 'post-cooking' area, they may become 'endemic' in auger systems, storage bins, centrifuges, presses, tallow tanks, etc., since they readily multiply in the warm moist product.

Full physical separation between cooked and uncooked material handling areas is desirable. However, this may not be wholly successful in preventing cross-contamination, since the *Salmonella* may be dust-borne, and hence other means of control must also be instituted (Bensink and Boland, 1979). Surface and, probably, internal contamination by insects in rendering plants has been demonstrated, while wild birds surrounding the premises may also acquire *Salmonella*, so their faeces may contaminate vehicles and storage equipment (Davies and Wray, 1994).

Liquid organic acid/acid salt products or combinations of formaldehyde and organic acids may be added to rendered products to reduce contamination and recontamination after cooking. Products providing the highest levels of free aldehydes and acids appear to be the most effective. It is also advisable not to include poultry proteins in poultry rations, because of the risk of recycling *S. enteritidis* infection, while, in some countries, ruminant protein may not be used in livestock rations, as part of the measures introduced to control BSE.

Fish-meal

This comprises either whole or parts of fish and shellfish that are unsuitable for human consumption and by-products from fish-oil manufacture. The production process for fish-meal is similar to that of meat and bone-meal in that the raw material is macerated and cooked under pressure (Miller and de Boer, 1988). The resulting oil and water is 'run off' and the remaining solids are pressed and centrifuged to remove the residual oil and the water. This is then passed through a concentrator to recover suspended solids, which are added to the fish-meal.

Fresh fish are considered free of natural *Salmonella* infection, but shellfish may accumulate the organisms from polluted waters. Contamination may occur during handling and storage of the fish on boats and dock sides, and much of the contamination originates from the faeces of sea birds. Fish-meal imported into the UK has a higher incidence of contamination than the domestic product.

Finished-feed Production

Home mixing

Feed ingredients may be fed to ruminants as 'straights', i.e. without milling and mixing, or with various levels of mixing before incorporation with farm-grown cereals or forages. Pelleting is rarely carried out during on-farm feed preparation, although some farms do pellet rations without heat treatment and large farming companies may have a feed mill where heat-treated rations are produced.

The number of ingredients used for on-farm mixed rations is usually more restricted and less variable than that used for compound feed manufacture. Ingredients are usually also purchased in bulk and remain on the farm for weeks or months, during which time the level and infectivity of any *Salmonella* present will diminish. In addition, the cereal ingredients used in the rations are usually home-produced. This limited range of ingredients and their relatively slow turnover mean that, although monitoring is less stringent than that used by the compound-feed industry, the chance of obtaining a sample of finished feed containing *Salmonella* is lower. In Denmark, the use of feed from compound-feed mills has been identified as an important risk factor for acquisition of *Salmonella* by pig herds, and farmers have been advised to consider home mixing as part of the national control strategy. In the UK, heat treatment of feed has been shown to be a protective factor against acquisition of *S. enteritidis* in poultry flocks, despite the rarity of isolation of that serovar from finished feeds or ingredients (S.J. Evans, CVL, Addlestone, personal communication, 1997). These observations suggest that acquisition of 'invasive' *Salmonella* from feed may occur more frequently than was previously thought.

Compound-feed production

A complete compound ration suitable for meeting the production requirements of livestock is formulated from a variety of basic ingredients to provide a nutritionally balanced ration. This is normally done on a least-cost basis, so that specific ingredients are chosen or substituted according to availability and price.

Several different types of processing methods are used in feed production. Cold processing includes crimping, cracking, rolling or crushing cereals for some ruminant and horse feeds and grinding in a hammer mill or grinder to various grist sizes for pig and poultry meals and all pelleted rations.

Most ruminant rations and 'coarse' pig and poultry meals are not heat-treated. However, hot processing increases the digestibility of the starch in cereals by partial cooking and also provides a variable degree of pasteurization. Pelleting or cubing is the most widely used processing method. The meal is conditioned by heating to 50–85°C by steam for a variable time and then forced by rollers through the rotating die of a pellet press. The friction involved in the process further raises the temperature by several degrees. The pelleting process increases feed-utilization efficacy and reduces waste during feeding, although there are concerns about the effect of very high temperatures on the levels of essential amino acids and vitamins (Andrews, 1991). A crumbed ration is produced by partially crushing pellets after heat treatment to produce a ration attractive for young birds.

Alternatives to pelleting include extrusion and expansion (Daw, 1991). Expanders produce conditioned meals by compressing a partially conditioned product through a variable-size orifice under pressure created by a compression screw. Extruders produce pellets or shapes through fixed dies from conditioned meal forced through by compression screws.

Expanders and anaerobic conditioning systems (Ekperigin *et al.*, 1990) may be used to produce heat-treated meals or, more commonly, their use may precede a conventional pelleting process.

Salmonella Contamination in the Compound-feed Mill

Salmonella contamination of lorries, both inside and out, may be demonstrated and this can be particularly important when the same vehicles are used for both ingredients and finished products. Moisture in delivery vehicles may allow low levels of *Salmonella* in ingredients to multiply, and contamination problems may be more prevalent in wet weather (Köhler, 1993). Ineffective disinfection may also accentuate *Salmonella* problems, and fogging lorries with low concentrations of disinfectant is contraindicated. Wet disinfection of feed-handling equipment should only be carried out with full agricultural-strength disinfectant and when there is enough time to allow complete drying of surfaces before reuse.

Feed ingredients delivered to a mill are deposited in the intake pit and contamination may be associated with faecal material carried by lorries, including their wheels, or contamination by wild animals. There may also be carry-over of material on grid supports and in dust, thus leading to cross-contamination between batches. Enclosed pits with automatic doors offer some protection from vermin and wild birds, but dust control in these systems must be effective, as movement of the door may lead to dust being dislodged into the pit.

Ingredients are then augered or blown, before or after grinding, from the intake point to storage bins. There may be limited carry-over of material in auger systems, particularly in the 'boots' of chutes or where the direction of flow changes, although there is little opportunity for significant multiplication of Salmonella to occur in high-throughput self-cleaning systems. Long-term contamination of ingredient storage bins is more likely. In older mills, open-topped bins may be used and the feed may be contaminated by droppings of wild birds that gain access to the loft space. In modern mills, all bins are closed, although there may still be problems either when low-risk ingredients, such as cereals, are stored after a high-risk ingredient, such as palm-kernel residues, or, more seriously, when finished products are stored in ingredient bins if there is a shortage of space. Multiplication of Salmonella may occur in bulk bins in warm, moist conditions and this is most likely to occur in the autumn, when warm days followed by cold nights result in condensation on the sides of bins (Köhler, 1993). Contamination in bulk bins can be reduced by a regular emptying and cleaning programme and installing appropriate insulation of the building.

Ingredients may be mixed either before or after grinding. Short-term blending bins and mixing and grinding equipment are rarely associated with long-term Salmonella contamination, but some carry-over of contamination between batches may occur. If a meal or mash ration is produced, then the feed is transferred to bulk finished-product bins or is bagged at this stage.

After grinding and mixing, feed to be pelleted is conditioned or expanded prior to pelleting. This process has a variable antibacterial effect on the feed, proportional to the temperature and conditioning time used (Mossel et al., 1967; Shackleford et al., 1987), and in practice some Salmonella may survive the pelleting process (Hacking et al., 1978; Cox et al., 1983; Blank et al., 1996). There is little current information on the effectiveness of heat treatment in modern, large-scale feed mills (Voeten and Van de Leest, 1989). Most studies on the effect of heat treat-

ment have been carried out using artificially contaminated samples, which do not necessarily mimic the commercial situation, since more heat is required to eliminate Salmonella from 'naturally contaminated materials (Williams, 1981). Conditioning and pelleting at 93°C for 90 s at 15% moisture has been shown to reduce Salmonella by 10,000-fold, which should pasteurize all but the most highly contaminated material (Himathongkhan et al., 1996). In practice, such high temperatures and conditioning times are rarely used, because of the cost of heat energy and the deleterious effect on pellet quality (McCapes et al., 1989; Blank et al., 1996). Heating to 85°C for 1 min is considered a good target requirement for Salmonella elimination (Liu et al., 1969), while temperatures of at least 80°C during conditioning, followed by pelleting, are likely to be successful in the majority of cases (Blankenship et al., 1985; Coven et al., 1985), since Salmonella numbers estimated by the most-probable-number method are usually less than 1 g^{-1} of feed (Taylor and McCoy, 1969). The biggest problems are likely to occur during the period immediately after production starts, when peak conditioning temperatures may not be reached. Older open-kettle and ripener systems operate at lower temperatures and are prone to Salmonella contamination and hence should not be used. Closed ripeners are better, but further heat treatment is usually necessary to ensure a Salmonella-free product.

Anaerobic pasteurization systems and expanders allow higher temperatures of up to 170°C to be achieved, but some types may be difficult to stabilize and there may be some degradation of proteins and amino acids at these higher temperatures (Jones et al., 1995).

Experimental work has shown that prior exposure of Salmonella to stresses, such as alkaline or acid conditions (Farber and Pagotto, 1992; Humphrey et al., 1993), prior heat shock (Bunning et al., 1990; Mackey and Derrick, 1990) or high mineral-salt concentrations (Palumbo et al., 1995), increases the heat tolerance of Salmonella. The most dramatic of these effects was found after S. typhimurium was habituated experimentally to extreme dehydration, after which it was able to survive 60 min at 100°C (Kirby and Davies, 1990). It is uncertain what relevance these observations have to Salmonella in a natural environment, but doubts about the consistency of heat treatment have led to calls for feed irradiation, where 10–40 kGy has been shown to control a wide range of feed contaminants (Leeson and Marcotte, 1993).

Heat-treated feed contains high levels of moisture and must be cooled as rapidly as possible before storage to prevent condensation and spoilage (Colburn, 1995). Simple vertical coolers, in which pellets pass slowly over perforated sheets that allow the passage of air, may be used in small mills. Larger mills which manufacture heat-treated mash rations and producers of vegetable protein may use much larger vertical coolers. Horizontal coolers, in which pellets in a moving bed are cooled by air from below, are generally being replaced by counter-flow cabinet coolers or rotary coolers, which occupy less space and cool the pellets more rapidly. Vertical coolers are more prone to long-term *Salmonella* contamination, because most are inaccessible for cleaning. Counter-flow coolers also often show a propensity for persistent contamination, despite reduced condensation and aggregate formation compared with other types. Horizontal coolers, despite the tendency for aggregates to accumulate rapidly, appear to become persistently contaminated with *Salmonella* relatively infrequently. This is likely to relate to the slow cooling of pellets, compared with more compact coolers, which maintains the conditioning heat within the mass of pellets for longer (Davies, 1992; Colburn, 1995).

Specific *Salmonella* serovars may persist inside individual coolers for years, producing intermittently contaminated batches of feed when fatty aggregates, which form around the intake pipe of the cooler, dislodge. There is considerable debate about the original source of cooler contamination. Contaminated dust is one suggested source, although in many cases there are just one or two contaminated coolers in a line of several adjacent machines. The contaminated coolers and post-cooling sieving equipment produce a great deal of contaminated dust locally, which should enter all the coolers if dust is the primary origin of the infection. Dust from poultry buildings has a similar distribution and, again, dust-borne cross-infection between adjacent buildings is rare (Schmidt and Hoy, 1996). Occasional failure of heat treatment and consequent passage of *Salmonella* into the cooler with incompletely treated feed are the more likely origin in most cases. This may be aggravated by structural work in the mill, which may alter condensation patterns within equipment.

Persistent contamination of equipment after the coolers is uncommon but carry-over of small amounts of feed and dust may occur in sieves, augers, blowers and hoppers. It is less common for persistent contamination to occur in bulk finished-product bins, because they are normally smaller than ingredient bins and are regularly emptied. However, open bins in old mills or bins in which the top hatches are not kept shut may become contaminated by faeces from wild birds. Wild birds are also a particular problem in the 'outloading' gantry of some feed mills, since they may nest or perch in the roof space and produce faeces, which may be shaken from beams into feed delivery lorries during the course of loading. Wild-bird faeces may also contaminate feed stored in open tote bins and the surfaces of feed bags. A similar problem may occur when vegetable ingredients and finished products are held in flat stores. Organic acids (Humphrey and Lanning, 1988) and formaldehyde-based treatments (Smyser and Snoeyenbos, 1979; Brown, 1996) have been used in poultry rations to reduce the effect of post-processing contamination in meals and occasionally in pelleted rations.

Decontamination of persistently infected feed mills and ingredient processors

Salmonella that become endemic residents in a feed plant may be difficult to eliminate. Birds can be excluded and deterred with sufficient investment in proofing and tidiness. Persistent contamination in bulk storage bins can be eliminated by thorough physical cleaning, followed by effective disinfection. Sufficient time must be allowed for the disinfectant to dry completely before refilling the bin and many mills do not have sufficient spare capacity to leave bins empty for more than a short time. In this situation, drying and the activity of disinfectants can be accelerated by using space heaters. If condensation is a regular problem, then appropriate building modifications, such as improved insulation and ventilation, may be required (Gabis, 1991). Augers, pipes and small-volume equipment can be decontaminated by multiple passage of a carrier material, ideally coarse bone-meal, as this has abrasive properties, which has been treated previously with a high concentration of a formaldehyde/organic acid product. If bone-meal cannot be used, for example because of BSE regulations, then wheat feed is a less effective alternative.

Persistent contamination of coolers may be difficult to eliminate because of their size and the inaccessibility of much of the equipment. Dry-cleaning inside and outside the coolers by brushing and the use of vacuum cleaners should be as thorough as possible before disinfectants are applied through a lance at medium pressure in order to soak all surfaces thoroughly. Formaldehyde, glutaraldehyde and quaternary ammonium compound combinations applied at high concentrations are the most suitable for elimination of *Salmonella* on surfaces that are difficult to clean. The spray disinfectant should be left to dry before the cooler is redisinfected using neat (40%) formaldehyde solution (formalin), applied with a fogging machine. After saturation of all surfaces, the cooler should be left closed for at least 24 h to maximize the penetrating effect of the formaldehyde vapour. An alternative to fogging is traditional fumigation, using formaldehyde vapour generated from formalin to which potassium permanganate has been added. Before this treatment is used, it is essential that arrangements are made to extract vapour safely via the venting system before the cooler is used again. All operators should use respirators designed for use with formaldehyde. It is also essential that the whole area surrounding the coolers is cleaned and disinfected to prevent possible recolonization of the disinfected surfaces via contaminated dust produced by the cooler before its decontamination. For this approach to be successful with large-volume coolers, several days total shut-down may be required to allow cleaning and disinfection to be carried out sufficiently thoroughly.

Auditing Feed Suppliers

A great deal of money is spent on testing feed ingredients and finished-feed samples with little hope of detecting contamination. A lorry load of feed has over 10^{13} potential particulate contamination sites for *Salmonella* but only one-millionth of them will be found in a standard-sized feed sample and even less as a proportion of a compos-

ite sample (Jones and Riche, 1994). Thus, the chance of finding *Salmonella* in feed increases with the size and number of samples (McChesney, 1995), although intensive sampling is not economically feasible. In addition, current *Salmonella* culture techniques are relatively time-consuming and the results only become available after the ingredients have been used or the finished feed has been consumed. Techniques are available which give a presumptive result within 24 h, although many of these are either laborious to use or lack sensitivity and hence are not widely used (D'Aoust *et al.*, 1995; Davies and Wray, 1996b).

Risk assessment and hazard-analysis critical control points (HACCP) systems are approaches that can be used to reduce *Salmonella* levels in the feed industry (Haselgrove, 1996). Development of HACCP plans must involve individual assessments of each mill, preferably backed up by identification of problem areas by the use of microbiological audits. Within a large feed company, there will be several mills, which can be compared by best-practice analysis (BPA) in order to identify those practices that are associated with low contamination rates and that can be adopted in other mills. Similarly, most major feed companies have drawn up detailed codes of practice for *Salmonella* control, although problems may still occur, despite their application. Error analysis can be used to identify the mistakes that can be made at each step of production and hence to alert staff of the dangers of failing to implement local operating procedures.

Feed-ingredient processors and feed mills receive ingredients from a number of sources and it will not normally be possible to audit all of these, especially as the original sources of ingredients may not be readily identifiable after materials have been mixed. Major farm suppliers and ingredient storage facilities, however, can be audited. Ideally, the premises should: (i) be more than 2 km from intensive livestock units, landfill sites, sewage-treatment plants, abattoirs, etc.; (ii) preferably have closed bulk-bin storage or no access of wild birds, cats or rodents to flat storage areas; and (iii) have dedicated handling equipment in the ingredient stores. Bacteriological monitoring of major suppliers by testing samples of spillage and dust from auger systems may also be undertaken. The presence of *Salmonella* suggests that the ingredients should only be used with caution. If microbiological auditing of the supplier's production premises is not practicable, then more intensive sampling of material delivered from potential suppliers may be of value initially, as those with a problem can be eliminated at an early stage.

The managers of feed mills can verify their own *Salmonella* situation, in part, by analysing the results obtained from testing ingredients and finished products, as specified by existing codes of practice. This can be supplemented by the more sensitive technique of testing material that has spilled or dust that has escaped from junctions in equipment in intake pits, ingredient bins, augers, grinders and mixers, coolers, finished-product bin augers and bagging plants/outloading gantries. This approach is helpful for identifying problem areas, since spillage and dust settlement tend to be relatively localized and hence deductions can be made about the likely source of contamination (Schmidt and Hoy, 1996). The presence of several serovars in the ingredient side of the mill suggests problems with ingredient suppliers. Contamination detected in ingredient-bin discharge augers, but not earlier in the plant, suggests persistent contamination in the bins. A high prevalence of contamination in cooler areas (including air filters) and beyond suggests persistent cooler contamination problems. This system of sampling can be used on a one-off basis, in order to establish the contamination status of the mill and to identify potential problem areas, by taking large numbers of samples throughout the mill. A less intensive approach, based on regular monitoring, may be of value, since it allows for fluctuations in the level of contamination of ingredients. When specific problems are identified, scrape samples can be taken from inside equipment to confirm the site of contamination. In the case of coolers, samples of feed taken, before and after cooling during the first 10 min after a shut-down period will often identify residual contamination. An HACCP-based monitoring system introduced in Sweden has resulted in improvements in the sensitivity of *Salmonella* detection and has identified many persistently infected mills, which were not identified by monitoring ingredients and finished product (Eld *et al.*, 1991; Malmqvist *et al.*, 1995). When the prevalence of *Salmonella* contamination is low, initial evidence of the efficiency of heat treatment can be obtained by assessing the reduction of other thermophilic enterobacteria after treatment (Veldman *et al.*, 1995).

Selection of Feed Sources by Customers

It is important to give *Salmonella*-free feed to grandparent and parent poultry flocks and primary pig breeders. These breeders may wish to carry out a risk assessment on potential new feed mill suppliers, based on the same criteria discussed for ingredient producers. In addition to this, prevention of the access of wild birds to intakes and gantries is important, as are the overall tidiness and lack of moisture in the mill. There is less risk of *Salmonella* contamination in specialist poultry mills or pig and poultry mills, since these do not produce ruminant rations, which frequently include a wider range of potentially contaminated vegetable proteins and are usually pelleted at lower temperatures. Mills that produce a high proportion of pelleted rations are less likely to be subject to cross-contamination from meals in finished-product storage and handling areas. Poultry companies may also prefer to obtain feed from mills that do not use poultry proteins in feeds.

The results of testing of ingredients and finished products should be examined and, in some cases, it may be worthwhile commissioning an independent detailed microbiological audit of the mill, in which every aspect of production, including wildlife on the site, is checked for *Salmonella* contamination.

Poultry-breeder customers should also ensure that separate vehicles and storage bins are always used for ingredients and finished products and that the level of heat treatment is sufficient to kill *Salmonella*. Organic-acid products added to the feed should also be chosen to optimize the level of active free acid in the finished ration, and details of the mill's policy for auditing ingredient suppliers should be scrutinized.

The Significance of Feed as a Source of Major Zoonotic *Salmonella* Serovars

Although the human population and livestock industry is responsible for the production of large quantities of faecal material contaminated with *S. enteritidis* and *S. typhimurium*, which is subsequently released into the environment, these serovars are rarely isolated during the monitoring of feed ingredients and finished feeds (Barrow,

1993; Bisping, 1993; Anon., 1996; Chief Veterinary Officer, 1997). When most-probable-number estimates of *Salmonella* levels are carried out on finished feeds, they are usually very low (Taylor and McCoy, 1969), but it is not certain whether it is an individual organism that is being counted or microcolonies 'attached' to small feed particles. It has been suggested that the low numbers of *Salmonella* found in finished feed are likely to be below the infective dose for all but newly hatched chicks, but it is common to find the same *Salmonella* serovars in poultry flocks as in the feed and the feed mill environment, so the occasional appearance of *S. enteritidis* and *S. typhimurium*, which are more virulent in animals than most other serovars, is likely to be significant. In the case of *S. typhimurium*, the contamination may be present in the form of discrete rodent and bird droppings, in which the *Salmonella* may have been 'primed' for enhanced invasion by passage through a wildlife host (Lax *et al.*, 1995). Low numbers of *Salmonella* may also cause infection in animals, including humans, when protected from the intestinal defence mechanisms by high concentrations of fat in the feed. *Salmonella* present in low numbers in feed may multiply in the warm, moist conditions that may develop in, for example, feed bins that are subject to condensation and ad lib feed hoppers, where there may be a build-up of feed aggregates moistened by saliva. The organism may also be passaged through rodents, with subsequent contamination of the environment from their faeces.

The Control of *Salmonella* in Feed

There are a number of ways by which the risks to farm livestock posed by the presence of *Salmonella* in animal feed can be either reduced in number or eliminated. However, it must be stressed that, since *Salmonella* can be transmitted to farm livestock by many routes, their control in feed must not be considered in isolation but, rather, as part of a broader strategy for the overall control of *Salmonella* infections (Oostrom, 1991).

Product management

It is essential that feed mills operate to appropriate standards of good management practice (GMP). This policy, which must involve the active cooperation of the whole workforce and may include the implementation of HACCP principles, will automatically take account of all relevant legislation and published codes of practice. The feed mill must have a workable layout so that the pre- and post-processing areas are clearly defined, and the employees and equipment they use must be assigned to either one place or the other.

The raw materials, additives and finished products must be kept separate and stored under hygienic conditions. The material should not be allowed to get wet, and all vertebrate and invertebrate pests should be controlled as far as possible.

Good ventilation and dust control are important and different sectors of the feed mill should have separate dust-filtration and extraction systems, so that recirculation of 'polluted' air, particularly to the 'clean' areas, is prevented. The fixed and movable equipment should be cleaned and, when necessary, decontaminated at regular intervals, although care must be taken to ensure that all surfaces coming into contact with feed are thoroughly dried after any wet cleaning, since the presence of moisture may allow residual *Salmonella* to multiply. In practice, most mills will limit regular decontamination to circulation of an organic acid or acid/formaldehyde product. Ingredient bins should be emptied and cleaned on a rotational basis, whenever possible. The frequency of cleaning will vary, depending on the degree of risk to the product.

Heat treatment

This subject has been referred to earlier, in the section entitled '*Salmonella* Contamination in the Compound-feed Mill'. To summarize, elimination of *Salmonella* and other pathogens from feed by heat treatment depends on several factors, including: (i) moisture content of feed; (ii) temperature; (iii) heating time; and (iv) initial levels of contamination. The heat used routinely during manufacture is often no more than sufficient to reduce levels of bacteria such as the coliforms and *Salmonella* by a few 100- or 1000-fold (e.g. Stott *et al.*, 1975; Cox *et al.*, 1983; Jones *et al.*, 1991; Davies, 1992; Veldman *et al.*, 1995) and hence sufficient *Salmonella* may persist after pelleting of highly contaminated rations for *Salmonella* to become established, for example, in chicks (Hinton and Linton, 1988).

Irradiation

Ingredients and finished feed have been treated with doses of gamma irradiation of up to 10 kGy without affecting nutritional quality (Williams, 1981; Leeson and Marcotte, 1993). Irradiation has been used to eliminate *Salmonella* from feed (Mossel *et al.*, 1967; Williams 1981; Leeson and Marcotte, 1993), although 10 kGy may prove ineffective *in vivo*, since this dose may only delay colonization of broiler chickens with *Salmonella* and not prevent it (Hinton *et al.*, 1987).

It has been suggested that a combination of improved sanitation at the mill, pelleting feed at the highest possible temperature and low-dose irradiation of the bagged product would ensure the destruction of all *Salmonella* and eliminate the problem of recontamination. However, it is unlikely that this method of decontamination will be adopted by the industry for commercial farm rations, because of: (i) the high costs of installing irradiation facilities in a feed mill; (ii) the logistics of transporting large amounts of feed to regional treatment centres; and (iii) the difficulties of dealing with a bagged feed system on large farms. Its use may be justified, however, for feed prepared for certain categories of specific-pathogen-free animals or those kept for experimental purposes, particularly laboratory rodents.

Chemical additives

Organic acids

Neither heat treatment nor irradiation will protect feed against recontamination during storage and subsequent distribution. An alternative technique, which does provide some protection against recontamination, is to add antimicrobial compounds, such as organic acids, to the feed (Hinton and Linton, 1988). These acids are effective in reducing the prevalence of *Salmonella* infections in poultry (Hinton and Linton, 1988; Humphrey and Lanning, 1988), although they only cause a small reduction in the number of *Salmonella* in the dry feed (Duncan and Adams, 1972; Vanderwal, 1979; Banton *et al.*, 1984;

Hinton and Linton, 1988; Humphrey and Lanning, 1988). This suggests that the acids are either inactive, or only slightly active, against bacteria in these circumstances. However, the feed becomes hydrated after it has been consumed by the bird and it is probably at this stage that the acids exert their antibacterial effect, since they are only effective when the water activity is sufficient to permit bacterial multiplication.

Acid-treated feed should be given throughout the rearing period, since the acids have no beneficial effect once the birds have become colonized (Hinton and Linton, 1988). Similarly, the acids are ineffective if the feed contains large numbers of organisms, although efficiency can be increased by increasing the concentration of the acid (Vanderwal, 1979; van Staden *et al.*, 1980; Hinton and Linton, 1988). This approach may be appropriate if a specific ingredient requires decontaminating but it may not be suitable for finished feed, on the grounds of cost and the effects of reduced palatability (Cave, 1984).

Formaldehyde

This compound has a high level of activity against *Salmonella* and is less likely to be inactivated by the presence of organic matter than most disinfectant groups. The action of formaldehyde is slow, however, and at least 3 h exposure is required. This can be achieved by adding liquid formaldehyde to the feed in a closed bin so that the vapour permeates the bulk of the ration. Previous work has shown superior decontamination of feed by formaldehyde, compared with acid treatment (Duncan and Adams, 1972; Smyser and Snoeyenbos, 1979; R.H. Davies, unpublished data), so the licensing of formaldehyde-based products to treat finished feed (Brown, 1996), as well as ingredients, should enhance control of *Salmonella* in feed as long as treatment is not used to mask ongoing mill-hygiene problems.

Sugars

The control of *Salmonella* colonization by the addition of lactose or mannose to the diet has generally proved inconclusive and may prove impracticable for a number of reasons, including cost. It has been suggested that lactose, which is not metabolized by the chick, may promote the growth of lactose-fermenting bacteria, which either compete directly with the *Salmonella* or produce volatile fatty acids that are toxic to them (Corrier *et al.*, 1990; Hollister *et al.*, 1994). *In vitro* studies involving mannose suggest that this sugar may reduce the ability of *Salmonella* to adhere to the cells of the intestinal tract (Lindquist *et al.*, 1987; Oyofo *et al.*, 1989).

Fermentation

Heat treatment of contaminated feed ingredients is expensive in terms of the energy required, and bacterial fermentation may be an alternative, particularly to render waste food and certain farm wastes fit for use as feedstuffs. The process can be made more effective by the addition of fermentable carbohydrates, such as molasses, and/or organic acids, such a propionic acid. The contradictory nature of reports on the effectiveness of fermentation, however, means that the process may not prove suitable for commercial use, since its success depends on an antimicrobial process that cannot be guaranteed (Evans and Smith, 1986).

Growth Promoters and *Salmonella* Shedding

Over the years, there has been much debate about whether the inclusion of growth-promoting antibiotics – agents that are generally active against Gram-positive bacterial genera – in feed favour the colonization of the intestinal tract with *Salmonella*. There has been relatively little published on this topic in recent years, particularly with respect to *Salmonella* carriage in birds reared on commercial farms, although the problems caused by a lack of veterinary supervision of the use of growth-promoting antibiotics has been highlighted recently (Richter *et al.*, 1996). Matthes (1985) reviewed the results of experimental studies and concluded that shedding was prolonged in some but not others. The majority of investigations referred to involved the study of broiler birds that were inoculated orally or were reared in contact with orally inoculated birds. There is relatively little information, however, concerning challenge via feed, an important potential source of *Salmonella* organisms for birds reared under commercial conditions. Barrow *et al.* (1984) reported that *Salmonella* shedding was

increased in birds given feed containing avoparcin and contaminated by the inclusion of unsterilized meat and bone-meal, while the results obtained by Hinton *et al.* (1986), who used naturally contaminated feed, were inconclusive. Further studies, involving young chicks given feed artificially contaminated with *Salmonella* and containing either avoparcin, virginiamycin or zinc bacitracin, have since been reported (Hinton *et al.*, 1992). The inclusion of the growth promoter in the feed was associated with an increase in the prevalence of *Salmonella* colonization of between 15 and 20%. However, the average number of *Salmonella* in caecal contents of the positive birds was increased by less that tenfold and hence the total numbers of *Salmonella* within a group of birds remained within the same order of magnitude. This means that, in microbiological terms, the growth promoter did not facilitate a 'super'-infection in the intestinal tract. Rather than ban their use on these grounds, it would be more appropriate to reduce the prevalence of *Salmonella* infections in poultry by other means, so that any slight detrimental effect that the growth-promoting antibiotic may have becomes irrelevant.

Antibiotic growth promoters may also select for resistance genes, including particularly those carried on plasmids. The risk that this may pose to the health of human and farm livestock has yet to be quantified, and it may never be possible to do so with any accuracy, since there is usually insufficient high-quality epidemiological information available. Nevertheless, the use of growth-enhancing antibiotics is now either totally or partially banned within the European Union and in certain other countries, because of the perception that the treatment of certain categories of bacterial diseases in human patients may be compromised by their continued use.

Conclusions

The presence of *Salmonella* in animal feedstuffs is a well-recognized phenomenon and there is documentary evidence to link contamination of feed given to farm livestock with the development of salmonellosis in humans consuming foods of animal origin. However, given the large quantities of animal feed that are consumed worldwide every year, the risk of either farm animals or humans developing clinical infections caused by strains originating from this source is small. On occasion, however, the consequences can be serious, since the *Salmonella* may become widely distributed in the farm-animal and human population, for example *S. agona,* which was originally isolated from fish-meal. The simultaneous appearance of *S. typhimurium* definitive type (DT) 104 in all livestock species also suggests the possibility of an as yet unidentified feed source (Davies and Wray, 1997).

The assessment or quantification of the risk from animal feed will not be easy and will inevitably involve some special considerations. For example, different categories of livestock will be on different feeding regimens and the 'dose' of *Salmonella* required to initiate infection will be dependent on a number of factors, including the species and age of the animals, the serovar, the concentration of the contaminated ingredient in the ration and the total amount of feed consumed. In addition, given appropriate conditions, particularly with respect to moisture (Allen, 1997), *Salmonella* can multiply readily at any time during the preparation, storage and distribution of the feed.

In theory, the application of heat is the simplest way of decontaminating animal feed. However, the process is expensive and its effectiveness is influenced by factors such as the heating time, the temperature and the moisture content of the feed. It also has no residual bene-fit, since the treated feed may become recontaminated after the process is complete (Wierup *et al.*, 1995). An alternative approach could involve several treatment processes or 'hurdles' – for example, the addition of antimicrobial agents, such as organic acids, and the application of heat and irradiation. A combination of organic-acid treatment and heat treatment is the only currently permissible method in the UK, and it is used by primary poultry breeders to ensure, as far as possible, *Salmonella*-free feed. Now that formaldehyde treatment has been approved for use in finished feed, it is likely that it will be more widely used.

It is essential that the strategy adopted for the control of *Salmonella* in feed forms part of a comprehensive coordinated control programme that covers the whole chain of production of foods of animal origin, from 'farm to fork'. Clearly, if this approach is to have any chance of success, it will require the active cooperation of many disparate organizations, including the primary producers, food processors, distributors, retailers and law-enforcement agencies. It is equally important that each of these partners' appreciates the other's problems, that the costs of the control measures are borne equitably and that one sector of the food industry is not expected to bear an unreasonable proportion of consequences of a failure, should the agreed control measures that are introduced prove unsuccessful.

References

Advisory Committee on the Microbial Safety of Food (1996) *Report on Poultry Meat.* London: HMSO.

Allen, V.M. (1997) *Salmonella* infections in broiler chickens: epidemiology and control during incubation and brooding. PhD thesis, University of Bristol, Bristol, UK.

Anderson, R.J., Walker, R.L., Hird, D.W. and Blanchard, P.C. (1997) Case–control study of an outbreak of clinical disease attributable to *Salmonella menhaden* infection in eight dairy herds. *Journal of the American Veterinary Medical Association* 210, 528.

Andrews, J. (1991) Pelleting: a review of why, how, value and standards. *Poultry Digest* 50, 64–71.

Anon. (1996) Salmonella *in Animal Feedingstuffs and Ingredients 1996 (Jan–Jun).* Animal Health (Disease Control) Division Report, MAFF, Tolworth, UK.

Banton, C.L., Parker, D. and Dunn, M. (1984) Chemical treatment of feed ingredients. *Journal of the Science of Food and Agriculture* 35, 637.

Barrow, P.A. (1993) *Salmonella* control – past, present and future. *Avian Pathology* 22, 651–669.

Barrow, P.A., Smith, H.W. and Tucker, J.F. (1984) The effect of feeding diets containing avoparcin on the excretion of *Salmonellas* experimentally infected with natural sources of *Salmonella* organisms. *Journal of Hygiene* 93, 97–99.

Baver, J. and Hormansdorfer, S. (1995) Salmonellosis in farm animals. *Fleischwirtschaft* 75, 952–960.

Bensink, J.C. and Boland, P.H. (1979) Possible pathways of contamination of meat and bone-meal with *Salmonella*. *Australian Veterinary Journal* 55, 521–524.

Bisping, W. (1993) *Salmonella* in feedstuffs. *Deutsche Tierärtzliche Wöchenschrift* 100, 262–263.

Blackman, J., Bowman, T., Chambers, J., Kisilenks, J., Parr, J., St Lavent, A.M. and Thompson, J. (1992) Controlling *Salmonella* in Livestock and Poultry Feeds. Report of Agriculture Canada and Canadian Feed Associates.

Blank, G., Savoie, S. and Campbell, L.D. (1996) Microbiological decontamination of poultry feed – evaluation of steam conditioning. *Journal of the Science of Food and Agriculture* 72, 299–305.

Blankenship, L.C., Shackelford, D.A., Cox, N.A., Burdick, D. and Bailey, J.S. (1985) Survival of *Salmonella* as a function of poultry feed processing conditions. In: Snoeyenbos, G.H. (ed.) *Proceedings of the International Symposium on Salmonella: New Orleans, 19–20 July, 1984.* American Association of Avian Pathology, Kennett Square, Pennsylvania, pp. 211–220.

Brown, R.H. (1996) FDA approves the use of formaldehyde in poultry feed. *Feedstuffs* 18, 15.

Bunning, V.K., Crawford, R.G., Tierney, J.T. and Peeler, J.T. (1990) Thermotolerance of *Listeria monocytogenes* and *Salmonella typhimurium* after sublethal heat shock. *Applied and Environmental Microbiology* 56, 3216–3217.

Cave, A.G. (1984) Effects of dietary propionic and lactic acids on feed intake by chicks. *Poultry Science* 63, 131–134.

Chief Veterinary Officer (1997) *Animal Health 1996.* HMSO, London.

Colburn, R. (1995) Building a new feedmill. *Feed Compounder* 15, 24–30.

Corrier, D.E., Hinton, A., Ziprin, R.L. and Deloach, J.R. (1990) Effects of dietary lactose on *Salmonella* colonization of broiler chicks. *Avian Diseases* 34, 668–676.

Coven M.S., Gary, J.T. and Binder, S.F. (1985) Reduction of standard plate counts, total coliform counts and *Salmonella* by pelleting animal feed. In: Snoeyenbos, G.H. (ed.) *Proceedings of the International Symposium on Salmonella: New Orleans, 19–20 July, 1984.* American Association of Avian Pathology, Kennett Square, Pennsylvania, pp. 221–231.

Cox, N.A., Bailey, J.S. and Thomson, J.E. (1983) *Salmonella* and other Enterobacteriaceae in commercial poultry feed. *Poultry Science* 62, 2169–2175.

D'Aoust, J.Y., Sewell, A.M., Greco, P., Mozola, M.A. and Colvin, R.E. (1995) Performance assessment of the Gene-Trak® colorimetric probe assay for the detection of food-borne *Salmonella* spp. *Journal of Food Protection* 58, 1069–1076.

Davies, R.H. (1992) *Salmonella* – the feedstuffs connection. In: *Proceedings of the Society of Veterinary Epidemiology and Preventive Medicine, Edinburgh, 1–3 April, 1992.* pp. 47–59.

Davies, R.H. and Wray, C. (1994) *Salmonella* pollution in poultry units and associated enterprises. In: ap Dewi, I., Axford, R.F.E., Fayez, I., Marai, M. and Omed, H. (eds) *Pollution and Livestock Systems.* CAB International, Wallingford, UK, pp. 137–169.

Davies, R.H. and Wray, C. (1996a) Persistence of *Salmonella enteritidis* in poultry units and poultry feed. *British Poultry Science* 37, 589–596.

Davies, R.H. and Wray, C. (1996b) Development and evaluation of a simple, one-step *Salmonella* isolation test. *Letters in Applied Microbiology* 22, 267–270.

Davies, R.H. and Wray, C. (1997) Distribution of *Salmonella* contamination in 10 animal feed mills. *Veterinary Microbiology* 51, 159–169.

Daw, D. (1991) Extrusion and expanders. *Feed Compounder* 11, 42–43.

Duncan, M.S. and Adams, A.W. (1972) Effects of chemical additive and of formaldehyde-gas fumigation of *Salmonella* in poultry feeds. *Poultry Science* 51, 797–802.

Ekperigin, H.E., McCapes, R.H., Redus, R., Ritchie, W.L., Cameron, W.J., Magaraja, K.V. and Noll, I.S. (1990) Microcidal effects of a new pelleting process. *Poultry Science* 69, 1595–1598.

El Boushy, A.R., van der Poel, A.F.O. and Walraven, O.E.D. (1990) Feather meal as a source of proteins. *World Poultry* 6, 70–73.

Eld, K., Gunnersson, A., Holmberg, T., Hurrell, B. and Wierup, M. (1991) *Salmonella* isolated from animals and feedstuffs in Sweden during 1983–1987. *Acta Veterinaria Scandinavica* 32, 261–277.

Evans, M.R. and Smith, M.P.W. (1986) Treatment of farm animal wastes. *Journal of Applied Bacteriology* 61 Symposium Suppl. 27S–41S.

Farber, J.M. and Pagotto, F. (1992) The effect of acid shock on the heat resistance of *Listeria monocytogenes*. *Letters in Applied Microbiology* 15, 197–201.

Gabis, D.A. (1991) Environmental factors affecting enteropathogens in feed and feedmills. In: Blankenship, L.C. (ed.) *Colonisation Control of Human Bacterial Enteropathogens in Poultry.* Academic Press, London, pp. 23–27.

Gareis, M. (1995) Salmonellae – a survey. *Fleischwirtschaft* 75, 954–957.

Hacking, W.C., Mitchell, W.R. and Carlson, H.C. (1978) *Salmonella* investigation in an Ontario feedmill. *Canadian Journal of Comparative Medicine* 42, 400–406.

Harris, I.T., Fedorka-Cray, P.J., Gray, J.T., Thomas, L.A. and Ferris, K. (1997) Prevalence of *Salmonella* organisms in swine feed. *Journal of the American Veterinary Medical Association* 210, 382–385.

Haselgrove, P. (1996) The increasing importance of HACCP techniques in mill management. *Feed Compounder* 16, 11–13.

Himathongkhan, S., des Gracas Pereira, M. and Riemanm, H. (1996) Heat destruction of *Salmonella* in poultry feed: effect of time, temperature and moisture. *Avian Diseases* 40, 72–77.

Hinton, M. (1993) Spoilage and pathogenic microorganisms in animal feed. *International Biodeterioration and Biodegredation* 32, 67–74.

Hinton, M. and Bale, M.J. (1990) Animal pathogens in feed. In: Wiseman, J. and Cole, D.J.A. (eds) *Feedstuff Evaluation.* Butterworths, London, pp. 429–444.

Hinton, M. and Linton, A.H. (1988) Control of *Salmonella* infections in broiler chickens by the acid treatment of their feed. *Veterinary Record* 123, 416–421.

Hinton, M. and Mead, G.C. (1990) The control of feed-borne bacterial and viral pathogens in farm animals. In: Haresign, W. and Cole, D.J.A. (eds) *Recent Advances in Animal Nutrition 1990.* Butterworths, London, pp. 31–46.

Hinton, M, Al-Chalaby, Z.A.M. and Linton, A.H. (1986) The influence of dietary protein and antimicrobial feed additives on *Salmonella* carriage by broiler chickens. *Veterinary Record* 119, 494–500.

Hinton, M., Al-Chalaby, Z.A.M. and Linton, A.H. (1987) Field and experimental investigations into the epidemiology of *Salmonella* infections in broiler chickens. In: Smulders, F.J.M. (ed.) *Elimination of Pathogenic Organisms from Meat and Poultry.* Elsevier, Amsterdam, pp. 27–36.

Hinton, M.H., Allen, V.M., and Wray, C. (1992) The influence of growth promoting antibiotics on the colonization of the caecum of young chicks following consumption of feed artificially contaminated with *Salmonellas.* In: Hinton, M.H. and Mulder, R.W.A.W. (eds) *Prevention and Control of Potentially Pathogenic Microorganisms in Poultry and Poultry Meat Processing. 7. The Role of Antibiotics in the Control of Pathogens.* Agricultural Research Department (DLO-NL), Beekbergen, pp. 69–75.

Hollister, A.G., Corrier, D.E., Nisbet, D.J., Bieir, R.C. and Deloach, J.R. (1994) Effects of cecal cultures lyophilised in skim milk or reagent 20 on *Salmonella* colonization of broiler chicks. *Poultry Science* 73, 1409–1416.

Humphrey, T.J. and Lanning, D.G. (1988) The vertical transmission of *Salmonella* and formic acid treatment of chicken feed. *Epidemiology and Infection* 100, 43–49.

Humphrey, T.J., Wallis, M., Hoad, M., Richardson, N.P. and Rowbury, R.J. (1993) Factors influencing alkali induced heat resistance in *Salmonella enteritidis* phage type 4. *Letters in Applied Microbiology* 16, 147–149.

Jones, F.T. and Richardson, K.E. (1996) Fallacies exist in current understanding of *Salmonella. Feedstuffs* 68, 22–25.

Jones, F.T. and Riche, S.C. (1994) Researchers propose HACCP plan for feedmills. *Feedstuffs* 66, 35–42.

Jones, F.T., Axtell, R.C., Rives, D.V., Scheideler, S.E., Tarver, F.R., Walker, R.L. and Wineland, M.J. (1991) A survey of *Salmonella* contamination in modern broiler production. *Journal of Food Protection* 54, 502–507.

Jones, F.T., Anderson. R.E. and Ferkel, P.R. (1995) Effect of extrusion on feed characteristics and broiler chicken performance. *Journal of Applied Poultry Science* 4, 300–319.

Jones, P.W., Collins, P., Brown, G.T.H. and Aitken, M. (1982) Transmission of *Salmonella mbandaka* to cattle from contaminated feed. *Journal of Hygiene* 88, 255–263.

Kirby, R.M. and Davies, R. (1990) Survival of dehydrated cells of *Salmonella typhimurium* LT2 at high temperatures. *Journal of Applied Bacteriology* 68, 241–246.

Köhler, B. (1993) Demonstration of dissemination and enrichment of *Salmonella* in the environment. *Deutsche Tierärtzliche Wöchenschrift* 100, 264–274.

Lax, A.J., Barrow, P.A., Jones, P.W. and Wallis, T.S. (1995) Current perspectives in salmonellosis. *British Veterinary Journal* 151, 351–377.

Leeson, S. and Marcotte, M. (1993) Irradiation of poultry feed. 1. Microbial status and bird response. *World's Poultry Science Journal* 49, 19–33.

Lindquist, B.L., Lebenthal, E., Lee, P.C., Stinson, M.W. and Merrick, J.M. (1987) Adherence of *Salmonella typhimurium* to small intestine enterocytes in the rat. *Infection and Immunity* 55, 3044–3050.

Liu, T.S., Snoeyenbos, G.H. and Carlson, V.L. (1969) Thermal resistance of *Salmonella senftenberg* 77W in dry animal feeds. *Avian Diseases* 13, 611–619.

McCapes, R.H., Ekperigin, U.E., Cameron, W.J., Richie, W.L., Slaughter, J., Strangelad, V. and Nagaraja, K.V. (1989) Effect of a new pelleting process on the level of contamination of poultry mash by *Escherichia coli* and *Salmonella. Avian Diseases* 33, 103–111.

McChesney, D.G. (1995) FDA survey results: *Salmonella* contamination of finished feed and the primary meal ingredient. In: *Proceedings of the 99th Annual Meeting of the USAHA, 2 November, Reno, Nevada,* pp. 174–175.

Mackenzie, M.A. and Bains, B.S. (1976) Dissemination of *Salmonella* serovars from raw feed ingredients to chicken carcasses. *Poultry Science* 55, 957–960.

Mackey, B.M. and Derrick, C. (1990) Heat shock protein synthesis and thermo-tolerance in *Salmonella typhimurium*. *Journal of Applied Bacteriology* 69, 373–383.

Malmqvist, M., Jacobsson, K.G., Haggblom, P., Cerenius, F., Sjoland, L. and Gunnarsson, A. (1995) *Salmonella* isolated from animals and feedstuffs in Sweden during 1988–1992. *Acta Veterinaria Scandinavica* 36, 21–39.

Matthes, S. (1985) Influence of antimicrobial agents on the ecology of the gut and *Salmonella* shedding. In: Helmuth, R. and Bulling, E. (eds) *Criteria and Methods for the Microbial Evaluation of Growth Promoters in Animal Feeds*. Institut für Veterinärmedizin des Bundesgesundheitsametes, Berlin, pp. 104–125.

Miller, E.L. and de Boer, F. (1988) By-products of animal origin. *Livestock Production Science* 19, 159–168.

Mossel, D.A.A., Van Schothorst, M. and Kampelmacher, E.H. (1967) Comparative study on decontamination of mixed feeds by radiation and by pelletisation. *Journal of Science and Food Agriculture* 18, 362–367.

Oostrom, J. (1991) Epidemiologic studies and proposed preventive measures in the fight against human salmonellosis. *International Journal of Food Microbiology* 12, 41–52.

Oyofo, B.A., Droleskey, R.E., Norman, J.O., Mollenhaur, H.H., Ziprin, R.L., Corrier, D.E. and Deloach, J.R. (1989) Inhibition by mannose of *in vitro* colonization of chicken small intestine by *Salmonella typhimurium*. *Poultry Science* 68, 1351–1356.

Palumbo, M.S., Beers, S.M., Bhaduri, S. and Palumbo, S.A. (1995) Thermal resistance of *Salmonella* spp. and *Listeria monocytogenes* in liquid egg yolk and egg yolk products. *Journal of Food Protection* 58, 960–966.

Pivnic, H. and Nurmi, E. (1982) The Nurmi concept and its role in the control of *Salmonella* in poultry. In: Davis, R. (ed.) *Developments in Food Microbiology*. Applied Science, London, pp. 41–70.

Richter, A., Loscher, W. and Witte, W. (1996) Feed additives with antimicrobial effects. *Praktische Tierärztliche* 77, 603.

Sadler, W.W., Brownell, J.R. and Fanelli, M.J. (1968) Influence of age and inoculum level on shed pattern of *Salmonella typhimurium* in chickens. *Avian Diseases* 13, 793–803.

Schmidt, R. and Hoy, S.E. (1996) Investigation on dust emission from chicken and layer houses. *Berliner Münchener Tierärztliche Wöchenschrift* 109, 95–100.

Shackleford, A.O., Blankenship, L.C., Charles, O.W. and Dickens, J.A. (1987) Experimental two stage pellet mill conditioner with paddle shaft steam injection. *Poultry Science* 66, 1737–1743.

Shapcott, R.C. (1984) Practical aspects of *Salmonella* control: progress report on a programme in a large broiler integration. In: Snoeyenbos, G.H. (ed.) *Proceedings of the International Symposium on Salmonella, New Orleans, 19–20 July, 1984* American Association of Avian Pathology, Kennett Square, Pennsylvania, pp. 109–114.

Smyser, C.F. and Snoeyenbos, G.H. (1979) Evaluation of organic acids and other compounds as *Salmonella* antagonists in meat and bone-meal. *Poultry Science* 58, 50–54.

Stott, J.A., Hodgson, J.E. and Chaney, J.C. (1975) Incidence of Salmonellae in animal feed and the effect of pelleting on the content of Enterobacteriaceae. *Journal of Applied Bacteriology* 39, 41–46.

Taylor, J. and McCoy, J.N. (1969) *Salmonella* and arizona infections. In: Riemann, H. (ed.) *Foodborne Infections and Intoxications*. Academic Press, New York, pp. 3–72.

Urlings, H.A.P., van Logtestijn, J.G. and Bijker, P.G.H. (1992) Slaughter by-products, preliminary research and possible solutions. *Veterinary Quarterly* 14, 34–38.

Vanderwal, P. (1979) *Salmonella* control of feedstuffs by pelleting or acid treatment. *World's Poultry Science Journal* 35, 70–78.

van Staden, J.J., Van der Made, H.N. and Jordaan, E. (1980) The control of bacterial contamination in carcass meal with propionic acid. *Onderstepoort Journal of Veterinary Research* 47, 77–82.

Veldman, A., Vahl, H.A., Borggreve, G.J. and Fuller, D.C. (1995) A survey of the incidence of *Salmonella* species and Enterobacteriaceae in poultry feeds and feed components. *Veterinary Record* 136, 169–172.

Voeten, A.C. and Van de Leest, L. (1989) Influence of the pelleting temperature used for feed in *Salmonella* infection in broilers. *Geflügelkunde* 53, 225–234.

Wierup, M., Engstrom, B., Engvall, A. and Wahlstrom, H. (1995) Control of *Salmonella enteritidis* in Sweden. *International Journal of Food Microbiology* 25, 219–226.

Williams, J.E. (1981) *Salmonella* in poultry feeds – a worldwide review. *World's Poultry Science Journal* 37, 97–105.

Wilson, S. (1989) Control of *Salmonella enteritidis* in poultry. *Veterinary Record* 125, 465.

Chapter 18
Competitive Exclusion

Carita Schneitz[1] and Geoffrey Mead[2]
[1]*Orion Corporation, Animal Health, PO Box 425, FIN-20101 Turku, Finland;*
[2]*Royal Veterinary College, University of London, Boltons Park, Hawkshead Road, Potters Bar, Hertfordshire EN6 1NB, UK*

Introduction

Despite increasing emphasis on hygiene control in the production and processing of poultry, problems caused by food-borne human infections from this source persist in many countries. *Salmonella* is known to be one of the most common human pathogens in poultry. This is due to modern methods of hatching and rearing the birds, which separate the generations from each other and leave the young vulnerable to invading pathogens (Baird-Parker, 1991; Bryan and Doyle, 1995).

It is well known that poultry can be successfully immunized against systemic infection by the host-specific serovar *S. gallinarum*, which has been virtually eradicated in developed countries. In contrast, colonization by most of the non-host-specific serovars that can infect poultry is largely confined to the gastrointestinal tract and is usually asymptomatic. It cannot be assumed that immunological control of these organisms can be obtained in the same way (Barrow, 1991). Although there is evidence that the immune system is involved in the normal clearance of food-poisoning *Salmonella* from the alimentary tract of chickens (Barrow *et al.*, 1988, 1990; Mead and Barrow, 1990), the main reason for adult birds being relatively resistant to *Salmonella* colonization is their complex intestinal microflora, which occupies the ecological niches sought by *Salmonella*. Newly hatched poultry, on the other hand, are susceptible to a variety of enteric dis-

ease agents, because their intestinal flora has yet to develop. In comparison, the flora of young mammals becomes fully established between 24 and 48 h of age and will remain reasonably stable thereafter (Ducluzeau, 1983).

Before intensive production systems were introduced, newly hatched birds acquired the full variety of intestinal bacteria during their first day or so of life. The organisms were obtained from the eggshells and by pecking at the faeces of adult birds. The bacteria rapidly colonized the intestine of the young bird, coating the intestinal mucosa and preventing the intrusion of *Salmonella* and other undesirable bacteria. The caeca are the main site for *Salmonella* colonization in poultry and the part of the gut that offers the most stable environment for microorganisms. The observation that the natural resistance of young birds to *Salmonella* infection increases with age was first reported by Milner and Shaffer (1952), but some 20 years then elapsed before both an explanation and a possible solution to the problem of early susceptibility to infection were provided by Nurmi and Rantala (1973). The phenomenon by which *Salmonella* is suppressed by the normal flora is known as 'competitive exclusion' (CE) and is sometimes described as the 'Nurmi concept', when referring to poultry (Pivnick and Nurmi, 1982).

Although the CE concept applies to different types of habitat in virtually all animals, this chapter will deal solely with the control of intestinal colonization by human enteropathogens in chickens and turkeys, a topic studied most intensively during the past quarter of a century.

The Intestinal Microflora

The intestinal microflora of the chick changes markedly with age. Initially, the gut contains bacteria belonging to only a few genera and these appear rapidly, derived, presumably, from the hatchery environment. Thus, within a few hours of hatching, genera of *Enterobacteriaceae* and species of *Enterococcus* and sometimes *Clostridium* may occur in the caeca and scattered throughout the remainder of the alimentary tract. Species of *Lactobacillus* become established by about the third day after the start of feeding. The native adult flora in the small intestine appears within the first 2 weeks of life, but the caecal flora can take more than 4 weeks to develop (Barnes *et al.*, 1972, 1979b; Mead and Adams, 1975).

The caecal microflora is ultimately dominated by obligate non-sporing anaerobic bacteria, including species of *Bacteroides*, *Fusobacterium*, *Peptostreptococcus*, anaerobic *Streptococcus*, *Eubacterium* and *Bifidobacterium*, and budding bacteria, such as *Gemmiger*. More than 40 different types of anaerobic Gram-negative and Gram-positive non-sporing rods and cocci have been isolated and characterized. At least 17 different species of *Clostridium* have also been isolated from the caecal contents (Barnes and Impey, 1970; Salanitro *et al.*, 1974, 1976; Barnes, 1979; Barnes *et al.*, 1979b; Croucher and Barnes, 1983). Apart from fusiforms, typical members of *Bacteroides* and *Bifidobacterium* did not appear as major components of the flora until 4 weeks of age (Barnes, 1977). It is easy to understand that the slow development of a mature caecal microflora makes chicks very susceptible to colonization by enteropathogens.

The Competitive Exclusion Concept

The CE concept may be defined as 'the early establishment of an adult intestinal microflora to prevent subsequent colonization by enteropathogens'. Tanami (1959) showed that *Lactobacillus acidophilus* or *Lactobacillus bifidus* given to germ-free guinea-pigs prior to the addition of *Escherichia coli* caused a decrease in the

number of *E. coli* found, and a great excess of *L. bifidus* eradicated the organism from the intestine of the guinea-pigs. The term 'competitive exclusion' was used for the first time by Greenberg in 1969. He showed that the exclusion of *S. typhimurium* from maggots of blowflies was so effective that the bacterium survived in the gut only if the normal microbiota was simplified or omitted. 'Colonization resistance' is an analagous term that was introduced by van der Waaij *et al.* (1971) in studying the intestinal populations in mice. The term 'competitive inhibition', which was used by Lloyd *et al.* (1974), would be more appropriate in the present context, because exclusion of *Salmonella* is rarely complete.

Microbial interactions and the mechanisms by which indigenous intestinal microorganisms inhibit colonization by invading pathogens are not yet fully understood. There is also a lack of information on the bacteria involved, although moderately effective mixtures of pure cultures have been developed (see below). Apparently, the only universally accepted fact concerning the mechanism of CE is that protection depends upon the oral administration of viable bacteria, especially anaerobes. Blanchfield *et al.* (1982) obtained good protection of chicks against *Salmonella* infection by oral administration of 10^{-5} g of faeces and similar protection was obtained with anaerobic broth cultures derived from 10^{-7} g of caecal or faecal material (Blanchfield *et al.*, 1984). As the quantity of inoculum was reduced from 10^{-7} g to 5×10^{-10} g, there was a gradual loss of protective activity. Excellent protection was evident with inocula of 10^{-7} g and even inocula of 10^{-8} g resulted in cultures that completely protected over 50% of challenged birds. Treatment cultures prepared with 10^{-10} g did not give any protection. This suggests that all the bacterial species needed for fully protective cultures were present only at a level of $10^7 - 10^8$ g^{-1} of caecal or faecal contents.

Different kinds of *in vivo* and *in vitro* models have been used to study the mechanism of CE. *In vivo* models include infant, antibiotic-treated and germ-free animals, while *in vitro* models, which are considered to be less relevant than experiments with animals, include batch, continuous-flow (CF) and agar cultures (Savage, 1977; Barnes *et al.*, 1979a; Freter *et al.*, 1983; Rolfe, 1991).

Among the mechanisms by which one or more bacterial species may inhibit proliferation or reduce the numbers of other bacterial types are the following (Rolfe, 1991).

1. Creation of a restrictive physiological environment.
2. Competition for bacterial receptor sites.
3. Elaboration of antibiotic-like substances.
4. Depletion of or competition for essential substrates.

These are discussed in the following sections.

Creation of a restrictive physiological environment

Volatile fatty acids (VFA), including acetic, propionic and butyric acids, which are produced by caecal anaerobes, are known to be inhibitory to *Salmonella*, especially in the undissociated state below pH 6.0. Meynell (1963), in studying the exclusion of *S. typhimurium* from the mouse gut, came to conclusion that *Salmonella* failed to multiply because of the combined inhibitory effects of a low oxidation–reduction potential (E_h) and short-chain fatty acids produced by the normal gut flora. He came to this conclusion after eliminating the caecal flora with streptomycin and finding that the antibiotic also abolished the mechanism responsible for removing *Salmonella* from the gut. At the same time, Meynell (1963) observed a decrease in concentration of fatty acids and a rise in pH and E_h.

The importance of VFA as part of the mechanism of CE has also been reported by others (Barnes *et al.*, 1979a; Nisbet *et al.*, 1993; Corrier *et al.*, 1995a, b). On the other hand, Seuna (1979) and Soerjadi *et al.* (1981a) found that protection against an oral *Salmonella* challenge starts to become apparent only 1–2 h after CE treatment. A similar observation was made by Mead *et al.* (1989b) when studying the effect of CE treatment on the transmission of *S. enteritidis* between chickens in delivery boxes. All these studies indicate that, initially, protection is predominantly a physical phenomenon rather than one involving synthesis of VFA or other metabolites.

Competition for bacterial receptor sites

The ability of bacteria to adhere is important in establishing or maintaining colonization of mucosal surfaces (Savage, 1977). The bacterial glycocalyx, which is considered to be any polysaccharide-containing component outside the cell wall (Costerton *et al.*, 1981), is thought to mediate adherence of protective bacteria to each other and to the intestinal epithelium of the chick (Soerjadi *et al.*, 1982b). Thus, a layer of cells is formed, and this blocks receptor sites for *Salmonella* attachment (Fuller and Turvey, 1971; Snoeyenbos *et al.*, 1979; Savage, 1983; Schleifer, 1985).

The ability to adhere to epithelial cells *in vitro* is a common property of *Lactobacillus*. Fuller and Brooker (1980) described surface fibrils and microcapsules as colonization determinants and Wadström *et al.* (1987) explored the role of surface hydrophobicity in the adherence of *Lactobacillus*. The proteinaceous nature of some determinants was demonstrated by Henriksson *et al.* (1991), while Conway and Adams (1989) showed that adherence of *Lactobacillus* is not mediated by extracellular polysaccharide. A proteinaceous surface layer (S layer) has been detected in many Gram-negative bacteria, e.g. in *Campylobacter fetus* (Fujimoto *et al.*, 1989), as well as in Gram-positive organisms, e.g. *Clostridium symbiosum* (Messner *et al.*, 1990), *Eubacterium* spp. (Sjögren *et al.*, 1988) and *Lactobacillus* spp. (Kawata, 1981; Schneitz *et al.*, 1993).

Loss of contact with the epithelial cell and its inductive influence may alter bacterial physiology. For example, Fujimoto *et al.* (1989) reported that a strain of *C. fetus* lost its S layer during cultivation on laboratory media but regained it after a single animal passage. According to Ray and Johnson (1986), the S layer can also be damaged by certain manipulations, e.g. freeze-drying. Schneitz *et al.* (1993) also noticed that changes in mode of growth and ultrastructure of *L. acidophilus*, which coincided with the loss of adherence to epithelial cells *in vitro*, resulted in simultaneous loss of CE capability *in vivo*.

Elaboration of antibiotic-like substances

The antibiotic-like metabolites elaborated by native gut organisms include a wide range of inhibitory substances of differing chemical composition and modes of action. Bacteriocins are the most comprehensively studied in this respect, but their significance as regulators of bacterial populations remains unclear (Rolfe, 1991).

Depletion of or competition for essential subtrates

The importance of competition for growth-limiting nutrients as a means of controlling microbial populations is difficult to evaluate *in vivo*, because of inhibitory factors in the intestinal environment (Rolfe, 1991).

Mechanisms discussed in the above two sections have been studied in relation to pathogens other than *Salmonella* and host systems other than poultry. Their role in the avian alimentary tract has yet to be investigated systematically. In addition to the mechanisms suggested above, there are a multitude of host-dependent factors that interact with the normal intestinal microflora and hence could influence the exclusion of pathogens. However, it is extremely unlikely that any single mechanism is responsible for excluding invading pathogens from the intestine.

Development of CE Preparations

Probiotics

The CE concept was first associated with probiotics, the use of which originates from the work of Elie Metchnikoff, a Russian biologist working at the Pasteur Institute at the turn of the century (Metchnikoff, 1908). He became convinced that fermented milks containing so-called 'lactic acid bacteria' were responsible for the longevity of Bulgarian farming families. It was assumed that the introduction of a benign non-toxigenic gut flora created conditions that favoured good health.

Probiotic products that are currently available are a heterogeneous group of preparations that contain microorganisms or microbial metabolites from various sources (Fuller, 1989; Juven *et al.*, 1991) and are used in agriculture to promote the growth of food animals and to control conditions such as scouring. Also, *Lactobacillus reuteri*, together with whey, has been shown to reduce *Salmonella* colonization in

turkeys (Edens *et al.*, 1991) and chick mortality associated with egg-borne *E. coli* was reduced by in-hatcher exposure to *L. reuteri* (Edens *et al.*, 1994). Otherwise, various probiotics tested in controlled chick assay trials have not been found to prevent or reduce *Salmonella* colonization in poultry (Hinton and Mead, 1991; Stavric *et al.*, 1992a).

The World Health Organization (WHO, 1994) suggested that undefined CE preparations of the kind now used commercially should be distinguished from live probiotics, which are preparations of only one or a few strains of microorganisms.

Preparations of native intestinal microorganisms

Undefined, mixed cultures

In their preliminary work, Nurmi and Rantala (1973) used diluted material from the crop and intestinal tract of adult chickens to protect newly hatched chicks against a challenge with *S. infantis*. Later, a similar degree of protection was obtained using the third daily subculture from an anaerobic broth culture of intestinal contents (Rantala and Nurmi, 1973). The results of this study have been confirmed and extended by several research groups around the world and reviewed by Pivnick and Nurmi (1982), Schleifer (1985), Mead and Impey (1987), Schneitz (1993) and Stavric and D'Aoust (1993).

Lloyd *et al.* (1977) increased the resistance of newly hatched chicks and turkey poults to colonization with *S. typhimurium* by oral pretreatment with filtered (glass-wool) caecal material from a mature bird. Material from other areas of the alimentary tract was less protective.

Most investigators who have used anaerobic broth cultures for protective purposes have utilized a modified Viande Levure (VL) medium (Barnes and Impey, 1971). The advantages of these cultures are that they can be subcultured many times without losing their effectiveness (Snoeyenbos *et al.*, 1978; Mead and Impey, 1986) and non-bacterial pathogens, such as viruses and protozoa, which are unable to proliferate in bacteriological culture media, are diluted out on subculture. The cultures can be tested directly to ensure the absence of known bacterial pathogens.

To minimize the risk of transferring any pathogens present from donor birds to recipients, Snoeyenbos *et al.* (1979) used faeces from a specific-pathogen-free (SPF) flock as the inoculum material. Birds in this flock had acquired their gut microflora by oral inoculation from a group of conventional birds, specially selected as one offering particularly good protection. This approach is said to avoid the problem of insufficient protection due to the retarded development of the intestinal microflora in SPF birds (Coloe *et al.*, 1984; Impey *et al.*, 1984; Mead and Impey, 1986).

Defined cultures

The main point in developing defined treatment preparations is to avoid the necessity of testing the product for pathogens. Thus, numerous attempts have been made to use a variety of pure cultures for protective purposes, but, despite some success, the results have generally been disappointing (Stavric and D'Aoust, 1993).

Newly hatched chicks were protected to some extent against colonization by *S. typhimurium* when given pure cultures of a *Clostridium* sp. (Rigby *et al.*, 1977) or *Enterococcus faecalis* (Soerjadi *et al.*, 1978). Protection was also obtained with a mixture of 23 bacterial strains, including species of *Lactobacillus* and other facultatively and strictly anaerobic strains originating from caecal material (Barnes *et al.*, 1980b). Impey *et al.* (1982, 1984) have reported protective activity equivalent to that of caecal cultures with mixtures of 48 and 65 organisms, in which *Bacteroides* were numerically predominant. However, neither of the two defined-culture mixtures from chicken provided any protection against *Salmonella* colonization in turkey poults (Impey *et al.*, 1984). Stavric *et al.* (1985) also succeeded in protecting chicks using a mixture of 50 pure cultures. Less protection was evident when the challenge dose was increased beyond 10^4 colony-forming units (cfu) per chick or when the number of cultures in the mixture was reduced. The proportion of strains from individual genera differed from that used by Impey *et al.* (1982). When newly hatched chicks were inoculated with a mixture of 11 pure cultures originating from a CF culture of caecal material, together with dietary lactose, caecal colonization by *S. typhimurium* was reduced by $\log_{10}$ 3.59 units compared with the control group

(Nisbet *et al.*, 1993). A mixture of 295 pure cultures of obligate anaerobes did not provide any protection against *S. infantis*, according to Goren *et al.* (1984a). On the other hand, a mixture of only three cultures of *E. coli*, isolated from sewage and from an abattoir, produced a substantial reduction in caecal populations of *S. typhimurium* over a period of 7 weeks (Barrow and Tucker, 1986). However, the strains were not equally effective against other *Salmonella* tested. Barrow *et al.* (1990) and Berchieri and Barrow (1990) reported a profound inhibition of *S. typhimurium* F98 induced by an avirulent rough-mutant strain of the same organism. Similar inhibition was obtained between two homologous strains of *Citrobacter* sp. and the same occurred with strains of *E. coli* (Barrow *et al.*, 1987).

The antagonistic effect of *Lactobacillus* on the ability of *Salmonella* to colonize the alimentary tract of young chicks has been studied extensively (reviewed by Juven *et al.*, 1991). Although evidence is conflicting, at least *L. acidophilus* (Schneitz *et al.*, 1993) and *L. reuteri* appear to be involved in CE of pathogens from the alimentary tract of both chickens and humans (Reid *et al.*, 1990; Edens *et al.*, 1991, 1994; Dunham *et al.*, 1994). Soerjadi *et al.* (1981b) showed that *Lactobacillus* reduced the number of *Salmonella* adhering to chicken crop epithelium by $\log_{10}$ 1–2 units. On the other hand, *Lactobacillus* given alone or together with a few other pure cultures had no effect in reducing *Salmonella* colonization of the caeca (Barnes *et al.*, 1980a; Weinack *et al.*, 1985b).

All the pure-culture preparations described to date have one feature in common (disregarding mutant strains of *Salmonella* and *E. coli*, which may be unacceptable from the public-health point of view): they give good or relatively good protection at first, but, when used successively over a long period of time, they tend to lose their effectiveness. There is no clear explanation for this, but the use of artificial laboratory media in isolating and cultivating the organisms may well change their physiology. Growing the strains together partially overcomes the problem (Stavric, 1992). Furthermore, Nisbet *et al.* (1996) have claimed that growing the organisms as a mixture under conditions of continuous culture is a way of stabilizing their protective activity.

Commercial CE products

At present, the commercially available CE products are AviFree, Aviguard, Broilact and Preempt (formerly CF3; DeLoach 29 in Japan). They are all essentially undefined, despite the claim that Preempt is a defined preparation. An *L. reuteri* culture is also marketed as a CE product, but appears to be, according to the results presented, relatively ineffective against enteropathogens.

AviFree

Evidently AviFree is an undefined mixed culture of whole caecal content from an adult chicken. It was developed by Alltech Ltd and launched in 1996. There is very little information concerning the composition or effectiveness of AviFree, but, according to Newman and Spring (1996), it was moderately protective against *S. typhimurium* 29E.

Aviguard

Aviguard is a lyophilized mixed culture developed by Life-Care Products Ltd in the UK and now marketed by Bayer AG. It was launched in 1993. Aviguard is a culture of caecal content from an adult SPF chicken. In small-scale trials, the product protected turkey poults against *S. typhimurium* (Cameron *et al.*, 1997) and *S. kedougou* (Ghazikhanian *et al.*, 1997) and chickens against *S. enteritidis* (Guillot *et al.*, 1997). A successful field trial was also carried out (Deruyttere *et al.*, 1997). Aviguard was successfully given to older birds after therapeutic doses of antibiotics to regenerate the intestinal microflora and prevent reinfection with *Salmonella* (Reynolds *et al.*, 1997).

Broilact

Broilact was the first commercial CE product and was developed by the Orion Corporation in Finland. It was launched in Finland and Sweden in 1987 and, since 1994, has been sold as a lyophilized preparation. Attempts have been made to characterize the principal microorganisms present. In total, 32 different types of bacteria were isolated, including 22 strictly anaerobic rods and cocci, representing five genera, and ten different facultatively anaerobic rods and cocci, representing three genera. Spore-formers were not found in this product.

An ability to associate with the intestinal epithelial surface is a common characteristic of microbes that colonize the alimentary tract (Rolfe, 1991; Schneitz *et al.*, 1993). Competition for adherence sites on the mucosa is one of the suggested mechanisms of CE, and Broilact is based on this hypothesis (Nurmi *et al.*, 1987). Most of the scientific investigations that have been carried out to date on commercial treatment preparations have been on Broilact. The work has shown that this product gives good protection in chicks, not only against caecal colonization, but also against subsequent invasion of specific organs (heart, liver and spleen) by *S. enteritidis* phage type (PT) 4 or *S. typhimurium* (Mead *et al.*, 1989b; Bolder *et al.*, 1992; Cameron and Carter, 1992; Nuotio *et al.*, 1992; Schneitz, 1992; Methner *et al.*, 1997). Broilact was also shown to be effective against *Salmonella* in the field (Wierup *et al.*, 1988, 1992; Bolder *et al.*, 1995; Palmu and Camelin, 1997). In one of these studies, Bolder *et al.* (1995) reported an additional protective effect against *Campylobacter* spp. In another study, protection was also obtained against both avian pathogenic *E. coli* and *E. coli* O157:H7 (Hakkinen and Schneitz, 1996). Furthermore, treatment of chicks with Broilact has been shown to decrease mortality due to necrotic enteritis and hepatitis and to reduce caecal levels of *Clostridium perfringens*, which is one of the causative agents of necrotic enteritis (Elwinger *et al.*, 1992; Kaldhusdal *et al.*, 1998). With older birds, Broilact has been used successfully following antibiotic therapy to regenerate the intestinal microflora (Johnson, 1992; Humbert *et al.*, 1997; Reynolds *et al.*, 1997). Field studies also indicate improvement in bird performance in terms of higher body weight and lower mortality in Broilact-treated birds (Bolder *et al.*, 1995; Palmu and Camelin, 1997). The results of a laboratory-scale study showed that the treatment decreased the viscosity of the ileal contents and increased the faecal dry-matter content. It also improved the metabolizable energy (ME) value of the feed, increased the concentration of propionic acid in the caeca and decreased that of butyric acid in the ileal contents (Schneitz *et al.*, 1998).

Preempt

Preempt, formerly called CF3 (Hume *et al.*, 1998) (DeLoach 29 in Japan), is a mixed culture developed by Corrier *et al.* (1995a), using a CF culture system and low pH of the culture medium to select for certain facultative and obligate anaerobes. The starting material was a homogenate of caecal tissue and contents prepared from 10-week-old broiler chickens (Corrier *et al.*, 1995b; Nisbet *et al.*, 1995, 1996). It has been shown to reduce *Salmonella* colonization in laboratory trials and in the field and to improve bird performance (Corrier *et al.*, 1995a,b, 1998).

Milk Specialties Co. has developed the commercial version of CF3 (Burns, 1995). Altogether, 29 different strains of bacteria have been isolated from CF3, including 14 strictly anaerobic rods and cocci, representing seven genera, and 15 facultatively anaerobic rods and cocci, representing seven genera. However, only one non-selective medium was used to isolate the anaerobes from CF3 and one non-selective and one selective medium were used to isolate the facultative anaerobes (Corrier *et al.*, 1995a). It is therefore unlikely that all the organisms present would have been recovered.

Other CE preparations

MSC

A preparation designated MSC (mucosal starter culture), which is identical to MCE (mucosal competitive exclusion), was developed by Stern *et al.* (1995). It was shown to be effective in both laboratory and field trials for controlling *Salmonella* in chickens (Blankenship *et al.*, 1993; Bailey *et al.*, 1997). There was also some activity against *Campylobacter* (Stern, 1994; Stern *et al.*, 1995). The preparation is of undefined composition, derived from scrapings of washed caeca or caecal sections incubated anaerobically in an appropriate culture medium (Stern, 1990). Continental Grain Co. holds a licence for product manufacture (Anon., 1995).

Lactobacillus reuteri

The protective effect of *L. reuteri* against enteropathogens is based on its ability to produce reuterin, a broad-spectrum antibiotic (Talarico *et al.*, 1988). Very little information is available concerning the use of this organism in poultry and its ability to prevent intestinal colonization with enteropathogens. However, it is claimed that *in ovo* treatment reduces chick mortality

caused by *Salmonella* (Dunham *et al.*, 1994). In another study, however, *L. reuteri* given *in ovo* to turkey poults had only a minor effect against *S. typhimurium* (Edens *et al*, 1997).

Saccharomyces boulardii

This is a yeast that is also used as a human probiotic. It was reported to reduce *Salmonella* infection of chicks when administered *in ovo* or via the feed, and has been given to the birds just before transportation from the farm to the processing plant (Line, 1997). Enteropathogens appear to adhere to the yeast surface and are removed from the bird when the yeast is voided with the faeces.

Vermicompost

The preparation known as vermicompost was produced by feeding the earthworm *Eisenia foetida* on a mixture of fresh chicken faeces and vegetable matter, the faeces being obtained from an SPF adult bird (Spencer and Garcia, 1995). This approach was seen as a means of producing large quantities of protective material easily and inexpensively. The material effectively protected chicks against a *Salmonella* challenge.

Applicability of Competitive Exclusion

Pathogen specificity

Studies carried out in several countries have shown that the CE concept applies to all serovars of *Salmonella* that are capable of intestinal colonization in the chick (Pivnick and Nurmi, 1982; Schleifer, 1985; Mead and Impey, 1987; Schneitz, 1993; Stavric and D'Aoust, 1993).

Although aimed originally at the control of *Salmonella* infections, it has been shown experimentally that CE treatment also protects chicks against pathogenic *E. coli* (Soerjadi *et al.*, 1981a; Weinack *et al.*, 1981, 1982, 1984; Stavric *et al.*, 1992b; Hakkinen and Schneitz, 1996), *Yersinia enterocolitica* (Soerjadi-Liem *et al.*, 1984b) and *Campylobacter jejuni* (Soerjadi *et al.*, 1982a; Soerjadi-Liem *et al.*, 1984a; Stern, 1994; Mead *et al.*, 1996). In addition, CE treatment decreases mortality due to necrotic enteritis and hepatitis and reduces levels of caecal *C. perfringens*, which is one of the causative factors in necrotic enteritis

(Barnes *et al.*, 1980b; Snoeyenbos *et al.*, 1983; Elwinger *et al.*, 1992; Kaldhusdal *et al.*, 1998). In a 7-day study, CE treatment also significantly reduced caecal colonization by *Listeria monocytogenes* (Hume *et al.*, 1998). On the other hand, Husu *et al.* (1990) reported that, in their study, most chicks eliminated *L. monocytogenes* from the body within 9 days after peroral inoculation without any treatment.

Host specificity

Protection of newly hatched chicks against *Salmonella* colonization using material from adult birds of the same species seems to be independent of the breed, strain or sex of bird, even though individual differences exist with respect to protective capability. Chickens can be protected to some extent by the microflora of a few other species of birds (Snoeyenbos *et al.*, 1979; Weinack *et al.*, 1982; Impey *et al.*, 1984), but material from other animals, e.g. horse and cow, has proved to be ineffective (Rantala and Nurmi, 1973). Weinack *et al.* (1982), Impey *et al.* (1984) and Schneitz and Nuotio (1992) showed that native chicken and turkey microflora provided reciprocal protection in chicks and turkey poults. However, defined cultures used experimentally appear to be completely specific for the avian species from which the organisms were obtained (Impey *et al.*, 1984).

Laboratory experience

The effectiveness of CE treatment has been evaluated and proved in numerous laboratory-scale trials involving small groups of birds (reviewed by Pivnick and Nurmi, 1982; Schleifer, 1985; Mead and Impey, 1987; Schneitz, 1993; Stavric and D'Aoust, 1993). In these trials, the chicks have been housed in cardboard boxes, cages or isolators with wire-mesh floors, or in wire-walled or solid-walled pens on concrete floors covered with litter. Three methods of administering the treatment material have been used: dosing of individual birds, treatment via drinking-water or spraying. The most frequent method of challenge is by dosing individual birds, although the seeder-bird technique and challenge via feed or water have also been used (Weinack *et al.*, 1979;

Linton *et al.*, 1985; Mead *et al.*, 1989b; Hinton, M. *et al.*, 1991; Schneitz *et al.*, 1991; Schneitz, 1992; Bailey *et al.*, 1998; Corrier *et al.*, 1998). Usually the trial lasts for 1 week. Both the rearing conditions and the method of administration may affect the results obtained. Chicks kept in cages with wire floors may be less prone to secondary challenge from their environment when any bird in the group starts to shed *Salmonella*. Dosing of individual birds is the most effective way to administer the treatment material, whereas spraying small groups of chicks effectively is difficult. The interval between treatment and challenge and the length of the rearing period may also affect the results. Experience has shown that the results are improved if the interval between treatment and challenge is prolonged. A longer rearing period has the effect of reducing *Salmonella* levels in the caeca, since *Salmonella* infections tend to be self-limiting in older birds.

In an attempt to standardize the methods used to evaluate different CE preparations, Mead *et al.* (1989a) described a recommended assay. Newly hatched chicks are treated orally on day 1, challenged orally with *Salmonella* 24 h later and examined 5 days after challenge to determine both the proportion of positive birds in treated and control groups and the levels of *Salmonella* carriage in infected individuals. The efficacy of the treatment is determined by the calculation of an infection factor (IF) value, which is the geometric mean of the number of *Salmonella* per gram of caecal content for all chicks in a particular group, and a protection factor, which is obtained by dividing the IF value for the control group by that for the treated group (Pivnick *et al.*, 1985).

Field experience

Carrying out reliable field trials is difficult because artificial challenge cannot be used. Some of the chicks may also be contaminated with *Salmonella* from the hatchery. Although there is evidence that levels of *Salmonella* infection in commercial broiler flocks can be reduced by the use of CE (Blankenship *et al.*, 1993; Bolder *et al.*, 1995; Palmu and Camelin, 1997), treatment must precede infection for maximum benefit (Seuna, 1979; Corrier *et al.*, 1998).

CE preparations have been used routinely in Finland since 1976 (Raevuori *et al.*, 1978) and now practically all Finnish broilers are given the treatment. The long-term use of CE preparations appears to have contributed significantly to the decline in *Salmonella* contamination of both flocks and carcasses (Hirn *et al.*, 1992) – a point that is often overlooked. The percentage of *Salmonella*-contaminated broiler flocks over this decade has averaged 2.0 (range 0.5–3.8%, Annual Reports of the National Veterinary and Food Research Institute).

In Sweden, CE treatment has been used since 1981, and an epidemiological study conducted by Wierup *et al.* (1988) during a period when *Salmonella* was being spread from contaminated feed also supported the view that CE treatment can be effective under field conditions. During 1981–1990, 179 flocks, involving 3.82 million chicks, were treated and only one of these flocks was found subsequently to be infected with *Salmonella* (Wierup *et al.*, 1992). The use of CE treatment in Sweden is now required by law, so that two consecutive broiler or turkey flocks following a *Salmonella*-infected flock must be given a CE preparation. The incidence of *Salmonella* in Swedish poultry is very low and poultry meat is currently marketed as 'Salmonella-free'. However, the industry in Sweden, as in Finland, is relatively small and the climate is colder than in many other countries. These factors may facilitate the control of *Salmonella* by CE treatment and other means.

CE preparations are normally given to newly hatched chicks or turkey poults as soon as possible after hatch, either in the hatchery or on the farm. Because the treatment is prophylactic rather than therapeutic, the chicks should be *Salmonella*-free prior to treatment. However, a reduction in *Salmonella* infection has been found in flocks that were *Salmonella*-positive in the hatchery (Blankenship *et al.*, 1993; Bolder *et al.*, 1995; Palmu and Camelin, 1997). The treatment has also been given successfully to older birds after antibiotic therapy to eliminate an existing *Salmonella* infection (e.g. Johnson, 1992; Reynolds *et al.*, 1997). The combined treatment may thus avoid the necessity of slaughtering breeding stock that is infected with invasive *Salmonella*, i.e. *S. enteritidis* and *S. typhimurium*, as required within the European Union (EU).

Both drinking-water administration and spraying in the hatchery are suitable for dosing birds under field conditions. Individual administration is restricted to treatment of valuable élite stock, where the number of birds to be treated is low.

Initially, the only way of administering CE preparations in the field was via the first drinking-water. This method was used successfully in Sweden by Wierup *et al.* (1988, 1992). However, the method has its disadvantages. Sometimes, some of the chicks fail to drink prior to feeding and protection spreads unevenly among the flock (Schneitz *et al.*, 1991). Also, the viability of the anaerobic organisms in the treatment preparation shows a rapid decline, especially in chlorinated water, and the product becomes ineffective before all the chicks have received an adequate dose (Seuna *et al.*, 1978). In addition, chicks may be exposed to *Salmonella* in the hatchery or during transportation to the farm and even earlier if there is vertical transmission from infected breeders. In the first two cases, treatment on the farm via drinking-water would be too late (Seuna, 1979). With respect to vertical transmission, CE treatment is likely to have little effect, though the spread of *Salmonella* among the chicks can be restricted by the use of CE (Mead *et al.*, 1989b; Bailey *et al.*, 1998; Corrier *et al.*, 1998).

The use of aerosols as a method of administering CE preparations was suggested by Pivnick and Nurmi (1982). Subsequently, Goren *et al.* (1984b; 1988) developed a method of spray application to treat newly hatched chicks in the hatchery, either in the hatchers themselves or in delivery boxes. Spraying in the hatcher, followed by drinking-water administration on the farm, was used by Blankenship *et al.* (1993) and shown to be effective in controlling *Salmonella* infection, but there is no evidence that a double treatment of this kind is necessary to obtain maximum protection.

Spray application, either manual (Schneitz *et al.*, 1990) or automated (Schneitz, 1992), enables chicks to be treated at the earliest possible point after hatching and ensures an even spread of the treatment material. Spray application does not have any adverse effects on the health or performance of the birds during growout (Corrier *et al.*, 1995b; Palmu and Camelin, 1997). The results of studies performed by several research groups and the 7-year field experience in Finland show that both manual and automated spray application are effective means of dosing newly hatched chicks.

The possibility of chicks becoming infected in the hatchers has encouraged researchers to look for a method of administration that would enable the birds to be treated prior to hatch (Cox *et al.*, 1990, 1991). Cox and Bailey (1993) developed an *in ovo* method in which the CE preparation was introduced into either the air cell or the amnion of the egg a few days before hatching. However, the use of a caecal culture containing highly proteolytic organisms and abundant gas-formers resulted in depressed hatchability when the material was introduced into the air cell (Cox *et al.*, 1992; Cox and Bailey, 1993). Introducing the preparation into the amnion prevented any of the chicks from hatching. Adverse effects on hatchability can be avoided by excluding strongly proteolytic and gas-forming organisms; also the timing of injection is crucial (C. Schneitz, unpublished data). Edens *et al.* (1997) showed that *L. reuteri* could be administered *in ovo* without loss of hatchability and the organism could be readily isolated from the resultant chicks. As mentioned previously, however, *L. reuteri* given *in ovo* to turkey poults had little effect on colonization with *S. typhimurium* (Edens *et al.*, 1997).

Other beneficial effects of CE

Claims have been made that CE treatment enhances growth and reduces mortality in treated birds. According to Goren *et al.* (1984b), an improvement in growth rate was observed in commercial broiler flocks sprayed with an undefined CE preparation. Corrier *et al.* (1995b) reported an improvement in the efficiency of feed utilization in broiler flocks that were given CE treatment on the day of hatch. An improvement in bird performance, in terms of higher body weight, better feed conversion and lower mortality, was reported by Abu-Ruwaida *et al.* (1995) and Palmu and Camelin (1997). Higher body weight and lower mortality in CE-treated flocks were also reported by Bolder *et al.* (1995).

In an attempt to explain the nutritional effects of CE treatment, a laboratory-scale study was conducted with the commercial CE product Broilact. The results showed that the treatment decreased the viscosity of the ileal contents and

increased the faecal dry-matter content. It also improved the ME value of the feed by 1.6%, increased the concentration of propionic acid in the caeca and decreased that of butyric acid in the ileal contents (Schneitz *et al.*, 1998).

Factors affecting the efficacy of CE

In addition to certain growth-promoting antibiotics and coccidiostats, which are discussed later, factors that can reduce the efficacy of CE treatment include stress and disease. Starving chicks for the first 24 h of life also has a negative effect (Goren *et al.*, 1984a), whereas, in older birds, the protective flora is more difficult to disrupt (Snoeyenbos *et al.*, 1985). With the day-old chick, physiological stress induced by high or low environmental temperatures or removal of feed and water either interfered with the colonization of protective organisms or reduced the protection provided by these organisms; however, there was no obvious effect at 2 weeks of age (Weinack *et al.*, 1985a).

Lafont *et al.* (1983) studied CE-treated chicks that were carrying low numbers of *Salmonella* in their intestines and administered oocysts of *Eimeria tenella* at a level known to produce caecal coccidiosis. The birds then shed large numbers of *Salmonella* for more than 2 weeks. Exposure of CE-treated chicks to aerosols of *Mycoplasma gallisepticum* and/or infectious bronchitis virus increased the number of birds shedding pathogenic *E. coli* or *S. typhimurium*, following a challenge 2 days after protective treatment (Weinack *et al.*, 1984).

Induced moulting of White Leghorn layers and subjecting market-age broilers to feed withdrawal have also been shown to increase both the numbers of *Salmonella* in the gastrointestinal tract and the proportion of infected individuals (Holt and Porter, 1993; Holt *et al.*, 1995; Macri *et al.*, 1997; Raminez *et al.*, 1997).

As discussed earlier, hatchery-acquired *Salmonella* contamination can substantially reduce the effectiveness of subsequent CE treatments to prevent *Salmonella* colonization of the young chicks (Bailey *et al.*, 1998), although treatment on the day of hatch can significantly help to reduce seeder establishment and the spread of *Salmonella* from seeders to highly susceptible contact birds (Mead *et al.*, 1989b; Schneitz, 1992; Corrier *et al.*, 1998).

Safety Requirements for Undefined CE Preparations

The criteria applicable to current undefined CE preparations are those described by Nurmi and Nuotio (1994).

1. A healthy donor bird from a regularly monitored flock, preferably SPF. This requires both ante- and post-mortem examination of the donor bird.
2. Good laboratory and manufacturing practices adopted throughout the production process.
3. Meticulous examination of primary inocula for human and avian pathogens in laboratories certified by licensing authorities.

Additional or supportive measures for the safety of an undefined CE product include the following.

1. Low incidence of contagious diseases in the country where the CE preparation is produced.
2. Series of consecutive cultivation steps in manufacture which provide a dilution of the original material of at least 1 in 100 million.
3. Media used for propagation of the organisms that do not support the proliferation of mycoplasmas or viruses.
4. Quality control of the final product by the detection of indicator organisms.

Other Means of Manipulating the Intestinal Microflora

Dietary additives

Antimicrobials

Poultry feeds generally contain antimicrobial agents at subtherapeutic levels, either as growth-promoters or to prevent infectious diseases caused by bacteria or protozoa (Bailey, 1987). Several growth-promoting agents, either alone or together with anticoccidials, have been examined for their effects on *Salmonella* shedding. Conflicting results have been obtained. According to Smith and Tucker (1978, 1980) and Barrow *et al.* (1984), chickens fed diets containing avoparcin remained infected for longer periods and shed higher numbers of *Salmonella* than those given diets without avoparcin. On the

other hand, trials conducted by Smith and Green (1980) and Gustafson and Kobland (1984) showed no such disadvantage. However, there is a tendency in many countries to restrict the use of antimicrobials at subtherapeutic levels because of the development of bacterial resistance, which may cause problems for human beings (Dupont and Steele, 1987). For example, vancomycin-resistant *Enterococcus faecium* strains have been isolated from production animals, and it has been suggested that this is due to the use of the vancomycin-like polypeptide avoparcin in the feed for growth promotion. This means that enterococci can now be resistant to all currently available antibiotics (Wegener *et al.*, 1997). In addition, Ginns *et al.* (1996) showed that the prophylactic use of certain antimicrobials could promote the persistence of many antimicrobial resistance genes in the *E. coli* populations of birds.

The compatibility of commonly used antimicrobial feed additives with CE treatment has been of interest (Fowler, 1992), but only a few reports describe the effects of such additives on the control of *Salmonella* colonization in newly hatched chicks given an adult gut microflora. Chicks receiving avoparcin in the feed following CE treatment had significantly more *Salmonella* in their caeca than birds given feed containing no antimicrobials, whereas chicks given a diet supplemented with flavomycin had levels similar to those of control birds with no dietary antimicrobials (Humbert *et al.*, 1991). Bailey *et al.* (1988) demonstrated that a combination of nicarbazin and bacitracin interfered with the protective effect of CE treatment. In contrast, treatment of newly hatched chicks for 5 days with either furazolidone (400 p.p.m. in the feed) or trimethoprim–methoxazole sulphate (0.02 g trimethoprim and 0.1 g sulphamethoxazole l^{-1} in drinking-water) did not affect the protective capability of Broilact (Bolder and Palmu, 1995).

Organic acids

Incorporation of organic acids, especially formic and propionic acids, in the feed has been shown to reduce the incidence of *Salmonella* infections in broiler chicks (Hinton *et al.*, 1985) and in both laying hens and their progeny (Humphrey and Lanning, 1988). Hinton and Linton (1988) showed that treatment of feed 1 week before adding *S. kedougou* to the feed prevented subsequent infection in chicks. However, Hinton, M. *et al.* (1991a) found that feed supplemented with the acids had no effect on caecal colonization by *S. enteritidis* PT4 when chicks were challenged via drinking-water. Differences in challenge organism, dose and method of challenge may have accounted for the conflicting results. In the UK, acidified feed is sometimes used for breeder flocks, but less so for broilers (Mead, 1990).

Carbohydrates

Different sugars added in feed or drinking-water have been tested in laboratory and pilot-scale trials for their ability to control *Salmonella* infections in broiler chicks and turkey poults. The results have varied considerably. Addition of D-mannose has been shown to prevent adherence of *Salmonella* to chicken-gut mucosa both *in vitro* (Oyofo *et al.*, 1989a) and *in vivo* (Oyofo *et al.*, 1989b), while dextrose, maltose and sucrose had only a minor effect. In contrast, results obtained by Izat *et al.* (1990) indicated that mannose reduced neither the incidence nor the numbers of *Salmonella* in the caeca and had no effect on contamination of pre-chill carcasses of market-age broilers. Mannose has the disadvantage of causing sticky droppings, which may lead to problems in the processing plant.

Lactose has also been reported to reduce *Salmonella* colonization in chicks (Oyofo *et al.* 1989c), and a combination of orally administered caecal microorganisms and lactose resulted in greater reductions in *Salmonella* colonization than either of the treatments used alone (Hinton, A. *et al.*, 1991). With turkey poults, Corrier *et al.* (1991) showed that protection against *Salmonella* colonization was similar for birds inoculated with anaerobic caecal cultures and those given dietary lactose, and a combined treatment resulted in a level of protection equal to or higher than that obtained when either of the two treatments was given separately. Nisbet *et al.* (1993) reported that dietary lactose and a CF culture of chicken caecal bacteria controlled *Salmonella* colonization in newly hatched broiler chicks. However, in a study conducted by Waldroup *et al.* (1992), no such advantage in the use of lactose could be shown. The use of lactose tends to cause slight scouring in the birds and therefore wet litter, which may lead to hock burn and breast blisters.

The influence of fructo-oligosaccharide (FOS) on *Salmonella* colonization in the chicken gut has been studied by Bailey *et al.* (1991) and Chambers *et al.* (1997). The results of both studies showed only a small reduction in levels of *Salmonella* in the caeca of birds fed FOS, when compared with control birds. However, when chickens were stressed by feed and water deprivation on day 13 and challenged the following day with a high dose of *Salmonella*, only 25% of the birds given FOS became infected, while the corresponding figure for control chicks was 92% (Bailey *et al.*, 1991).

Under some conditions, dietary carbohydrates may cause a change in the intestinal flora of chicks and favour that part of the flora that is detrimental to *Salmonella*.

Lactoperoxidase system

The antibacterial activity of bovine colostrum and post-colostral milk was first observed in the 1920s and was intensively studied by Reiter and his co-workers between 1964 and 1985. In addition to specific antibodies, colostrum and post-colostral milk contain non-specific antibacterial factors, such as lysozyme, lactoferrin and the lactoperoxidase system (LPS) (Reiter and Härnulv, 1984). The LPS consists of three components: the enzyme lactoperoxidase (LP), thiocyanate (SCN$^-$) and hydrogen peroxide (H$_2$O$_2$). LP reacts with H$_2$O$_2$ and oxidizes SCN$^-$ to a short-lived intermediary oxidation product, hypothiocyanite, which is antibacterial. The end-products of the oxidation (sulphate, carbon dioxide and ammonia) are inert (Reiter *et al.*, 1976). Besides milk, the LPS occurs in, for example, saliva. The bactericidal activity of LPS against *E. coli*, *S. typhimurium* and *Pseudomonas aeruginosa* was found to be greatest at pH 5.5 and below (Reiter *et al.*, 1976). On the other hand, pH 5.5 and lower is known to be inhibitory to Gram-negative, facultatively anaerobic bacteria. Wray and McLaren (1987) found that when calves were fed on fresh milk containing the LPS and challenged with high doses of *S. typhimurium*, the *Salmonella* excretion patterns and clinical findings were similar to those of control calves fed on heated milk. A growth-promoting effect of LPS was shown in calves by Reiter *et al.* (1981). However, the effect of LPS administration on enteropathogen colonization in poultry does not appear to have been studied.

Bacteriophage

Elimination of *Salmonella* from the chicken gut with lytic bacteriophage was studied by Berchieri *et al.* (1991). A lytic phage, isolated from human sewage, was used to inoculate *S. typhimurium*-infected chicks via the feed. The phage took longer to establish in the caeca than did the *Salmonella* and it disappeared when caecal levels of *S. typhimurium* fell to 10^6 cfu ml^{-1}. No neutralizing antibodies to the phage were detected in the serum of these birds. In a second trial, Berchieri *et al.* (1991) inoculated five of 30 infected chicks with the phage. Within 3 days, the phage was isolated from 72% of the contact (non-inoculated) birds. A second phage, isolated from sewage, when inoculated into newly hatched chicks at the same time as any one of three strains of *S. typhimurium*, produced a considerable reduction in mortality in the birds. This effect was only produced by using high concentrations of phage (> 10^{10} plaque forming units ml^{-1}). However, the *Salmonella*-reducing effect of the phage lasted for no more than 12 h in the crop, small intestine and caeca, with smaller effects in the liver after 24 and 48 h.

Used litter

The impact on *Salmonella* colonization of dosing chicks with used-litter preparations was studied by Rigby and Pettit (1980) and Corrier *et al.* (1992). In three separate trials carried out by the former, groups of 180–200 newly hatched broiler chicks were treated, respectively, with a lyophilized extract of breeder-flock nest litter, an anaerobic culture of this extract and an anaerobic culture of adult chicken faeces. Both treated and control chicks were challenged with *S. typhimurium* at 3 days of age and reared to 7–8 weeks. Cultures of litter samples and of intestines from chicks that died or were killed throughout the growing period showed that the incidence of infection at market age was significantly lower in all treated groups than in untreated controls.

Corrier *et al.* (1992) placed newly hatched chicks on fresh, unused litter or on litter that had been collected and stored for 1, 4 or 50 days before the start of the trial. At 3 days of age, all chicks were challenged orally with *S. typhimurium*. *Salmonella* levels in caecal contents and the proportion of positive chicks were significantly lower in the birds reared on used litter of any age than in those on new litter. The chicks reared on used litter also showed higher concentrations of caecal VFA than birds kept on new litter.

These two studies suggest that the types of bacteria involved in controlling *Salmonella* colonization in poultry are relatively aerotolerant.

Conclusions

There is increasing interest worldwide in measures to reduce the symptomless carriage of human enteropathogens in food animals and thereby minimize the risk of consumer illness from contaminated products. The prospects for developing successful control programmes appear to be best for intensively reared animals that are kept indoors in controlled-environment housing, and poultry is a prime example. In this situation, the newly hatched chick should be regarded as a critical control point, because of its extreme susceptibility to colonization with enteropathogens, especially *Salmonella*, a fact that can be attributed to the slow rate of development of a competing microflora under modern commercial conditions. Both the reason for chick susceptibility and a possible solution to the problem were highlighted in the seminal publication of Nurmi and Rantala (1973).

The use of CE preparations for *Salmonella* control has a number of advantages. The treatment is relatively cheap and easy to apply, and it appears to be effective against all non-host-specific *Salmonella* that can colonize the alimentary tract of the bird. Treatment efficacy is unaffected by the breed, strain or sex of the bird or by most antimicrobial feed additives. There are also indications that current treatment products may be of value in controlling infections with other enteropathogens, including *C. perfringens* and *E. coli*, should the need arise. Control of *C. jejuni* is also possible by this means, but may involve different protective bacteria, because of the specific niche in the caecum occupied by campylobacters (Beery *et al.*, 1988).

Treatment of chicks is essentially a preventive measure and must be supported by effective biosecurity on the rearing site. It also requires the use of good husbandry practices to minimize bird stress, otherwise the effectiveness is reduced (Weinack *et al.*, 1985a). Because of its preventive nature, successful use of CE treatment is dependent upon a consistent supply of *Salmonella*-free chicks. Failure to meet one or other of these requirements may explain why field trials have sometimes given poor results. Thus, CE treatment can only be a part of any programme aimed at reducing enteropathogen infections in poultry. However, it can be readily combined with other measures, including, where appropriate, vaccination and the use of acid-treated feed. There is also an important application to adult breeder birds that are infected with invasive *Salmonella* and would otherwise need to be slaughtered under current EU regulations. Loss of valuable stock may be avoided, under some circumstances, by use of antibiotic therapy followed by CE treatment to repair the antibiotic-damaged gut microflora and prevent reinfection.

The CE concept exploits a natural phenomenon relating to microbial competition in the alimentary tract and the effect on any *Salmonella* ingested by the chick is to prevent their multiplication (Impey and Mead, 1989). Thus, invading *Salmonella* are unable to colonize and eventually they are flushed out of the gut. Most of the currently available treatment products are relatively complex in composition and apparently undefined, although at least two of them have been partly characterized. No fully effective defined preparations have yet been developed and progress in this area is hampered by a lack of knowledge of the precise mechanism of protection and of the types of bacteria involved.

Undefined treatment products must be carefully screened for the presence of any avian or human pathogens before being used commercially. This is done for some of the available products, and commercial application has never revealed any safety problems. Nor have there been any adverse effects on bird health or growth performance. Under some conditions, there may even be an improvement in bird growth and feed utilization and a reduction in the mortality of chicks (Goren *et al.*, 1984b; Abu-Ruwaida *et al.*, 1995; Bolder *et al.*, 1995; Corrier *et al.*, 1995b; Palmu and Camelin, 1997). More work is needed to define the conditions under which such benefits can be obtained.

Despite the long usage of undefined CE preparations in Scandinavia and the absence of any indication of a safety hazard to humans or birds, there is reluctance on the part of some regulatory authorities to allow commercial use of these products. This is one reason for the slow rate of market penetration in certain countries. Since CE preparations cannot be defined in the same manner as either a vaccine or a veterinary medical product, the World Health Organization has proposed a special product category ('normal gut flora') which should be distinguished from conventional probiotics. It remains to be seen whether undefined CE products will eventually gain widespread acceptance among regulators and how long it will be before the apparent food-safety benefits will lead to increased use of CE treatment in those countries where this approach has already been sanctioned. In particular, regular use in broiler flocks would seem to represent a significant step forward in relation to *Salmonella* control.

References

Abu-Ruwaida, A.S., Husseini, M. and Banat, I.M. (1995) *Salmonella* exclusion in broiler chicks by the competitive action of adult gut microflora. *Microbios* 83, 59–69.

Anon. (1995) Continental Grain licenses natural *Salmonella* blocker. *Broiler Industry* July, 18.

Bailey, J.S. (1987) Factors affecting microbial competitive exclusion in poultry. *Food Technology* July, 88–92.

Bailey, J.S., Blankenship, L.C., Stern, N.J., Cox, N.A. and McHan, F. (1988) Effect of anticoccidial and antimicrobial feed additives on prevention of *Salmonella* colonization of chicks treated with anaerobic cultures of chicken feces. *Avian Diseases* 32, 324–329.

Bailey, J.S., Blankenship, L.C. and Cox, N.A. (1991) Effect of fructooligosaccharide on *Salmonella* colonization of the chicken intestine. *Poultry Science* 70, 2433–2438.

Bailey, J.S., Stern, N.J. and Cox, N.A. (1997) Control of salmonellae in broiler chickens using different application methods and dosage levels of mucosal starter culture. In: *Proceedings of the* Salmonella *and Salmonellosis Symposium.* Zoopôle Developpement, Ploufragan, France, pp. 487–491.

Bailey, J.S., Cason, J.A. and Cox, N.A. (1998) Effect of *Salmonella* in young chicks on competitive exclusion treatment. *Poultry Science* 77, 394–399.

Baird-Parker, A.C. (1991) Food borne salmonellosis. *Lancet* 336, 1231–1235.

Barnes, E.M. (1977) Ecological concepts of the anaerobic flora in the avian intestine. *American Journal of Clinical Nutrition* 30, 1793–1798.

Barnes, E.M. (1979) The intestinal microflora of poultry and game birds during life and after storage. *Journal of Applied Bacteriology* 46, 407–419.

Barnes, E.M. and Impey, C.S. (1970) The isolation and properties of the predominant anaerobic bacteria in the caeca of chickens and turkeys. *British Poultry Science* 11, 467–481.

Barnes, E.M. and Impey, C.S. (1971) The isolation of anaerobic bacteria from chicken caeca with particular reference to members of the family *Bacteroidaceae*. In: Shapton, D.A. and Board, R.G. (eds) *Isolation of Anaerobes*. Academic Press, London, pp. 115–123.

Barnes, E.M., Mead, G.C., Barnum, D.A. and Harry, E.G. (1972) The intestinal flora of the chicken in the period 2 to 6 weeks of age, with particular reference to the anaerobic bacteria. *British Poultry Science* 13, 311–326.

Barnes, E.M., Impey, C.S. and Stevens, J.H. (1979a) Factors affecting the incidence and anti-*Salmonella* activity of the anaerobic caecal flora of the young chick. *Journal of Hygiene, Cambridge* 82, 263–283.

Barnes, E.M., Mead, G.C., Impey, C.S. and Adams, B.W. (1979b) Analysis of the avian intestinal flora. In: Lovelock, D.W. and Davies, R. (eds) *Techniques for the Study of Mixed Populations*. Technical Series 11, Society for Applied Bacteriology, Academic Press, London, pp. 89–105.

Barnes, E.M., Impey, C.S. and Cooper, D.M. (1980a) Competitive exclusion of salmonellas from the newly hatched chick. *Veterinary Record* 106, 61.

Barnes, E.M., Impey, C.S. and Cooper, D.M. (1980b) Manipulation of the crop and intestinal flora of the newly hatched chick. *American Journal of Clinical Nutrition* 33, 2426–2433.

Barrow, P.A. (1991) Immunological control of *Salmonella* in poultry. In: Blankenship, L.C. (ed.) *Colonization Control of Human Bacterial Enteropathogens in Poultry*. Academic Press, San Diego, pp. 199–217.

Barrow, P.A. and Tucker, J.F. (1986) Inhibition of colonization of the chicken caecum with *Salmonella typhimurium* by pre-treatment with strains of *Escherichia coli*. *Journal of Hygiene, Cambridge* 96, 161–169.

Barrow, P.A., Smith, H.W. and Tucker, J.F. (1984) The effect of feeding diets containing avoparcin on the excretion of salmonellas by chickens experimentally infected with natural sources of *Salmonella* organisms. *Journal of Hygiene, Cambridge* 93, 439–444.

Barrow, P.A., Tucker, J.F. and Simpson, J.M. (1987) Inhibition of colonization of the chicken alimentary tract with *Salmonella typhimurium* Gram-negative facultatively anaerobic bacteria. *Epidemiology and Infection* 98, 311–322.

Barrow, P.A., Simpson, J.M. and Lovell, M.A. (1988) Intestinal colonization in the chicken by food-poisoning *Salmonella* serovars: microbial characteristics associated with faecal excretion. *Avian Pathology* 17, 571–588.

Barrow, P.A., Hassan, J.O. and Berchieri, A., Jr (1990) Reduction in faecal excretion of *Salmonella typhimurium* strain F98 by chickens by vaccination with live and killed *S. typhimurium* organisms. *Epidemiology and Infection* 104, 413–426.

Beery, J.T., Hugdahl, M.B. and Doyle, M.P. (1988) Colonization of gastrointestinal tracts of chicks by *Campylobacter jejuni*. *Applied and Environmental Microbiology* 54, 2365–2370.

Berchieri, A., Jr and Barrow, P.A. (1990) Further studies on the inhibition of colonization of the chicken alimentary tract with *Salmonella typhimurium* by pre-colonization with an avirulent mutant. *Epidemiology and Infection* 104, 427–441.

Berchieri, A., Jr, Barrow, P.A. and Lovell, M.A. (1991) The activity in the chicken alimentary tract of bacteriophages lytic for *Salmonella typhimurium*. *Research in Microbiology* 142, 541–549.

Blanchfield, B., Gardiner, M.A. and Pivnick, H. (1982) Nurmi concept for preventing infection of chicks by *Salmonella*: comparison of fecal suspensions and fecal cultures administered into the crop and in drinking water. *Journal of Food Protection* 45, 345–347.

Blanchfield, B., Stavric, S., Gleeson, T. and Pivnick, H. (1984) Minimum intestinal inoculum for Nurmi cultures and a new method for determining competitive exclusion of *Salmonella* from chicks. *Journal of Food Protection* 47, 542–545.

Blankenship, L.C., Bailey, J.S., Cox, N.A., Stern, N.J., Brewer, R. and Williams, O. (1993) Two-step mucosal competitive exclusion flora to diminish salmonellae in commercial broiler chickens. *Poultry Science* 72, 1667–1672.

Bolder, N.M. and Palmu, L. (1995) Effect of antibiotic treatment on competitive exclusion against *Salmonella enteritidis* PT4 in broilers. *Veterinary Record* 137, 350–351.

Bolder, N.M., van Lith, L.A.J.T., Putirulan, F.F., Jacobs-Reitsma, W.F. and Mulder, R.W.A.W. (1992) Prevention of colonization by *Salmonella enteritidis* PT4 in broiler chickens. *International Journal of Food Microbiology* 15, 313–317.

Bolder, N.M., Vereijken, P.F.G., Putirulan, F.F. and Mulder, R.W.A.W. (1995) The effect of competitive exclusion on the *Salmonella* contamination of broilers (a field study). In: Briz, R.C. (ed.) *Proceedings of the 2nd Annual Meeting of EC COST Working Group No. 2*. Graficas Imprinter, Zaragoza, Spain, pp. 89–97.

Bryan, F. and Doyle, M. (1995) Health risks and consequences of *Salmonella* and *Campylobacter jejuni* in raw poultry. *Journal of Food Protection* 58, 326–344.

Burns, R. (1995) Commercial competitive exclusion bacterial mix ready soon. *Feedstuffs*, 13 March, 27.

Cameron, D.M. and Carter, J.N. (1992) Evaluation of the efficacy of BROILACT in preventing infection of broiler chicks with *S. enteritidis* PT4. *International Journal of Food Microbiology* 15, 319–326.

Cameron, D.M., Carter, J.N., Mansell, P. and Redgrave, V.A. (1997) Floor-pen efficacy study with Aviguard against *Salmonella typhimurium* DT 104 colonization in turkeys. In: *Proceedings of the Salmonella and Salmonellosis Symposium*. Zoopôle Developpement, Ploufragan, France, pp. 481–485.

Chambers, J.R., Spencer, J.L. and Modler, H.W. (1997) The influence of complex carbohydrates on *Salmonella typhimurium* colonization, pH, and density of broiler ceca. *Poultry Science* 76, 445–451.

Coloe, P.J., Bagust, T.J. and Ireland, L. (1984) Development of the normal gastrointestinal microflora of specific pathogen-free chickens. *Journal of Hygiene, Cambridge* 92, 79–87.

Conway, P.L. and Adams, R.F. (1989) Role of erythosine in the inhibition of adhesion of *Lactobacillus fermentum* strain 737 to mouse stomach tissue. *Journal of General Microbiology* 135, 1167–1173.

Corrier, D.E., Hinton, A., Jr, Kubena, L.F., Ziprin, R.L. and DeLoach, J.R. (1991) Decreased *Salmonella* colonization in turkey poults inoculated with anaerobic cecal microflora and provided dietary lactose. *Poultry Science* 70, 1345–1350.

Corrier, D.E., Hinton, A., Jr, Hargis, B. and DeLoach, J.R. (1992) Effect of used litter from floor pens of adult broilers on *Salmonella* colonization of broiler chicks. *Avian Diseases* 36, 897–902.

Corrier, D.E., Nisbet, D.J., Scanlan, C.M., Hollister, A.G. and Deloach, J.R. (1995a) Control of *Salmonella typhimurium* colonization in broiler chicks with continuous-flow characterized mixed culture of cecal bacteria. *Poultry Science* 74, 916–924.

Corrier, D.E., Nisbet, D.J., Scanlan, C.M., Hollister, A.G., Caldwell, D.J., Thomas, L.A., Hargis, B.M., Tomkins, T. and Deloach, J.R. (1995b) Treatment of commercial broiler chickens with a characterized culture of cecal bacteria to reduce salmonellae colonization. *Poultry Science* 74, 1093–1101.

Corrier, D.E., Byrd, J.A., II, Hume, M.E., Nisbet, D.J. and Stanker, L.H. (1998) Effect of simultaneous or delayed competitive exclusion treatment on the spread of *Salmonella* in chicks. *Journal of Applied Poultry Research* 7, 132–137.

Costerton, J.W., Irvin, R.T. and Cheng, K.-J. (1981) The bacterial glycocalyx in nature and disease. *Annual Review of Microbiology* 35, 299–324.

Cox, N.A. and Bailey, J.S. (1993) Introduction of bacteria *in ovo*. US Patent 5,206,015.

Cox, N.A., Bailey, J.S., Mauldin, J.M. and Blankenship, L.C. (1990) Presence and impact of salmonellae contamination in commercial broiler hatcheries. *Poultry Science* 69, 1606–1609.

Cox, N.A., Bailey, J.S., Mauldin, J.M. and Blankenship, L.C. (1991) Extent of salmonellae contamination in breeder hatcheries. *Poultry Science* 70, 416–418.

Cox, N.A., Bailey, J.S., Mauldin, J.M. and Blankenship, L.C. (1992) *In ovo* administration of a competitive exclusion treatment to broiler embryos. *Poultry Science* 71, 1781–1784.

Croucher, S.C. and Barnes, E.M. (1983) The occurrence and properties of *Gemmiger formicilis* and related budding bacteria in the avian caecum. *Journal of Applied Bacteriology* 54, 7–22.

Deruyttere, L., Klaasen, J., Froyman, R. and Day, C.A. (1997) Field study to demonstrate the efficacy of Aviguard against intestinal *Salmonella* colonization in broilers. In: *Proceedings of the Salmonella and Salmonellosis Symposium*. Zoopôle Developpement, Ploufragan, France, pp. 523–525.

Ducluzeau, R. (1983) Implantation and development of the gut flora in the newborn animal. *Annales de Recherches Vétérinaires* 14, 354–359.

Dunham, H.J., Edens, F.W., Casas, I.A. and Dobrogosz, W.J. (1994) Efficacy of *Lactobacillus reuteri* as a probiotic for chickens and turkeys. *Microbial Ecology in Health and Disease* 7, 52–53.

Dupont, H.L. and Steele, J.H. (1987) Use of antimicrobial agents in animal feeds: implications for human health. *Review of Infectious Diseases* 9, 447–460.

Edens, F.W., Parkhurst, C.R. and Casas, I.A. (1991) *Lactobacillus reuteri* and whey reduce *Salmonella* colonization in the ceca of turkey poults. *Poultry Science* 70 (Suppl. 1), 158.

Edens, F.W., Casas, I, A, Parkhurst, C.R. and Joyce, K. (1994) Reduction of egg-borne *E. coli*-associated chick mortality by in-hatcher exposure to *Lactobacillus reuteri*. *Poultry Science* 73 (Suppl. 1), 79.

Edens, F.W., Parkhurst, C.R. and Casas, I.A. (1997) Principles of *ex ovo* competitive exclusion and *in ovo* adminis-tration of *Lactobacillus reuteri*. *Poultry Science* 76, 179–196.

Elwinger, K., Schneitz, C., Berndtson, E., Fossum, O., Teglöf, B. and Engström, B. (1992) Factors affecting the incidence of necrotic enteritis, caecal carriage of *Clostridium perfringens* and bird performance in broiler chicks. *Acta Veterinaria Scandinavica* 33, 369–378.

Fowler, N.G. (1992) Antimicrobials and competitive exclusion. *International Journal of Food Microbiology* 15, 277–279.

Freter, R., Stauffer, E., Cleven, D., Holdeman, L.V. and Moore, W.E.C. (1983) Continuous-flow cultures as *in vitro* models of the ecology of large intestinal flora. *Infection and Immunity* 39, 666–675.

Fujimoto, S., Umeda, A., Takade, A., Murata, K. and Amako, K. (1989) Hexagonal surface layer of *Campylobacter fetus* isolated from humans. *Infection and Immunity* 57, 2563–2565.

Fuller, R. (1989) Probiotics in man and animals: a review. *Journal of Applied Bacteriology* 66, 365–378.

Fuller, R. and Brooker, B.E. (1980) The attachment of bacteria to the squamous epithelial cell and its importance in the microecology of the intestine. In: Berkeley, R.C.W., Lynch, J.M., Melling, J., Rutter, P.R. and Vincent, B. (eds) *Microbial Adherence to Surfaces*. Ellis Horwood, Chichester, UK, pp. 495–507.

Fuller, R. and Turvey, A. (1971) Bacteria associated with the intestinal wall of the fowl (*Gallus domesticus*). *Journal of Applied Bacteriology* 34, 617–622.

Ghazikhanian, G.Y., Bland, M.C., Hofacre, C.L. and Froyman, R. (1997) Floor-pen study to determine the effect of Aviguard application to day-old turkey poults in reduction of clinical disease and intestinal colonization caused by *S. kedougou* infection. In: *Proceedings of the* Salmonella *and* Salmonellosis *Symposium*. Zoopôle Developpement, Ploufragan, France, pp. 531–533.

Ginns, C.A., Browning, G.F., Benham, M.L., Anderson, G.A. and Whithear, K.G. (1996) Antimicrobial resis-tance and epidemiology of *Escherichia coli* in broiler chickens. *Avian Pathology* 25, 591–605.

Goren, E., de Jong, W.A., Doornenbal, P., Koopman, J.P. and Kennis, H.M. (1984a) Protection of chicks against *Salmonella infantis* infection induced by strict anaerobically cultured intestinal microflora. *Veterinary Quarterly* 6, 22–26.

Goren, E., de Jong, W.A., Doornenbal, P., Koopman, J.P. and Kennis, H.M. (1984b) Protection of chicks against *Salmonella* infection induced by spray application of intestinal microflora in the hatchery. *Veterinary Quarterly* 6, 73–79.

Goren, E., de Jong, W.A., Doornenbal, P., Bolder, N.M., Mulder, R.W.A.W. and Jansen, A. (1988) Reduction of *Salmonella* infection of broilers by spray application of intestinal microflora: a longitudinal study. *Veterinary Quarterly* 10, 249–255.

Greenberg, B. (1969) *Salmonella* suppression by known populations of bacteria in flies. *Journal of Bacteriology* 99, 629–635.

Guillot, J.F., Salmon, A., Mouline, C., Delaporte, J. and Magnin, M. (1997) Effect of a gut microflora (Aviguard) against controlled *Salmonella enteritidis* contamination in chickens. In: *Proceedings of the* Salmonella *and* Salmonellosis *Symposium*. Zoopôle Developpement, Ploufragan, France, p. 521.

Gustafson, R.H. and Kobland, J.D. (1984) Factors influencing *Salmonella* shedding in broiler chickens. *Journal of Hygiene, Cambridge* 92, 385–394.

Hakkinen, M. and Schneitz, C. (1996) Efficacy of a commercial competitive exclusion product against chicken pathogenic *Escherichia coli* and *E. coli* O157:H7. *Veterinary Record* 139, 139–141.

Henriksson, A., Szewzyk, R. and Conway, P.L. (1991) Characteristics of the adhesive determinants of *Lactobacillus fermentum* 104. *Applied and Environmental Microbiology* 57, 499–502.

Hinton, A., Jr, Corrier, D.E., Ziprin, R.L., Spates, G.E. and DeLoach, J.R. (1991) Comparison of the efficacy of cultures of cecal anaerobes as inocula to reduce *Salmonella typhimurium* colonization in chicks with or without dietary lactose. *Poultry Science* 70, 67–73.

Hinton, M. and Linton, A.H. (1988) Control of *Salmonella* infections in broiler chickens by the acid treatment of their feed. *Veterinary Record* 123, 416–421.

Hinton, M. and Mead, G.C. (1991) *Salmonella* control in poultry: the need for the satisfactory evaluation of probi-otics for this purpose. *Letters in Applied Microbiology* 13, 49–50.

Hinton, M., Linton, A.H. and Perry, F.G. (1985) Control of *Salmonella* by acid disinfection of chick's food. *Veterinary Record* 116, 502.

Hinton, M., Mead, G.C. and Impey, C.S. (1991) Protection of chicks against environmental challenge with *Salmonella enteritidis* by 'competitive exclusion' and acid-treated feed. *Letters in Applied Microbiology* 12, 69–71.

Hirn, J., Nurmi, E., Johansson, T. and Nuotio, L. (1992) Long-term experience with competitive exclusion and sal-monellas in Finland. *International Journal of Food Microbiology* 15, 281–285.

Holt, P.S. and Porter, R.E., Jr (1993) Effect of induced molting on the recurrence of a previous *Salmonella enteritidis* infection. *Poultry Science* 72, 2069–2078.

Holt, P.S., Macri, P. and Porter, R.E., Jr (1995) Microbial analysis of the early *Salmonella enteritidis* infection in molted and unmolted hens. *Avian Diseases* 39, 55–63.

Humbert, F., Lalande, F., L'Hospitalier, R., Salvat, G. and Bennejean, G. (1991) Effect of four antibiotic additives on the *Salmonella* contamination of chicks protected by an adult caecal flora. *Avian Pathology* 20, 577–584.

Humbert, F., Carraminana, F., Lalande, F. and Salvat, G. (1997) Bacteriological monitoring of *Salmonella enteritidis* carrier birds after decontamination using enrofloxacin, competitive exclusion and movement of birds. *Veterinary Record* 141, 297–299.

Hume, M.E., Byrd, J.A., Stanker, L.H. and Ziprin, R.L. (1998) Reduction of caecal *Listeria monocytogenes* in Leghorn chicks following treatment with a competitive exclusion culture (PEEMPT™). *Letters in Applied Microbiology* 26, 432–436.

Humphrey, T.J. and Lanning, D.G. (1988) The vertical transmission of salmonellas and formic acid treatment of chicken feed: a possible strategy for control. *Epidemiology and Infection* 100, 43–49.

Husu, J.R., Beery, J.T., Nurmi, E. and Doyle, M.P. (1990) Fate of *Listeria monocytogenes* in orally dosed chicks. *International Journal of Food Microbiology* 11, 259–269.

Impey, C.S. and Mead, G.C. (1989) Fate of salmonellas in the alimentary tract of chicks pretreated with a mature caecal microflora to increase colonization resistance. *Journal of Applied Bacteriology* 66, 469–475.

Impey, C.S., Mead, G.C. and George, S.M. (1982) Competitive exclusion of salmonellas from the chick caecum using a defined mixture of bacterial isolates from the caecal microflora of an adult bird. *Journal of Hygiene, Cambridge* 89, 479–490.

Impey, C.S., Mead, G.C. and George, S.M. (1984) Evaluation of treatment with defined and undefined mixtures of gut microorganisms for preventing *Salmonella* colonization in chicks and turkey poults. *Food Microbiology* 1, 143–147.

Izat, A.L., Hierholzer, R.E., Kopek, J.M., Adams, M.H., Reiber, M.A. and McGinnis, J.P. (1990) Research note: effects of D-mannose on incidence and levels of salmonellae in ceca and carcass samples of market age broilers. *Poultry Science* 69, 2244–2247.

Johnson, C.T. (1992) The use of an antimicrobial and competitive exclusion combination in *Salmonella*-infected pullet flocks. *International Journal of Food Microbiology* 15, 293–298.

Juven, B.J., Meinersman, R.J. and Stern, N.J. (1991) Antagonistic effects of lactobacilli and pediococci to control intestinal colonization by human enteropathogens in live poultry. *Journal of Applied Bacteriology* 70, 95–103.

Kaldhusdal, M., Schneitz, C., Hofshagen, M. and Skjerve, E. (1998) Broilact® reduces the incidence of necrotic enteritis in broiler chickens. In: Hakkinen, M., Nuotio, L., Nurmi, E. and Mead, G.C. (eds) *Proceedings of Cost Action 97, Working Group 1*, Helsinki, Finland, Office for Official Publications of the European Communities, Luxembourg, pp. 85–90.

Kawata, T. (1981) Electron microscopic studies on surface structures of bacterial cells with special reference to regular arrays on bacterial surfaces. *Electron Microscopy* 15, 115–122.

Lafont, J.P., Brée, A., Naciri, M., Yvoré, P., Guillot, J.F. and Chaslus-Dancla, E. (1983) Experimental study of some factors limiting 'competitive exclusion' of *Salmonella* in chickens. *Research in Veterinary Science* 34, 16–20.

Line, J.E. (1997) Administration of *Saccharomyces boulardii* to reduce salmonellae colonization in poultry. In: *Proceedings of the* Salmonella *and Salmonellosis Symposium*. Zoopôle Developpement, Ploufragan, France, pp. 475–479.

Linton, A.H., Al-Chalaby, Z.A.M. and Hinton, M.H. (1985) Natural subclinical *Salmonella* infection in chickens: a potential model for testing the effects of various procedures on *Salmonella* shedding. *Veterinary Record* 116, 361–364.

Lloyd, A.B., Cumming, R.B. and Kent, R.D. (1974) Competitive exclusion as exemplified by *Salmonella typhimurium*. In: *Proceedings of the Australian Poultry Science Convention*. Hobart, Tasmania, pp. 185–186.

Lloyd, A.B., Cumming, R.B. and Kent, R.D. (1977) Prevention of *Salmonella typhimurium* infection in poultry by pretreatment of chickens and poults with intestinal extracts. *Australian Veterinary Journal* 53, 82–87.

Macri, N.P., Porter, R.E., Jr and Holt, P.S. (1997) The effects of induced molting on the severity of acute intestinal inflammation caused by *Salmonella enteritidis*. *Avian Diseases* 41, 117–124.

Mead, G.C. (1990) Food poisoning salmonellas in the poultry-meat industry. *British Food Journal* 92, 32–36.

Mead, G.C. and Adams, B.W. (1975) Some observations on the caecal microflora of the chick during the first two weeks of life. *British Poultry Science* 16, 169–176.

Mead, G.C. and Impey, C.S. (1986) Current progress in reducing salmonella colonization of poultry by 'competitive exclusion'. *Journal of Applied Bacteriology* Symposium Suppl. 61, 65S–75S.

Mead, G.C. and Impey, C.S. (1987) The present status of the Nurmi concept for reducing carriage of food-poisoning salmonellae and other pathogens in live poultry. In: Smulders, F.J.M. (ed.) *Elimination of Pathogenic Organisms from Meat and Poultry*. Elsevier Science Publishers, Amsterdam, the Netherlands, pp. 57–77.

Mead, G.C. and Barrow, P.A. (1990) *Salmonella* control in poultry by 'competitive exclusion' or immunization. *Letters in Applied Microbiology* 10, 221–227.

Mead, G.C., Barrow, P.A., Hinton, M.H., Humbert, F., Impey, C.S., Lahellec, C., Mulder, R.W.A.W., Stavric, S. and Stern, N.J. (1989a) Recommended assay for treatment of chicks to prevent *Salmonella* colonization by competitive exclusion. *Journal of Food Protection* 52, 500–502.

Mead, G.C., Schneitz, C.E., Nuotio, L.O. and Nurmi, E.V. (1989b) Treatment of chicks using competitive exclusion to prevent transmission of *Salmonella enteritidis* in delivery boxes. In: *IXth International Congress of the World Veterinary Poultry Association*. Brighton, UK, p. 115 (abstract).

Mead, G.C., Scott, M.J., Humphrey, T.J. and McAlpine, K. (1996) Observations on the control of *Campylobacter jejuni* infection of poultry by 'competitive exclusion'. *Avian Pathology* 25, 69–79.

Messner, P., Bock, K., Christian, R., Schulz, G. and Sleytr, U.B. (1990) Characterization of the surface layer glycoprotein of *Clostridium symbiosum* HB25. *Journal of Bacteriology* 172, 2576–2583.

Metchnikoff, E. (1908) *Prolongation of Life*. G.P. Putnam and Sons, New York, USA.

Methner, U., Barrow, P.A., Martin, G. and Meyer, H. (1997) Comparative study of the protective effect against *Salmonella* colonization in newly-hatched Spf chickens using live, attenuated *Salmonella* vaccine strains, wild-type *Salmonella* strains or a competitive-exclusion product. *International Journal of Food Microbiology* 35, 223–230.

Meynell, G.G. (1963) Antibacterial mechanisms of the mouse gut. II. The role of E_h and volatile fatty acids in the normal gut. *British Journal of Experimental Pathology* 44, 209–219.

Milner, K.C. and Shaffer, M.F. (1952) Bacteriologic studies of experimental *Salmonella* infections in chick. *Journal of Infectious Diseases* 90, 81.

Newman, K.E. and Spring, P. (1996) Effect of a commercial competitive exclusion culture (AviFree) on *Salmonella typhimurium* concentration in broiler chicks. In: *12th Annual Symposium on Biotechnology in the Feed Industry*. Enclosure Code AVI 2.1 (poster).

Nisbet, D.J., Corrier, D.E., Scanlan, C.M., Hollister, A.G., Beier, R.C. and Deloach, J.R. (1993) Effect of a defined continuous-flow derived bacterial culture and dietary lactose on *Salmonella typhimurium* colonization in broiler chickens. *Avian Diseases* 37, 1017–1025.

Nisbet, D.J., Corrier, D.E. and DeLoach, J.R. (1995) Probiotic for control of *Salmonella*. United States Patent 5,478,557.

Nisbet, D.J., Corrier, D.E., Ricke, S.C., Hume, M.E., Byrd, J.A., II and DeLoach, J.R. (1996) Maintenance of the biological efficacy in chicks of a cecal competitive-exclusion culture against *Salmonella* by continuous-flow fermentation. *Journal of Food Protection* 59, 1279–1283.

Nuotio, L., Schneitz, C., Halonen, U. and Nurmi, E. (1992) Use of competitive exclusion to protect newly-hatched chicks against intestinal colonization and invasion by *Salmonella enteritidis* PT4. *British Poultry Science* 33, 775–779.

Nurmi, E. and Nuotio, L. (1994) Safety requirements for commercial competitive exclusion products. In: *Proceedings of the 9th European Poultry Conference*, Vol. II. The World's Poultry Science Association, Glasgow, pp. 90–93.

Nurmi, E.V. and Rantala, M. (1973) New aspects of *Salmonella* infection in broiler production. *Nature* 241, 210.

Nurmi, E.V., Schneitz, C.E. and Mäkelä, P.H. (1987) Process for the production of a bacterial preparation for the prophylaxis of intestinal disturbances in poultry. US Patent 4,689,226.

Oyofo, B.A., Droleskey, R.E., Norman, J.O., Mollenhauer, H.H., Ziprin, R.L., Corrier, D.E. and DeLoach, J.R. (1989a) Inhibition by mannose of *in vitro* colonization of chicken small intestine by *Salmonella typhimurium*. *Poultry Science* 68, 1351–1356.

Oyofo, B.A., DeLoach, J.R., Corrier, D.E., Norman, J.O., Ziprin, R.L. and Mollenhauer, H.H. (1989b) Prevention of *Salmonella typhimurium* colonization of broilers with D-mannose. *Poultry Science* 68, 1357–1360.

Oyofo, B.A., DeLoach, J.R., Corrier, D.E., Norman, J.O., Ziprin, R.L. and Mollenhauer, H.H. (1989c) Effect of carbohydrates on *Salmonella typhimurium* colonization in broiler chickens. *Avian Diseases* 33, 531–534.

Palmu, L. and Camelin, I. (1997) The use of competitive exclusion in broilers to reduce the level of *Salmonella* contamination on the farm and at the processing plant. *Poultry Science* 76, 1501–1505.

Pivnick, H. and Nurmi, E. (1982) The Nurmi concept and its role in the control of *Salmonella* in poultry. In: Davies, R. (ed.) *Developments in Food Microbiology – 1*. Applied Science Publishers, Barking, pp. 41–70.

Pivnick, H., Barnum, D., Stavric, S., Gleeson, T. and Blanchfield, B. (1985) Investigations on the use of competitive exclusion to control *Salmonella* in poultry. In: Barnum, D.A. (ed.) *Proceedings of International Symposium on* Salmonella *and Prospects for Control*. University of Guelph, Ontario, p. 263.

Raevuori, M., Seuna, E. and Nurmi, E. (1978) An epidemic of *Salmonella infantis* infection in Finnish broiler chickens in 1975. *Acta Veterinaria Scandinavica* 19, 317–330.

Raminez, G.A., Sarlin L.L., Caldwell, D.J., Yezak, C.R., Hume, D.E., Corrier, D.E., DeLoach, J.R. and Hargis, B.M. (1997) Effect of feed withdrawal on the incidence of *Salmonella* in the crops and ceca of market age broiler-chickens. *Poultry Science* 76, 654–656.

Rantala, M. and Nurmi, E. (1973) Prevention of the growth of *Salmonella infantis* in chicks by the flora of the alimentary tract of chickens. *British Poultry Science* 14, 627–630.

Ray, B. and Johnson, M.C. (1986) Freeze-drying injury of surface layer protein and its protection in *Lactobacillus acidophilus*. *Cryo-Letters* 7, 210–217.

Reid, G., Bruce, A.W., McGroarty, J.A., Cheng, K.-J. and Costerton, J.W. (1990) Is there a role for lactobacilli in prevention of urogenital and intestinal infections? *Clinical Microbiology Reviews* 3, 335–344.

Reiter, B. and Härnulv, G. (1984) Lactoperoxidase antibacterial system: natural occurrence, biological functions and practical applications. *Journal of Food Protection* 47, 724–732.

Reiter, B., Marshall, M.E., Björck, L. and Rosén, C.-G. (1976) Nonspecific bactericidal activity of the lactoperoxidase–thiocyanate–hydrogen peroxide system of milk against *Escherichia coli* and some Gram-negative pathogens. *Infection and Immunity* 13, 800–807.

Reiter, B., Fulford, R.J., Marshall, V.M., Yarrow, N. and Ducker, M.J. (1981) An evaluation of the growth promoting effect of the lactoperoxidase system in newborn calves. *Animal Production* 32, 297–306.

Reynolds, D.J., Davies, R.H., Richards, M. and Wray, C. (1997) Evaluation of combined antibiotic and competitive exclusion treatment in broiler breeder flocks infected with *Salmonella enterica* serovar Enteritidis. *Avian Pathology* 26, 83–95.

Rigby, C.E. and Pettit, J.R. (1980) Observations on competitive exclusion for preventing *Salmonella typhimurium* infection of broiler chickens. *Avian Diseases* 24, 604–615.

Rigby, C.E., Pettit, J. and Robertson, A. (1977) The effects of normal intestinal flora on the *Salmonella* carrier state in poultry with special reference to *S. thompson* and *S. typhimurium*. In: Barnum, D.A. (ed.) *Proceedings of the International Symposium on* Salmonella *and Prospects for Control*. University of Guelph, Ontario, p. 263.

Rolfe, R.D. (1991) Population dynamics of the intestinal tract. In: Blankenship, L.C. (ed.) *Colonization Control of Human Bacterial Enteropathogens in Poultry*. Academic Press, San Diego, pp. 59–75.

Salanitro, J.P., Fairchilds, I.G. and Zornicki, Y.D. (1974) Isolation, culture characteristics, and identification of anaerobic bacteria from the chicken caecum. *Applied Microbiology* 27, 678–687.

Salanitro, J.P., Muirhead, P.A. and Goodman, J.R. (1976) Morphological and physiological characteristics of *Gemmiger formicilis* isolates from chicken ceca. *Applied and Environmental Microbiology* 32, 623–632.

Savage, D.C. (1977) Microbial ecology of the gastrointestinal tract. *Annual Review of Microbiology* 31, 107–133.

Savage, D.C. (1983) Mechanisms by which indigenous microorganisms colonize gastrointestinal epithelial surfaces. *Progress in Food and Nutrition Science* 7, 65–74.

Schleifer, J.H. (1985) A review of the efficacy and mechanism of competitive exclusion for the control of *Salmonella* in poultry. *World's Poultry Science Journal* 41, 72–83.

Schneitz, C. (1992) Automated droplet application of a competitive exclusion preparation. *Poultry Science* 71, 2125–2128.

Schneitz, C. (1993) Development and evaluation of a competitive exclusion product for poultry. PhD thesis, Department of Veterinary Medicine, University of Helsinki, Helsinki, Finland.

Schneitz, C. and Nuotio, L. (1992) Efficacy of different microbial preparations for controlling salmonella colonization in chicks and turkey poults by competitive exclusion. *British Poultry Science* 33, 207–211.

Schneitz, C., Hakkinen, M., Nuotio, L., Nurmi, E. and Mead, G. (1990) Droplet application for protecting chicks against *Salmonella* colonization by competitive exclusion. *Veterinary Record* 126, 510.

Schneitz, C., Nuotio, L., Kiiskinen, T. and Nurmi, E. (1991) Pilot-scale testing of the competitive exclusion method in chickens. *British Poultry Science* 32, 877–880.

Schneitz, C., Nuotio, L. and Lounatmaa, K. (1993) Adhesion of *Lactobacillus acidophilus* to avian intestinal epithelial cells mediated by the crystalline bacterial cell surface layer (S-layer). *Journal of Applied Bacteriology* 74, 290–294.

Schneitz, C., Kiiskinen, T., Toivonen, V. and Näsi, M. (1998) Effect of Broilact® on the physicochemical conditions and nutrient digestibility in the gastrointestinal tract of broilers. *Poultry Science* 77, 426–432.

Seuna, E. (1979) Sensitivity of young chickens to *Salmonella typhimurium* var. *copenhagen* and *S. infantis* infection and the preventive effect of cultured intestinal microflora. *Avian Diseases* 23, 392–400.

Seuna, E., Raevuori, M. and Nurmi, E. (1978) An epizootic of *Salmonella typhimurium* var. *copenhagen* in broilers and the use of cultured chicken intestinal flora for its control. *British Poultry Science* 19, 309–314.

Sjögren, A., Wang, D.N., Hovmöller, S., Haapasalo, M., Ranra, H., Kerosuo, E., Jousimies-Somer, H. and Lounatmaa, K. (1988) The three-dimensional structures of S-layers of two novel *Eubacterium* species isolated from inflammatory human processes. *Molecular Microbiology* 2, 81–87.

Smith, H. and Green, S.I. (1980) Effect of feed additives on the incidence of naturally acquired *Salmonella* in turkeys. *Veterinary Record* 107, 289.

Smith, H.W. and Tucker, J.F. (1978) The effect of antimicrobial feed additives on the colonization of the alimentary tract of chickens by *Salmonella typhimurium*. *Journal of Hygiene, Cambridge* 80, 217–231.

Smith, H.W. and Tucker, J.F. (1980) Further observations on the effect of feeding diets containing avoparcin, bacitracin and sodium arsenilate on the colonization of the alimentary tract of poultry by *Salmonella* organisms. *Journal of Hygiene, Cambridge* 84, 137–150.

Snoeyenbos, G.H., Weinack, O.M. and Smyser, C.F. (1978) Protecting chicks and poults from salmonellae by oral administration of 'normal' gut microflora. *Avian Diseases* 22, 273–286.

Snoeyenbos, G.H., Weinack, O.M. and Smyser, C.F. (1979) Further studies on competitive exclusion for controlling salmonellas in chickens. *Avian Diseases* 24, 904–914.

Snoeyenbos, G.H., Weinack, O.M. and Soerjadi, A.S. (1983) Our current understanding of the role of native microflora in limiting some bacterial pathogens of chickens and turkeys. In: *Australian Veterinary Poultry Association and International Union of Immunological Societies, Proceedings No. 66, Disease Prevention and Control in Poultry Production*, Sydney, Australia. pp. 45–51.

Snoeyenbos, G.H., Weinack, O.M., Soerjadi-Liem, A.S., Miller, B.M., Woodward, D.E. and Weston, C.R. (1985) Large-scale trials to study competitive exclusion of *Salmonella* in chickens. *Avian Diseases* 29, 1004–1011.

Soerjadi, A.S., Lloyd, A.B. and Cumming, R.B. (1978) *Streptococcus faecalis*, a bacterial isolate which protects young chickens from enteric invasion by salmonellae. *Australian Veterinary Journal* 54, 549–550.

Soerjadi, A.S., Stehman, S.M., Snoeyenbos, G.H., Weinack, O.M. and Smyser, C.F. (1981a) Some measurements of protection against paratyphoid *Salmonella* and *Escherichia coli* by competitive exclusion in chickens. *Avian Diseases* 24, 706–712.

Soerjadi, A.S., Stehman, S.M., Snoeyenbos, G.H., Weinack, O.M. and Smyser, C.F. (1981b) The influence of lactobacilli on the competitive exclusion of paratyphoid salmonellae in chickens. *Avian Diseases* 25, 1027–1033.

Soerjadi, A.S., Snoeyenbos, G.H. and Weinack, O.M. (1982a) Intestinal colonization and competitive exclusion of *Campylobacter fetus* subsp. *jejuni* in young chicks. *Avian Diseases* 26, 520–524.

Soerjadi, A.S., Rufner, R., Snoeyenbos, G.H. and Weinack, O.M. (1982b) Adherence of salmonellae and native gut microflora to the gastrointestinal mucosa of chicks. *Avian Diseases* 26, 576–584.

Soerjadi-Liem, A.S., Snoeyenbos, G.H. and Weinack, O.M. (1984a) Comparative studies on competitive exclusion of three isolates of *Campylobacter fetus* subsp. *jejuni* in chickens by native gut microflora. *Avian Diseases* 28, 139–146.

Soerjadi-Liem, A.S., Snoeyenbos, G.H. and Weinack, O.M. (1984b) Establishment and competitive exclusion of *Yersinia enterocolitica* in the gut of monoxenic and holoxenic chicks. *Avian Diseases* 28, 256–260.

Spencer, J.L. and Garcia, M.M. (1995) Resistance of chicks and poults fed vermicompost to caecal colonization by *Salmonella*. *Avian Pathology* 24, 157–170.

Stavric, S. (1992) Defined cultures and prospects. *International Journal of Food Microbiology* 15, 245–263.

Stavric, S. and D'Aoust, J.-Y. (1993) Undefined and defined bacterial preparations for the competitive exclusion of *Salmonella* in poultry – a review. *Journal of Food Protection* 56, 173–180.

Stavric, S., Gleeson, T.M., Blanchfield, B. and Pivnick, H. (1985) Competitive exclusion of *Salmonella* from newly hatched chicks by mixtures of pure bacterial cultures isolated from fecal and cecal contents of adult birds. *Journal of Food Protection* 48, 778–782.

Stavric, S., Gleeson, T.M., Buchanan, B. and Blanchfield, B. (1992a) Experience of the use of probiotics for *Salmonellae* control in poultry. *Letters in Applied Microbiology* 14, 69–71.

Stavric, S., Buchanan, B. and Gleeson, T.M. (1992b) Competitive exclusion of *Escherichia coli* O157:H7 from chicks with anaerobic cultures of faecal microflora. *Letters in Applied Microbiology* 14, 191–193.

Stern, N.J. (1990) Influence of competitive exclusion on chicken cecal colonization by *Campylobacter jejuni*. *Poultry Science* 69 (Suppl. 1), 130.

Stern, N.J. (1994) Mucosal competitive exclusion to diminish colonization of chickens by *Campylobacter jejuni*. *Poultry Science* 73, 402–407.

Stern, N.J., Bailey, J.S., Cox, N.A. and Blankenship, L.C. (1995) Mucosal competitive exclusion flora. United States Patent 5,451,400.

Talarico, T.L., Casas, I.A., Chung, T.C. and Dobrogosz, W.J. (1988) Production and isolation of reuterin, a growth inhibitor produced by *Lactobacillus reuteri*. *Antimicrobial Agents and Chemotherapy* 32, 1854–1858.

Tanami, J. (1959) Studies on germfree animals. *Journal of Chiba Medical Society* 35, 1–24. Reprinted in: Luckey, T.D. (ed.) (1963) *Germ-free Life and Gnotobiology*. Academic Press, New York, p. 411.

van der Waaij, D., Berghuis-de Vries, J.M. and Lekkerkerk-van der Wees, J.E.C. (1971) Colonization resistance of the digestive tract in conventional and antibiotic-treated mice. *Journal of Hygiene* 69, 405–511.

Wadström, T., Andersson, K., Sydow, M., Axelsson, L., Lindgren, S. and Gullmar, B. (1987) Surface properties of lactobacilli isolated from the small intestine of pigs. *Journal of Applied Bacteriology* 62, 513–520.

Waldroup, A.L., Yamaguchi, W., Skinner, J.T. and Waldroup, P.W. (1992) Effects of dietary lactose on incidence and levels of salmonellae on carcasses of broiler chickens grown to market age. *Poultry Science* 71, 288–295.

Wegener, H.C., Madsen, M., Nielsen, N. and Aarestrup, F.M. (1997) Isolation of vancomycin resistant *Enterococcus faecium* from food. *International Journal of Food Microbiology* 35, 57–66.

Weinack, O.M., Snoeyenbos, G.H. and Smyser, C.F. (1979) A supplemental test system to measure competitive exclusion of salmonellae by native microflora in the chicken gut. *Avian Diseases* 23, 1019–1030.

Weinack, O.M., Snoeyenbos, G.H., Smyser, C.F. and Soerjadi, A.S. (1981) Competitive exclusion of intestinal colonization of *Escherichia coli* in chicks. *Avian Diseases* 25, 696–705.

Weinack, O.M., Snoeyenbos, G.H., Smyser, C.F. and Soerjadi, A.S. (1982) Reciprocal competitive exclusion of *Salmonella* and *Escherichia coli* by native intestinal microflora of the chicken and turkey. *Avian Diseases* 26, 585–595.

Weinack, O.M., Snoeyenbos, G.H., Smyser, C.F. and Soerjadi-Liem, A.S. (1984) Influence of *Mycoplasma gallisepticum*, infectious bronchitis, and cyclophosphamide on chickens protected by native intestinal microflora against *Salmonella typhimurium* or *Escherichia coli*. *Avian Diseases* 28, 416–425.

Weinack, O.M., Snoeyenbos, G.H., Soerjadi-Liem, A.S. and Smyser, C.F. (1985a) Influence of temperature, social and dietary stress on development and stability of protective microflora in chickens against *S. typhimurium*. *Avian Diseases* 29, 1177–1183.

Weinack, O.M., Snoeyenbos, G.H. and Soerjadi-Liem, A.S. (1985b) Further studies on competitive exclusion of *Salmonella typhimurium* by lactobacilli in chickens. *Avian Diseases* 29, 1273–1277.

WHO (1994) Unpublished report of the WHO-FEDESA-FEP workshop on competitive exclusion, vaccination and antimicrobials in *Salmonella* control in poultry. WHO/CDS/VPH/94. 134 WHO, Geneva.

Wierup, M., Wold-Troell, M., Nurmi, E. and Hakkinen, M. (1988) Epidemiological evaluation of the *Salmonella*-controlling effect of a nationwide use of a competitive exclusion culture in poultry. *Poultry Science* 67, 1026–1033.

Wierup, M., Wahlström, H. and Engström, B. (1992) Experience of a 10-year use of competitive exclusion treatment as part of the *Salmonella* control programme in Sweden. *International Journal of Food Microbiology* 5, 287–291.

Wray, C. and McLaren, I. (1987) A note on the effect of the lactoperoxidase systems on salmonellas *in vitro* and *in vivo*. *Journal of Applied Bacteriology* 62, 115–118.

Chapter 19

Vaccination against *Salmonella* Infections in Food Animals: Rationale, Theoretical Basis and Practical Application

Paul A. Barrow and Timothy S. Wallis

Institute for Animal Health, Compton Laboratory, Compton, Newbury, Berkshire RG20 7NN, UK

Introduction

Over the last decade, the animal and public-health problems associated with *Salmonella* in poultry have increased to such an extent that they have become major political issues, of which the general public has become very aware. *S. enteritidis*, in particular, has become a worldwide problem, arising mainly from poultry (Rodrigue *et al.*, 1990). In many countries, individual phage types of this serovar have replaced *S. typhimurium* as the most dominant type in poultry and humans. Control in poultry has become a major issue and immunity, whether acquired or, more speculatively, innate, is seen as one of the possible means of containing the problem. In the UK cattle industry, clinical salmonellosis in its acute and chronic forms, together with subclinical infection, remains a major economic, welfare and public-health problem. While the incidence of acute clinical salmonellosis in pigs is less of a problem compared with that of cattle, subclinical *Salmonella* infections that results in carcass contamination lead to the introduction of the organism into the food-chain. Widespread use of antibiotics has led to the emergence of multiple antibiotic-resistant bacteria, especially *S. typhimurium*. These problems have indicated to the industry and government agencies an increasing requirement for effective vaccines to control this important zoonotic infection.

This review discusses the reasons for the relatively poor success in immunizing food animals against those non-host-specific *Salmonella* serovars that usually produce food poisoning, compared with the success obtained with the small number of serovars that more typically produce systemic 'typhoid-like' diseases in a restricted range of host species. Most of our understanding of immunity to salmonellosis arises from experimental work with typhoid-like diseases, usually *S. typhimurium* infection in mice. Such work may not be entirely relevant to the, often disease-free, colonization by most *Salmonella* serovars. Whereas live, attenuated vaccines against host-specific serovars are highly protective, similarly developed vaccine strains have traditionally been less effective in protecting chickens, calves and pigs against intestinal colonization. Newer methods of attenuation are being developed, but their exploitation and success will depend on appropriate attenuation, delivery, their use for the types of infection that have been shown to be amenable to immune control and their effectiveness under field conditions. However, from the point of view of consumer safety there is a school of thought that considers inactivated or subunit vaccines to be the safest. The benefits of developing effective killed or subunit vaccines over the use of live vaccines are enormous. Recently, there have been significant advances in the development of adjuvants, for example microspheres, which are capable of potent immunostimulation, targeting different arms of the immune system (Morein *et al.*, 1996). The exploitation of such technology in conjunction with the ongoing developments in identifying key *Salmonella* virulence determinants should form the next generation of *Salmonella* subunit vaccines for the control of this important group of pathogens.

Pathogenesis of Salmonellosis

The development of an effective vaccine is dependent on an understanding of how *Salmonella* organisms infect their hosts and the host response to infection. A major problem with this goal is that *Salmonella* pathogenicity is both serovar- and host-dependent and that the factors influencing serovar–host specificity are not known (see Chapter 4).

The pathogenesis of *Salmonella* infections can be divided into two major groups. One group typically produces systemic disease and is rarely involved in human food poisoning, while the other typically produces food poisoning and only produces systemic disease under particular circumstances, such as during parturition, during lay, in very young or old animals or after some viral or parasite infections. A comparison of the biological aspects of the two groups might explain our success in immunological control of the former group and our limited success in the control of the latter. It might also contribute to our understanding of the immune responses to infection and it is obviously central to any appraisal of experimental work on pathogenesis and immunity that like is compared with like.

Our current understanding of *Salmonella* pathogenesis and immunity is largely derived from experimental infection studies, usually with *S. typhimurium*, in inbred laboratory strains of mice. This type of infection is characteristic of a small number of serovars that produce severe systemic disease, initially involving the reticuloendothelial system in a restricted number of host species. Following inoculation of mice with an infective dose, a systemic infection occurs without enteritis. Thus mice are only a useful model for studying the systemic form of disease. The severity of the infection is dependent on the virulence of the *Salmonella* strain, the route of infection, the innate resistance of the mouse strain used and its immune status. Similar information

on *S. dublin* in cattle or *S. gallinarum* in poultry is much less complete.

The second group comprises the remaining 2400 or more serovars. They are not restricted to particular host species and their epidemiology can therefore be complex. Most are able to colonize the alimentary tract of animals without production of disease.

Some strains of particular serovars, most notably *S. typhimurium* and *S. enteritidis*, are capable of producing clinical disease under certain circumstances. This is particularly the case for young animals or during or after a period of physiological stress, as mentioned above. In these cases, a systemic disease, again initially involving the reticuloendothelial system, occurs in addition to faecal excretion. The extent of disease and mortality varies according to the strain, but with serovars such as these there is always some invasion of the intestinal mucosa and reticuloendothelial system.

Current Understanding of Immunity to *Salmonella* Infection

In comparison with the increasingly detailed information becoming available on the cellular and humoral responses to *Salmonella* infection in mice, very little is known about how the host responds to the different types of infection seen in cattle, pigs and poultry (see Chapter 5). Recent reviews of avian immunology include those by Davison *et al.* (1996) and on the increasing information relating to the secretory immune system by Lillehoj and Chung (1991) and Schat (1994). Similar reviews on bovine immunology have also been compiled (Morrison, 1986).

As described previously, at least two types of infection need to be considered: (i) essentially an infection of the reticuloendothelial system, with little or no initial intestinal involvement; and (ii) an extensive intestinal infection by an organism with varying degrees of systemic dissemination. *S. gallinarum* infection in immunologically mature birds is likely to be essentially similar to *S. typhimurium* infection in mice. In many cases, *S. dublin* infection in adult cattle and *S. cholerae-suis* in adult pigs are likely to be similar. Full protection can be obtained at the level of the reticuloendothelial system in these infections. In the field the oral median lethal dose (LD_{50}) may be *c.* 10^4 colony-forming units (cfu), although the inoculum received may vary considerably. It may therefore be that, with large oral doses, challenge organisms may penetrate to the reticuloendothelial system, whereas smaller challenge doses are eliminated at the level of the gut. There is some experimental evidence for this (P.A. Barrow and B.A.D. Stocker, unpublished results).

Studies with typhoid-like infections in mice suggest that both humoral and cell-mediated immune (CMI) responses are involved in acquired resistance to *Salmonella* infection (Collins, 1971; Hochadel and Keller, 1977; Eisenstein and Sultzer, 1983; Hsu 1989; Michetti *et al.*, 1992; Guilloteau *et al.*, 1993). However, the majority of studies assessing *Salmonella* vaccine efficacy have used highly susceptible inbred mouse strains, in which both types of the immunity are simultaneously required for protection (Mastroeni *et al.*, 1993b). Live *Salmonella* vaccines are particularly effective in eliciting both humoral and CMI responses and therefore considerable efforts have been made in developing this class of vaccine.

Knowledge of the mechanisms whereby these two arms of the immune system operate and cooperate has resulted from work on *Leishmania* and other infections in mice. In all cases, T cells are involved. Thus, organisms (most notably viruses) inside non-professional phagocytic cells may release antigens, which can be presented at the cell surface by major histocompatibility complex (MHC) class I molecules. These are recognized by surface receptors on CD8+ T cells, which are thus stimulated to divide and to release chemicals, which kill the infected cell. Microorganisms inside macrophages may be destroyed or may release antigens, which can be degraded to peptides that are presented at the cell surface by MHC class II molecules. These are recognized in a specific way by CD4+ T (helper) cells, which are thus stimulated to release cytokines, particularly interferon-γ (IFNγ). These stimulate the macrophage to produce tumour necrosis factor alpha (TNF-α), nitric oxide and oxygen radicals, which enhance microbial killing. In these cases, the recognition event between presented antigen and T cell is supplemented by an essential signal indicating that the presenting cell is infected. This is thought to be mediated by the B7 mole-cule, which interacts with CD28 on the surface of the T cell.

The outcome of infection of mice with *S. typhimurium* depends on a number of factors, including the dose and route, the virulence of the strain and the host genetic background. When mice are inoculated orally with *S. typhimurium*, the number of organisms in the spleen gradually increases (Dunlap *et al.*, 1991). Many studies are carried out with parenteral, particularly intravenous, inoculation, so that bacteria reach the reticuloendothelial system almost immediately. The pattern of bacterial multiplication can be divided into four phases (Hormaeche, 1979).

1. A short phase of early inactivation taking place within 24 h of inoculation. This is potentiated by factors that activate macrophages and by antibody administration (Collins, 1974). Accumulation of polymorphonuclear leucocytes is also a major feature of early infection, although its significance has not yet been fully assessed (van Furth *et al.*, 1993).

2. A phase of exponential growth. This is unaffected by antibody, which may therefore delay but not suppress it (Collins, 1974; Maskell *et al.*, 1987). It is strongly influenced by bacterial virulence and by genetically defined resistance, particularly the Ity phenotype expressed in macrophages (Lissner *et al.*, 1985; Harrington and Hormaeche, 1986). If bacterial multiplication continues, the outcome is fatal, induced by the effects of endotoxin and possibly other unidentified toxins.

3. In sublethal infections, the growth phase reaches a plateau, characterized by an influx of bone-marrow-derived monocytes. These cells increase the rate of killing already carried out by Kupffer cells, and the recruitment of polymorphonuclear leucocytes may also play a role (Campbell, 1986). The effect is also TNF-α-dependent and this factor may be involved throughout the plateau phase (Mastroeni *et al.*, 1993a), although the mechanism is unknown.

4. Finally organisms are slowly cleared from the reticuloendothelial system at a variable rate, which is regulated by genes inside and outside the MHC complex (O'Brien *et al.*, 1981; Hormaeche *et al.*, 1985; Nauciel *et al.*, 1988).

Vaccination with a suitable live, attenuated vaccine can be construed as stimulating the immune system in the same way that occurs following a sublethal infection with a wild-type strain.

Many genetically defined attenuating mutations have been described (for review, see Curtiss *et al.*, 1991); the *aroA* (Hoiseth and Stocker, 1981) and the *cya crp* mutations (Curtiss and Kelly, 1987) have been most widely exploited. Following oral dosing of mice with such mutants of *S. typhimurium*, invasion of the intestinal epithelium and gut-associated lymphoid tissue occurs and the vaccine strains become disseminated to mesenteric lymph nodes, liver and spleen, from which they are cleared over the following 4–5 weeks. Both the natural route of infection and persistence *in vivo* are thought to be crucial in eliciting local and systemic humoral and CMI responses which can protect mice against up to 10,000 LD_{50} of the virulent parental strain. However, not all mutants that colonize, invade and persist induce protection in this typhoid model (O'Callaghan *et al.*, 1988).

It is generally accepted that live vaccines confer better protection than killed vaccines, probably because the former stimulate both cellular and humoral immunity, while the latter stimulate strong antibody production only (Collins, 1974). This is undoubtedly essential to protection against the typhoid-like nature of murine salmonellosis and probably related infections, such as fowl typhoid (Collins and Mackaness, 1968; Collins, 1974; Villareal *et al.*, 1992). Genetically resistant mice can be protected by vaccination with killed vaccines (Eisenstein *et al.*, 1984), whereas genetically susceptible mice, such as Balb/c, require vaccination with live vaccines (Killar and Eisenstein, 1985; Hormaeche *et al.*, 1991). However, not all attenuated vaccines are equally protective, although the reasons for this are unclear.

T cells are essential for the specific recall of immunity after the vaccine strain has been cleared from the reticuloendothelial system. T cell-depleted mice were unable to control a challenge infection, even though substantial levels of circulating antibodies were present (Mastroeni *et al.*, 1992). Depletion of CD4+ and CD8+ T cells removed protection and the former cell type may be more important than the latter, although this is not clear-cut (Mastroeni *et al.*, 1992, 1993b). Administration of anti-IFN-α or anti-TNF-α

antibodies also impaired cellular immunity, and granulomas and mononuclear infiltration were absent and necrosis was more extensive than in normally immunized mice.

Delayed-type hypersensitivity (DTH) is not an accurate indicator of the immune status of an animal (Hormaeche *et al.*, 1991; Killar and Eisenstein, 1984, 1986). One of the reasons for this is the confusion caused by Arthus reactivity and additional hypersensitivity reactions, possibly induced by the lipopolysaccharide (LPS) content of the crude preparations most frequently used (Mastroeni *et al.*, 1994). This leads to greater sensitivity to TNF-α (Matsuura and Galanos, 1990).

LPS is undoubtedly important in the specificity of protection in mice (Hormaeche *et al.*, 1991), although it is not the only bacterial factor involved. Thus *S. typhimurium* (04) does not protect fully against *S. enteritidis* (09). That other antigens are involved is also indicated by strong interleukin-2 (IL-2) production in cell-proliferation assays in response to whole-cell, lysed, bacterial preparations but not to LPS on its own.

Outer-membrane proteins (OMPs) have also been used as vaccines. Their use arose from the realization that earlier good results with ribosomal vaccines resulted from contamination with membrane proteins (Smith and Bigley, 1972; Johnson, 1993). Such proteins can be protective in mice (Udhayakumar and Muthukkaruppan, 1987) and also stimulate DTH (Hassan *et al.*, 1991), although, again, contamination with LPS cannot be ruled out.

Until studies are carried out using natural routes of infection and the relative contribution of protection at the intestinal level, in addition to protection in the reticuloendothelial system, is assessed, it will be impossible to assess fully the role of humoral and cellular immunity.

The relative roles of the CMI and humoral immunity in clearance of *S. typhimurium* from chickens are unclear. Lee *et al.* (1981, 1983) found that clearance correlated with CMI responses rather than humoral responses. It seems likely that CMI responses are responsible for tissue clearance, but how such responses could be responsible for intestinal clearance remains unclear. It would seem far more logical for secretory immunoglobulin A (IgA) to be a major factor (Slauch *et al.*, 1993). There is no correlation between colonization ability and tissue invasion *per se* and colonization does not necessarily involve epithelial association of any sort (Barrow *et al.*, 1988). However, in *S. typhimurium*, more virulent strains are eventually eliminated more quickly than less virulent strains, suggesting that immunity may be involved in intestinal clearance (Barrow *et al.*, 1988). Neither B nor T cells are involved in gut colonization (Corrier *et al.*, 1991). Following invasion, rapid heterophil responses are thought to be central to the relatively high resistance of chickens to *S. enteritidis*, which behaves in a similar way to *S. typhimurium*. Treatment with 5-fluorouracil reduced heterophil populations without affecting mononuclear cells, which correlated with a much greater susceptibility to parenteral infection and, to a lesser extent, to oral infection (Kogut *et al.*, 1993, 1994).

A virulent organism such as *Salmonella*, which infects the reticuloendothelial system, may have indirect effects on the ability of the immune system to respond to other infections. Immunosuppression is known to occur following viral infections, e.g. infectious bursal disease of chickens, and also after *S. typhimurium* infection in mice (Eisenstein *et al.*, 1988). It remains to be seen how far infection with an organism such as *S. typhimurium* would affect responses to other antigens. There is some evidence to this effect (Hassan and Curtiss, 1994), in the same way as has been reported in mice (Eisenstein *et al.*, 1984).

Vaccines for Poultry

Infection of poultry with its host-species-specific serovar *S. gallinarum–pullorum* induces a strong protective immunity against reinfection. Live, attenuated vaccines have been developed that can be administered parenterally and which are protective, because the disease is primarily systemic. These vaccines are used extensively. The 9R *S. gallinarum* vaccine (Smith, 1956) is a rough strain and therefore it does not stimulate the production of antibodies to the somatic antigen. This is advantageous, since the vaccine's use will not interfere with serological tests to detect natural infection. A second vaccine, 9S, which is a smooth strain, is more protective but slightly more virulent (Smith, 1956). The 9R vaccine has been assessed extensively (Gordon and Luke,

1959; Gordon *et al.*, 1959; Silva *et al.*, 1981), indicating its usefulness, although some residual virulence exists for highly susceptible breeds.

More recently, a number of other methods of attenuation for *S. gallinarum* have been used. *S. gallinarum* strain 9, cured of its virulence plasmid, is greatly attenuated for chickens of more than a few weeks old (Barrow *et al.*, 1987). Chickens that had received a parenteral inoculation with such a strain were very resistant to reinfection with the parent strain (Barrow, 1990). However, the resistance to oral challenge obtained by immunization with either this strain or a rough, phage-resistant derivative was not as great as that induced by the 9R strain, given either parenterally or orally (Barrow, 1990). An *aro*A mutant of strain 9, was generated by genetic deletion and, since the growth of these mutants requires *p*-aminobenzoic acid, which is not freely available in mammalian and avian tissues, such mutants are attenuated. The mutant was completely avirulent for 2-week-old chickens (increase in LD_{50} of 10^6). This strain was an effective vaccine, but, as with the virulence-plasmid-cured derivative, the degree of protection produced was not as great as that produced by 9R. However, it was suggested that such more attenuated, less protective vaccine strains might be used in more susceptible breeds, followed by a second vaccination with the 9R mutant (Griffin and Barrow, 1993).

Because of the paucity of information on colonization and immunity relating to the *Salmonella* serovars that are usually associated with food poisoning, the development of vaccines for use with poultry has, until recently, been almost exclusively empirical. Various types of non-living vaccines have been used experimentally and in the field. Killed vaccines may be inactivated with formalin, acetone, glutaraldehyde or heat treatment. *S. enteritidis* grown under conditions of iron restriction has been used to prepare an inactivated vaccine, which has been licensed in a number of European countries. The vaccine has been shown to reduce the level of flock infection (S.B. Houghton, Milton Keynes, personal communication, 1998). Otherwise, oil adjuvants, aluminium hydroxide and other immune-stimulating compounds have been used. They certainly generate an immune response, but they produce a poor or inconsistent protective effect. Some of this work is described below.

In several experiments, Truscott immunized chickens with heated bacterial sonicates, prepared from different combinations of serovars and incorporated into the feed (Truscott, 1981). This was followed some weeks later by oral challenge. The extent of protection in the different experiments, measured by isolation of the challenge strain from cloacal swabs, varied from very good to poor.

Other groups (Bisping *et al.*, 1971; Thain *et al.*, 1984) have found that heated, whole-cell bacterins have little effect on faecal shedding, either in vaccinated birds or when their newly hatched progeny were challenged. Some reductions in mortality have, however, been demonstrated (McCapes *et al.*, 1967; Truscott and Friars, 1972), but the biological significance of these findings in relation to faecal shedding and subsequent carcass contamination is unknown.

Some live, attenuated vaccines have been adapted for use with poultry to mimic natural infection, but whether they behave in the same way as the parent or field strains in stimulating immunity is not known.

A mutant of a mammalian *S. typhimurium* defective in the enzyme UDP-galactose epimerase (*galE*), which persists for only a short time *in vivo*, was originally developed in mice for use with humans (Tagliagbue *et al.*, 1986) and cattle (Wray *et al.*, 1977), but it has also been tested in poultry. Pritchard *et al.* (1978) vaccinated 1-day-old chickens with this strain and challenged 2 weeks later with an avian *S. typhimurium* strain. Small reductions were obtained in faecal excretion and in the number of isolations from the viscera and also when chickens were vaccinated twice, at 4 or 6 weeks of age, followed by challenge 2 weeks later. Surprisingly, protection was better when chickens were inoculated intramuscularly rather than orally (Subhabphant *et al.*, 1983). One drawback of the use of any *galE* mutant is the question of attenuation, because some mutants are known to retain some virulence for the host (Nnalue and Stocker, 1986; Hone *et al.*, 1988).

Others have used a murine streptomycin-dependent *S. typhimurium* strain (Schimmel *et al.*, 1974). After vaccination of very young chicks and challenge 8 days later, protection was poor and inconclusive. The value of such experiments, in which animals are challenged when still immature, so soon after vaccination and with extremely high challenge doses (10^{10} cfu), must be questioned.

Some vaccine strains are excreted in the faeces for longer periods than are others (Smith and Tucker, 1980; Barrow *et al.*, 1988). A general inverse correlation between virulence for chickens and the duration of faecal excretion can be found. More invasive strains may therefore stimulate a systemic as well as a local immune response, which should combine to clear such strains from the gut more quickly than might occur following colonization by less invasive strains. However, it might also follow, to our advantage, that, after immunization with an invasive strain, clearance of both invasive and less invasive challenge strains might occur to the same extent.

Oral inoculation of 4-day-old chickens with a virulent avian *S. typhimurium* strain, F98, resulted in faecal shedding for several weeks. When these birds were rechallenged orally after they had virtually ceased to shed the organism, the challenge strain was excreted in smaller numbers and for a shorter period of time than in a previously uninfected control group, demonstrating the induction of protective immunity controlling intestinal colonization and faecal shedding. Inoculation with large numbers of killed organisms in the feed or intramuscularly had virtually no protective effect. An *aroA* mutant was produced from this strain and in addition a rough strain was selected by its resistance to virulent bacteriophage. When vaccinated intramuscularly, both mutants produced initially good reductions in faecal excretion of the challenge strain, but this did not persist. By oral inoculation, the *aroA* mutant produced little protection, while the rough mutant was still very protective (Barrow *et al.*, 1990a). Oral ingestion of the mutants by human volunteers did not result in illness and the strains did not persist in their faeces for longer than 3 days.

An *aroA* mutant of *S. enteritidis* has been tested in chickens. This also produces a good initial reduction in faecal excretion of a challenge *S. enteritidis* strain, which, however, is of limited duration (Cooper *et al.*, 1990). However, when challenge was carried out by contact with orally infected birds, considerably better protection was obtained (Cooper *et al.*, 1992). Cross-protection against challenge with *S. typhimurium* was not obtained.

Vaccines developed for use against host-adapted serovars are of less value in reducing faecal excretion by non-host-specific serovars. The *S. dublin* 51 vaccine has little effect in reducing excretion of *S. typhimurium* in chickens, although it reduces systemic multiplication of the challenge strain (Knivett and Stevens, 1971; Knivett and Tucker, 1972). Such vaccines, however, might be profitably used to reduce systemic infection with, for example, *S. enteritidis*, which belongs to the same serogroup. An *aroA* strain of *S. enteritidis* phage type 4, also made rough to reduce the titres of anti-LPS antibodies produced, did not protect layers from challenge, whereas the *S. gallinarum* 9R vaccine did, although this latter vaccine strain was still present in the ovaries 1 month later at the time of challenge. Isolations of the challenge strain from the ovary were reduced to zero by vaccination with 9R (Barrow *et al.*, 1990b). There was also a reduction in the number of isolations from the liver and spleen. The effect on isolations from the gut was less clear. Both vaccines produced a reduction in the number of contaminated eggs, although this is probably confused by eggs being infected by both ovarian and cloacal routes.

Whether the effects described above would be sufficient to improve the public-health problem remains to be seen. However, it does show that some current problems can still be tackled empirically with some success.

Other attenuations have been used with varying success (Curtiss *et al.*, 1991) and auxotrophic mutants, including *pur* and *thy* mutations, have been produced, though some have not yet been tested thoroughly (O'Callaghan *et al.*, 1988; Sigwart *et al.*, 1989). However, a purine-requiring mutant is being used extensively in the field in Germany (Meyer *et al.*, 1992). Transposon and deletion mutants have also been produced which are defective in adenylate cyclase activity (*cya*) or cyclic AMP receptor protein (*crp*) and which have been tested for their protective value in chickens (Hassan and Curtiss, 1994). A double-deletion mutant (Δ*cya*Δ*crp*) was an effective vaccine when birds were challenged with homologous or heterologous serovars 2 weeks after the last vaccination. It was unclear whether the vaccine strain was still present at the time of challenge. Mutants defective in expression of OMPs have been found to be attenuated for mice (Dorman *et al.*,

1989), but they have not been assessed in birds. Administration of purified OMPs has, however, been shown to protect birds against *S. gallinarum* infection (Bouzouba *et al.*, 1987; Nagaraja *et al.*, 1991).

Like competitive exclusion, immunity is likely to have little protective effect against infections that are extant at the time of immunization. Following vaccination, a protective immunity takes several days to develop. This delay could be overcome by using a live strain, which, in newly hatched birds, shows the colonization-blocking effect, a form of competitive exclusion that occurs between closely related enteric bacteria (Barrow *et al.*, 1987; Berchieri and Barrow, 1990). This might mean that an appropriate live, attenuated strain could protect against infection acquired in the hatchery during the first few days of life, followed by the development of true immunity. Thus, it should also be active in immunologically naïve animals in the presence of maternal antibody.

Vaccines for Cattle

Several small studies show significant variation in the efficacy of vaccines based on dead *Salmonella* organisms. A heat-inactivated *S. dublin* vaccine administered intradermally was shown to be effective when the cattle were challenged by the intravenous route (Aitkin *et al.*, 1982). Whether this type of vaccine would protect cattle against oral challenge was not determined. Acid-hydrolysed *S. typhimurium* coated with alkali-hydrolysed LPS or LPS alone was ineffective in protecting cattle against challenge with this serovar (Anderson, 1991). Since calves may become infected with *Salmonella* within the first few days after birth (Jones *et al.*, 1983) and the peak of mortality occurs between 3 and 4 weeks of age, passive protection by vaccination of the dam is one approach that can be adopted. Immunization of the dam 7 and 2 weeks prior to parturition, using formalin-killed, log-phase *S. typhimurium* was shown to be safe and effective for preventing illness in the calf (Jones *et al.*, 1988a).

Live, attenuated vaccine strains have been assessed in cattle, with varying success rates. Many publications attest to the efficacy of live vaccines in experimental trials, but few are available commercially. In the former East Germany, however, the use of live auxotrophic mutants of S. *typhimurium* and S. *dublin* and a rough mutant of S. *choleraesuis* has had a profound effect on the incidence of salmonellosis in food-production animals (Meyer et al., 1992). However, the precise nature of the attenuating mutations in these particular vaccine strains is unclear and therefore questions arise as to the stability of such strains in the field, although reversion to virulence has not been detected after administration of over 1 million doses (Meyer et al., 1992). Clearly, live vaccines should ideally carry defined, multiple, attenuating mutations and specific genetic markers to distinguish them from wild-type strains. Three vaccine strains of S. *typhimurium* carrying defined mutations in the *aroA* gene were tested for safety and efficacy in 2- to 3-week-old calves. All three strains were attenuated, although only one strain, SL1479, was protective after either oral or subcutaneous administration (Smith et al., 1984a). Of considerable concern, however, is that, when the same strain was assessed in 5-day-old calves, a lethal infection resulted (Mikula et al., 1989). A live S. *dublin aroA* vaccine has also been assessed in calves. Calves suffered a transient pyrexia and mild diarrhoea following intramuscular vaccination. When challenged with either wild-type S. *dublin* or S. *typhimurium*, some enteritis was seen, although all animals survived, demonstrating that cross-protection between serovars in large animals is possible (Smith et al., 1984b), as had previously been reported with the live S. *dublin* vaccine that was then commercially available (Rankin et al., 1967). However, in both of these studies, animals were challenged shortly after immunization and therefore protection may have be mediated by a short-lived, non-specific immune stimulation. Such vaccines would nevertheless be useful if this mechanism protected animals when at their most susceptible. The possibility of reversion to virulence of the vaccine strains can be substantially reduced by the introduction of a second attenuating mutation in a different area of the chromosome. However, if overly attenuated, vaccine strains will become less immunogenic. Nevertheless, an S. *typhimurim aroA aroD* strain was shown to protect calves from challenge with virulent S. *typhimurium* (Jones et al., 1991).

Vaccines for Pigs

Live, attenuated vaccine strains have also been assessed in pigs. An *aroA* vaccine strain of S. *typhimurium* was shown to be attenuated and it significantly reduced faecal shedding of a virulent strain following challenge, though in this study the challenge was performed only 1 week after vaccination; thus the duration of protection is not clear (Lumsden et al., 1991). An S. *typhimurium cya crp* vaccine strain was also shown to be protective in pigs; however, this strain caused significant pyrexia for 4 days post-vaccination (Coe and Wood, 1992). In the USA, where there is a continuing problem with S. *choleraesuis*, significant progress has been made with a live, attenuated vaccine strain of S. *choleraesuis* (Roof and Doitchinoff, 1995). The strain was generated by repeated passage of a virulent strain through porcine neutrophils in vitro and was cured of the 50 kb virulence plasmid. Following oral vaccination, the vaccine was well tolerated and it provided significant protection for up to 20 weeks after vaccination. One slight concern with a vaccine of this type is the reacquisition by the vaccine strain of the virulence plasmid from other *Salmonella* that may be present within the herd. This could well result in reversion of the vaccine strain to a virulent phenotype. Surprisingly, this same vaccine strain has also been shown to be protective in cattle against challenge with S. *dublin* (Fox et al., 1997), though, again, in this study, protection was assessed 2 weeks after immunization and thus protection may have resulted from non-specific stimulation of the immune system after vaccination. Furthermore, the vaccine was not as effective in cattle as it was in pigs and induced a significant pyrexia. This latter observation is not altogether surprising, as plasmid-free strains of S. *dublin* and S. *typhimurium* can induce a fatal enteritis in calves (Jones et al., 1988b; Wallis et al., 1995). With the recent increases in frequency of isolation of multiple antibiotic-resistant strains of S. *typhimurium* definitive type (DT) 104 from pigs, the development of effective vaccines for this serovar is an urgent requirement.

Vaccines for Sheep

A number of vaccines have been developed to control S. *abortusovis* and S. *typhimurium* infection in sheep. An *S. abortusovis* strain RV6, which represents a non-dependent reverse mutant selected from a streptomycin-dependent strain, has been evaluated in thousands of sheep in different parts of France (Sanchis and Pardon, 1984; Pardon *et al.*, 1990). A metabolic-drift mutant of *S. typhimurium* has been produced by Linde *et al.* (1992) and shown to protect against *S. abortusovis* infection.

GalE and AroA mutants of *S. typhimurium* have been shown to protect sheep against experimental *Salmonella* infections. Although *S. abortusovis* infection is a serious problem in some countries, most sheep are kept under free-range conditions and *Salmonella* infection is only a problem when sheep are stressed, e.g. under adverse weather conditions or when transported. Following an outbreak of salmonellosis, sheep are generally more resistant the following year and this should be borne in mind when evaluating vaccines.

Problems Associated with Killed Vaccines

Vaccines based on dead *Salmonella* bacteria have long been known to protect experimental animals and those in the field against salmonellosis, but, as indicated earlier, there are many contradictions in the scientific literature relating to their actual efficacy. In part, this is due to the huge variety of specific formulations, adjuvants, methods of bacterial inactivation and animal models, which make the different studies very difficult to compare. Parenteral administration of dead vaccines generally results in a potent serum antibody response, though often in the absence of secretory antibody and CMI responses. Widespread laboratory studies with genetically susceptible mouse strains have demonstrated that, in such animals, dead vaccines are not very effective. However, several studies, frequently using outbred mice or genetically more resistant inbred mouse strains, have shown killed vaccines to be highly effective (Lindberg *et al.*, 1974; Kuusi *et al.*, 1981; Eisenstein *et al.*, 1984). Furthermore, a recent review was carried out comparing the efficacy of different types of *S. typhi* vaccine in over 1.8 million humans. Whole-cell, killed vaccines were more effective than a live, attenuated vaccine and a vaccine based on Vi capsular polysaccharide, although the killed vaccine was associated with more adverse side-effects

(Engels *et al.*, 1998). Obviously, typhoid fever in humans represents a different form of infection from acute enteric salmonellosis in cattle and sub-clinical intestinal carriage in pigs, but killed vaccines should clearly not be dismissed on the basis of experimental laboratory studies using highly susceptible mouse strains.

There are three major potential problems with dead *Salmonella* vaccines. First, they only contain antigens induced by the environmental conditions *in vitro* in which they were grown, although this may be partially overcome by simulated *in vivo* conditions *in vitro* by, for example, growing organisms under iron or other nutrient restriction. Secondly, they fail to elicit CMI immune responses, which are considered important for long-term protection. This too may be overcome with the use of an appropriate adjuvant. Finally, they generally fail to elicit secretory IgA responses, which are potentially important in protecting mucosal surfaces. However, some dead vaccines have been shown to induce immune responses at mucosal surfaces. Baljer *et al.* (1986) protected calves by oral vaccination with repeated doses of heat-inactivated and disrupted cells, and found that anti-O antibody was produced in intestinal secretions. In the UK, at present only inactivated vaccines are licensed and therefore widely available for cattle, and these are claimed to be effective in combating outbreaks of disease in cattle. However, information relating to their actual efficacy is scarce.

Residual Virulence of Live Vaccines: Addressing the Problem and Criteria for an Effective Vaccine

A common problem with the use of live vaccine strains in calves is the mild scouring and associated shedding of the vaccine strain that often follows oral vaccination with an effective dose. This is likely to be a reason for concern over their use in all food-production animals. Not only is residual virulence for the host animal undesirable, but effects on the human population are possible, since the vaccine strains are likely to enter the human food-chain. Recent developments in our understanding of the virulence factors mediating *Salmonella*-induced enteritis are of potential value in further modifying the pre-existing live vaccine strains, in order to reduce these adverse side-effects. A virulence locus on the *Salmonella* chromosome, named *Salmonella* pathogenicity island 1 (SPI-1), encodes a protein secretory apparatus that translocates proteins into target eukaryotic cells (for review, see Galan, 1996). Mutation of genes in SPI-1 significantly reduces intestinal invasion of *S. typhimurium* and *S. dublin* in cattle (Watson *et al.*, 1995; Galyov *et al.*, 1997) and significantly attenuated *S. typhimurium* in cattle, although mild pyrexia and scours still occurred (Watson *et al.*, 1998). Further studies in our laboratory have identified an effector protein, named SopB, which is secreted and translocated by SPI-1 and is directly involved in the induction of enteritis. SopB has inositol phosphate phosphatase activity (Norris *et al.*, 1998), which influences intracellular signalling and thus electrolyte transport in eukaryotic cells. Disruption of the expression of SopB significantly reduces the intestinal fluid secretion and inflammation induced by *S. dublin*, without influencing the invasive phenotype (Galyov *et al.*, 1997). Another secreted protein, SopD, appears to act in concert with SopB in enteropathogenesis (Jones *et al.*, 1998), although the precise function of SopD is not yet known. Neither of these proteins is involved in mediating intestinal invasion and thus such mutants should be fully immunogenic. *SopB* maps to another pathogenicity island, called SPI-5, which encodes for four other proteins, all of which influence enteropathogenesis in cattle but not systemic pathogenesis in mice (Wood *et al.*, 1998). These observations suggest that SPI-5 is a cluster of genes acquired relatively recently in the evolutionary history of *Salmonella* which bestow the enteropathogenic phenotype. The full potential of these new developments for vaccine design have not yet been investigated. However, the use of mutations that specifically reduce enteritis, in conjunction with mutations that specifically prevent the systemic form of disease, has obvious potential benefits over the existing live, attenuated vaccine strains. What is surprising is that these major virulence determinants are conserved between different *Salmonella* serovars and antibodies to them can be found in convalescent sera (Galyov *et al.*, 1997). However, the protective immune response operating in the gut, elicited in vaccinated and convalescent animals, is very serovar-specific (Villarreal *et al.*, 1997). This may be a consequence of these effector proteins

being directly injected into cells by *Salmonella* and therefore avoiding immune defence mechanisms acting at the mucosal surface.

There are several additional problems with the current use of live vaccines. First, there is a problem with the formulation. Live vaccines have to be prepared in the lyophilized form and therefore require storage at 4°C and appropriate rehydration on the farm before administration to animals. Second, and most important, is the acceptability to the public of the use of genetically manipulated organisms in food-producing animals. Finally, there is the time required for development and registration of new live vaccines for use in the field. With the emergence and decline of different serovars, as typified by *S. enteritidis* phage type 4, by the time the vaccine is licensed, the *Salmonella* problem for which the vaccine was developed may no longer exist. The use of dead or subunit vaccines overcomes these particular problems, in that they are stable, are relatively quick to produce and do not contaminate the environment with genetically modified microorganisms.

It should thus eventually be possible to delete from the *Salmonella* chromosome those genes responsible for fluid secretion (enterotoxigenicity), thereby producing a mutant that should behave like the parent strain in the alimentary tract of poultry but which would be avirulent for humans. The criteria in selecting an ideal vaccine have been discussed previously (Pritchard *et al.*, 1978; Barrow, 1991). These are outlined below.

Strong protection against intestinal and systemic infection is required. An additional requirement is stable avirulence for humans (see discussion above). In view of the current increased public awareness of *Salmonella* food poisoning, it is unclear whether the greatest fear will be of salmonellosis, thereby increasing the chance of acceptance of the use of live vaccines, or of the use of a live vaccine itself. There may always be some resistance to the use of a vaccine that is *Salmonella*-derived, unless it can be shown that the vaccine is no more virulent for humans than other microorganisms, which are also present on the carcass at slaughter. The virulence of some of these candidate vaccines for humans has been assessed (Barrow *et al.*, 1990b), whereas the virulence of the *aroA S. enteritidis* strains for humans has not yet been studied. Ideally, the vaccine strains should not enter the human food-chain and, for this and reasons associated with the release of genetically modified organisms, the strain should not contaminate the environment.

The ideal route of administration to poultry would be orally via the drinking-water or food or by spray. For cattle and pigs, this should be achieved with a single dose. However, parenteral administration may be an additional requirement for maximum protection. Although the ideal vaccine should be avirulent for chickens, oral vaccination may require the use of an invasive strain to stimulate maximum immunity, because immunogenicity may be correlated with invasiveness. It seems likely that a strong secretory IgA response will be required for this. Whichever route of inoculation is used, some residual virulence may result in vertical transmission, as occurs occasionally with the *S. gallinarum* 9R vaccine. This should present few problems, providing that the vaccine produces no disease in the progeny. More importantly, the vaccine must not affect animal productivity.

Protection should obviously last as long as possible. Protection of broilers is required for a matter of weeks, although *Salmonella* control in broilers is not required under European Union (EU) legislation. However, control in breeders is an integral part of the European control programme, and here protection is needed for many months. Protection against *S. gallinarum* induced either by the 9R vaccine (Smith, 1956) or by an *aroA* vaccine (P.A. Barrow and B.A.D. Stocker, unpublished results) lasts for between 3 and 6 months. Protection is therefore not likely to be any longer than this. How long protection against faecal excretion will last is completely unknown.

It seems likely that, when the vaccine strain has been completely eliminated, little cross-protection against other serovars occurs. With the small number of major serovars involved, this should not be a major problem. Whether it will be possible in the future to vaccinate against different serovars with a single vaccine remains to be seen.

Legislation in many countries requires that all isolations of *Salmonella* from poultry must be reported. Thus wild-type strains must be easily differentiated from vaccine strains. This can be achieved positively or negatively by antigenic markers or their absence, by auxotrophy or by molecular differentiation, many of which techniques would be beyond the competence of many laboratories. The use of antibiotic resistance is unlikely to be adopted, because it is so widespread, but resistance to heavy metals might be used. If monitoring is carried out serologically, the vaccine must not express the antigens responsible for stimulating antibodies detected in the test used.

Vaccination against salmonellosis can produce a degree of protection over and above that possessed by unimmunized chickens with a fully mature intestinal flora. Vaccination should therefore be compatible with the use of competitive exclusion (early application or replacement of gut flora). No problems should arise that are similar to those resulting from overuse of antibiotics. Vaccination should also be compatible with the use of growth-promoting antibiotics.

Cross-protection Between Serovars

It is known that different serovars of *Salmonella*, either as virulent parent strains or when attenuated, induce different degrees of protection in the mouse typhoid model (Collins, 1974; Hormaeche et al., 1991). The role of O antigen in eliciting protective responses is unclear. Hybrid vaccine strains expressing O antigen common to both *S. typhimurium* and *S. dublin* induced protection against both serovars in mice (Lindberg et al., 1993), but not in cattle (Segall and Lindberg, 1993), suggesting that O antigen is important in conferring protection in systemic salmonellosis in mice, but not in enteritis in larger animals. In contrast, Hormaeche et al. (1991) found no key role for O antigen in protection in mice, and proposed a role for other undefined antigens. Studies in poultry have been confusing. Cooper and his group (1993) have suggested that there is no cross-protection between *S. typhimurium* and *S. enteritidis*. However, more recently, Hassan and Curtiss (1994) indicated a considerable degree of cross-protection between a variety of serovars. It is assumed, however, that cross-protection between serovars does not generally occur and that protection lasts no longer than 6 months.

The problem of protecting animal herds against a broad spectrum of different *Salmonella* serovars remains unsolved. Whether or not it will ever be achievable is uncertain. No study to date has demonstrated good cross-protection between different *Salmonella* serovars for any significant duration after vaccination. We performed a study assessing cross-protection in calves immunized with a live *S. typhimurium aroA* or *S. dublin aroA* and challenged with the homologous serovar to check (and boost) the protective immune response. When wild-type *S. typhimurium* and *S. dublin* strains were used to infect ligated ileal loops constructed in these immune animals 3 weeks after challenge with the virulent organism, clear serovar-specific protection was found. In animals vaccinated with *S. typhimurium*, only *S. dublin* strains invaded intestinal mucosa and elicited enteropathogenic responses and vice versa in *S. dublin* immunized cattle (Wallis *et al.*, 1997), despite these two serovars sharing common enteropathogenic virulence mechanisms

(Wood *et al.*, 1998). This being the case, it is difficult to see how effective cross-protection between different serovars can be achieved in the future. One approach is to include different serovars in the vaccine preparation, be it live or dead. However, preliminary observations with dual, live *S. typhimurium aroA* and *S. enteritidis aroA* vaccination regimens in poultry suggest that the vaccine strains interfere with one another, reducing their efficacy compared with a monovalent regimen (M. Woodward, personal communication). It would appear that, at present, the only really effective means of controlling outbreaks of uncommon serovars is through the use of herd-specific killed vaccines, which have been shown to be protective in acute outbreaks (Bauer, 1986) and very effective in reducing faecal shedding (Aitkin *et al.*, 1982; Weber, 1993).

References

Aitken, M.M., Jones, P.W. and Brown, G.T. (1982) Protection of cattle against experimentally induced salmonellosis by intradermal injection of heat-killed *Salmonella dublin*. *Research in Veterinary Science* 32, 368–373.

Anderson, J., Smith, B.P. and Ulrich, J.T. (1991) Vaccination of calves with a modified bacterin or oil-in-water emulsion containing alkali-detoxified *Salmonella typhimurium* lipopolysaccharide. *American Journal of Veterinary Research* 52, 596–601.

Baljer, G., Hoerstke, M., Dirksen, G., Sailer, J. and Mayr, A. (1986) Efficacy of local and/or parenteral vaccination against calf salmonellosis with inactivated vaccines. *Journal of Veterinary Medicine* 33, 206–212.

Barrow, P.A. (1990) Immunity to experimental fowl typhoid in chickens induced by a virulence plasmid-cured derivative of *Salmonella gallinarum*. *Infection and Immunity* 58, 2283–2288.

Barrow, P.A. (1991) Immunological control of *Salmonella* in poultry. In: Blankenship, L.C. (ed.) *Colonization Control of Human Bacterial Enteropathogens in Poultry*. Academic Press, pp.199–217.

Barrow, P.A., Simpson, J.M., Lovell, M.A. and Binns, M.M. (1987) Contribution of *Salmonella gallinarum* large plasmid towards virulence in fowl typhoid. *Infection and Immunity* 35, 388–392.

Barrow, P.A., Simpson, J.M. and Lovell, M.A. (1988) Intestinal colonization in the chicken by food-poisoning *Salmonella* serovars: microbial characteristics associated with faecal excretion. *Avian Pathology* 17, 571–588.

Barrow, P.A., Hassan, J.O. and Berchieri, A. (1990a) Reduction in faecal excretion of *Salmonella typhimurium* strain F98 in chickens vaccinated with live and killed *S. typhimurium* organisms. *Epidemiology and Infection* 104, 413–426.

Barrow, P.A., Hassan, J.O. and Berchieri, A. (1990b) Immunisation of laying hens against *Salmonella enteritidis* phage type 4 with live, attenuated vaccines. *Veterinary Record* 126, 241–242.

Bauer, K. (1986) Experiences with local administration of herd-specific vaccines. *Tierazẗliche Praxis* 14, 51–54.

Berchieri, A. and Barrow, P.A. (1990) Further studies on the inhibition of colonization of the chicken alimentary tract with *Salmonella typhimurium* by pre-colonization with an avirulent mutant. *Epidemiology and Infection* 104, 427–441.

Bisping, W., Dimitriadis, I. and Seippel, M. (1971) Versuche zur oralen Immunisierung von Huhnen mit hitzeinaktivierter *Salmonella*-Vakzine 1. Mitteilung; Impf-und Infektionsversuche an Huhnerkuken. *Zentralblatt für Veterinärmedizin* B 18, 337–346.

Bouzouba, K., Nagaraja, K.V., Newman, J.A. and Pomeroy, B.S. (1987) Use of membrane proteins from *Salmonella gallinarum* for prevention of fowl typhoid infection in chickens. *Avian Diseases* 31, 699–674.

Campbell, P.A. (1986) Are inflammatory phagocytes responsible for resistance to intracellular bacteria? *Immunology Today* 7, 70–72.

Coe, N.E. and Wood, R.L. (1992) The effect of exposure to a Δ*cya* Δ*crp* mutant *of Salmonella typhimurium* on the subsequent colonization of swine by the wild-type parent strain. *Veterinary Microbiology* 31, 207–220.

Collins, F.M. (1971) Mechanisms of antimicrobial immunity. *Journal of the Reticuloendothelial Society* 10, 59–99.

Collins, F.M. (1974) Vaccines and cell-mediated immunity. *Bacteriological Reviews* 38, 371–402.

Collins, F.M. and Mackaness, G.B. (1968) Delayed hypersensitivity and Arthus reactivity in relation to host resistance in *Salmonella*-infected mice. *Journal of Immunology* 101, 830–845.

Cooper, G.L., Nicholas, R.A.J., Cullen, G.A. and Hormaeche, C.E. (1990) Vaccination of chickens with a *Salmonella enteritidis aroA* live oral *Salmonella* vaccine. *Microbial Pathogenesis* 9, 255–265.

Cooper, G.L., Venables, L.M., Nicholas, R.A.J. and Cullen, G.A. (1993) Further studies of the application of live *Salmonella* enteritidis aroA vaccines in chickens. *Veterinary Record* 133, 31–36.

Cooper, G.L., Venables, L.M., Nicholas, R.A.J., Cullen, G.A. and Hormaeche, C.E. (1992) Vaccination of chickens with chicken-derived *Salmonella enteritidis* phage type 4 *aroA* live oral *Salmonella* vaccines. *Vaccine* 10, 247–254.

Corrier, D.E., Elissalde, M.H., Ziprin, R.L. and Deloach, J.R. (1991) Effect of immunosuppression with cyclophosphamide, cyclosporin or dexamethasone on *Salmonella* colonization of broiler chicks. *Avian Diseases* 35, 40–45.

Curtiss, R., III and Kelly, S.M. (1987) *Salmonella typhimurium* deletion mutants lacking adenylate cyclase and cyclic AMP receptor protein are avirulent and immunogenic. *Infection and Immunity* 55, 3035–3043.

Curtiss, R., Porter, S.B., Munson, M., Tinge, S.A., Hassan, J.O., Gentry-Weeks, C. and Kelly, S.M. (1991) Non-recombinant and recombinant *Salmonella* live vaccines for poultry. In: Blankenship, L.C. (ed.) *Colonization Control of Human Bacterial Enteropathogens in Poultry*. Academic Press, pp. 169–198.

Davison, T.F., Morris, T.R. and Payne, L.N. (eds) (1996) *Poultry Immunology*. Poultry Science Symposium Series Vol. 24, Carfax Publishing, Abingdon, UK.

Dorman, C.J., Chatfield, S., Higgins, C.F., Hayward, C. and Dougan, G. (1989) Characterization of porin and *ompR* mutants of a virulent strain of *Salmonella typhimurium*: *ompR* mutants are attenuated *in vivo*. *Infection and Immunity* 57, 2136–2140.

Dunlap, N.E., Benjamin, W.N., McCall, R.D., Tilden, A.B. and Briles, D.E. (1991) A safe site for *Salmonella typhimurium* is within splenic cells during the early phase of infection in mice. *Microbial Pathogenesis* 10, 297–310.

Eisenstein, T.K. and Sultzer, B.M. (1983). Immunity to *Salmonella* infection. *Advances in Experimental Medicine and Biology* 162, 261–296.

Eisenstein, T.K., Killar, L.M. and Sultzer, B.M. (1984) Immunity to infection with *Salmonella typhimurium*: mouse-strain differences in vaccine- and serum-mediated protection. *Journal of Infectious Diseases* 150, 425–435.

Eisenstein, T.K., Dalal, N., Killar, L., Lee, J. and Schafer, R. (1988) Paradoxes of immunity and immunosuppression in *Salmonella* infection. *Advances in Experimental Biology and Medicine* 239, 353–366.

Engels, E.A., Falagas, M.E., Lau, J. and Bennish, M.L. (1998) Typhoid fever vaccines: a meta-analysis of studies on efficacy and toxicity. *British Medical Journal* 316, 110–116.

Fox, B.C., Roof, M.B., Carter, D.P., Kesl, L.D. and Roth, J.A. (1997) Safety and efficacy of an avirulent live *Salmonella choleraesuis* vaccine for protection of calves against *S. dublin* infection. *American Journal of Veterinary Research* 58, 265–271.

Galán, J.E. (1996) Molecular genetic bases of *Salmonella* entry into host cells. *Molecular Microbiology* 20, 263–271.

Galyov, E.E., Wood, M.W., Rosqvist, R., Mullen, P.B., Watson, P.R., Hedges, S. and Wallis, T.S. (1997) A secreted effector protein of *Salmonella dublin* is translocated into eukaryotic cells and mediates inflammation and fluid secretion in infected ileal mucosa. *Molecular Microbiology* 25, 903–912.

Gordon, R.F. and Luke, D. (1959) A note on the use of the 9R fowl typhoid vaccine in poultry breeding flocks. *Veterinary Record* 71, 926–927.

Gordon, R.F., Garside, J.S. and Tucker, J.F. (1959) The use of living attenuated vaccines in the control of fowl typhoid. *Veterinary Record* 71, 300–305.

Griffin, H.G. and Barrow, P.A. (1993) Construction of an *aroA* mutant of *Salmonella* serovar Gallinarum: its effectiveness against experimental fowl typhoid. *Vaccine* 11, 457–462.

Guilloteau, L., Buzoni-Gatel., D., Bernard, F., Lantier, I. and Lantier, F. (1993) *Salmonella abortusovis* infection in susceptible balb/c mice – importance of Lyt-2+ and L3T4+ T-cells in acquired immunity and granuloma formation. *Microbial Pathogenesis* 14, 45–55.

Harrington, K.A. and Hormaeche, C.E. (1986) Expression of the innate resistance gene *Ity* in mouse Kupffer cells infected with *Salmonella typhimurium in vitro*. *Microbial Pathogenesis* 1, 269–274.

Hassan, J.O. and Curtiss, R. (1994) Development and evaluation of an experimental vaccination program using a live–avirulent *Salmonella typhimurium* strain to protect immunized chickens against challenge with homologous and heterologous *Salmonella* serovars. *Infection and Immunity* 62, 5519–5527.

Hassan, J.O., Mockett, A.P.A., Catty, D. and Barrow, P.A. (1991) Infection and reinfection of chickens with *Salmonella typhimurium*: bacteriology and immune responses. *Avian Diseases* 35, 809–819.

Hochadel, J.F. and Keller, K.F. (1977) Protective effects of passively transferred immune T- or B-lymphocytes in mice infected with *Salmonella typhimurium*. *Journal of Infectious Diseases* 135, 813–823.

Hone, D.M., Attridge, S.R., Forrest, B., Morona, R., Daniels, D., La Brooy, J.L., Bartholomeusz, R.C.A., Shearman, D.J.C. and Hackett, J. (1988) A *galE Via* (Vi-antigen negative) mutant of *Salmonella typhi* Ty2 retains virulence in humans. *Infection and Immunity* 56, 1326–1333.

Hormaeche, C.E. (1979) Natural resistance to *Salmonella typhimurium* in different inbred mouse strains. *Immunology* 37, 311–318.

Hormaeche, C.E., Harrington, K.A. and Joysey, H.S. (1985) Natural resistance to Salmonellae in mice: control by genes within the major histocompatibility complex. *Journal of Infectious Diseases* 5, 1050–1056.

Hormaeche, C.E., Joysey, H.S. DeSilva, L., Izhar, M. and Stocker, B.A.D. (1991) Immunity conferred by aro⁻ *Salmonella* live vaccines. *Microbial Pathogenesis* 10, 149–158.

Hosieth, S.F. and Stocker, B.A.D. (1981) Aromatic-dependent *Salmonella typhimurium* are non-virulent and effective as live vaccines. *Nature* 291, 238–239.

Hsu, H.S. (1989) Pathogens and immunity in murine salmonellosis. *Microbiological Reviews* 53, 390–409.

Johnson, W. (1993) Ribosomal vaccines. II. Specificity of the immune response to ribosomal ribonucleic acid and protein isolated from *Salmonella typhimurium*. *Infection and Immunity* 8, 395–400.

Jones, M.A., Wood, M.W., Mullen, P.B. Watson, P.R., Wallis, T.S. and Galyov, E.E. (1998) Secreted effector proteins of *Salmonella dublin* act in concert to induce enteritis. *Infection and Immunity* 66, 5799–5804.

Jones, P.W., Collins, P., Brown, G.T. and Aitken, M.M. (1983) *Salmonella saint-paul* infection in two dairy herds. *Journal of Hygiene, Cambridge* 91, 243–257.

Jones, P.W., Collins, P. and Aitken, M.M. (1988a) Passive protection of calves against experimental infection with *Salmonella typhimurium*. *Veterinary Record* 123, 536–541.

Jones, P.W., Collins, P. and Lax, A. (1988b) The role of large plasmids in the pathogenesis of salmonellosis in cattle [abstract]. *Journal of Medical Microbiology* 27, x.

Jones, P.W., Dougan, G., Hayward, C., Mackenzie, N., Collins, P. and Chatfield, S.N. (1991) Oral vaccination of calves against experimental salmonellosis using a double *aro* mutant of *Salmonella typhimurium*. *Vaccine* 9, 29–34.

Killar, L.M. and Eisenstein, T.K. (1984) Differences in delayed-type hypersensitivity responses in various mouse strains in the C3H lineage infected with *Salmonella typhimurium* strain SL3235. *Journal of Immunology* 133, 1190–1196.

Killar, L.M. and Eisenstein, T.K. (1986) Delayed-type hypersensitivity and immunity to *Salmonella typhimurium*. *Infection and Immunity* 52, 503–508.

Knivett, V.A. and Stevens, W.K. (1971) The evaluation of a live *Salmonella* vaccine in mice and chickens. *Journal of Hygiene, Cambridge* 69, 233–245.

Knivett, V.A. and Tucker, J.F. (1972) Comparison of oral vaccination or furazolidone prophylaxis for *Salmonella typhimurium* infection in chicks. *British Veterinary Journal* 128, 24–34.

Kogut, M.H., Tellez, G., Hargis, B.M., Corrier, D.E. and Deloach, J.R. (1993) The effect of 5-fluorouracil treatment of chicks: a cell depletion model for the study of avian polymorphonuclear leucocytes and natural host defences. *Poultry Science* 72, 1873–1880.

Kogut, M.H., Tellez, G.I., McGruder, E.D., Hargis, B.M., Williams, J.D., Corrier, D.E. and Deloach, J.R. (1994) Heterophils are decisive components in the early responses of chickens to *Salmonella enteritidis* infections. *Microbial Pathogenesis* 16, 141–151.

Kuusi, N., Nurminen, N., Saxen, H. and Makela, P.H. (1981) Immunization with outer membrane protein (Porin) preparations in experimental murine salmonellosis: effect of lipopolysaccharide. *Infection and Immunity* 34, 328–332.

Lee, G.M., Jackson, G.D.F. and Cooper, G.N. (1981) The role of serum and biliary antibodies and cell-mediated immunity in the clearance of *S. typhimurium* from chickens. *Veterinary Immunology and Immunopathology* 2, 233–252.

Lee, G.M., Jackson, G.D.F. and Cooper, G.N. (1983) Infection and immune responses in chickens exposed to *Salmonella typhimurium*. *Avian Diseases* 27, 577–583.

Lillehoj, H.S. and Chung, K.S. (1991) Intestinal immunity and genetic factors influencing colonization of microbes in the gut. In: Blankenship, L.C. (ed.) *Colonization Control of Human Bacterial Enteropathogens in Poultry*. Academic Press, pp. 219–241.

Lindberg, A.A., Rosenberg, L.T., Ljunggren, A., Ganegg, P.J., Svensson, S. and Wallin, N.H. (1974) Effect of synthetic disaccharide–protein conjugate as immunogen in *Salmonella* infection in mice. *Infection and Immunity* 10, 541–545.

Lindberg, A.A., Segall, T., Weintraub, A. and Stocker, B.A.D. (1993) Antibody response and protection against challenge in mice vaccinated intraperitoneally with a live *aroA* 04-09 hybrid *Salmonella dublin* strain. *Infection and Immunity* 61, 1211–1221.

Linde, K., Bondarenko, V. and Sviridenko, V. (1992) Prophylaxis of *Salmonella abortus ovis*-induced abortion of sheep by a *Salmonella typhimurium* live vaccine. *Vaccine* 10, 337–340.

Lissner, C.R., Weinstein, D.R. and O'Brien, A.D. (1985) Mouse chromosome 1 *Ity* locus regulates microbicidal activity of isolated peritoneal macrophages against a diverse group of intracellular and extracellular bacteria. *Journal of Immunology* 135, 544–547.

Lumsden, J.S., Wilkie, B.N. and Clarke, R.C. (1991) Resistance to fecal shedding of salmonellae in pigs and chickens vaccinated with an aromatic-dependent mutant of *Salmonella typhimurium*. *American Journal of Veterinary Research* 52, 1784–1787.

McCapes, R.H., Coffland, R.T. and Christie, L.E. (1967) Challenge of turkey poults originating from hens vaccinated with *Salmonella typhimurium* bacterins. *Avian Diseases* 11, 15–24.

Maskell, D.J., Hormaeche, C.E., Harrington, K.A., Joysey, H.S. and Liew, F.Y. (1987) The initial suppression of bacterial growth in a *Salmonella* infection is mediated by a localised rather than a systemic response. *Microbial Pathogenesis* 2, 295–302.

Mastroeni, P., Villarreal-Ramos, B. and Hormaeche, C.E. (1992) Role of T cells, TNFα and IFNα in recall of immunity to oral challenge with virulent *Salmonellae* in mice vaccinated with live attenuated *aro⁻* *Salmonella* vaccines. *Microbial Pathogenesis* 13, 477–491.

Mastroeni, P., Villarreal-Ramos, B., Demarco de Hormaeche, D. and Hormaeche, C.E. (1993a) Delayed (footpad) hypersensitivity and Arthus reactivity using protein-rich antigens and LPS in mice immunized with live attenuated *aroA* *Salmonella* vaccines. *Microbial Pathogenesis* 14, 369–379.

Mastroeni, P., Villarreal Ramos, B. and Hormaeche, C.E. (1993b) Adoptive transfer of immunity to oral challenge with virulent *Salmonella* in innately susceptible balb/c mice requires both immune serum and T-cells. *Infection and Immunity* 61, 3981–3984.

Mastroeni, P., Villarreal-Ramos, B., Harrison, J.A., Demarco de Hormaeche, R. and Hormaeche, C.E. (1994) Toxicity of lipopolysaccharide and of soluble extracts of *Salmonella typhimurium* in mice immunized with a live attenuated *aroA* salmonella vaccine. *Infection and Immunity* 62, 2285–2288.

Matsuura, M. and Galanos, C. (1990) Induction of hypersensitivity to endotoxin and tumour necrosis factor by sublethal infection with *Salmonella typhimurium*. *Infection and Immunity* 58, 935–937.

Meyer, H., Barrow, P. and Pardon, P. (1992) *Salmonella* immunization in animals. In: *Proceedings of the International Symposium on* Salmonella *and Salmonellosis*. Ploufragan/St Brieuc, France, pp. 345–374.

Michetti, P., Mahan, M.J., Slauch, J.M., Mekalanos, J.J. and Neutpa, M.R. (1992) Monoclonal secretary immunoglobulin-a protects mice against oral challenge with the invasive pathogen *Salmonella typhimurium*. *Infection and Immunity* 60, 1786–1792.

Mikula, I., Rosocha, J. and Pilipcinec, E. (1989) Immunization of calves with live and inactivated whole-cell vaccines against *Salmonella typhimurium* infection. *Acta Veterinaria Hungary* 37, 219–226.

Morein, B., Villacres-Eriksson, M., Sjolander, A. and Bengtsson, K.L. (1996) Novel adjuvants and vaccine delivery systems. *Veterinary Immunology and Immunopathology* 54, 373–384.

Morrison, W.I. (1986) *The Ruminant Immune System in Health and Disease*. Cambridge University Press, Cambridge.

Nagaraja, K.V., Kim, C.J., Kumar, M.C. and Pomeroy, B.S. (1991) Is vaccination a feasible approach for the control of *Salmonella*? In: Blankenship, L.C. (ed.) *Colonization Control of Human Bacterial Enteropathogens in Poultry*. Academic Press, San Diego, pp. 243–256.

Nauciel, C., Ronco, E., Guenet, J.L. and Pla, M. (1988) Role of H-2 and non-H-2 genes in control of bacterial clearance from the spleen in *Salmonella typhimurium* infected mice. *Infection and Immunity* 56, 2407–2411.

Nnalue, N.A. and Stocker, B.A.D. (1986) Some *galE* mutants of *Salmonella cholerae-suis* retain virulence. *Infection and Immunity* 54, 635–640.

Norris, F.A., Wilson, M.P., Wallis, T.S., Galyov, E.E. and Majerus, P. (1998) SopB, a protein required for virulence of *Salmonella dublin*, is an inositol phosphate phosphatase. *Proceedings of the National Academy of Sciences USA* 95, 14057–14059.

O'Brien, A.D., Scher, I. and Metcalf, E.S. (1981) Genetically conferred defect in anti-*Salmonella* antibody formation renders CBA/N mice innately susceptible to *Salmonella typhimurium* infection. *Journal of Immunology* 126, 1368–1373.

O'Callaghan, D., Maskell, D., Liew, F.Y., Easmon, C.S.F. and Dougan, G. (1988) Characterization of aromatic-aro purine dependent *Salmonella typhimurium*: attenuation, persistence and ability to induce protective immunity in Balb/C mice. *Infection and Immunity* 56, 419–423.

Pardon, P., Sanchis, R., Marly, J., Lantier, F., Guilloteau, L., Buzoni-Gatel, D., Oswald, I.P., Pepin, M., Kaeffer, B. and Berthon, P. (1990) Experimental ovine salmonellosis (*Salmonella abortusovis*): pathogenesis and vaccination. *Research in Microbiology* 141, 945–953.

Pritchard, D.G., Nivas, S.C., York, M.D. and Pomeroy, B.S. (1978) Effect of Gal-E mutant of *Salmonella typhimurium* on experimental salmonellosis in chickens. *Avian Diseases* 22, 562–575.

Rankin, J.D., Taylor, R.J. and Neaman, G. (1967) The protection of calves against infection with *Salmonella typhimurium* by means of a vaccine prepared from *Salmonella dublin* (strain 51). *Veterinary Record* 80, 720–726.

Rodrigue, D.C., Tauxe, R.V. and Rowe, B. (1990) International increase in *Salmonella enteritidis*: a new pandemic? *Epidemiology and Infection* 105, 21–27.

Roof, M.B. and Doitchinoff, D.D. (1995) Safety, efficacy and duration of immunity induced in swine by use of an avirulent live *Salmonella choleraesuis*-containing vaccine. *American Journal of Veterinary Research* 56, 39–44.

Sanchis, R., Pardon, P. and Abadie, G. (1991) Abortion and serological reaction of ewes after conjunctival instillation of *Salmonella enterica* subsp. *enterica* ser. abortusovis. *Annales de Recherches Veterinaire* 22, 59–64.

Schat, K.A. (1994) Cell-mediated immune effector functions in chickens. *Poultry Science* 73, 1077–1081.

Schimmel, D., Linde, K., Marx, G. and Ziedller, K. (1974) Zum Einsatz einer Smd-*Salmonella typhimurium* Mutante bei Kucken. *Archiv für Experimentelle Veterinärmedizin* 28, 551–558.

Segall, T. and Lindberg, A.A. (1993) Oral vaccination of calves with an aromatic-dependent *Salmonella dublin* (09, 12) hybrid expressing 04, 12 protects against *S. dublin* (09, 12) but not against *Salmonella typhimurium* (04, 5, 12). *Infection and Immunity* 61, 1222–1231.

Sigwart, D., Stocker, B.A.D. and Clements, J.D. (1989) Effect of a *purA* mutation on efficacy of *Salmonella* live vaccine vectors. *Infection and Immunity* 57, 1858–1861.

Silva, E.N., Snoeyenbos, G.H., Weinack, O.M. and Smyser, C.F. (1981) Studies on the use of 9R strain of *Salmonella gallinarum* as a vaccine in chickens. *Avian Diseases* 25, 38–52.

Slauch, J.M., Mahan, M.J., Michetti, P., Neutra, M.R. and Mekalanos, J.J. (1993) Mucosal immunity: the role of secretory immunoglobulin A in protection against the invasive pathogen *Salmonella typhimurium*. In: Cabello, F., Hormaeche, C., Mastroeni, P. and Bonina, L. (eds) *Biology of Salmonella*. NATO ASI Series A245, Plenum Press, New York, pp. 289–298.

Smith, B.P., Reina-Guerra, M., Hosieth, S.K. *et al.* (1984a) Aromatic-dependent *Salmonella typhimurium* as modified live vaccine for calves. *American Journal of Veterinary Research* 45, 59–66.

Smith, B.P., Reina-Guerra, M., Stocker, B.A.D., Hosieth, S.R and Johnson, E. (1984b) Aromatic-dependent *Salmonella dublin* as a parenteral modified live vaccine for calves. *American Journal of Veterinary Research* 45, 2231–2235.

Smith, H.W. (1956) The use of live vaccines in experimental *Salmonella gallinarum* infection in chickens with observations on their interference effect. *Journal of Hygiene, Cambridge* 54, 419–432.

Smith, H.W. and Tucker, J.F. (1980) The virulence of *Salmonella* strains for chickens: their excretion by infected chickens. *Journal of Hygiene, Cambridge* 84, 479–488.

Smith, R.A. and Bigley, N.J. (1972) Ribonucleic acid-protein fractions of virulent *Salmonella typhimurium* as protective immunogens. *Infection and Immunity* 6, 377–383.

Subhabphant, W., York, M.D. and Pomeroy, B.S. (1983) Use of two vaccines (live G30D or killed RW16) in the prevention of *Salmonella typhimurium* infections in chickens. *Avian Diseases* 27, 602–615.

Tagliabue, A., Villa, L., De Magistris, M.T., Romano, M., Silvestri, S., Boraschi, D. and Nencioni, L. (1986) IgA-driven T cell-mediated anti-bacterial immunity in man after live oral Ty 21a vaccine. *Journal of Immunology* 137, 1504–1510.

Thain, J.A., Baxter-Jones, C., Wilding, G.P. and Cullen, G.A. (1984) Serological response of turkey hens to vaccination with *Salmonella hadar* and its effect on their subsequently challenged embryos and poults. *Research in Veterinary Science* 36, 320–325.

Truscott, R.B. (1981) Oral *Salmonella* antigens for the control of *Salmonella* in chickens. *Avian Diseases* 25, 810–820.

Truscott, R.B. and Friars, G.W. (1972) The transfer of endotoxin induced immunity from hens to poults. *Canadian Journal of Comparative Medical and Veterinary Science* 36, 64–68.

Udhayakumar, V. and Muthukkaruppan, V.R. (1987) Protective immunity induced by outer membrane proteins of *Salmonella typhimurium* in mice. *Infection and Immunity* 55, 816–821.

van Furth, R., van Dissel, J.T. and Langermans, J.A.M. (1993) The role of granulocytes, naïve and activated macrophages in the host resistance against *Salmonella typhimurium*. In: Cabello, F., Hormaeche, C., Mastroeni, P. and Bonina, L. (eds) *The Biology of Salmonella*. NATO ASI Series A245, Plenum Press, London, pp. 249–253.

Villaread, B., Mastroeni, P., Demarco de Hormaeche, R. and Hormaeche, C. (1992) Proliferative and T-cell specific interleukin (IL-2/IL-4) production responses in spleen cells from mice vaccinated with *aroA* live attenuated *Salmonella* vaccines. *Microbial Pathogenesis* 13, 305–315.

Villarreal, B., Paulin, S.M., Watson, P.R., Manser, J.M., Jones, P.W. and Wallis, T.S. (1997). Analysis of the specificity of protection induced by *Salmonella* in cattle. In: *Proceedings of the 2nd International Symposium on* Salmonella *and Salmonellosis, Ploufragran, France.* pp. 471–473.

Wallis, T.S., Paulin, S.M., Plested, J.S., Watson, P.R. and Jones, P.W. (1995) The *Salmonella dublin* virulence plasmid mediates systemic but not enteric phases of salmonellosis in cattle. *Infection and Immunity* 63, 2755–2761.

Watson, P.R., Paulin, S.M., Jones, P.W. and Wallis, T.S. (1995) Characterisation of intestinal invasion by *Salmonella typhimurium* and *Salmonella dublin* and effect of a mutation in the *invH* gene. *Infection and Immunity* 63, 2743–2754.

Watson, P. R., Galyov, E.E., Paulin, S.M., Jones, P.W. and Wallis, T.S. (1998) *invH*, but not *stn*, influences *Salmonella*-induced enteritis in cattle. *Infection and Immunity* 66(4), 1432–1438.

Weber, A., Bernt, C., Bauer, K. and Mayr, A. (1993) The control of bovine salmonellosis under field conditions using herd-specific vaccines. *Tieraztliche Praxis* 21, 511–516.

Wood, M.W., Jones, M.A., Watson, P.R, Hedges, S., Wallis, T.S. and Galyov, E.E. (1998) Identification of a pathogenicity island required for *Salmonella* enteropathogenicity. *Molecular Microbiology* 29, 883–891.

Wray, C., Sojka, W.J., Morris, J.A. and Brinley-Morgan, W.J. (1977) The immunization of mice and calves with *galE* mutants of *Salmonella typhimurium*. *Journal of Hygiene, Cambridge* 79, 17–24.

Chapter 20

Epidemiology and Salmonellosis

Kathy Hollinger

FDA CVM, Division of Epidemiology and Surveillance, 7500 Standish Place, Rockville, Maryland 20855, USA

Epidemiology is defined as the study of the distribution and determinants of health-related states or events in specified populations, and the application of this study to the prevention and control of health problems (Last, 1995). This study in animal populations alone is termed epizootiology. Because the diseases are often transferred between human and animal populations, the distinction between the two populations has been considered semantic. Thus the term epidemiology is commonly used to describe the investigation of health conditions in human and animal populations (Thrusfield, 1995).

The earliest structured approach to studying disease causality was focused solely on the agent, and involved the three Henle–Koch postulates:

1. The organism is present in all cases of disease,
2. The organism does not occur in cases of another disease, and
3. The organism is isolated in pure culture from an animal and induces the same disease in other experimental animals.

Over time, scientists recognized that the presence of the organism in animals did not appear to be the main determinant of disease rates. Factors other than exposure to the agent were recognized as important in the development of clinical disease. Later, the concept of necessary and sufficient causes of disease was introduced. *Necessary* causes were those causal factors that were required elements in the disease process. For example, the causative agent or microorganism was a necessary cause. *Sufficient* causes included those factors that were predisposing, enabling, precipitating or reinforcing. Most disease can be attributed to a combination of the necessary and sufficient causes, and can be explained by interactions between agent, host and environment. A schematic representation of the complex interactions between agent, host and environment for *Salmonella* is demonstrated in Fig. 20.1. The veterinary epidemiologist is primarily interested in those determinants or factors that can be controlled to reduce or to prevent disease and in animal populations, these controllable factors often include management and husbandry practices. In the study of infectious diseases, the investigation of subclinical carriage of a potential pathogen, and cases of clinical disease may be equally important to the epidemiologist, and are essential in the design of effective prevention and control programmes. This chapter is an introduction to epidemiology and its vocabulary. For more complete coverage of the topics, excellent sources are available and are referenced at the end of this chapter.

Goals of Epidemiological Investigation

The objectives of epidemiological investigation include: detecting the origin of disease, determining factors associated with the expression of disease, controlling or preventing disease,

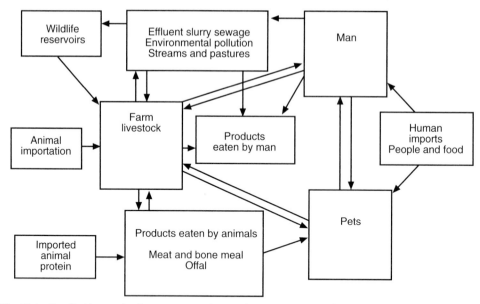

Fig. 20.1. Detailed interaction between agent, host and environment in salmonellosis.

understanding the ecology of infectious agents, assessing the economic consequences of disease and of interventions to control or prevent disease (Thrusfield, 1995). Techniques used to achieve these objectives are based on statistical and other mathematical tools.

The Use of Rates in Epidemiology

Epidemiology involves the study of health-related events in populations as opposed to studying individual cases of disease. One of the earliest recorded epidemiological studies was conducted by John Snow, a London obstetrician in the mid 1800s, utilizing the distribution of disease rates in populations. Snow had observed three cholera outbreaks prior to 1850, and proposed a theory that water was the source of the disease. During a later epidemic, in 1854, Snow carried out a systematic study of the population, meticulously recording cases and their sources of drinking water. He then compared the rates of disease based on drinking water source. He identified one of two water companies, in particular one water pump, the Broad Street pump, as the source of the disease. After removal of the pump handle the epidemic was rapidly contained.

Evaluating the health of populations is based upon the determination of rates and ratios of infection, disease, or other outcomes of interest in specified subgroups and comparison of rates to another defined population. The use of rates as the basis for comparison implies that disease is not randomly distributed in populations, but instead is distributed based on determinants of disease present in subgroups of the population. Disease determinants are factors that are directly or indirectly associated with the disease. By comparing disease rates in subgroups of a specified population, epidemiological associations between disease and potential exposure factors can be derived. Identification of these factors can be used to generate hypotheses that are the basis for further investigations to determine the nature of the association and investigate potential causality between the determinant and disease.

Rates are a measure of the frequency of occurrence of an event. The numerator represents the number of reported incidents during a defined time period, and the denominator represents the population during the same time period. Some epidemiological rates measuring disease occurrence include incidence, prevalence, attack rate, and case–fatality rate. Incidence represents the absolute risk of develop-

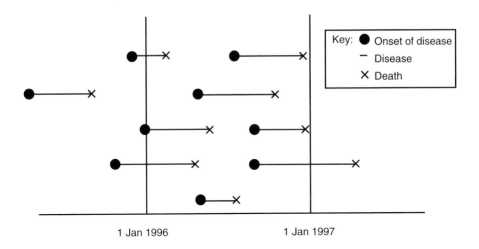

Key: ● Onset of disease
      – Disease
      ✕ Death

1 Jan 1996   1 Jan 1997

Incidence (1 Jan 1996–31 Dec 1996) 6/8 (75%)   Period prevalence (1 Jan 1996–31 Dec 1996) 8/8 (100%)
Point prevalence (1 Jan 1996) 3/8 (38%)   Total population = 9 (100.0%)

Fig. 20.2. Relationship between incidence, point prevalence and period prevalence.

ing disease in a population and is defined as the number of new cases in a specified period of time, divided by the size of the defined population during the same period. Prevalence is the number of current cases present in a population divided by the size of the population at risk, or susceptible to disease. It can be measured at a point in time, in which case it is a 'point prevalence,' or during a period of time, when it is termed a 'period prevalence' (see Fig. 20.2). The relationship between incidence (I) and prevalence (P) is represented by the formula, $P = I \times D$, where D=duration of disease. As an example of the impact of duration of disease on incidence and prevalence, introduction of a new therapy that would prolong the life of a diseased animal would result in increased prevalence while incidence remained constant. Alternatively, an increase in the incidence of a disease exhibiting increased virulence, with a shortened course of illness and a high death rate may result in a decrease in prevalence. Prevalence data are useful for evaluating the economic impact of interventions and other changes introduced to manage disease, while incidence data give an estimate of the proportion of the population affected.

The population at risk, representing the denominator, in herds and flocks of animals is often not a static population. The calculation of the population at risk on the farm or in other animal settings can be difficult, as animals may leave and enter the herd during the period of interest. The population at risk may be calculated at the beginning, middle or end of the period or, alternatively, by using an average of the two periods. At times it may not be feasible to obtain good denominators representative of the population at risk. In calculating the incidence of clinical disease caused by *S. typhimurium* DT104 in cattle herds in Great Britain (GB), one denominator, available from agricultural census data, would have overestimated the true number of herds, because mixed beef and dairy herds were counted twice. This would have resulted in a calculation of incidence that was lower than the actual incidence. An alternative approach involved selecting a sample population and determining the number of new cases occurring during a given period. The sample-derived incidence gave an estimate that was closer to the true incidence. The extrapolation of rates obtained from sample populations to another population is limited by the representativeness of the sample to the population of interest. If the composition of the sample population was not similar to cattle herds in GB, the extrapolation of this estimate would not have been valid.

Attack rates are defined as the cumulative number of cases over the period of an outbreak divided by the population at risk at the

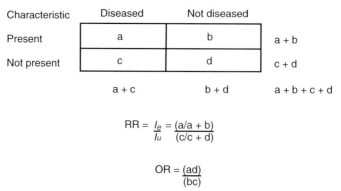

Fig. 20.3. 2 × 2 Table for determination of the strength of the association and chi-squared test.

beginning of the outbreak. Attack rates are commonly used to describe outbreak incidence. Other descriptive rates are used to describe the impact of disease on populations, such as morbidity, mortality or case–fatality rates. Morbidity is the proportion of individuals affected by the disease during a specified period. The mortality rate is the number of deaths in the population at risk during a defined period. Case–fatality rates are a measure of the severity of disease and are the number of deaths due to a specific disease divided by the number of cases of that disease during the period.

Disease rates in a population can be expressed as crude rates, such as those described above, specific rates or as adjusted rates. Crude rates are usually overall population or herd/flock rates, and are the average rates, weighted by the characteristics of the specified population. Comparison of crude rates is appropriate if the populations being compared share similar disease characteristics that make them susceptible to disease. Specific rates are rates that are stratified into categories or characteristics, such as age, geographical location or herd size. Rates can also be adjusted to allow comparison of two populations that differ in composition. Two methods of adjustment are commonly used; direct and indirect adjustment. Direct adjustment applies the rates in the population of interest to a standard population and can be used to compare two populations directly. Indirect adjustment may be used when the rates in the population of interest are not known. It involves applying the rates from a standard population to the population of interest; the comparison between two populations is limited to comparison of each independently with the standard population.

Associations between Determinants and Outcome

The determination of rates of disease in a population is the basis of descriptive epidemiology. The distribution of the study population into categories, based on disease status and exposure, is a means of calculating the strength of association. Two measures of the strength of an association are the relative risk (RR) and the odds ratio (OR). The relative risk of an exposure can be calculated mathematically by determining the incidence of disease in exposed (Ie), and unexposed (Iu) populations. The ratio of these two incidence rates represents the relative magnitude of the difference between the exposed and unexposed groups. The formula for RR is given in Fig. 20.3. An OR is a second RR measuring the probability of an event occurring over the probability the event will not occur. When using incident cases, the OR is considered a good estimate of relative risk and is derived from the simplified formula ad/bc (Fig. 20.3). If the exposure to a putative risk factor is not related to the disease, the OR and RR will equal 1. An OR or RR > 1 indicates a positive association between the determinant and disease, and a negative or 'protective' association will result in an OR or RR < 1. If the rate of disease in the population is not low, ORs can tend to overestimate and exaggerate the RR. Rates can be compared to determine absolute, as well as relative differences. An absolute difference, termed the attributable risk, between exposed and unexposed populations is calculated by subtracting the incidence in the unexposed population from the incidence in the exposed population. This results in a value that represents the risk due to the single exposure. Often multiple factors play a role in the occurrence of a disease. These multiple factors can not only affect the outcome of interest, but also affect and may be affected by other determinants. Statistical methods used to determine the independent effect of a determinant in the presence of these other factors involve multivariable analysis. Statistical methods for multivariable analysis include linear and logistic regression and are facilitated by the use of statistical packages that simplify these calculations.

Types of Association

Types of associations made in epidemiological studies can be artefactual, indirect or directly causal in nature. Artefactual associations may occur by chance alone or due to bias, and are not linked to the outcome of interest. Bias is the deviation from a true association. Biases can arise during sample selection, study design, data collection, data analysis or interpretation of results. Bias can be either systematic or random in nature. Systematic bias is most problematic because it results in the distortion of the relationship between the exposure and the outcome and may result in erroneous associations. Random bias is less likely to alter the association between the determinant and outcome as it is multi-directional. Some types of biases are selection bias, misclassification bias and recall bias. Selection bias occurs when participants in a study possesses characteristics that differ from the reference population and distorts the relationship between exposure and disease. Misclassification bias occurs when cases are not accurately identified in a study. Recall bias is the lack of accurate memory of events and may impair the collection of accurate exposure histories. Artefactual associations may also be introduced by confounders. Confounders have three characteristics: (i) they are risk factors for the outcome of interest; (ii) they are not intermediate between exposure and disease; and (iii) they are associated with the exposure of interest. An example of a confounder may be herd size. In a study of *Salmonella* in cattle herds, larger herds had higher rates of disease than the smaller herds. The rate of introduction of newly acquired animals was greater in large compared with small herds due to replacement of culled animals. Large herds purchased cattle from higher risk sources; dealers and markets, compared with the small herds that tended to acquire cattle from private farm sources. Therefore, herd size was a risk factor for the disease, was not intermediate between exposure and causality and was associated with the introduction of newly acquired animals.

The effect of confounders in a study can be controlled either in the study design, by matching cases and controls on the confounding effect, or during the analysis of the data by use of adjustment procedures, either stratification in univariate analysis or inclusion of important confounders in multivariable analysis. The impact of bias in distorting an association emphasizes the importance, in the planning stages of a study, of reducing or eliminating the potential biases in order to allow the best determination of the true association.

Valid associations between the risk determinant and the disease may be either direct or indirect. Indirect associations are classified into two types: those which are due to the factor and the disease sharing a common underlying condition, or those due to an intermediate factor modulating the effect of the factor on the disease. Direct causal associations, linking exposure and disease, are best determined within the construct of controlled experiments, but it is often not possible to conduct these studies. Formal criteria used to distinguish causal from non-causal epidemiological associations are the demonstration of: (i) the strength of an association; (ii) a dose–response association; (iii) consistency or repeatability of an association; (iv) a temporally plausible association; (v) the specificity of an association; (vi) use of additional supporting experimental evidence; and (vii) the biological plausibility of the association. Many associations may initially appear to be causal, but are later found to be indirectly associated with disease as the association is refined by further studies.

Tests of Association

Once an association has been made, statistical methods can be used to estimate the probability of the observed degree of association between the event and the exposure of interest. The chi-squared test for association is commonly used in

$$\chi^2 = \text{sum of } \frac{\text{(observed – expected)}^2}{\text{expected}} \text{ for each cell in the array}$$

Fig. 20.4. Formula for chi-squared statistic (χ^2) calculation. The numbers for the expected cell values are derived by multiplying column totals by row totals and dividing by the grand total $(a + b + c + d)$ (from Fig. 20.3).

disease investigations and determines the probability of the association occurring by chance alone. This test is a mathematical comparison between observed and expected disease rates or proportions. The larger the chi-squared statistic, the larger the difference between observed and expected rates and the less likely that the observed cell values would have occurred by chance alone. The chi-squared test statistic is generated from the data (see Fig. 20.4), and a P value derived from a table of chi-squared statistics. The derived P value is compared to the previously determined significance level. The level of significance is often defined as $P \leq 0.05$, representing a rare event, and is the point at which the event is unlikely to occur by chance alone. If the calculated chi-squared statistic defines a P value of less than 0.05, then the event is considered a statistically 'significant' association, as it has a chance of occurring less than one out of 20 times.

In addition to the chi-squared test for significance, the Fisher exact test and others are used to calculate the significance of an association. The Fisher exact test is used when 20% or more of the expected observations contain less than five observations per cell. Statistically significant associations should be interpreted with caution, as the biological plausibility of an association should be determined prior to the acceptance of statistically significant results. Software programs are available to calculate the significance of concurrent multivariable associations in logistic regression and other analyses.

Confidence intervals are another method used to define an association between a determinant and an outcome variable. A confidence interval simultaneously describes the magnitude of an effect and the variability in the sample estimate, in contrast with the limited information obtained from P values which simply assesses the likelihood of the association.

Other Common Epidemiological Definitions: Sensitivity, Specificity, Validity and Reliability

Rates are ideally determined based on case definitions that are sensitive and specific to the condition of interest and appropriate to the goals of the investigation. Sensitivity is the probability of correctly identifying a case and specificity is the probability of correctly identifying a non-diseased individual. The relationship between sensitivity and specificity is typically inverse; as the ability to correctly identify a case increases there is a loss of specificity resulting in an increase in falsely identified cases. Reliability and validity are also important epidemiological terms. Reliability is a measure of the repeatability or the precision of the results, and validity is a measure of the truth of an estimate.

Types of Epidemiological Studies

Epidemiological studies are divided into broad categories termed interventional or observational. Interventional studies are controlled investigational studies involving an intentional change to test a hypothesized relationship between a factor and the rate of disease in the study population. Some examples of interventional studies are clinical trials, such as those commonly used to determine the efficacy of drugs or vaccines, and controlled experiments that may test specific husbandry practices. In contrast, observational studies involve the collection of information from a natural setting, where the exposure to a defined characteristic is determined and disease rates in the population are compared, without investigator intervention. Three types of observational studies are: (i) case–control, (ii) cohort and (iii) prevalence or cross-sectional studies. Case–control studies are commonly used in outbreak investigations or when cases are rare. A clear case definition is established by the investigator that is appropriate to the goals of the investigation, and highly specific to assure identification of true cases. Controls are selected, without the disease of interest, and may be matched to the cases based on potentially confounding factors such as age or seasonality of illness. Once cases are matched to controls based on a factor, the effect of that factor can no longer be assessed in the analysis. For example, in a case–control study where cases and controls were selected and matched on age at sample submission, factors associated with age could not be assessed, as there were no differences between cases and controls. In a case–control study, the exposure to hypothesized risk determinants is typically assessed retrospectively, after the outcome of interest is identified. A limitation of retrospective assessment of exposure can be the introduction of recall bias because the respondent may not remember accurate exposure histories, unless the exposure was well documented while it occurred. Cohort studies involve the selection of a study population in which exposure data can be assessed over time. This population is followed either prospectively or records are examined or histories taken retrospectively, often over a long period, to determine disease outcome and relevant exposures. Exposure in cohort studies is recorded prior to the onset or determination of disease in prospective or concurrent cohort studies and the rates of disease in the exposed and unexposed groups are compared to identify exposure–disease associations. Exposure may also be assessed from records in which case the study is termed a non-concurrent or retrospective cohort study. The incidence of a health event in the population can be calculated from cohort studies as the cases occurring during the period of observation are new cases of disease. Cross-sectional studies involve the concurrent collection of disease and exposure data; therefore, one of the criteria for establishing causality, the temporal relationship between exposure and disease, cannot be established.

Each study type has advantages and disadvantages associated with cost, time required to conduct the study, the suitability to the study topic, sample size and comparison group availability among others (Thrusfield, 1995). These studies can provide data for the modelling and prediction of epidemics using computer-assisted techniques, and the data can be used for policy making, decision analysis, cost–benefit assess-

ments, and the evaluation of interventions. Epidemiologic data describing the transmission of *S. dublin* were used to model the introduction and spread of the currently exotic organism into New Zealand, to investigate the effects of a reduction in funding for national animal health surveillance. It was evident from the model that a reduction in the funding would result in a longer time to diagnosis and potentially allow spread of disease to a greater number of herds (Sanson and Thornton, 1997). Epidemiologic studies that are well designed and constructed can provide valuable information for the prevention and control of disease by veterinarians and farmers as well as provide data for decision analysis and policy setting by public health officials.

Sample Size

The determination of an appropriate sample size during the planning phase of a study is an important component of a well designed study. An inadequate sample size may not allow the investigator to detect a difference between two comparison groups whereas an excessively large sample can be wasteful of resources. Factors influencing the size of the sample are:

1. The prevalence or incidence of the disease or outcome of interest (although this is often unknown, a pilot study or estimate obtained from previously published literature may be helpful).
2. The specified minimum difference in disease prevalence or incidence between comparison groups that is thought to be of interest.
3. The probability of a type I error, which is termed the alpha (α) and is the level of significance set by the investigator. This represents the probability or chance of statistically detecting a difference in disease prevalence or incidence between cases and controls when there is no difference.
4. The probability of a type II error (β), which represents the chance of not detecting a difference when one actually exists. The power of the study to detect a specified effect or difference between the two groups is defined as $1-\beta$. By convention, β is often set at 0.20 with a resultant power of 0.80.

Table 20.1 provides a tabulation of the

Table 20.1. Sample size for detection of *Salmonella* in a flock of 10,000–100,000 birds at a confidence level of 95% (from WHO, 1994).

Detectable prevalence level (%)	Sample size	Detectable prevalence level (%)	Sample size	Detectable prevalence level (%)	Sample size
1	298	11	26	55	4
2	149	12	24	60	4
3	99	15	19	65	4
4	74	20	14	70	3
5	59	25	11	75	3
6	49	30	9	80	3
7	42	35	8	85	3
8	36	40	6	90	2
9	32	45	6	95	2
10	29	50	5	100	1

estimated sample size needed to detect the presence of disease in a finite population, at different prevalence levels. At times there are a limited number of cases available for a study. If the number of cases is limited, the investigator can increase the power, or ability to detect a difference, by increasing the ratio of controls to cases. However, the effect of increasing controls to greater than four controls per case results in diminishing incremental increases in the power and may not be cost effective. When the number of available cases is limited, the power of the study should be calculated to determine whether performing the study is a worthwhile endeavour. The factors to be considered for the determination of sample size are not restricted to statistical considerations. Sample size will depend on such pragmatic and technical considerations as laboratory capacity and personnel availability as well as statistically based reasons for the final number of subjects. The investigator will balance these statistical and practical considerations during the design phase of a study.

Outbreak Investigation in Animal Populations

An outbreak is a localized increase in the number of new cases of a disease. Outbreak investigations in animal populations are commonly conducted by veterinarians. Often, the determination of the agent causing the outbreak is not the objective of interest, as it is easy to identify the cause. It is usually more difficult to identify the source of the

agent, or factors associated with the appearance of clinical disease in the animals. An outbreak investigation can be conducted using a structured or non-structured approach. A non-structured investigation relies on the investigator's experience and interests, and can be limited by these biases, resulting in inconclusive or false associations. A structured approach to an investigation reduces the impact of investigator biases and is generally broader in scope. Once an approach has been determined, the investigation should be planned prior to arrival on site. A structured approach begins with confirmation of the outbreak as the first step of the outbreak investigation. This begins with verification of the clinical diagnosis with laboratory results and establishing the existence of an epidemic by comparing previous disease rates with current disease incidence. After the diagnosis and epidemic status have been established, the goals of the investigation must be defined. Often, the goals of the investigation are to reduce the rate of disease and to prevent recurrence in the herd or flock. The goals of the investigation should be directed to focus on those factors in the causal pathway of disease that can be altered and are controllable in the management of the affected animal population. Focusing on the agent of infection does not take into account the multifactorial nature of disease expression, and will limit the information gathered during the investigation. In many cases, the organism is endemic to the affected farms, and the determinants of interest are factors associated with the progression from infection to disease, see Box 20.1 for a brief overview of outbreak investigation procedures. For a more detailed description of techniques, see Thrusfield (1995).

Box 20.1. Outline of the outbreak investigation process.

1. Confirm clinical findings and diagnosis with laboratory results and verify an increase in the rate of disease to establish the existence of an epidemic.

2. Define the cases precisely to distinguish disease from illness that is attributable to other causes.

3. Plan the investigation by first defining the goals of the investigation, focusing on those risk factors associated with the disease, with an emphasis on management factors that are alterable. Collect the data using quantitative or defined qualitative measures.

4. Describe the outbreak by time, place and animal characteristics. Describe the temporal pattern of the epidemic, creating an epidemic curve from the number of cases per unit of time. From the epidemic curve, determine the time of exposure, the incubation period and the curve type, describing the source of transmission, i.e. common-source or propagated epidemic curve. Describe the spatial distribution of the epidemic by mapping cases and correlating with the temporal pattern of distribution. Describe the animal pattern of disease by describing the characteristics of affected and unaffected animal groups. Descriptive characteristics may include: age, breed, sex, production stage, diet, vaccination status and any other factors that may be relevant to the population.

5. Data analysis: factor-specific attack rates, case–fatality rates, relative risk and attributable risk are calculated. Statistical tests, such as the chi-squared test for independence, are used to determine the statistical significance of risk determinants.

6. Formulation and testing of hypothesis to determine the cause(s), the source(s) and the mode(s) of transmission.

7. Further investigation and analysis.

8. Report of findings of the investigation.

Surveillance and Monitoring: Definitions and Use of Data

Monitoring and surveillance of health events involves the continuous collection of data on cases in a specified population. Monitoring is defined as the performance and analysis of routine measurements, aimed at detecting changes in the environment or health status of populations. Monitoring may imply interventions in light of the observations made during the analysis of the data. Surveillance is defined as the ongoing systematic collection, analysis and interpretation of health data in the process of describing and monitoring a health event and the timely dissemination of this data to those who need to know (Last, 1995). Mechanisms for the collection of surveillance data can be active, passive or sentinel. Passive data collection refers to data that is reported by the veterinarian, laboratory or other party without solicitation or contact from the surveillance system coordinators. Active surveillance involves the solicitation of data from health care providers or laboratories. This type of system is more costly, but typically results in data that is more complete and

accurate than that collected in passive surveillance systems. Sentinel surveillance systems rely upon reported cases of disease and typically alert health officials that control and prevention methods are inadequate. Some examples of diseases often monitored by sentinel surveillance systems are rabies, salmonella and anthrax.

Surveillance data can be applied to strategies for the control and prevention of disease as well as understanding the disease by describing the time, place and population characteristics of the reported cases. An understanding of biases in the surveillance system, and knowledge of the structure, purpose and objectives of the system allow informed use of surveillance and monitoring data. Biases in surveillance data often occur, due to the circumstances inherent in the sampling and reporting process. A pyramidal model demonstrating stepwise reduction in reported cases of salmonellosis in a surveillance system can be seen in Fig. 20.5. Examples of selection biases in *Salmonella* surveillance systems can include those that arise due to sampling only those animals exhibiting clinical disease or sampling only those animals that are of economic value to the producer.

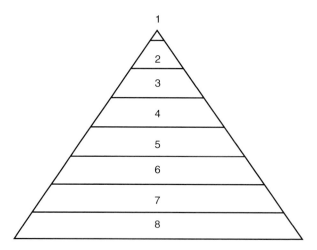

Fig. 20.5. Pyramidal scheme of *Salmonella* reporting. 1, Reporting of the isolation. 2, Isolation of organism. 3, Animal shedding. 4, Veterinary sampling of animal. 5, Veterinary visit to farm. 6, Clinically apparent infection. 7, *Salmonella* infection. 8, *Salmonella* exposure.

Surveillance data are often used to describe temporal trends in epidemics. Temporal trends are demonstrated by epidemic curves that are graphed by counting the number of cases by time of onset of illness. They can represent short-term epidemics such as outbreaks, seasonal, long-term or secular and cyclical epidemics. The shape of an epidemic curve can provide clues about the type or source of exposure. Two types of epidemic curves are common source and propagated epidemics. In common source epidemics a group of individuals is exposed to an agent from a single source. The exposure may be a single, repeated multiple, or continued exposure over a period of time. If the exposure is simultaneous or limited to a short period all cases of disease develop within one incubation period of the disease. Propagated epidemics involve the serial transfer of the disease to other susceptible individuals in the population.

Other temporal trends in the distribution of disease are seasonal or secular in nature. Data obtained from the Zoonoses Order database indicate that the *S. enterica* serovar Typhimurium DT104 epidemic in cattle herds in GB shows a seasonal trend that peaks during the calving season. The seasonal increase in the incidence of DT104 infections reflects the increased susceptibility of adult cows during the immediate postpartum period and the increased number of susceptible calves. Seasonal determinants may be

a direct result of agent survivability due to climatic or environmental conditions, population changes reflecting changes in host susceptibility, reservoir status, diet and many other factors that may influence exposure or expression of disease in the population.

Secular trends are changes over a long period of time, involving years or decades. Secular trends may show variability in disease occurrence due to changes in the population of susceptible individuals over time, interventions that have had an impact on the disease and random effects. Other techniques used in temporal trend analysis, such as the establishment of correlation, time series and cluster analysis are covered in brief in epidemiology texts (Thrusfield, 1995) and in detail by statistical texts and are not addressed here.

Geographical representation of surveillance data can be achieved by the mapping of cases, rates or proportions. Mapping can include the addition of landscape features, such as streams and other natural characteristics, which can result in a sophisticated representation of natural barriers and modes of transmission.

Characteristics of *Salmonella* of Epidemiological Significance

Salmonellae can infect a diverse range of animal hosts, from insects, reptiles and birds to mammals, including man, and can be present and persist in the environment. All *Salmonella* serovars are considered potential pathogens in most animal species. However, the pathogenicity of some serovars appears to be limited to a narrow range of animal host species and are considered 'host adapted', such as *S. dublin* in cattle. Other serovars such as *S. typhimurium* can affect many animal species, including humans and are termed 'non-host adapted.' *Salmonella* can cause a spectrum of disease expression and outcomes, ranging from sub-apparent infection, impaired productivity, or clinical disease that may range from mild to fatal in severity of infection. The organism can be shed for a variable period after recovery; for example *S. typhimurium* DT104 has been detected in samples taken up to 18 months after clinical disease in a herd of dairy cattle (Evans and Davies, 1996). Isolation of the organism from animals may be impaired by the intermittent shedding of the organism in faeces that is characteristic of infected animals. It may be difficult to detect *Salmonella* when it is present in low numbers due to the competitive overgrowth of other *Enterobacteriaceae*. Distinct patterns of distribution, virulence, pathogenesis, and resistance can be ascribed to specific serovars or even sub-serotypic classifications of *Salmonella*, based on phage type, plasmid profile, antimicrobial susceptibility profile or other methods of differentiation.

Serological determination of the presence of infection also has limitations. The lag period between exposure and antibody response, interference by passively transferred immunoglobulins, the inability to identify infecting serovar and cross-reactivity with antibodies due to immunization are a few of the limitations. Therefore, the determination of an infection rate, using isolation techniques or serology, can be difficult to assess in affected populations.

The risk factors known to be associated with isolation of *Salmonella* and clinical expression of salmonellosis in cattle are diverse. In a recent case–control study of *S. typhimurium* DT104 in GB, factors associated with increased risk of clinical disease in cattle herds were: cattle dealer premises, access to feed stores by wild birds, the presence of feral cats on the farm, the housing of cattle, a lack of isolation facilities, and the introduction of newly purchased cattle (Evans, 1996).

Some determinants of *Salmonella* prevalence are specific to the serovar and affected animal host; for example, the rate of isolation of *S. enteritidis* PT4 from the gut contents of laying hens was shown to be increased 24 h after withdrawal of feed compared with a group of fed birds (Humphrey *et al.*, 1993). Differences in disease determinants are also evident in subsets of a population based on age, breed, or gender differences in disease susceptibility.

In addition to these factors, some *Salmonella* serovars can be persistent in the environment, *S. typhimurium* DT204c was detected on a calf-rearer's premises up to 2.5 years after the first isolation on a farm with a mean period of persistence of 14 months on five farms (McLaren and Wray, 1991). Environmental persistence of the organism facilitates the dissemination of infection, reflected by rates of disease in recently purchased calves or cattle and the facility with which *Salmonella* are isolated from transport vehicles and market calf pens (Wray *et al.*, 1991). Other potential sources of environmental exposures are water courses, sewage, septic tank effluents, some domestic pets, wild birds and rodents.

An understanding of the organism may be gained from the literature, based on the serovar and specific population of interest. Construction of a list of previously determined associations with the disease will provide the investigator with some of the information needed to begin to define potential exposures for a proposed investigation.

Molecular Epidemiology and the Differentiation of *Salmonella* Strains

Typing of strains into groups below the level of serovar may be necessary for the purposes of an epidemiological investigation. Typing may allow investigators to determine the relatedness of outbreak-associated strains, trace the spread of an epidemic and make the distinction between a relapse and a newly introduced infection. The typing of isolates for an epidemiological investigation assumes that clonal isolates are epidemiologically related and represent the amplification and dissemination of a single isolate. This assumption is used cautiously as it can be simplistic. Bacterial populations are constantly changing by mutation, transfer of genetic determinants and many other mechanisms for change that create the diversity observed in microbial populations in nature.

Typing systems are broadly categorized into phenotypic techniques that identify the expressed molecular characteristics of an organism and genotypic techniques that identify the chromosomal or extra-chromosomal DNA composition of an organism. Phenotypic techniques include biotyping, antimicrobial susceptibility testing, serotyping, bacteriophage typing and bacteriocin typing. Phenotypic classification of strains is limited by the immense capacity of organisms to alter the expression of these characteristics. Genotypic strain typing techniques include plasmid profile analyses, restriction endonuclease analysis (REA) of chromosomal DNA, Southern blot of restriction length polymorphisms (RFLPs), pulsed-field gel electrophoresis (PFGE) and nucleotide sequence analysis. Interpretation of the typing results can be difficult, as epidemiologically related isolates may demonstrate typing differences in phenotype, plasmid content, phage reactivity or sequence mutations. Designation of degree of difference between similar isolates into indistinguishable, closely related, possibly related and different strains may facilitate the interpretation of typing results for epidemiological study (Arbeit, 1999).

Conclusion

Investigation of salmonellosis using epidemiological techniques can present unique challenges to the investigator. It is important to define the goals and objectives before planning any studies or outbreak investigations. Clear understanding of the limitations of the data are also important in the analysis and interpretation of data from surveillance and other sources. The use of a well constructed epidemiological approach to the study of *Salmonella* in animals can be used to effectively target prevention and control practices that can reduce the impact of this zoonotic enteric pathogen on animal and human populations.

References and Further Reading

Arbeit, R.D. (1995) Laboratory procedures for the epidemiological analysis of microorganisms. In: Murray, P.R., Baron, E.J., Pfaller, M.A., Tenover, F.C. and Yolken, R.H. (eds) *Manual of Clinical Microbiology*, 6th edn. ASM Press, Washington, DC, pp. 190–208.

Evans, S.J. and Davies, R. (1996) Case control study of multiple resistant *Salmonella typhimurium* DT 104 infection of cattle in Great Britain. *Veterinary Record* 139, 557–558.

Evans, S.J. (1996) A case control study of multiple Resistant *Salmonella typhimurium* DT104 infection of cattle in Great Britain. *Cattle Practice* 4, 259–266.

Hancock, D.D. and Wikse, S.E. (1988) Investigation planning and data gathering. In: Lessard, P.R. and Perry, B.D. (eds) *Investigation of Disease Outbreaks and Impaired Productivity*. Veterinary Clinics of North America – Food Animal Practice, March. W.B. Saunders, Philadelphia, pp. 1–16.

Hueston, W.D. (1988) Evaluating risk factors in disease outbreak. In: Lessard, P.R. and Perry, B.D. (eds) *Investigation of Disease Outbreaks and Impaired Productivity*. Veterinary Clinics of North America – Food Animal Practice, March. W.B. Saunders, Philadelphia, pp. 79–96.

Humphrey, T.J., Baskerville, A., Whitehead, A., Rowe, B. and Henley, A. (1993) Influence of feeding patterns on the artificial infection of laying hens with *Salmonella enteritidis* phage type 4. *Veterinary Record* 132, 407–409.

Last, J.M. (1995) *A Dictionary of Epidemiology*, 3rd edn. Oxford University Press, New York.

Lessard, P.R. (1988) The characterization of disease outbreaks. In: Lessard, P.R. and Perry, B.D. (eds) *Investigation of Disease Outbreaks and Impaired Productivity*. Veterinary Clinics of North America – Food Animal Practice, March. W.B. Saunders, Philadelphia, pp. 17–32.

Lillenfeld, D.E., Stolley, P.D. (1994) *Foundations of Epidemiology*, 3rd edn. Oxford University Press, Oxford.

McLaren, I.M. and Wray, C. (1991) Epidemiology of *Salmonella typhimurium* infection in calves: persistence of *Salmonella* on calf units. *Veterinary Record* 129, 461–462.

Mausner, J.S. and Kramer, S. (1985) *Mauser and Bahn Epidemiology – an Introductory Text*, 2nd edn. W.B. Saunders, Philadelphia.

Osterholm, M.T., Hedberg, C.W. and MacDonald, K.L. (1995) Epidemiology of infectious diseases. In: Mandell, G.L., Bennet, J.E. and Dolin, R. (eds) *Principles and Practice of Infectious Diseases*, 4th edn. Churchill Livingstone, New York, pp. 159–168.

Rosen, G.A. (1993) *History of Public Health*, expanded edn. Johns Hopkins University Press, Baltimore, Maryland, pp. 261–264.

Rothman, K.J. (1986) *Modern Epidemiology*. Little, Brown, Boston.

Sanson, R.L. and Thornton, R.N. (1997) A modeling approach to the quantification of benefits of a national surveillance programme. *Preventive Veterinary Medicine* 30, 37–47.

Spangler, L. (1988) The investigation of outbreaks of infectious disease. In: Lessard, P.R. and Perry, B.D. (eds) *Investigation of Disease Outbreaks and Impaired Productivity*. Veterinary Clinics of North America – Food Animal Practice, March. W.B. Saunders, Philadelphia, pp. 159–169.

Thrusfield, M. (1995) *Veterinary Epidemiology*, 2nd edn. Blackwell Science, Oxford.

WHO (1994) *Guidelines on Detection and Monitoring of* Salmonella *Infected Poultry Flocks with Particular Reference to* Salmonella enteritidis (edited by Wray, C. and Davies, R.H.). WHO/Zoon./94.173.

Wray, C., Todd, N., McLaren, I.M., Beedell, Y.E. and Rowe, B. (1990) The epidemiology of *Salmonella* infection of calves: the role of dealers. *Epidemiology and Infection* 105, 295–305.

Wray, C., Todd, N., McLaren, I. and Beedell, Y.E. (1991) The epidemiology of *Salmonella* in calves: the role of markets and vehicles. *Epidemiology and Infection* 107, 521–525.

Chapter 21

Methods for the Cultural Isolation of *Salmonella*

W. Douglas Waltman

Georgia Poultry Laboratory, Oakwood, GA 30566, USA

Introduction

There is a tremendous amount of information in the scientific literature concerning the media and methods for the isolation of *Salmonella*; however, it is often confusing and contradictory. This results from the multiplicity of media and methods that are available; perhaps more than for any other single species of bacteria. Some of the variables that can result in differences in recovery, include: (i) the type, amount and sources of samples; (ii) whether the samples are artificially or naturally contaminated; (iii) the use of pre-enrichment and the type used; (iv) the type and formulation of selective enrichment used; (v) the enrichment incubation temperature; (vi) the enrichment incubation time; (vii) the plating media used; and (viii) the number of colonies screened from the plating media.

Historically, the need to isolate *Salmonella* originated in the late 1800s, when *Salmonella typhi* was found to be the aetiological agent of typhoid fever. The resulting media and methods were specifically designed for culturing *S. typhi*. In the early 1900s, it became apparent that other *Salmonella* serovars were also responsible for clinical disease. Even though these *Salmonella* were characteristically different from *S. typhi*, the same methods were generally used for isolation. In many cases, the origin of these infections was found to be food-borne. The existing cultural procedures were applied to the isolation of *Salmonella*

from these foods and, as these foods were tested, it became apparent that a significant proportion of the *Salmonella* outbreaks could be traced to meat and food-producing animals. It therefore became necessary to test and monitor food-producing animals for *Salmonella*. Environmental sampling has been found to be one of the best ways to monitor for *Salmonella* on the farm. This progression of inherently different sample types and sources, from clinical disease to food to environment, has important consequences on the methods by which *Salmonella* may be isolated. The media and methods that are best with one particular sample type may not necessarily be optimal for other samples. Therefore particular procedures need to be evaluated for different types of samples.

An early survey of European laboratories found variations in the methodology used to isolate *Salmonella*, which resulted in significant differences in the detection of *Salmonella* (Edel and Kampelmacher, 1968). A nationwide survey in the USA of laboratories primarily involved in culturing *Salmonella* from poultry and poultry environments also showed a large variation in the methodology used by different laboratories (Waltman and Mallinson, 1995). These authors recommended a re-evaluation of the methods for *Salmonella* isolation based on comparative studies to determine the best procedures. A similar recommendation had been made earlier by the World Health Organization (1994).

Several review articles have been published discussing the isolation of *Salmonella* and may be referred to for more detailed information (Jameson, 1962; McCoy, 1962; Fagerberg and Avens, 1976; Harvey and Price, 1979; D'Aoust, 1981; Moats, 1981; Fricker, 1987; Busse, 1995; Andrews, 1996).

Regardless of the sample source, the isolation of *Salmonella* generally includes: (i) direct culture; (ii) non-selective pre-enrichment; (iii) selective enrichment; (iv) inoculation of plating media; (v) screening suspect colonies; and (vi) biochemical and serological confirmation.

Direct Culture

The process of directly inoculating tissues or other samples on to plating media, except in the case of acute infections, is usually non-productive. Typically, in chronic infections, carrier animals or environmental samples, the numbers of *Salmonella* are low, especially relative to the high numbers of other bacteria. These samples should be inoculated into selective enrichment media for optimal recovery of *Salmonella*.

In clinical-disease situations, whether systemic or enteric, samples may be inoculated directly on to plating media. Samples of internal organs, which are normally sterile, should be inoculated on to non-selective (e.g. blood agar, nutrient agar) or at least weakly selective media (e.g. MacConkey agar), in addition to the more selective plating media (e.g. brilliant green agar, xylose lysine desoxycholate agar). This is a good practice for those rare cases when a *Salmonella* strain may be inhibited by the selective agents in the plating media.

Some studies have investigated the use of direct culture for testing eggs for *Salmonella*. Gast (1993) and Gast and Holt (1995) found that direct culture was not as sensitive as selective or non-selective enrichment. However, it was found that isolation by direct culture could be improved by incubating the egg mixture with additional iron or culture media (Reissbrodt and Rabsch, 1993; Gast and Holt, 1995).

Pre-enrichment

Early studies showed that direct plating and direct selective enrichment of certain types of samples were often unsuccessful for the detection of *Salmonella*. Typically, samples that have been dried, heated, irradiated or otherwise processed require the use of non-selective pre-enrichment for optimal recovery of *Salmonella*. In these samples, *Salmonella* may be present but are 'damaged' or 'sublethally injured'. These organisms, although still viable and able to cause disease under the right conditions, are easily killed if placed into the harsh environment of most selective-enrichment broths, especially when incubated at higher temperatures (Corry *et al.*, 1969). Non-selective pre-enrichment broth allows these injured cells to resuscitate prior to transfer into the selective enrichment media.

Several pre-enrichment media have been advocated (Edel and Kampelmacher, 1973; Gabis and Silliker, 1974; D'Aoust, 1981). Lactose broth (LB) was perhaps the first to receive widespread use. But there has been concern that LB, because of the fermentation of lactose and resulting acidity, would allow the pH to fall to a level that is inhibitory or lethal to *Salmonella* (Hilker, 1975).

The recovery of *Salmonella* is not dependent on the nutritional value of the pre-enrichment media (D'Aoust, 1981). Other pre-enrichment media have been formulated without a fermentable sugar and with greater buffering capacities, e.g. buffered peptone water (BPW), M9 and universal pre-enrichment (Bailey and Cox, 1992).

Several studies have found that BPW was better than LB for isolating *Salmonella* (Thomason *et al.*, 1977; Thomason and Dodd, 1978; van Leusden *et al.*, 1982; Fricker, 1987). Juven *et al.* (1984) compared LB, LB with tergitol, M9 and BPW and found that both BPW and M9 were better than LB. The pH of the LB cultures after incubation ranged from 4.8 to 5.5, whereas the ranges for BPW and M9 were 5.8–6.4 and 5.9–6.2, respectively. BPW has been the pre-enrichment broth of choice for use in conjunction with Rappaport–Vassiliadis (RV) enrichment media.

Likewise, it has been shown that pre-enrichment is useful for isolating *Salmonella* from faeces and environmental samples (Edel and Kampelmacher, 1973; van Schothorst and Renaud, 1983; Vassiliadis, 1983; Fricker *et al.*, 1985; Tate *et al.*, 1990; Schlundt and Munch, 1993). For many years, pre-enrichment of these samples was not advocated, because it was thought that non-selective enrichment would allow overgrowth of organisms other than *Salmonella* and give false-negative results. In environmental samples, *Salmonella* may become stressed by the level of available water, pH, temperature or other inhibitory effects of the microecosystem in which they are trying to survive.

Selective Enrichment

Selective-enrichment broths are formulated to selectively inhibit other bacteria while allowing *Salmonella* to multiply to levels that may be detected after plating. There are currently three major types of selective-enrichment media: tetrathionate, selenite and RV. There are also different formulations within each type of enrichment.

A survey of laboratories (Waltman and Mallinson, 1995) found that 13, 17 and 15 different enrichments or combinations of enrichment media were being used for isolating *Salmonella* from poultry tissues, the environment and large animal samples, respectively. RV was not being used for poultry or poultry environmental samples and it was used in only 2% of laboratories for large animal samples. In Europe, however, RV is widely used and is one of the recommended media for monitoring poultry flocks, as laid down in the EC directive 92/117.

Generally, as the number of selective-enrichment media is increased, the number of positive samples also increases. This is also true for incubation conditions and plating media. This potential to increase the numbers of *Salmonella* isolations by using multiple media must, however, be compared with the additional labour and expense involved.

Tetrathionate enrichment media

Muller (1923) described a selective enrichment broth that contained iodine and sodium thiosulphate, which combined to form tetrathionate. Kauffmann (1930, 1935) modified the enrichment broth of Muller by adding ox bile and brilliant green (BG) (tetrathionate brilliant green (TBG)). The mode of action of tetrathionate broth on Gram-negative bacteria is poorly understood; however, it is postulated to involve inactivation of sulphydryl groups of enzymes (Palumbo and Alford, 1970).

Jeffries (1959) modified the tetrathionate enrichment by adding 40 µg ml^{-1} novobiocin to suppress the growth of *Proteus* and *Providencia* spp. Hajna and Damon (1956) also modified the conventional tetrathionate enrichment broth by adding yeast extract, glucose and mannitol, decreasing the bile salts concentration and increasing the sodium thiosulphate concentration. The resulting medium is referred to as tetrathionate Hajna. It is important to recognize that modification of tetrathionate enrichment broths, especially the change in the concentration of sodium thiosulphate, will alter the concentration of the iodine/potassium iodide solution that is added. Failure to use the correct iodine solution will affect the tetrathionate concentration of the enrichment and detrimentally alter its effectiveness.

Different formulations of tetrathionate enrichment have contributed to the often confusing comparative studies. The relative concentrations of sodium thiosulphate and iodine have a dramatic influence on the molar concentration of tetrathionate, which changes the selectivity and specificity of the enrichment media. It also alters the temperature at which the enrichment will be effective and the growth of some *Salmonella* serovars is inhibited in tetrathionate at higher temperatures.

Waltman *et al.* (1995) compared three tetrathionate formulations for the recovery of *Salmonella* from poultry environmental samples. They found no significant differences in the three media. Several studies have shown that tetrathionate enrichment is better than selenite enrichment (Smyser *et al.*, 1970; Carlson and Snoeyenbos, 1972, 1974; D'Aoust *et al.*, 1992; Waltman *et al.*, 1995).

Selenite enrichment media

Leifson (1936) formulated the first selenite enrichment, commonly known as selenite F (SF). Selenite is reduced by bacteria, which increases the pH and reduces the toxicity of selenite. Therefore, a fermentable sugar, usually lactose, which is fermented by enterococci and coliforms, is added to keep the pH within the acidic range. The mode of action of selenite is not well known; however, it has been postulated that selenite or selenium reacts with sulphydryl groups or is incorporated into analogues of sulphur compounds of bacteria (Weiss *et al.*, 1965). North and Bartram (1953) modified SF by adding cystine (selenite cystine (SC)). It had been reported by Leifson (1936) that selenite selectivity was enhanced under reduced conditions. SC reportedly performs better than SF in the presence of organic material.

Stokes and Osborne (1955) modified SF by changing the carbohydrate source from lactose to mannitol and adding sodium taurocholate and brilliant green (selenite brilliant green (SBG)). The formulation was changed to suppress the overgrowth of *Proteus* spp. and coliforms. After it was found that SBG did not work well with egg samples, sulphapyridine was added (SBGS) and the inhibitory properties of the enrichment were restored.

Harvey and Price (1975) found that SBG was not effective in isolating *S. dublin* or in isolating other *Salmonella* when incubated at 43°C. Selenite enrichment medium has been reported to be inhibitory for *S. choleraesuis* (Leifson, 1936; Smith, 1952; Harvey and Price, 1975). Greenfield and Bigland (1970) found that *S. pullorum* and *S. gallinarum* did not grow well in SF. Carlson and Snoeyenbos (1974) found that SBGS was unsatisfactory, because the number of *Salmonella* decreased between 24 and 48 h incubation, especially when incubated at 43°C. Yamamoto *et al.* (1961) also found that SBGS was too selective, and less effective than SC and tetrathionate. In a comparative study, Waltman *et al.* (1995) compared SF, SC, and SBG with tetrathionate and RV enrichment broths. The tetrathionate and RV enrichment results were better than the results from the use of selenite enrichments.

Other disadvantages that have resulted in the reduced use of selenite enrichment media include their shorter shelf-life relative to tetrathionate or RV enrichments. Sodium acid selenite is reduced to the toxic heavy metal selenium, which has been shown to reduce fertility and produce congenital defects, is considered to be embryotoxic and teratogenic and has produced an increase in hepatocellular carcinomas and adenomas in animals (Andrews, 1996; Goyer, 1996). Because selenium is considered a hazardous chemical, selenite enrichment media must be disposed of as hazardous waste.

Rappaport–Vassiliadis enrichment media

Rappaport *et al.* (1956) described an enrichment medium based on the ability of *Salmonella* to: (i) survive relatively high osmotic pressures (achieved using magnesium chloride); (ii) multiply at relatively low pH (pH 5.2); (iii) survive in malachite green (106 mg l^{-1}); and (iv) grow with minimal nutritional requirements (5 g peptone l^{-1}). Later, Vassiliadis *et al.* (1970) modified the medium by reducing the concentration of malachite green (medium referred to as R25). Further modifications by Vassiliadis *et al.* (1976) reduced the malachite green concentration to 36 mg l^{-1}, which made the medium suitable for incubation at 43°C (medium referred to as R10 or RV).

Pre-enrichment of the sample, regardless of the type or source, is advocated with RV broth. The inoculation ratio commonly used for tetrathionate and selenite enrichment broths is 1 : 10; with RV broth, however, it is 1 : 100 (Rappaport *et al.*, 1956; Vassiliadis, 1983; Fricker *et al.*, 1985). Some investigators have even advocated a ratio of 1 : 1000. The recommended incubation temperature is 41.5°C; however, some studies advocate temperatures as high as 43°C (Vassiliadis *et al.*, 1976), although the higher temperature has been found to be inhibitory to some *Salmonella*, including *S. dublin* (Peterz *et al.*, 1989). The enrichment is normally plated after 24 h incubation, although Fricker *et al.* (1985) found that isolation of *Salmonella* was best after 48 h incubation.

Several studies have shown that RV enrichment was better than either tetrathionate or selenite enrichment broths (Bager and Peterson, 1991; Oboegbulem, 1993; Schlundt and Munch, 1993; Waltman *et al.*, 1995). RV enrichment medium has been approved for isolating *Salmonella* from raw, highly contaminated food and poultry feeds, replacing the use of selenite enrichment media (June *et al.*, 1996).

A number of studies have shown that tetrathionate and selenite enrichment media were unsatisfactory for the isolation of *S. choleraesuis* (Smith, 1952; Rappaport *et al.*, 1956; Smith, 1959; Sharma and Packer, 1969). Conflicting reports have been published concerning their effectiveness with *S. dublin* (Harvey and Price, 1975; Schlundt and Munch, 1993).

Modified semi-solid Rappaport–Vassiliadis media

Goossens *et al.* (1984) developed a semi-solid medium based on RV selective enrichment broth (modified semi-solid RV (MSRV)). A few years later, De Smedt and Bolderdijk (1987) developed the commercial MSRV as a selective enrichment medium in a plate format. The selective properties of MSRV differ from RV in having increased nutrients and buffering capacity, decreased magnesium chloride concentration and added novobiocin. The MSRV plate is incubated at 41.5°C and the motility of *Salmonella* further selects and differentiates *Salmonella* from other microorganisms. Davies and Wray (1994a) modified the use of MSRV media by adding a disc soaked in polyvalent H antisera on to the surface of the media and observing a zone of inhibition around the disc when motile *Salmonella* were present.

Several studies have shown the advantages of the MSRV culture method (Aspinall *et al.*, 1992; Davies and Wray, 1994a; Read *et al.*, 1994). The MSRV culture method cannot be used for isolating non-motile *Salmonella*, such as *S. pullorum* and *S. gallinarum*, unless the growth at the point of inoculation is sub-cultured, which may result in a large number of false positive samples.

Gram-negative enrichment broth

Gram-negative (GN) broth, which was developed by Hajna (1956) for the culture of Gram-negative bacteria, contains sodium citrate and sodium desoxycholate as selective agents. Cox *et al.* (1972) found that GN was less effective than RV and SC for isolating *Salmonella* from egg, meat, chicken faeces and turkey rolls. Taylor and Schelhart (1971) compared GN and SF enrichment for the isolation of *Salmonella* from 1597 stool samples. The GN enrichment was as sensitive as SF, but not as specific. Gram-negative enrichment may be useful if other enteric organisms, e.g. *Shigella*, need to be isolated.

Two recent pig and cattle surveys conducted in the USA used direct inoculation of tetrathionate and GN enrichment broths with faecal and feed samples (Fedorka-Cray *et al.*, 1995). After 24 h, the GN broth was subcultured into RV broth. The combination of GN followed by transfer into RV broth was not as effective as the combination of tetrathionate followed by transfer into RV broth.

Incubation Conditions

Pre-enrichment

Since the pre-enrichment step increases the time required for the isolation of *Salmonella*, several investigators have attempted to shorten the pre-enrichment time. Most of these studies have found that incubating for less than 18 h resulted in reduced sensitivity (D'Aoust and Maishment, 1979; D'Aoust, 1981; Andrews, 1986; D'Aoust *et al.*, 1990). Therefore, incubating the pre-enrichment for 18–24 h is recommended to allow resuscitation of *Salmonella* before transfer into selective-enrichment media.

The recommended incubation temperature for pre-enrichment is 35–37°C. Since the purpose of pre-enrichment is the resuscitation of damaged *Salmonella*, a higher temperature should be avoided. After incubation, an aliquot of the pre-enrichment broth in the volumes described previously is transferred into 10 ml of selective enrichment broth (1 : 10 ratio), except with RV enrichment broth, where a ratio of 1 : 100 is recommended.

Selective enrichment

Incubation temperature
Generally, samples, such as internal organs or tissues, having low levels of background flora are incubated at 35–37°C. A higher temperature is not necessary to suppress contaminants in these samples. Moreover, some of the host-adapted *Salmonella* (e.g. *S. pullorum* and *S. gallinarum* in poultry and *S. dublin* in cattle) may be inhibited at higher temperatures. Intestinal and environmental samples, which generally have higher levels of competing bacteria, may be incubated at higher temperatures (40–43°C), because *Salmonella* are more tolerant to the higher temperature. The optimal growth temperature of *Salmonella* is about 37°C; however, with naturally contaminated samples the selective benefits of incubation at higher temperatures may offset the reduced growth potential.

Harvey and Thompson (1953) were the first to report that incubation of selenite broth at 43°C was better than incubation at 37°C with human faecal samples. Others have also shown the benefits of incubation at 40°C or higher (Dixon, 1961; Carlson et al., 1967; Harvey and Price, 1968; Smyser and Snoeyenbos, 1969; Banffer, 1971; Carlson and Snoeyenbos, 1972; D'Aoust, 1981; Dusch and Altwegg, 1995). Jameson (1962) suggested that, when selenite enrichment was incubated at 37°C, its concentration of selenite was less than optimal. This suggests that the concentration of some selective agent that has a bacteriostatic effect on a competitor at 37°C may exert a bactericidal effect at a higher temperature. Conversely, a concentration of some selective agent that is effective at 37°C may become inhibitory to _Salmonella_ at higher temperatures.

Some investigators have found that incubation at higher temperatures, especially 43°C, can be inhibitory or even lethal to some _Salmonella_ (McCoy, 1962; Carlson and Snoeyenbos, 1974; Harvey and Price, 1979). Certainly, any deviation above 43°C may be lethal (Busse, 1995). Harvey and Thompson (1953) found that _S. typhi_ and _S. pullorum_ could not be isolated satisfactorily at 43°C. Therefore, a selective enrichment cannot be arbitrarily incubated at any temperature without proper validation. Generally, a temperature of 41–42°C is advocated for the high incubation temperature to allow for some variation in incubator temperature. The temperature of incubators should be routinely monitored and checked at more than one area to determine if any 'hot spots' exist. Another precaution involves the use of large volumes of cold enrichment media. Since both tetrathionate and RV enrichment broths can be stored and refrigerated, these must be allowed to warm up to at least room temperature before addition of the sample and incubation. Cold media require at least 4–8 h to allow the sample to reach the appropriate temperature. If the normal incubation period before plating is 18–24 h, this dramatically reduces the incubation time and may reduce the ability to isolate _Salmonella_.

Incubation time

The US laboratory survey found that 51%, 48% and 65% of the laboratories culturing poultry tissues and environmental and large-animal samples,

respectively, were incubating the selective-enrichment broths for only 24 h. Another 28%, 25% and 30% of laboratories, respectively, were incubating the selective-enrichment broths for another 24 h and replating (Waltman and Mallinson, 1995).

Sharma and Packer (1969) subcultured enrichment broth cultures after 6, 9, 12, 15, 18, 24, 30, 36 and 48 h and were unable to isolate _Salmonella_ before 12 h and only small numbers until 24 h. Likewise, Grunnet (1975) found that enrichment times less than 24 h were not effective. After 24 h, 60% of the samples were positive; subculture after 48, 72 and 96 h increased the isolation rates by 23%, 11% and 6%, respectively. Using pure cultures, Carlson and Snoeyenbos (1972) examined the population dynamics of pure cultures in SBGS and tetrathionate and found that the maximum number of cells was not reached until about 32 h.

Several investigators have recommended a second subculture after 48 h incubation (Galton et al., 1968; Edel and Kampelmacher, 1974; D'Aoust et al., 1992). No further advantage of extending the incubation time beyond 48 h was found; in fact, other bacteria will usually overgrow _Salmonella_ (Carlson et al., 1967). However, with some enrichment media, especially when incubated at higher temperatures, _Salmonella_ begin to die between 24 and 48 h (Osborne and Stokes, 1955; Carlson and Snoeyenbos, 1974).

Delayed secondary enrichment

Delayed secondary enrichment (DSE) is the process whereby the original selective-enrichment broth (usually tetrathionate) is held at room temperature after the initial 24 h incubation and subsequent culture. If the initial subculture is negative for _Salmonella_, the enrichment broth is left at room temperature for 5–7 days; then 0.5–1.0 ml of the original selective-enrichment broth is transferred into 10 ml of fresh selective-enrichment broth, which is incubated at 37°C for 18–24 h and then subcultured on to plating media. The difference between DSE and using extended incubation times, which often showed a decrease in isolation of _Salmonella_ after 48 h, is that DSE cultures are held at room temperature, not incubation temperatures of 35°C or higher.

Several studies have shown increased isolation of _Salmonella_ using DSE (Pourciau and Springer, 1978; Rigby and Pettit, 1980; Waltman et al., 1991). In a study of over 1400 clinical and environmental samples, the number from which *Salmonella* were isolated after 24 h, after 48 h and following DSE was 107, 137 and 187, respectively (Waltman *et al.*, 1993). If the selective enrichments had only been plated after 24 h, then 45% of the total *Salmonella* isolated would not have been detected, and subculture after 48 h still failed to detect 29% of the *Salmonella*.

In a follow-up study, Waltman *et al.* (1993) compared incubation for 24 h and 48 h and with a 3-day and 5-day DSE. The number from which *Salmonella* were isolated was 32, 48, 58 and 65, respectively. Incubation times of 24 and 48 h would have failed to detect 55% and 32% of the total *Salmonella,* respectively. This study also showed that the longer, 5-day, DSE was better than the shorter, 3-day, DSE.

Plating Media

Principles

The enrichment process is designed to increase the number of *Salmonella* in the culture to a level that may be detected on plating media. Various plating media have been developed for the isolation of *Salmonella* by using the principles of selectivity and differentiation. Selectivity involves the incorporation of inhibitory substances into the media that selectively inhibit other bacteria. The differential characteristics of a plating media involve the addition of substances that differentiate *Salmonella* colonies from other bacteria. This often results from the production of hydrogen sulphide (H_2S) or the production of acid from some sugar, e.g. lactose.

Many different plating media are available for the isolation of *Salmonella*. The US laboratory survey showed that 14 different plating media were being used with poultry, environmental and large-animal samples (Waltman and Mallinson, 1995). It is recommended that at least two plating media be used, with different selective and differential properties. For example, the combination of brilliant green with novobiocin (BGN) and xylose lysine tergitol 4 (XLT4) agars has been found to be very effective for the isolation of *Salmonella* (Mallinson, 1990; Waltman *et al.*, 1995).

Plating media should be judged not only on their ability to selectively grow *Salmonella*, but also on their ability to differentiate colonies of *Salmonella* from those of other bacteria. The inability to distinguish *Salmonella* colonies may result in false positives, which increase the time, labour and expense involved in culturing the samples. One of the main problems is the presence of *Proteus* spp., but the addition of novobiocin to plating media has essentially eliminated their presence and resulted in a decreased number of false-positive samples (Moats, 1978; Komatsu and Restaino, 1981; Waltman *et al.*, 1995). In these studies, the addition of novobiocin not only increased the specificity of the medium, but it also increased its sensitivity. Novobiocin prevents the overgrowth of other bacteria, which may mask the presence of *Salmonella* either physically or by altering the pH of the media, resulting in atypical colonies, and it also reduces the presence of a high number of colonies similar to *Salmonella*, which might prevent the detection of *Salmonella* unless several colonies are screened.

It is important that laboratory personnel be aware of atypical *Salmonella*, for example, Waltman *et al.* (1995) reported that 13% of *Salmonella* isolated from poultry and poultry environmental samples were not H_2S-positive on plating media. Also, *S. pullorum*, *S. gallinarum* and some *S. choleraesuis* do not produce H_2S on most media. Other atypical reactions include lysine decarboxylase-negative or lactose- or sucrose-positive isolates. It is therefore recommended that at least two plating media are used, with different selective agents and differential characteristics.

Plating media are incubated at 35–37°C for 20–24 h and observed for suspected *Salmonella*. Some of the host-adapted *Salmonella*, especially *S. pullorum* and *S. abortusovis*, grow slowly. Therefore, the plates should be incubated for an additional 24 h before discarding as negative. Two exceptions to this 35–37°C incubation temperature are the incubation of a Rambach agar plate at 41.5°C when it is subcultured from MSRV (Davies and Wray, 1994a) and the incubation of the EF-18 agar plate at 42°C (Entis and Boleszczuk, 1991).

Characteristics of plating media

Bismuth sulphite agar

Bismuth sulphite (BS) agar was developed by Wilson and Blair (1927) for the isolation of

S. typhi. The selective properties are conferred by BS and BG. The differential properties of BS agar differ from those of other media in that carbohydrate fermentation is not used. An H_2S indicator system results in typical *Salmonella* colonies appearing black with a characteristic black sheen.

BS agar has been advocated for the detection of *S. typhi* and *Salmonella* that ferment lactose. A major disadvantage of BS agar involves its stability, which is the subject of controversy. Some investigators recommend using BS agar within a few days of preparation, while others recommend 'ageing' the medium (McCoy, 1962; Fagerberg and Avens, 1976). Moats (1981) found that inhibitory properties of BS agar change during storage. *Salmonella* other than *S. typhi* have been shown to grow better on BS agar if it has been aged in the refrigerator for a few days (Cook, 1952; McCoy, 1962; Hobbs, 1963). McCoy (1962) also found that 'ageing' was necessary for the development of characteristic black colonies surrounded by a clear translucent periphery.

Another difficulty with BS agar is the differentiation of *Salmonella* colonies from those of other bacteria (Montford and Thatcher, 1961; Erdman, 1974). It is often difficult to distinguish dark brown and black colonies on BS, especially when the colonies are poorly separated. Some *Salmonella*, especially *S. pullorum*, *S. gallinarum* and *S. choleraesuis*, do not produce characteristic colonies (McCoy, 1962).

Brilliant green agar

Brilliant green agar (BGA) was developed by Kristensen *et al.* (1925) and later modified by Kauffmann (1935). The selectivity of the agar derives from the presence of the BG dye and the presence of lactose and sucrose, which is the basis for the differential capabilities of the media. Almost all *Salmonella* fail to ferment either lactose or sucrose and their colonies appear pink to red, with a reddening of the media.

Two major modifications of BGA have been developed primarily to prevent *Proteus* spp. overgrowth. The first was the addition of sulphadiazine (Galton *et al.*, 1954) or sulphapyridine (Osborne and Stokes, 1955) (BGS) and the other was the addition of novobiocin (Tate and Miller, 1990) (BGN).

The growth of *Salmonella* on BG and BGS agars was compared with growth on nutrient or trypticase soy agar and very little if any inhibition of *Salmonella* was found (Moats and Kinner, 1974; Moats, 1981). Several investigators have advocated the use of BG (Smyser *et al.*, 1963; Pourciau and Springer, 1978), BGS (Ellis *et al.*, 1976) and BGN agars (Mallinson, 1990; Waltman *et al.*, 1995; National Poultry Improvement Plan, 1996). *S. dublin* may either fail to grow or may produce small colonies on BGA (Harvey and Price, 1975).

Deoxycholate citrate agar

Deoxycholate citrate agar (DCA) was developed by Leifson (1935) and later modified by Hynes (1942) as a medium to isolate *Salmonella* and other enteric pathogens, such as *Shigella*. The selective ability of DCA involves sodium deoxycholate and, to a lesser extent, sodium citrate. The presence of lactose and an H_2S indicator system provides the differential characteristics of the agar. Since *Salmonella* usually do not ferment lactose and do produce H_2S, their colonies are clear with black centres.

Although some studies have shown favourable results with DCA (Oboegbulem, 1993), a major disadvantage of this medium is the high number of false-positive colonies, primarily due to *Proteus* spp., which are not easily differentiated from those of *Salmonella* (Jeffries, 1959; Fagerberg and Avens, 1976).

EF-18

EF-18 is a plating media that is predominantly used in conjunction with hydrophobic membrane filtration (Entis and Boleszczuk, 1991); however, a few studies have been conducted using it as a standard plating media from enrichment broth. EF-18 is highly selective and contains bile salts, crystal violet, sulphapyridine and novobiocin. It is also incubated at 42°C. The differential properties are conferred by the presence of sucrose and lysine, and *Salmonella* colonies appear blue-green. Warburton *et al.* (1994b) encountered problems with the use of this technique for the isolation of *Salmonella*, because the colonies were reduced in size and overgrown by other bacteria.

Hektoen enteric agar

Hektoen enteric (HE) agar was formulated by King and Metzger (1968) for the isolation of *Salmonella* and *Shigella*, while inhibiting normal intestinal flora. The selective agent in HE agar is bile salts. The differential capabilities involve the sugars lactose, sucrose and salicin and an H$_2$S indicator system. Most *Salmonella* do not ferment the three sugars, but produce H$_2$S, which results in bluish-green colonies with black centres.

A major disadvantage of HE is the number of false-positive colonies that may be present, primarily *Proteus* spp. (Poisson *et al.*, 1992; Dusch and Altwegg, 1995). Several investigators have added novobiocin to the medium (HEN) to overcome this problem (Hoben *et al.*, 1973; Restaino *et al.*, 1977, 1982). Each study showed that the addition of novobiocin increased the sensitivity and specificity of the medium.

MacConkey agar

One of the first selective media, MacConkey (MAC) agar, was described by MacConkey (1905) for the selective isolation of Gram-negative enteric bacteria. MAC agar uses bile salts and crystal violet as selective agents; however, the concentration of bile salts is lower than that in other *Salmonella* isolation media. This produces very little selectivity, although it is useful for isolating *Salmonella* strains that may be inhibited on more selective media. The differential ability of MAC agar lies in the fermentation of lactose. Most *Salmonella* do not ferment lactose and, as a consequence, the colonies are clear. MacConkey and other similar media, e.g. violet red bile and eosin methylene-blue agar, are excellent for isolating enteric bacteria from clinical specimens; however, they lack sufficient selectivity and differential properties for isolating *Salmonella* from intestinal and environmental sources, either directly or from enrichment broths.

Mannitol lysine crystal violet brilliant green agar

Mannitol lysine crystal violet brilliant green (MLCB) agar was developed by Inoue *et al.* (1968) and modified by van Schothorst *et al.* (1987). The media contains BG and crystal violet as selective agents. The differential characteristics are conferred by lysine, mannitol and an H$_2$S detection system. Typical *Salmonella* produce large bluish-purple colonies that may have black centres or be almost entirely black depending on the amount of H$_2$S produced. *Proteus* spp. produce similar colonies.

Modified lysine iron agar

Modified lysine iron agar (MLIA) was developed by Rappold and Bolderdijk (1979) and modified to its commercial form by Bailey *et al.* (1988).

They took advantage of the differential capabilities of lysine iron agar and added lactose, sucrose and novobiocin. The combination of glucose, lactose, sucrose, lysine and an H$_2$S indicator system enables *Salmonella* to be differentiated. Typical *Salmonella* are blue, with black centres. The presence of bile salts and novobiocin selectively inhibits other bacteria, including *Proteus* spp. Bailey *et al.* (1988) found MLIA to be better than BGS and xylose lysine desoxycholate with novobiocin (XLDN) agar for isolating *Salmonella* from turkey, cured chicken and broiler carcasses. Waltman *et al.* (1995) found MLIA to be as effective as BGN, HEN and XLDN agars.

Novobiocin brilliant green glycerol lactose agar

Novobiocin brilliant green glycerol lactose (NBGL) agar was developed by Poisson (1992). The selective agents include BG and novobiocin. The differential characteristics are conferred by lactose, glycerol and an H$_2$S indication system. Glycerol was added to help differentiate *Salmonella* from *Citrobacter*. Typical *Salmonella* produce colonies with black centres.

Rambach agar

One of the more recently formulated media, Rambach (RAM) agar was developed based on the finding that *Salmonella* produce acid from propylene glycol (Rambach, 1990). The selective ability results from the presence of bile salts. The differential characteristics involve the fermentation of propylene glycol, typical *Salmonella* appearing as crimson-red colonies, and a chromogenic detection system for the production of β-D-galactosidase by other *Enterobacteriaceae*, which, after reaction with the chromogenic substance, X-Gal (5-bromo-4-chloro-3-indole-β-D-galactopyranoside), results in the formation of blue-green colonies. *Salmonella* belonging to the subspecies IIIa, IIIb and V produce β-D-galactosidase and would also appear as blue-green colonies. Some serovars do not produce the characteristic red colonies, e.g. *S. pullorum*, *S. gallinarum*, *S. choleraesuis*, *S. abortussuis*, *S. abortusequi*, and *S. dublin* (Kuhn *et al.*, 1994; Pignato *et al.*, 1995a, b). Likewise, *S. typhi* and *S. paratyphi* A and B fail to produce acid and appear as colourless colonies.

Salmonella–Shigella agar

Closely related to DCA, *Salmonella–Shigella* (SS) agar was formulated to inhibit coliforms, while allowing the recovery of *Salmonella* and *Shigella*. The selective agents include, bile salts, BG and, to a lesser degree, sodium citrate. The differential agents are lactose fermentation and the presence of an H_2S indicator system. Typical *Salmonella* are colourless with black centres. A major disadvantage of SS agar is the number of false positives, primarily due to *Proteus* spp. (Yamamota *et al.*, 1961; Pollock and Dahlgren, 1974; Fagerberg and Avens, 1976).

Salmonella *identification agar*

Salmonella identification (SM-ID) agar is a chromogenic medium similar to RAM agar; however, it can be used for the isolation of *S. typhi* and *S. paratyphi*. Differential characteristics involve fermentation of D-glucuronate and a β-galactosidase indicator. *Salmonella*, including *S. typhi* and *S. paratyphi*, are typically red. Davies and Wray (1994b) found the sensitivity and specificity of SM-ID to be as good as or better than BG, RAM or xylose lysine desoxycholate (XLD) agars.

Xylose lysine desoxycholate *agar*

XLD agar was developed by Taylor (1965). The selective agent is sodium desoxycholate and the differential ability comes from the combination of the sugars xylose, lactose and sucrose, the amino acid lysine and an H_2S indicator system. Most *Salmonella* ferment xylose, but not lactose and sucrose, decarboxylate lysine and produce H_2S, and typical *Salmonella* colonies are red with black centres.

Although an effective isolation medium, the major disadvantage of XLD is the high false-positive rate primarily due to *Proteus* spp. and non-H_2S-producing *Salmonella*. The addition of novobiocin (XLDN) has been shown to increase the specificity and the sensitivity (Restaino *et al.*, 1977, 1982; Komatsu and Restaino, 1981; Waltman *et al.*, 1995).

Xylose lysine tergitol 4 *agar*

Miller *et al.* (1991) modified XLD agar by substituting tergitol 4 (Niaproof), which is inhibitory to *Proteus*, *Pseudomonas* and *Providencia*, as the selective agent. The differential characteristics are the same as for XLD; however, additional peptone is added to enhance the weak H_2S reactions (Miller *et al.*, 1995). Most *Salmonella* colonies are black.

Table 21.1 shows the appearance of *Salmonella* and other organisms on the more commonly used selective agars.

Comparison of plating media

Several studies have compared the various plating media; however, no single study has compared all the media (Tables 21.2 and 21.3). Studies have

Table 21.1. Appearance of *Salmonella* and other organisms on selective agars (after WHO/Zoon./94.173).

Medium*	Salmonella	Proteus spp.		Coliforms		Pseudomonas	
		Growth	Appearance	Growth	Appearance	Growth	Appearance
BS	Black metallic sheen	1	Black	3	Brown-green	–	–
BGA	Red	3	Red	2	Yellow-green	2	Red
BGS	Red	3	Red	4	Yellow-green	2	Red
BGN	Red	4	Red	3	Yellow-green	3	Red
DCA	Colourless BC	1	Colourless BC	2	Pink	–	–
Grassner	Yellow	1	Yellow	1	Black	1	Yellow
HE	Blue-green BC	1	Blue-green BC	3	Pink	–	–
MAC	Colourless	1	Colourless	1	Red–pink	1	Colourless
RAM	Crimson with pale borders	2	Colourless	2	Blue–violet	3	Dull orange
SS	Colourless BC	2	Colourless BC	2	Pink	–	–
XLD	Red BC	1	Yellow BC	2	Yellow	–	–
XLT4	Red BC	4	–	3	Yellow	3	–

*See text for descriptions of plating media.
Growth: 1, good; 2, fair to good; 3, poor; 4, absent to poor. BC, black centre due to H_2S production.

Table 21.2. Results of studies comparing the sensitivity of various plating media for isolating *Salmonella*.

Study						Percentage *Salmonella* isolated on each plating medium*								
	BG	BGS	BGN	HE	HEN	XLD	XLDN	XLT4	MAC	SS	BS	DCA	RAM	NBGL
A	-	-	-	80	-	83	-	-	42	74	-	-	-	-
B	-	-	-	52	58	68	84	-	-	-	-	-	-	-
C	-	65	91	75	85	50	82	-	-	-	-	-	-	-
D	82	72	85	20	-	33	81	98	12	-	-	-	-	-
E	71	-	79	-	-	30	84	98	-	-	-	-	-	-
F	-	-	-	45	78	44	78	83	-	-	-	-	73	-
G	-	93	-	95	-	89	-	-	-	-	91	-	87	91
H	-	-	-	-	-	-	-	-	-	92	91	-	-	-
I TT37	-	-	-	100	-	-	-	97	-	-	-	-	100	81
TT42	-	-	-	100	-	-	-	100	-	-	-	-	100	97
J	-	-	-	83	-	-	-	-	-	84	-	-	-	92
K	-	-	-	89	-	-	-	-	-	84	-	-	87	94

* See text for descriptions of plating media.
Study: A, Taylor and Schelhart, 1971, stool samples; B, Moats, 1978, beef and turkey samples; C, Komatsu and Restaino, 1981, meat products; D, Mallinson, 1990, drag swabs; E, Miller *et al.*, 1991, drag swabs; F, Waltman *et al.*, 1995, environmental samples; G, Warburton *et al.*, 1994a, food and environmental samples; H, Ruiz *et al.*, 1996, stool samples; I, Dusch and Altwegg, 1995, stool samples; J, Poisson *et al.*, 1992, stool samples; K, Poisson *et al.*, 1993, stool samples.

Table 21.3. Results of studies comparing the specificity (false positives) of various plating media for isolating *Salmonella*.

Study						Percentage false-positive samples from each plating medium*								
	BG	BGS	BGN	HE	HEN	XLD	XLDN	XLT4	MAC	SS	BS	DCA	RAM	NBGL
A	-	-	-	62	-	25	-	-	21	52	-	-	-	-
B	-	16	-	45	19	14	2	-	-	-	-	-	-	-
C	-	-	3	50	16	38	5	0	-	-	-	-	6	5
D	-	-	-	32	7	35	1	-	-	39	17	-	-	-
E	-	-	-	-	-	-	-	-	-	-	-	-	-	-
F TT37	-	-	-	39	-	-	-	0	-	-	-	-	3	6
TT42	-	-	-	19	-	-	-	0	-	-	-	-	0	0
G	20	24	-	-	-	-	-	-	-	58	82	61	-	-
H	-	-	-	16	-	-	-	-	-	16	-	-	-	10
I	-	-	-	10	-	-	-	-	-	-	-	-	1	1

*See text for description of plating media.
Study: A, Taylor and Schelhart, 1971, stool samples; B, Moats, 1978, beef and turkey samples; C, Komatsu and Restaino, 1981, meat products; D, Waltman *et al.*, 1995, environmental samples; E, Ruiz *et al.*, 1996, stool samples; F, Dusch and Altwegg, 1995, stool samples; G, Montford and Thatcher, 1961, egg products; H, Poisson *et al.*, 1992, stool samples; I, Poisson *et al.*, 1993, stool samples.

shown a correlation between sensitivity and specificity. Generally, as the specificity increases, i.e. false-positives decrease, the sensitivity of the media also increases. The addition of novobiocin dramatically increases the sensitivity of the media by inhibiting the competing flora, especially *Proteus*. Other media that have also been shown to be good isolation media typically have a low percentage of false-positive reactions.

Selection of Presumptive *Salmonella* Colonies

Following incubation, plating media are observed for typical *Salmonella* colony formation and morphology, which will differ according to the plating media used. Therefore, individuals need to be trained in the appearance of *Salmonella* on various plating media and to be aware of atypical colonies. Depending on the selectivity of the respective plating media, there may be almost a pure culture of *Salmonella* or there may be several species of bacteria, including *Salmonella*, growing on the plates. Herein lies the usefulness of the more selective plating media, which will reduce the level of colonies similar to *Salmonella* and the labour involved in screening them. The use of more than one plating medium will help to detect atypical strains.

Generally, suspect *Salmonella* colonies are selected and inoculated into tubes of composite media, e.g. triple sugar iron (TSI) and lysine iron (LI) agar. These tubes are incubated overnight at 37°C and the resulting biochemical reactions observed. The typical reactions are shown in Table 21.4.

Three to five suspect colonies should be routinely screened. This allows a reasonable probability of detecting *Salmonella* and the possible presence of more than one serovar. Additional colonies may be screened in situations where more confidence is desired, to ensure the presence or absence of a particular serovar.

Harvey and Price (1979) found that plating enrichment broths at different times was as efficient for the isolation of different serovars as screening a large number of colonies. In samples that are prone to have multiple serovars, it may be best to use more than one selective-enrichment broth and two or more different plating media.

Other methods include the use of miniaturized test kits, automated and semi-automated systems (see Chapter 22) and latex agglutination tests. A commercially available kit is available for specifically screening for group D *Salmonella* (Lamichhane *et al.*, 1995). A membrane is placed over the isolation plate and then processed through a series of enzyme-linked immunosorbent assay (ELISA)-like steps. Any positive colonies may then be picked from the agar and confirmed.

Serotyping

The genus *Salmonella* has perhaps the most elaborate and certainly the largest serological typing system among bacteria. Serovars are very specifically identified on the basis of their respective somatic and flagellar antigens, and this is dealt with in Chapter 23.

Biochemical Identification

It is often necessary to biochemically identify or at least screen isolates. Occasionally, other bacterial species will react serologically with the *Salmonella* antisera, especially somatic sera. Also, there are some *Salmonella* biotypes that can only be determined based on biochemical reactions. For example, the separation of *S. pullorum* from *S. gallinarum* in poultry and the separation of *S. choleraesuis* from *S. choleraesuis* var. Kunzendorf in pigs require biochemical testing. 'Typical' biochemical reactions of *Salmonella* and some other commonly encountered bacteria are shown in Table 21.4.

Table 21.4. Typical biochemical reactions of *Salmonella* and other commonly encountered bacteria (from Ewing, 1986; Farmer, 1995).

Bacteria	TSI	LIA	ONPG	LAC	MAN	SAL	SUC	XYL	LYS	MLN	URE
					Biochemical test results						
Citrobacter freundii	A/AG,H₂S	K/A	+	+	+	–	–	+	–	–	–
Enterobacter cloacae	A/AG	K/A	+	+	+	+	+	+	–	–	+/–
Escherichia coli	A/AG	K/K	+	+	+	+/–	+/–	+	+	–	–
Klebsiella pneumoniae	A/AG	K/K	+	+	+	+	+	+	+	+	+
Morganella morganii	K/AG	R/A	–	–	–	–	–	–	–	–	+
Proteus mirabilis	K/AG,H₂S	R/A	–	–	–	+/–	+	+	–	–	+
Proteus vulgaris	A/AG,H₂S	R/A	–	–	–	+/–	+	+	–	–	+
Providencia rettgeri	K/A	R/A	–	–	+	+/–	+/–	–	–	–	+
'Typical' *Salmonella*	K/AG,H₂S	K/K	–	–	+	–	–	+	+	–	–
Salmonella arizonae	K/AG,H₂S	K/K	+	+	+	–	–	+	+	+	–

+, > 75% positive reaction; +/–, 25–75% positive reaction; –, < 25% positive reaction. Tests: TSI, triple sugar iron agar; LIA, lysine iron agar; ONPG, ortho-nitro-phenyl-galactopyranoside (beta-galactosidase); LAC, lactose fermentation; MAN, mannitol fermentation; SAL, salicin fermentation; SUC, sucrose fermentation; XYL, xylose fermentation; LYS, lysine decarboxylase; MLN, malonate utilization; URE, urease. TSI: A, acid (yellow); K, alkaline (red); G, gas (bubbles); H₂S, hydrogen sulphide (black). LIA: A, acid (yellow); K, alkaline (purple); R, lysine deaminase (red).

References

Andrews, W.H. (1986) Resuscitation of injured *Salmonella* spp. and coliforms from foods. *Journal of Food Protection* 49, 62–75.

Andrews, W.H. (1996) Evolution of methods for the detection of *Salmonella* in foods. *Journal of AOAC International* 79, 4–12.

Aspinall, S.T., Hindle, M.A. and Hutchinson, D.N. (1992) Improved isolation of salmonellae from faeces using a semisolid Rappaport–Vassiliadis medium. *European Journal Clinical Microbiology and Infectious Disease* 11, 936–939.

Bager, F. and Petersen, J. (1991) Sensitivity and specificity of different methods for the isolation of *Salmonella* from pigs. *Acta Veterinarian Scandinavica* 32, 473–481.

Bailey, J.S. and Cox, N.A. (1992) Universal preenrichment broth for the simultaneous detection of *Salmonella* and *Listeria* in foods. *Journal of Food Protection* 55, 256–259.

Bailey, J.S., Chi, J.Y., Cox, N.A. and Johnson, R.W. (1988) Improved selective procedure for detection of salmonellae from poultry and sausage products. *Journal of Food Protection* 51, 391–396.

Banffer, J.R.J. (1971) Comparison of the isolation of salmonellae from human faeces at 37°C and 43°C. *Zentralblatt für Bakteriologie, Parasitenkunde, Infectionskrankheiten und Hygiene I Abteilung Originale* 217, 35–40.

Busse, M. (1995) Media for *Salmonella*. *International Journal of Food Microbiology* 26, 117–131.

Carlson, V.L. and Snoeyenbos, G.H. (1972) Relationship of population kinetics of *Salmonella typhimurium* and cultural methodology. *American Journal of Veterinary Research* 33, 177–184.

Carlson, V.L. and Snoeyenbos, G.H. (1974) Comparative efficacies of selenite and tetrathionate enrichment broths for the isolation of *Salmonella* serovars. *American Journal of Veterinary Research* 35, 711–719.

Carlson, V.L., Snoeyenbos, G.H., McKie, B.A. and Smyser, C.F. (1967) A comparison of incubation time and temperature for the isolation of *Salmonella*. *Avian Diseases* 11, 217–225.

Cook, C.T. (1952) Comparison of two modifications of bismuth sulphite agar for the isolation and growth of *Salmonella typhi* and *Salmonella typhimurium*. *Journal of Pathology and Bacteriology* 64, 559–566.

Corry, J.E.L., Kitchell, A.G. and Roberts, T.A. (1969) Interaction in the recovery of *Salmonella typhimurium* damaged by heat or gamma radiation. *Journal of Applied Bacteriology* 32, 415–511.

Cox, N.A., Davis, B.H., Kendall, J.H., Watts, A.B. and Colmer, A.R. (1972) *Salmonella* in the laying hen. 3. A comparison of various enrichment broths and plating media for the isolation of *Salmonella* from poultry faeces and poultry food products. *Poultry Science* 51, 1312–1316.

D'Aoust, J.Y. (1981) Update on preenrichment and selective enrichment conditions for detection of *Salmonella* in foods. *Journal of Food Protection* 44, 369–374.

D'Aoust, J.Y. and Maishment, C. (1979) Pre-enrichment conditions for effective recovery of *Salmonella* in foods and feed ingredients. *Journal of Food Protection* 42, 153–157.

D'Aoust, J.Y., Daley, E. and Sewell, A.M. (1990) Performance of the microplate BacTrac ELISA technique for detection of food-borne *Salmonella*. *Journal of Food Protection* 53, 841–845.

D'Aoust, J.Y., Sewell, A.M. and Jean, A. (1992) Efficacy of prolonged (48 h) selective enrichment for the detection of food-borne *Salmonella*. *International Journal of Food Microbiology* 15, 121–130.

Davies, R.H. and Wray, C. (1994a) Evaluation of a rapid cultural method for identification of salmonellas in naturally contaminated veterinary samples. *Journal of Applied Bacteriology* 77, 237–241.

Davies, R.H. and Wray, C. (1994b) Evaluation of SMID agar for identification of *Salmonella* in naturally contaminated veterinary samples. *Letters in Applied Microbiology* 18, 15–17.

De Smedt, J.M. and Bolderdijk, R.F. (1987) Dynamics of *Salmonella* isolation with modified semi-solid Rappaport–Vassiliadis medium. *Journal of Food Protection* 50, 658–661.

Dixon, J.M.S. (1961) Rapid isolation of salmonellae from faeces. *Journal of Clinical Pathology* 14, 397–399.

Dusch, H. and Altwegg, M. (1995) Evaluation of five new plating media for isolation of *Salmonella* species. *Journal of Clinical Microbiology* 33, 802–804.

Edel, W. and Kampelmacher, E.H. (1968) Comparative studies on *Salmonella*-isolation in eight European laboratories. *Bulletin of the World Health Organization* 39, 487–491.

Edel, W. and Kampelmacher, E.H. (1973) Comparative studies on the isolation of 'sublethally injured' salmonellae in nine European laboratories. *Bulletin of the World Health Organization* 48, 167–174.

Edel, W. and Kampelmacher, E.H. (1974) Comparative studies on *Salmonella* isolations from feeds in ten laboratories. *Bulletin of the World Health Organization* 50, 421–426.

Ellis, E.M., Williams, J.E., Mallinson, E.T., Snoeyenbos, G.H. and Martin, W.J. (1976) *Culture Methods for the Detection of Animal Salmonellosis and Arizonosis*. A manual of the American Association of Veterinary Laboratory Diagnosticians, Iowa State University Press, Ames, Iowa, 87 pp.

Entis, P. and Boleszczuk, P. (1991) Rapid detection of *Salmonella* in foods using EF-18 agar in conjunction with the hydrophobic grid membrane filter. *Journal of Food Protection* 54, 930–934.

Erdman, E.E. (1974) ICMSF methods studies. IV. International collaborative assay for the detection of *Salmonella* in raw meat. *Canadian Journal of Microbiology* 20, 715–720.

Ewing, W.H. (1986) *Edwards and Ewing's Identification of Enterobacteriaceae*, 4th edn. Elsevier Science, New York, 536 pp.

Fagerberg, D.J. and Avens, J.S. (1976) Enrichment and plating methodology for *Salmonella* detection in food: a review. *Journal of Milk and Food Technology* 39, 628–646.

Farmer, J.J., III (1995) Enterobacteriaceae: introduction and identification. In: Murray, P.R. (ed. in chief) *Manual of Clinical Microbiology*, 6th edn. American Society of Microbiology, Washington, DC, pp. 438–449.

Fedorka-Cray, P.J., Gray, J.T., and Thomas, L.A. (1995) Comparison of culture media for the isolation of *Salmonella*. In: *Proceedings of Symposium on the Diagnosis of* Salmonella *Infections*. United States Animal Health Association and American Association of Laboratory Veterinary Diagnosticians, Reno, Nevada, pp. 116–123.

Fricker, C.R. (1987) The isolation of salmonellas and campylobacters: a review. *Journal of Applied Bacteriology* 63, 99–116.

Fricker, C.R., Quail, E., McGibbon, L. and Girdwood, R.W.A. (1985) An evaluation of commercially dehydrated Rappaport–Vassiliadis medium for the isolation of salmonellae from poultry. *Journal of Hygiene, Cambridge* 95, 337–344.

Gabis, D.A. and Silliker, J.H. (1974) ICMSF methodology studies. II. Comparison of analytical schemes for detection of *Salmonella* in high moisture foods. *Canadian Journal of Microbiology* 20, 663–669.

Galton, M.M., Lowery, W.D. and Hardy, A.V. (1954) *Salmonella* in fresh and smoked pork sausage. *Journal of Infectious Diseases* 95, 232–235.

Galton, M.M., Morris, G.K. and Martin, W.T. (1968) *Salmonellae in Foods and Feeds. Review of Isolation Methods and Recommended Procedures*. United States Department of Health, Education, and Welfare/Public Health Service. Centers for Disease Control, Atlanta, Georgia.

Gast, R.K. (1993) Research note: evaluation of direct plating for detecting *Salmonella enteritidis* in pools of egg contents. *Poultry Science* 72, 1611–1614.

Gast, R.K. and Holt, P.S. (1995) Iron supplementation to enhance the recovery of *Salmonella enteritidis* from pools of egg contents. *Journal of Food Protection* 58, 268–272.

Goossens, H., Wauters, G., De Boeck, M., Janssens, M. and Butzler, J. (1984) Semisolid selective-motility enrichment medium for isolation of salmonellae from fecal specimens. *Journal of Clinical Microbiology* 19, 940–941.

Goyer, R.A. (1996) Toxic effects of metals. In: Klaassen, C.D., Amdur, M.O. and Doull, J. (eds) *Casarett and Doull's Toxicology: the Basic Science of Poisons*, 5th edn. McGraw Hill, New York, pp. 691–736.

Greenfield, J. and Bigland, C.H. (1970) Selective inhibition of certain enteric bacteria by selenite media incubated at 35 and 43°C. *Canadian Journal of Microbiology* 16, 1267–1271.

Grunnet, K. (1975) Development of a standard method for isolation *Salmonella* from sewage and receiving waters. In: *Salmonella in Sewage and Receiving Waters*. FADL's Forlag, Copenhagen, pp. 69–78.

Hajna, A.A. (1956) A new enrichment broth medium for Gram negative organisms of the intestinal group. *Public Health Laboratory* 13, 83–89.

Hajna, A.A. and Damon, S.R. (1956) New enrichment and plating media for the isolation of *Salmonella* and *Shigella* organisms. *Applied Microbiology* 4, 341–345.

Harvey, R.W.S. and Price, T.H. (1968) Elevated temperature of incubation of enrichment media for the isolation of salmonellas from heavily contaminated materials. *Journal of Hygiene, Cambridge* 66, 377–381.

Harvey, R.W.S. and Price, T.H. (1975) Studies on the isolation of *Salmonella dublin*. *Journal of Hygiene, Cambridge* 74, 369–374.

Harvey, R.W.S. and Price, T.H. (1979) A review: principles of *Salmonella* isolation. *Journal of Applied Bacteriology* 46, 27–56.

Harvey, R.W.S. and Thompson, S. (1953) Optimum temperature of incubation for isolation of salmonellae. *Monthly Bulletin of the Ministry of Health and the Public Health Laboratory Service* 12, 149–150.

Hilker, J.S. (1975) Enrichment serology and fluorescent antibody procedures to detect salmonellae in foods. *Journal of Milk and Food Technology* 38, 227–231.

Hobbs, B.C. (1963) Techniques for the isolation of salmonellae from eggs and egg products. *Annales de l'Institut Pasteur* 104, 621–637.

Hoben, D.A., Ashton, D.H. and Peterson, A.C. (1973) Some observations on the incorporation of novobiocin into Hektoen enteric agar for improved *Salmonella* isolation. *Applied Microbiology* 26, 126–127.

Hynes, M. (1942) The isolation of intestinal pathogens by selective media. *Journal of Pathology and Bacteriology* 54, 193–207.

Inoue, T., Takagi, S, Ohnishi, A., Tamura, K. and Suzuki, A. (1968) Food-borne disease *Salmonella* isolation medium (MLCB). Paper presented at the 66th Annual Meeting of the Society of Veterinary Science of Japan.

Jameson, J.E. (1962) A discussion of the dynamics of *Salmonella* enrichment. *Journal of Hygiene, Cambridge* 60, 193–207.

Jeffries, L. (1959) Novobiocin–tetrathionate broth: a medium of improved selectivity for the isolation of salmonellae from faeces. *Journal of Clinical Pathology* 12, 568–571.

June, G.A., Sherrod, P.A., Hammack, T.S., Amaguana, R.M. and Andrews, W.A. (1996) Relative effectiveness of selenite cystine broth, tetrathionate broth, and Rappaport–Vassiliadis medium for recovery of *Salmonella* spp. from raw flesh, high contaminated foods, and poultry feed: collaborative study. *Journal of AOAC International* 79, 1307–1323.

Juven, B.J., Cox, N.A., Bailey, J.S., Thomson, J.E., Charles, O.W. and Schutze, J.V. (1984) Recovery of *Salmonella* from artificially contaminated poultry feeds in non-selective and selective broth media. *Journal of Food Protection* 47, 299–302.

Kauffmann, F. (1930) Die technik der Typhenbestimmung in der Typhus-paratyphus-gruppe. *Zentralblatt für Bakteriologie, Parasitenkunde und Infektionskrankheiten I. Abteilung Originale* 119, 152–160.

Kauffmann, F. (1935) Weitere Erfahrungen mit der Kombinierten anreicherungsverfahren für *Salmonella* Bazillen. *Zeitschroft für Hygiene Infektionskrankheiten* 117, 26–32.

King, S. and Metzger, W.I. (1968) A new plating medium for the isolation of enteric pathogens. I. Hektoen enteric agar. *Applied Microbiology* 16, 577–578.

Komatsu, K.K. and Restaino, L. (1981) Determination of the effectiveness of novobiocin added to two agar plating media for the isolation of *Salmonella* from fresh meat products. *Journal of Food Safety* 3, 183–192.

Kristensen, M., Lester, V. and Jurgens, A. (1925) Use of trypsinized casein, brom-thymol blue, brom-cresol-purple, phenol-red and brilliant green for bacteriological nutrient media. *British Journal of Experimental Pathology* 6, 291–299.

Kuhn, H., Wonde, B., Rabsch, W., and Reissbrodt, R. (1994) Evaluation of Rambach agar for detection of *Salmonella* subspecies I to VI. *Applied and Environmental Microbiology* 60, 749–751.

Lamichhane, C.M., Joseph, S.W., Waltman, W.D., Secott, T., Odor, E.M., DeGraft-Hanson, J., Mallinson, E.T., Vo, V. and Blankford, M. (1995) Rapid detection of *Salmonella* in poultry using the colony lift immunoassay. Presented at the 16th meeting of the Southern Poultry Science Society.

Leifson, E. (1935) New culture media based on sodium desoxycholate for the isolation of intestinal pathogens and for the enumeration of colon bacilli in milk and water. *Journal of Pathology and Bacteriology* 40, 581–599.

Leifson, E. (1936) New selenite enrichment media for the isolation of typhoid and paratyphoid (*Salmonella*) bacilli. *American Journal of Hygiene* 24, 423–432.

MacConkey, A. (1905) Lactose fermenting bacteria in faeces. *Journal of Hygiene* 5, 333–379.

McCoy, J.H. (1962) The isolation of salmonellae. *Journal of Applied Bacteriology* 25, 213–224.

Mallinson, E.T. (1990) *Salmonella* monitoring system simplifies evaluation of farms. *Poultry Digest September,* 46–47.

Miller, R.G, Tate, C.R., Mallinson, E.T. and Scherrer, J.A. (1991) Xylose-lysine-tergitol 4: an improved selective agar medium for the isolation of *Salmonella*. *Poultry Science* 70, 2429–2432.

Miller, R.G., Tate, C.R. and Mallinson, E.T. (1995) Improved XLT4 agar: small addition of peptone to promote stronger production of hydrogen-sulfide by salmonellae. *Journal of Food Protection* 58, 115–119.

Moats, W.A. (1978) Comparison of four agar plating media with and without added novobiocin for isolation of salmonellae from beef and deboned poultry meat. *Applied and Environmental Microbiology* 36, 747–751.

Moats, W.A. (1981) Update on *Salmonella* in foods: selective plating media and other diagnostic media. *Journal of Food Protection* 44, 375–380.

Moats, W.A. and Kinner, J.A. (1974) Factors affecting the selectivity of brilliant green-phenol red agar. *Applied Microbiology* 27, 118–123.

Montford, J. and Thatcher, F.S. (1961) Comparison of four methods of isolating salmonellae form foods, and elaboration of a preferred procedure. *Journal of Food Science* 26, 510–517.

Muller, L. (1923) Un nouveau milieu d'enrichissement pour la recherche du bacille typhique et des paratyphiques. *Comptes Rendus des Séances de la Société de Biologie et de ses Filiales* 89, 434–437.

National Poultry Improvement Plan and Auxiliary Provisions (1996) APHIS 91–55–031, April, United States Department of Agriculture, Animal Plant Health Inspection Service, pp. 73–84.

North, W.R. and Bartram, W.T. (1953) The efficiency of selenite broth of different compositions in the isolation of *Salmonella*. *Applied Microbiology* 1, 130–134.

Oboegbulem, S.I. (1993) Comparison of two enrichment media and three selective media for isolation of salmonellae from fresh chicken carcass rinse fluids and sewer swabs. *International Journal of Food Microbiology* 18, 167–170.

Osborne, W.W. and Stokes, J.L. (1955) A modified selenite brilliant-green medium for the isolation of *Salmonella* from eggs. *Applied Microbiology* 3, 295–299.

Palumbo, S.A. and Alford, J.A. (1970) Inhibitory action of tetrathionate enrichment broth. *Applied Microbiology* 20, 970–976.

Peterz, M., Wiberg, C. and Norberg, P. (1989) The effect of incubation temperature and magnesium chloride concentration on growth of *Salmonella* in homemade and in commercially available dehydrated Rappaport–Vassiliadis broths. *Journal of Applied Bacteriology* 66, 523–528.

Pignato, S., Giammanco, G. and Giammanco, S. (1995a) Rambach agar and SM-ID medium sensitivity for presumptive identification of *Salmonella* subspecies I–VI. *Journal of Medical Microbiology* 43, 68–71.

Pignato, S., Marino, A.M., Emanuele, M.C., Iannotta, V., Caracappa, S. and Giammanco, G. (1995b) Evaluation of new culture media for rapid detection and isolation of salmonellae in foods. *Applied and Environmental Microbiology* 61, 1996–1999.

Poisson, D.M. (1992) Novobiocin, brilliant green, glycerol, lactose agar: a new medium for the isolation of *Salmonella* strains. *Research in Microbiology* 143, 211–216.

Poisson, D.M., Niocel, B., Florence, S. and Imbault, D. (1992) Comparison of hektoen and *Salmonella–Shigella* agar on 6033 stools of human origin submitted for routine isolation of *Salmonella* sp. and *Shigella* sp. *Pathology and Biology* 40, 21–24.

Poisson, D.M., Nugier, J.P. and Rousseau, P. (1993) Study of Rambach and NBGL agar on 4037 stools of human origin and 584 veterinary samples submitted for isolation of salmonellae. *Pathology and Biology* 41, 543–546.

Pollock, H.M. and Dahlgren, B.J. (1974) Clinical evaluation of enteric media in the primary isolation of *Salmonella* and *Shigella*. *Applied Microbiology* 27, 197–201.

Pourciau, S.S. and Springer, W.T. (1978) Evaluation of secondary enrichment for detecting salmonellae in bobwhite quail. *Avian Diseases* 22, 42–45.

Rambach, A. (1990) New plate medium for facilitated differentiation of *Salmonella* spp. from *Proteus* spp. and other enteric bacteria. *Applied and Environmental Microbiology* 56, 301–303.

Rappaport, F., Konforti, N. and Navon, B. (1956) A new enrichment medium for certain salmonellae. *Journal of Clinical Pathology* 9, 261–266.

Rappold, H. and Bolderdijk, R.F. (1979) Modified lysine iron agar for isolation of *Salmonella* from food. *Applied and Environmental Microbiology* 38, 162–163.

Read, S.C., Irwin, R.J., Poppe, C. and Harris, J. (1994) A comparison of two methods for isolation of *Salmonella* from poultry litter samples. *Poultry Science* 73, 1617–1621.

Reissbrodt, R.K. and Rabsch, W. (1993) Selective pre-enrichment of *Salmonellae* from eggs by siderophore supplements. *Zentralblatt für Bakteriologie* 279, 344–353.

Restaino, L., Grauman, G.S., McCall, W.A. and Hill, W.M. (1977) Effects of varying concentrations of novobiocin incorporated into two *Salmonella* plating media on the recovery of four Enterobacteriaceae. *Applied and Environmental Microbiology* 33, 585–589.

Restaino, L., Komatsu, K.K. and Syracuse, M.J. (1982) A note on novobiocin in XLD and HE agars: the optimum levels required in two commercial sources of media to improve isolation of salmonellas. *Journal of Applied Bacteriology* 53, 285–288.

Rigby, C.E. and Pettit, J.R. (1980) Delayed secondary enrichment for the isolation of salmonellae from broiler chickens and their environment. *Applied and Environmental Microbiology* 40, 783–786.

Ruiz, J., Nunez, M., Diaz, J., Lorente, I., Perez, J. and Gomez, J. (1996) Comparison of five plating media for isolation of *Salmonella* species from human stools. *Journal of Clinical Microbiology* 34, 686–688.

Schlundt, J. and Munch, B. (1993) A comparison of the efficiency of Rappaport Vassiliadis, tetrathionate and selenite broths with and without pre enrichment for the isolation of *Salmonella* in animal waste biogas plants. *Zentralblatt für Bakteriologie* 279, 336–343.

Sharma, R.M. and Packer, R.A. (1969) Evaluation of culture media for isolation of salmonellae from faeces. *Applied Microbiology* 18, 589–595.

Smith, H.G. (1959) Observations on the isolation of salmonellae from selenite broth. *Journal of Applied Bacteriology* 22, 116–124.

Smith, H.W. (1952) The evaluation of culture media for the isolation of salmonellae from faeces. *Journal of Hygiene, Cambridge* 50, 21–36.

Smyser, C.F. and Snoeyenbos, G.H. (1969) Evaluation of several methods of isolating salmonellae from poultry litter and animal feedstuffs. *Avian Diseases* 13, 134–141.

Smyser, C.F., Bacharz, J. and Van Roekel, H. (1963) Detection of *Salmonella typhimurium* from artificially contaminated poultry feed and animal by-products. *Avian Diseases* 7, 423–434.

Smyser, C.F., Snoeyenbos, G.H. and McKie, B. (1970) Isolation of salmonellae from rendered by-products and poultry litter cultured in enrichment media incubated at elevated temperature. *Avian Diseases* 14, 248–254.

Stokes, J.L. and Osborne, W.W. (1955) A selenite brilliant green medium for the isolation of *Salmonella*. *Applied Microbiology* 3, 217–220.

Tate, C.R. and Miller, R.G. (1990) Modification of brilliant green agar by adding sodium novobiocin to increase selectivity for *Salmonella*. *Maryland Poultryman* April, 7–10.

Tate, C.R., Miller, R.G., Mallinson, E.T., Douglass, L.W. and Johnson, J.W. (1990) The isolation of salmonellae from poultry environmental samples by several enrichment procedures using plating media with and without novobiocin. *Poultry Science* 69, 721–726.

Taylor, W.F. (1965) Isolation of shigellae. I. Xylose-lysine agars: new media for the isolation of enteric pathogens. *American Journal of Clinical Pathology* 44, 471–475.

Taylor, W.I. and Schelhart, D. (1971) Isolation of shigellae. VIII. Comparison of xylose lysine desoxycholate agar, Hektoen enteric agar, *Salmonella–Shigella* agar, and eosin methylene blue agar with stool specimens. *Applied Microbiology* 21, 32–37.

Thomason, B.M. and Dodd, D.J. (1978) Enrichment procedures for isolating salmonellae from raw meat and poultry. *Applied and Environmental Microbiology* 36, 627–628.

Thomason, B.M., Dodd, D.J. and Cherry, W.B. (1977) Increased recovery of salmonellae from environmental samples enriched with buffered peptone water. *Applied and Environmental Microbiology* 34, 270–273.

van Leusden, F.M., Van Schothorst, M. and Beckers, H.J. (1982) The standard *Salmonella* isolation method. In: Coffy, J.E.L., Roberts, D. and Skinner, F.A. (eds) *Isolation and Identification Methods for Food Poisoning Organisms*. SAB Technical Series, Academic Press, London, pp. 35–49.

van Schothorst, M. and Renaud, A.M. (1983) Dynamics of *Salmonella* isolation with modified Rappaport's medium (R10). *Journal of Applied Bacteriology* 54, 209–215.

van Schothorst, Renaud, M. and van Beek, C. (1987) *Salmonella* isolation using RVS broth and MLCB agar. *Food Microbiology* 4, 11–18.

Vassiliadis, P. (1983) The Rappaport–Vassiliadis (RV) enrichment medium for the isolation of salmonellas: an overview. *Journal of Applied Bacteriology* 54, 69–76.

Vassiliadis, P., Trichopoulos, D., Papadakis, G. and Politi, G. (1970) *Salmonella* isolations in abattoirs in Greece. *Journal of Hygiene, Cambridge* 68, 601–609.

Vassiliadis, P., Pateraki, E., Papaiconomou, N. and Papadakis, J.A. and Trichopoulos, D. (1976) Nouveau procédé d'enrichissiment de *Salmonella*. *Annales de Microbiologie* (*Institut Pasteur*) 127B, 195–200.

Waltman, W.D. and Mallinson, E.T. (1995) Isolation of *Salmonella* from poultry tissue and environmental samples: a nationwide survey. *Avian Diseases* 39, 45–54.

Waltman, W.D., Horne, A.M., Pirkle, C. and Dickson, T.G. (1991) Use of delayed secondary enrichment for the isolation of *Salmonella* in poultry and poultry environments. *Avian Diseases* 35, 88–92.

Waltman, W.D., Horne, A.M. and Pirkle, C. (1993) Influence of enrichment incubation time on the isolation of *Salmonella*. *Avian Diseases* 37, 884–887.

Waltman, W.D, Horne, A.M. and Pirkle, C. (1995) Comparative analysis of media and methods for isolating *Salmonella* from poultry and environmental samples. In: *Proceedings of Symposium on the Diagnosis of Salmonella Infections*. United States Animal Health Association and American Association of Laboratory Veterinary Diagnosticians, Reno, Nevada, pp. 1–14.

Warburton, D.W., Bowen, B., Konkle, A., Crawford, C., Durzi, S., Foster, R., Fox, C., Gour, L., Krohn, G., La Casse, P., Lamontagne, G., McDonagh, S., Arling, V., Mackenzie, J., Todd, E.C.D., Oggel, J., Plante, R., Shaw, S., Tiwari, N.P., Trottier, Y. and Wheeler, B.D. (1994a) A comparison of six different plating media used in the isolation of *Salmonella*. *International Journal of Food Microbiology* 22, 277–289.

Warburton, D.W., Arling, V., Worobec, S., Mackenzie, J., Todd, E.C.D., Lacasse, P., Lamontagne, G., Plante, R., Shaw, S., Bowen, B. and Konkle, A. (1994b) A comparison study of the EF-18 agar/hydrophobic grid membrane filter (HGMF) method and the enzyme linked antibody (ELA)/HGMF method to the HPB standard method in the isolation of *Salmonella*. *International Journal of Food Microbiology* 23, 89–98.

Weiss, K.F., Ayres, J.C. and Kraft, A.A. (1965) Inhibitory action of selenite on E. coli, P. vulgaris, S. thompson. *Journal of Bacteriology* 90, 857–862.

Wilson, W.J. and Blair, E.M.McV. (1927) Use of a glucose bismuth sulphite iron medium for the isolation of B. typhosus and B. proteus. *Journal of Hygiene* 26, 374–391.

World Health Organization Consultation (1994) Control of *Salmonella* infections in animals and prevention of human food-borne *Salmonella* infections. *Bulletin of the World Health Organization* 72, 831–833.

Wray, C. and Davies, R.H. (1994) (eds) *Guidelines on Detection and Monitoring of Salmonella-infected Poultry Flocks with Particular Reference to Salmonella enteritidis*. WHO, Graz, Austria, p. 46. (WHO/Zoon/94.173.)

Yamamoto, R., Sadler, W.W., Adler, H.E. and Stewart, G.F. (1961) Comparison of media and methods for recovering *Salmonella typhimurium* from turkeys. *Applied Microbiology* 9, 76–80.

Chapter 22

Methods for the Rapid Detection of *Salmonella*

Henk van der Zee[1] and Jos H.J. Huis in't Veld[2]

[1]Inspectorate for Health Protection, Regional Service East, De Stoven 22, 7206 AX Zutphen, The Netherlands; [2]Department of Food of Animal Origin, Utrecht University, PO Box 80.175, 3508 TD Utrecht, The Netherlands

Introduction

Conventional isolation of *Salmonella* is accomplished by using cultural methods. These cultural procedures generally have four distinct phases: (i) pre-enrichment in a non-selective medium, to allow resuscitation of any injured cells and multiplication of the target organism and others present in the sample; (ii) selective enrichment, to allow the survival or growth of the *Salmonella*, while inhibiting other accompanying organisms in the selective broth; (iii) isolation, using selective agar media that restrict growth of bacteria other than *Salmonella*, in order to produce presumptive isolates; and (iv) confirmation, where isolates are subjected to a variety of biochemical and serological tests to confirm that they are *Salmonella* and to determine their serovar. In general, each individual step needs at least 16 h, up to a maximum of 48 h. Thus the whole procedure takes 4–7 days to complete and is therefore laborious and requires substantial available staff.

The detection of *Salmonella* in some specific veterinary samples can also be accomplished by omitting one or more of these steps. This depends on the type of sample, the infection level and the reason for the investigation. Direct incubation in a selective broth or direct plating on selective agar can be used for clinical material, such as faeces, post-mortem or organ samples (intestine/liver/spleen/gonads) and rectal and/or cloacal swabs, which are expected to contain high numbers of *Salmonella*.

In contrast, those samples that generally harbour low levels of *Salmonella* spp., such as animal feed, litter, environmental swabs, eggs and egg material, require all four phases of the cultural procedure to be performed. In these samples the *Salmonella* can also be distributed unevenly and/or have undergone some sort of injury (e.g. water activity (A_w), pH, presence of antimicrobial substances). Therefore special attention should be paid to sampling procedures.

This double-edged approach implies that different alternative methods can also be used to replace, partially or in total, the conventional clinical or cultural methods to obtain a reliable result in as short a time as possible and preferentially in a less laborious way. The time span that is required to obtain a negative result for clinical methods is at least 24 h and can be delayed up to 96 h when the whole cultural method has to be performed. Presumptive positive results, serotyping not included, require a minimum analysis time of 48–96 h for the clinical and cultural methods, dependent on the method used.

Introduction of alternative methods aims not only reduce the analytical time but also at savings in staff time and media requirements to improve overall performance of the *Salmonella* detection procedure by enhancing the sensitivity of the detection procedures. When large numbers of samples have to be investigated, the need for alternative, more rapid and more convenient methods than the current culture methods becomes especially obvious.

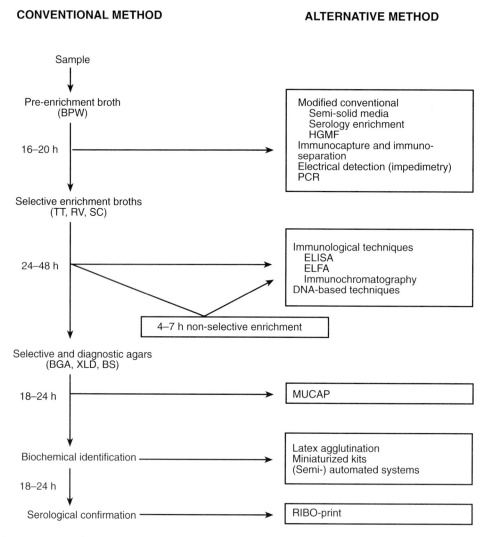

Fig. 22.1. Points of application within classical cultural techniques of alternative methods for *Salmonella* detection. BPW, buffered peptone water; RV, Rappaport–Vassiliadis; SC, selenite cystine; BGA, brilliant green agar; XLD, xylose-lysine-deoxycholate; BS, bismuth sulphite; HGMF, hydrophobic grid membrane filter; PCR, polymerase chain reaction; ELISA, enzyme-linked immunosorbent assay; ELFA, enzyme-linked fluorescent assay; MUCAP, 4-methylumbelliferyl caprylate.

Alternative systems and test kits are derived from different basic technologies, including isolation techniques, which are modifications of conventional methods and non-isolation methods, such as: (i) electrical conductance and impedance; (ii) immunological techniques; (iii) nucleic acid-based assays; and (iv) the polymerase chain reaction (PCR) (Blackburn, 1993; Giese, 1995; Feng, 1996; Patel, 1997; van der Zee and Huis in't Veld, 1997). These methods generally take

24–52 h to screen for or detect and identify *Salmonella* in veterinary samples. The points of application of the alternative methods within conventional cultural techniques are shown in Fig. 22.1.

The majority of these tests, particularly the enzyme-linked immunological techniques, also need 10^5 cells ml^{-1} for reliable results (Blackburn, 1993; Giese, 1995). Accordingly, all these test protocols involve a pre-enrichment stage, and some also involve a selective enrichment and some sort of post-enrichment culture, each based on appropriate culture media, before the alternative technique can be used.

Modified Conventional Methods

Modification of conventional methods can be applied to all four phases of the cultural methods, and can include addition of specific substances to conventional media or the use of specific characteristics of *Salmonella*, e.g. motility, to facilitate isolation and/or identification.

Pre-enrichment

In food microbiology, a number of media for pre-enrichment have been developed for various foods, but for the majority of these products, as well as for most veterinary samples, buffered peptone water (BPW) is the medium of choice. In specific cases, e.g. protocols for transport of environmental swabs and for investigation of eggs and egg products for *S. enteritidis*, modified pre-enrichment procedures are needed (van der Zee, 1994; Davison *et al.*, 1995).

Environmental swabs
The validity of environmental monitoring requires determination of the best sites for sampling, and therefore one must have confidence in the isolation procedures used. A variety of transport media, which can be considered as primary enrichment media, have been recommended to ensure the isolation of *Salmonella* from the environmental samples. For this purpose, BPW (Thomason *et al.*, 1977), double-strength skim milk (Mallinson, 1990; Opara *et al.*, 1992, 1994), asparagine and double-distilled water (Davison *et al.*, 1995) can be used.

Pre-enrichment for eggs and egg products
The addition of ferric-ammonium-citrate to BPW made it possible to isolate *S. enteritidis* from the yolk of fresh hen's eggs, while its omission led to negative results (Dolzinsky and Kruse, 1992). Pre-enrichment of 20 ml pooled egg contents in 180 ml trypticase soy broth (TSB), supplemented with 35 mg l^{-1} ferrous sulphate to ensure adequate iron availability to support bacterial growth in the

presence of the iron-chelating albumen protein ovotransferrin, is also recommended (Gast, 1993; Gast and Holt, 1995). For the same reason, supplementation of BPW with the siderophore ferrioxamine E, 1 μg ml^{-1}, was recommended (Reissbrodt and Rabsch, 1993). It remains unclear whether the addition of the above-mentioned supplements favour exclusively the outgrowth of *S. enteritidis* or also that of other serovars.

It is obvious that the pre-enrichment phase will become a more important field of future research, because many of the alternative novel detection systems for *Salmonella* will depend on pre-enrichment. As a consequence, novel ways to enhance the numbers of specific pathogens in the presence of background microbial contamination in primary enrichment broths have to be investigated further (Patel, 1997).

Selective enrichment

The objective of selective enrichment is to select *Salmonella* from all the other organisms present in the pre-enriched sample. The ideal selective enrichment should repress competing organisms and allow *Salmonella* to multiply unrestrictedly. Three 'families' of selective enrichment are in common use: tetrathionate, selenite and Rappaport. These media and their different modifications are discussed more fully in Chapter 21. Tetrathionate was formulated by Mueller (Mueller, 1923) and modified by Kauffmann, who added brilliant green dye and bile salts as selective agents (Kauffmann, 1935) and it was further improved by adding yeast extract as growth stimulant (Hajna, 1956). Other modifications, such as addition of novobiocin (Jeffries, 1959), have been reported, although their use has not become widely accepted.

For selenite media, several modifications to the original medium, described by Leifson (1936), have been proposed. Well-known modifications include selenite F, selenite cystine, dulcitol selenite and selenite brilliant green. The performance of these media has been investigated, but with no overall agreement (Carlson and Snoeyenbos, 1974; Fagerburg and Avens, 1976). Selenite enrichment can have a slight advantage over other selective-enrichment media for the isolation of *S. pullorum* or *S. gallinarum*. It is frequently used for direct enrichment of faeces and post-mortem samples.

The original Rappaport medium contains malachite green as a selective agent (Rappaport et al., 1956) and was modified by Vassiliadis (Vassiliadis et al., 1981). This medium has become known as Rappaport–Vassiliadis (RV) medium and is now widely used in Europe, in preference to selenite and tetrathionate media (Beckers et al., 1987). In the USA, RV medium was recommended recently for analysis of raw flesh foods and poultry feed (June et al., 1996).

Modification of conventional enrichment methods includes the use of 'motility enrichment media' such as the modified semi-solid Rappaport–Vassiliadis medium (MSRV), diagnostic Salmonella medium (DIASALM) and test kits based on this principle (DeSmedt et al., 1986; Holbrook et al., 1989).

For the detection of Salmonella, the use of motility enrichment on MSRV and DIASALM is widely recognized as an effective procedure for the identification of contaminated products (van der Zee and van Netten, 1992; Davis and Wray, 1994b; Wiberg and Norberg, 1996). This method, based on the motility of Salmonella, indicates the organism as a swarm zone after inoculation with pre-enrichment cultures and incubation at 41.5–42°C for only 18–24 h in a Petri dish.

It was observed that the use of semi-solid agar as a selective enrichment seems to favour the isolation of S. enteritidis (Svastova et al., 1984; Perales and Erkiaga, 1991; van Netten et al., 1991; Poppe et al., 1992). The reason for this phenomenon when using this type of medium is still not clear.

Addition of 0.0015% nitrofurantoin to DIASALM (DIASALM-N) also resulted in specific suppression of serovars other than S. enteritidis in both low- and high-moisture foods, and resulted in a two- to threefold increase in recovery of S. enteritidis (van der Zee and van Netten, 1992). Rampling et al. (1990) found that 98% of S. enteritidis phage type (PT) 4 strains were resistant to nitrofurantoin with minimal inhibitory concentrations (MICs) $\leq$16 mg l^{-1}. Addition of nitrofurantoin to selective media could thus offer a much better chance for isolating S. enteritidis from samples that also contain other serovars (Ruzickova et al., 1996).

Other commercial systems based on motility enrichment include combinations of selective, elective and indicator media in U-shaped tubes and jars. In the case of the oxoid Salmonella rapid test (OSRT), as described by Holbrook et al. (1989), the presumptive detection of Salmonella is indicated by the use of motility migration media (RV broth and a brilliant green agar), combined with indicator media (lysine iron agar and lysine-iron-cystine-neutral red agar). In the Salmonella 1–2 test, used for poultry feeds (Oggel et al., 1995), this is realized by incubation in modified tetrathionate, followed by a visible immunoprecipitation reaction in motility agar containing anti-flagella antibodies. The 1–2 test gave reliable results for Salmonella analysis of poultry feeds and poultry-feed supplements, poultry products, egg products and environmental samples (Warburton et al., 1994a; Oggel et al, 1995).

Selective plating

Selective-plating agars rely on different selective agents, indicator systems or differential characteristics (such as carbohydrate fermentations and hydrogen sulphide production) to distinguish Salmonella from many related enteric bacteria. Because none of the plating agars is ideal when used singly, it has been recommended to use at least two media. Each should contain different selective agents and indicator systems or differential characteristics, in order to detect a wide range of Salmonella serovars from different situations (Fricker, 1987; Andrews, 1996).

Many different plating agars have been developed and their characters and the different modifications are discussed in Chapter 21. Plating media should be incubated at 35–37°C for 18–24 h, unless specified otherwise.

Recently, some new selective and differential plating media, based on enhanced selectivity and other abilities of Salmonella to produce specific identifying characteristics, have been developed. Xylose-lysine-tergitol 4 (XLT-4) is a modification of xylose-lysine-deoxycholate (XLD) agar. The surfactant tergitol 4 (7-ethyl-2-methyl-4-undecanol hydrogen sulphate sodium salt) is added as a selective inhibitor for Proteus spp. and other Enterobacteriaceae to the xylose-lysine agar base (Miller et al., 1991, 1995).

Other modifications include media in which chromogenic substances are incorporated in order to create a more distinct differentiation (e.g. red colonies for *Salmonella*) from accompanying bacteria on agar media. Rambach agar uses acid formation from propylene glycol (PG) by *Salmonella*, which results in red colonies and β-D-galactosidase production by other *Enterobacteriaceae*, which, after reaction with the chromogenic substance 5-bromo-4-chloro-3-indole-β-D-galactopyranoside (X-Gal) results in the formation of blue-green colonies (Rambach, 1990). However, all strains of *Salmonella* subspecies IIIa, IIIb and V produce β-D-galactosidase and therefore also appear as blue-green colonies on this medium (Kühn *et al.*, 1994). The possibility of the presence of these particular subspecies in samples from sheep, turkeys and cold-blooded animals cannot be excluded. On the other hand, *S. typhi* and *S. paratyphi* A and B failed to produce acid from PG, resulting in colourless colonies on Rambach agar.

Salmonella identification (SM-ID) agar is based on the metabolism of glucuronate by *Salmonella* on this medium. Combined with the absence of β-galactosidase activity, this also results in specific pink/red colonies (Dusch and Altwegg, 1993). Due to the fact that *S. typhi* and *S. paratyphi* B also form acid from glucuronate, specific colonies belonging to these serovars can also be detected on the SM-ID medium.

Although both media are not very selective, they may perform better after selective enrichment (Dusch and Altwegg, 1993) and they have been successfully used in differentiating *Salmonella* from other enteric bacteria in clinical isolates (Freydiere and Gille, 1991; Gruenewald *et al.*, 1991; Davies and Wray, 1994a,b).

Another modification of conventional plating methods is the hydrophobic grid membrane filter (HGMF) technique. For *Salmonella* detection, this test uses conventional pre-enrichment, followed by 6 h selective enrichment. A portion of the selective enrichment is filtered through an HGMF and incubated on EF-18 agar (Entis and Boleszczuk, 1991). Negative results are obtained in 42 h. Presumptive positive results (green colonies on EF-18 agar) need an additional 24 h for confirmation. The performance of the test was evaluated for egg products against the US Department of Agriculture *Salmonella* method. Both methods performed identically in frozen-, liquid- and dried-egg products (Entis, 1996). In another study, which included poultry products and feed samples,

problems with the method resulted from the inability to isolate colonies of *Salmonella* on the HGMF, due to small colony size, abnormal colony coloration and overgrowth by competitors (Warburton *et al.*, 1994b).

Confirmation

Biochemical screening

Typical colonies on selective agar media are screened biochemically and have to undergo a battery of differential tests, depending on the prescribed (standardized) method that is used. Before this procedure is started, as a first screening step, colonies resembling *Salmonella* on selective agar plates can be submitted to the methylumbelliferyl caprylate fluorescence (MUCAP) test. This test is an easy and rapid method, which allows the presumptive detection of *Salmonella* directly on media such as brilliant green (BG), Hektoen enteric (HE) and XLD agar. One drop (c. 5 µl) of reagent added to each colony produces a strong blue fluorescence when observed under ultraviolet (UV) light (wavelength 366 nm) after 3 min. This reaction is due to the release of 4-methylumbelliferone from the fluorogenic substrate 4-methylumbelliferyl-caprilate in the reagent, after this is cleaved by the enzyme C_8-esterase, which is present in *Salmonella*. Although there have also been reports that some strains of *Pseudomonas* spp., *Aeromonas* spp., *Proteus* spp., *Serratia* spp. and *Enterobacter cloaca* gave false positive results (Olson *et al.*, 1991; Ruiz *et al.*, 1991; Manafi and Sommer, 1992), the MUCAP test can be considered as an easy, rapid method for screening colonies suspected of being *Salmonella*. This will considerably reduce the number of isolates that have to undergo an extensive array of biochemical confirmation tests.

Alternative systems to perform or replace biochemical screening procedures are available, such as: (i) miniaturized test kits; (ii) automated and semi-automated systems; and (iii) latex agglutination tests.

MINIATURIZED TEST KITS. Miniaturized biochemical test kits, such as API, Micro-ID and Enterotube II Biotest ID-GNI, can generally provide results after 4–24 h. The biochemical patterns of the reactions are translated into profile numbers, which lead, upon comparison with a database, to genus or species identifications with a stated probability of correct identification. It should be remembered, however, that the databases are produced predominantly with human isolates, and atypical *Salmonella*, e.g. *S. pullorum*, may be misidentified as another bacterial species.

AUTOMATED AND SEMI-AUTOMATED SYSTEMS. There are automated and semi-automated systems, using different technologies. The AutoMicrobic System GNI uses a diversity of biochemical tests, interpreted by an optical reader and with a computer to generate identification profiles (Knight *et al.*, 1990). Other commercially available automated systems are Microbial-ID, producing identification profiles based on gas chromatography of cellular fatty acids (Dziezak, 1990), and the Biolog Identification System, in which oxidation of different carbon sources is measured to generate metabolic fingerprints for identification (Miller and Rhoden, 1991).

LATEX AGGLUTINATION TESTS. Material from suspected colonies can be submitted directly to latex agglutination tests. In these test kits, *Salmonella*-specific antibodies have been coupled to latex particles, to improve the sensitivity and visualization of the agglutination reaction. Depending on the manufacturer of the test, antisera to flagellar, somatic and fimbrial antigens are used. Performances of kits from different manufacturers can vary greatly (D'Aoust, 1992), and therefore these agglutination tests can only be used in this presumptive stage of the confirmation procedure and cannot replace conventional serotyping.

Serotyping

Conventional serotyping is a labour-intensive procedure, involving serological identification of antigens located on flagella (H antigens) and the somatic antigens (O antigens) in the cell wall. The combined identification of flagellar and somatic antigens comprises the antigenic profile. This procedure is therefore performed only in centralized or reference laboratories. There have been attempts to automate the conventional procedure, but rapid techniques have only been used recently. For enhanced flagellar-antigen phase inversion of diphasic *Salmonella*, the immunomagnetic separation technique (explained later)

can be used (Davies and Wray, 1997).

Recently, an automated system, the Ribo-Printer, based on DNA fingerprinting technology, has been introduced (Bruce, 1996). In this system, the DNA extracted from cells from colonies on agar media is fragmented and then separated by electrophoresis. These fragments are relocated to a membrane and the pattern is then recognized by using a DNA probe, which, in combination with a chemiluminescent reaction, results in emission of light from the fragments. This is then captured by a camera, and can be compared with other patterns in a database. This system can be a valuable tool for rapid epidemiological investigation and tracing sources of *Salmonella* contamination (Bailey *et al.*, 1996).

As far as isolation techniques are concerned, it can be stated that modified conventional methods are generally reliable and produce accurate results, comparable with those of standard or conventional methods. The modifications, aimed at more expedient reporting time, as well as savings in materials and analytical time, do not affect their reliability. The detection limit will at least be equal to that of conventional methods (van der Zee and van Netten, 1992). In most cases, the specificity will also be superior to that of conventional methods (Oggel *et al.*, 1990). In general, the costs for materials will be equal to or lower than those of conventional methods. In some cases, however, when expensive substances like X-Gal are needed, costs may be higher. The methods are as easy to carry out as the 'original' conventional method – or even easier.

The use of a reliable conventional cultural method is important in assuring cross-comparisons with alternative techniques, especially when non-isolation techniques are involved. Many studies have compared non-isolation assays for *Salmonella* detection with isolation by culture to evaluate the non-isolation technique. However, these evaluations employed a variety of media, which appeared to be less sensitive than the non-isolation assays being assessed. Consequently, in naturally contaminated samples, a false-positive result obtained with a non-isolation technique cannot be differentiated from a false-negative isolation result (Tate *et al.*, 1992). These considerations have to be kept in mind when results of studies are mentioned in the reviews of the non-isolation techniques mentioned subsequently.

Non-isolation Techniques

Electrical conductance and impedance

When microbial growth and metabolism take place in a culture medium, changes in conductance of the medium will occur. The time necessary for these changes to reach a threshold value, the detection time, is inversely proportional to the initial inoculum. Until there is a large microbial population (10^5–10^6 colony-forming units (cfu) ml^{-1}) of actively metabolizing microorganisms, the instrument cannot recognize these changes (Firstenberg-Eden and Eden, 1984). Impedimetric automated systems for direct and indirect measurement of changes in electrical properties of culture media are widely used in the food and feed industry for detection and estimation of a range of organisms (Bolton and Gibson, 1994). These systems have gained in significance for the rapid detection of *Salmonella*, are easy to apply and able to examine 120–640 samples at the same time.

Three systems (Malthus, Rabit and BacTrac) are actively marketing their own tests and media for *Salmonella* detection, whilst the fourth system (Bactometer) can be used with media of one's own choice (Bullock and Frodsham, 1988; Di Falco *et al.*, 1993).

When detecting pathogens with impedimetry, reliability largely depends on the performance of the selective medium used in the assay. For the detection of *Salmonella*, Malthus assays, with media based on the metabolism of trimethylamine oxide (TMAO) to trimethylamine (TMA), were tested extensively (Gibson, 1987; Ogden, 1988). Assay modifications involved the inclusion of an antibiotic mixture in the pre-enrichment step, followed by subculture into two selenite-based media, held in disposable cells, one containing TMAO and dulcitol and the other containing lysine (Bolton and Gibson, 1994). Users of this method have reported false negatives (up to 12%) and false positives (up to 10%) with raw foods, including meat, feed and environmental samples (D'Aoust, 1992; Quinn *et al.*, 1995). Other electrical *Salmonella* assays give the possibility of carrying out the assays in RV medium by the indirect conductance method (Bolton and Gibson, 1994).

The BacTrac 4100 system operates according to the so-called impedance-splitting method, which allows the simultaneous and separate determination of changes of the impedance of the culture medium (M value) and of the electrode impedance (E value). A volume of 0.1 ml pre-enrichment culture has to be transferred to 9.9 ml impedance-splitting *Salmonella* medium, which consist of magnesium chloride, malachite green oxalate, novobiocin, phosphate buffer, mannitol, peptone and yeast extract. The impedance changes are recorded automatically in the system during incubation at 40°C for up to 22 h. In this way, it can be used as a screening test, in which negative results can be obtained within 38 h. In a study involving 250 food samples, a false-negative rate of 7.4% and a false-positive rate of 4.9% were found (Pless *et al.*, 1994).

The fact that at least 10^5–10^6 cfu ml^{-1} actively metabolizing *Salmonella* are needed to register a preliminary detection means that a time span of at least 20 h is required when only low levels of *Salmonella* are present. This is a serious drawback to the use of this technique for the specific detection of pathogens. A strong point is, of course, the fact that these systems are very easy to handle and fully automated and can analyse large numbers of samples at the same time. The instruments have proved to be very useful when used for some defined samples (Silley and Forsythe, 1996). When working with these systems, one should, however, realize that large investments are necessary. Additional costs for consumables and/or reagents for each individual test also have to be considered.

Immunological techniques

Immunological techniques comprise the largest group of alternative methods and are currently available in many test kits for the detection of *Salmonella*. The techniques used in these test kits include immunodiffusion (see also modified conventional techniques), enzyme-linked immunosorbent assays (ELISA) and enzyme-linked immunofluorescent assays (ELFA), based on the 'sandwich' principle, and immunoagglutination, immunocapture, immunoprecipitation and combinations of these techniques.

Enzyme-linked immunoassays, based on the 'sandwich' principle (Fig. 22.2) that are in use for *Salmonella* have detection limits that range from 10^3–10^5 cfu ml^{-1}, and therefore pre-enrichment and/or selective enrichment for at least 16–24 h

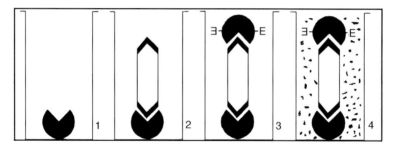

Fig. 22.2. Principles of the enzyme-linked immunosorbent assay (ELISA) in sandwich configuration.
1. Microwells coated with antibodies to *Salmonella* antigen. 2. *Salmonella* antigen from the sample and the antibodies bound to the microwell form immune complexes. 3. Enzyme-labelled antibodies (conjugate) are added and bind to the antigen part of the complex (antibody–antigen–antibody 'sandwich' complex). 4. A substrate is added on which the complex-bound enzyme works to produce a colour reaction. If complex bound enzyme is not present, no colour reaction will follow.

is required (Blackburn, 1993). Numerous commercial test kits are available. As far as reliability is concerned, the false-positive rates for environmental, feed and poultry carcass samples ranged from 2.3% to 5.8% and false-negative rates from 0.4% to 22.0% (Tate *et al.*, 1992; Quinn *et al.*, 1995). In general, the assays that have to be operated manually (e.g. *Salmonella* Tek, TECRA *Salmonella*, *Salmonella* EIA) are not as easy to perform as conventional, modified conventional or impedimetry-based methods, although fully automated systems (TECRA OPUS) are marketed now (Patel, 1997).

The fully automated Vitek immunodiagnostic assay system (VIDAS) based on ELFA, can also be used for detection of *Salmonella*. Capture antibodies are coated on the internal surface of a pipette-like solid-phase receptacle (SPR), which is used to draw up the necessary reagents from a pre-prepared test strip (Fig. 22.3). The VIDAS

SLM assay detects motile and non-motile *Salmonella* through the incorporation of somatic and flagellar antibodies. The SLM test has a detection limit of 10^5 cfu ml^{-1}. This test also needs pre-enrichment, selective enrichment and post-enrichment in M broth. The assay time after the 42 h enrichment protocol is 45 min. The VIDAS SLM *Salmonella* kit gave good agreement with conventional cultural methods, but false-positive reactions have also been reported, in some cases, for food samples (Blackburn *et al.*, 1994) and raw turkey (Curiale *et al.*, 1997). Additional investments for the instrument also have to be made and the costs of the consumables needed for the tests are similar to those of other commercially available ELISA kits.

Immunocapture, separation and concentration techniques result from the requirement to detect very low numbers of pathogens in foods. This technology achieves its goal by immuno-

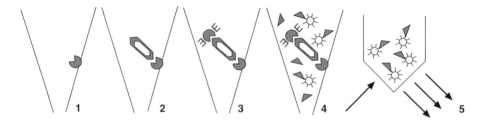

Fig. 22.3. Principles of the VIDAS automated enzyme-linked fluorescent assay (ELFA).
1. Antibody-coated pipette-tip-like SPR. 2. *Salmonella* antigen is bound by the antibody. 3. The bound immunocomplex is detected by the anti-*Salmonella* antibody conjugated to alkaline phosphatase. 4. The bound enzyme catalyses the added substrate. 5. The product is released in the test strip and the fluorescence reaction is measured by the VIDAS scanner.

logical binding (capture), eventually followed by physical separation of target organisms from a mixed enrichment culture, resulting in concentration of the target organism. Immunocapture, separation and concentration techniques can be applied after pre-enrichment and/or selective enrichment, which can vary in time from 6 to 24 h. A 5–7 h post-enrichment period is needed before the final ELISA, ELFA or immunochromatographic test is performed (Blackburn, 1993; Giese, 1995). After pre-enrichment and/or post-enrichment, at least 10^2–10^3 cfu ml^{-1} of the organism need to be present to enable *Salmonella* detection (Skjerve and Olsvik, 1991; Vermunt *et al.*, 1992).

At this time, three systems, using different solid phases (magnetic polystyrene-based particles, dipstick and the internal surface of a hollow spur) on which the *Salmonella* antibody is coated, are available.

An antibody-coated dipstick is used in the TECRA immunocapture system, developed for use with the TECRA *Salmonella* visual immunoassay, as well as in the TECRA UNIQUE, both 24 h *Salmonella* test kits. The dipstick is used to capture *Salmonella* selectively from the incubated pre-enrichment broth. The dipstick-bound *Salmonella* are then reincubated in M broth for at least 5 h before performing an ELISA (visual immunoassay) or immunoprecipitation test (TECRA UNIQUE). The UNIQUE test was compared with MSRV medium in a study involving 100 samples of chicken-carcass rinses, chicken portions and dust samples from chicken houses. The UNIQUE *Salmonella* test had a sensitivity of only 59.3% and a specificity of 97.6%, compared with the MSRV method (Poppe and Duncan, 1996).

In the VIDAS ICS system, a pipette-like SPR, coated with specific antibodies, is the basis of a fully automated test for the detection of *Salmonella* within 24 h. The interior of the SPR has antibodies against *Salmonella* adsorbed on its surface. An aliquot of the pre-enrichment broth is placed into a reagent strip, which contains washing and releasing solutions, and the sample is cycled in and out of the SPR for a specific length of time, performed automatically by the VIDAS instrument. *Salmonella* present in the broth will bind to the *Salmonella* antibodies. Unbound sample components are then washed away. The captured *Salmonella* are then released in a well in the reagent strip and this immunoconcentrated solution can be used for final detection, either by VIDAS SLM assay or plate inoculation (Fig. 22.4).

Dynal and Vicam provide magnetizable particles coated with specific antibodies against *Salmonella* (Giese, 1995). The target organisms in the sample bind to the immunomagnetic beads, which are then isolated from the other sample material and microorganisms in a magnetic field. The *Salmonella* can then be either detected by subculture on plating media or used in combination with other techniques, such as ELISA or PCR (Fig. 22.5).

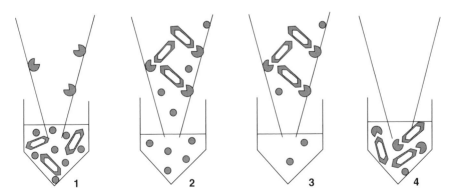

Fig. 22.4. Principles of the VIDAS immunoconcentration *Salmonella* test.
1. *Salmonella* and other organisms in the pre-enrichment broth in a well in the test strip + antibody coated SPR. 2. The *Salmonella* are captured by the antibody coated SPR. 3. Washing procedures eliminate most of the other organisms. 4. An immunoconcentrated solution of *Salmonella* is released in a well in the test strip. This solution can be used for final detection by either VIDAS SLM assay or selective plating.

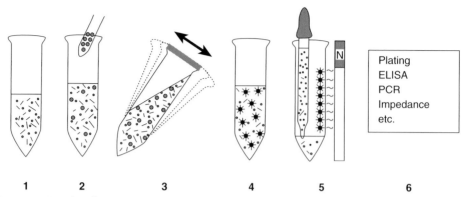

Plating
ELISA
PCR
Impedance
etc.

1   2   3   4   5   6

Fig. 22.5. Principles of immunomagnetic separation (IMS).
1. Incubated pre-enrichment broth containing all microorganisms present in the sample. 2. Magnetizable beads coated with anti-*Salmonella* antibodies are added. 3. Pre-enrichment broth with latex beads is mixed for 20 min. 4. *Salmonella* are bound to the surface of the latex particles. 5. A magnetic field is applied, and the supernatant aspirated. 6. The beads with the immobilized *Salmonella* can be used to perform the final test(s).

The semi-automated EIAFOSS system combines the immunomagnetic separation (IMS) technique with ELISA. It is a 48 h screening test for qualitative determination of *Salmonella* in feed. The concept includes a three-stage *Salmonella* enrichment procedure, with primary (19–28 h), selective (18–26 h) and post- (2.5 h) enrichment steps. The *Salmonella* are immuno-captured by the sensitized magnetic particles (Dyna beads) and heated for 15 min. A second antibody, conjugated with β-galactosidase, is then added, which binds the immunocomplex. With the addition of 4-MUG substrate, a fluorescent compound is formed, which is quantified by a fluorescence detector. The samples have to be verified using conventional cultural methods.

The method gave good results in studies that included animal feed, broiler carcasses and raw egg products (Krusell and Skovgaard, 1993; Fierens and Huygebaert, 1996).

The PATH-Stik, a dipstick-based immunochromatography assay, is also available for detection of *Salmonella*. After pre-, selective and post-enrichment, cell lysis is performed and subsequently enzyme-labelled antibodies against *Salmonella* are added. The resulting complex is captured with an antibody-coated dipstick. A positive detection shows as a coloured line on the dipstick after incubation in a chromogenic substrate (Fig. 22.6).

The PATH-Stik *Salmonella* assay has a detection limit of 10^6 cfu ml^{-1} after cultural

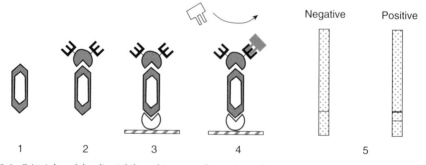

Negative   Positive

1   2   3   4   5

Fig. 22.6. Principles of the dipstick-based immunochromatographic assay.
1. *Salmonella* antigen in post-enrichment broth. 2. Enzyme-conjugated anti-*Salmonella* antibody is added and forms immunocomplex with the antigen. 3. The dipstick, coated with *Salmonella* antibody, is added and formation of a 'sandwich complex' takes place. 4. A chromogenic substrate is added and will be converted from colourless to purple–blue in the presence of complex bound enzyme. 5. This reaction is presented as a positive result by the appearance of two blue–black lines on the stick.

enrichment and the assay takes only 20 min (Blackburn *et al.*, 1994). The method was evaluated in comparison with the International Organization for Standardization (ISO) 6579 method (Brinkman *et al.*, 1995). For products with low A_w (animal feed, soya powder, egg powder), the sensitivity was 88.2% and the specificity 97.9%. Products with a high A_w (raw-egg product, raw poultry) resulted in a specificity of 92.3% and a sensitivity of 96.2%.

Nucleic acid-based assays

Nucleic acid-based assays include DNA-probe technology. Only one commercial assay based on DNA probes with tests for *Salmonella* is currently available. The colorimetric Gene Trak *Salmonella* assay employs *Salmonella*-specific DNA probes that target ribosomal RNA in the bacteria and a colorimetric system for detection. For *Salmonella* detection, a 44 h culture period, consisting of pre-, selective and post-enrichment, is needed before the probe assay itself can be completed in 2.5 h. This results in a total analytical time of about 48 h. The commercially available probe assay requires the presence of at least 10^3–10^5 target organisms ml^{-1} (Giese, 1995) to yield positive results. The performance of this assay has been studied but with no overall agreement. It proved to be comparable to conventional bacteriology in a study for the identification of *Salmonella* in cloacal samples (Cotter *et al.*, 1995). False-negative rates for this assay of 1.25–13.4% and false positives of 0.57–4.9% have been reported (St Clair and Klenk, 1990; Tate *et al.*, 1992; Quinn *et al.*, 1995) for raw chicken, poultry feed and environmental samples. In practice, the colorimetric *Salmonella* assay was more labour-intensive than commercially available ELISA tests and thus far more labour-intensive than modified conventional or automated methods; additional costs for the instruments are needed also.

Polymerase chain reaction

The PCR amplifies a targeted region of the *Salmonella* genome by repetition of a three-step process: (i) denaturing double-stranded DNA into single strands by heating; (ii) annealing specific, complementary, oligonucleotide primers to

the single-stranded DNA by cooling; and (iii) enzymatically extending the primers to produce an exact copy of the original double-stranded target sequence (Fig. 22.7).

The entire process, in which an amount of DNA equivalent to an 'overnight culture' is obtained, can be completed within 2 h. The technique was evaluated for detection of *Salmonella* in eggs (McElroy *et al.*, 1996).

Three manufacturers (Sanofi, Dupont and Perkin-Elmer) have released PCR-based test kits for *Salmonella*. All three now provide a kit in which the primer-mediated enzymatic amplification of target DNA is relatively easy to perform, because they use pre-dispensed PCR mixtures, which reduces the possibility of contamination and pipetting errors when preparing DNA polymerase, nucleotide triphosphates and buffer for the reaction. These kits also include a lysis buffer, which eliminates the need for DNA extraction by organic solvents or isolation columns. The assays are available as 24 h *Salmonella* assays, and they focus on detection of *Salmonella* in pre-enrichment culture with at least 10^2–10^3 target cells ml^{-1} (manufacturers' product information). Special instruments to perform the DNA amplification and final tests are also needed for these assays. The final detection step of amplified material differs in all three products. Gel electrophoresis is used in the Dupont BAX test, a colorimetric ELISA procedure by the Sanofi PROBELIA *Salmonella* PCR and an automated fluorescence immunoassay by Perkin-Elmer.

Application of rapid methods and test kits for detection of *Salmonella*

To stimulate the application of alternative methods, *Salmonella* detection tests with at least three main characteristics are needed. The tests should be: (i) highly efficient at accurately and reliably detecting low levels of *Salmonella*; (ii) cheap, rapid, simple and consumer-friendly, with the possibility of computerization; and (iii) internationally accepted.

EFFICIENCY. For this criterion, it can be said that the amount of false-negative and/or false-positive results is decisive. False-negative results are, in principle, unacceptable, although sometimes inevitable. False positives will cause a false alarm

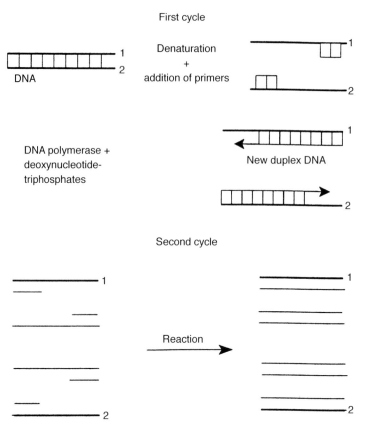

Fig. 22.7. Principle of the polymerase chain reaction (PCR). Small specific DNA sequences (primers) are allowed to bind to heat-denatured (separate) DNA strands. A heat stable enzyme (DNA polymerase) extends the primers to new complementary strands, after which the cycle is repeated.

and, as a consequence, extra work. Accordingly, the amount of false-positive results has to be limited to acceptable ranges.

LOW DETECTION LIMIT. This criterion is essential, because some stringent product requirements exist, in which absence of *Salmonella* in a 25 g sample is required.

COSTS. The initial costs of purchasing automated instruments have to be considered. Reagent supply and operational costs per test can be high. The volume of tests and the effectiveness of use of the instrument are essential issues in recovering the initial investment.

RAPIDITY. This can sometimes be the main reason for using these methods. Methods that can be used in 'on-line' processes and that are suit-

able for use in hazard-analysis critical control points (HACCP) schemes would be ideal.

EASY TO HANDLE. Some of the methods require highly skilled personnel, which may not always be present in quality-control laboratories in the food industry.

AUTOMATION. This allows the possibility of examining large numbers of samples at the same time and the possibility of computerization. This is a must nowadays for the modern food microbiology laboratories.

INTERNATIONAL ACCEPTABILITY. This is a requirement of great importance to international trade.

A summary of the ways in which alternative methods for pathogens generally meet some of

Table 22.1. Alternative screening and detection methods for *Salmonella* and their characteristics.

Method	Accuracy	Detection limit[1]	Analysis time (h)	Ease of handling
Modified conventional	**	10^1–10^2	24–60	+
Impedimetry	*	10^5–10^6	24–60	+
Immunological techniques				
ELISA	**	10^5–10^6	48–60	+/±
IMS/IC	**	10^2–10^3	24–48	+/±
DNA techniques	**	10^3–10^5	22–60	±
PCR	**	10^2–10^3	22–24	±

[1] Detection limit is expressed as the number of cells ml^{-1} to yield detection in the substrate that is tested in the test kit or instrument.
*, Accuracy good for some defined groups of samples; **, high. Ease of handling: +, high; ±, intermediate.

the previously mentioned criteria is given in Table 22.1.

Problems with the application of novel techniques

Testing for *Salmonella* is needed at different stages in the animal husbandry 'production chain'. Also the need for quick results and required sensitivity or specificity of the tests may vary. Consequently, different approaches can be developed and applied. As shown previously the potential methodology is available but, unfortunately, it must be admitted that not many of the newly developed techniques have been used in routine work. Apparently, there is a gap between what is offered and what is suitable for implementation. The reasons for this include the following observations.

- Some of the methods require highly skilled personnel, who are not always available in routine laboratories.
- The isolation of the target organism from food, feed, organ or faecal samples is a difficult step. This is not usually accounted for in the instructions delivered with the instrument or test kit.
- In some non-isolation techniques there are no objective results. Subjective judgements must be made by highly trained personnel.
- Operational costs may be high. Alternative methods are considerably more expensive than conventional methods as far as utensils and reagents are concerned. Therefore, the benefits of reduced labour and time, low detection limits and the possibility of computerization should be weighed against these higher costs.

- A barrier to acceptance of new alternative tests, especially in the food and feed industry, and also the fact that there is no worldwide standardization for validation and approval of these tests. For this reason, accredited laboratories might be reluctant to introduce new techniques. New methods for detection of *Salmonella* from different samples have to be evaluated against standard methods before being introduced. At the international level, in the USA, the evaluation programmes of the Association of American Chemists (AOAC) International are the most widely accepted. In France, the Association Française de Normalisation (AFNOR) has a similar position. To deal with these problems, an initiative was started by the European Community. The so-called MICROVAL project, started in 1993, aims at the inception of a European validation procedure, which should take no longer than 9 months to validate a new alternative method.

Future Perspectives and Concluding Remarks

In developing new techniques for *Salmonella* detection, the problems – as mentioned above – have to be considered. As a first step, creative combinations of immunoassays, DNA technology and automated instrumental techniques hold promise and can form the basis of future microbiological monitoring systems. There are a few areas where advances are being made to develop new techniques for microbial analysis in general and *Salmonella* detection in particular. Some of these areas are described below:

More efficient cultural techniques

Novel ways to rapidly enhance the multiplication of *Salmonella* in the presence of background contamination and interference from food components in pre-enrichment broth are being investigated. The use of ferrioxamine (Reissbrodt and Rabsch, 1993) or oxyrase enzyme (Fung *et al.*, 1994) in these broths is an example.

Bacteriophage-based assays

lux *genes*

Recombinant bacteriophages specific for *Salmonella* and containing bacterial luciferase can be constructed. The phage will cause the host cell to luminesce when it is mixed with *Salmonella*. Luminescence can be detected using a so-called photon-counting charge-couple device camera, a luminometer or an X-ray film (Chen and Griffiths, 1996).

Bacterial ice-nucleation diagnostics (BIND)

The assay uses genetically engineered bacteriophage containing a so-called 'ice-nucleation activity' (*ina*) gene (Wolber and Green, 1990). When delivered to *Salmonella*, the gene induces production of a protein that increases the freezing-point of the solution (Fig. 22.8). If *Salmonella* bacteria are present in a sample, they will receive the *ina* gene from the bacteriophage in the BIND assay, causing the cell to produce ice-nucleation proteins. When the temperature of the sample is then lowered to −8.5°C, positive samples freeze and a freezing indicator dye, which is yellow-green and fluorescent in the liquid state, will turn to orange; negative samples remain yellow-green. The BIND assay can identify *Salmonella* in only 22 h.

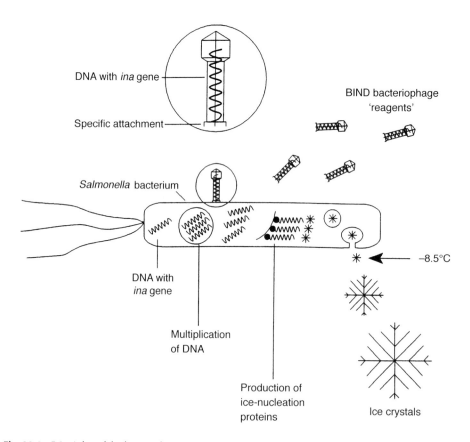

Fig. 22.8. Principles of the bacterial ice nucleation diagnostics (BIND) test.

Biosensors

Other powerful tools for monitoring *Salmonella* infections are biosensors, in particular immunosensors based on surface plasmon resonance (SPR). The technique offers a number of potential advantages over conventional immunoassays (Severs and Schasfoort, 1993). The detection is quick and in real time, while the immunochemical reaction takes place. Moreover, the system is easy to use, does not need highly sophisticated environments and can easily be automated for decentralized testing. The SPR immunosensor is based on changes in the reflection of laser light at a metal–liquid surface, due to a change in refractive index. Local refractive-index changes, produced by binding of (serum) antibodies to immobilized antigens, disturb this field and the plasmons arise at a different angle of incidence. This shift can be detected by measuring a calibrated change in intensity of the reflected beam at a fixed angle of incidence with a pin diode. The sensitivity of the SPR immunosensor normally depends on the molecular weight of the ligand to be detected. Antibody titres in serum can be detected directly after binding to the antigen-coated slide. Biosensors based on SPR certainly hold promise for future monitoring of *Salmonella* infections in animal husbandry. The availability of automated instrumentation (BIAcore system) indicates a further likely increase of the future applications of this technology.

In particular, research should aim at selection of the appropriate methodology, the problems of sample preparation and organism separation and the specificity of results, all working from sample matrix towards analytical end points, rather than the other way around.

It has been proved that the accuracy and efficacy of alternative methods for *Salmonella* detection are equal to or even better than that of conventional standard methods.

The greatest benefit of alternative tests for *Salmonella* is savings in time in comparison with conventional culture methods, although all these techniques are dependent on some form of culture enrichment to acquire a sufficient concentration of the target organism to achieve detection. Accordingly, the name 'rapid methods' may, in some cases, seem to be a little misleading, because some of these methods only replace the materials used in the isolation and confirmation stage, whereas the time needed for pre-enrichment and/or selective enrichment procedures remains the same. In some cases, however, a rapid screening of enrichment cultures before starting the complicated conventional isolation and confirmation procedure can save considerable work.

If the problem of standardization of validation can be solved satisfactorily and easy-to-perform tests with even shorter reporting times than the 22–24 h now achieved can be developed, then the use of alternative methods will increase substantially in the near future.

Acknowledgements

The authors thank Colin Bubeneck and Peter Jansen for providing the illustrations.

References

Andrews, W.H. (1996) Evolution of methods for the detection of *Salmonella* in foods. *Journal of the AOAC International* 79, 4–12.

Bailey, J.S., Cox, N.A., Bruce, J. and Fritschell, S. (1996) Comparison of automated ribotyping to conventional serotyping of *Salmonella*. In: *Abstracts of the Annual Meeting, 1996*. American Society for Microbiology, Washington, DC, p. 25.

Beckers, H.J., Roberts, D., Pietsch, O., van Schothorst, M., Vassiliadis, P. and Kampelmacher, E.H. (1987) Replacement of Mueller–Kauffmann's tetrathionate brilliant green bile broth by Rappaport–Vassiliadis' magnesium chloride malachite green broth in the standard method for detection of *Salmonella*. *International Journal of Food Microbiology* 4, 59–64.

Blackburn, C. de W. (1993) Rapid and alternative methods for the detection of *Salmonella* in foods. *Journal of Applied Bacteriology* 75, 199–214.

Blackburn, C. de W., Curtis, L.M., Humpheson, L. and Petit, S.B. (1994) Evaluation of the Vitek immunodiagnostic assay system (VIDAS) for the detection of *Salmonella* in foods. *Letters in Applied Microbiology* 19, 32–36.

Bolton, F.J. and Gibson, D.M. (1994) Automated electrical techniques in microbiological analysis. In: Patel, P.D. (ed.) *Rapid Analysis Techniques in Food Microbiology*. Blackie Academic, Glasgow, pp. 131–167.

Brinkman, E., Van Beurden, R., Mackintosh, R. and Beumer, R. (1995) Evaluation of a new dipstick test for the rapid detection of *Salmonella* in food. *Journal of Food Protection* 58, 1023–1027.

Bruce, J. (1996) Automated system rapidly identifies and characterizes microorganisms in food. *Food Technology* January, 77–81.

Bullock, R.D. and Frodsham, D. (1988) Rapid impedance detection of *Salmonella* in confectionery using modified LICNR broth. *Journal of Applied Bacteriology* 66, 385–391.

Carlson, V.L. and Snoeyenbos, G.H. (1974) Comparative efficiencies of selenite and tetrathionate enrichment broths for the isolation of *Salmonella* serovars. *American Journal of Veterinary Research* 35, 711–718.

Chen, J. and Griffiths, M.W. (1996) *Salmonella* detection in eggs, using Lux+ bacteriophages. *Journal of Food Protection* 59, 908–914.

Cotter, P.F., Murphy, J.E., Klinger, J.D. and Taylor, R.L. (1995) Identification of *Salmonella enteritidis* from experimentally infected hens using a colorimetric DNA hybridization method. *Avian Diseases* 39, 873–878.

Curiale, M.S., Gangar, V. and Gravens, C. (1997) VIDAS enzyme linked immunoassay for detection of *Salmonella* in foods: collaborative study. *Journal of AOAC International* 80, 491–504.

D'Aoust, J.Y. (1992) Commercial diagnostic kits for the detection of food-borne *Salmonella* In: *Reports and Communications Symposium* Salmonella *and Salmonellosis*. Ploufragan, France, pp. 9–19.

Davies, R.H. and Wray, C. (1994a) Evaluation of SMID agar for identification of *Salmonella* in naturally contaminated veterinary samples. *Letters in Applied Microbiology* 18, 15–17.

Davies, R.H. and Wray, C. (1994b) Evaluation of a rapid cultural method for identification of *Salmonella* in naturally contaminated veterinary samples. *Journal of Applied Bacteriology* 77, 237–241.

Davies, R.H. and Wray, C. (1997) Immunomagnetic separation for enhanced flagellar antigen phase inversion in *Salmonella*. *Letters in Applied Microbiology* 24, 217–220.

Davison, S., Benson, C.E. and Eckroade, R.J. (1995) Comparison of environmental monitoring protocols for the detection of *Salmonella* in poultry houses. *Avian Diseases* 39, 475–479.

DeSmedt, J.M., Bolderdijk, R.F., Rappold, H.F. and Lautenschlager, D. (1986) Rapid *Salmonella* detection in foods by motility enrichment on a modified semi-solid Rappaport–Vassiliadis medium. *Journal of Food Protection* 49, 510–514.

Di Falco, G., Giaccone, V., Amerio, G.P. and Parisi, E. (1993) A modified impedance method to detect *Salmonella* spp. in fresh meat. *Food Microbiology* 10, 421–429.

Dolzinsky, B. and Kruse, K. (1992) Zur Isolierung von Salmonellen unter Verwendung Ammoniumeisen(III)-citrat-haltiger Medien. *Archiv für Lebensmittelhygiene*. 43, 124–125.

Dusch, H. and Altwegg, M. (1993) Comparison of Rambach agar, SM-ID medium. and Hektoen enteric agar for primary isolation of non-typhi *Salmonella* from stool samples. *Journal of Clinical Microbiology* 31, 410–412.

Dziezak, J.D. (1990) New analyzer automates *Salmonella* testing. *Food Technology* 44, 116.

Entis, P. (1996) Validation of the ISO-GRID 2-day rapid screening method for detection of *Salmonella* spp. in egg products. *Journal of Food Protection* 59, 555–558.

Entis, P. and Boleszczuk, P. (1991) Rapid detection of *Salmonella* in foods using EF-18 agar in conjunction with the hydrophobic grid membrane filter. *Journal of Food Protection* 54, 930–934.

Fagerburg, D.J. and Avens, J.S. (1976) Enrichment and plating methodology for *Salmonella* detection in food: a review. *Journal of Milk and Food Technology* 39, 628–646.

Feng, P. (1996) Emergence of rapid methods for identifying microbial pathogens in foods. *Journal of AOAC International* 79, 809–812.

Fierens, H. and Huygebaert, A. (1996) Screening of *Salmonella* in naturally contaminated feeds with rapid methods. *International Journal of Food Microbiology* 31, 301–309.

Firstenberg-Eden, R. and Eden, G. (1984) What is impedance microbiology? In: Sharpe, A.N. (ed.) *Impedance Microbiology*. Innovation in Microbiology Series, Research Studies Press, Letchworth, UK, pp. 7–21.

Freydiere, A.M. and Gille, Y. (1991) Detection of *Salmonella* by using Rambach agar and by a C8 esterase spot test. *Journal of Clinical Microbiology* 29, 2357–2359.

Fricker, C.R. (1987) The isolation of *Salmonella* and campylobacters. *Journal of Applied Bacteriology* 63, 99–116.

Fung, D.Y.C., Niroomand, F. and Tuitemwong, K. (1994) Novel methods to stimulate growth of food pathogens by oxyrase and related membrane fractions. In: Spencer, R.C., Wright, E.P. and Newsom, S.W.B. (eds) *Rapid Methods and Automation in Microbiology and Immunology*. Intercept, Andover, UK, pp. 313–318.

Gast, R.K. (1993) Recovery of *Salmonella enteritidis* from inoculated pools of egg contents. *Journal of Food Protection* 56, 21–24.

Gast, R.K., and Holt, P.S. (1995) Iron supplementation to enhance recovery of *Salmonella enteritidis* from pools of egg contents. *Journal of Food Protection* 58, 268–272.

Gibson, D.M. (1987) Some modification to the media for rapid automated detection of *Salmonella* by conductance measurement. *Journal of Applied Bacteriology* 63, 299–304.

Giese, J. (1995) Rapid microbiological testing, kits and instruments. *Food Technology* 49, 64–71.

Gruenewald, R., Henderson, R.W. and Yappow, S. (1991) Use of rambach propylene glycol containing agar for identification of *Salmonella* spp. *Journal of Clinical Microbiology* 29, 2354–2356.

Hajna, A.A. (1956) A new enrichment broth for gram negative organisms of the intestinal group. *Public Health Laboratory* 13, 83–89.

Holbrook, R., Andersen, J.M., Baird-Parker, A.G., Doods, L.M., Sawhney, A., Stuchbury, S.H. and Swaine, D. (1989) Rapid detection of *Salmonella* in foods, a convenient two-day procedure. *Letters in Applied Micobiology* 8, 139–142.

Jeffries, L. (1959) Novobiocin–tetrathionate broth: a medium of improved selectivity for the isolation of *Salmonella* from faeces. *Journal of Clinical Pathology* 12, 568–571.

June, G.A., Sherrod, P.S., Hammack, T.S., Amagua, R.M. and Andrews, W.H. (1996) Relative effectiveness of selnite cystine broth, tetrathionate broth and Rappaport–Vassiliadis medium for recovery of *Salmonella* spp. from raw flesh, highly contaminated foods, and poultry feed: collaborative study. *Journal of AOAC International* 79, 1307–1322.

Kauffmann, F. (1935) Weitere Erfahrungen mit dem kombinierten Anreicherungsverfahren für *Salmonella* Bazillen. *Zeitschrift für Hygiene und Infectionskrankheiten* 117, 26–32. Cited by Fricker, C.R. (1987) The isolation of *Salmonella* and campylobacters. *Journal of Applied Bacteriology* 63, 99–116.

Knight, M.T., Wood, D.W., Black, J.F., Gosney, G., Rigney, R.O., Agin, J.R., Gravens, C.K. and Farnam, S.M. (1990) Gram-negative identification card for identification of *Salmonella*, *Escherichia coli* and other Enterobacteriaceae isolated from foods. *Journal of AOAC International* 73, 729–733.

Krusell, L. and Skovgaard, N. (1993) Evaluation of a new semi-automated screening method for the detection of *Salmonella* in foods within 24 h. *International Journal of Food Microbiology* 20, 123–130.

Kühn, H., Wonde, B., Rabsch, W. and Reissbrodt, R. (1994) Evaluation of Rambach agar for detection of *Salmonella* subspecies I to VI. *Applied and Environmental Microbiology* 60, 749–751.

Leifson, E. (1936) New selenite enrichment media for the isolation of typhoid and paratyphoid (*Salmonella*) bacilli. *American Journal of Hygiene* 24, 423–432. Cited by Fricker, C.R. (1987) The isolation of *Salmonella* and campylobacters. *Journal of Applied Bacteriology* 63, 99–116.

McElroy, A.P., Cohen, N.D. and Hargis, B.M. (1996) Evaluation of the polymerase chain reaction for the detection of *Salmonella enteritidis* in experimentally inoculated eggs and eggs from experimentally challenged hens. *Journal of Food Protection* 59, 1273–1278.

Mallinson, E.T. (1990) *Salmonella* monitoring system simplifies evaluation of farms. *Poultry Digest* 46–48.

Manafi, M. and Sommer, R. (1992) Comparison of three rapid screening methods for *Salmonella* spp.: MUCAP test, microscreen latex and Rambach agar. *Letters in Applied Microbiology* 14, 163–166.

Miller, J.M. and Rhoden, D.L. (1991) Preliminary evaluation of Biolog, a carbon source utilization method for bacterial identification. *Journal of Clinical Microbiology* 29, 1143–1147.

Miller, R.G., Tate, C.R., Mallinson, E.T. and Scherrer, J.A. (1991) Xylose-lysine-tergitol 4: an improved selective agar medium for the isolation of *Salmonella*. *Poultry Science* 70, 2429–2432.

Miller, R.G., Tate, C.R. and Mallinson, E.T. (1995) Improved XLT4 agar: small addition of peptone to promote stronger production of hydrogen-sulfide by *Salmonella*. *Journal of Food Protection* 58, 115–119.

Mueller, L. (1923) Un nouveau milieu d'enrichissement pour la recherche du bacile typhique et des paratyphiqe. *Comptes Rendus des Séances de la Société de Biologie et de ses Filiales* 89, 434–437. Cited by Fricker, C.R. (1987) The isolation of *Salmonella* and campylobacters. *Journal of Applied Bacteriology* 63, 99–116.

Ogden, I.D. (1988) A conductance medium to distinguish between *Salmonella* and *Citrobacter* spp. *International Journal of Food Microbiology* 7, 287–297.

Oggel, J.J., Nundy, D.D. and Randall, C.J. (1990) Modified 1–2 test system as a rapid screening method for detection of *Salmonella* in food and feed. *Journal of Food Protection* 8, 656–658.

Oggel, J.J., Nundy, C., Zebuck, P.A. and Shaw, S.J. (1995) Reliability of the modified semi-solid Rappaport–Vassiliadis agar and the modified 1–2 test system for detection of *Salmonella* in poultry feeds. *Journal of Food Protection* 58, 98–101.

Olson, M., Syk, A. and Wollin, R. (1991) Identification of *Salmonella* with the 4- methylumbilliferyl caprilate fluorescence test. *Journal of Clinical Microbiology* 29, 2631–2632.

Opara, O.O., Mallinson, E.T., Tate, L.E., Miller, R.G., Steward, L., Kelleher, C., Johnston, R.W. and Joseph, S.W. (1992) The effect of exposure, storage, times, and types of holding media on the drag-swab monitoring technique for *Salmonella*. *Avian Diseases* 36, 63–68.

Opara, O.O., Mallinson, E.T., Tate, L.E., Miller, R.G., Steward, L. and Joseph, S.W. (1994) Evaluation of possible alternatives to double strength skim milk used to saturate drag swabs for *Salmonella* detection. *Avian Diseases* 38, 293–296.

Patel, P. (1997) Recent advances in microbiological methods in food control laboratories. In: *The Second Symposium on Food Safety, Organised by Ministry of Public Health, Preventive Health Department, Doha, Qatar, 29 April–1 May, 1997.*

Perales, I. and Erkiaga, E. (1991) Comparison between semisolid Rappaport and modified semisolid Rappaport–Vassiliadis media for the isolation of *Salmonella* species from foods and feed. *International Journal of Food Microbiology* 14, 51–57.

Pless, P., Futchick, K. and Schopf, E. (1994) Rapid detection of *Salmonella* by means of a new impedance-splitting method. *Journal of Food Protection* 57, 369–376.

Poppe, C. and Duncan, C.L. (1996) Comparison of detection of *Salmonella* by the Tecra Unique *Salmonella* test and the modified Rappaport Vassiliadis medium. *Food Microbiology* 13, 75–81.

Poppe, C., Johnson, R.P., Forsberg, C.M. and Irwin, R.J. (1992) *Salmonella enteritidis* and other *Salmonella* in laying hens and eggs from flocks with *Salmonella* in their environment. *Canadian Journal of Veterinary Research* 56, 226–232.

Quinn, C., Ward, J., Griffin, M., Yearsley, D. and Egan, J. (1995) A comparison of conventional culture and three rapid methods for the detection of *Salmonella* in poultry feeds and environmental samples. *Letters in Applied Microbiology* 20, 89–91.

Rambach, A. (1990) New plating medium for facilitated differentiation of *Salmonella* spp. from *Proteus* spp. and other enteric bacteria. *Applied Environmental Microbiology* 56, 127–130.

Rampling, A., Upson, R. and Brown, D.F.J. (1990) Nitrofurantoin resistance in isolates of *Salmonella enteritidis* phage type 4 from poultry and humans. *Journal of Antimicrobial Chemotherapy* 25, 285–290.

Rappaport, F., Konforti, N. and Navon, B. (1956) A new enrichment medium for certain *Salmonella*. *Journal of Clinical Pathology* 9, 261–266.

Reissbrodt, R.L. and Rabsch, W. (1993) Selective pre-enrichment of *Salmonella* from eggs by siderophore supplements. *Zentralblatt für Bakteriologie* 279, 344–353.

Ruiz, J., Valera, M.C., Sempere, M.A., Lopez, M.L., Gomez, J. and Oliva, J. (1991) Presumptive identification of *Salmonella enterica* using two rapid tests. *European Journal of Clinical Microbiology and Infectious Diseases* 10, 649–651.

Ruzickova, V., Karpiskova, R. and Pakrova, E. (1996) Semi-solid medium diasalm for the isolation of *Salmonella enteritidis*. *Veterinarni Medicina* 41, 283–288.

Severs, A.H. and Schasfoort, R.B.M. (1993) Enhanced surface plasmon resonance inhibition test (ESPRIT) using latex particles. *Biosensors and Bioelectronics* 8, 365–370.

St Clair, V.J. and Klenk, M.M. (1990) Performance of three methods for the rapid identification of *Salmonella* in naturally contaminated foods and feed. *Journal of Food Protection* 53, 961–964.

Silley, P. and Forsythe, S. (1996) Impedance microbiology – a rapid change for microbiologists. *Journal of Applied Microbiology* 80, 233–243.

Skjerve, E. and Olsvik, O. (1991) Immunomagnetic separation of *Salmonella* from foods. *International Journal of Food Microbiology* 14, 11–17.

Svastova, A., Skalka, B. and Smola, J. (1984) A modified medium for *Salmonella* isolation by the selective motility test. *Zentralblatt für Veterinaer Medizin (Serie B)* 31, 396–399.

Tate, C.R., Miller, R.G. and Mallison, E.T. (1992) Evaluation of two nonisolation methods for detecting naturally occuring *Salmonella* from broiler flock environmental drag-swab samples. *Journal of Food Protection* 55, 964–967.

Thomason, B.M., Dodd, D.J. and Cherry, W.B. (1977) Increased recovery of *Salmonella* from environmental samples enriched with buffered peptone water. *Applied Environmental Microbiology* 34, 270–273.

van der Zee, H. (1994) Conventional methods for the detection and isolation of *Salmonella enteritidis*. *International Journal of Food Microbiology* 21, 41–46.

van der Zee, H. and Huis in't Veld, J.H.J. (1997) Rapid and alternative screening methods for microbiological analysis. *Journal of AOAC International* 80, 934–940.

van der Zee, H. and van Netten, P. (1992) Diagnostic selective semisolid media based on Rappaport–Vassiliadis broth for the detection of *Salmonella* spp. and *Salmonella enteritidis* in foods. In: *Proceedings Symposium Salmonella and Salmonellosis, Ploufragan: Reports and Communications*, pp. 69–77.

van Netten, P., van der Zee, H. and van Moosdijk, A. (1991) The use of a diagnostic semisolid medium for the isolation of *Salmonella enteritidis* from poultry. In: *Quality of Poultry Products. III Safety and Marketing Aspects.* Spelderholt Jubilee Symposia, Doorwerth, pp. 59–66.

Vassiliadis, P., Kalapothaki, V., Trichopoulos, D., Mavrommati, C. and Sérié, C. (1981) Improved isolation of *Salmonella* from naturally contaminated meat products by use of Rappaport–Vassiliadis enrichment broth. *Applied Environmental Microbiology* 42, 615–618.

Vermunt, A.E.M., Franken, A.A.J.M. and Beumer, R.R. (1992) Isolation of *Salmonella* by immunomagnetic separation. *Journal of Applied Microbiology* 72, 112–118.

Warburton, D.W., Oggel, J., Bowen, B., Crawford, C., Durzi, S., Gibson, E., Foster, R., Fox, C., Gour, L., Krohn, G., McDonah, S., MacKenzie, J., Todd, E.C.D., Shaw, S., Tiwari, N.P., Trottier, Y. and Wheeler, B.D. (1994a) A comparison study of the modified 1–2 test and the HPB standard method in the isolation of *Salmonella*. *Food Microbiology* 11, 253–263.

Warburton, D.W., Arling, V., Worobec, S., MacKenzie, J., Todd, E.C.D., Lacasse, P., Lamontagne, G., Plante, R., Shaw, S., Bowen, B. and Konkle, A. (1994b) A comparison study of the EF-18/hydrophobic grid membrane filter (HGMF) method and the enzyme linked antibody (ELA)/HGMF method to the HPB standard method in the isolation of *Salmonella*. *International Journal of Food Microbiology* 23, 89–98.

Wiberg, C. and Norberg, P. (1996) Comparison between a cultural procedure using Rappaport–Vassiliadis broth and motility enrichment on modified semisolid Rappaport–Vassiliadis medium for *Salmonella* detection from food and feed. *International Journal of Food Microbiology* 29, 353–360.

Wolber, P.K. and Green, R.L. (1990) Detection of bacteria by transduction of ice nucleation genes. *Trends in Biotechnology* 8, 276–279.

Chapter 23

Laboratory Aspects of *Salmonella*

Yvette E. Jones, Ian M. McLaren and Clifford Wray

Veterinary Laboratories Agency (Weybridge), New Haw, Addlestone, Surrey KT15 3NB, UK

Biochemical Identification

Selective agar media are examined for typical *Salmonella* colony formation and morphology, which will differ according to the media used. Laboratory personnel need to be aware of the appearance of *Salmonella* on the different plating media and also to be aware of atypical colonies (see Chapter 21). Depending on the selectivity of the respective plating media, there may be almost a pure culture of *Salmonella* or the presence of several different bacterial species in addition to *Salmonella*: hence the use of the more selective plating media, which will reduce the level of colonies similar to *Salmonella* and the labour involved in screening them. The use of more than one plating medium will help to detect atypical strains, e.g. non-hydrogen sulphide (H_2S) producers.

The methylumbelliferyl caprylate fluorescence (MUCAP) test is a useful screening test (Chapter 22). One drop (c. 5 µl) of the reagent is added to each colony and observed under ultraviolet (UV) light (wavelength 366 nm), when a strong blue fluorescence is produced after 3 min. However, other species of bacteria, e.g. pseudomonads, may occasionally give false-positive results.

For a fuller account of the biochemical characteristics of *Salmonella* the reader is referred to Ewing (1986). The genus *Salmonella* consists of two species, *S. enterica* and *S. bongori* (Chapter 1). *S. enterica* consists of six subspecies, which

can be differentiated by the tests indicated in Table 23.1. Most isolates of *Salmonella* from domestic animals belong to subspecies *enterica* and the behaviour of these organisms is, with few exceptions, regular and may be classified as typical for the subspecies.

The principal biochemical tests by which *Salmonella* can be identified are given in Table 23.2. It should, however, be pointed out that variable test results are likely, e.g. lactose-fermenting *Salmonella* of subspecies 1 have been described, and reference to Ewing (1986) shows that, with few exceptions, no test is 100% positive or negative when a wide range of strains are examined. Since most of the isolates belong to subspecies 1 (*enterica*), these can be readily confirmed with a simplified range of biochemical tests, usually the fermentation of glucose, mannitol and dulcitol, inability to ferment sucrose, salicin and lactose, inability to hydrolyse urea, O-nitrophenyl-β-D-galactopyranoside (ONPG)-positive and production of H_2S.

However, it is probably more normal to use composite media such as triple sugar iron agar (TSI). This contains glucose, lactose and sucrose, an H_2S detection system and an indicator. The medium is inoculated by stabbing into the centre of the butt and continuing down to the base and then streaking the inoculum on to the slope, with incubation at 37°C. Organisms capable of fermenting glucose, but not lactose or sucrose, will show an initial acid (yellow) slant in a short period as the small quantity of glucose is utilized.

Table 23.1. Differential characteristics of *Salmonella* species and subspecies (from Le Minor *et al.*, 1982, 1986).

| Characters | S. enterica | | | | | | S. bongori |
	subsp. *enterica*	subsp. *salamae*	subsp. *arizonae*	subsp. *diarizona*	subsp. *houtenae*	subsp. *indica*	
Dulcitol	+	+	−	−	−	d	+
ONPG (2 h)	−	−	+	+	−	d	+
Malonate	−	+	+	+	−	−	−
Gelatinase	−	+	+	+	+	+	−
Sorbitol	+	+	+	+	+	−	+
Culture with KCN	−	−	−	−	+	−	+
L(+)-tartrate*	+	−	−	−	−	−	−
Galacturonate	−	+	−	+	+	+	+
γ-Glutamyltransferase	+†	+	−	+	+	+	+
β-Glucuronidase	−	d	−	+	−	d	−
Mucate	+	+	+	−(70%)	−	+	+
Salicin	−	−	−	−	+	−	−
Lactose	−	−	−(75%)	+(75%)	−	d	−
Lysis by phage 01	+	+	−	+	−	+	d
Usual habitat	Warm-blooded animals		Cold-blooded animals and environment				

* d-tartrate. † Typhimurium d, Dublin −. +, 90% or more positive reactions; −, 90% or more negative reactions. d, different reactions given by different serovars; ONPG, *o*-nitropenyl-β-D-galactopyranoside; KCN, potassium cyanide.

However, the reaction under the aerobic conditions on the slant reverts and becomes alkaline (red) because of protein breakdown in the medium. Under anaerobic conditions in the butt of the tube, the medium remains acid (yellow) and H_2S production is characterized by a blackening of the medium. The tube should be lightly stoppered; otherwise all the medium may become black. Some *Proteus* and other species may give a similar reaction, but can readily be distinguished by their ability to hydrolyse urea, and urea agar should always be used in parallel. Typical reactions in TSI are shown in Table 23.3.

Other composite media include Kliger's iron agar and Kohn's two-tube media. Alternative systems, such as miniaturized biochemical test kits, e.g. API, Micro-ID and Enterotube, may also be used. The results are translated into a numbered biochemical profile, which can be compared with a database for genus or species identification with a probability of correct identification. It should be remem-

bered, however, that the databases are produced predominantly with human isolates and atypical *Salmonella*, e.g. *S. pullorum* may be misidentified as another bacterial species, e.g. *hafnia* and also that *S. arizonae* O18:z_4,z_{32} and O18:z_4,z_{23} are important pathogens of poultry, and O61:k:1, 5, 7 is a frequent isolate from sheep and *S.* II *sofia* is widespread in Australian poultry.

The biochemical reactions of the 'host-adapted' serovars, such as *S. choloraesuis*, *S. gallinarum/pullorum* and, especially *S. abortusovis*, are variable (Table 23.4) and are auxotrophic microorganisms, which require one or more growth factors. Thus *S. choloraesuis* fails to ferment trehalose, which differentiates it from *S. paratyphi* C and *S. typhisuis*. The subdivision of *S. paratyphi* B and *S. java* (a frequent isolate from terrapins) is based on the ability of the former to produce a slime wall and its inability to utilize D-tartrate. *S. pullorum* and *S. gallinarum* can be distinguished by the inability of the former to ferment maltose and dulcitol.

Table 23.2. Biochemical reactions of typical *Salmonella enterica* subsp. *enterica* strains.

Test	Reaction of typical strains	Test	Reaction of typical strains
Motility	+	Fermentation of:	
Nitrate reduction	+	Glucose	+
Oxidase	−	Mannitol	+
O/F	F	Maltose	+
Urea hydrolysis	−	Lactose	−
Indole	−	Sucrose	−
H₂S production	+	Salicin	−
Citrate utilization	+	Adonitol	−
Sodium malonate	−	Dulcitol	+
Growth on KCN	−	Lysine decarboxylase	+
MR	+	Arginine dihydrolase	+
VP	−	Ornithine decarboxylase	+
Gelatin liquefaction	−	Deamination of phenylalanine	−
ONPG	−		

KCN, potassium cyanide; MR, methyl red; O/F, oxidation/fermentation; VP, Voges Proskauer.

Biotyping

In its broadest sense, the term biotyping can be used to describe members of a group of living entities which resemble one another but which can be distinguished from one another on the basis of one or more biological parameters. Bacterial biotyping predates the age of genetic technology and historically these parameters have been phenotypic rather than genomic in nature. This is based on an assessment of the physical constitution of the organism as determined by interaction with its environment, as opposed to genetic content. Until recent times, such phenotypic characters were relied upon for the purpose of bacterial taxonomy and identification. The knowledge from phenotypic characterization was also applied in the development of improved isolation media and protocols. An account of bacterial biotyping could include bacteriophage susceptibility, antimicrobial susceptibility, resistance to heavy metals and dyes and production of bacteriocins. However, the present chapter will only deal with the subdivision of *Salmonella* serovars/phage types on the basis of substrate utilization.

Table 23.3. Reactions on triple sugar iron agar.

Organism	Butt	Slope	H₂S
Enterobacter aerogenes	AG	A	−
Enterobacter cloacae	AG	A	−
Escherichia coli	AG	A	−
Proteus vulgaris	AG	A	+
Morganella morgianii	A or AG	NC or ALK	−
Shigella dysenteriae	A	NC or ALK	−
Salmonella typhi	A	NC or ALK	−
Salmonella paratyphi	AG	NC or ALK	−
Salmonella typhimurium (and other typical *Salmonella*)	AG	NC or ALK	+

AG, acid (yellow) and gas formation; A, acid (yellow); NC, no change; ALK, alkaline (red); +, hydrogen sulphide (black); −, no hydrogen sulphide (no blackening).

Table 23.4. Biochemical reactions of 'host-adapted' *Salmonella enterica* serovars.

	Typical Salmonella	S. choleraesuis	S. pullorum	S. gallinarum	S. typhi	S. typhisuis	S. paratyphi A
H₂S	+	− (var. *kunzendorf*) +	±	±	+ (weak)	+	−
Citrate	+	+*	+	+	−	−	−
Gas from glucose	+	+	+	±	−	+	+
Dulcitol	+	−	−	+	−	(+)	+
Mucate	+	−	−	+	−	−	−
Maltose	+	+	−	+	+	+	+
Trehalose	+	−	+	+	+	+	+

* Simmon's citrate-negative, Christensen citrate-positive.

Biotyping may be applied to *Salmonella* for a variety of reasons. It may, for example, be used to determine the extent of changes in the phenotypic activity of mutant strains created for research purposes. This would be especially applicable where a deletion or insertion might affect more than the intended target character. Biotyping has been applied in many instances to improve our understanding of the spread of epidemic strains and also to phylogenetic studies which aim to determine how groups of bacteria may be related. Irrespective of the application, a good biotyping scheme should satisfy three criteria. First, it should have the desired discriminatory power to subdivide the existing group. Secondly, it needs to be reliable. This refers to both reproducibility of replicate tests on the same strain and stability of the typing character during the course of the epidemic spread. Thirdly, it needs to be relatively cheap and easy to use. This is particularly important for epidemiological studies, which may involve testing many strains.

A scheme with these qualities was devised for *S. typhimurium* and described by Duguid *et al.* (1975). It consists of five primary tests (xylose utilization, fermentation of meso-inositol, rhamnose and D-tartrate and meso-tartrate resistance) and ten supplementary tests (fimbriation, motility, fermentation at 37°C of L-tartrate, xylose, trehalose, glycerol and inositol at 25°C, rhamnose utilization, nicotinamide requirement and cysteine). Using the scheme, they were able to suggest possible mutational routes by which various groups of *S. typhimurium* had developed. In another study, in combination with phage typing, Anderson *et al.* (1978) gained an improved understanding of the phylogeny and spread of epidemic strains by distinguishing clones of different biotypes within the same phage type.

The same biotyping scheme with slight modifications has been successfully applied to studies on *S. agona* (Barker *et al.*, 1982), *S. paratyphi* B (Barker *et al.*, 1988), *S. montevideo* (Reilly *et al.*, 1985), *S. senftenberg* (Tuchili *et al.*, 1991) and *S. livingstone* (Odongo *et al.*, 1990).

General considerations for devising a biotyping scheme

Strains
A panel of strains representing the group of interest should be assembled from as wide a geographic, temporal and host range as possible. In this way, the likelihood of obtaining different biotypes will be maximized.

Inoculum
This should be dense and prepared just before use from freshly grown subcultures. The use of old material is strongly discouraged.

Substrates
Conventional biochemical test media are suitable; however, the substrates commonly listed in identification tables are likely to be of limited value for the purposes of subdividing serovars and phage types. This is because they will have been selected to ensure shared characteristics of all members of that group. More likely to be of value are substrates excluded from these lists or those recorded as giving 'variable' results.

Table 23.5. Indicator pH range and colour change.

Indicator	pH range	Colour
Methyl red	4.2–6.3	Red–yellow
Andrades	5.0–8.0	Pink–yellow
Bromocresol purple	5.2–6.8	Yellow–purple
Bromothymol blue	6.0–7.6	Yellow–blue
Neutral red	6.8–8.0	Red–yellow
Phenol red	6.8–8.4	Yellow–red

Incubation time and temperature

For the sake of good reproducibility, the standardization of these criteria is vital. In addition, it should be borne in mind that the incubation period selected must be sufficiently long to allow positive results to appear but at the same time not so long that mutational fermentations arise. Generally, the incubation temperature should be between 35°C and 37°C. However, some enzyme systems are temperature-sensitive and differentiation of some strains is possible with incubation at alternative temperatures, e.g. meso-inositol at 25°C and 37°C in the scheme of Duguid *et al.* (1975).

Reading of test results

Tests can be read as growth/turbidity. This can be done by eye alone or in a more quantified way using an optical device, such as a spectrophotometer. Alternatively, an indicator can be included in the medium to show pH change. The choice of indicator may influence the results, as each has a particular pH range over which the colour change takes place (Table 23.5).

Factors influencing the pH and hence the result but which are external to the organism under test should be minimized. This includes the precise pH of the medium, the degree of gaseous exchange and even possibly the cleanliness of the glassware used. It is also important to standardize the reading criteria as much as possible, particularly in the intermediate area between fully positive and fully negative.

Quality control of media

Care needs to be exercised to limit the effect of potential variation between batches of media or ingredients. New batches should be tested with a panel of control strains representing positive, negative and, if possible, intermediate reactions.

Other formats

There are numerous identification kits for *Enterobacteriaceae*, which may be of value for biotyping *Salmonella*. They are mostly in micro format, e.g. API 20E (Biomerieux), Micro ID (Organon Teknika), etc. Notwithstanding the earlier comment regarding the choice of substrate, such kits are generally intended for identification of genus/species rather than for subdividing serovars. Generally, the more substrates included, the greater the potential for subdivision of existing groups. Perhaps the greatest number of substrates offered in a single test format is that of Biolog. Their Gram-negative (GN) plate presents 95 different carbon sources in a single microplate format. Incubation times are shortened considerably because the 'signal' is read not as growth but as a change in the colour of tetrazolium violet, caused by variation in the redox potential within the substrate through bacterial respiration.

Biotyping has been a valuable tool for the bacteriologist and should continue to be seen as such. With care and attention to detail, it affords a reliable, relatively cheap and easy means of characterizing strains of *Salmonella*.

Salmonella Serotyping

Salmonella are identified by characterization of their surface antigens, as described in the White–Kauffmann–Le Minor (WKL) scheme. The identity, known as the *Salmonella* serovar, is designated by the identification of two sets of antigens. The somatic antigens, known as the 'O' antigens, are present on the main body of the organism. They are polysaccharide antigens, identified by agglutination with specific agglutinating antibody in serum, usually prepared in rabbits, to give the serogroup, e.g. *S. typhimurium* is a group B *Salmonella* due to the presence of the O4 somatic antigen. Two phases of antigen (termed H) are present on the protein flagellar structures that extend from the surface of the bacterium and such antigens are described as diphasic. Some serovars possess only one phase of flagellar antigen and are described as monophasic. Two serovars, *S. pullorum* and *S. gallinarum*, have no flagella and are non-motile strains. However, the majority of *Salmonella* possess two phases of flagellar antigens, which are occasionally present together and may be identi-fied by agglutination with specific antisera. It is more usual, however, to find one set of 'H' anti-gens at any one time. To change the 'H' phase expressed, the organism is placed in an environ-ment surrounded by antibody to the 'H' antigen already identified to induce phase inversion. The agglutination tests are then repeated to identify the remaining 'H' antigens.

Agglutination tests should be carried out with a range of absorbed *Salmonella* antisera. Commercial supplies of specific agglutinating antisera for slide testing are available in most countries. For accurate serotyping, good-quality antisera are essential. Serotyping should take place only when a culture has been biochemi-cally confirmed as *Salmonella*. This is due to the presence of similar antigens on other members of the *Enterobacteriaceae*, which may agglutinate with *Salmonella* antisera, e.g. *Citrobacter* species with group C antisera.

Agglutination tests can be carried out in three ways. The method of choice is dependent on the number of cultures being serotyped and resources available. All three methods utilize the specific recognition of the antigen by homolo-gous antibody. When a culture of *Salmonella* is mixed with specific antisera, agglutination occurs, which can be seen macroscopically as 'cells clumping', when homologous antibody and antigen are present. Repeating this process with a range of specific antisera will characterize which antigens are present.

When serotyping *Salmonella*, always plate out the organism on an agar plate to obtain sin-gle colonies. It is vital to conduct all tests from one colony (a clone from a single cell), i.e. grow up antigen from one colony to eliminate the pos-sibility of serotyping a mixed culture of *Salmonella*, which will give false-positive results. If a mixed culture is suspected, inoculate individ-ual colonies into separate nutrient broths and grow as individual cultures for serotyping. Agglutination techniques are described below.

Slide agglutination test

This test requires a high degree of skill. The den-sity and smoothness of the cell suspension placed on the slide will affect the result. Culture taken directly from inhibitory or differential culture medium may adversely affect the end result and basic culture medium, such as nutrient agar, is best. Care should be taken whilst working with live cultures. To obtain inocula suitable for both somatic and flagellar agglutination tests, a nutri-ent agar slope is inoculated with a few drops from an overnight culture of one *Salmonella* colony grown in nutrient broth. Ensure that the inocu-lum comes into contact with the sloped agar sur-face and that approximately 3 mm depth of liquid culture is visible at the base of the slope. Incubate overnight at approximately 37°C.

Identification of somatic O antigens

Place a loopful of normal saline on to a clean glass, microscope slide. This will act as a negative control and identify roughness in the organism. (If agglutination is seen in the saline control when the culture is added, the culture is autoag-glutinating, cannot be typed further and is assigned as a 'rough strain'.)

Place a second loopful of saline next to the first.

Take half a loopful of culture from the growth on the agar slope and mix equal portions into the two saline drops until a fairly thick sus-pension is achieved. If the suspension is too thick, more than one drop of antisera will be needed to give agglutination. If the suspension is too weak, it may be difficult to see agglutination.

Take a loopful of polyvalent 'O' antisera (this usually represent O groups A–I or A–S) and add to one of the suspension drops. Mix this with the loop (or by carefully tilting the slide back and forth) for at least 60 s or until agglutination is seen. Agglutination is identified by the forma-tion of white granular lumps in the suspension. If no agglutination occurs, the culture is unlikely to be *Salmonella* and should be examined further biochemically. If agglutination does occur, repeat the process with group-specific antisera until the somatic antigens are identified.

Proceed through the range of group specific antisera, as in Box 23.1.

It is important to note that the expression of certain O antigens is controlled by the presence or absence of lysogenic phage, e.g. O3, 10 lysoge-nized by phage ε_{15} will become O3, 15 and, if lyso-genized by phage ε_{15} and ε_{34}, it will become O3, 15, 34; these were previously known as E_2 and E_3

Box 23.1. Identification of somatic O antigens.

O2-positive ⇒ test O12 ⇒ (group A antigens identified)
⇓
Negative
⇓
O4-positive ⇒ test O5 and O27 ⇒ (group B antigens identified)
⇓
Negative
⇓
O6, 7-positive ⇒ test O8 and O14 ⇒ (group C_1 and C_2 antigens identified)
⇓
Negative
⇓
O8-positive ⇒ test O20 ⇒ (group C_3 antigens identified)
⇓
Negative
⇓
O9-positive ⇒ test O12 ⇒ (group D_1 antigens identified)
⇓   ⇓
Negative   if negative test O46 ⇒ (group D_2 antigens identified)
⇓
O3, 10, 15, 19-positive ⇒ test O10, O15, O19, O34 (group E antigens identified)
⇓
Negative
⇓
O13, 22, 23-positive ⇒ test O22 and O23 ⇒ (group G antigens identified)
⇓
Then test O11 and work through remaining group sera up to O67. Not all group sera are available commercially.

O subgroups, respectively. Members of the subgroups were formerly named *Salmonella* serovars, but current practice in the WKL scheme is to call the former *S. binza* : *S. orion* (15+) and the former *S. thomasville* : *S. orion* (15+, 34+). From the taxonomic viewpoint this is correct, but for the epidemiologist it can be confusing.

Identification of flagellar H antigens

Flagellar agglutination is best observed under low-power magnification, preferably with a darkened background. Agglutination can be observed by eye, but is finer in appearance than somatic agglutination and weak positive reactions may easily be missed.

Place two loopfuls of inocula taken from the liquid culture at the base of the agar slope next to one another on a clean glass slide. One of these drops will act as the negative control and identify autoagglutination. Add one loopful of polyvalent H antisera to one drop on the slide and mix. Observe for agglutination whilst mixing.

If positive, repeat the test with specific *Salmonella* 'H' antisera. Most commercial companies make polyvalent mixes or rapid sera for this purpose, where antisera have been combined to give a range of five or six sera to test. This eliminates a large number of possible antisera and reduces the amount of testing. Once a positive polyvalent serum has been found, it remains to test the individual constituents of that serum in order to identify which flagellar antigens are present.

Phase inversion – also called phase induction and phase reversal

Some *Salmonella*, e.g. *S. enteritidis* (9,12:gm:–) are monophasic, having only one flagellar phase. However, most serovars are diphasic and possess two flagellar phases.

Once the first-phase flagellar antigens of a monophasic *Salmonella* have been identified, the culture is identified as a serovar by reference to the WKL scheme. For diphasic *Salmonella*, the other flagellar phase must be induced. This can be done in several ways, e.g. the method of Svengard, where antibody is incorporated into semi-solid agar in a Petri dish and inoculated in the centre with the *Salmonella* culture. The plate is incubated and the culture at the edge of the radiating growth is tested by slide agglutination for the presence of flagellar antigens. At the Veterinary Laboratories Agency, the Craigie tube method is used. In this method, semi-solid culture medium (0.27% agar in nutrient broth) in a universal bottle contains an open-ended glass tube of approximately 5 mm diameter. The agar is melted by placing the universal bottle in a water-bath at 100°C for 10–15 min. The agar is then kept in the water-bath until its temperature has reduced to 56°C. At this point, approximately 30 μl of *Salmonella* antiserum is added to the agar. The antiserum chosen for addition to the medium is that which has already identified one phase of flagellar antigen, e.g. if the first set of flagella slide tests identified Hi, then i antiserum is incorporated into the semi-solid agar. Allow the agar to set at room temperature.

Inoculate a 1 μl loop of *Salmonella* culture taken from the liquid at the base of the nutrient agar slope into the glass tube, until the bottom of the loop is approximately 0.5 cm below the surface of the agar. Incubate the medium overnight at 37°C and observe for growth on the surface of the agar. This growth should normally appear overnight, but it may take 2–3 days' incubation, dependent on the culture and agar content. The *Salmonella* culture will 'swim', using flagella to propel through the medium, down to the base of the tube, out into the surrounding medium and finally upward toward the surface of the agar. The growth on the surface, which has a milky appearance, can be used to conduct a repeat of the serotyping steps described above to identify the other flagellar phase. If the original antigens identified are still present, phase inversion must be repeated by passage of the surface growth until the phase changes and the remaining phase antigens have been identified. If the antigen structure is complete, e.g. somatic and flagellar antigens fully identified, the 'name' can be found by reference to the WKL scheme. Over 90% of *Salmonella* strains isolated

are of the subspecies *enterica*. This subspecies were originally given a serovar 'name', which can still be found in the WKL scheme.

Some flagellar antigens are grouped into 'complexes' according to shared antigens and consequent relationships they have with one another. Thus:

1 complex contains flagellar antigens 1,2; 1,5;
    1,6; 1,7
E complex: eh; enx; enz_{15}
L complex: lw; lv; lz_{13}; lz_{28}
G complex: gp; gm; p; u; s; m; t; f; q; z_{51}

Flagellar antigens not included in complexes include: a; b; c; d; i; k; r; y; z; z_{39}; z_{41}; (z_6 may be found included with 1 complex; it is known to cross-react with 1,5).

Microagglutination test (MAT)

This method was devised at the Laboratory of Enteric Pathogens by Rowe and Hall and their co-workers and it is the most useful when serotyping large numbers of cultures.

Before commencing this technique, it is important to titrate the antisera to be used against a wide bank of positive and negative *Salmonella* cultures in order to determine the optimum dilution for each antiserum so as to identify agglutination. Most commercial antisera can be diluted with sterile saline for use in the MAT. The titre of diluted antisera drops over a period of time. It is therefore recommended that diluted antisera be stored at 4°C for a maximum of 10 days. The MAT has several advantages. It requires less skill than slide testing, many more cultures can be processed and it is safer, as the cultures being tested are killed. Microscopic examination of agglutination is not necessary.

Once the optimal titres have been determined, 96-well 'U-bottomed' microtitration plates can be prepared to screen for the antigens present. One plate should be prepared for O identification and another for H. The most common serovars found locally will dictate which antisera are contained in the primary screening plates for O and H antigens.

Preparation of antigens for MAT
From a single colony, inoculate two 3 ml volumes of heart infusion broth (Difco) with 1% added glucose (HIB). Incubate these for 4–5 h at 37°C.

PREPARATION OF MAT SOMATIC ANTIGEN. Place one HIB culture in a water-bath at 100°C for 1 h. Allow the culture to cool before use.

PREPARATION OF MAT FLAGELLAR ANTIGEN. To the other culture, add 3 ml of 1% formal saline. Leave at room temperature overnight before dispensing into the microtitration plate.

Somatic MAT

It is suggested that the O screen plate should contain 25 µl of the following somatic antisera: polyvalent O; O4; O5; O27; O6,7; O8; O9; O12; O3,10,15,19; O11; O13,22,23; saline (negative control). Then 25 µl of killed antigen is added to each test well. Cover with cling film or a plastic plate cover and incubate in a water-bath at 51°C for approximately 2 h. After this time, the MATs can be moved into a refrigerator for a further 18 h.

For reading, the plates should be placed on a microtitration plate reader in an overhead-lit area, preferably within a dark background. When agglutination does not occur, the bacterial cells fall to form a tight button in the bottom of the well. Agglutination is identified by a distortion of the button of cells or a loose button or sometimes clumps of cells are visible.

Flagellar MAT

Prepare a microtitration plate by adding 25 µl of optimally diluted antisera to individual wells. Add 25 µl of killed antigen to each well, cover with cling film or a plastic cover and incubate in a water-bath at 51°C for 2 h. The plates should have their bottoms just on the water-line. After incubation, read the test results under the lighting conditions described previously. Negative results will appear as a cloudy suspension. Positive results are identified by the presence of floccular agglutination, accompanied by a clearing of the background suspension.

A suggested set of antisera to include in the screening H plate is: polyvalent H; 1 complex; E complex; L complex; G complex; b; i; r; y; z; z10; saline (negative control).

Follow-up MAT

Where a polyvalent serum has given a positive result, repeat the MAT with the individual constituents of the serum.

Phase inversion should be carried out, as previously described, to identify both flagellar phases. The milky liquid growth on the surface of the Craigie medium can be inoculated into HIB broth and tested in the same way as for the previously identified phase.

Tube agglutination test

This method is very similar to MAT, except that it is carried out with larger volumes in tubes suspended in a rack and is consequently more time-consuming.

The test method will vary, depending on the serum used. It is therefore advisable to follow the instructions from the serum manufacturer.

Preparation of somatic antigen for tube agglutination test

A culture inoculated on to a nutrient agar slope is most suitable for this method. After overnight incubation at 37°C, wash the growth off the agar slope with approximately 2 ml of absolute alcohol. Use a Pasteur pipette to loosen growth from the agar if necessary. Transfer the suspension of cells to another container and leave at room temperature for at least 3 h. Centrifuge the suspension to gently deposit cells to the bottom of the container. Discard the supernatant. Resuspend the cells in enough normal saline to give the equivalent of a Brown's tube 2 (McFarland 3) suspension.

Preparation of flagellar antigen for tube agglutination test

Inoculate 3 ml of nutrient broth with approximately one-quarter of a single colony and incubate overnight at 37°C.

The culture is killed by the addition of 1 ml of 10% formalin and leaving at room temperature for at least 15 min before testing. NB: when the formalin is disposed of, ensure that it is not likely to come into contact with substances that can form toxic gases, e.g. bleach.

The 'virulence' antigen, commonly called Vi, is present on some isolates of some *Salmonella* serovars, e.g. *S. typhi*, *S. paratyphi* A and *S. dublin*, and *Citrobacter*. Its presence is identified by agglutination with Vi antisera. If present, any O antigens will not be identifiable until the culture is boiled for approximately 30 min, when any of the procedures described previously will identify the presence of O antigen.

Table 23.6. Method of recording degrees of lysis on enteric phage-typing plates.

Plaque size	No. of plaques/lysis
Large (L)	0–5
Normal (N)	± 6–20
Small (S)	± 21–40
Minute (M)	+± 41–60
Visible only with	++ 61–80
hand-lens (µ)	++± 81–100
	+++ > 120
	Semi-confluent lysis (SCL)
	Confluent lysis (CL)
	< SCL (intermediate degrees
	< CL of lysis)
	Confluent 'opaque' lysis (opacity
	due to heavy secondary growth)
	(OL)

Plaques should be described as follows: large (L), normal (N), small (S), minute (M) and mu (µ). The latter are only visible with a 10 × magnification hand lens.

Phage Typing

Phage typing has a long history with regard to the differentiation of *Salmonella* serovars of human and animal origin (Craigie and Yen, 1938a,b; Felix and Callow, 1943; Anderson *et al.*, 1977; Ward *et al.*, 1987). Since the discovery of bacteriophages by Twort/D'Herelle in 1915/1917, many different typing schemes have been developed. Most of the systems are based on virulent phages, and more detailed technical and historical information will be found in Guinée and van Leeuwen (1978). Phage-typing schemes for *S. typhi*, *S. paratyphi* A and B, *S. typhimurium* and *S. enteritidis* are used in most reference laboratories and a number of schemes have been developed for other *Salmonella* serovars, which may be of more local importance.

Many phages occur in sewage, sewage-contaminated water and faeces and, once isolated, they have to be purified by repeated single-plaque isolation. Preliminary identification consists of establishing the host range and plaque morphology. Typing-phage preparations can be produced by growing equal concentrations of purified phage particles and host cells in their logarithmic phase of growth in pre-warmed nutrient broth at 37°C

Table 23.7. Phage-typing schemes in common use.

	No. of phage preparations	No. of phage types
S. typhi	110	110
S. paratyphi A	9	15
S. paratyphi B		
S. paratyphi B (var. Java)	12	53
S. enteritidis	16	65
S. typhimurium	37	260+
S. hadar	9	62
S. virchow	15	68
S. thompson	10	38
S. pullorum	5	13
S. agona	9	38

for as long as lysis visibly continues and usually no longer that 7 h. The mixture is then heated at 60°C for 30 min to kill the host cells; most typing phages for *Salmonella* survive such treatment. If the phage is not thermostable, the host cells may be killed by the addition of a few drops of chloroform. The lysates are then centrifuged to remove the killed bacteria. It is sometimes impossible to prepare a phage suspension of sufficiently high titre in liquid media and the agar-layer technique should be used (Guineé and van Leeuwen, 1978). The phage preparation is then titrated on its host strain to determine the routine test dilution (RTD), which is defined as the highest dilution that produces confluent or, in some cases, semi-confluent lysis of the host strain that was used for propagation.

The *Salmonella* to be typed is grown in 5 ml nutrient broth in an incubator at 37°C. The period of time and conditions depend on the serovar, e.g.105 min with shaking at 60 r.p.m. for *S. enteritidis*. Nutrient agar plates that have been dried at 37°C for 180 min are flooded with the *Salmonella* culture, so that a thin film results. The quality and thickness of the agar may affect the results; in general, better results are obtained with a thicker agar. The plates are dried again for a maximum of 15 min at 37°C and the phage preparations added at their RTD. Various mechanical devices, such as the Lidwell apparatus, may be used to spot the phage on to the *Salmonella* lawn. The plates are then read after being incubated overnight at 37°C. The different degrees of lysis are shown in Table 23.6.

In addition to phage-typing schemes for *S. typhi*, *S. paratyphi* A and B and *S. typhimurium*,

Table 23.8. Phage-typing schemes for other *Salmonella* serovars.

S. adelaide	Atkinson *et al.* (1952)
S. anatum	Gershman (1974b)
S. bareilly	Majumdar and Singh (1973), Jayasheela *et al.* (1987)
S. blockley	Sechter and Gerichter (1969)
S. bovismorbificans	Atkinson *et al.* (1952), Atkinson (1956)
S. braenderup	Sechter and Gerichter (1968)
S. dublin	Williams Smith (1951b), Lilleengen (1950), Vieu *et al.* (1990)
S. pullorum/gallinarum	Lilleengen (1952)
S. newport	Gershman (1974a), Petrow and Kasatiya (1974)
S. panama	Guinée and Scholtens (1967)
S. thompson	Gershman (1972), Williams Smith (1951a)
S. virchow	Velaudapillai (1959); Chambers *et al.* (1987)
S. weltevreden	Garg and Singh (1971), Kaliannan and Mallick (1973)
S. hadar	De Sa *et al.* (1980)
S. oranienberg	Bordini and Trefouel (1970)
S. choleraesuis	Lalko (1971)
S. enteritidis	Macierewicz *et al.* (1968), Ward *et al.* (1987), Lilleengen (1950)
S. good	Kawanishi (1972)
S. montevideo	von Thal (1957), Vieu *et al.* (1981)
S. infantis	Kasatiya *et al.* (1978)

schemes have been developed for many other *Salmonella* serovars. More than one typing scheme may have been developed for a serovar, e.g. *S. typhimurium*, and, as a consequence, the results of typing may not be comparable. The number of phage preparations and different phage types for the more commonly used typing schemes are shown in Table 23.7. Other phage-typing schemes that have been developed for some of the other *Salmonella* serovars are shown in Table 23.8.

References

Anderson, E.S., Ward, L.R., De Saxe, M.J. and De Sa, J.D.H. (1977) Bacteriophage typing designations of *S. typhimurium*. *Journal of Hygiene, Cambridge* 78, 297–300.

Anderson, E.S., Ward, L.R., De Saxe, M.J., Old, D.C., Barker, R. and Duguid, J.P. (1978) Correlation of phage-type biotype and source in strains of *Salmonella typhimurium*. *Journal of Hygiene, Cambridge* 81, 203–217.

Atkinson, N. (1956) Lysogenicity and lysis patterns in the salmonellas: a bacteriophage grouping scheme for *S. bovis-morbificans*. *Australian Journal of Experimental Biology and Medical Science* 34, 231–234.

Atkinson, N., Geutenbeek, H., Swann, M.C. and Wollaston, J.M. (1952) Lysogenicity and lysis patterns in the salmonellas: a bacteriophage grouping scheme for *S. adelaide*, *S. waycross* and *S. bovis morbificans*. *Australian Journal of Experimental Biology and Medical Science* 30, 333–340.

Barker, R.M., Old, D.C. and Tye, Z. (1982) Differential typing of *Salmonella agona*: type divergence in a new serovar. *Journal of Hygiene, Cambridge* 88, 413–423.

Barker, R.M., Kearney, G.M., Nicholson, P., Blair, A.L., Porter, R.C. and Crichton, P.B. (1988) Types of *Salmonella paratyphi* B and their phylogenetic significance. *Journal of Medical Microbiology* 26, 285–293.

Bordini, A. and Trefouel, J. (1970) Bactériophages et lysotypie de *S. oranienburg*. *Comptes Rendus Académie des Sciences (Paris)* 270D, 567–570.

Chambers, R.M., McAdam, R., De Sa, J.D.H., Ward, L.R. and Rowe, B. (1987) A phage typing scheme for *Salmonella virchow*. *FEMS Microbiology Letters* 40, 155–157.

Craigie, J. and Yen, C.H. (1938a) The demonstration of types of *B. typhosus* by means of preparations of type II Vi phage Part 1. *Canadian Public Health Journal* 29, 448–484.

Craigie, J. and Yen, C.H. (1938b) The demonstration of types of *B. typhosus* by means of preparations of type II Vi phage Part 2. *Canadian Public Health Journal* 29, 484–496

De Sa, J.D.H., Ward, L.R. and Rowe, B. (1980) A scheme for the phage-typing of *Salmonella hadar*. *FEMS Microbiology Letters* 9, 175–177.

Duguid, J.P., Anderson, E.S., Alfresson, G.A., Barker, R. and Old, D.C. (1975) A new biotyping scheme for *Salmonella typhimurium* and its phylogenetic significance. *Journal of Medical Microbiology* 8, 149–166.

Ewing, W.H. (1986) *Edwards and Ewing's Identification of Enterobacteriaceae*, 4th edn. Elsevier Science, New York.

Felix, A. and Callow, B.R. (1943) Typing of paratyphoid B bacilli by means of Vi-bacteriophage. *British Medical Journal* ii, 127.

Garg, D.N. and Singh, I.P. (1971) A phage typing scheme for *S. weltevreden*. *Annales de l'Institut Pasteur* 121, 751–762.

Gershman, M. (1972) Preliminary report: a system for typing *S. thompson*. *Applied Microbiology* 23, 831–832.

Gershman, M. (1974a) A phage typing scheme for *S. newport*. *Canadian Journal of Bacteriology* 20, 769–771.

Gershman, M. (1974b) A phage typing system for *S. anatum*. *Avian Diseases* 18, 565–568.

Guinée, P.A.M. and Scholtens, R.T. (1967) Phage typing of *S. panama* according to the method of Craigie and Yen. *Antonie van Leeuwenhoek* 33, 25–29.

Guinée, P.A.M. and van Leeuwen, W.J. (1978) Phage typing of *Salmonella*. In: *Methods in Microbiology*, Vol. 11. Academic Press, New York, pp. 158–190.

Guinée, P.A.M., van Leeuwen, W.J. and Pruys, D. (1974) Phage typing of *S. typhimurium* in the Netherlands. *Zentalblatt für Bakteriologie I Abt. Orig.* 226, 194–200.

Jayasheela, M., Singh, G., Sharma, N.C. and Saxena, S.N. (1987) A new scheme for phage-typing *Salmonella bareilly* and characterization of typing phages. *Journal of Applied Bacteriology* 62, 429–432.

Kaliannan, K. and Mallick, B.B. (1973) Phage typing scheme for *S. weltevreden*. *Indian Journal of Animal Science* 43, 24–26.

Kasatiya, S., Caprioli, T. and Champoux, S. (1978) Bacteriophage typing scheme for *Salmonella infantis*. *Journal of Clinical Microbiology* 10, 637–640.

Kawanishi, T. (1972) On the isolation of bacteriophage from *S. good* and phage typing. *Bulletin National Institute Hygiene Science Tokyo* 47, 75–80.

Lalko, Y. (1971) Differentiation of *S. cholerae-suis* by means of bacteriophages. In: *Bulgariya Proceedings 2nd National Conference*, pp. 175–178.

Le Minor, L., Veron, M. and Popoff, M.Y. (1982a) Taxonomie des *Salmonella*. *Annales de Microbiologie* 133B, 223–243.

Le Minor, L., Veron, M. and Popoff, M.Y. (1982b) Proposal of *Salmonella* nomenclature. *Annales de Microbiologie* 133B, 245–254.

Le Minor, L., Popoff, M.Y., Laurent, B. and Hermant, D. (1986) Individualisation d'une septième sous-espèce de *Salmonella: S. choleraesuis* subsp. *indica* subsp. nov. *Annales de l'Institut Pasteur/Microbiologie* 137B, 211–217.

Lilleengen, K. (1950) Typing of *S. dublin* and *S. enteritidis* by means of bacteriophage. *Acta Pathologica Microbiologica Scandinavica* 27, 625–640.

Lilleengen, K. (1952) Typing of *S. gallinarum* and *S. pullorum* by means of bacteriophage. *Acta Pathologica Microbiologica Scandinavica* 30, 194–202.

Macierewicz, M., Kaluzewski, S. and Lalko, J. (1968) Differentiation of strains of *S. enteritidis*. *Experimental Medical Microbiology* 20, 138–146.

Majumdar, A.K. and Singh, S.P. (1973) A phage typing scheme for *S. bareilly*. *Indian Veterinary Journal* 50, 1161–1166.

Odongo, M.O., McLaren, I.M., Smith, J.E. and Wray, C. (1990) A biotyping scheme for *Salmonella livingstone*. *British Veterinary Journal* 146, 75–79.

Petrow, S. and Kasatiya, S.S. (1974) A phage typing scheme for *S. newport*. *Annales de l'Institut Pasteur* 125A, 433–445.

Reilly, W.J., Old, D.C., Munro, D.S. and Sharp, J.C.M. (1985) An epidemiological study of *Salmonella montevideo* by biotyping. *Journal of Hygiene, Cambridge* 95, 23–28.

Sechter, I. and Gerichter, C.B. (1968) Phage typing scheme for *S. braenderup*. *Applied Microbiology* 16, 1708–1712.

Sechter, I. and Gerichter, C.B. (1969) A phage typing scheme for *S. blockley*. *Annales de l'Institut Pasteur* 116, 190–199.

Tuchili, L.M, McLaren, I.M., Smith, J.E. and Wray, C. (1991) Differentiation of *Salmonella seftenberg* into biogroups. *Veterinary Record* 129, 530–531.

Velaudapillai, T. (1959) Phage typing of *S. virchow*. *Zeitschrift für Hygiene* 146, 84–88.

Vieu, J.F., Hassan-Massoud, B., Klien, B. and Leherissey, M. (1981) Bacteriophage typing and biotyping of *Salmonella montevideo*. In: FEMS *Symposium on Salmonella and Salmonellosis. Istanbul, 15–17 September, 1981*.

Vieu, J.F., Jeanjean, S., Tournier, B. and Klein, B. (1990) Application d'une série unique de bactériophages à la lysotypie de *Salmonella* sérovar Dublin et de *Salmonella* sérovar Enteritidis. *Médecine et Maladies Infectieuses* 20, 229–233.

von Thal, E. (1957) Observations on variability within the group of S. *montevideo*. *Nordiske Veterinar Medicin* 9, 831–838.

Ward, L.R., De Sa, J.D.H. and Rowe, B. (1987) A phage typing scheme for S. *enteritidis*. *Epidemiology and Infection* 99, 291–294.

Williams Smith, H. (1951a) The typing of S. *thompson* by means of bacteriophage. *Journal of General Microbiology* 5, 472–479.

Williams Smith, H. (1951b) The typing of S. *dublin* by means of bacteriophage. *Journal of General Microbiology* 5, 919–925.

Chapter 24

Serological Diagnosis of *Salmonella* by ELISA and Other Tests

Paul A. Barrow

Institute for Animal Health, Compton Laboratory, Compton, Newbury, Berkshire RG20 7NN, UK

Introduction

The public and professional interest in *Salmonella* and salmonellosis in relation to human food poisoning, in addition to the ever-present economic problems associated with animal disease, has created an incentive to produce accurate data on the prevalence of infected flocks and herds, and of *Salmonella*-infected animals within flocks and herds. Bacteriological methods have been the traditional means of obtaining such data (Statutory Instruments, 1989; Commission of the European Communities, 1992).

A major advantage of serological over bacteriological methods is that, whereas *Salmonella* organisms are excreted intermittently (Richardson, 1973a; Williams and Whittemore, 1976), serum immunoglobulin G (IgG) concentrations are generally persistent. The logistical problems of frequent sampling of large numbers of animals are thus reduced. One major disadvantage is that, soon after initial infection, serum IgG concentrations will be low (although increasing), whereas bacterial excretion may be at a maximum. Also, serovars generally regarded as non-invasive (although perhaps less significant numerically, from the public-health point of view) may not be detected serologically. Such tests are therefore probably mainly applicable for the initial screening of stock for invasive serovars, such as *S. dublin* in cattle, *S. choleraesuis* in pigs, *S. enteritidis*, *S. pullorum* and *S. gallinarum* in poultry and *S. typhimurium* in all species. However,

many *Salmonella* serovars are invasive in poultry and mice (Barrow *et al.*, 1994) and serovars considered to be poorly invasive, such as *S. infantis* and *S. bredeney*, which produce little or no systemic disease in animals, nevertheless may produce infections that are detectable serologically (Smith *et al.*, 1972; Nielsen *et al.*, 1995b).

A number of serological tests were developed earlier for detecting invasive serovars, the most successful being slide agglutination, using either serum or whole blood, for the detection of poultry flocks infected with either *S. gallinarum* or *S. pullorum*. This test has been applied successfully for more than 50 years in the control and eradication of *S. pullorum*. It is easy to carry out on the farm and, at the time of its development, saved time and money in comparison with other serological assays then available. The advantages and disadvantages of this assay, and also the frequency of false reactions, are discussed in detail by Gordon and Brander (1942).

Although relatively crude, it was successful in largely eradicating pullorum disease from the intensive sectors of the poultry industry in many countries. The test can be used for other serovars, but generally with limited success. Tube and microagglutination tests and the more sensitive microantiglobulin tests (MAGs) have been applied to experimental and field infections with serogroups B, C and D. However, these tests are cumbersome and do not lend themselves readily to extensive use for large-scale herd or flock screening.

The detection systems required by recent European Community legislation for poultry (Commission of the European Communities, 1992) are based on bacterial culture. However, the option is available for serological screening, providing that those assays used give the same guarantees that hatchery investigations provide. Serological assays, mainly the enzyme-linked immunosorbent assay (ELISA), are now currently used in this context in a number of countries.

Recent advances in ELISA technology, including miniaturization and mechanization, together with the availability of commercially prepared high-quality reagents, have resulted in its application to the detection of infection caused by several pathogens in humans, poultry, pigs and cattle. ELISAs based on lipopolysaccharide (LPS) and virulence (Vi) antigens have been used to detect *S. typhimurium* and *S. typhi* infection in rodents and humans respectively (Carlsson *et al.*, 1972; Lentsch *et al.*, 1981; Engleberg *et al.*, 1983). They have also been used to monitor experimental *Salmonella* infections in calves (Robertsson *et al.*, 1982a) and various avian pathogens, including infectious-bronchitis virus (Mockett and Darbyshire, 1981) and *Pasteurella multocida* (Briggs and Skeeles, 1983). However, it is only more recently that these assays have been applied to detecting *Salmonella* infections in poultry, cattle and pigs on a large scale.

Serological Response to *Salmonella* Infections

Work with experimental *Salmonella* infections can assist in understanding the parameters that may affect the performance of the ELISA and other assays in detecting infection. Most of these studies have been carried out with *S. typhimurium*, but there is no reason to believe that *S. enteritidis* or *S. dublin* would produce essentially different results. However, the pathogenesis of these infections is dissimilar. In addition, many serovars are considerably less invasive than those mentioned above and, despite the detection of antibody in poultry infections caused by *S. bredeney* and *S. virchow* (Smith *et al.*, 1972), it must be borne in mind that the amount of circulating specific IgG may be less in such infections.

In poultry, a number of studies have been carried out with different types of assays, using a standard indirect ELISA, with either LPS, flagella, outer-membrane proteins (OMPs) or a sonicated whole-cell protein extract as detecting antigen. IgG, IgM and IgA were detected following infection of specified pathogen-free (SPF) chickens. Serum IgG titres rose 2 weeks after infection, reached a peak at 5 weeks and remained high for a further 4 weeks. Serum IgM titres appeared before IgG, as expected, and disappeared earlier. In bile and gut washings, IgA was the predominant isotype. The responses to all antigens were similar (Hassan *et al.*, 1991). High IgG titres persist for many weeks following *S. enteritidis* (Barrow and Lovell, 1991; Kim *et al.*, 1991) or *S. typhimurium* (Barrow, 1992) infection, in the latter case high titres still being present when the experiment was terminated at 45 weeks after infection. This has practical consequences, since high IgG titres may persist in the absence of infection.

Not all those investigating immune responses have detected typical IgM responses. Using commercial reagents, IgM was not detected in commercial chickens naturally infected with *S. enteritidis* (Chart *et al.*, 1990c), although the time interval between sampling and infection of these birds was not known. Neither IgM nor IgG was detected in commercial layers experimentally infected with *S. enteritidis* (Humphrey *et al.*, 1989). The apparent absence of detectable IgM by ELISA was puzzling, since it contributes considerably to the tube agglutination reaction (Nicholas and Cullen, 1991). However, later studies by the same group indicated that considerable quantities of specific IgM were detectable, apparently for many weeks after infection. In these cases, the concentrations measured were often greater than those of IgG (Humphrey *et al.*, 1991a,b).

In some ways, immunoblotting has additional sensitivity, because the response, albeit semi-quantitative, to individual antigens can be discerned (Chart *et al.*, 1990c). Variable responses to LPS, flagella and OMPs can be observed. Non-specific reactions occur with chicken serum, which increase with the age of the bird from which the sera were taken; these must be treated with caution.

Hassan *et al.* (1991) also detected delayed-type hypersensitivity responses in chickens to a number of antigens, including LPS and OMPs by intradermal inoculation and examination of skin swelling.

Cellular and humoral immune responses in calves to attenuated vaccine strains and after challenge with a virulent *S. typhimurium* organism have been studied (Robertsson *et al.*, 1982a,b; Lindberg and Robertsson, 1983; Robertsson, 1984). Following vaccination of 4- to 5-week-old calves, specific serum IgG and IgM titres against LPS and porins were found to have increased after 3 weeks. Similar titres to both antigen types were obtained by subcutaneous vaccination with a heat-killed strain, whereas the anti-porin antibody titre was greater with oral vaccination with a live attenuated strain. Three weeks after challenge, the titres to the two antigens were similar. There was no response to heterologous LPS. Lymphocyte stimulation by LPS and porins from an antigenically homologous but not a heterologous organism were also observed. Skin reactivity was directed primarily at LPS as a macromolecule. High specific serum IgG titres and, to a lesser extent, IgM titres have been found in experimentally infected cattle for periods of up to 3 months. Milk and serum titres remained high for up to 1 year in mammary-gland carrier animals (Smith *et al.*, 1989).

Specific circulating IgG responding to LPS has been observed to rise in titre between 1 and 3 weeks after experimental inoculation of pigs with *S. typhimurium* or *S. infantis*. High concentrations, indicated by high (OD) values, persisted for several weeks (Nielsen *et al.*, 1995b).

A number of host factors affect humoral response to salmonellosis, including the genetic background of the animal (Barrow, 1992) and its age. In the latter case, older animals generally produce stronger responses to a variety of antigens than very young animals (Williams and Whittemore, 1975; Hassan *et al.*, 1990; House *et al.*, 1993). The response of young animals to LPS may be poorer than to flagella (Roden *et al.*, 1992). In addition, animals may differ in their response to different LPS types (Roden *et al.*, 1992).

Salmonella Diagnosis Using Agglutination-based Serology and Other Systems

A number of assays based on agglutination or complement fixation have been used over the years. In some cases, the sensitivity of these has been increased by using antiglobulins (Coombs *et al.*, 1945; Cunningham, 1968). As will be seen below, in all cases, the systems either suffer from relatively poor sensitivity or, in the process of increasing sensitivity, become cumbersome and less specific.

The crudest of tests, based on slide agglutination, has been used very successfully with poultry (see above) to eliminate carriers of *S. pullorum*. The use of rapid agglutination was introduced by Runnells *et al.* (1927) and a stained antigen–whole-blood combination was developed by Schaffer *et al.* (1931). This was introduced in the UK by Gordon and Brander (1942) and proved to be of great value in eliminating carrier birds. Retesting at 2- to 4-weekly intervals is recommended until two consecutive negative results are obtained with the whole flock (Anon., 1984). False-positive reactions can occasionally cause problems and the whole-blood test (WBT) has been found to be unreliable in turkeys and ducks. These problems may occur because of infection with other *Salmonella*, but in many cases it has not been possible to discover the cause. Other bacteria, including members of the *Enterobacteriaceae* and enterococci, have also been reported to produce cross-reactions (Garrard *et al.*, 1948). Such problems are likely to be more significant as the incidence of true positive reactions falls.

The test has a number of other disadvantages besides cross-reactions. The results vary, depending on antigen quality, which may vary with the strain of *S. pullorum* and is dependent on operator expertise. Gast (1997) found that there were slight differences in results between different stained antigen preparations, although it was unlikely that birds infected with the different variants of *S. pullorum* would be missed. Whether it is cheaper for large-scale screening than microtests, such as the ELISA, is questionable, because it is labour-intensive, but one advantage is that the seropositive birds can be culled on the spot.

The slide-agglutination WBT for *S. pullorum* has been adapted for use with *S. enteritidis*, using saponin-containing antigen to lyse the blood and produce a clearer result (Chart *et al.*, 1990a,b). Chart *et al.* (1990c) found reasonably good correlation between the results obtained by slide agglutination and immunoblotting and those obtained with an indirect ELISA for sera from an *S. enteritidis*-infected flock. However, exceptions were observed, in that some non-agglutinating sera produced relatively high OD values while two agglutinating sera produced low ODs. Further differences between the results produced by slide agglutination and ELISA were found in a subsequent study (Chart et al., 1990d), where, in five flocks infected with S. enteritidis, a poor correlation was found for individual sera, although high and low OD values in the ELISA and positive and negative agglutinations were found in each of the five flocks.

The relative merits of slide agglutination and ELISA for detecting S. enteritidis-infected flocks have been discussed extensively in published correspondence (Chart et al., 1990a,b, 1991; Barrow, 1991; Cullen and Nicholas, 1991). It is also possible that slide agglutination would preferentially identify IgM-producing birds. The proportion of birds producing IgM will be relatively low and this isotype does not persist at high titre, but it may assist in identifying current infections.

Agglutination-based assays may be more sensitive in the early stages of infection, when IgM titres are highest. For poultry, tube agglutination tests have been found to be unreliable with S. bredeney and S. virchow (Smith et al., 1972) and titres against other serovars were found to vary widely (Smyser et al., 1966; Cooper et al., 1989; Nicholas and Cullen, 1991). In contrast, the indirect haemagglutination (IHA) test is more sensitive and specific antibody titres against LPS may persist for several months at titres in excess of 1 : 100 (Smith et al., 1972).

After the discovery of antibodies against Salmonella O and H antigens in cattle by Lovell (1934), tube agglutination with these antigens has been used to attempt to detect cows that are carriers of S. dublin. The titres obtained by Field (1948) and Richardson (1973b) for both antigens were considered to be of diagnostic value but they were unable to identify some animals known to be latent carriers of S. dublin. Wray et al. (1975) found IHA, using both alkaline and ether extracts, to be no more sensitive than tube agglutination. In addition to false-negative reactions, which might occur early in infection, false-positive reactions also occurred. Titres to flagellar antigens in the serum agglutination test (SAT) were higher and persisted for longer periods than those detected in the IHA test. Serum agglutination has been made more sensitive by the antiglobulin test. For S. dublin in cattle, Wray et al. (1981) found that false-positive and false-negative results still occurred at different levels of sensitivity. However, although it was possible to differentiate naturally infected animals from those vaccinated with a rough strain by comparing agglutination and antiglobulin titres, it was not possible to differentiate carrier animals from convalescent, non-infected animals using the antiglobulin test. This would be desirable for S. dublin control. The MAG has also been applied to Salmonella infections in poultry (Williams and Whittemore, 1972, 1975) and has been found to be more reliable than agglutination assays. The MAG was more sensitive and high titres were found to persist for several months. It can also be used with serovars other than S. typhimurium, including S. thompson and S. infantis (Williams and Whittemore, 1979; Thain et al., 1980). Results obtained with this test correlated well with the ELISA, both in birds infected experimentally with S. enteritidis and in the field (Cooper et al., 1989; Nicholas and Cullen, 1991). However, as these authors point out, the practical difficulties in using this test for large-scale screening outweigh any potential advantages.

Similar problems were found by Wray and Sojka (1976) with the complement fixation test (CFT). They found that serum agglutinins against flagellar antigens remained at a diagnostic titre for longer than did complement-fixing antibodies using a number of somatic antigen preparations. Non-specific reactions also occurred and a considerable number of infected cattle reacted negatively with both the CFT and agglutination test. Hinton (1973) adapted the Brucella milk-ring test to the detection of S. dublin infections following abortion. Use of whey for agglutination of flagellar antigens was practical and nearly as sensitive as using serum. The use of paired samples, the second taken 2–3 weeks after the first, assisted in diagnosis. He drew attention to the problems associated with infection with related serovars causing rises in antibody titres. There was little correlation between agglutinin titre and the results of the milk-ring test.

Thus agglutination-based assays suffer from limited sensitivity and a tendency to produce both false-positive and false-negative results.

Given the sensitivity and specificity of ELISAs, combined with their ease of use for large-scale screening, it is not surprising that these assays have largely overtaken others. A few comparisons have been made between different systems for *S. enteritidis* infections in poultry. Cooper *et al.* (1989) and Nicholas and Cullen (1991) found that agglutination-based systems were less sensitive than other tests and identified fewer infected birds. Similar results were found by Timoney *et al.* (1990). Previous workers have also found slide agglutination to be too unreliable and insensitive for *S. typhimurium* (Nagaraja *et al.*, 1991). On the positive side, use of the slide-agglutination test with *S. pullorum* antigen resulted in fewer false-positive reactions in *S. enteritidis*-infected flocks (Gast and Beard, 1990).

Agglutination-based systems have also been developed for the rapid identification of *S. enteritidis* cultures. SEF14 fimbriae, found only in *S. enteritidis*, a proportion of *S. dublin* strains and the obscure, related serovars *S. blegdam* and *S. moscow* (Thorns *et al.*, 1990), have been used as the basis of a latex agglutination test using monoclonal antibodies (Thorns *et al.*, 1992). However, this has been tested in different laboratories to confirm cultures of *S. enteritidis*, rather than for serological purposes (McLaren *et al.*, 1992).

A few investigations have additionally been made into the possibility of using delayed-type hypersensitivity for detecting cattle infected with *S. dublin*. Aitken *et al.* (1978) showed that intradermal inoculation of killed *S. dublin* cells evoked a typical delayed response in most (25/28) intravenously inoculated cattle. There was a poor correlation with O and H agglutinin titres, but the reactions could be evoked up to 493 days post-infection. On retesting, reduced reactions were observed in a considerable proportion of the animals. Robertsson *et al.* (1982b) also demonstrated similar responses by inoculation of crude supernatant fractions.

Because of the high sensitivity and specificity, together with their ease of use for large-scale screening and the fact that little capital equipment is necessarily required, ELISA technology has largely superseded other systems for herd and flock monitoring. Subsequent sections of this review concentrate almost solely on this system.

ELISAs Available for Serological Diagnosis of *Salmonella*

There are two basic systems available, the indirect ELISA (Barrow, 1992) and the competitive

'sandwich-type' ELISA (van Zijderveld *et al.*, 1992). However, the definition is not completely clear, as will be seen.

The indirect ELISA involves the use of a detection antigen coated on to the wells of a microtitre plate. After the application of a blocking reagent to reduce non-specific binding, test samples are applied to the wells. Specifically bound antibody in the sample is detected by an antibody–enzyme conjugate. A variety of antigens, including LPS, flagella, SEF14 fimbriae and cruder antigen preparations, have been used. Selective or improved binding of the antigen may be obtained by precoating with a monoclonal antibody, although, in this case, there is no competitive aspect to the assay.

The competitive-sandwich ELISA employs a specific mechanism of coating antigen to wells, namely monoclonal antibody. This is then followed by a pure or crude antigen preparation. Test samples are applied followed by conjugated monoclonal antibody, which will not bind to the antigen if the test sample contains antibodies. The assay can be shortened by adding both test sample and conjugate together. Monoclonal antibodies have been prepared for flagella, LPS and SEF14 for *S. enteritidis* (van Zijderveld *et al.*, 1992; Thorns *et al.*, 1994). Standard indirect assays may also be used in a competitive way by using the blocking effect of an unknown serum on a standard positive serum (Hoorfar *et al.*, 1994).

There are advantages and disadvantages to both basic systems. The indirect assay is simpler and reagents are available that are applicable to all *Salmonella* serovars in chickens, turkeys and ducks. The competitive ELISA can be applied to all animal species and, in general, shows higher specificity. However, reagents are not commercially available for all serovars. There are also some affinity problems and it may be less sensitive than the indirect assays. In the field, both systems have been known to produce false-positive reactions.

ELISA – choice of antigen

LPS is the most frequently used discriminatory detecting antigen. Sera obtained from chickens infected either with *S. typhimurium* (Barrow, 1992) or with *S. gallinarum* (Barrow *et al.*, 1992) yielded much higher titres with homologous than with heterologous LPS as detecting antigen. Similar specificity can be demonstrated using groups B and D LPS tested with sera raised in chickens against *Salmonella* strains from different serogroups, such as B, C_1, C_4 (now also grouped within group C_1), D, E_1, E_4 and O35 (Chart *et al.*, 1990c; Hassan *et al.*, 1990). Differentiation between infections caused by *S. enteritidis*, *S. gallinarum* and *S. pullorum* using LPS has been claimed (Cooper *et al.*, 1989), but there have been no further reports of this. However, varying degrees of cross-reactions between groups B and D LPS have been demonstrated with sera from chickens infected with *S. typhimurium* (Nicholas and Cullen, 1991; Barrow, 1992; Barrow *et al.*, 1992). The cross-reactions were caused by antibodies against the common 12 antigens present in groups B and D, as was demonstrated by immunoblotting (Chart *et al.*, 1990c). The problem may be reduced by adjustment of the dilution of the test sample (Barrow *et al.*, 1992). There is also some evidence that periodate treatment of group D LPS can destroy O1 and O12 cross-reacting epitopes, while preserving the O9 and O4/5 specificities (House *et al.*, 1993; van der Heijden, 1994). A third possibility is to use an O antigen, such as that from group D_2, containing 9,46 specificity. However, in experimental infection, this antigen did not react with sera from *S. enteritidis*-infected chickens (Huis in't Veld *et al.*, 1994). The specificity of reactions may also be checked by blocking the reaction with a monoclonal antibody specific for the reactive epitope present on the detecting antigen. Thus a monoclonal antibody specific for O9 LPS has been used to confirm infection of cattle with a group D (O9) serovar, which, with bacteriological evidence, indicates *S. dublin* infection (Hoorfar *et al.*, 1994, 1996).

Experimental infection of chickens with other members of the *Enterobacteriaceae* does not induce the production of high titres of cross-reacting antibodies (Nicholas and Cullen, 1991; Barrow, 1992; Barrow *et al.*, 1992). Organisms tested include avian pathogenic *E. coli* serovars and strains of *Citrobacter*, *Klebsiella* and *Proteus*.

Flagella antigens have also been used to avoid this problem of differentiating serovars possessing cross-reacting LPS. Cross-reactions caused by different serovars sharing the same flagella antigens may conceivably pose a problem,

and no information is currently available on the expression *in vivo* of phase 1 and phase 2 antigens. Under experimental conditions, using *S. enteritidis* infections (Baay and Huis in't Veld, 1993) or *S. typhimurium* infections (Hassan *et al.*, 1990), specific immune responses can be demonstrated. However, in the field, confusing cross-reactions seem to occur. The former authors also report that flagella-specific IgG is not as persistent as LPS-specific IgG, detectable quantities disappearing within 4 months, and that the titres peak before those of LPS-specific antibody.

In contrast, in early work, Humphrey and colleagues (Humphrey *et al.*, 1989) and Chart *et al.* (1990c) were unable to detect flagella-specific IgG or IgM in chickens infected experimentally or naturally with *S. enteritidis*. One interesting feature found by Timoney *et al.* (1990) is that not all chickens experimentally infected with *S. enteritidis* develop high flagella-specific antibody titres; in the majority of the birds, no raised titres were observed. R. Nicholas (Weybridge, personal communication, 1999) found that flagella responses may be poorer than those against LPS if the strain is poorly invasive.

One area where flagella antigen can already be used with success is to differentiate infection caused by flagellate and non-flagellate serovars, particularly *S. enteritidis* from *S. gallinarum* and *S. pullorum*. Timoney *et al.* (1990) were able to differentiate these infections in this way. Barrow *et al.* (1992) found similar results by ELISA and by immunoblotting. Birds in a Brazilian flock, in which clinical fowl typhoid was observed, were identified by high IgG titres to group D LPS and by low gm-H-specific IgG in contrast to control sera from *S. enteritidis*-infected chickens.

One study has been carried out on SEF14 antigen by Thorns *et al.* (1994), using both indirect and capture ELISAs. Specific IgG was detectable by 1 week post-infection and persisted for at least 8 weeks when the experiment was terminated. No cross-reactions were observed with sera from *S. typhimurium*- or *S. gallinarum*-infected birds. A capture ELISA was used to examine field sera and was compared with an indirect ELISA, using flagella or heat-extract antigen. The SEF14 antigen ELISA was as sensitive as the heat-extract antigen ELISA and as specific as the flagella ELISA.

Other antigens of varying specificity have been used and some claims for specificity have been made. They include surface-protein antigens, some of which are said to be specific (Nagaraja *et al.*, 1984, 1986; Dadrast *et al.*, 1990; Kim *et al.*, 1991), sonicated whole-cell soluble-protein antigen (Hassan *et al.*, 1990; Steinbach *et al.*, 1994). In the latter case, calves infected with group B, C or D serovars produced stronger absorbency values using LPS as detecting antigen with *S. dublin* or *S. typhimurium* infection. High-molecular-weight proteins produced the highest absorbence values, but these were not very specific. Poor responses were obtained with *S. agona* and *S. infantis* although the animals were only bled once at 2 weeks after inoculation.

ELISA – choice of sample

Because of its high IgG concentrations, egg yolk (Rose and Orlans, 1981; Kaspers *et al.*, 1991), rather than serum, has already been applied to the diagnosis of other avian infections with pathogens such as Newcastle disease virus, infectious bronchitis virus (Piela *et al.*, 1984) and *Mycoplasma* (Piela *et al.*, 1984; Mohammed *et al.*, 1986). It has also been applied to *S. enteritidis*, *S. typhimurium* and *S. gallinarum* (Dadrast *et al.*, 1990; Nicholas *et al.*, 1990; McLeod and Barrow, 1991; Nicholas and Andrews, 1991; Barrow, 1992; Barrow *et al.*, 1992). In chickens experimentally infected with either *S. typhimurium* or *S. gallinarum*, specific IgG titres in egg yolk were found (after an initial lag) to be very similar to serum titres (Barrow, 1992; Barrow *et al.*, 1992). Fewer data are available from testing egg yolk for *S. enteritidis*-specific IgG. Dadrast *et al.* (1990) indicated in a pilot study that 60% of egg yolks from an infected flock had high LPS-specific IgG titres. Nicholas and Andrews (1991) studied three flocks infected with *S. enteritidis* and one infected with *S. typhimurium*. The infection rates in the first three flocks determined by ELISA with group-D LPS and with a threshold determined with SPF sera were 20/100, 99/180 and 55/180. Using a heat-extracted antigen the rates were higher. None of the yolks from the *S. typhimurium*-infected flock was considered as indicating infection using *S. enteritidis*-derived detecting antigens in the ELISA. McLeod and Barrow (1991) examined egg yolks from chickens in 17 houses. From four houses, 159/300 yolk samples had ODs greater than 0.2 and 58/300

greater than 0.8 and *S. enteritidis* was isolated from birds in these houses. In contrast, 177/450 yolk samples taken from the 13 uninfected houses produced values of less than 0.1, 416/450 less than 0.2 and all 450 less than 0.4.

There is some indication that a standard indirect ELISA might be more appropriate for analysing egg yolk than a competitive (blocking) sandwich ELISA. The low dilution of the samples used in the latter ELISA may lead to interference and to reduced sensitivity and false-negative reactions (van de Giessen *et al.*, 1994).

Any of the tissue fluids are likely to contain concentrations of IgG equivalent to those present in serum. Meat juice, obtained from butchered meat, has been used routinely in Denmark to trace *Salmonella*-infected pig herds (Nielsen *et al.*, 1995a).

In addition to producing systemic disease in cattle, *S. dublin* may become localized in the mammary gland and be shed in the milk, which may give rise to the production of local specific IgG$_1$ immunoglobulins (Watson, 1980). As described previously, these have been used to detect infection using agglutination-based assays. They may also be used in ELISAs for monitoring infection with this serovar, using LPS as detecting antigen (Robertsson, 1984; Smith *et al.*, 1989; Spier *et al.*, 1990; Hoorfar *et al.*, 1995a,b). Milk IgG and IgM titres from individual samples were considerably lower than those in the sera of experimentally infected cows and known carrier animals. In the former group, they were only marginally higher than those in uninoculated control animals. These authors (Smith *et al.*, 1989; Spier *et al.*, 1990) concluded that, although milk samples were convenient, two serum IgG samples had more predictive value of carrier status than two milk samples taken 60 days apart. Since a large proportion of milk IgG$_1$ is derived from serum antibody, milk may also be used to detect systemic infection in the absence of localized mammary infection. It has been used in this regard for *S. dublin* (see above references) and *S. typhimurium* (Hoorfar and Wedderkopp, 1995).

Meat juice, released from a sample, of standard size, cut from a carcass and frozen and thawed, has been increasingly used for epidemiological purposes in Denmark. This method has been applied particularly in monitoring pig herds (Nielsen *et al.*, 1996, 1997, 1998) but has also been applied to predicting the infection status of cattle (Hoorfar et al., 1997). Not surprisingly, pig meat from the diaphragm or sternomastoid muscle yielded juice with similar ODs. Considerable variation was observed in the volume of juice released per gram of tissue, which Nielsen et al., (1998) ascribe to variations in residual blood content of the muscle, the level of pre-slaughter stress and the associated effect of glycogen content and pH post-slaughter. Apart from a slightly lower titre seen with meat juice in comparison with sera, the former samples behaved in a similar manner to sera on dilution (Nielsen et al., 1998).

ELISA – factors affecting immunoglobulin titres and practical aspects

With a serovar such as S. typhimurium, the extent of invasiveness appears to have little effect on the IgG titres produced in chickens (Hassan et al., 1990). However, the extent of multiplication of the bacteria in the tissues did appear to affect the immune response (Bumstead and Barrow, 1988; Barrow, 1992). There is also some evidence that non-specific factors, including handling the chickens (Barrow, 1992) and the onset of lay (Barrow et al., 1992), may affect the IgG response.

Reinfection of chickens with the same organism does not seem to boost the serum IgG titres greatly (Williams and Whittemore, 1976; Hassan et al., 1991; Barrow, 1992). The IgG titre obtained is, to some extent, dose-dependent and is related to the extent of faecal excretion, but not necessarily so for individual birds (Humphrey et al., 1991a; Barrow, 1992).

However, slightly different results were found in cattle when challenged with different doses of Salmonella. Lindberg and Robertsson (1983) found that vaccination of calves with live or killed S. typhimurium organisms produced a rise in IgG and IgM, which was not potentiated by challenge if the vaccine was killed, but a further increase occurred if the vaccine was a live, attenuated strain.

The ELISA is relatively robust and can accommodate poor-quality sera. Antibody titres do not deteriorate if serum or blood is dried on to absorbent paper (Hassan et al., 1990; Minga and Wray, 1990). The dried serum on paper can be

stored at room temperature for several weeks without a loss of titre. Likewise it can be treated with some disinfectants, such as phenol vapour or chloroform with no effect on antibody titre. However, ethylene oxide, formalin vapour and gamma-irradiation all reduce titres to varying degrees. Heat inactivation of serum (56°C, 30 min) has no effect on antibody titre (Hassan et al., 1990). Kyvsgaard et al. (1996) found that electron-beam irradiation (β particles) reduced titres against S. dublin in bovine sera when the sera were liquid but not when the sera were irradiated when frozen with dry ice. The same authors showed that addition of 5 or 10 mM binary ethylenimine had no effect on the titres obtained from liquid sera. Such techniques might allow transfer of serum, blood and diluted egg yolk between laboratories in different countries without risk of introducing other pathogens.

The effect of antibiotic therapy on antibody levels might be considered to be limited; however, the evidence is mixed. One group (Goren, 1992) has reported that not only might a 10–14 day treatment of Salmonella-infected poultry with enrofloxacin, followed by competitive exclusion, virtually eliminate infection and prevent reinfection but also the birds treated in the field become serologically negative. However, insufficient data were presented to assess this properly. Experimental work with S. enteritidis indicated that enrofloxacin treatment started 13 days post-infection, followed by competitive exclusion, might reduce antibody titres significantly but not eliminate them (Desmidt et al., 1992). Antibiotic therapy starting at 3 days post-infection had a greater effect. The effect of therapy on a well-established infection was not examined (Desmidt et al., 1992).

The problem of cross-reactions induced by vaccination may initially seem more intractable. Immunization with live, attenuated (aroA) S. enteritidis strains that are smooth produces detectable but lower titres in LPS-specific ELISAs (Cooper et al., 1990). This has also been confirmed by immunization with pur mutants (G. Steinbach, unpublished results). There may be two ways to avoid this problem more completely. Immunization with reduced doses of vaccine may induce lower titres of IgG (C. Wray, Weybridge, personal communication, 1999). Alternatively, mutants may be produced that are devoid of the antigens used for detecting IgG in ELISAs or, less satisfactorily, marker antigens may be introduced. The little experimental evidence available indicates that the antibody response to LPS or flagella antigen is poor when chickens are infected with strains not possessing these (Barrow, 1992).

Similarly, in cattle, vaccination with a live *aroA S. typhimurium* strain also generated detectable, specific, circulating IgG (Lindberg and Robertsson, 1983). Slight rises in temperature followed oral vaccination, indicating invasion of the tissues and resultant febrile response.

Application of ELISA to field infection in poultry

Several ELISAs have been used to study field infections caused by S. *enteritidis*, S. *gallinarum* and S. *pullorum*. Little work has been reported on its use in studying S. *typhimurium* infections. Both serum and egg yolk have been used as samples. In most cases, the logistics of flock sampling require that the absorbance value obtained with a single serum dilution, rather than titration, is used as an estimation of antibody concentration. Cooper *et al.* (1989) found high O group B LPS-specific titres (log$_2$ 7.8–13.6) in 25 sera taken from three laying and breeder flocks infected with S. *enteritidis*. In contrast, only one of ten sera taken from a flock infected with S. *typhimurium* showed a high titre (log$_2$ 8.96) when tested against group D LPS. The cut-off point used to decide whether an absorbance indicated infection was based on OD values obtained by testing sera from uninfected SPF chickens, but sera from no other *Salmonella*-free chickens were examined.

Using the same ELISA, Nicholas and Cullen (1991) found high absorbance values determined with LPS and heat-extract antigen as detection antigens, with 25/40 and 40/40 sera respectively from a flock of layers, 10/40 of which were identified bacteriologically as infected. Absorbance values obtained with 40 sera from an S. *enteritidis*-free flock were all less than the threshold value.

Chart *et al.* (1990c) examined 29 laying hens, ten of which were found to harbour S. *enteritidis*. A wide range of absorbance values was obtained with sera from these birds. No attempt was made to classify chickens as infected on the basis of the ELISA results. A second study involved six different flocks, all of which had been shown to be infected with S. *enteritidis* (Chart *et al.*, 1990d). Absorbance values from 99 sera from one flock were very low, perhaps because of recent infection. From the remaining five flocks, a wide range of absorbance values was obtained in each case, the authors concluding that S. *enteritidis*-infected flocks probably vary considerably in the levels of LPS-specific IgG detectable by their ELISA.

Using a flagella antigen prepared from an S. *enteritidis* strain in an ELISA, Timoney *et al.* (1990) examined an S. *enteritidis*-infected breeder flock and an uninfected layer flock. Of 151 serum samples from the uninfected flock, all produced absorbance values of less than 0.5 and 147 were less than 0.25. Of the 47 sera from the infected flock, 32 were greater than 0.50 and 14 were greater than 1.0.

Fewer data are available from testing egg yolk for S. *enteritidis*-specific IgG. The work of Dadrast *et al.* (1990), Nicholas and Andrews (1991) and McLeod and Barrow (1991) has already been dealt with in detail earlier in this chapter.

Barrow *et al.* (1989) examined sera from seven flocks, four of which had bacteriological evidence of S. *typhimurium* infection. High antibody titres detected by the whole-cell soluble-protein sonicate antigen were found in three of these flocks and in none of the three flocks considered uninfected. Nicholas and Andrews (1991) found that 6/29 egg-yolk samples from a flock of free-range layers from which S. *typhimurium* had previously been isolated were deemed to be positive by ELISA using group B LPS and one was positive using group D LPS.

Timoney and colleagues (1990) used flagella prepared from S. *enteritidis* to differentiate field infections caused by this serovar from those caused by S. *pullorum*, since both serovars mentioned above belong to group D (O1, 9, 12). All of eight sera from a breeder flock and ten from a layer flock, both infected with S. *pullorum*, produced absorbance values of less than 0.25, a similar profile to an uninfected flock examined simultaneously. In contrast, high absorbance values of greater than 0.5 were seen in over 40% of chickens either experimentally or naturally infected with S. *enteritidis*. Berchieri *et al.* (1995) examined Brazilian flocks, some of which showed evidence of either fowl typhoid or pullorum disease. Sera from the uninfected flocks gave relatively low absorbance values when tested with group D LPS as detection antigen (all less than 0.4), whereas between 9 and 42% of the sera from the infected flock gave absorbances greater than 1.0. As expected, absorbance values obtained with flagella prepared from *S. enteritidis* (H, gm) were uniformly low, indicating absence of infection with this serovar.

Van de Giessen *et al.* (1994) analysed two pools each of five egg yolks from 170 laying flocks and found that 15.8% of them had elevated titres. Five of six flocks examined bacteriologically yielded *S. enteritidis*.

De Jong (1994) analysed breeding flocks as part of the Dutch national *S. enteritidis* eradication programme. In this, an LPS-based indirect and double antibody sandwich blocking ELISA based on flagella were compared. Of 1148 serum samples considered negative by the blocking ELISA, 1038 were also considered to be so by the indirect ELISA, 110 being considered positive by the indirect ELISA. A small number of flocks giving discrepancies were the results of vaccination against *S. typhimurium*. Five flocks were considered positive by both assays. Following cultural examinations, the author considered the blocking ELISA to be more specific.

Nielsen *et al.* (1994) also developed a blocking ELISA, using monoclonal antibodies produced against the O9 antigen. Little evidence was presented, but the indication again was that this assay was less sensitive but more specific than the indirect ELISA, since sera from *S. typhimurium*-infected flocks did not give positive reactions using the former assay.

Desmidt *et al.* (1994) analysed sera from 33 different types of flocks using an LPS-based indirect ELISA. There was good correlation between the number of sera giving high ODs (> 0.65 – >1.00) and cultural evidence of infection. Of sera from flocks known to be infected with *S. typhimurium*, only a small number gave high values and those flocks infected with serovars from groups C_1, C_2 and E_3 generally never gave such high values. Analysis of egg yolk from six selected flocks produced similar results.

Results obtained by Thorns *et al.* (1994) using SEF14 antigen have already been presented.

As can be seen, most work has involved salmonellosis in chickens. Assays have been successfully developed for *S. arizonae* infection in turkeys (Nagaraja *et al.*, 1984, 1986) and also for salmonellosis in ducks (P.A. Barrow and R.M. Hatfield, unpublished results).

Applications of ELISA to field infections in cattle

A small number of studies have shown the value of using bovine serum or milk in ELISAs for monitoring *S. dublin* or *S. typhimurium* infection in cattle.

Smith *et al.* (1989) developed an indirect assay with LPS as detecting antigen. In addition to cattle experimentally infected with *S. dublin*, they examined seven naturally occurring carrier cows, six uninfected, randomly selected, control cows and six normal, uninfected, *S. dublin*-negative cows in the same herd from which the carrier animals were selected. The mean serum ODs in the random control cows and the uninfected animals from the *S. dublin*-infected herd were 14% and 8%, respectively of the value of the positive control for serum samples and −1.0% and 0.6%, respectively, for milk. The mean values for serum and milk from the carriers were 85.2% and 70.6%, respectively. This compared with a mean serum value of 17.7% obtained from experimentally infected cows. It was suggested that specific serum IgG concentrations might be useful as an indicator of carrier animals. The serum IgG in experimentally infected animals rose to a maximum approximately 60 days after infection and thereafter diminished, whereas serum and milk titres in carrier animals remained consistently high for more than 1 year (the entire period of observation).

The same authors (Spier *et al.*, 1990) monitored milk and serum IgG and IgM specific for *S. dublin* LPS in similar groups of cows to those studied by the group earlier (Smith *et al.*, 1989). Again, serum IgG specific for *S. dublin* was the most indicative parameter predictive of carrier status. Two serum IgG titres obtained 2 months apart were a better indicator than the IgG : IgM ratio from a single sample. In milk, carrying out two ELISAs 60 days apart and analysis of IgG and IgM levels was useful but was not as accurate as serum immunoglobulin assays.

The group studied a milking herd and, as a result of screening milk and serum samples, again concluded that specific serum IgG could be used to identify *S. dublin* carrier animals, later identified at necropsy, as infected (House *et al.*, 1993).

Similar ELISAs have been used in Denmark for the detection of both *S. dublin* and *S. typhimurium*. Using a standard indirect LPS-based ELISA and confirmation of LPS specificity with a blocking assay, Hoorfar *et al.* (1994) showed the value of serological examination of cattle as a herd test for *S. dublin*. High positive ODs were always obtained from herds where infection was endemic but were never obtained from herds on islands where salmonellosis never occurred. Similar results were found using milk in the ELISA.

Hoorfar *et al.* (1997) have also compared meat juice with sera for the detection of *Salmonella* in Danish cattle. Whereas all cattle from the *Salmonella*-free Island of Bornholm produced sera with OD values of much less than the cut-off point (0.5), 19% (28/144) of cattle from an area where *S. dublin* was endemic produced ELISA-positive sera and meat-juice samples. In contrast, 2% (3/144) were positive for muscle fluid only and 1/144 was positive with serum only. This produced sensitivity and specificity values of 97%. Similar correlations were also produced with an indirect *S. typhimurium* ELISA and in an O9 blocking ELISA. Highly significant correlation was found between the results produced with serum and meat juice.

One or two high OD values were found in uninfected cattle, which were possibly the result of infection with a different serovar, but this was not pursued (Hoorfar *et al.*, 1995a,b). This is an important finding, because it is well known that cross-reactions occur with LPS as detecting antigens with sera from poultry infected with group B or D serovars. This might be avoided by using group D-specific (periodate treated) LPS (House *et al.*, 1993; van der Heijden, 1994), but it is important to know the extent to which this might occur. Use of a blocking ELISA with the homologous LPS alone does not show this.

Similar results have been found with *S. typhimurium* infections in cattle (Hoorfar and Wedderkopp, 1995). The problem of specificity between group D and B was appreciated (Hoorfar and Bitsch, 1995) and it was considered that group LPS was reasonably specific. The specificity was not total and was useful to the point that the test was valuable as a herd test (Hoorfar

et al., 1996). B.P. Smith and colleagues (personal communication) have also used LPS ELISAs to determine the level of herd exposure to *Salmonella* groups C_1, C_2, C_3 and E_1.

Applications of ELISA to field infections in pigs

Increasing information is being published on the use of ELISAs for diagnosis of *Salmonella* infection in pigs by workers in Denmark (Nielsen *et al.*, 1995a,b, 1996, 1997, 1998). In that country, the standard indirect ELISA is included in a comprehensive scheme for controlling the incidence of salmonellosis in pigs, intended as a means to control zoonotic infection (Nielsen *et al.*, 1997). Thus serum ELISAs using standard indirect assays with several O antigens (1, 4, 5, 6, 7) are used for monitoring infection in breeder and multiplier herds. For slaughter herds, all herds producing more than 100 finishers per year are screened by use of the meat-juice ELISA at slaughter. Based on the results, slaughter herds are divided into three groups, one with no or an acceptably low incidence of reactors, a second in which this incidence has increased and a third in which the level is unacceptably high. Restrictions are placed on these last two groups to attempt to reduce the size of the problem. All samples are analysed centrally and results are sent directly to farmers.

Initial screening of farms known to be infected with different serovars indicated that considerable variation occurred between the ODs obtained with meat juice from individual animals from herds known to be infected with individual serovars. ODs of greater than 10% of a positive control were regarded as positive and were obtained from many herds, including some that were culture-negative. However, the mean OD% values were of greater value, again showing the value as a herd test (Nielsen *et al.*, 1996). The use of LPS from *S. typhimurium* and *S. choleraesuis* (mix ELISA) was found to be of value and was used by this group in later studies (Nielsen *et al.*, 1998). It is estimated that 800,000 samples are analysed annually in this way in a totally automated system at a low price (US 20 cents per finisher). As a herd test, the assay has great predictive value, whether or not the control scheme is a long-term success.

Determination of cut-off points and standardization

For all biological parameters, normal variation occurs, and it is essential to take this into account in determining the titre or OD above which sera will be considered to have been taken from *Salmonella*-infected animals. In assays such as these, it may be necessary to adjust the cut-off point according to the disease prevalence in order to maximize the financial advantages of screening (Ridge and Vizard, 1993). However, this could create confusion by allowing different cut-off points for different countries. Other problems of interpretation are currently being tackled for *Salmonella* ELISAs.

The cut-off point can be measured by screening a set of serum samples from animals known to be free of infection followed by a calculation of the mean value, with a number of standard deviations added. For poultry, SPF birds have been used for this purpose (Nicholas and Cullen, 1991). Commercial birds maintained under high security conditions may also be used. These generally produce a higher cut-off than SPF sera (Timoney *et al.*, 1990; McLeod and Barrow, 1991), possibly because of a wider range of antigen exposure, including vaccination against other pathogens. This has also been done for pig sera (Nielsen *et al.*, 1995b).

The use of a cut-off point derived from a mean plus standard deviations means that the more samples that are tested the more the chance that the cut-off point will be exceeded. Even the addition of four standard deviations will be exceeded in approximately one in 60,000 tests.

It may, therefore, be appropriate to look at other ways of interpretation, which would avoid this problem and the problem caused by variation in interpretation as a result of differing disease incidence. Analysis of the profiles of ODs reveals that, in most infected flocks and cattle herds, sera producing ODs well above the cut-off point can be observed (Timoney *et al.*, 1990; McLeod and Barrow, 1991; Desmidt *et al.*, 1994; Hoorfar *et al.*, 1994). It is possible to utilize this fact by introducing a cut-off point well above that calculated on the basis of clean sera, milk or yolk. In the case of poultry, flocks have been identified as infected by the presence of sera or egg yolk producing high ODs of 0.65 or greater

(Desmidt *et al.*, 1994). Sensitivity and specificities for an ELISA were estimated for the calculation of a cut-off on a herd basis (Hoorfar *et al.*, 1994, 1995a). In this case, an OD value of 0.3 gave a sensitivity of 1.0 and a specificity of 0.9. This has also been used for bulk tank milk samples. For sera, this value resulted in no *Salmonella*-free herds reacting positively in the ELISA. The cost of large-scale testing following identification of false positives must be borne in mind. However, ELISAs have also been used as a predictive assay for individual cows (Spier *et al.*, 1990).

The value of sampling bulk milk raises the question over pooling samples. Because of the convenience, combined with a predictive value, this has been recommended by Hoorfar and his colleagues as a prelude to testing individual animals (Hoorfar *et al.*, 1995a). However, pooling of samples such as egg yolk or sera may lead to a reduction in the sensitivity of the assay.

Many assays are currently in use and national and international authorities are under pressure to adopt them for screening, and it is therefore desirable that some degree of standardization is obtained. One such exercise has been carried out under the aegis of the World Health Organization (WHO) (Barrow *et al.*, 1996). A set of sera from infected and uninfected SPF and commercial chickens was provided for several groups. The results produced indicate that, with some exceptions, the different assays behaved similarly and that there was some consensus in interpretation of results. This has also been carried out with bovine antisera, with very similar results (Dilling and Smith, 1992). Given that serological monitoring of poultry for *Salmonella* infections is likely to be used increasingly frequently, the idea has been raised that a standard assay system should be recommended. However, more than one basic type of assay exists. In addition, national laboratories already use assays that they themselves have developed. The provision of a bank of antisera against which assays may be calibrated might be the most appropriate approach to standardization. This could be organized by an organization such as the WHO, OIE (Office International des Epizooties) or, at a regional level, by a European Union reference laboratory.

In the case of cattle, bulk milk or serum from all animals in a herd will be sampled. This might not be possible for poultry. However, for the latter species, recommended sampling sizes of 60 per flock are less than adequate and samples of 300 per house are more useful, since this increases the sensitivity to allow detection of an infection rate of 1% rather than 5%.

Conclusion

The above discussion indicates that, for poultry, the ELISA must be regarded as a flock test. For *S. dublin*, it is possible to use it for individual animals to predict carrier status. For *S. typhimurium* infections in cattle and pigs, where environmental contamination may be more extensive, it is only of value as a herd test. For organisms such as *S. dublin* and *S. pullorum*, where carriage may occur in the absence of extensive excretion, ELISA serology might be used on its own for prediction of carriage. However, for other serovars in all animal species, ELISAs must be followed up by bacteriological examination. However, when carried out at intervals, it may become as sensitive as or more sensitive than bacteriology.

The ELISA appears to be the most widely used and accepted serological assay, although there may still be a place for other systems as these are refined. The ELISA has a number of advantages, including ease of operation, cost, sensitivity and specificity. There are areas where improvements are required and more information is needed. These include a reduction in the cross-reactions between groups B and D LPS. This may be less important in poultry than it seems, because some current legislation (Commission of the European Communities, 1992) involves control of both *S. typhimurium* and *S. enteritidis* at breeder level. The use of a test with a combination of detecting antigens may therefore be more appropriate. More information is required on the potential value of flagella and SEF14 antigen for screening purposes. Further work is required on the effects of chemotherapy and vaccination on serological status. Further discussions are also required on the choice of 'cut-off' values and the interpretation of data, together with the potential value of a set of reference sera against which assays could be calibrated.

On the basis of the experimental and field data acquired so far, the ELISA appears to satisfy most of the requirements for its adoption as a serological test to detect infections produced by

S. dublin, *S. enteritidis*, *S. typhimurium*, *S. gallinarum* and *S. pullorum*. Although sera may be used, it would seem sensible for egg-yolk samples to be used to minimize disruption and milk to facilitate initial screening. Some consideration must also be given to increasing the number of samples taken at each time. The replacement of 60 samples by 300 to increase the sensitivity of monitoring of large herds or flocks is desirable. Whether this could be carried out at parent level remains to be seen, but it should certainly be adopted for grandparent and élite stock. Similarly, there would seem to be no reason why the rapid slide-agglutination test for *S. pullorum* and *S. gallinarum* should not now be replaced by ELISA.

Recommended Protocols for Immunological Assays

Sample collection

Samples should be sent to the laboratory the same day or stored at 4°C before centrifugation to aid separation of serum. Serum may be stored for several days at 4°C or indefinitely at −20°C before testing. Sodium merthiolate (0.02% w/v) may be added as a preservative.

Whole blood may be dried on filter paper discs and sent to the laboratory by post.

Egg yolk should be diluted 1 : 10 in phosphate-buffered saline containing 0.1% Tween 20 (PBST) before storage.

Milk samples should be stored at −20°C immediately on arrival in the laboratory or centrifuged at 2000 *g* at 4°C to remove the fat layer prior to storage. Samples may be added to wells undiluted or diluted in PBST.

It is recommended (Nielsen *et al.*, 1988; Hoorfar *et al.*, 1997) that meat juice be prepared in a standard manner. For this, a piece of meat 3 cm × 1 cm × 1 cm is recommended for pigs (Nielsen *et al.*, 1998) or 10 g for cattle (Hoorfar *et al.*, 1997). This should be taken from the sternomastoid or the diaphragm muscle. It is transferred to a muscle fluid collection system (patent pending in Europe (application no. 95610056) and the USA), which consists of a polyethylene funnel connected to a 10 ml polyethylene tube. The funnel is covered and placed in a −20°C freezer overnight. The next morning, the system is incubated at 20–22°C (room temperature) for 5 h to allow thawing to take place. The fluid collected should be stored at −20°C until analysed.

Whole-blood test

Characteristics
The WBT provides a rapid test that can be used on the farm. It is a cheap test to perform and requires little equipment. For use in the laboratory, serum may be separated and tested.

The sensitivity of the WBT is low and, in inexperienced hands, false-positive and negative results may be recorded. In experienced hands, however, the results are comparable to those of the SAT.

The stained pullorum antigen will only detect the somatic antibody response and there is, as yet, no commercially available reagent to detect flagellar antibodies.

Fimbrial antigens may cause false-positive results.

Blood, 0.02 ml, is taken from the wing vein of a bird, after pricking with a needle, and placed on a white ceramic tile, using a loop of the appropriate size.

Polyvalent crystal violet-stained S. pullorum antigen (e.g. stained S. pullorum antigen, PA0191, Central Veterinary Laboratory, UK) or other source, 0.04 ml, is added to the blood and mixed with it.

The tile is rocked gently for 2 min, after which time the test is read.

Positive and negative control sera should be included in each batch that is tested.

The test is unreliable for turkeys and ducks.

Rapid slide-agglutination test

The test components must reach room temperature before being used.

Serum, 0.02 ml, is mixed with 0.02 ml polyvalent crystal violet-stained antigen.

The tile should be rocked gently for 2 min, after which time the test is read.

Test sera should be free from contamination and haemolysis. It may be helpful to centrifuge serum samples that have been stored for any period. In the case of turkey sera, it may be necessary to extend the mixing period to 3 min.

Due to transient, non-specific reactions, positive/suspicious sera should be retested after heat inactivation at 56°C for 30 min.

Serum agglutination test

Characteristics
Relatively insensitive and many older birds have low levels of agglutinins in their sera caused by enterobacteria other than Salmonella.

Single samples are of little value; paired samples are needed as the minimal requirement.

The test is relatively cheap; the antigens can be readily prepared and expensive equipment is not necessary. It can be adapted to the microtitre format.

It can be readily used to determine somatic and flagellar titres.

It is desirable to have standard sera for quality-control SAT (antigen preparation).

Preparation of somatic antigen
1. Plate out the Salmonella culture from the appropriate stock culture on to a blood agar base (BAB) plate, or other suitable medium, for single-colony growth. Incubate overnight at 37°C (± 2°C).
2. Select a smooth colony and carry out a slide agglutination test to ensure that the required somatic antigen is present.
3. Using a sterile loop, inoculate a nutrient agar slope in a universal container with the selected colony.
4. Incubate the culture for 8–12 h at 37°C (± 2°C).
5. After incubation, wash off the culture, preferably inside a safety cabinet, with approximately 2 ml of absolute alcohol, and transfer, using a Pasteur pipette, into a sterile universal container.
6. Leave the antigen for 4–6 h at room temperature to enable the alcohol to kill the bacteria and detach flagella.
7. Spin the universal container in a bench-top centrifuge for 5 min at 3600 r.p.m. Pour off the liquid and add enough phenol saline to give the antigen an opacity equivalent to Brown's tube 2 (c. 10^8 ml^{-1}) or another appropriate standard.
8. Carry out standard titration with known serum to ensure that the antigen is positive for

the required factor.

9. Store in a refrigerator at 4°C until required.

Preparation of flagellar antigens

1. Plate out the appropriate *Salmonella* stock culture on to a BAB plate, or other appropriate medium. Incubate overnight at 37°C (± 2°C).

2. Passage in semi-solid agar (c. 0.3%) in a Craigie's tube or other suitable container to induce optimum expression of the appropriate flagellar antigen. If the serovar is biphasic, H antiserum corresponding to the phase to be suppressed is added to the agar.

3. Use slide agglutination to check that the *Salmonella* is in the required phase. If this is correct, inoculate a loopful of culture into 20 ml nutrient broth. Incubate for 12–18 h at 37°C (± 2°C) for optimum growth. (If the phase is incorrect, passage through semi-solid agar again.)

4. Pipette 250 µl of 40% formaldehyde into the antigen suspension. (Use gloves and preferably work in a safety cabinet.)

5. Test the antigen by SAT, using the appropriate typing serum.

Realization of SAT

1. It is easiest to screen the sera at a dilution of 1 : 20. Antigen (0.25 ml) is added to 0.25 ml serum pre-diluted to 1 : 10 in normal saline.

2. The tests are incubated in a water-bath at 50°C, for 24 h in the case of somatic antigens and for 4 h for the flagellar antigens.

3. Sera that give a positive reaction are then diluted 1 : 20 to 1 : 320 and tested again with the appropriate antigen. Titres of 1/160 to 1/1320 (cattle) or greater than 1/40 (poultry) are considered to indicate infection.

Enzyme-linked immunosorbent assays of immunoglobulins specific for S. *enteritidis*

Two main basic systems are available for detection of IgG (IgY), the indirect ELISA and the competitive sandwich-type ELISA.

The indirect ELISA involves the use of a detecting antigen coated on to the wells of a microtitre plate. After the application of a blocking reagent to reduce non-specific binding, test samples are applied to the wells. Specifically bound antibody in the sample is detected by an antibody–enzyme conjugate. A variety of antigens, including LPS, flagella, SEF14 fimbriae, OMPs and cruder antigen preparations, have been used.

The competitive sandwich ELISA employs a specific reagent for coating antigen to wells, namely monoclonal antibody. This is then followed by a pure or crude antigen preparation. Test samples are applied, followed by conjugated monoclonal antibody, which will not bind to the antigen if the sample contains specific antibodies. The assay time can be shortened by adding both test sample and conjugate together. Monoclonal antibodies have been prepared for LPS, flagella and SEF14 for S. *enteritidis*.

There are advantages and disadvantages to both systems. The indirect assay is simpler and reagents are available to all *Salmonella* serovars for chickens, turkeys, ducks and mammalian hosts. The competitive ELISA can be applied to all animal species and, in general, shows higher specificity. However, reagents are not commercially available for all serovars. There are also some affinity problems and it may be less sensitive than the indirect assays. In the field, both systems have produced false-positive reactions.

Both types of assay may be used with serum, yolk or reconstituted dried blood, which may be treated with some antibacterial agents without loss of antibody titre. Some differentiation can be made between infections produced by *Salmonella* serovars from different serogroups. Some cross-reaction occurs between groups B and D, which, ideally, should be reduced. The optimal method for choosing a cut-off absorbance value above which sera are designated as having come from an S. *enteritidis*-infected flock has not yet been decided.

ELISAs are readily adapted to automation and hence large-scale testing programmes. A major problem is that expensive equipment is necessary and many of the reagents are also expensive. However, they may be incubated at ambient temperatures and read by eye, provided that control samples are used.

S. enteritidis lipopolysaccharide preparation

Although LPS to some *Salmonella* 'O' groups can be purchased commercially, it is relatively easy to prepare LPS.

1. Two litres (4 × 500 ml in 4 × 1 l flasks) of nutrient broth are inoculated with a smooth *S. enteritidis* phage type 4, strain P125589 or other suitable strain. Cultures are incubated in a shaking incubator at 37°C.

2. Cultures are centrifuged in Sorvall RC5B centrifuge (7000 r.p.m., 15 min, 4°C).

3. Pellets are resuspended in a total of 40 ml (2 × 20 ml) distilled H_2O and centrifuged (7000 r.p.m., 10 min, 4°C).

4. Both pellets are resuspended in 2 × 20 ml H_2O and mixed well.

5. Acetone, 400 ml, is cooled to −20°C with alcohol and frozen CO_2.

6. The *S. enteritidis* suspension (40 ml) is placed in a 500 ml flask with 360 ml acetone cooled to −20°C. This mixture is kept at approximately −20°C for 60 min.

7. The mixture is filtered on to filter-paper. This is washed with 200 ml of fresh acetone, also at −20°C.

8. The filter-paper and bacterial deposit are allowed to dry on the bench for 30 min and then dried overnight in a desiccator or in a vacuum (lyophilizing) machine for 5 min.

9. The bacterial deposit is carefully scraped off the paper and ground to a fine powder with a pestle and mortar.

10. The ground bacterial powder is resuspended in distilled water to 6% w/v and warmed to 65°C in a water-bath.

11. Phenol, to a final concentration of 90% and at a temperature of 65°C is added. The mixture is incubated at 65°C for 5 min and then put on ice at 0°C for 30 min.

12. The mixture is centrifuged at 8000 g for 20 min at 2°C.

13. The top layer is removed carefully and dialysed, either against running water or against several changes of water, for 48 h.

ELISA procedure used at Institute for Animal Health, Compton, Newbury, UK

Flexible polyvinyl chloride micro-ELISA plates (Dynatech Laboratories) are used. Wells along the sides of the plates may give slightly higher readings. The ELISA comprises five stages, each using 100 μl of the appropriate reagent, incubated for 30 min at 37°C unless otherwise stated. After each stage, the reagents are removed by aspiration and the wells washed three times in phosphate-buffered saline containing 0.1% Tween-20 using an ELISA washer (Skatron A/S, PO Box 8, 3401 Lier, Norway).

The stages are as follows.

1. LPS antigen, prepared as described previously and diluted to 1/500 (c. 70 g ml^{-1}) in 50 mM carbonate buffer, pH 9.6 (15 mM Na_2CO_3, 35 mM $NaHCO_3$ in freshly distilled water), is added to the wells for 1 h at 37°C.

2. Bovine serum albumin, 3% (Fraction V, Sigma), in carbonate buffer pH 9.6 is added for 30 min.

3. The test serum, diluted at 1/500 in PBST, is added for 30 min. Positive and negative control sera must always be used.

4. Rabbit anti-chicken IgG alkaline phosphatase conjugate (Sigma), diluted 1/1000 in PBST, is added for 30 min.

5. Substrate – *p*-nitrophenyl phosphate (1 mg ml^{-1}) in diethanolamine buffer pH 9.8 – is added for 30 min. The buffer is made up by adding 19.4 ml diethanolamine and 20 mg $MgCl_2$ to 160 ml distilled water. The pH is adjusted to 9.8 with 1 M HCl and the volume made up to 200 ml.

6. The reaction is stopped by the addition of 50 μl 3 M NaOH.

The absorbance is read at 405 nm, using an automatic micro-ELISA reader (Titertek Multiskan MCC/340, Flow Labs).

If washing equipment and ELISA readers are not available, the plates may be washed with PBST by gravity flow and the assays may be read by eye, comparing the test sera with the positive and negative control sera included on each plate.

Incubation may also be carried out at ambient temperature.

Acknowledgements

I would like to thank Prof. B.P. Smith and Dr P. Lind for reading the manuscript critically and Miss J. Howard for typing the manuscript.

References

Aitken, M.M., Hall, G.A. and Jones, P.W. (1978) Investigation of a cutaneous delayed hypersensitivity response as a means of detecting *Salmonella dublin* infection in cattle. *Research in Veterinary Science* 24, 370–374.

Anon. (1984) *The National Poultry Improvement Plan and Auxillary Provisions*. United States Department of Agriculture, Washington, DC.

Baay, M.F.D. and Huis in't Veld, J.H.J. (1993) Alternative antigens reduce cross-reactions in an ELISA for the detection of *Salmonella enteritidis* in poultry. *Journal of Applied Bacteriology* 74, 243–247.

Barrow, P.A. (1991) Serological analysis for antibodies to *S. enteritidis*. *Veterinary Record* 128, 43–44.

Barrow, P.A. (1992) Further observations on the serological response to experimental *Salmonella typhimurium* infection in chickens measured by ELISA. *Epidemiology and Infection* 108, 231–241.

Barrow, P.A. and Lovell, M.A. (1991) Experimental infection of egg-laying hens with *Salmonella enteritidis* phage type 4. *Avian Pathology* 20, 335–348.

Barrow, P.A., Hassan, J.O., Mockett, A.P.A. and McLeod, S. (1989) Serological analysis of chicken flocks for antibodies to *Salmonella enteritidis*. *Veterinary Record* 127, 501–502.

Barrow, P.A., Berchieri, A. and Al-Haddad, O. (1992) The serological response of chickens to infection with *Salmonella gallinarum–pullorum* detected by ELISA. *Avian Disease* 36, 227–236.

Barrow, P.A., Huggins, M.B. and Lovell, M.A. (1994) Host specificity of *Salmonella* infection in chickens and mice is expressed *in vivo* primarily at the level of the reticuloendothelial system. *Infection and Immunity* 62, 4602–4610.

Barrow, P.A., Desmidt, M., Ducatelle, R., Guittet, M., van der Heijden, H.M.J.F., Holt, P.S., Huis in't Velt, J.H.J., McDonough, P., Nagaraja, K.V., Porter, R.E., Proux, K., Sisak, F., Staak, C., Steinbach, G., Thorns, C.J., Wray, C. and van Zijderveld, F. (1996) World Health Organisation – supervised interlaboratory comparison of ELISAs for the serological detection of *Salmonella enterica* serovar Enteritidis in chickens. *Epidemiology and Infection* 117, 69–77.

Berchieri, A., Iba, A.M. and Barrow, P.A. (1995) Examination by ELISA of sera obtained from chicken breeder and layer flocks showing evidence of fowl typhoid or pullorum disease. *Avian Pathology* 24, 411–420.

Briggs, D.J. and Skeeles, J.K. (1983) An enzyme-linked immunosorbent assay for detecting antibodies to *Pasteurella multocida* in chickens. *Avian Diseases* 28, 208–215.

Bumstead, N. and Barrow, P.A. (1988) Genetics of resistance to *Salmonella typhimurium* in newly hatched chicks. *British Poultry Science* 29, 521–529.

Carlsson, H.E., Lindberg, A.A. and Hammerstrom, S. (1972) Titration of antibodies to *Salmonella* O antigens by enzyme-linked immunosorbent assay. *Infection and Immunity* 6, 703–708.

Chart, H., Rowe, B., Baskerville, A. and Humphrey, T.J. (1990a) Serological tests for *Salmonella enteritidis* in chickens. *Veterinary Record* 126, 20.

Chart, H., Rowe, B., Baskerville, A. and Humphrey, T.J. (1990b) Serological tests for *Salmonella enteritidis* in chickens. *Veterinary Record* 126, 92.

Chart, H., Rowe, B., Baskerville, A. and Humphrey, T.J. (1990c) Serological response of chickens to *Salmonella enteritidis* infection. *Epidemiology and Infection* 104, 63–71.

Chart, H., Rowe, B., Baskerville, A. and Humphrey, T.J. (1990d) Serological analysis of chicken flocks for antibodies to *Salmonella enteritidis*. *Veterinary Record* 127, 501–502.

Chart, H., Rowe, B., Baskerville, A. and Humphrey, T.J. (1991) Serological analysis for antibodies to *S. enteritidis*. *Veterinary Record* 218, 215.

Commission of the European Communities (1992) *Concerning Means for Protection against Specified Zoonoses and Specified Zoonotic Agents in Animals and Products of Animal Origin in order to Prevent Outbreaks of Food-borne Infections and Intoxications*. Council Directorate 92/117/EEC.

Coombs, R.R.A., Mourant, A.E. and Race, R.R. (1945) A new test for the detection of weak and 'incomplete' RH agglutinins. *British Journal of Experimental Pathology* 26, 255–266.

Cooper, G.L., Nicholas, R.A.J. and Bracewell, C.D. (1989) Serological and bacteriological investigations of chickens from flocks naturally infected with *Salmonella enteritidis*. *Veterinary Record* 125, 567–572.

Cullen, G.A. and Nicholas, R.A.J. (1991) Serological analysis for antibodies to *S. enteritidis*. *Veterinary Record* 128, 387–388.

Cunningham, B. (1968) The control and eradication of brucellosis I. Serological responses in cattle following vaccination with S 19 and killed *Brucella* 45/20 adjuvant vaccine. *Veterinary Record* 82, 7–10.

Dadrast, H., Hesketh, R. and Taylor, D.J. (1990) Egg-yolk antibody detection in identification of *Salmonella* infected poultry. *Veterinary Record* 126, 219.

De Jong, W. A. (1994) *Salmonella enteritidis* eradication programme in the Netherlands. In: *Proceedings of the EC*

Workshop on ELISAs for Serological Detection of Salmonella *in Poultry, Brussels, 8–9 June 1993*. Directorate General for Agriculture, Commission of the European Communities, Brussels, pp. 83–86.

Desmidt, M., Uyttebroek, E., de Groot, P.A., Ducatelle, R. and Haesebrouck, F. (1992) Lipopolysaccharide versus whole germ ELISA and possible consequences of antibiotic treatment on seroconversion to *Salmonella enteritidis*. In: Hinton, M.H. and Mulder, R.W.A.W. (eds) *The Role of Antibiotics in the Control of Food-Borne Pathogens*. FLAIR no. 6. *The Prevention and Control of Potentially Pathogenic Microorganisms in Poultry and Poultry Meat Processing*. DLO Centre for Poultry Research and Information Services, Beekbergen, The Netherlands, pp. 103–110.

Desmidt, M., De Groot, P. A., Ducatelle, R., Haesebrouck, F., Bale, J., Allen, V. and Hinton, M. (1994) Determination of a workable cut-off value in a lipopolysaccharide based ELISA for the detection of antibodies against *Salmonella enteritidis* and *Salmonella typhimurium* in chicken sera and eggs. In: *Proceedings of the EC Workshop on ELISAs for Serological Diagnosis of* Salmonella *Infection in Poultry, Brussels, 8–9 June, 1993*. Directorate General for Agriculture, Commission of the European Communities, Brussels, pp. 141–150.

Dilling, G.W. and Smith, B.P. (1992) International standardization of ELISAs for serological detection of *Salmonella dublin* carrier cows. In: *Proceedings of the International Symposium on* Salmonella *and Salmonellosis – Posters*. Ploufragan, France, pp. 87–88.

Engleberg, N.C., Barrett, T.J., Fisher, H., Porter, B., Hurtado, E. and Hyghes, J.M. (1983) Identification of a carrier by using Vi enzyme-linked immunosorbent assay serology in an outbreak of typhoid fever on an Indian reservation. *Journal of Clinical Microbiology* 18, 1320–1322.

Field, H.I. (1948) A survey of bovine salmonellosis in mid- and west-Wales. *Veterinary Journal* 104, 251–266.

Garrard, E.H., Burton, W.A. and Carpenter, J.A. (1948) Non-pullorum agglutination reactions. In: *Proceedings of the 8th World's Poultry Congress*, pp. 626–631.

Gast, R.K. (1997) Detecting infections of chickens with recent S. *pullorum* isolates using standard serological methods. *Poultry Science* 76, 17–23.

Gast, R.K. and Beard, C.W. (1990) Serological detection of experimental *Salmonella enteritidis* infection in laying hens. *Avian Diseases* 34, 721–728.

Gordon, R.F. and Brander, G.C. (1942) The value of the rapid whole-blood stained antigen agglutination test in the eradication of pullorum disease. *Veterinary Record* 54, 275–280.

Goren, E. (1992) [Untitled letter]. In: *World Health Organisation on National and Local Schemes of Salmonella Control in Poultry, Ploufragan, France, 18–19 September, 1992*. WHO/CDS/ VPH/92.110, World Health Organization, Geneva.

Hassan, J.O., Barrow, P.A., Mockett, A.P.A. and McLeod, S. (1990) Antibody response to experimental *Salmonella typhimurium* infection in chickens measured by ELISA. *Veterinary Record* 126, 519–522.

Hassan, J.O., Mockett, A.P.A., Catty, D. and Barrow, P.A. (1991) Infection and reinfection of chickens with *Salmonella typhimurium*: bacteriology and immune response. *Avian Diseases* 35, 809–819.

Hinton, M.H. (1973) *Salmonella dublin* abortion in cattle II. Observations in the whey agglutination test and the milk ring test. *Journal of Hygiene, Cambridge* 71, 471–479.

Hoorfar, J. and Bitsch, V. (1995) Evaluation of an O-antigen ELISA for screening cattle herds for *Salmonella typhimurium*. *Veterinary Record* 137, 374–379.

Hoorfar, J. and Wedderkopp, A. (1995) Enzyme-linked immunosorbent assay for screening of milk samples for *Salmonella typhimurium* in dairy herds. *American Journal of Veterinary Research* 56, 1549–1554.

Hoorfar, J., Feld, N.C., Schirmer, A.L., Bitsch, V. and Lind, P. (1994) Serodiagnosis of *Salmonella dublin* infection in Danish dairy herds using O-antigen based enzyme linked immunosorbent assay. *Canadian Journal of Veterinary Research* 58, 268–274.

Hoorfar, J., Lind, P. and Bitsch, V. (1995a) Evaluation of an O antigen enzyme-linked immunosorbent assay for screening of milk samples for *Salmonella dublin* infection in dairy herds. *Canadian Journal of Veterinary Research* 59, 142–148.

Hoorfar, J., Wedderkopp, A., Lind, P. and Bitsch, V. (1995b) Anvendelse af ELISA til Pavisning af antistoffer mod *Salmonella dublin* i maelk. *Dansk Veterinaertidsskrift* 78, 491–494.

Hoorfar, J., Wedderkopp, A. and Lind, P. (1996) Comparison between persisting anti-lipopolysaccharide antibodies and culture at post-mortem in *Salmonella*-infected cattle herds. *Veterinary Microbiology* 50, 81–94.

Hoorfar, J., Wedderkopp, A. and Lind, P. (1997) Detection of antibodies to *Salmonella* lipopolysaccharide in muscle fluid from cattle. *American Journal of Veterinary Research* 58, 334–337.

House, J.K., Smith, B.P., Dilling, G.W. and Roden, L.D. (1993) Enzyme-linked immunosorbent assay for serological detection of *Salmonella dublin* carriers in a large dairy. *American Journal of Veterinary Research* 54, 1391–1399.

Huis in't Veld, J.H.J., van Asten, A.J.A.M., Baay, M.F.D., Kusters, J.G. and van der Zeijst, B.A.M. (1994) Selection of alternative antigens for the development of a specific ELISA for the detection of *Salmonella enteritidis* in poultry. In: *Proceedings of the EC Workshop on ELISAs for Serological Diagnosis of* Salmonella *Infection in Poultry, Brussels, 8–9 June, 1993*. Directorate General for Agriculture, Commission of the European Communities, Brussels, pp. 111–118.

Humphrey, T.J., Baskerville, A., Chart, H. and Rowe, B. (1989) Infection of egg-laying hens with *Salmonella enteritidis* PT4 by oral inoculation. *Veterinary Record* 125, 531–532.

Humphrey, T.J., Baskerville, A., Chart, H., Rowe, B. and Whitehead, A. (1991a) *Salmonella enteritidis* PT4 infection in specific pathogen free hens: influence of infecting dose. *Veterinary Record* 129, 482–485.

Humphrey, T.J., Chart, H., Baskerville, A. and Rowe, B. (1991b) The influence of age on the response of SPF hens to infection with *Salmonella enteritidis* PT4. *Epidemiology and Infection* 106, 33–43.

Kaspers, B., Schranner, I. and Lösch, U. (1991) Distribution of immunoglobulins during embryogenesis in the chicken. *Journal of Veterinary Medicine* A38, 73–79.

Kim, C.J., Nagaraja, K.V. and Pomeroy, B.S. (1991) Enzyme-linked immunosorbent assay for the detection of *Salmonella enteritidis* infection in chickens. *American Journal of Veterinary Research* 52, 1069–1074.

Kyvsgaard, N.C., Lind, P., Preuss, T., Kamstrup, S., Lei, J.C., Bøgh, H.O. and Nansen, P. (1996) Activity of antibodies against *Salmonella dublin, Toxoplasma gondii*, or *Actinobacillus pleuropneumoniae* in sera after treatment with electron beam irradiation or binary ethylenimine. *Clinical and Diagnostic Laboratory Immunology* 3, 628–634.

Lentsch, R.H., Batema, R.P. and Wagner, J.E. (1981) Detection of *Salmonella* infections by polyvalent enzyme-linked immunosorbent assay. *Journal of Clinical Microbiology* 14, 281–287.

Lindberg, A.A. and Robertsson, J.A. (1983) *Salmonella typhimurium* infection in calves: cell mediated and humoral immune reactions before and after challenge with live virulent bacteria in calves given live or inactivated vaccines. *Infection and Immunity* 41, 751–757.

Lovell, R. (1934) The presence and significance of agglutinins for some members of the *Salmonella* group occurring in the sera of normal animals. *Journal of Comparative Pathology and Therapeutics* 47, 107–124.

McLaren, I.M., Sojka, M.G., Thorns, C.J. and Wray, C. (1992) An interlaboratory trial of a latex agglutination test for rapid identification of *Salmonella enteritidis*. *Veterinary Record* 131, 235–236.

McLeod, S. and Barrow, P.A. (1991) Lipopolysaccharide-specific IgG in egg yolk from two chicken flocks infected with *Salmonella enteritids*. *Letters in Applied Microbiology* 13, 294–297.

Minga, U. and Wray, C. (1990) Serological tests for *Salmonella enteritidis* in chickens. *Veterinary Record* 126, 20–21.

Mockett, A.P.A. and Darbyshire, J.H. (1981) Comparative studies with an enzyme-linked immunosorbent assay (ELISA) for antibodies to avian infectious bronchitis virus. *Avian Pathology* 10, 1–10.

Mohammed, H.O., Yamamoto, R., Carpenter, T.E. and Ortmayer, H.B. (1986) A statistical model to optimise enzyme-linked immunosorbent assay parameters for detection of *Mycoplasma gallisepticum* and M. *synoviae* antibodies in egg yolk. *Avian Diseases* 30, 389–397.

Nagaraja, K.V., Emery, D.A., Sherlock, L.F., Newman, J.S. and Pomeroy, B.S. (1984) Detection of *Salmonella arizonae* in turkey flocks by ELISA. In: *Proceedings of the 27th Annual Meeting of the American Association of Veterinary Laboratory Diagnosticians*, pp. 185–204.

Nagaraja, K.V., Ausherman, L., Emery, D.A. and Pomeroy, B.S. (1986) Update on enzyme-linked immunosorbent assay for its field application in the detection of *Salmonella arizonae* infection in breeder flocks of turkeys. In: *Proceedings of the 29th Annual Meeting of the American Association of Veterinary Laboratory Diagnosticians*, pp. 347–356.

Nagaraja, K.V., Pomeroy, B.S. and Williams, J.E. (1991) Paratyphoid infections. In: Calnek, B.W., Barnes, H.J., Beard, C.W., Reid, W.M. and Yoder, H.W. (eds) *Diseases of Poultry*, Iowa State University Press, Ames, Iowa, pp. 99–130.

Nicholas, R.A.J. and Andrews, S.J. (1991) Detection of antibody to *Salmonella enteritids* and S. *typhimurium* in the yolk of hens' eggs. *Veterinary Record* 128, 98–100.

Nicholas, R.A.J. and Cullen, G.A. (1991) Development and application of an ELISA for detecting antibodies to *Salmonella enteritidis* in chicken flocks. *Veterinary Record* 128, 74–76.

Nicholas, R.A.J., Cullen, G.A. and Duff, P. (1990) Detection of salmonellas. *Veterinary Record* 126, 147.

Nielsen, B., Feld, N.C., Hoorfar, J. and Lind, P. (1994) ELISAs for detection of *Salmonella* infections in poultry, pigs and cattle. In: *Proceedings of the EEC Workshop on ELISAs for Serological Diagnosis of* Salmonella *Infection in Poultry, Brussels, 8–9 June, 1993*. Directorate General for Agriculture, Commission of the European Communities, Brussels, pp. 81–82.

Nielsen, B., Bager, F., Mousing, J., Dahl, J., Halgaard, C. and Christensen, H. (1995a) Danish perspective on the implementation of HACCP in the swine industry. In: *Proceedings of a Symposium on Hazard Analysis and Critical Control Point (HACCP): 75th Animal Meeting of the Conference of Research Workers in Animal Diseases*. Chicago, Illinois.

Nielsen, B., Baggesen, D., Bager, F., Haugergaard, J. and Lind, P. (1995b) The serological response to *Salmonella* serovars *typhimurium* and *infantis* in experimentally infected pigs: the time course followed with an indirect anti-LPS ELISA and bacteriological examinations. *Veterinary Microbiology* 47, 205–218.

Nielsen, B., Baggessen, D., Lind, P., Feld, N. and Wingstrand, A. (1996) Serological surveillance of *Salmonella* infections in swine herds by use of an indirect LPS ELISA. In: *Proceedings of the 14th International Pig Veterinary Society Congress*. Bologna, p. 169.

Nielsen, B., Sørensen, L.L. and Emborg, H.-D. (1997) The Danish *Salmonella* surveillance programme for pork. In: *Proceedings of the Second International Symposium on* Salmonella *and Salmonellosis*. Ploufragan, France, pp. 619–625.

Nielsen, B., Ekeroth, L., Bager, F. and Lind, P. (1998) Use of muscle juice as a source of antibodies for serological detection of *Salmonella* infection in slaughter pig herds. *Journal of Veterinary Diagnostic Investigation* 10, 158–163.

Piela, T.H., Gulka, G.M., Yates, V.J. and Chang, P.W. (1984) Use of egg yolk in serological tests (ELISA and HI) to detect antibody to Newcastle disease, infectious bronchitis and *Mycoplasma gallisepticum*. *Avian Diseases* 28, 877–883.

Richardson, A. (1973a) Serological responses of *Salmonella dublin* carrier cows. *British Veterinary Journal* 129, Liii–Lv.

Richardson, A. (1973b) The transmission of *Salmonella dublin* to calves from adult carrier cows. *Veterinary Record* 92, 112–115.

Ridge, S.E. and Vizard, A.L. (1993) Determination of the optimal cut-off value for a serological assay: an example using the Johne's absorbed ELA. *Journal of Clinical Microbiology* 31, 1256–1261.

Robertsson, J.A. (1984) Humoral antibody responses to experimental and spontaneous *Salmonella* infections in cattle measured by ELISA. *Zentralblatt für Veterinärmedizin* 31, 367–380.

Robertsson, J.A., Fossum, C., Svenson, S.B. and Lindberg, A.A. (1982a) *Salmonella typhimurium* infection in calves: specific immune reactivity against O-antigenic polysaccharide detectable in *in vitro* assays. *Infection and Immunity* 37, 728–736.

Robertsson, J.A., Svenson, S.B. and Lindberg, A.A. (1982b) *Salmonella typhimurium* infection in calves: delayed specific skin reactions directed against the O-antigenic polysaccharide chain. *Infection and Immunity* 37, 737–748.

Roden, L.D., Smith, B.P., Spier, S.J. and Dilling, G.W. (1992) Effect of calf age and type of *Salmonella* bacterin type on ability to produce immunoglobulins directed at *Salmonella* whole cells or lipopolysaccharide. *American Journal of Veterinary Research* 53, 1895–1899.

Rose, M.E. and Orlans, E. (1981) Immunoglobulins in the egg, embryo and young chick. *Developmental and Comparative Immunology* 5, 15–20.

Runnels, R.A., Coon, C.J., Farley, H. and Thorp, F. (1927) An application of the rapid method agglutination test in the diagnosis of bacillary white diarrhoea infection. *Journal of the American Veterinary Medical Association* 70, 660–667.

Schaffer, J.M., McDonald, A.D., Hall, W.J. and Bunyea, H. (1931) A stained antigen for the rapid whole blood test for pullorum disease. *Journal of the American Veterinary Medical Association* 79, 236–240.

Smith, B.P., Oliver, D.G., Singh, P., Dilling, G., Marvin, P.A., Ram, B.P., Jang, L.S., Sharkov, N., Osborn, J.S. and Jackett, K. (1989) Detection of *Salmonella dublin* mammary gland infection in carrier cows, using an enzyme-linked immunosorbent assay for antibody in milk or serum. *American Journal of Veterinary Research* 50, 1352–1360.

Smith, P.J., Larkin, M. and Brooksbank, N.H. (1972) Bacteriological and serological diagnosis of salmonellosis of fowls. *Research in Veterinary Science* 13, 460–467.

Smyser, C.F., Adinarayanan, N., van Roekel, H. and Snoeyenbos, G.H. (1966) Field and laboratory observations on *Salmonella heidelberg* infections in three chicken breeding flocks. *Avian Diseases* 10, 314–329.

Spier, S.J., Smith, B.P., Tyler, J.W., Cullar, J.S. and Dilling, G.W. (1990) Use of ELISA for detection of immunoglobulins G and M that recognize *Salmonella dublin* lipopolysaccharide for production of carrier status in cattle. *American Journal of Veterinary Science* 51, 1900–1904.

Statutory Instruments (1989) *The Poultry Breeding Flocks and Hatcheries (Registration and Testing) Order No. 1963*. Her Majesty's Stationery Office, London.

Steinbach, G., Dinjus, U., Gottschaldt, J., Kreutzer, B. and Staak, C. (1994) Studies on the serovar specificity of antibodies produced during experimental *Salmonella* infection. In: *Proceedings of the EC Workshop on ELISAs for Serological Diagnosis of* Salmonella *Infections in Poultry, Brussels, 8–9 June, 1993*. Directorate General for Agriculture, Commission of the European Communities, Brussels, pp. 87–96.

Thain, J.A. (1980) An evaluation of the microantiglobulin test in monitoring experimental *Salmonella* group C infections in chickens. *Research in Veterinary Science* 28, 212–216.

Thorns, C.J., Sojka, M. and Chasey, D. (1990) Detection of a novel fimbrial structure on the surface of *Salmonella enteritidis* by using a monoclonal antibody. *Journal of Clinical Microbiology* 28, 2409–2414.

Thorns, C.J., Sojka, M.G., McLaren, I.M. and Dibb-Fuller, M. (1992) Characterisation of monoclonal antibodies against a fimbrial structure of *Salmonella enteritidis* and certain other serogroup D salmonellae and their application as serotyping reagents. *Research in Veterinary Science* 53, 300–308.

Thorns, C.J., Nicholas, R.A.J., Bell, M.M. and Chisnall, S.C. (1994) Preliminary results on the use of SEF14 fimbrial antigen in an ELISA for the serodiagnosis of *Salmonella enteritidis* infection in chickens. In: *Proceedings of the EC Workshop on ELISAs for Serological Diagnosis of* Salmonella *Infection in Poultry, Brussels, 8–9 June, 1993*. Directorate General for Agriculture, Commission of the European Communities, Brussels, pp. 119–124.

Timoney, J.F., Sikora, N., Shivaprasad, H.L. and Opitz, M. (1990) Detection of antibody to *Salmonella enteritidis* by a gm flagellin-based ELISA. *Veterinary Record* 127, 168–169.

van de Giessen, A.W., Dufrenne, J.B. Ritmeester, W.S., Van Zijderveld, F.G. and Notermans, S.H.W. (1994) Identification of *Salmonella enteritidis* infected poultry flocks by detection of egg yolk antibody. In: *Proceedings of the EC Workshop on ELISAs for Serological Diagnosis of* Salmonella *Infection in Poultry, Brussels, 8–9 June, 1993*. Directorate General for Agriculture, Commission of the European Communities, Brussels, pp. 133–140.

van der Heijden, H.M.J.F. (1994) Effect of periodate oxidation of LPS on the specificity of *Salmonella*-ELISAs in poultry. In: *Proceedings of the EC Workshop on ELISAs for Serological Diagnosis of* Salmonella *Infection in Poultry, Brussels, 8–9 June, 1993*. Directorate General for Agriculture, Commission of the European Communities, Brussels, pp. 107–110.

van Zijderveld, F.G., van Zijderveld-van Bemmel, A.M. and Anakotta, J. (1992) Comparison of four different enzyme-linked immunosorbent assays for serological diagnosis of *Salmonella enteritidis* infections in experimentally infected chickens. *Journal of Clinical Microbiology* 30, 2560–2566.

Watson, D.L. (1980) Immunological function of the mammary gland and its secretion – comparative review. *Australian Journal of Biological Science* 33, 403–422.

Williams, J.E. and Whittemore, A.D. (1972) Microantiglobulin test for detecting *Salmonella typhimurium* agglutinins. *Applied Microbiology* 23, 931–937.

Williams, J.E. and Whittemore, A.D. (1975) Influence of age on the serological response of chickens to *Salmonella typhimurium* infection. *Avian Diseases* 19, 745–760.

Williams, J.E. and Whittemore, A.D. (1976) Comparison of six methods of detecting *Salmonella typhimurium* infection in chickens. *Avian Diseases* 20, 728–734.

Williams, J.E. and Whittemore, A.D. (1979) Serological response of chickens to *Salmonella thompson* and *Salmonella pullorum* infections. *Journal of Clinical Microbiology* 9, 108–114.

Wray, C., Morris, J.A. and Sojka, W.J. (1975) A comparison of indirect haemagglutination tests and serum agglutination tests for the serological diagnosis of *Salmonella dublin* in cattle. *British Veterinary Journal* 131, 727–737.

Wray, C., Callow, R.J., Sojka, W.J. and Jones, P.C. (1981) A study of the antiglobulin test for the diagnosis of *Salmonella dublin* infection of cattle. *British Veterinary Journal* 137, 53–59.

Wray, C. and Sojka, W.J. (1976) A study of the complement fixation test in *Salmonella dublin* infection. *Research in Veterinary Science* 21, 184–189.

Chapter 25

Molecular Typing of *Salmonella*

John E. Olsen

Department of Veterinary Microbiology, The Royal Veterinary and Agricultural University, Stigbøjlen 4, DK-1870 Frederiksberg C. Denmark

Introduction

Zoonotic *Salmonella* serovars are a major cause of food-borne disease in humans in most Western countries (Fisher, 1997). Several serovars also regularly cause disease in animals and humans by transmission from one individual to another; notable examples are the host-specific and host-adapted serovars, characterized by their ability to cause the systemic type of salmonellosis (Bäumler *et al.*, 1998).

Investigations into the epidemiology of animal and human salmonellosis are often performed by use of bacterial typing methods. Outbreak investigations and the tracing of *Salmonella* from livestock through the food-chain to humans are important examples of situations where such methods are used, but, as seen from Table 25.1, several reasons exist for the application of typing methods to studies of salmonellosis.

At present, typing investigations of *Salmonella* begin with serotyping according to the White–Kaufmann scheme (Kauffmann, 1961), and additional methods would normally only be applied if this method indicates relationships between isolates. With the more commonly isolated serovars, phage typing is then often applied to obtain further discrimination, e.g. typing of *S. typhimurium* according to the method of Anderson *et al.* (1977) and *S. enteritidis* according to Ward *et al.* (1987). The use of these methods has been discussed in previous

chapters and will not be dealt with here. However, both serotyping and phage typing usually show limited discriminatory power when isolates from a narrow geographical region or isolates obtained during a limited time period are typed. Typing by other phenotypic methods, such as biotyping (Odongo *et al.*, 1990; Katouli *et al.*, 1992) and antibiogram typing (Threlfall *et al.*, 1983; Wray *et al.*, 1987), may improve discrimination in certain situations, but usually a much more detailed typing result can be obtained by the use of DNA-based, molecular typing methods.

In this chapter, the use of molecular typing methods for studies of *Salmonella* epidemiology is reviewed. The methodology will be presented, and the methods will be discussed in the context of their application to both salmonellosis among livestock and the transmission of *Salmonella* from animals to humans.

Molecular Methods for Typing *Salmonella*

Numerous genotyping methods have been applied to the typing of *Salmonella*. Two of the main advantages of these methods are that they do not depend on the expression of phenotypic properties and that all strains of *Salmonella* can be typed by the same methodology, with only minor modifications for use with the individual serovar. The genotypic typing methods have

Table 25.1. Reasons for application of (molecular) typing to strains of *Salmonella*.

Typing situation	Purpose of typing
Surveillance	Obtain baseline information and react on changing patterns; estimate the attribution of animal reservoirs to human cases
Outbreak	Identification of common source
Endemic	Identify factors that contribute to persistence and spread
Treatment	Evaluation of treatment failure (recurrence or new infection)
Diagnosis	Identification of virulent subtypes
Food production	Monitor critical point for cross-contamination

normally been applied in a comparative way, i.e. only strains analysed on the same gel are fully comparable, but, with the development of fully automated gel-scanning systems and well-designed gel analysis systems, this is changing. A few methods have been used intensively and will be discussed here. No single method can at pre-

sent be recommended for all typing purposes within *Salmonella*. The choice of methods will depend on the specific circumstances, and often more than one method is applied to improve the quality of the typing.

Plasmid profiling

Methodology

Plasmid profiling was the first genotypic method used for strain separation within *Enterobacteriaceae* (Shaberg *et al.*, 1981). When the bacterium divides, copies of resident plasmids will be distributed between the two daughter cells, and members of the same clonal line are expected to carry the same plasmids. Strains of *Salmonella* often carry plasmids ranging in size from 2 to 150 kb (for an example, see Fig. 25.1), but frequencies and size distributions vary between serovars.

Table 25.2 reports baseline information on a number of common serovars, based on investigations of a diverse collection of strains of each serovar.

Plasmid profiling of strains of *Salmonella* can be done by simple methods, and useful results

(a) 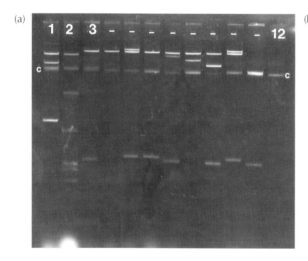   (b)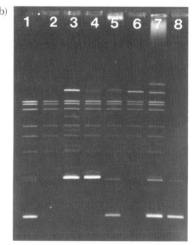

Fig. 25.1. An example of demonstration of plasmids and plasmid restriction enzyme profiles in strains of *Salmonella*. (a) Lanes 1 and 2 are plasmid size markers in *E. coli* strains 39R861 (147 kb, 63 kb, 36 kb and 7 kb plasmids) and V517 (54 kb, 7.4 kb, 5.6 kb, 5.1 kb, 4.4 kb, 3.3 kb, 2.7 kb and 2.1 kb plasmids). Lanes 3–12 show the plasmid content of ten Danish *S. dublin* strains. The strains in lanes 3–11 differ with respect to the plasmid profile. The strain in lane 12 does not carry any plasmids. (b) The *Hind*III restriction endonuclease profile of the plasmids shown in lanes 3–9 and lane 11 in part (a) are demonstrated in the order mentioned. C, chromosomal band.

Table 25.2. Examples of baseline studies on plasmid content in strains from serovars of zoonotic *Salmonella*.

Serovar	Source of strains	Reference
S. berta	Denmark, poultry	Sørensen *et al.* (1991)
S. dublin	Denmark, cattle and humans	Olsen *et al.* (1990)
	USA, clinical isolates	Ferris *et al.* (1992)
	Universal strain collection	Browning *et al.* (1995)
S. gallinarum/pullorum	Universal strains collection	Christensen *et al.* (1992)
S. enteritidis	UK, clinical isolates	Threlfall *et al.* (1989)
	USA, selected strains	Rodrigue *et al.* (1992)
	Phage-type reference strains	Brown *et al.* (1993)
	Danish strains	Brown *et al.* (1994)
S. goldcoast	UK and France	Threlfall *et al.* (1986a)
S. muenster	Canada, humans, animals and food	Bezanson *et al.* (1983)
S. typhimurium	Japan, animals	Nakamura *et al.* (1986)
	Scotland, different sources	Brown *et al.* (1986)
	UK, animal strains of phage type 204c	Wray *et al.* (1987)
	UK, strains of phage type 49	Threlfall *et al.* (1990)

can be obtained in less than 2 h from an overnight broth culture or colonies on an agar plate. The methods described by Birnboim and Doly (1979) and Kado and Liu (1981) in particular have been very popular. A slightly modified version of the Kado and Liu protocol that is in use in our laboratory is shown in Box 25.1.

Two plasmids of the same molecular size but with different DNA sequences will look identical in a plasmid profile. Confidence in the results can therefore be improved by subjecting the plasmids to digestion with restriction endonuclease enzymes (Thompson *et al.*, 1974; Platt *et al.*, 1986, 1987), producing a characteristic pattern of bands for each true plasmid profile (see Box 25.1). Extended versions of the Birnboim and Doly (1979) and Kado and Liu (1981) protocols can produce DNA of sufficient quality, but usually other plasmid isolation protocols are used for this purpose. A method developed in our laboratory (Olsen, 1990), and which we routinely use for restriction analysis of plasmids in wild-type strains of *Salmonella*, is shown in Box 25.1.

The interpretation of plasmid profiles should be done with an appreciation of the possible mistakes that can be made. Some plasmids are transferable by conjugation and/or mobilization, and, in animal flocks that have a high population density, several isolates have to be typed to identify the 'flock type', as major and minor plasmid profile types will be present (Brown *et*

al., 1992). Certain resistance (R) plasmids in *S. typhimurium* have been shown to undergo rearrangement or deletions if the selection pressure is not in favour of maintaining the R factors (Brown *et al.*, 1991), and the way the cultures are treated prior to analysis may influence the plasmid profiles, because plasmid loss may occur in stab cultures during storage (Olsen *et al.*, 1994a). In general, however, plasmid profiles of epidemic strains of *Salmonella* seem to be stably maintained and, in situations where there is doubt, reproducibility can be improved by accepting minor differences in plasmid profiles and still allocating isolates into the same group. It is also important to acknowledge that the assumption that strains containing the same plasmids belong to the same clonal line is not always valid. In both *S. dublin* and *S. gallinarum/pullorum*, identical plasmid profiles have been observed in strains allocated to separate clonal lines based on chromosomal typing (Christensen *et al.*, 1992, 1993; Olsen and Skov, 1994). In a similar manner, *S. enteritidis* and *S. typhimurium* virulence plasmids are prevalent and indistinguishable in several different phage types (clones) of these serovars (Brown *et al.*, 1986, 1993).

Application of plasmid profiling to typing of Salmonella

Plasmid profiling and plasmid restriction analysis have been used intensively for typing *Salmonella*, often to fine-tune conclusions based on phage-

Box 25.1. Methods suitable for isolation of plasmids from *Salmonella* for plasmid profiling (A) and restriction enzyme digestion (B). Method (A) is a slightly modified version of Kado and Liu (1981) and method (B) is according to Olsen (1990). EDTA, ethylenediamine tetra-acetic acid.

A	B
Transfer 0.5–1 ml of an overnight broth (O/N) culture or 10–15 colonies suspended in 1 ml 50 mM Tris, 10 mM EDTA, pH 8 (TE 50 : 10) to an Eppendorf tube	Transfer 1–1.5 ml of an O/N culture or 15–20 colonies suspended in 1 ml TE (50 : 10) to an Eppendorf tube
Centrifuge for 2 min at 13,000 r.p.m. in a microcentrifuge. Decant supernatant and remove remaining supernatant carefully with a micropipette	As Method A
Resuspend cells in 20 µl TE 50 : 10. Vortex until cells are completely resuspended	Resuspend cells in 40 µl TE (50 : 10). Vortex until cells are completely resuspended
Add 100 µl 3% SDS, 50 mM Tris, pH 12.46 (lysis buffer). Mix gently by inverting the tube ten times. Solution should turn clear	Add 400 µl lysis-buffer. Mix gently by inverting the tube ten times
Incubate at 56°C for 30 min	As Method A
Add 100 µl unbuffered phenol : chloroform : isoamyl-alcohol (25 : 24 : 1) and vortex until the suspension turns homogeneous milk-white	Add 300 µl 1.5 M KAc pH 5.2 and mix by inverting the tube 50 times. Incubate on ice for 1 h
Centrifuge 15 min at 13,000 r.p.m. in a microcentrifuge	Centrifuge at 13,000 r.p.m. for 15 min. Decant supernatant through a sterile layer of gauze into an Eppendof tube containing 600 µl isopropanol. Mix gently and leave on bench for 15 min
Carefully remove 40 µl of the upper phase into a clean Eppendorf tube containing 15 µl 30% glycerol, 1 mM EDTA, 1 p.p.m. bromophenol blue	Harvest the precipitated DNA by centrifugation at 13,000 r.p.m. for 15 min. Decant supernatant and wash pellet in ice-cold 70% ethanol. Maintain pellet by a short centrifugation
Load 15–20 µl on a 0.8% agarose gel, and separate for 1–1.5 h at 100 V constant current. Plasmids in *E. coli* 39R861 (Threlfall *et al.*, 1986b) and V517 (Macrina *et al.*, 1978) can be run alongside the samples to provide size marker plasmids	Air-dry pellet and resuspend in 50–100 µl TE (10 : 1)

typing results. One of the first examples of outbreak investigations using this technique linked the transmission of S. *muenchen* to contaminated marijuana (Taylor *et al.*, 1982). Other examples are numerous, and only a few will be highlighted here to illustrate the potentials for typing.

In the environment of calf dealers in the UK, Wray *et al.* (1990) used plasmid profiling in combination with phage typing and biotyping to demonstrate routes of transmission between farms. Different *Salmonella* serovars were isolated and, by typing of strains of S. *typhimurium* defini-

tive type (DT) 204c, persistent infection was demonstrated in only one out of 12 units. Based on the analysis of three separate epidemics, the central role of calf dealers in the dissemination of salmonellosis among cattle was clearly demonstrated.

Brown *et al.* (1992) showed that a particular line (plasmid profile) of S. *berta* persisted through three generations of birds in a poultry farm, despite the fact that proper hygiene measures were enforced to clean houses between generations. The same study illustrated the need for baseline information to ensure the correct interpretation of results. The *S. berta* isolates from other chicken houses in the same production unit carried a 5.7; 2.0 kb plasmid profile, which was demonstrated in 66% of all isolates from that year (Olsen *et al.*, 1992). Its value as an epidemiological marker was consequently of little significance. The inadequacy of cleaning and disinfection routines to prevent persistent *Salmonella* infection has also been illustrated in calf-rearing units. McLaren and Wray (1991) monitored four farms for several years and, by demonstrating persistence of particular types on the same location, persistence of *Salmonella* in the environment of calf-rearing units was shown be an important factor in the epidemiology of the disease.

Persistent infection of *S. typhimurium* DT110 due to infestation problems with the beetle *Alphitobius diaperinus* has been demonstrated in poultry-rearing flocks, using phage typing and plasmid profiling. By restriction analysis of the virulence-associated 90 kb plasmid, concurrent epidemics with different subclones of the same phage type were implicated at two locations in this study (Baggesen *et al.*, 1992), clearly demonstrating the potential for high discrimination by plasmid profiling combined with restriction analysis.

Plasmid profiling has played an important role in studies of the zoonotic aspects of salmonellosis. For example, outbreaks associated with *Salmonella* in milk (Ryan *et al.*, 1987), hamburgers (Ryan and Steele, 1987), precooked roast beef (Riley *et al.*, 1983), and imported chicken (Threlfall *et al.*, 1992) and direct transmission from animals to humans (Olsvik *et al.*, 1985) have been studied by use of this technique.

Plasmid profiling has also played a role in basic studies of *Salmonella* biology. An important example is the observation that several of the commoner serovars, such as *S. dublin*, *S. typhimurium*, *S. choleraesuis* and *S. enteritidis*, carry virulence-associated plasmids, which are essential for systemic infection in the mouse model (Helmuth *et al.*, 1985). This observation paved the way for the many studies into the so-called *spv* (*Salmonella* plasmid virulence genes) operon (reviewed by Gulig *et al.*, 1993).

Genotyping based on restriction analysis of total DNA

Methods that use direct restriction analysis

RESTRICTION ENZYME ANALYSIS. *Salmonella* genomes can be compared by isolation of total DNA and electrophoretic separation of DNA fragments generated by digestion with restriction endonuclease enzymes, using the same principle as for the restriction analysis of plasmids.

In the first application to typing of *Salmonella*, the DNA fragments were compared directly after separation by traditional agarose gel electrophoresis restriction enzyme analysis and serovar-specific patterns were obtained (Tompkins *et al.*, 1986). Results from this approach are difficult to reproduce and to interpret, as several hundred fragments have to be compared. In an attempt to overcome this problem, polyacrylamide gels, which have a higher sieving effect were introduced (Kapperud *et al.*, 1989), but they were never used extensively.

PULSED-FIELD GEL ELECTROPHORESIS. By using a 'rare cutter', i.e. a restriction enzyme that has only relatively few recognition sites in the bacterial genome, much fewer and larger restriction fragments can be produced. These can in turn be separated using pulsed-field gel electrophoresis (PFGE) (Schwartz *et al.*, 1983), where the polarity of the current through the gel is switched at intervals throughout the run, producing clear and easily scored restriction patterns (see Fig. 25.2).

An essential part of the methodology is isolation of very large pieces of DNA. A flow diagram of a PFGE analysis is shown in Box 25.2, based on the isolation procedure of Cameron *et al.* (1994) and with the electrophoretic conditions suggested by Olsen and Skov (1994).

Like plasmid profiling, PFGE patterns should be interpreted with care. A useful guideline to handling PFGE profiles has been produced by Tenover *et al.* (1995). While it is not directly applicable to all typing situations, it gives a very good introduction to the genetic events that influence PFGE profiles and recommendations for which enzyme to use with which bacteria.

PFGE is both time- and labour-intensive, but it has proved valuable in outbreak investiga-

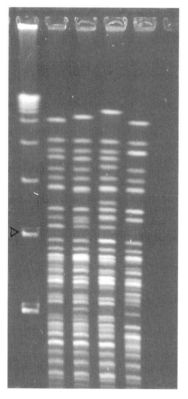

Fig. 25.2. Pulsed-field gel electrophoretic (PFGE) patterns of *Not*I-digested DNA of four strains of *S. dublin* (from Olsen and Skov, 1994). Lane 1 contains a 50-kb base ladder. The arrow points to the fragment of 100 kb. Lanes 2–5 are four strains of *S. dublin*, each showing a different PFGE pattern.

tions, often in situations where other typing methods have failed to separate strains. A pork-associated outbreak of salmonellosis caused by *S. infantis* occurred in Denmark during spring 1993. A particular meat plant and a meat retail market were implicated as the source of the human infections, based on a unique PFGE profile, which was not demonstrated among control isolates of the same serovar, including isolates from patients not involved in the outbreak (Wegener and Baggesen, 1996).

In an attempt to link Swiss isolates of the previously uncommon serovar *S. brandenburg*, PFGE was more discriminatory than ribotyping and insertion sequence (IS) 200 typing. The outbreak strains were characterized by a particular PFGE profile, and multiple food vehicles were

implicated as the source of the outbreak (Baquar *et al.*, 1994).

Similarly, Weide-Botjes *et al.* (1998) found PFGE to be the most discriminatory method for typing *S. hadar* and they used this method, in combination with plasmid profiling, for epidemiological investigations in poultry parent layer flocks in Germany. Recurrent infections, rather than persistence of clones, were demonstrated. The high discriminatory power of PFGE has been indicated by the fact that a strain-specific identification of the vaccine strain Zoosaloral H was obtained with either of the enzymes *Xba*I and *Spe*I, compared with other strains of *S. typhimurium* (Schwarz and Liebisch, 1994).

It has been noted that PFGE may be too discriminatory when comparing strains that have been isolated over a longer time period. Christensen *et al.* (1994) investigated the source of the first *S. gallinarum* outbreak among commercial poultry in Denmark in recent times. By plasmid profiling and ribotyping, using the highly discriminatory *Hind*III enzyme, the outbreak strains were similar to historical isolates from a neighbouring country suspected of being the origin of the infection. Transport of animals had taken place between the particular poultry production unit and the suspected country of origin and transport crates had been reused. By PFGE analysis, however, the outbreak strains had an additional band compared with the historical control strains. Several genetic events could have resulted in the appearance of this band, but its presence weakened the conclusions from the study.

Methods that use Southern analysis to simplify results

RESTRICTION FRAGMENT LENGTH POLYMORPHISMS. Restriction fragment length polymorphism (RFLP) analysis encompasses all methods that use probes to highlight polymorphism in specific DNA fragments. To simplify the reading of results of REA, DNA fragments are transferred to hybridization membranes and hybridized with a labelled hybridization probe. Thereafter, strain comparison is only based on restriction fragments that show homology to the probe. Figure 25.3 outlines the working protocol of this kind of technique. The commonest methods have been given names of their own, such as ribotyping and

Box 25.2. A method suitable for PFGE typing of *Salmonella*. The DNA isolation protocol is according to Cameron *et al.* (1994) and was developed for typing strains of *Vibrio cholerae*. The electrophoretic conditions are as recommended by Olsen and Skov (1994). LB, lactose broth; OD, optical density; EDTA, ethylenediamine tetra-acetic acid; TE, Tris (10 mM)–EDTA (1 mM); UV, ultraviolet.

A

Transfer three or four colonies to LB and incubate until OD_{600} reaches 0.1. Wash cells once in 1 M sodium chloride, 10 mM Tris (pH 8), 100 mM EDTA (STE-buffer). Resuspend in 1 ml STE and mix with an equal volume of 1% chromosome-grade agarose

Lyse in STE with 0.1 mg ml^{-1} lysozyme 0.05% sarcosyl, 0.02% sodium deoxycholate, 2 µg ml^{-1} RNase for 1 h at 37°C. Remove lysis buffer and incubate O/N in ESP buffer (0.5 M EDTA, 1% sarcosyl, 1 mg ml^{-1} Proteinase K)

Rinse plugs in water. Wash twice in TE 10 : 1 containing 30 µl 0.1 M phenyl-methylsulphonyl fluoride for 30 min each time. Rinse in TE 10 : 1

Incubate a slice of the agarose in the appropriate restriction enzyme buffer for 1 h, replace with fresh buffer and perform restriction digestion for 4 h. The enzymes *Not*I and *Xba*I are commonly used for analysis of strains of *Salmonella*

B

Perform agarose gel electrophoresis in 1% agarose and 0.5 TBE buffer

Suitable running conditions are 12 V cm^{-1} at 14°C for 22 h
The pulse time is increased by stepping as follows: 2 s for 5 h; 5 s for 6 h; 9 s in 6 h, 12 s in 5 h
Use polymerized phage lambda as size marker

Stain gel in 2 µg ml^{-1} aqueous ethidium bromide for 15 min, destain in distilled water for 15 min and photograph under 254 nm UV light

---

IS200 typing, and this section will be mainly devoted to these methods.

RFLP analysis was first used to analyse strains of *S. typhimurium*, *S. dublin* and *S. enteritidis*, using two randomly cloned chromosomal fragments of *S. enteritidis*. Strains that grouped together by phage typing, antibiograms and plasmid profiling also showed the same hybridization pattern by RFLP (Tompkins *et al.*, 1986). However, no major epidemiological studies have been undertaken using random cloned fragments as probes.

RIBOTYPING. Hybridization with 16S and/or 23S ribosomal RNA sequences was first proposed as a means for the study of taxonomic relationships between strains (Grimont and Grimont, 1986) and, within *Salmonella*, several studies of interrelationships within and between serovars have been performed using this technique (Chowdry *et al.*, 1993; Olsen and Skov, 1994; Olsen *et al.*, 1994b, 1997; Stanley and Baquer, 1994). Probes

have consisted of cloned rRNA subfragments (Martinetti and Altwegg, 1990), cDNAs of 16S and 23 S rRNA (Grimont and Grimont, 1986) or polymerase chain reaction (PCR)-generated internal fragments of 16S rRNA only (Chowdry *et al.*, 1993). Based on typing of comparable but not identical strain collections of *S. dublin* with the two latter probe systems (Chowdry *et al.*, 1993; Olsen and Skov, 1994), the 16S/23S cDNA probe would appear to give the higher discriminatory power, which is not surprising, since more restriction sites are analysed.

In general, ribotyping seems to highlight restriction polymorphism in the areas surrounding the seven rRNA operons in *Salmonella*, as opposed to sequence variation within the genes themselves. This can be inferred from comparison of 16S and 23S rRNA gene sequences in *Salmonella*, where, for example, some of the prominent serovars of *S. enterica* subspecies I show identical sequences (Christensen *et al.*, 1998).

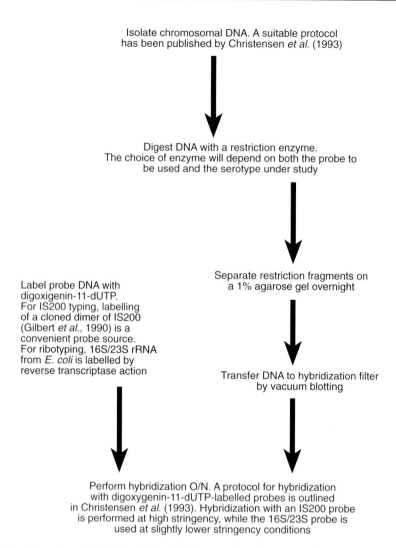

Fig. 25.3. Working protocol for RFLP methods. The diagram shows the principal steps of an IS200 typing and a 16S/23S ribotyping.

Ribotyping has a relatively low discriminatory power and, for that reason, it has not been commonly used for outbreak investigations. In a Danish outbreak with *S. saintpaul*, ribotyping was applied, together with serofactor subtyping and plasmid profiling, in the outbreak investigation. The presence of the O:5 factor and the lack of plasmid suggested that pork was the major source of the outbreak, while ribotyping indicated that Danish turkey was the main source. It was concluded that a third, unrecognized source could be the true origin of the outbreak (Baggesen *et al.*, 1996).

IS200 TYPING. Lam and Roth (1983) identified a *Salmonella*-specific insertion sequence, which they termed IS200. The element was demonstrated to be present in all serovars analysed apart from *S. agona* (Gibert *et al.*, 1990). Later, the absence of IS200 was also noted in strains of *S. daressalam* (Olsen and Skov, 1994) and *S. hadar* (Weide-Botjes *et al.*, 1998). Gilbert *et al.* (1990) cloned an internal *Hind*III–*Hind*III fragment of this element, and this has become the universally applied IS200 probe. An example of an IS200 profile is shown in Fig. 25.4 together with a 16S/23S ribotype.

(a)

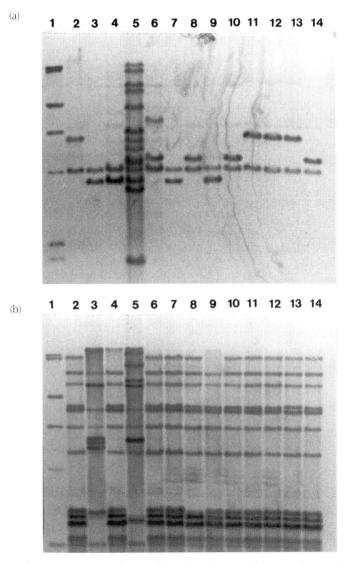

Fig. 25.4. Examples of IS200 (a) and ribotype (b) profiles of strains of *Salmonella*. In both figures, lane 1 is the size marker lambda *Hind*III, and lanes 2–14 show *Pst*I IS200 profiles and *Sma*I ribotype pattens of strains of *S. enteritidis*.

Stanley *et al.* (1991) were among the first to demonstrate the use of this probe for *Salmonella* typing by showing that strains of *S. enteritidis* could be grouped into possible evolutionary lines, based on the number and location of IS200 elements on *Pst*I- and *Pvu*II-digested DNA. Demonstration of IS200 has since been applied to the typing of many different serovars. *S. infantis* from Finland were typed using *Ban*I-fragmented DNA. Eleven profiles were demon-

strated, some of which could be subdivided by 16S rRNA typing. Based on other epidemiological indications, broilers were believed to be a major source of human infections, but surprisingly no profiles were shared between broilers and human isolates (Pelkonen *et al.*, 1994).

Also focusing on poultry, Millemann *et al.* (1995) compared typing methods for *S. typhimurium* and *S. enteritidis* from different flocks in France and found that, while isolates of *S.*

typhimurium from within a flock were generally characterized by a common plasmid profile, the same ribotype and a particular IS200 profile, a small number of strains were outliers. The highest discrimination was obtained using a combined reading of IS200 typing and ribotyping. All strains of *S. enteritidis* showed the same IS200 type but differed with respect to ribotype, showing that no universal conclusions on choice of typing method can be drawn across serovars. Unfortunately, phage-typing results were not reported, which makes comparison with other studies difficult. As an important observation, both ribotype and IS200 patterns remained stable during a 15-week rearing period where artificially challenged chickens were sampled and isolates of *S. typhimurium* typed on a regularly basis. During a national outbreak of *S. berta*, we also demonstrated that the chromosomal markers were stable during the epidemic (Olsen *et al.*, 1996a).

In a comparison of typing methods, IS200 typing showed the highest discriminatory power (D = 0.78 for *Pvu*II and *Eco*RI digests) for 30 strains of *S. panama*, compared with plasmid profiling (D = 0.50) and ribotyping (D < 0.5 for all four enzymes used). PFGE was not applicable to this serovar, as DNA was degraded in the agarose blocks (Stanley *et al.*, 1995). In some serovars where typing schemes have not been developed extensively, IS200 typing has become a standard typing method. For example, for typing the sheep-adapted serovar *S. abortusovis*, IS200 typing is recommended as the method of choice (Schiaffino *et al.*, 1996).

Like ribotyping, IS200 typing has often been applied to more basic taxonomic studies of relationships within and between *Salmonella* serovars, for example *S. heidelberg* (Stanley *et al.*, 1992), *S. typhimurium* (Olsen *et al.*, 1997), *S. dublin* (Chowdry *et al.*, 1993; Olsen and Skov, 1994), *S. paratyphi* B/S Java (Ezquerra *et al.*, 1993) and *S. enteritidis* (Stanley *et al.*, 1991; Olsen *et al.*, 1994b).

PCR-based typing

By use of PCR method, specific DNA fragments can be amplified to detectable amounts in a short time. Typing based on this technique is rapid and cheap in terms of equipment and chemicals, but it has drawn surprisingly little attention for the

typing of *Salmonella*. This is possibly because the typing schemes are already well developed, and it has been difficult to obtain a higher discrimination than is possible by the more traditional methods.

Random amplification of polymorphic DNA
The specificity of the PCR reaction depends mostly on the sequence of the short oligonucleotides used to prime the polymerase enzyme and the temperature used to anneal the primers. When an arbitrary oligonucleotide with many possible sites for hybridization is used for priming, a number of DNA fragments will be amplified and, by comparing the sizes and numbers of fragments produced, typing of strains can be performed (Williams *et al.*, 1990). This technique is known as RAPD and was shown to give serovar-specific patterns in an investigation of a modest number of serovars (Hilton *et al.*, 1996). The investigation demonstrated that phage typing was more discriminatory than RAPD for *S. typhimurium* and *S. enteritidis* under the conditions used for RAPD in that study. Laconcha *et al.* (1998) also reported poor discrimination between strains of *S. enteritidis* by RAPD, but, by a careful selection of primers among 65 candidates, Lin *et al.* (1996) selected six oligonucleotides, which, when used in parallel in RAPD, produced better discrimination than other typing methods. Strains of *S. enteritidis* have been shown to maintain a stable RAPD profile after carrying out an infection in chicken (Fadl and Khan, 1997), but, in general, reproducibility has been problematic with this method. Careful standardization and optimization may overcome this problem (Tyler *et al.*, 1997).

PCR–RFLP
A different application of PCR and an altogether more reproducible approach, PCR–RFLP, involves PCR amplification of a specific fragment and restriction analysis with restriction enzymes of the amplified DNA (Kilger and Grimont, 1993). In a study based on the amplification of the flagella gene, *fliC*, and digestion with *Taq*I and *Sca*I, the patterns could separate H-antigen groups b, i, d, j, l, v, and z_{10} from each other and from the group of g types (g,m, g,p or g,m,s), which could not be separated individually. Flagellar types r and e,h could be distinguished from these groups but not from each other.

Serovar *S. gallinarum*, which does not express flagella, has a cryptic *fliC* gene, and this serovar was shown to have the g type of amplified fragment length polymorphism (AFLP) profile. The first practical application of the system was to tentatively identify two non-motile mutants as strains of *S. typhimurium* and *S. panama*. These strains were untypable by serotyping, and the authors foresee a system whereby serotyping of *Salmonella* is performed by PCR-based typing methods. In a follow-up to this investigation, Dauga *et al.* (1998) amplified both phase 1 and phase 2 flagellin genes of 264 serovars and compared patterns obtained with two restriction enzymes. One hundred and sixteen patterns were obtained, 80% of which were specifically associated with one antigen. Flagellin-antigen RFLP did not precisely match the diversity evidenced by flagellar agglutination.

Other fragments used to type *Salmonella* by this approach include primers deduced from the *rfb* genes encoding the oligosaccharide repeating units of lipopolysaccharide (LPS) (Luk *et al.*, 1993) and the rRNA genes (Shah and Romick, 1997). The 16S and 23S rRNA genes are highly conserved and present in seven copies per *Salmonella* cell (Christensen *et al.*, 2000). The intergenic region between these genes (internal transcribed spacer; ITS) shows variation between operons, and a specific amplification of these regions will produce a pattern that can be used for typing. When the amplicons are separated by non-denaturing polyacrylamide gel electrophoresis, serovar-specific patterns can be produced (Jensen and Hubner, 1996). Using this electrophoretic system, the patterns are difficult to reproduce. The use of denaturing gel electrophoresis results in more reproducible patterns, but, in this case, some serovars, such as *S. enteritidis*, *S. typhimurium*, *S. gallinarum* and *S. dublin*, share profiles. With few exceptions, serovar-specific patterns can be obtained by restriction analysis, but no efficient separation below serovar level is obtained (Christensen *et al.*, 2000).

Amplified fragment length polymorphism
High-resolution typing of *Salmonella* has been performed by a technique in which PCR is used to amplify restriction fragments produced by standard restriction digestion (AFLP) (Vos *et al.*, 1995). An example of the type of results obtained is demonstrated in Fig. 25.5.

Aarts *et al.* (1998) used the method to type 78 strains of 62 serovars and produced serovar-specific patterns, each consisting of approximately 50 bands in the recommended set-up. No separation below phage-type level was obtained with strains of *S. enteritidis*. AFLP patterns are normally scored by automatic reading of gels and direct transfer into suitable software programs, which can store and compare patterns. More than 200 bands can be compared for each strain. The patterns of strains of *Salmonella* are highly reproducible and, with further experience, the technique may form the basis for a new, definitive typing scheme, which could effectively provide a substitute for traditional serotyping.

Clonal Structure of *Salmonella*

Typing methods are used to compare isolates and to allocate strains with identical typing results into the same group. Although not often taken into consideration, the stability of the population structure is important for the strength of the conclusions that can be drawn from typing results, particularly in studies covering a protracted period of time. The currently accepted division of the genus *Salmonella* into two species, *S. enterica*, with six underlying subspecies (Le Minor and Popoff, 1987), and *S. bongori* (Reeves *et al.*, 1989), can be recognized, irrespective of whether grouping is based on comparison of housekeeping genes (Selander *et al.*, 1990; Christensen and Olsen, 1998), virulence genes (Nelson *et al.*, 1991; Nelson and Selander, 1992) or ribosomal gene sequences (Christensen *et al.*, 1998). Many serovars have been shown to be composed of several evolutionary lines (subclones), which can be distinguished by multilocus enzyme electrophoresis (Beltran *et al.*, 1988) or by suitable DNA markers, such as the copy number and location of insertion sequence IS200 (Stanley *et al.*, 1991; Chowdry *et al.*, 1993; Olsen *et al.*, 1994b, 1996b), and which, in general, also seem to have a clonal structure. Strains of phage types DT49, DT110, DT120 and DT135 of *S. typhimurium* have been shown by molecular typing to cluster together and to be separate from strains of other phage types, indicating that these phage types are related and belong to a particular clonal lineage of *S. typhimurium* (Olsen *et al.*, 1997).

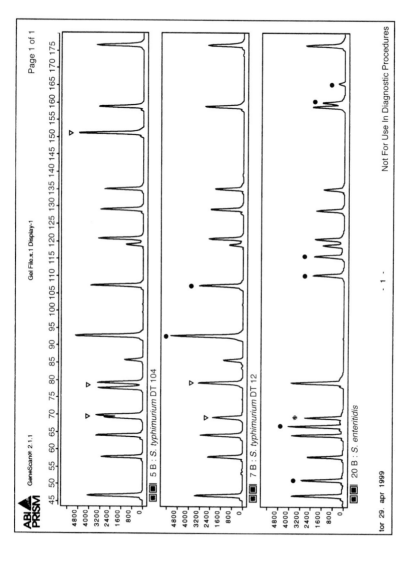

Fig. 25.5. Demonstration of AFLP electropherograms of strains of *Salmonella*. The figure shows a window of the electropherograms of two strains of *S. typhimurium* of different phage type and one strain of *S. enteritidis*. Arrows point to differences between the two strains of *S. typhimurium*, and dots indicate serovar differences. The bands were produced by simultaneous digestion of purified DNA with *BglII* and *MfeI* and by use of non-selective primers. Patterns were detected on an ABI 377 automated DNA sequencer. Data collection, pre-processing, fragment sizing and pattern analysis were done using GENES-SCAN fragment analysis software. In the figure below, the analysis of the two strains of *S. typhimurium* has been repeated to demonstrate the high reproducibility. Electropherograms are by courtesy of B. Kokotovic, Danish Veterinary Laboratory, Copenhagen, Denmark.

The fundamental basis for the successful application of many typing techniques to *Salmonella* is therefore to determine the truly clonal structure of the genus at all levels. Evidence for recombination between clonal lines is scarce in most parts of the genome (Smith *et al.*, 1993), and even relatively small genomic differences observed between types may therefore normally be taken as indicative of a different origin. The high stability of the different clones is, however, also the reason for some of the major failures to obtain useful results in the investigation of some *Salmonella*. When applying typing techniques to those highly successful clones that have spread to many countries and to different livestock populations, it has often been extremely difficult to draw any suitable conclusions. An excellent example of this failure has been the inability to satisfactorily subtype isolates involved in the global spread of *S. enteritidis* phage type (PT) 4 (reviewed by Helmuth and Schroeter, 1994) and it may be feared that a similar situation is happening with the more recent spread of *S. typhimurium* DT104.

phage typing, where applicable, are the most obvious choices for continuous surveillance of *Salmonella*, as these methods are cheap and definitive. Apart from the limitation of being restricted to specialized reference laboratories, they are laborious systems to maintain and may show untypable isolates. Due to technical problems with the scoring of profiles produced by molecular typing methods, these have mainly been used in a comparative fashion. With the arrival of automated gel electrophoresis and the improvements in analytical software for normalizing and storing profiles, this is changing. Results can now be regarded as definitive – that is to say, strains typed years apart or in different laboratories can be compared. PCR-based typing methods, especially AFLP, may have the potential to allocate strains into relevant groups more quickly than traditional typing methods and without the requirement for additional manpower. If this can be achieved, it may become the reference method that creates the clonal architecture in which outbreak strains can be related to each other.

Future Prospects

Control of zoonotic *Salmonella* has a high priority. This control involves extended surveillance of the possible reservoirs and follow-up on incidents in both livestock and the human population. To fully benefit from this and to be able to react to changing patterns, typing must be performed continuously. At present, serotyping and

Acknowledgement

My colleague Derek Brown is thanked for a critical reading of this manuscript and for suggestions for improvements. His input has been highly appreciated. Branko Kotovic of the Danish Veterinary Laboratory is thanked for supplying the AFLP patterns shown in Fig. 25.5.

References

Aarts, H.J.M., van Lith, L.A.J.T. and Keijer, J. (1998) High-resolution genotyping of *Salmonella* strains by AFLP-fingerprinting. *Letters in Applied Microbiology* 26, 131–135.

Anderson, E.S., Ward, L.R., de Saxe, M.J. and de Sa, J.D.H. (1977) Bacteriophage-typing designations of *Salmonella typhimurium*. *Journal of Hygiene, Cambridge* 78, 297–300.

Baggesen, D.L., Olsen, J.E. and Bisgaard, M. (1992) Plasmid profiles and phage types of *Salmonella typhimurium* isolated from successive flocks of chickens on three parent stock farms. *Avian Pathology* 21, 569–579.

Baggesen, D.L., Wegener, H.C. and Christensen, J.P. (1996) Typing of *Salmonella enterica* serovar Saintpaul: an outbreak investigation. *APMIS* 104, 411–418.

Baquar, N., Burnens, A. and Stanley, J. (1994) Comparative evaluation of molecular typing of strains of a national epidemic due to *Salmonella brandenburg* by rRNA gene and IS200 probes and pulsed-field gel electrophoresis. *Journal of Clinical Microbiology* 32, 1876–1880.

Bäumler, A.J., Tsolis, R.M., Ficht, T.A. and Adams, L.G. (1998) Evolution of host adaptation in *Salmonella enterica*. *Infection and Immunity* 66, 4579–4587.

Beltran, P., Musser, J.M., Helmuth, R., Farmer, J.J., Frerichs, W.M., Wachsmuth, I.K., Ferris, K., McWhorter, A.C., Wells, J.G., Cravioto, A. and Selander, R.K. (1988) Toward a population genetic analysis of *Salmonella*: genetic diversity and relationships among strains of serovars *S. choleraesuis*, *S. derby*, *S. dublin*, *S. enteritidis*, *S. heidelberg*, *S. infantis*, *S. newport* and *S. typhimurium*. *Proceedings of the National Academy of Sciences, USA* 85, 7753–7757.

Bezanson, G.S., Kharkhria, R. and Pagnutti, D. (1983) Plasmid profiles of value in differentiating *Salmonella muenster* isolates. *Journal of Clinical Microbiology* 17, 1159–1160.

Birnboim, H.C. and Doly, J.A. (1979) A rapid alkaline extraction procedure for screening recombinant plasmid DNA. *Nucleic Acids Research* 7, 1513–1523.

Brown, D.J., Munro, D.S. and Platt, D.J. (1986) Recognition of the cryptic plasmid pSLT by restriction fingerprinting and a study of its incidence in Scottish *Salmonella* isolates. *Journal of Hygiene, Cambridge* 97, 193–197.

Brown, D.J., Threlfall, E.J. and Rowe, B. (1991) Instability of multiple drug resistance plasmids in *Salmonella typhimurium* isolated from poultry. *Epidemiology and Infection* 106, 247–257.

Brown, D.J., Olsen, J.E. and Bisgaard, M. (1992) *Salmonella enterica*: infection, cross infection and persistence within the environment of a broiler parent stock unit in Denmark. *Zentralblatt für Bakteriologie* 277, 129–138.

Brown, D.J., Threlfall, E.J., Hampton, M.D. and Rowe, B. (1993) Molecular characterization of plasmids in *Salmonella enterica* phage types. *Epidemiology and Infection* 110, 209–216.

Brown, D.J., Baggesen, D.L., Hansen, H.B., Hansen, H.C. and Bisgaard, M. (1994) The characterization of Danish isolates of *Salmonella enterica* serovar Enteritidis by phage typing and plasmid profiling: 1980–90. *APMIS* 102, 208–214.

Browning, L.M., Wray, C. and Platt, D.J. (1995) Diversity and molecular variation among plasmids in *Salmonella enterica* serovar Dublin based on restriction enzyme fragmentation pattern analysis. *Epidemiology and Infection* 114, 237–248.

Cameron, D.N., Khambaty, F.H., Wachsmuth, I.K., Tauxe, R.V. and Barrett, T.J. (1994) Molecular characterization of *Vibrio cholerae* O1 strains by pulsed-field gel electrophoresis. *Journal of Clinical Microbiology* 32, 1685–1690.

Chowdry, N., Threlfall, E.J., Rowe, B. and Stanley, J. (1993) Genotype analysis of faecal and blood isolates of *Salmonella dublin* from humans in England and Wales. *Epidemiology and Infection* 110, 217–225.

Christensen, H.C. and Olsen, J.E. (1998) Phylogenetic relationships of *Salmonella* based on DNA sequence comparison of *atpD* encoding the ATPase-subunit of APT synthase. *FEMS Microbiology Letters* 161, 89–96.

Christensen, H.C., Nordentoft, S. and Olsen, J.E. (1998) Phylogenetic relationships of *Salmonella* based on rRNA sequences. *International Journal of Systematic Bacteriology* 38, 605–610.

Christensen, J.P., Olsen, J.E., Hansen, H.C. and Bisgaard, M. (1992) Characterization of *Salmonella enterica* serovar Gallinarum biovars *gallinarum* and *pullorum* by plasmid profiling and biochemical analysis. *Avian Pathology* 21, 461–470.

Christensen, J.P., Olsen, J.E. and Bisgaard, M. (1993) Ribotypes of *Salmonella enterica* serovar Gallinarum biovars *gallinarum* and *pullorum*. *Avian Pathology* 22, 725–738.

Christensen, J.P., Skov, M.N., Hinz, K.H. and Bisgaard, M. (1994) *Salmonella enterica* serovar Gallinarum biovar gallinarum in layers: epidemiological investigations of a recent outbreak in Denmark. *Avian Pathology* 23, 489–501.

Christensen, H.C., Møller, P.L., Vogensen, F.Q. and Olsen, J.E. (2000). Sequence variation of the 16S to 23S rRNA spacer region in *Salmonella enterica*. *Research in Microbiology* (in press).

Dauga, C., Zabrovskaia, A., Grimont, P.A.D. (1998) Restriction fragment length polymorphism analysis of some flagellin genes of *Salmonella enterica*. *Journal of Clinical Microbiology* 36, 2835–2843.

Ezquerra, E., Burnes, A., Jones, C. and Stanley, J. (1993). Genotypic typing and phylogenetic analysis of *Salmonella paratyphi* B and *S. java* with IS200. *Journal of General Microbiology* 139, 2409–2414.

Fadl, A.M. and Khan, M.I. (1997) Genotypic evaluation of *Salmonella enteritidis* isolates of known types by arbitrarily primed polymerase chain reaction. *Avian Diseases* 41, 732–737.

Ferris, K.E., Andrews, R.E., Thoen, C.O. and Blackburn, B.O. (1992) Plasmid analysis, phage typing, and antibiotic sensitivity of *Salmonella dublin* from clinical isolates in the United States. *Veterinary Microbiology* 32, 51–52.

Fisher, I. (1997) Salm/Enter-Net records a resurgence in *Salmonella enteritidis* infection throughout the European Union. *Eurosurveillance Weekly* 26 June. www.outbreak.org.uk/1997/270626.html

Gibert, I., Barbe, J. and Casadesús, J. (1990) Distribution of insertion sequence IS200 in *Salmonella* and *Shigella*. *Journal of General Microbiology* 136, 2555–2560.

Grimont, F. and Grimont, P.A.D. (1986) Ribosomal ribonucleic acid gene restriction patterns as potential taxonomic tools. *Annales Institut Pasteur/Microbiologie* 137B, 165–175.

Gulig, P.A., Danbara, H., Guiney, D.G., Lax, A., Norel, F. and Rhen, M. (1993) Molecular analysis of the *spv* virulence genes of *Salmonella* virulence plasmids. *Molecular Microbiology* 7, 825–830.

Helmuth, R. and Schroeter, A. (1994) Molecular typing methods for *S. enteritidis*. *International Journal of Food Microbiology* 21, 69–77.

Helmuth, R., Stephan, R., Bunge, C., Hoog, B., Steinbeck, A. and Bulling, E. (1985) Epidemiology of virulence-associated plasmids and outer membrane protein patterns within seven common *Salmonella* serovars. *Infection and Immunity* 48, 175–182.

Hilton, A.C., Banks, J.G. and Penn, C.W. (1996) Random amplification of polymorphic DNA (RAPD) of *Salmonella*: strain differentiation and characterization of amplified sequences. *Journal of Applied Bacteriology* 81, 575–584.

Jensen, M.A. and Hubner, R.J. (1996) Use of homoduplex ribosomal DNA spacer amplification products and heteroduplex cross-hybridization products in the identification of *Salmonella* serovars. *Applied and Environmental Microbiology* 59, 945–952.

Kado, C.I. and Liu, S.T. (1981) Rapid procedure for detection and isolation of large and small plasmids. *Journal of Bacteriology* 145, 1365–1373.

Kapperud, G., Lassen, J., Dommarsnes, K., Kristiansen, B.E., Caugant, D.A., Ask, E. and Jahkola, M. (1989) Comparison of epidemiology marker methods for identification of *Salmonella typhimurium* isolates from an outbreak caused by contaminated chocolate. *Journal of Clinical Microbiology* 27, 2019–2024.

Katouli, M., Kühn, I., Wollin, R. and Möllby, R. (1992) Evaluation of the PhP system for biochemical-fingerprint typing of strains of *Salmonella* of serovar Typhimurium. *Journal of Medical Microbiology* 37, 245–251.

Kauffmann, F. (1961) The species definition in the Enterobacteriaceae. *International Bulletin of Bacteriological Nomenclature and Taxonomy* 11, 5–6.

Kilger, G. and Grimont, P.A.D. (1993) Differentiation of *Salmonella* phase 1 flagellar antigen types by restriction of the amplified *fliC* gene. *Journal of Clinical Microbiology* 31, 1108–1110.

Laconcha, I., López-Molina, N., Rementeria, A., Audicana, A., Perales, I. and Garaizar, J. (1998) Phage typing combined with pulsed-field gel electrophoresis and random amplified polymorphic DNA increases discrimination in the epidemiological analysis of *Salmonella enteritidis* strains. *International Journal of Food Microbiology* 40, 27–34.

Lam, S. and Roth, J.R. (1983) IS200, a *Salmonella*-specific insertion sequence. *Cell* 34, 951–960.

Le Minor, L. and Popoff, M.Y. (1987) Request for an opinion. Designation of *Salmonella enterica* sp. nov., nom. rev., as the type and only species of the genus *Salmonella*. *International Journal Systematic Bacteriology* 37, 465–468.

Lin, A.W., Usera, M.A., Barrett, T.J. and Glodsby, R.A. (1996) Application of random amplified polymorphic DNA analysis to differentiate strains of *Salmonella enteritidis*. *Journal of Clinical Microbiology* 34, 870–876.

Luk, J.M.C., Kongmuang, U., Reeves, P.R. and Lindberg, A.A. (1993) Selective amplification of abequose and paratose synthase genes (*rfb*) by polymerase chain reaction for identification of *Salmonella* major serogroups (A, B, C2, and D). *Journal of Clinical Microbiology* 31, 2118–2123.

McLaren, I.M. and Wray, C. (1991) Epidemiology of *Salmonella typhimurium* infection in calves: persistence of *Salmonella* on calf units. *Veterinary Record* 129, 461–462.

Macrina, F.L., Kopecko, D.J., Jones, K.R., Ayers, D.J. and McCowen, S.M. (1978) A multiple plasmid-containing *Escherichia coli* strain: convenient source of size reference plasmid molecules. *Plasmid* 1, 417–420.

Martinetti, G. and Altwegg, M. (1990) rRNA gene restriction patterns and plasmid analysis as a tool for typing *Salmonella enteritidis*. *Research in Microbiology* 141, 1151–1162.

Millemann, Y., Lesage, M.-C., Chaslus-Dancla, E. and Lafont, J.-P. (1995) Value of plasmid profiling, ribotyping, and detection of IS200 for tracing isolates of *Salmonella typhimurium* and *S. enteritidis*. *Journal of Clinical Microbiology* 33, 173–179.

Nakamura, M., Sato, S., Ohya, T., Suzuki, S. and Ikeda, S. (1986) Plasmid profile analysis in epidemiological studies of animal *Salmonella typhimurium* infection in Japan. *Journal of Clinical Microbiology* 23, 360–365.

Nelson, K. and Selander, R.K. (1992) Evolutionary genetics of the proline permease gene (*putP*) and the control regions of the proline utilization operon in populations of *Salmonella* and *Escherichia coli*. *Journal of Bacteriology* 174, 6886–6895.

Nelson, K., Whittam, T.S. and Selander, R.K. (1991) Nucleotide polymorphisms and the evolution in the glyceraldehyde-3-phosphate dehydrogenase gene (*gapA*) in natural populations of *Salmonella* and *Escherichia coli*. *Proceedings of the National Academy of Sciences, USA* 88, 6667–6671.

Odongo, M.O., McLaren, I.M., Smith, J.E. and Wray, C. (1990) A biotyping scheme for *Salmonella livingstone*. *British Veterinary Journal* 146, 75–79.

Olsen, J.E. (1990) An improved method for isolation of plasmid DNA from wild type Gram negative bacteria for plasmid restriction profile. *Letters in Applied Microbiology* 10, 209–212.

Olsen, J.E. and Skov, M.N. (1994) Genomic lineage of *Salmonella dublin*. *Veterinary Microbiology* 40, 271–282.

Olsen, J.E., Baggesen, D.L., Nielsen, B.B. and Larsen, H.E. (1990) The prevalence of plasmids in Danish bovine and human isolates of *Salmonella dublin*. *APMIS* 98, 735–740.

Olsen, J.E., Sørensen, M., Brown, D.J., Gaarslev, K. and Bisgaard, M. (1992) Plasmid profiles as an epidemiological marker in *Salmonella enterica* serovar Berta infections: comparison of isolates obtained from humans and poultry. *APMIS* 100, 221–228.

Olsen, J.E., Brown, D.J., Skov, M.N. and Christensen, J.P. (1993) Bacterial typing methods suitable for epidemiological investigations: applications in investigations of salmonellosis among livestock. *Veterinary Quarterly* 15, 125–134.

Olsen, J.E., Brown, D.J., Baggesen, D.L. and Bisgaard, M. (1994a) Stability of plasmids maintained in stab cultures at different temperatures. *Journal of Applied Bacteriology* 77, 155–159.

Olsen, J. E., Skov, M., Threlfall, E.J. and Brown, D.J. (1994b) Clonal lines of *Salmonella enterica* serovar Enteritidis documented by IS200, ribotyping, pulsed field gel electrophoresis, and RFLP typing. *Journal of Medical Microbiology* 40, 15–22.

Olsen, J.E., Skov, M.N., Brown, D.J., Christensen, J.P. and Bisgaard, M. (1996a) Virulence and genotype stability of *Salmonella enterica* serovar Berta during a natural outbreak. *Epidemiology and Infection* 116, 267–274.

Olsen, J.E., Skov, M.N. and Christensen, J.P.. (1996b) Genomic lineage of *Salmonella enteritidis* serovar Gallinarum. *Journal of Medical Microbiology* 45, 413–418.

Olsen, J.E., Skov, M., Angen, Ø., Threlfall, E.J. and Bisgaard, M. (1997) Genomic relationships between selected phage types of *Salmonella enterica* serovar Typhimurium defined by ribotyping, IS200 typing and PFGE. *Microbiology* 143, 1471–1479.

Olsvik, Ø., Sørum, H., Birkeness, K., Wachsmuth, K., Fjølstad, M., Larsen, J., Fossum, K. and Feeley, J.C. (1985) Plasmid characterization of *Salmonella typhimurium* transmitted from animals to humans. *Journal of Clinical Microbiology* 22, 336–338.

Pelkonen, S., Romppanen, E.-L., Sitonen, A. and Pelkonen, J. (1994) Differentiation of *Salmonella* serovar Infantis isolates from human and animal sources by fingerprinting IS200 and 16rrn loci. *Journal of Clinical Microbiology* 32, 2128–2133.

Platt, D.J., Chesham, J.S., Brown, D.J., Kraft, C.A. and Taggart, J. (1986) Restriction enzyme fingerprinting of enterobacterial plasmids: a simple strategy with wide application. *Journal of Hygiene, Cambridge* 97, 205–210.

Platt, D.J., Brown, D.J., Old, D.C., Barker, R.M., Munro, D.S. and Taylor, J. (1987) Old and new techniques together resolve a problem of identification by *Salmonella typhimurium*. *Epidemiology and Infection* 99, 137–142.

Reeves, M.W., Evins, G.M., Heiba, A.A., Plikaytis, B.D. and Farmer, J.J., III (1989) Clonal nature of *Salmonella typhi* and its genetic relatedness to other *Salmonella* as shown by multilocus enzyme electrophoresis, and proposal of *Salmonella bongori* comb. nov. *Journal of Clinical Microbiology* 27, 313–320.

Riley, L.W., DiFerdianado, G.T., Jr, DeMelfi, T.M. and Cohen, M.L. (1983) Evaluation of isolated cases of salmonellosis by plasmid profile analysis: introduction and transmission of a bacterial clone by precooked roast beef. *Journal of Infectious Diseases* 148, 12–17.

Rodrigue, D.C., Cameron, D.N., Puhr, N.D., Brenner, F.W., St Louis, M.E., Wachsmuth, I.K. and Tauxe, R.V. (1992) Comparison of plasmid profiles, phage types, and antimicrobial resistance patterns of *Salmonella enteritidis* isolated in the United States. *Journal of Clinical Microbiology* 30, 854–857.

Ryan, C.P. and Steele, J.H. (1987) Chloramphenicol-resistant *Salmonella newport* traced through hamburgers to dairy farms. *New England Journal of Medicine* 317, 632.

Ryan, C.A., Nickels, M.K., Hargrett-Bean, N., Potter, M.E., Endo, T., Mayer, L., Langkop, C.W., Gibson, C., McDonald, R.C., Kenney, R.T., Puhr, N.D., McDonnell, P.J., Cohen, M.L. and Blake, P.A. (1987) Massive outbreak of antimicrobial resistant salmonellosis traced to pasteurized milk. *Journal of the American Medical Association* 258, 3269–3274.

Schiaffino, A., Beuzon, C.R., Uzzau, S., Leori, G., Cappuccinelli, P., Casadesús, S. and Rubino, S. (1996). Strain typing with IS200 fingerprints in *Salmonella abortusovis*. *Applied and Environmental Microbiology* 62, 2375–2380.

Schwartz, D.C., Saffran, W., Welsh, J., Haas, R., Goldenberg, M. and Cantor, C.R. (1983) New techniques for purifying large DNAs and studying their properties and packaging. *Cold Spring Harbor Symposium on Quantitative Biology* 47, 190–195.

Schwarz, S. and Liebisch, B. (1994) Pulsed field gel electrophoretic identification of *Salmonella enterica* serovar Typhimurium live vaccine strain Zoosaloral H. *Letters in Applied Microbiology* 19, 469–472.

Selander, R.K., Beltran, P., Smith, N.H., Helmuth, R., Rubin, F.A., Kopecko, D., Ferris, K., Tall, B.D., Cravioto, A. and Musser, J.M. (1990) Evolutionary genetic relationships of clones of *Salmonella* serovars that cause human typhoid and other enteric fevers. *Infection and Immunity* 58, 2262–2275.

Shaberg, D.R., Tompkins, L.S. and Falkow, S. (1981) Use of agarose gel electrophoresis of plasmid deoxyribonucleic acid to fingerprint Gram-negative bacilli. *Journal of Clinical Microbiology* 13, 1105–1108.

Shah, S.A. and Romick, T.L. (1997) Subspecies differentiation of *Salmonella* by PCR-RFLP of the ribosomal operon using universal primers. *Letters in Applied Microbiology* 25, 54–57.

Smith, J.M., Smith, N.H., O'Rourke, M. and Spratt, B.G. (1993) How clonal are bacteria? *Proceedings of the National Academy of Sciences, USA* 90, 4384–4388.

Sørensen, M., Brown, D.J., Bisgaard, M., Hansen, H.C. and Olsen, J.E. (1991) Plasmid profiles of *Salmonella enterica* serovar Berta isolated from broilers in Denmark. *APMIS* 99, 609–614.

Stanley, J. and Baquar, N. (1994) Phylogenetics of *Salmonella enteritidis*. *International Journal of Food Microbiology* 21, 79–87.

Stanley, J., Jones, C. and Threlfall, E.J. (1991) Evolutionary lines among *Salmonella enteritidis* phage types are identified by insertions sequence IS200 distribution. *FEMS Microbiology Letters* 82, 83–90.

Stanley, J., Burnens, A., Powell, N., Chowdry, N. and Jones, C. (1992) The insertion sequence IS200 fingerprints chromosomal genotypes and epidemiological relationships in *Salmonella heidelberg*. *Journal of General Microbiology* 138, 2329–2336.

Stanley, J., Baquar, N. and Burnens, A. (1995) Molecular subtyping scheme for *Salmonella panama*. *Journal of Clinical Microbiology* 33, 1206–1211.

Taylor, D.N., Wachsmuth, I.K., Shangkuan, Y.-H., Schmidt, E.V., Barrett, T.J., Shchrader, J.S., Scherach, C.S., McGee, H.B., Feldman, R.A. and Brenner, D.J. (1982) Salmonellosis associated with marijuana: a multistate outbreak traced by plasmid fingerprinting. *New England Journal of Medicine* 306, 1249–1253.

Tenover, F.C., Arbeit, R.D., Goering, R.V., Mickelsen, P.A., Murray, B.E., Persing, D.H. and Swaminathan, B. (1995) Interpreting chromosomal DNA restriction patterns produced by pulsed-field gel electrophoresis: criteria for bacterial strain typing. *Journal of Clinical Microbiology* 33, 2233–2239.

Thompson, R., Hughes, S.G. and Broda, P. (1974) Plasmid identification using specific endonucleases. *Molecular and General Genetics* 133, 141–149.

Threlfall, E.J., Ward, L.R. and Rowe, B. (1983) The use of phage typing and plasmid characterization in studying the epidemiology of multiresistant *Salmonella typhimurium*. In: Russel, A. and Quesnel, L. (eds) *Antibiotics: Assessment of Antimicrobial Activity and Resistance*. Society for Applied Bacteriology, pp. 285–297.

Threlfall, E.J., Hall, M.L.M. and Rowe, B. (1986a) *Salmonella gold-coast* from outbreaks of food-poisoning in the British Isles can be differentiated by plasmid profiles. *Journal of Hygiene, Cambridge* 97, 115–122.

Threlfall, E.J., Rowe, B., Ferguson, J.L. and Ward, L.R. (1986b) Characterization of plasmids conferring resistance to gentamicin and apramycin in strains of *Salmonella typhimurium* phage type 204c isolated in Britain. *Journal of Hygiene, Cambridge* 97, 419–426.

Threlfall, E.J., Rowe, B. and Ward, L.R. (1989) Subdivision of *Salmonella enteritidis* phage types by plasmid profile typing. *Epidemiology and Infection* 102, 459–465.

Threlfall, E.J., Frost, J.A., Ward, L.R. and Rowe, B. (1990) Plasmid profile typing can be used to subdivide phage type 49 of *Salmonella typhimurium* in outbreak investigations. *Epidemiology and Infection* 104, 243–251.

Threlfall, E.J., Hall, M.L.M., Ward, L.R. and Rowe, B. (1992) Plasmid profiles demonstrate that an upsurge in *Salmonella berta* in humans in England and Wales is associated with imported poultry meat. *European Journal of Epidemiology* 8, 27–33.

Tompkins, L.S., Troup, N., Labigne-Roussel, A. and Cohen, M.L. (1986) Cloned, random chromosomal sequences as probes to identify *Salmonella* species. *Journal of Infectious Diseases* 154, 156–162.

Tyler, K.D., Wang, G., Tyler, S.D. and Johnson, W.M. (1997) Factors affecting reliability and reproducibility of amplification-based DNA fingerprinting of representative bacterial pathogens. *Journal of Clinical Microbiology* 35, 339–346.

Vos, P., Hogers, R., Bleeker, M., Qeigans, M., van de Lee, T., Homes, M., Frijters, A., Pot, S., Peleman, D., Kuiper, M. and Zabeau, M. (1995) AFLP: a new technique for DNA fingerprinting. *Nucleic Acids Research* 23, 4407–4414.

Ward, L.R., de Sa, J.D.H. and Rowe, B. (1987) A phage-typing scheme for *Salmonella enteritidis*. *Epidemiology and Infection* 99, 291–294.

Weide-Botjes, M., Kobe, B., Lange, C. and Schwarz, S. (1998) Molecular typing of *Salmonella enterica* subspecies *enterica* serovar Hadar: evaluation and application of different typing methods. *Veterinary Microbiology* 61, 215–227.

Wegener, H.C. and Baggesen, D.L. (1996) Investigation of an outbreak of human salmonellosis caused by *Salmonella enterica* ssp. *enterica* serovar Infantis by use of pulsed field gel electrophoresis. *International Journal of Food Microbiology* 32, 125–131.

Williams, J.G.K., Kubelik, A.R., Livak, K.J., Rafalski, J.A. and Tingey, S.V. (1990) DNA polymorphisms amplified by arbitrary primers useful as genetic markers. *Nucleic Acids Research* 18, 6531–6535.

Wray, C., McLaren, I., Parkinson, N.M. and Beedell, Y. (1987) Differentiation of *Salmonella typhimurium* DT204c by plasmid profile and biotyping. *Veterinary Record* 121, 514–516.

Wray, C., Todd, N., McLaren, I., Beedell, Y. and Rowe, B. (1990) The epidemiology of *Salmonella* infection of calves: the role of the dealers. *Epidemiology and Infection* 105, 295–305.

Index

Figures in **bold** indicate major references.
Figures in *italic* refer to diagrams, photographs and tables.